Autodesk® Inventor® 2027 Essentials Plus

Daniel T. Banach, Shawna Lockhart & Sheila Markazi

SDC PUBLICATIONS

SDC Publications
P.O. Box 1334
Mission, KS 66222
913-262-2664
www.SDCpublications.com
Publisher: Stephen Schroff

ISBN-13: 978-1-63057-803-9
ISBN-10: 1-63057-803-7

Printed and bound in the United States of America.

CONTENTS

2 Sketching, Constraining, and Dimensioning 35

3 Creating & Editing Sketched Features 85

4 Creating Placed Features 125

5 Creating & Editing Drawing Views 187

6 Creating & Documenting Assemblies 267

INTRODUCTION

Welcome to the Autodesk Inventor 2027 Essentials Plus manual. This manual provides a thorough coverage of the features and functionalities offered in Autodesk Inventor.

Each chapter in this manual is organized with the following elements:

Objectives. Describes the content and learning objectives.

Topic Coverage. Presents a concise, thorough review of the topic.

Exercises. Presents the workflow for a specific command or process through illustrated, step-by-step instructions.

Checking Your Skills. Tests your understanding of the material using True/False and multiple-choice questions.

Note to the Learner

Autodesk Inventor is designed for easy learning. Autodesk Inventor's help system provides you with ongoing support as well as access to online documentation. Any time in this manual the word Inventor is used, it refers to Autodesk Inventor, the software by the parent company.

As described above, each chapter in this manual has the same instructional design, making it easy to follow and understand. Each exercise is task-oriented and based on real-world mechanical engineering examples.

Who Should Use This Manual?

The manual is designed to be used in instructor-led courses, although you may also find it helpful as a self-paced learning tool.

Recommended Course Duration

Four days (32 hours) to seven days (56 hours) are recommended, or used through a semester.

User Prerequisites

It is recommended that you have a working knowledge of Microsoft® Windows 10 as well as a working knowledge of mechanical design principles.

Manual Objectives

The primary objective of this manual is to provide instruction on how to create part and assembly models, document those designs with drawing views, and automate the design process.

Upon completion of all chapters in this manual, you will be proficient in the following tasks:

- Basic and advanced part modeling techniques
- Drawing view creation techniques
- Assembly modeling techniques

While working through these materials, we encourage you to make use of the Autodesk Inventor help system, where you may find solutions to additional design problems that are not addressed specifically in this manual.

Book Description

This book provides the foundation for a hands-on course that covers basic and advanced Autodesk Inventor features used to create, edit, document, and print parts and assemblies. You learn about the part and assembly modeling through the real-world exercises in this manual.

Exercise Files

The files that are used in the exercises can be downloaded from:

https://www.sdcpublications.com/Downloads/978-1-63057-803-9

Projects

Most designers and engineers work on several projects at a time, with each project consisting of many files. To accommodate this, Autodesk Inventor uses projects to help organize related files and maintain links between files.

Each project has a project file that stores the paths to all files related to the project. When you attempt to open a file, Autodesk Inventor uses the paths in the current project file to locate other necessary files.

For convenience, a project file is provided for the exercises.

Using the Project File

Before starting the exercise, you must complete the following steps:

1. Start Autodesk Inventor.
2. Click the File tab > Manage > Projects.
3. In the Projects window, select Browse. Navigate to the folder where you placed the Essentials Exercises (the default location is C:\Inv Ess Plus) and double-click the file "*Inv Ess Plus.ipj*".

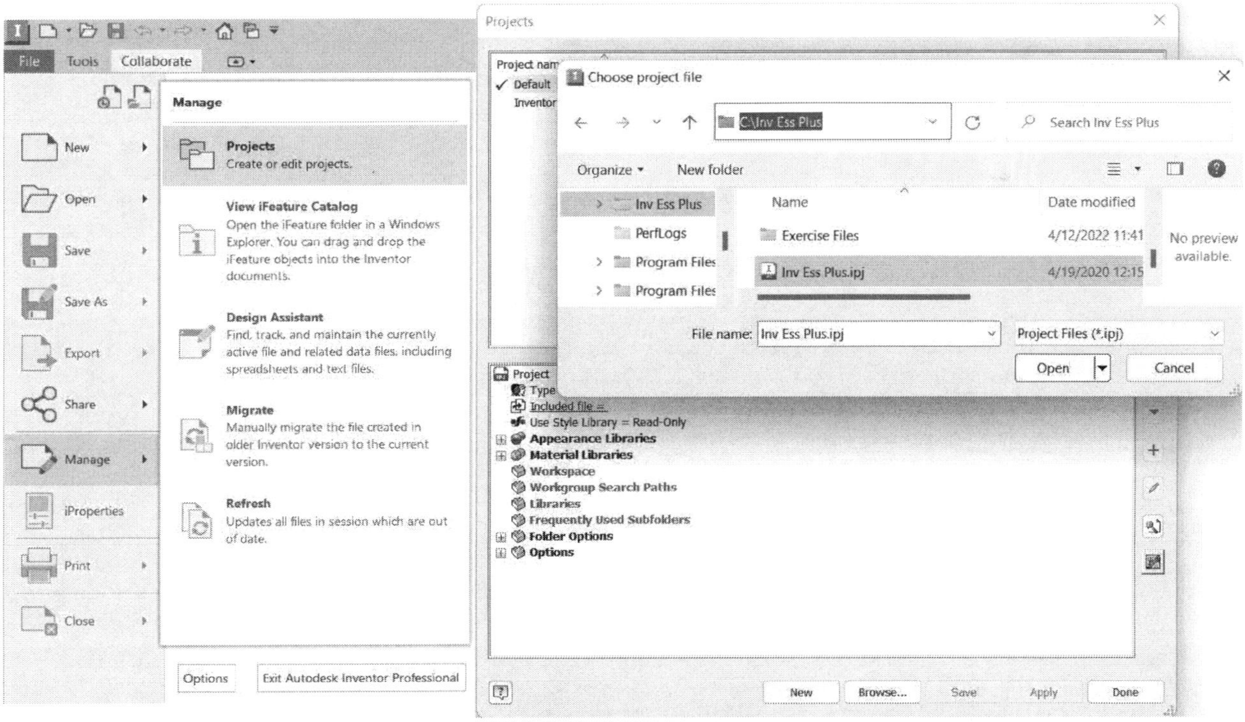

4. Click Open, so that the "Inv Ess Plus" project becomes the current project.
5. Click Done in the Projects dialog box.
6. You can now start doing the exercises.

Acknowledgements

The authors would like to thank the Autodesk Inventor team for their help and insight, and the SDC Publications staff for their expertise, knowledge, and attention to detail, as they have added a great deal to this book.

1 Getting Started

INTRODUCTION

In this chapter, you learn about the following: Inventor's user interface, application options that control how Inventor looks and acts, how to start commands, create and control projects, and how to change your viewpoint to view parts from different perspectives. The knowledge that you learn in this chapter will lay a strong foundation for you to master Inventor.

OBJECTIVES

After completing this chapter, you will be able to do the following:

- List the main areas of Inventor's user interface
- Open files
- Create new files
- List the file types used in Autodesk Inventor
- Explain how to use the three save commands
- Use the Application Options to control how Inventor looks
- Start commands
- Describe how to use Inventor's help system
- Describe the purposes of a project file
- Create a project file for a single user
- Use navigation commands to change how you view a part

GETTING STARTED WITH AUTODESK INVENTOR

Autodesk Inventor's Home screen looks like that shown in Figure 1.1. From here you can create a new file, open an existing file, see a list of recent files, and set the current project.

You can return to the home screen by clicking the Home icon at the bottom-left corner of the screen.

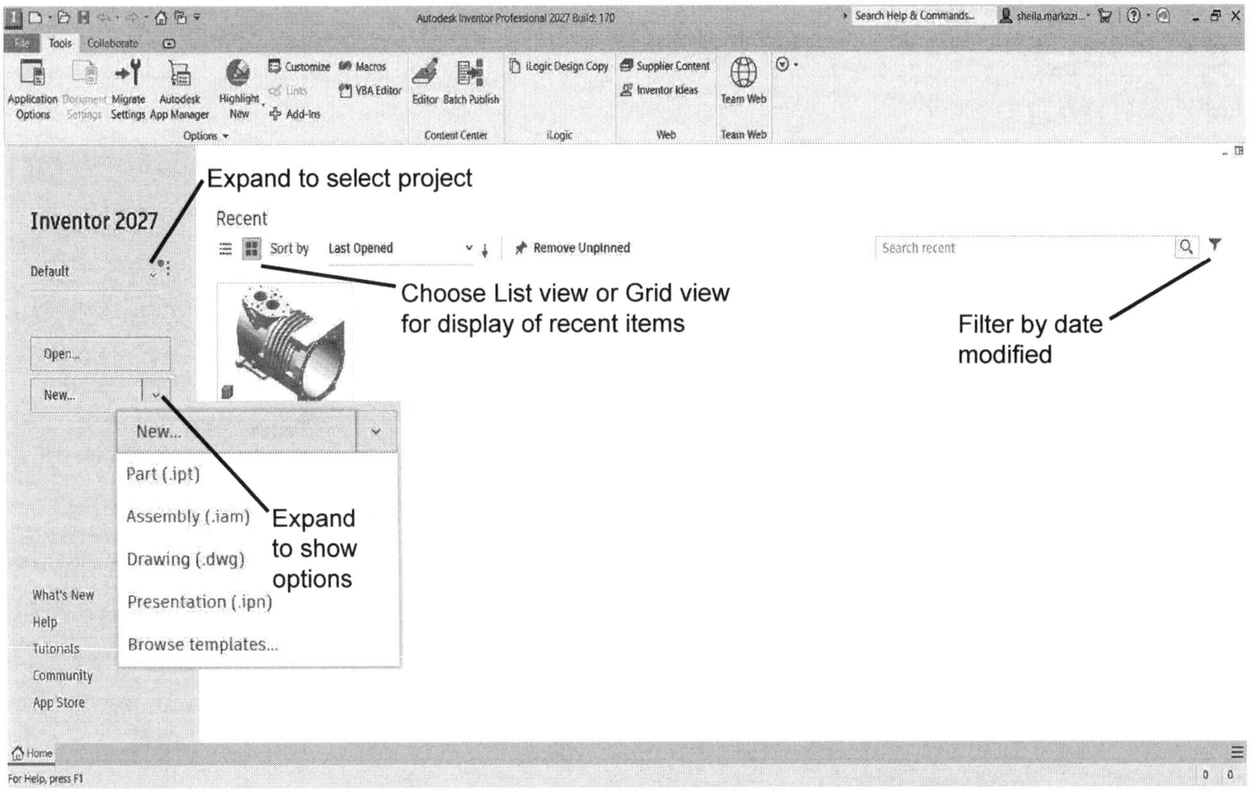

Figure 1.1

USER INTERFACE

The default sketch environment of a part (.ipt) is shown in the following image.

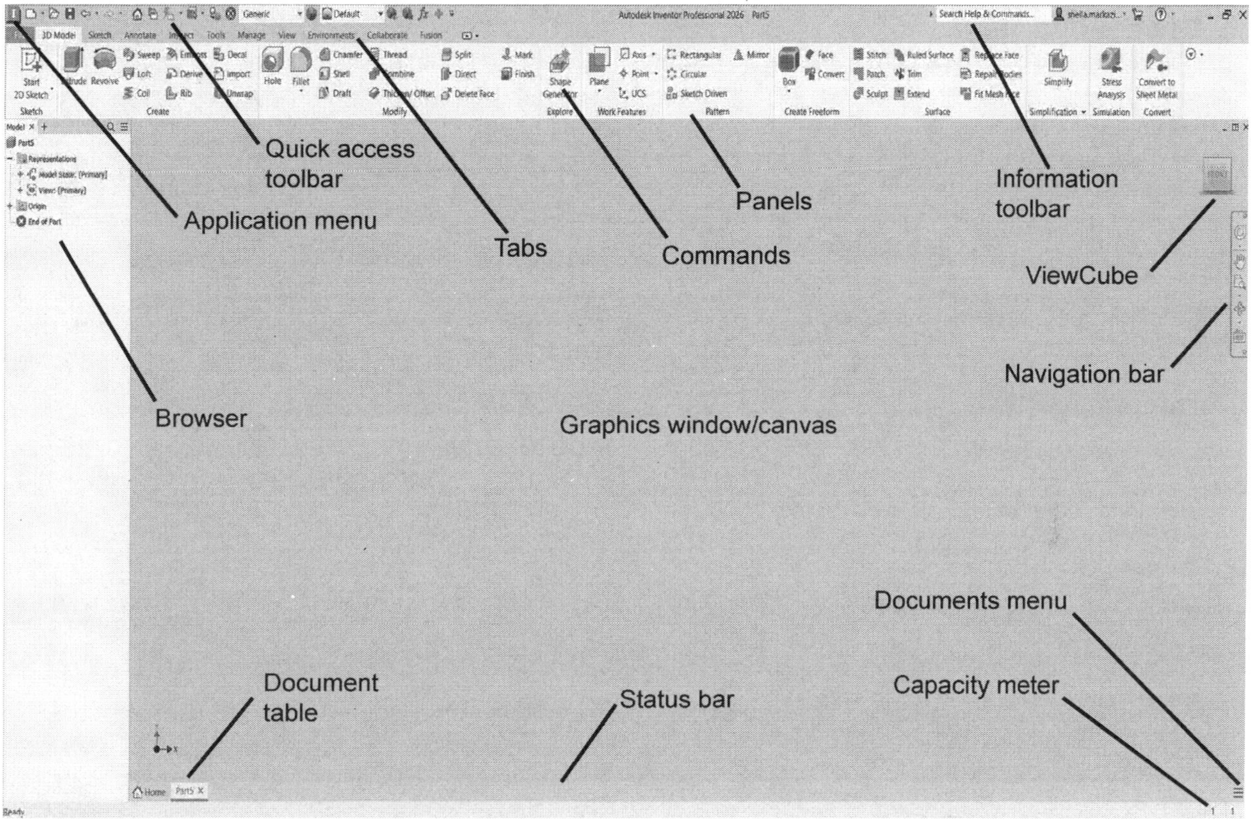

Figure 1.2

The screen is divided into the following areas:

Application Menu

Contains commands to minimize, maximize, and close Inventor.

Quick Access Toolbar

Access common commands as well as commands that can be added or removed.

Tabs

Change available commands by clicking on a tab.

Panels

The panels will change to show available commands for the active tab; click on a tab to display a set of new panels and commands.

Commands

Access basic Windows and Autodesk Inventor commands. The set of commands will change to reflect the environment in which you are working.

Information Toolbar

Displays common help commands as well as Subscription services.

ViewCube

Displays the current viewpoint and allows you to change the orientation of the view.

Navigation Bar
Displays common viewing commands. Viewing commands can be added by clicking the bottom drop arrow.

Documents Menu
Control how to cascade and tile windows, and view the active model in a separate window.

Capacity Meter
Displays how many occurrences (parts) are in the active document, the number of open documents in the current session, and how much memory is being used. Note: The capacity meter that shows memory usage is available only on 32-bit computers.

Status Bar
Displays text messages about the current process and lists common commands for setting how Inventor operates.

Document Tabs
Each open file will be displayed on its own tab, and the Home tab takes you to the Home screen if it has not been closed. If you have multiple files open and if you have a second monitor, you can drag a tab to the second monitor and it will open on that monitor.

Browser
Shows the history of how the contents in the file were created. The browser can also be used to edit features and components.

Graphics Window/Canvas
Displays the graphics of the current file.

FILE TAB

Besides selecting commands for working with files, you can control how the recent or open documents are listed in the menu. The following image shows the functionality available from the File tab.

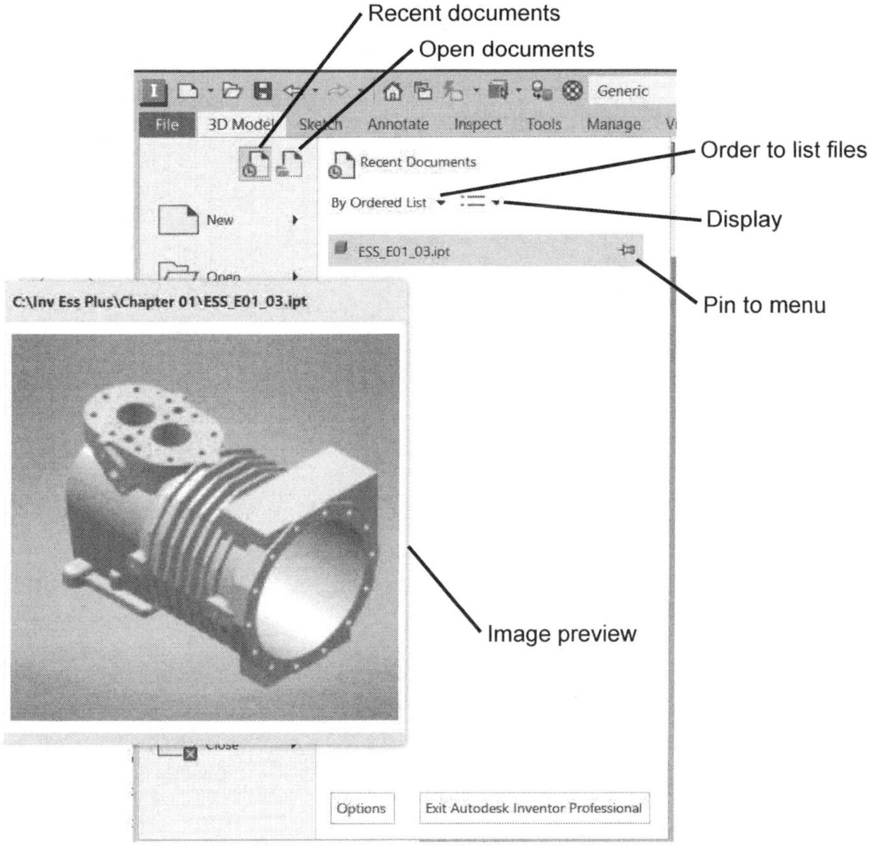

Figure 1.3

Recent Documents

Displays documents that were previously opened.

Open Documents

Displays documents that are opened.

Display

Controls what is displayed in the list: icons or images and what size.

Order to List Files

Controls the order that the files are listed: by Ordered List or by Type.

Pin to Menu

Click the push pin to keep the file in the list no matter when it was last opened.

Preview Image

Hover the cursor over a file to see a larger image of the file when it was last saved.

TIP: Double-click on the Inventor Application button, and a dialog box will appear asking you to save each unsaved document and then Inventor will close.

RIBBON

The ribbon displays commands that are relevant to the selected tab. The currently selected tab is highlighted. The commands are arranged by panels. Commands that cannot be used with the current selections appear "grayed out." Common commands often have larger icons, while less used commands are smaller and positioned to the right of the larger command. Commands may also be available in the drop list of a command or in the name of the panel. For example, there is a drop list available under the Circle icon located in the Create panel, as shown in the following image.

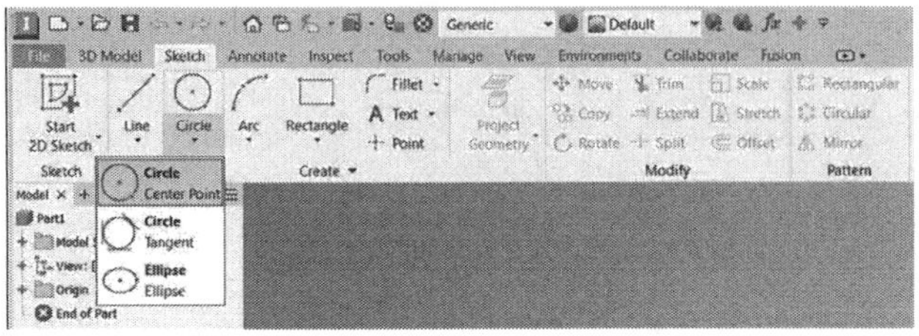

Figure 1.4

The ribbon can be modified by right-clicking on the panel; the following image on the left shows the main options or you can click the down arrow to the right of the panel to turn panels on / off as shown in the following image on the right. Note that you must exit the sketch environment to turn panels on / off.

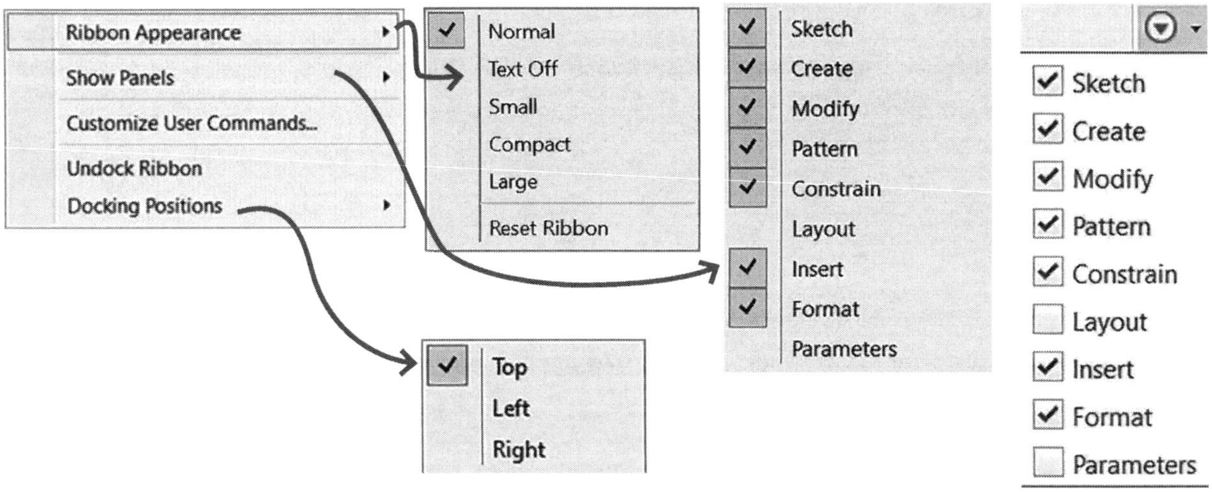

Figure 1.5

- **Ribbon Appearance** – Changes how the Ribbon is displayed, turns text off, and changes the size of the commands
- **Show Panels** – Adds and removes Panels
- **Customize User Commands** – Adds commands to a User Panel
- **Undock Ribbon** – Allows the ribbon to be freely moved
- **Docking Positions** – Changes the location of the Ribbon

QUICK ACCESS TOOLBAR

The Quick Access toolbar is located at the top left corner of the screen and is used to access commonly used commands. Commands can be added or removed from the Quick Access toolbar.

To add a command to the Quick Access toolbar, follow these steps:

1 Move the cursor over a command that you want to add to the Quick Access toolbar and right-click.

2 Click Add to Quick Access Toolbar from the menu as shown in the following image.

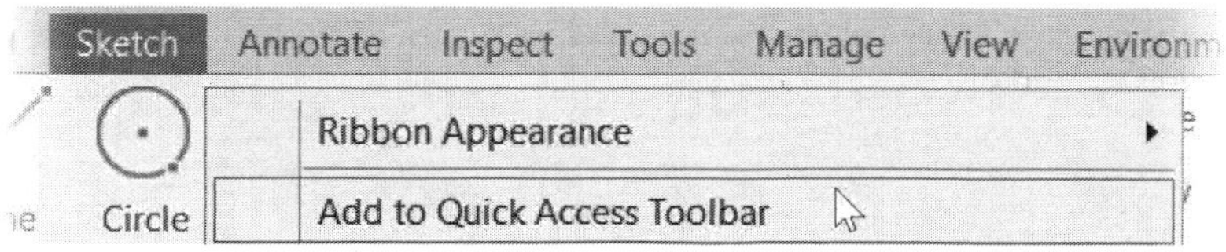

Figure 1.6

To remove a command from the Quick Access Toolbar, follow these steps:

1 Move the cursor over the command on the Quick Access Toolbar and right-click.

2 Click Remove from Quick Access Toolbar as shown in the following image.

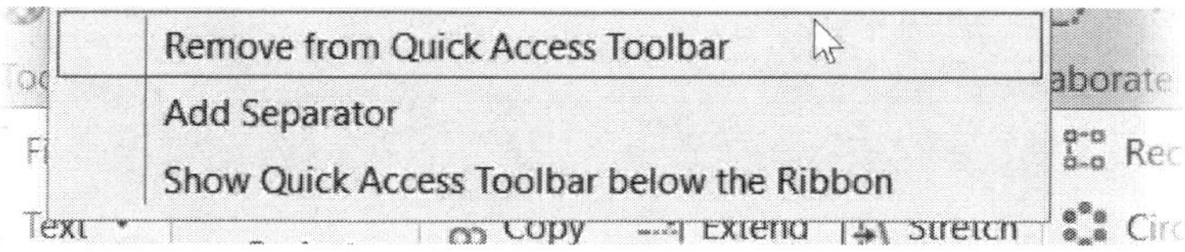

Figure 1.7

OPEN FILES

To open files follow one of these techniques:

- Click the File tab in the top-left corner of the screen and click Open.
- Click the Open command in the Quick Access toolbar.
- Press CTRL + O.

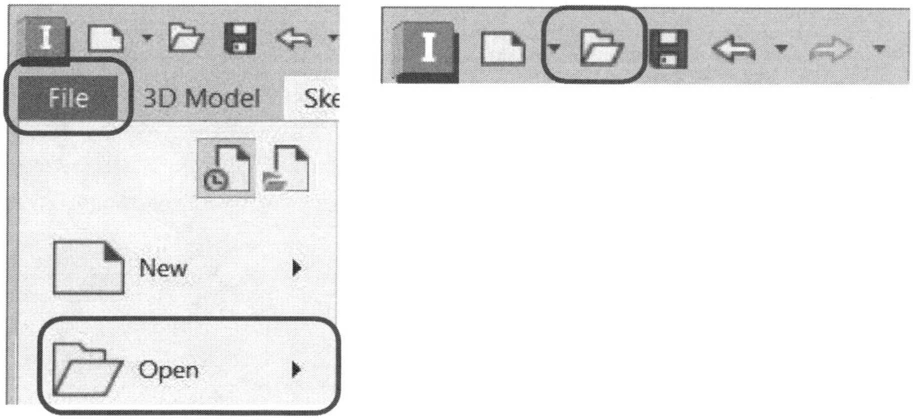

Figure 1.8

The Open dialog box appears as shown in Figure 1.9. The default directory to open is set by the current project file. You can open files from other directories that are not defined in the current project file, but this is **not** recommended. Part, drawing, and assembly relationships may not be resolved when you reopen an assembly that contains components outside the locations defined in the current project file. Projects are covered later in this chapter.

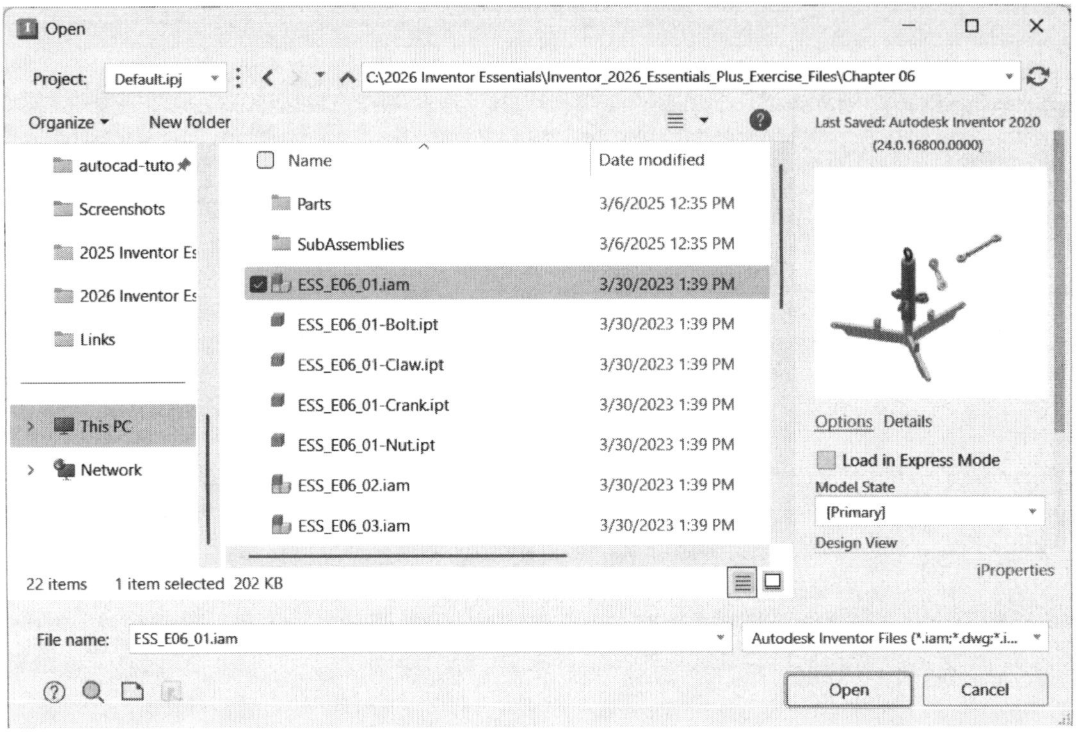

Figure 1.9

Opening Multiple Documents

You can open multiple Inventor files at the same time by holding down the CTRL key and selecting the files to open as shown in Figure 1.10. Each file opens in its own window in a single Inventor session. The files can be arranged to fit the screen or to appear cascaded.

Only one file from those open can be active at a time. Click a file to activate it.

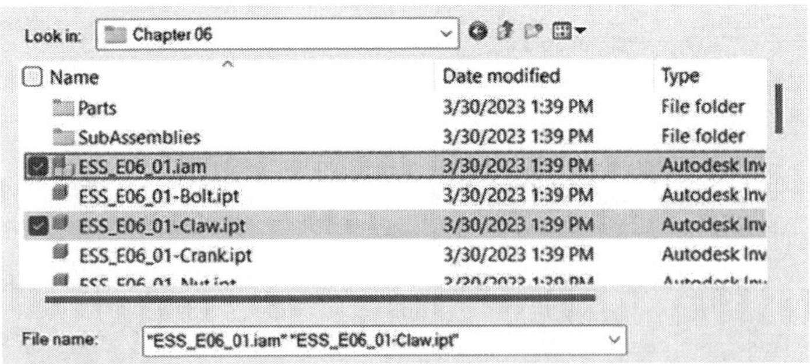

Figure 1.10

Document Tabs

When multiple documents are open in Inventor, each document appears in a tab in the lower left corner of the graphics window. The current document is represented with an "x" to the right of the file name and the tab has a white background. You can see a preview of an open document by hovering the cursor over the tab as shown in Figure 1.11. You can also click the Documents Menu on the lower-right corner of the screen to control how to arrange, tile horizontally, tile vertically, or close the document tabs.

TIP: If you have multiple files open and if you have a second monitor, you can drag a tab to the second monitor and it will open on that monitor.

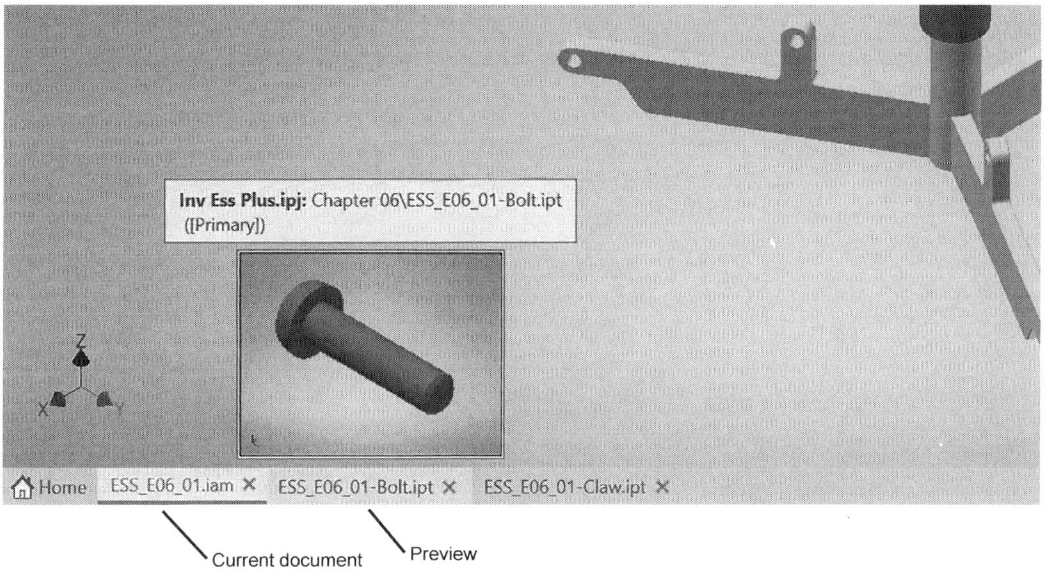

Figure 1.11

NEW FILES

Like the Open command there are many ways to create a new file. To create a new Inventor file, use one of these methods:

- Click the New command in the Quick Access toolbar.
- Click the File tab in the top-left corner of the screen and click New.
- Use a default templates from the down arrow next to the Quick Access toolbar New icon.
- Click New on the Home page.
- Press CTRL + N.

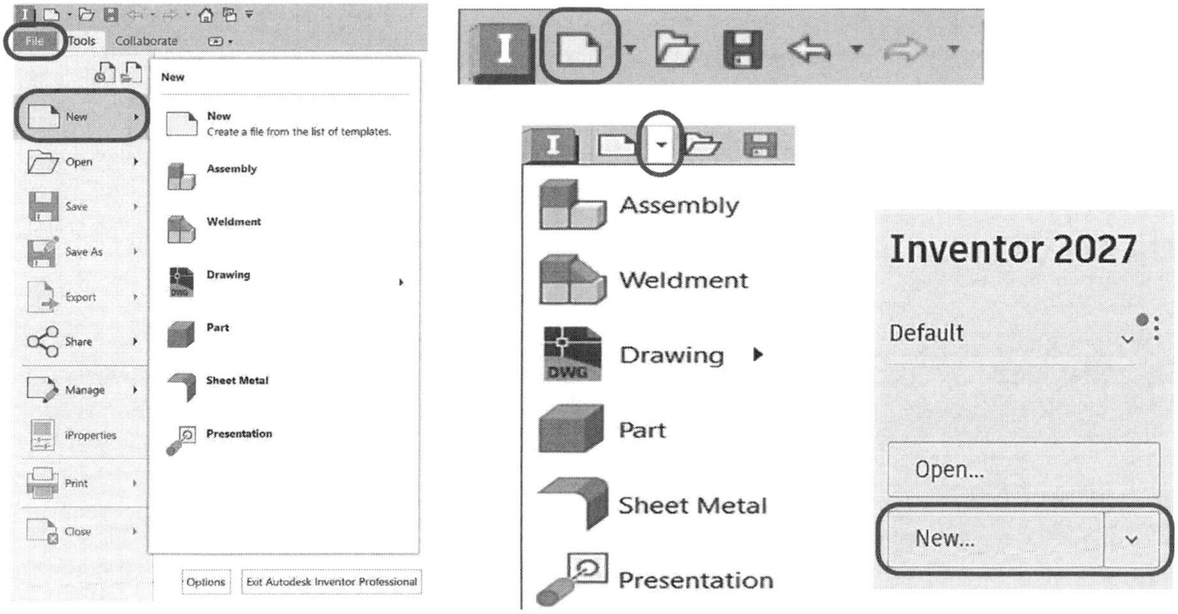

Figure 1.12

The Create New File dialog box appears as shown in Figure 1.13. Begin by selecting the type of file to create or one of the folders from the left column, and then click on a template from the Part, Assembly, Drawing or Presentation section. A preview and a description will appear on the right side of the dialog box. To start a new file, double-click on a template or after a template is selected, click Create on the bottom-right of the dialog box. If Autodesk Inventor options are installed, a Mold Design folder will exist in the list of template folders.

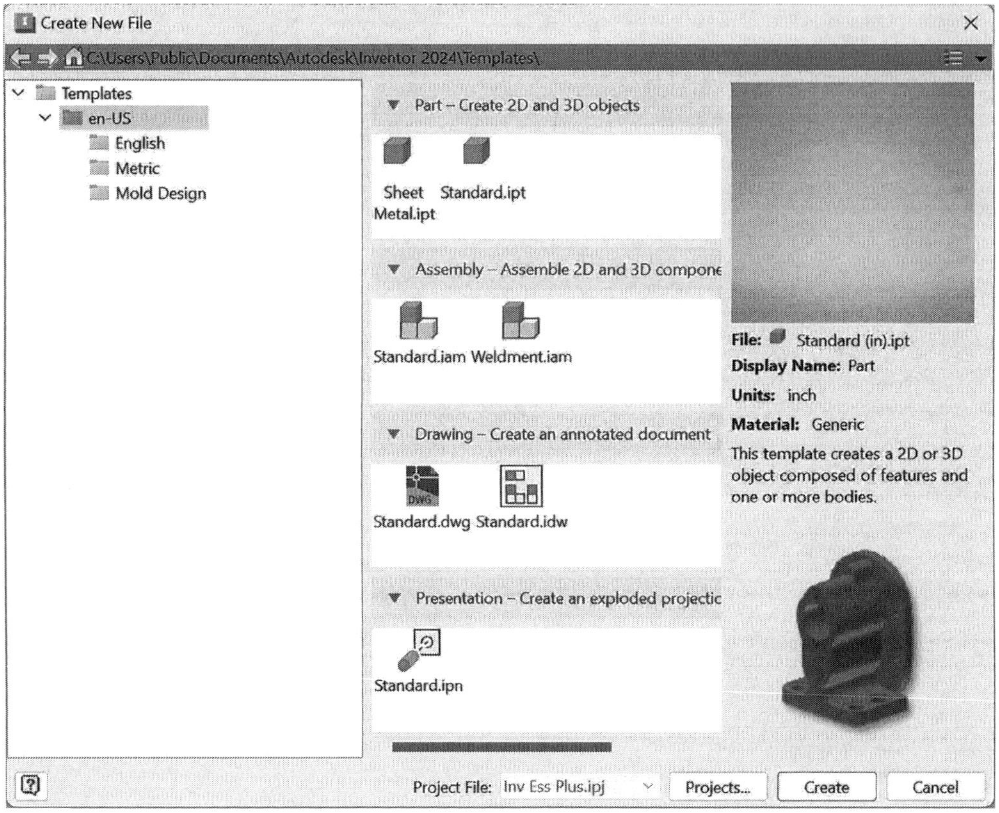

Figure 1.13

FILE INFORMATION

While creating parts, assemblies, presentation files, and drawing views, data is stored in separate files with different file extensions. This section describes the different file types and the options for creating them.

File Types

The following section describes the main file types that you can create in Autodesk Inventor, their file extensions, and descriptions of their uses.

Part (.ipt)

Part files contain only one part, which can be either 2D or 3D.

Assembly (.iam)

Assembly files can consist of a single part, multiple parts, or subassemblies. The parts themselves are saved to their own part file and are referenced (linked) in the assembly file. See Chapter 6 for more information about assemblies.

Presentation (.ipn)

Presentation files show parts of an assembly exploded in different states. A presentation file is associated with an assembly, and any changes made to the assembly will be updated in the presentation file. A presentation file can be animated, showing how parts are assembled or disassembled. The presentation file extension is ipn, but you save animations as an AVI or WMV file. See Chapter 6 for more information about presentation files.

Sheet Metal (.ipt)

Sheet metal files are part files that have the sheet metal environment loaded. In the sheet metal environment, you can create sheet metal parts and flat patterns. You can create a sheet metal part while in a regular part. This requires that you load the sheet metal environment manually.

Drawing (.dwg and .idw)

Drawing files can contain 2D projected drawing views of parts, assemblies, and/or presentation files. You can add dimensions and annotations to drawing views. The parts and assemblies in drawing files are linked, like the parts and assemblies in assembly and presentation files. See Chapter 5 for more information about drawing views.

Project (.ipj)

Project files are structured XML files that contain search paths to locations of all the files in the project. The search paths are used to find the files in a project.

iFeature (.ide)

iFeature files can contain one or more 3D features or 2D sketches that can be inserted into a part file. You can place size limits and ranges on iFeatures to enhance their functionality.

SAVE FILE OPTIONS

There are three options on the File tab for saving your files: Save, Save Copy As, and Save All as shown in the following image.

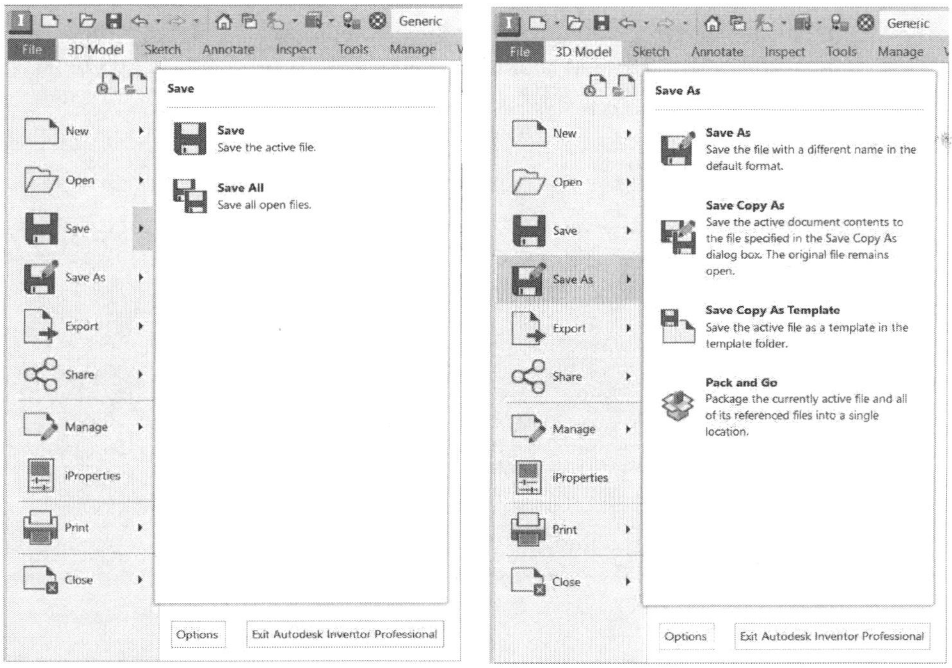

Figure 1.14

Save

The Save command saves the current document, with the same name, to the location where you created it. If this is the first time that a new file is saved, you are prompted for a file name and file location.

TIP: To use the Save command, click the Save icon on the Quick Access toolbar, use the *shortcut keys CTRL + S, or click Save on the File tab.*

Save All

Use the Save All command to save all open documents and their dependents. The files are saved with the same name to the location where you created them. The first time that a new file is saved, you will be prompted for a file name and file location.

Save As

Use the Save As command to save the active document with a new name and location, if required. A new file is created and is made active.

Save Copy As

Use the Save Copy As command to save the active document with a new name and location, if required. A new file is created but is not made active. You can also save the current file as different file formats that other CAD systems can open.

Save Copy As Template

Use the Save Copy As Template command to save the current file to the template folder. New files can be based on the template file. Templates can be saved in the existing folders, or you can create a subdirectory in the Autodesk\Inventor (version number)\ templates directory, and add a file to it. A new template tab, with the same name as the subdirectory, is created automatically when a file is added to the new folder.

Pack and Go

Use the Pack and Go command to copy all the files that are used to create the current file to a specified location.

Save Reminder

You can have Inventor remind you to save a file. Inventor will NOT automatically save the file. After a predetermined amount of time has expired without saving the file, a notification bubble appears in the upper right corner of the screen as shown on the left of the following image. The time can be adjusted via the Application Options Save tab as shown on the right of the following image. The time can be adjusted from 1 minute to 9999 minutes, or uncheck this option to turn off the notification.

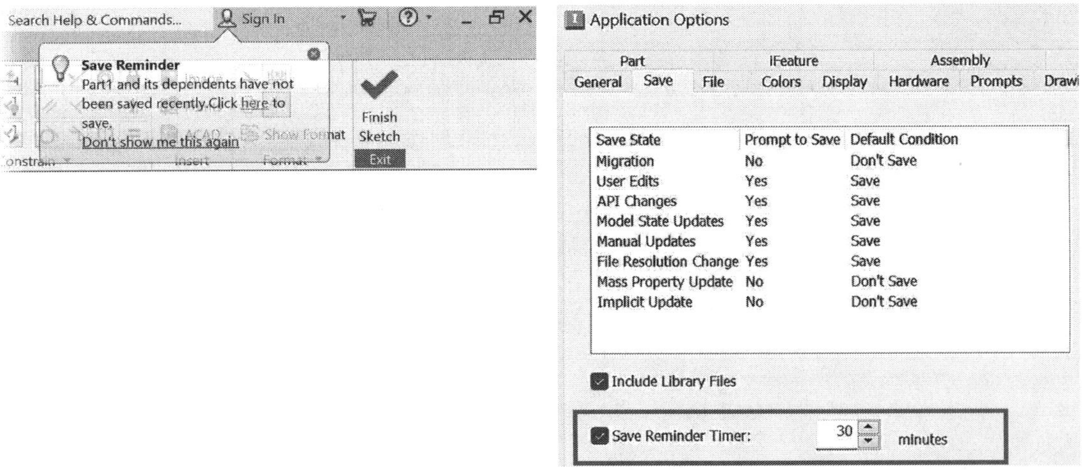

Figure 1.15

APPLICATION OPTIONS

Autodesk Inventor can be customized to your preferences. On the File tab, click Options, or from the Tools tab > Option panel > Application Options, to open the Application Options dialog box as shown in Figure 1.16.

You set options from each of the tabs to control specific actions in the Inventor software. The application options affect all Inventor documents that are open or will be created.

Each section is covered in more detail in the pertinent sections throughout this book. For more information about application options, see the Help system.

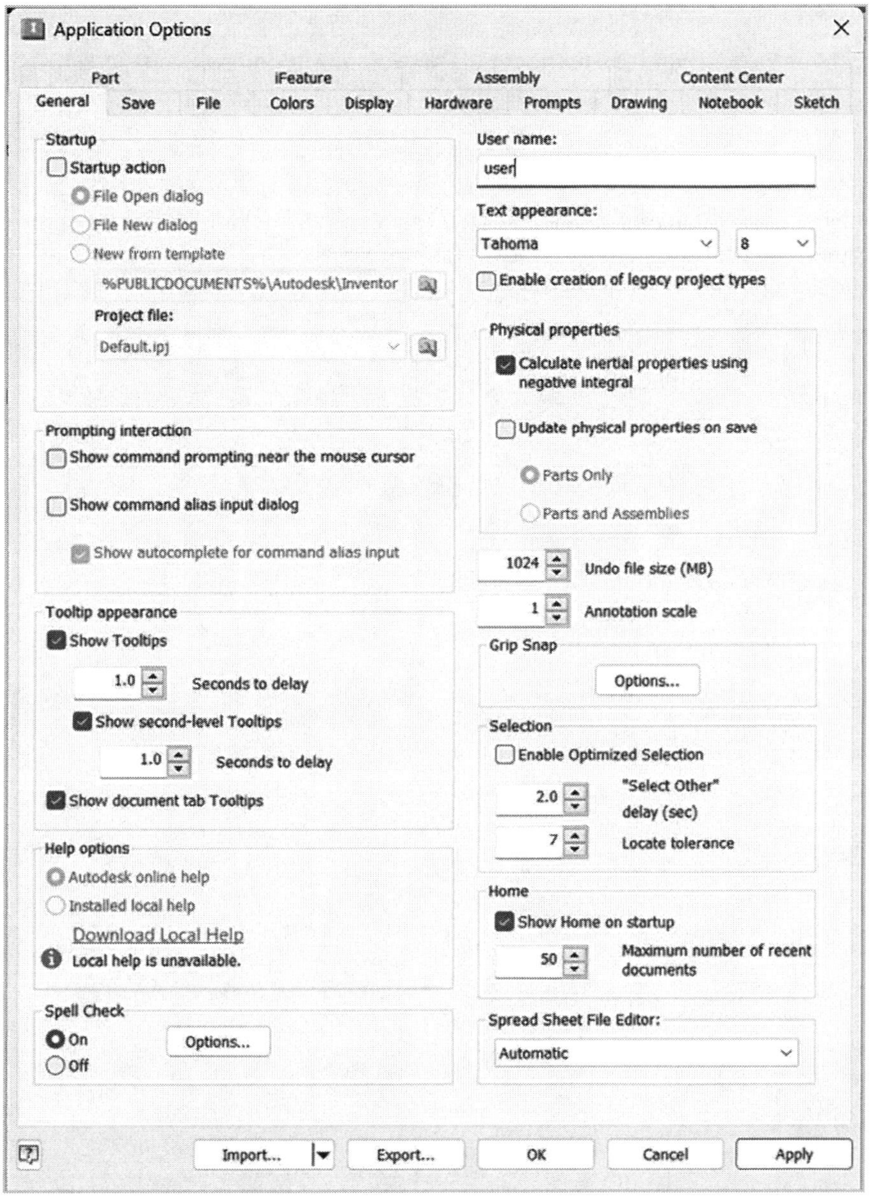

Figure 1.16

General

Set general options for how Autodesk Inventor operates.

Save

Set how files are saved.

File

Set where files are located.

Colors

Change the color scheme and color of the background on your screen. Determine if reflections and textures will be displayed.

Display
Adjust how parts look. Your video card and your requirements affect the appearance of parts on your screen. Experiment with different settings to achieve maximum video performance.

Hardware
Adjust the interaction between your video card and the Inventor software. The software is dependent upon your video card. Take time to make sure that you are running a supported video card and the recommended video drivers. If you experience video-related issues, experiment with the options on the Hardware tab. Consult the help system for more information about video cards and drivers.

Prompts
Modify the response given to messages that are displayed.

Drawing
Specify the way that drawings are created and displayed.

Notebook
Specify how the Engineer's Notebook is displayed.

Sketch
Modify how sketch data is created and displayed.

Part
Change how parts are created.

iFeature
Adjust where iFeature's data is stored.

Assembly
Specify how assemblies are controlled and behave.

Content Center
Specify the preferences for using the Content Center.

EXERCISE 1-1: USER INTERFACE

In this exercise, you change the user interface by moving the Ribbon, Quick Access toolbar, and add and remove commands to the Quick Access toolbar. Use Inventor and follow along with the steps listed.

1 Click the New command from the Quick Access toolbar (upper-left corner of the screen).

2 In the Create New File dialog box click the English folder on the left column, and in the Part section double-click Standard (in).ipt as shown.

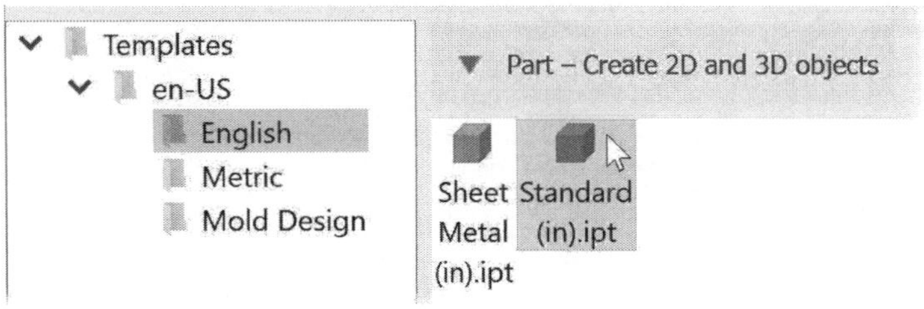

Figure 1.17

3 Move the Ribbon to different locations.

4 Move the cursor anywhere over the Ribbon and right-click.
From the menu click Docking Positions > Left as shown in Figure 1.18.

5 Repeat the process to move the Ribbon to the Right side.

6 Undock the Ribbon by right-clicking on the Ribbon, then
from the menu click Undock Ribbon as shown in Figure 1.18 (right).

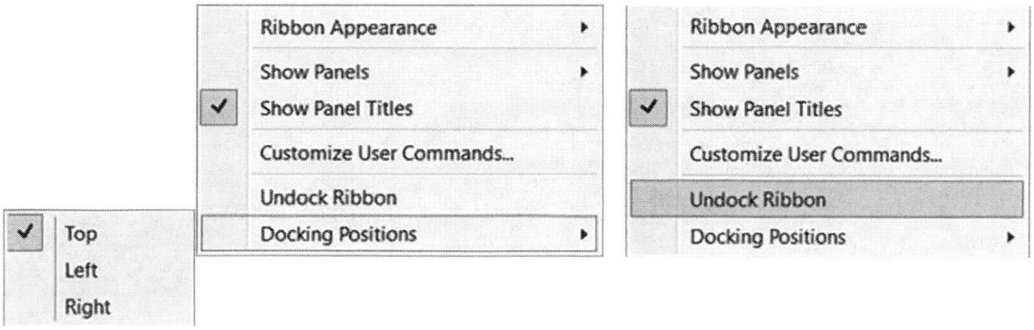

Figure 1.18

7 Move the Ribbon. Try several different locations.

8 Move the Ribbon back to its original top position by right-clicking on the Ribbon, and using the menu to click Docking Positions > Top as shown in Figure 1.18 left.

9 Change the appearance of the Ribbon by right-clicking on the Ribbon, and from the menu choose Ribbon Appearance. Try the different options to change the Ribbon's appearance.

10 Reset the Ribbon back to its original state by clicking Reset Ribbon from the menu as shown in Figure 1.19. Click Yes when prompted to reset the ribbon.

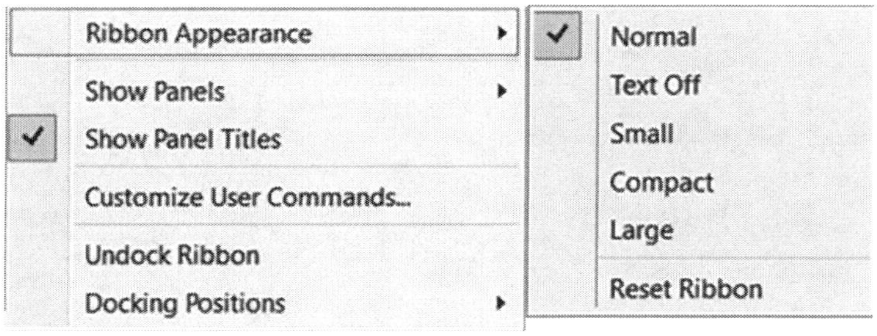

Figure 1.19

11 Change the background color of the graphics screen.
Click the Application Options command, by clicking the File tab, and then click Options at the bottom of the menu.

12 Click the Colors tab, and from the Color scheme area select a scheme.
Click Apply to see the change.

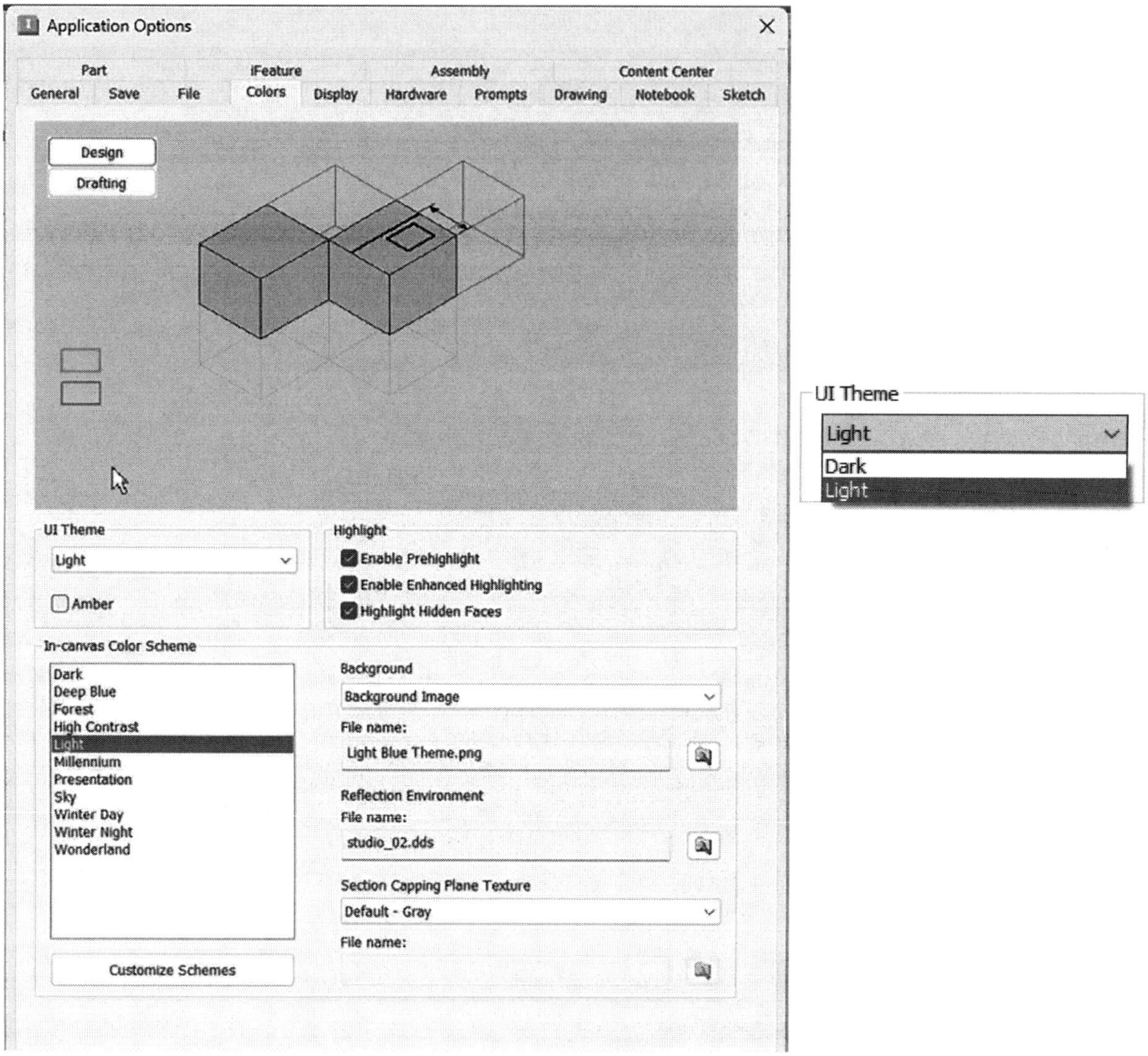

Figure 1.20

13 Experiment with the Background and UI Theme options. See Figure 1.20.

14 Add a command to the Quick Access toolbar by clicking the Tools tab > Options panel and right-clicking on the Application Options command. Click Add to Quick Access toolbar as shown on the left in Figure 1.21.

15 If desired, remove the Application Options command from the Quick Access toolbar. Move the cursor over the Application Options command in the Quick Access toolbar, right-click, and click Remove from Quick Access Toolbar as shown in Figure 1.21 right.

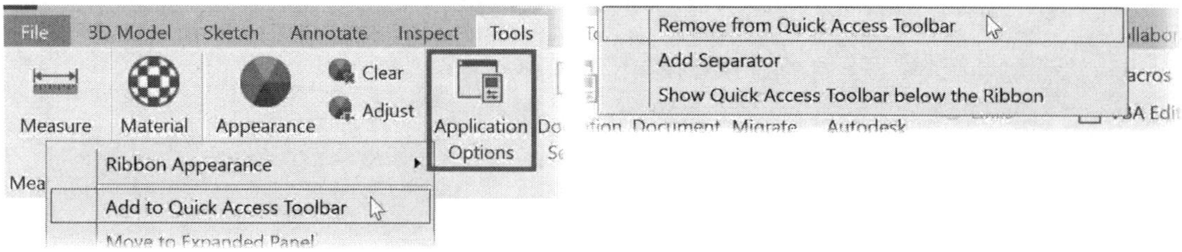

Figure 1.21

16 As you work with Inventor, adjust the user interface to meet your requirements.

17 Close the file. Do not save changes.

End of exercise.

COMMAND ENTRY

There are several methods to issue commands in Inventor. In the following sections, you will learn how to start a command. There are no right or wrong methods for starting a command, and with experience, you will develop your own preference.

To stop a command, either press the **ESC** key, right-click, and click Done from the menu or select another icon.

Tooltips

In the last section, you learned how to control the appearance of the Ribbon. The main function of the Ribbon is to arrange the commands in a logical fashion by dividing them into panels. To start a command from a panel, move the cursor over the desired icon. A command tip appears with the name of the command when you hover the cursor over the command icon.

You can control the tooltip from the Application Options under the General tab as shown in Figure 1.22 left. The first level tooltip displays an abbreviated command description as shown in the middle image. If the cursor hovers over the icon longer, a more detailed description appears as shown on the right.

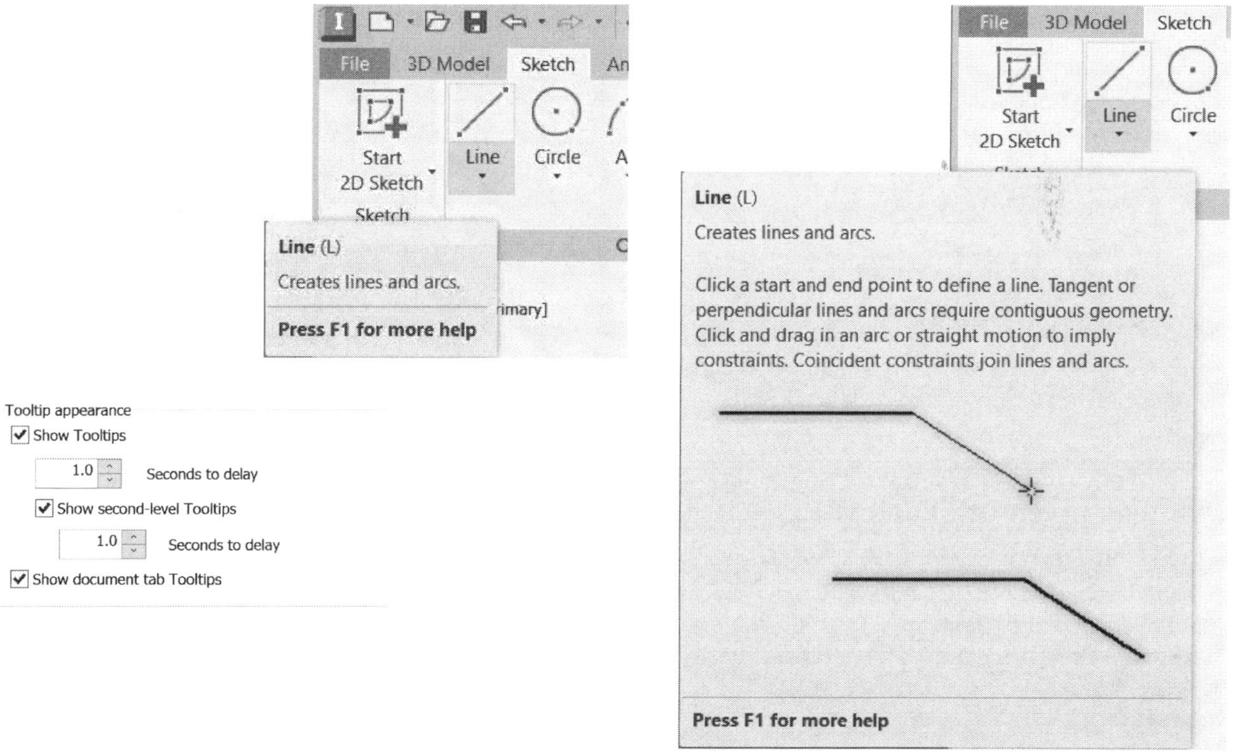

Figure 1.22

Some of the icons in the Panel have a small down arrow in the right side. Select the arrow to see additional commands. To activate a command, move the cursor over a command icon and click. The command that is selected will appear first in the list replacing the previous command.

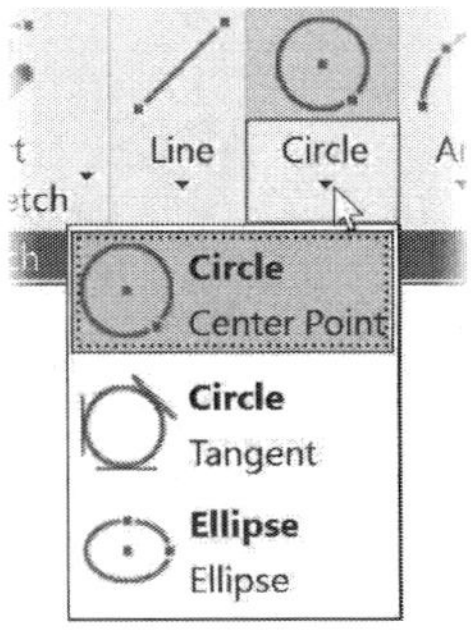

Figure 1.23

Marking Menus and Context (Overflow) Menus

Autodesk Inventor also uses marking menus and context menus, also referred to as overflow menus. These menus appear when you press the right mouse button. The marking menu consists of commands that appear around the center of the cursor and consist of commands that are commonly performed for the current environment. The context (overflow) menus appear below the marking menu and contain options that are relevant to the current task. A marking menu only appears when you right-click in the graphics window. The following image on the left shows the marking menu that appears while in the Line command. As you gain experience with Inventor, you can start a command from the top portion of the marking menu by right-clicking and moving the cursor (before the menu appears within 250 milliseconds) in the direction of the command and releasing the mouse button as shown on the right in the following image. This technique is referred to as gesturing.

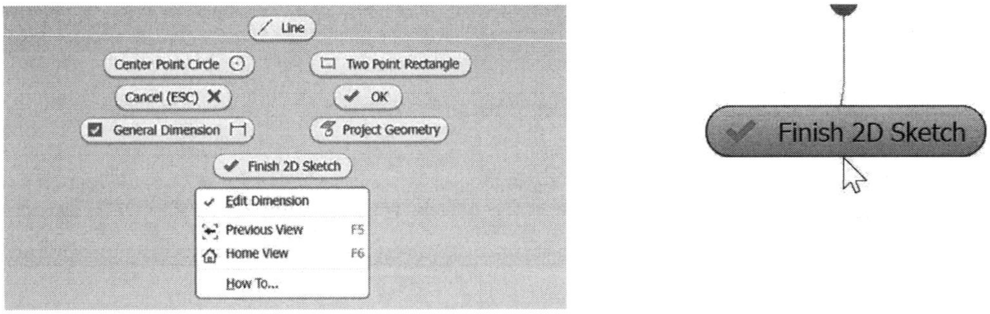

Figure 1.24

Autodesk Inventor Shortcut Keys

Autodesk Inventor has keystrokes called shortcut keys that are preprogrammed. While in a command, the tooltip displays the shortcut key in parentheses if a shortcut key is available. To start a command via a shortcut key, press the desired preprogrammed key(s). The keys can be reprogrammed by clicking the Tools tab and clicking Customize.

Repeat Last Command

To restart a command without reselecting the command in a panel bar, either press ENTER or the spacebar, or right-click and click the top entry in the menu, and Repeat "the last command." The following image shows the Line command being restarted.

Figure 1.25

Undo and Redo

You may want to undo an action that you just performed, or undo an undo. The Undo command backs up Autodesk Inventor one function at a time. If you undo too far, you can use the Redo command to move forward one step at a time. The Zoom, Orbit, and Pan commands do not affect the Undo and Redo commands. To start the commands, select the command from the Quick Access toolbar as shown in the following image. The Undo command is to the left, and the Redo command is to the right. The shortcut keys are CTRL Z for Undo and CTRL Y for Redo.

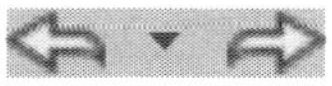

Figure 1.26

HELP SYSTEM

The commands in the Information Toolbar on the top-right corner of the screen will assist you while you design. To get help on a topic, enter a keyword in the area entitled "Type a keyword or phrase." To see the other help mechanisms that make up the Help System, click the drop arrow to the right of the question mark as shown in Figure 1.27. Ask the Autodesk Assistant for help by clicking the chat icon to the far right.

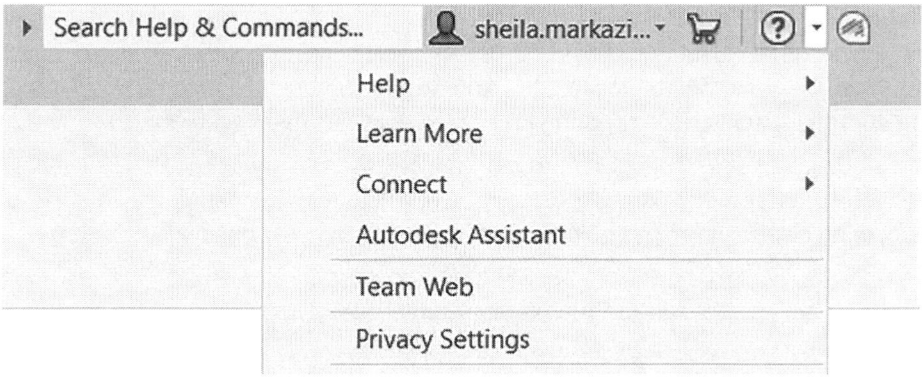

Figure 1.27

Other options to access the Help system include the following methods:

- Press the F1 key, and the Help system assists you with the active operation.
- Click an option on the Help menu.
- Click a Help option on the Information Toolbar.
- In any dialog box, click the icon and then click Help.
- Click an option on the Get Started tab.
- Right-click and click on Help Topics.

PROJECTS IN AUTODESK INVENTOR

Almost every design that you create in Autodesk Inventor involves more than a single file. Each part, assembly, presentation, and drawing created is stored in a separate file. Each of these files has a unique file extension. There are many times when a design will reference other files. An assembly file, for example, will reference many individual part files and/or additional subassemblies. When you open the parent or top-level assembly, it must contain information that allows Autodesk Inventor to locate each of the referenced files. Autodesk Inventor uses a project file to organize and manage these file-location relationships. There is no limit to the number of projects you can create, but only one project can be active at any given time.

You can structure the file locations for a design project in many ways. A single person design shop has different needs from a large manufacturing company or a design team with multiple designers working on the same project. In addition to project files, Autodesk Inventor includes a program called Autodesk Vault that controls basic check-out and check-in file-reservation mechanisms; these control file access for multiuser design teams. Autodesk Inventor always has a project named Default. Specifically, if all the files defining a design are in a single folder or in a folder tree where each referenced part is located with its parent or in a subfolder underneath the parent, the Default project may be all that is required.

TIP: It is recommended that files in different folders never have the same name to avoid the possibility of Autodesk Inventor resolving a reference to a file of the same name but in a different folder.

Project Setup

To reduce the possibility of file resolution problems later in the design process, always plan your project folder structure before you start a design. A typical project might consist of parts and assemblies unique to the project; standard components that are unique to your company; and off-the-shelf components such as fasteners, fittings, or electrical components.

Project File Search Options

Before you create a project, you need to understand how Autodesk Inventor stores cross-file reference information and how it resolves that information to find the referenced file. Autodesk Inventor stores the file name, a subfolder path (if present) to the file, and a library name (optionally) as the three fundamental pieces of information about the referenced file.

When you use the Default project file, the subfolder path is located relative to the folder containing the referencing file. It may be empty or may go deeper in the subfolder hierarchy, but it can never be located at a level above the parent folder.

When you create a project file, you do not need to add subfolder(s) as search paths. The subfolder(s) are automatically searched and do not need to be added to the project file.

Creating Projects

To create a new project or edit an existing project, use the Autodesk Inventor Project File Editor. The Project File Editor displays a list of shortcuts to previously active projects. A project file has an .ipj file extension and typically is stored in the home folder for the design-specific documents, while a shortcut to the project file is stored in the Projects Folder. The Projects Folder is specified on the File tab of the Application Options dialog box as shown in the following image. All projects with a shortcut in the Projects Folder are listed in the top pane of the Project File Editor.

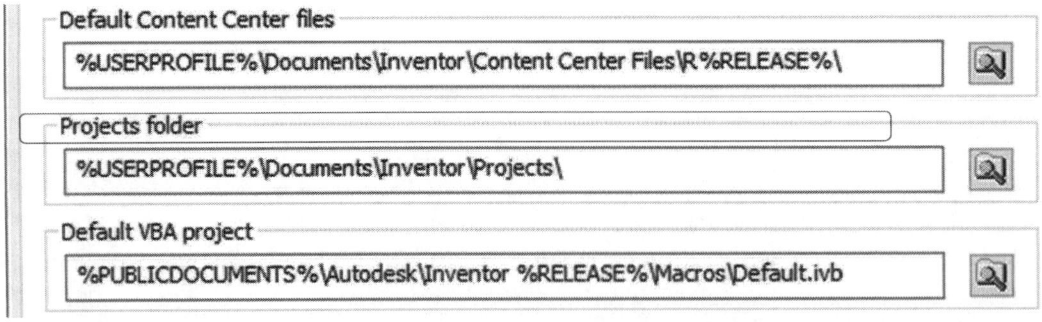

Figure 1.28

You create or edit a project file by clicking the Projects button in the New or Open dialog box, or from the Quick Access Toolbar, or from File tab > Manage > Projects as shown in the following image.

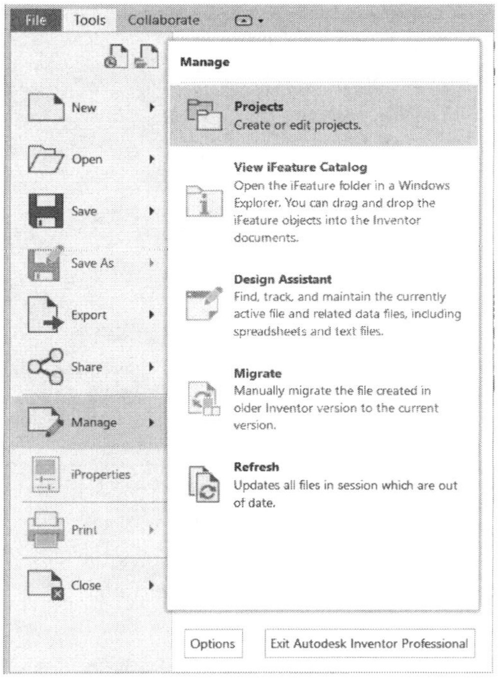

Figure 1.29

The Projects dialog box will appear as shown in the following image. The Projects dialog box is divided into two panes. The top pane lists shortcuts to the project files that have been active previously. Double-click on a project's name to make it the active project. All Inventor files must be closed before making a project active. Only one project file can be active in Autodesk Inventor at a time. The bottom pane reflects information about the project selected in the top pane. If a project file already exists, click on the Browse button on the bottom of the dialog box, then navigate to and select the project file.

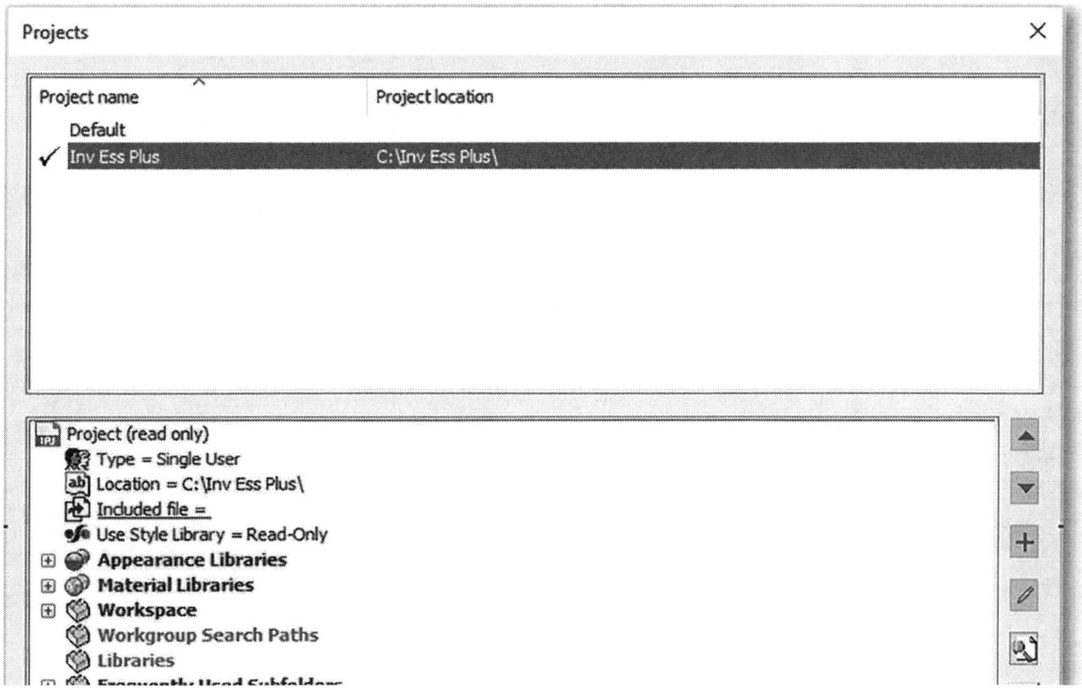

Figure 1.30

TIP: When defining a path to a folder on a network, it is recommended to define a Universal Naming Convention (UNC) path starting with the server name (*Server*\...) and not to use shared (mapped) network drives.

To create a new project, follow these steps:

1 In the Projects dialog box, click the New button at the bottom to initiate the Inventor project wizard.

2 In the Inventor project wizard, follow the prompts to the following questions.

What Type of Project Are You Creating?

If Autodesk Vault is installed, you will be prompted to create a New Vault project or a New Single-User project. If Autodesk Vault is not installed, only a New Single-User project type will appear in the list.

New Single-User Project

This is the default project type, which is used when only one user will reference Autodesk Inventor files. It creates one workspace where Autodesk Inventor files are stored and any needed library location(s), and it sets the Project Type to Single User. No workgroup is defined but can be defined later. The next section covers the steps for creating a new single-user project. For more information on projects, consult the online Help system.

New Vault Project

This project type is used with Autodesk Vault and is not available until you install Autodesk Vault. It creates a project with one workspace and any needed library location(s), and it sets the multiuser mode to Vault. More information about Autodesk Vault appears later in this section.

Creating a New Single-User Project

Click on New Single User Project as shown in the following image. If Autodesk Vault is not installed, New Single User Project will be the only available option.

Figure 1.31

Name

Click the Next button, and specify the project file name and location on the second page of the Inventor project wizard. Enter a descriptive project file name in the Name field as shown in the following image. The project file will use this name with an .ipj file extension.

Project (Workspace) Folder

This specifies the path to the home or top-level folder for the project. You can accept the suggested path, enter a path, or click the Browse button (...) to manually locate the path. The default home folder is a subfolder under My Documents, or wherever you last browsed, that is named to match the project file name.

Project File to Be Created

The full path name of the project file is displayed below the Location field.

Figure 1.32

TIP: In the workspace folder it is recommended that the project file (.ipj) be the only Inventor file stored in this folder, then create subfolders under the workspace folder.

Click the Next button at the bottom of the Inventor project wizard, and specify the project library search paths.

You can add library search paths from existing project files to this new project file. The library search paths from every project with a shortcut in your Projects Folder are listed on the left in the Inventor project wizard dialog box as shown in the following image.

You can add and remove libraries from the New Project area by clicking on their names in either the All Projects area or the New Project area and then clicking the arrows in the middle of the dialog box. The libraries listed by default in the All Projects list will match those in the project file that you selected prior to starting the New Project process.

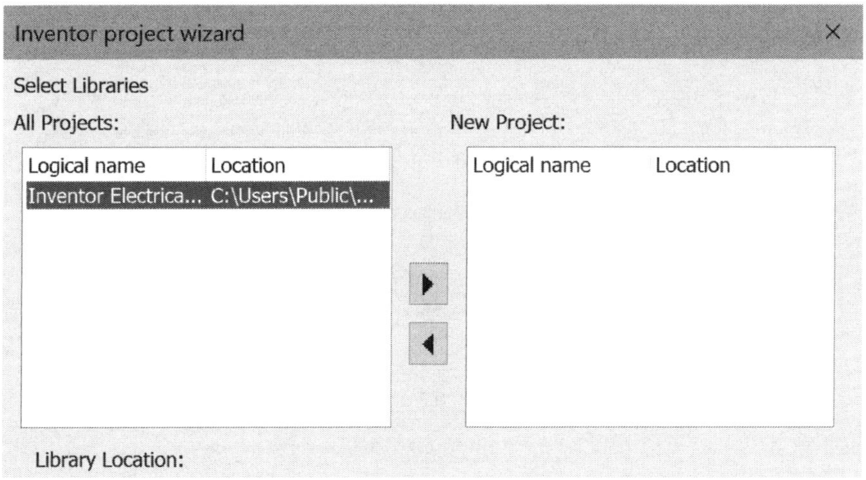

Figure 1.33

Click the Finish button to create the project. If a new directory will be created, click OK in the Inventor Project Editor dialog box. The new project will appear in the Open dialog box. Double-click on a project's name in the Projects dialog box to make it the active project. A checkmark will appear to the left of the active project as shown in Figure 1.34.

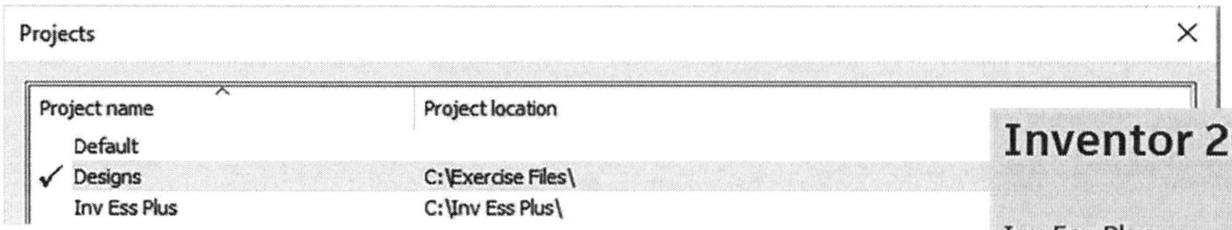

Figure 1.34

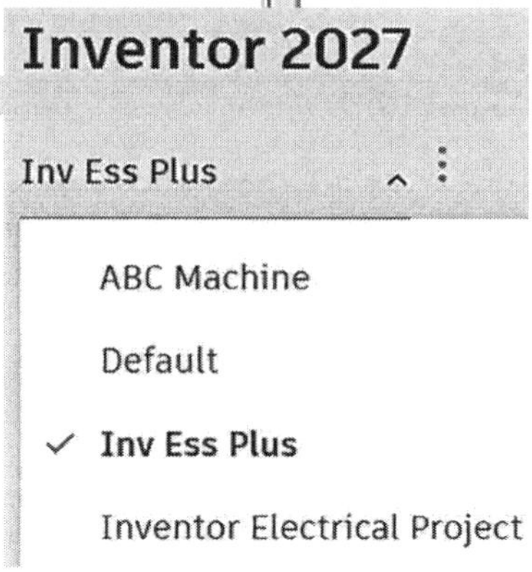

TIP: You can also select to make a project active from the Home screen.

AUTODESK VAULT

Autodesk Vault is available when Inventor is installed. Autodesk Vault enhances the data management process by managing more than just Autodesk Inventor files and by tracking file versions as well as team member access. Controlling access to data, tracking modifications, and communicating the design history are important aspects of managing collaborative data. When working with a vault project, your data files are stored in a central repository that records the entire development history of the design. The vault manages Inventor and non-Inventor files alike. To modify a file, it first must be checked out of the vault. When the file is checked back into the vault, the modifications are stored as the most recent version for the project, and the previous version is sequentially indexed as part of the living history of the design.

EXERCISE 1-2: PROJECTS

In this exercise, you create a project file for a single-user project, open existing files, delete the project file, and make an existing project file current.

1 Prior to creating a new project file, close all Inventor files and ensure that the exercise files have been copied to your computer. The default folder is *C:\Inv Ess Plus.* To download the exercise, see the Essentials Exercise Files section in the Introduction section in the front of the book.

2 From File tab > Manage, click Projects.

3 You now create a new project file. Click the New button at the bottom of the Projects dialog box.

4 For the type of project to create, click New Single User Project, and then click Next.

5 In the next dialog box of the Inventor project wizard, enter ABC Machine in the Name field.

6 In the Project (Workspace) Folder field type *C:\My First Project* as shown in Figure 1.35.

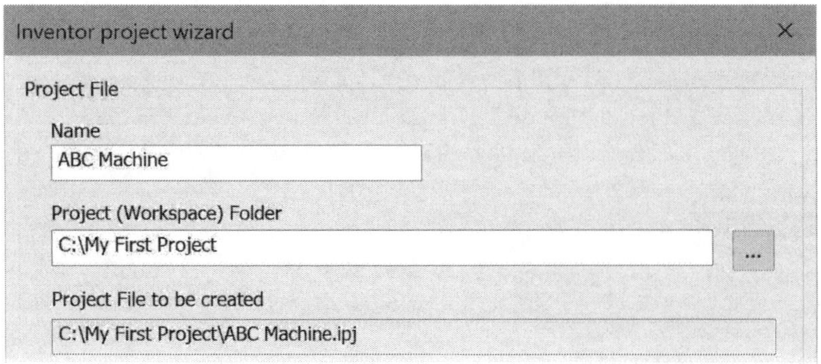

Figure 1.35

TIP: It is a good practice to place files such as parts, assemblies, and drawings in sub-folders below the top-level folder, where the project file is located.

7 Click Next to see a list of Libraries that are used in other projects' library search paths. The New Project list on the right side should be blank. For this project, no libraries will be added.

8 Click Finish and then click OK to create the new project path. The new project file is highlighted in the upper pane of the Project File Editor and should have a check mark to the left of its name indicating that it is the active project.

9 The options for the active project are listed in the lower pane. In the lower pane expand Workspace and notice that the Workspace search path is listed as a period "." as shown in Figure 1.36. The "." denotes that the workspace location is relative to the location where the project file is saved.

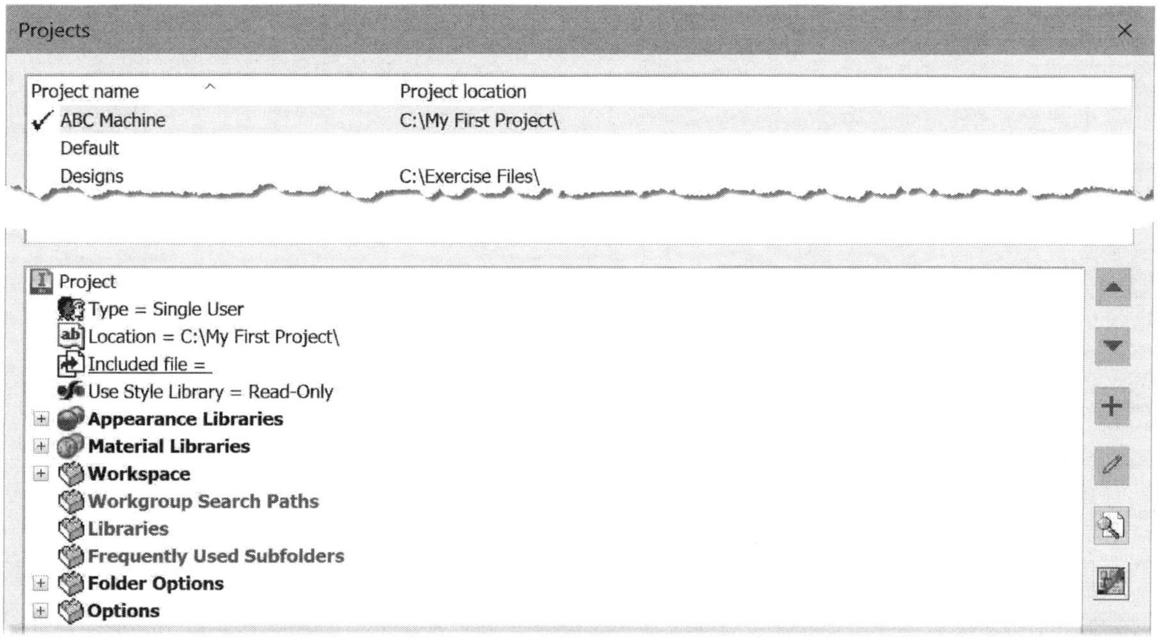

Figure 1.36

TIP: To make it easier to use subfolders below the workspace, add them by right-clicking on Frequently Used Subfolders, then click Add Paths from Directory as shown. The project file that you will use for the remaining exercises has the frequently used subfolders added.

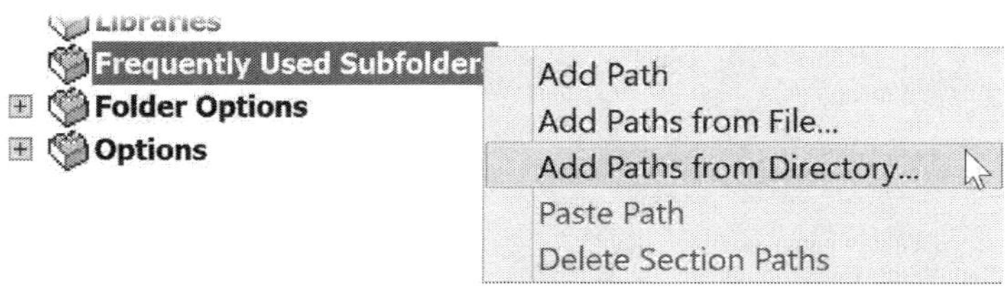

Figure 1.37

10 In the bottom of the Projects dialog box, click Done.

11 Click the Open icon on the Quick Access toolbar. Notice that the address bar is showing the My First Project folder and the project file is ABC Machine as shown in the following image.

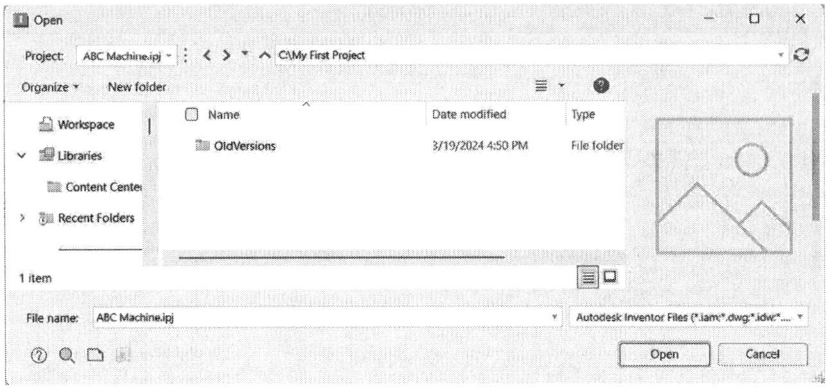

Figure 1.38

12 Close the Open dialog box by clicking the Cancel button.

13 You must change the active project file to successfully complete the remaining exercises. Close all files currently open in Autodesk Inventor.

14 Click File tab then click Manage > Projects.

15 To add an existing project file to the current list, click the Browse ... button at the bottom of the dialog box. Navigate to and select *C:\Inv Ess Plus\Inv Ess Plus.ipj*. This project should now be the current project. This project will be used for the remaining exercises. Expand the Frequently Used Subfolders section to verify that each chapter has its own subfolder.

16 Right-click on the project ABC Machine in the upper pane and select delete as shown in the following image on the left.

17 You can also see which project is current or make a project current in the Home screen as shown in the following image on the right.

TIP: You can only make a different project current if all Inventor files are closed.

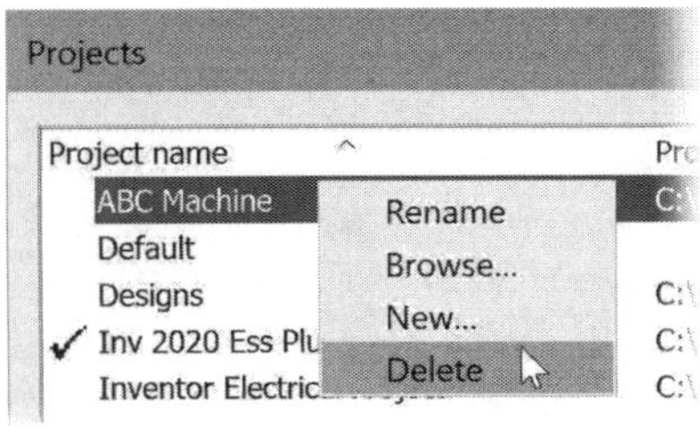

Figure 1.39

18 Close the Projects dialog box by clicking Done.

End of exercise.

CHANGING YOUR VIEWPOINT

When you work on a 2D sketch, the default view is looking straight down at the XY plane, which is often referred to as a plan view. When you work in 3D, it is helpful to view objects from a different viewpoint and to zoom in and out or pan the objects on the screen. The next section guides you through the most common methods for viewing objects from different perspectives and viewpoints. As you use these commands, the physical objects remain unmoved. Your perspective or viewpoint of the objects is what changes, not the location of the part on the coordinate system. If you are performing an operation during a viewing command, the operation resumes after the transition to the new view is completed.

Home (Isometric) View

Change to an isometric viewpoint by pressing the F6 key, by clicking the home icon above the ViewCube as shown in Figure 1.40 left, or by right-clicking in the graphics window and selecting Home View from the menu as shown in Figure 1.40 right. The view on the screen changes to a predetermined home view. You can redefine the home view with a ViewCube option. The ViewCube is explained later in this section.

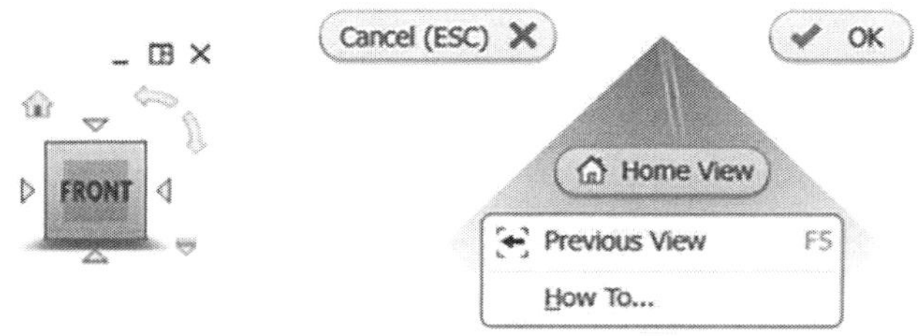

Figure 1.40

Navigation Bar

The Navigation bar, as shown in the following image, contains commands that will allow you to zoom, pan, and rotate the geometry on the screen. The default location for the Navigation bar is on the right side of the graphics window, but it can be repositioned. Commands are also available below the displayed commands. When commands are selected, they will become the top-level command. Descriptions of the viewing commands follow.

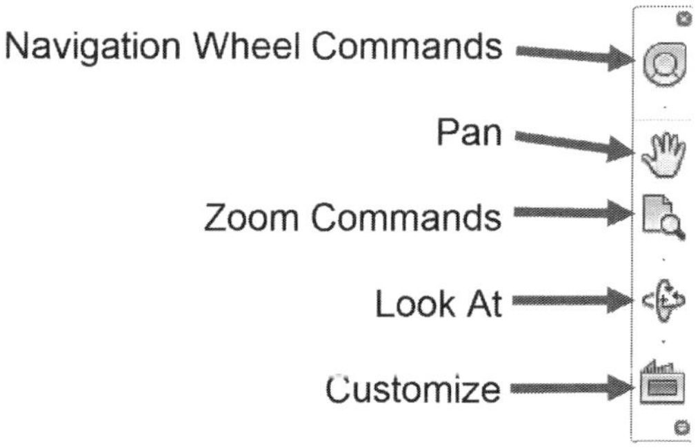

Figure 1.41

TIP: The navigation commands are also available on the View tab > Navigate panel.

Navigation Wheel

Click the Navigation Wheel icon to turn on the steering wheel, which contains eight viewing options.

Pan

Moves the view to a new location. Issue the Pan command, or press and hold the F2 key. Press and hold the left mouse button, and the screen moves in the same direction that the cursor moves. If you have a mouse with a wheel, hold down the wheel, and the screen moves in the same direction that the cursor is moved.

Zoom Commands

- **Zoom:** Zooms in or out from the parts. Issue the Zoom command, or press and hold the F3 key. Then, in the graphics window, press and hold the left mouse key. Move the mouse toward you to make the parts appear larger, and away from you to make the parts appear smaller. Another method is to roll the wheel on the mouse toward you, and the parts appear larger; roll the wheel away from you, and the parts appear smaller.
- **Zoom All:** Maximizes the screen with all parts that are in the current file. The screen transitions to the new view.
- **Zoom Window:** Zooms in on an area that is designated by two points. Issue the Zoom Window command and select the first point. With the mouse button depressed, move the cursor to the second point. A rectangle representing the window appears. When the correct window is displayed on the screen, release the mouse button, and the view transitions to it.
- **Zoom Selected:** Fills the screen with the maximum size of a selected face, faces, or a part. Either select the face or faces and then issue the Zoom Selected command, or press the END key, or launch the Zoom Selected command. Once the command is started, select the face or faces to which you wish to zoom.

Orbit Commands

- **Free Orbit:** Rotates your viewpoint dynamically. Issue the Free Orbit command; an Orbit symbol which is a circular image with lines at the quadrants and a center point appears. To rotate your viewpoint, click a point inside the circle, and keep the mouse button pressed as you move the cursor. Your viewpoint rotates in the direction of the cursor movement. When you release the mouse button, the viewpoint stops rotating. To accept the view orientation, press the ESC key or right-click and select Done from the menu.

  Click the outside of the circle to rotate the viewpoint about the center of the circle. To rotate the viewpoint about the vertical axis, click one of the horizontal lines on the circle and, with the mouse button pressed, move the cursor sideways. To rotate the viewpoint about the horizontal axis, click one of the vertical lines on the circle and, with the mouse button pressed, move the cursor upward or downward. The Constrained Orbit command is available below the Orbit command.

  - A shortcut to start the Free Orbit command is to press and hold down the F4 key and a circular image appears with lines at the quadrants and center. With the F4 key depressed, rotate the viewpoint. When you finish rotating the viewpoint, release the F4 key. If you are performing an operation while the F4 key is pressed, that operation will resume after you release the F4 key.
  - Another option to quickly rotate your viewpoint is to hold down the Shift key and press down the middle mouse button or wheel on the mouse. Once the rotate glyph appears on the screen, you can release the Shift key. While holding down the wheel, move the mouse, and your viewpoint will rotate. With this option, no other options are available. Release the wheel to complete the operation.
- **Constrained Orbit:** Use the Constrained Orbit command to rotate the model about the horizontal screen axes like a turntable. Click one of the horizontal lines on the circle and, with the mouse button pressed, move the cursor sideways. The model is rotated about the model space center point set in the Navigation Wheel CENTER command.

Look At

Changes your viewpoint so that you are looking perpendicular to a plane or looking horizontal to an edge. Issue the Look At command or press the PAGE UP key, then select a plane or edge. The Look At command can also be issued by selecting a plane or edge, right-clicking while the cursor is in the graphics window, and selecting Look At from the menu.

Customize the Navigation Bar

To modify the Navigation bar, click the down arrow on the bottom of the Navigation Toolbar as shown in the following image. Click a command to add or remove it from the Navigation bar. You can also reposition the toolbar by clicking the options under Docking positions.

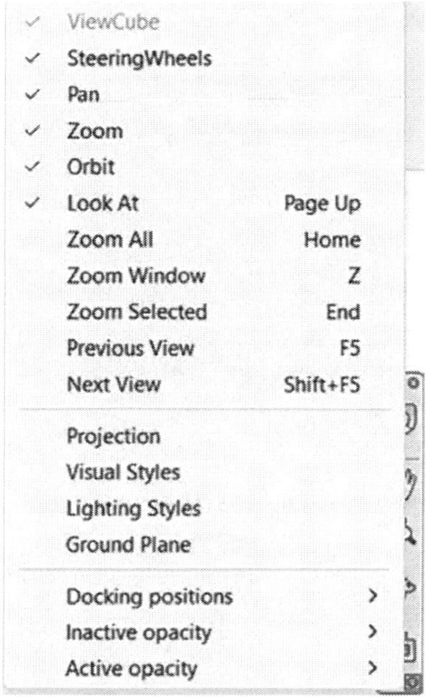

Figure 1.42

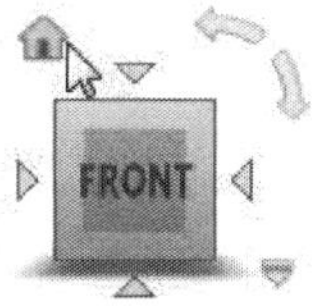

Figure 1.43

ViewCube

The ViewCube allows you to quickly change the viewpoint of the screen. With the ViewCube turned on, move the cursor over the ViewCube. Move the cursor over the home icon in the View-Cube to return to the default home (isometric) view as shown in the following image. Change the viewpoint by using one of the following techniques.

Isometric

Change to a different isometric view by clicking a corner on the ViewCube (Figure 1.44 A.)

Face

Change the viewpoint so it looks directly at a plane by clicking a plane on the ViewCube (Figure 1.44 B.)

Rotate in 90 Degree Increments

When looking at the plane of a ViewCube, you can rotate the view 90 degrees by clicking one of the arcs with arrows or one of the four inside facing triangles (Figure 1.44 C.)

Edge

You can also orient the viewpoint to an edge. Click an edge on the ViewCube (Figure 1.44 D.)

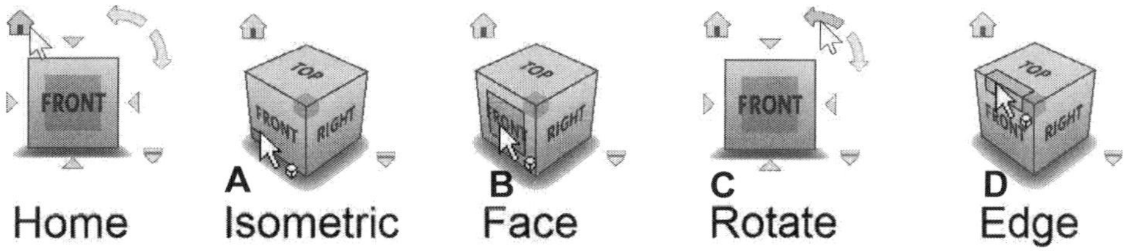

Figure 1.44

Dynamic Rotate via ViewCube

To dynamically rotate your viewpoint, click on the ViewCube and keep the mouse button pressed as you move the cursor. Your viewpoint rotates in the direction of the cursor movement as shown in Figure 1.45. When you release the mouse button, the viewpoint stops rotating. To smooth the rotation, right-click on the ViewCube and click Options and uncheck Snap to closest view.

Figure 1.45

Dynamically Rotate Viewpoint Shortcuts

While working, press and hold down the F4 key. The circular image appears with lines at the quadrants and center. With the F4 key still depressed, rotate the viewpoint. When you finish rotating the viewpoint, release the F4 key. If you are performing an operation while the F4 key is pressed, that operation will resume after you release the F4 key.

Another option to quickly rotate your viewpoint is to hold down the Shift key and press the middle mouse button (or mouse wheel). Once the rotate glyph appears on the screen, you can release the Shift key and while holding down the wheel, move the mouse, and your viewpoint will rotate. With this option, no other options are available. The model(s) are rotated about the center of the model(s). Release the wheel to complete the operation.

EXERCISE 1-3: VIEWING A MODEL

In this exercise, you use Navigation commands that make it easier to work on your designs.

1 Open *C:\Inv Ess Plus\Chapter 01\ESS_E01_03.ipt*. Click the Chapter 01 subfolder from the Frequently Used Subfolders area on the left side of the dialog box, and then double-click on the file name as shown in the following image.

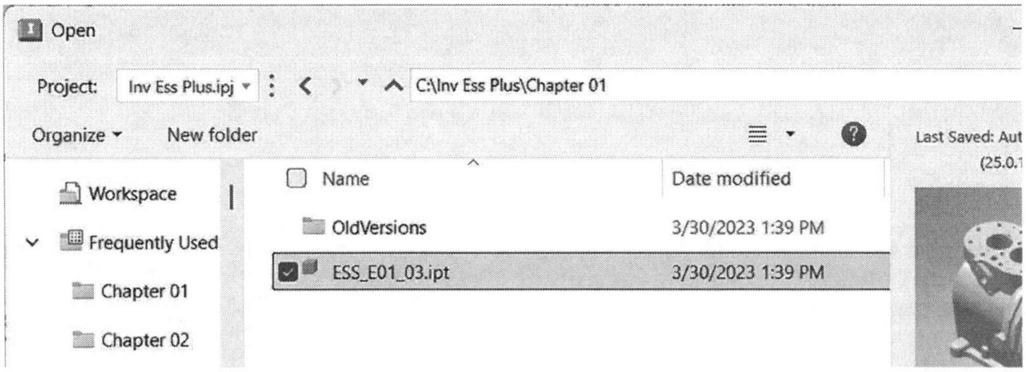

Figure 1.46

2 Move the cursor over the lower left foot of the motor housing and spin the mouse wheel toward you. Inventor will zoom into the location of the cursor. This will fill the screen as shown in Figure 1.47.

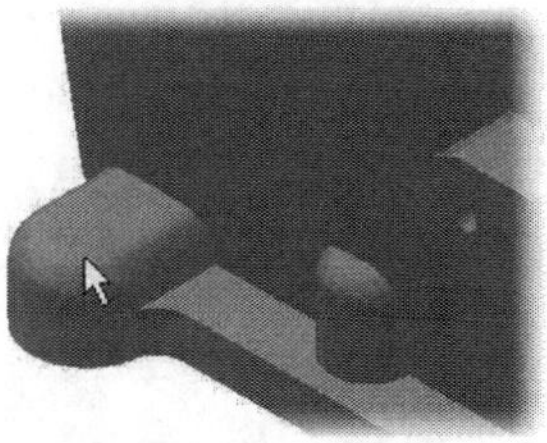

Figure 1.47

3 Press F5 a few times to change the viewpoint to the previous views.

4 Press the F6 key to change the viewpoint to the home view.

5 Next rotate the viewpoint; from the Navigation bar, click the Free Orbit command. The Orbit symbol appears in the graphics screen as shown in Figure 1.48.

Figure 1.48

6 Move the cursor inside the circle of the Orbit symbol; notice how the cursor (glyph) changes as the cursor moves inside the orbit's circle.

TIP: While the Free Orbit command is active, you can re-center the point in which the model rotates by clicking inside or near the Orbit symbol (the circle).

7 Click and drag the cursor to rotate the model.

8 To return to the Home View, press F6; the Free Orbit command will still be active.

9 Place the cursor on one of the horizontal handles of the Orbit symbol, noting the cursor display.

10 Drag the cursor right or left to rotate the viewpoint about the Y axis. As you rotate the viewpoint, you will see the bottom of the assembly.

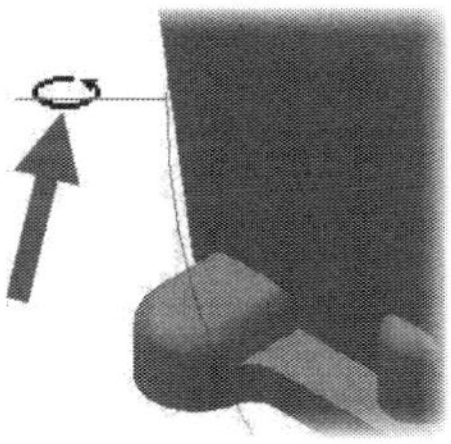

Figure 1.49

11 Cancel the command by pressing the ESC key.

12 Next you use the same Free Orbit command but use the shortcut key, press and hold down the F4 key, move the cursor inside the orbit's circle and click and hold down the left mouse button, and move the mouse to rotate the viewpoint. As soon as you release the F4 key the command will stop.

13 Press and hold down the Shift key and click and hold down the wheel on the mouse. Move the mouse to rotate the viewpoint.

14 Return to the home view by clicking the Home symbol above the ViewCube.

15 To change the viewpoint to predetermined isometric views, or directly at a plane, use the ViewCube. Click the Top-Front left corner of the ViewCube as shown on the left of the following image. Continue clicking the other corners of the ViewCube as shown in the middle two images.

16 Change to the front view by clicking the Front plane on the ViewCube as shown on the right of the following image.

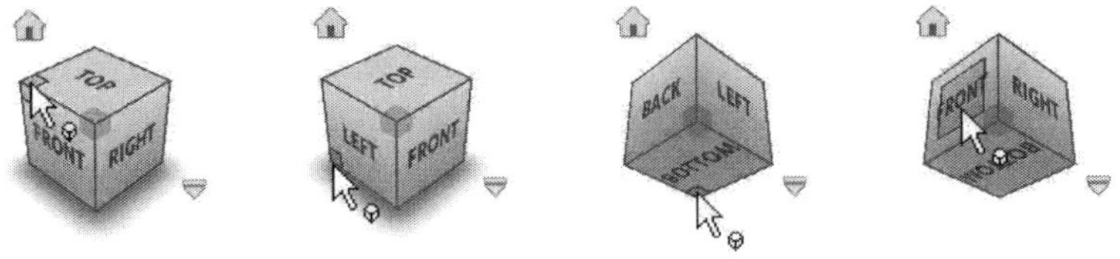

Figure 1.50

17 Rotate the view 90 degrees by clicking the arc with an arrow in the ViewCube as shown on the left of the following image.

18 To look at the Top view that is also rotated by 90 degrees, click the left arrow as shown on the right of the following image.

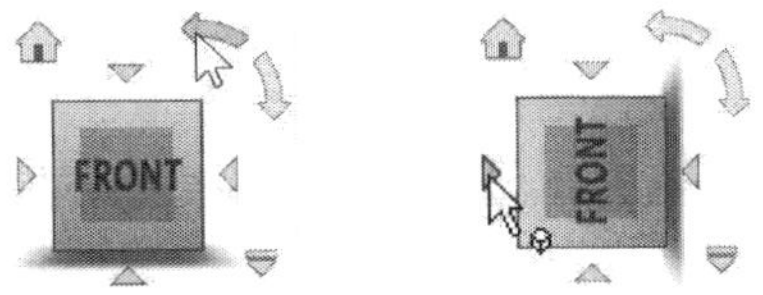

Figure 1.51

19 Practice changing your viewpoint by clicking the corners, faces, and edges of the ViewCube.

20 Click the Look At command from the Navigation bar as shown in the following image on the left. Select the planar face on the part as shown in the following middle image. The viewpoint will change so you are looking perpendicular to the selected plane as shown in the following image on the right.

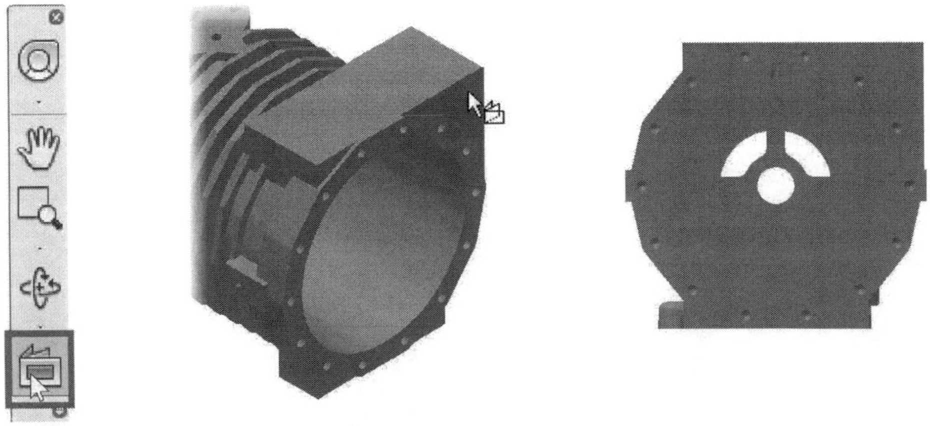

Figure 1.52

21 Continue to practice using the navigation commands.

22 Close the file. Do not save changes to the model.

End of exercise.

CHECKING YOUR SKILLS

Use these questions to test your knowledge of the material covered in this chapter.

1. Explain the reasons why a project file is used.
2. True__ False__ Only one project can be active at any time.
3. True__ False__ For files to be found, subfolders need to be added to the project file in order.
4. True__ False__ Autodesk Inventor stores the part, assembly information, and related drawing views in the same file.
5. True__ False__ Press and hold down the F4 key to dynamically rotate the viewpoint.
6. True__ False__ The Save Copy As command saves the active document with a new name, and then makes it current.
7. List four ways to access the Help system.
8. Explain how to change the location of the Ribbon.
9. True__ False__ Commands can be added to the Quick Access toolbar.
10. True__ False__ The ViewCube is used to rotate the active part or assembly file about the X axis.
11. True__ False__ Marking menus are used to add annotations to a part file.
12. True__ False__ The settings in the Application Options are global and affect all open and new Inventor files.
13. List three methods to repeat the last command.
14. Explain how to start a command from the marking menu without displaying the marking menu first.
15. True__ False__ By default, Inventor will automatically save the current file every 10 minutes.

TIP: Use the Help system to find the Keyboard Shortcut Reference page as shown below. Use shortcut keys to save time as you work.

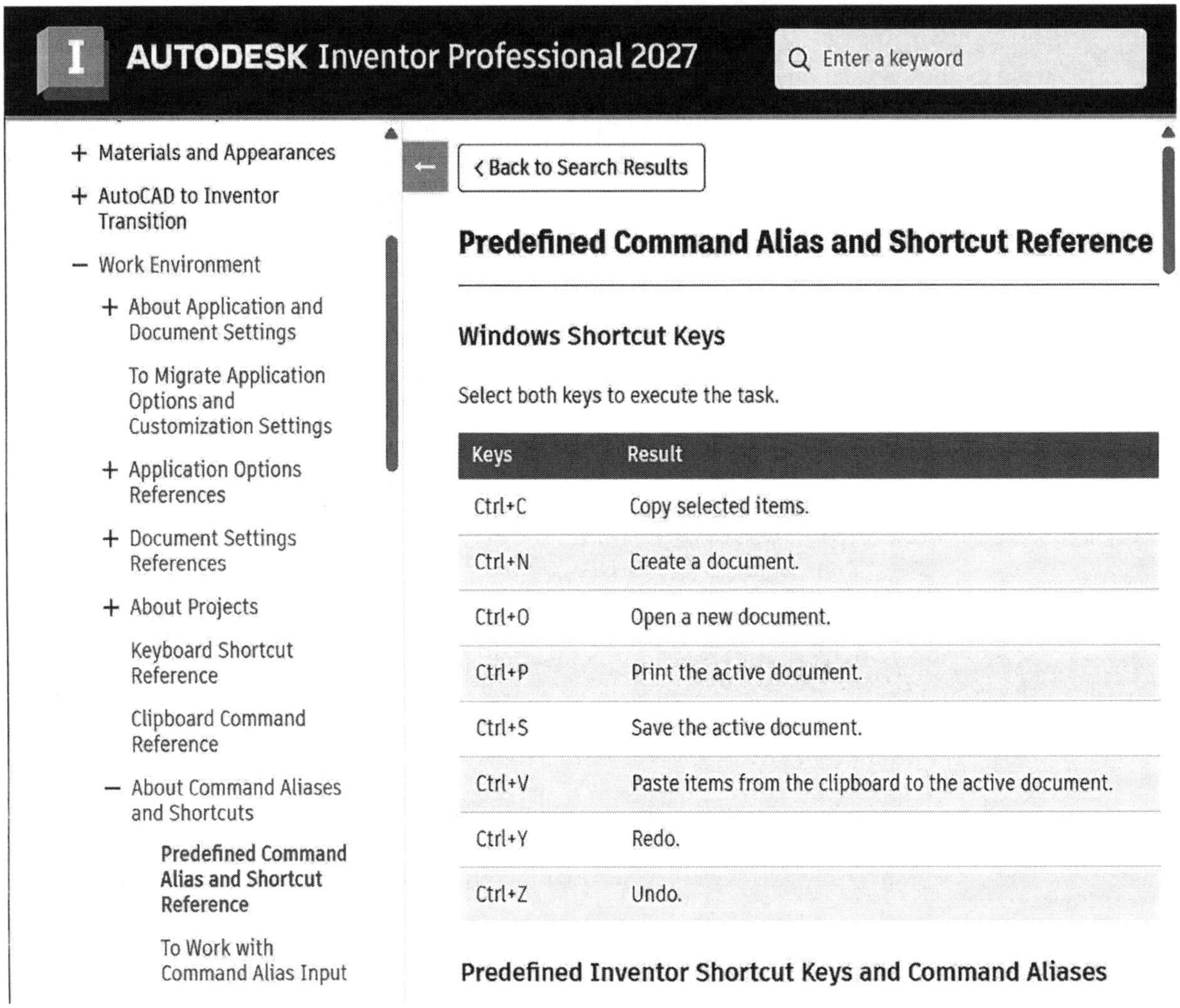

2 Sketching, Constraining, and Dimensioning

INTRODUCTION

Most 3D parts in Autodesk Inventor start from a 2D sketch. This chapter first provides a look at the application options for creating a part file and sketching. It then covers the three steps in creating a 2D parametric sketch: sketching a rough 2D outline of a part, applying geometric constraints, and then adding parametric dimensions. Lastly, you learn how to use 2D AutoCAD data in a sketch.

OBJECTIVES

After completing this chapter, you will be able to do the following:

- Change the part and sketch Application Options to meet your needs
- Sketch an outline of a part
- Create geometric constraints to a sketch to control design intent
- Use construction geometry to help constrain a sketch
- Dimension a sketch
- Change a dimension's value in a sketch
- Insert AutoCAD DWG data into a part's sketch

PART AND SKETCH APPLICATION OPTIONS

Before you start a new part, examine the part and sketch options in Autodesk Inventor that will affect how the part file will be created and how the sketching environment will look and act. While learning Autodesk Inventor, refer to these option settings to determine which ones work best for you—there are no right or wrong settings.

Part Options

You can customize Inventor Part options to your preferences. Click the File tab > Options button, and click on the Part tab, as shown in the following image. Descriptions of a couple of the most common Part options follow. For more information about the Application Options consult the help system. These settings are global—they will affect all active and new Inventor documents.

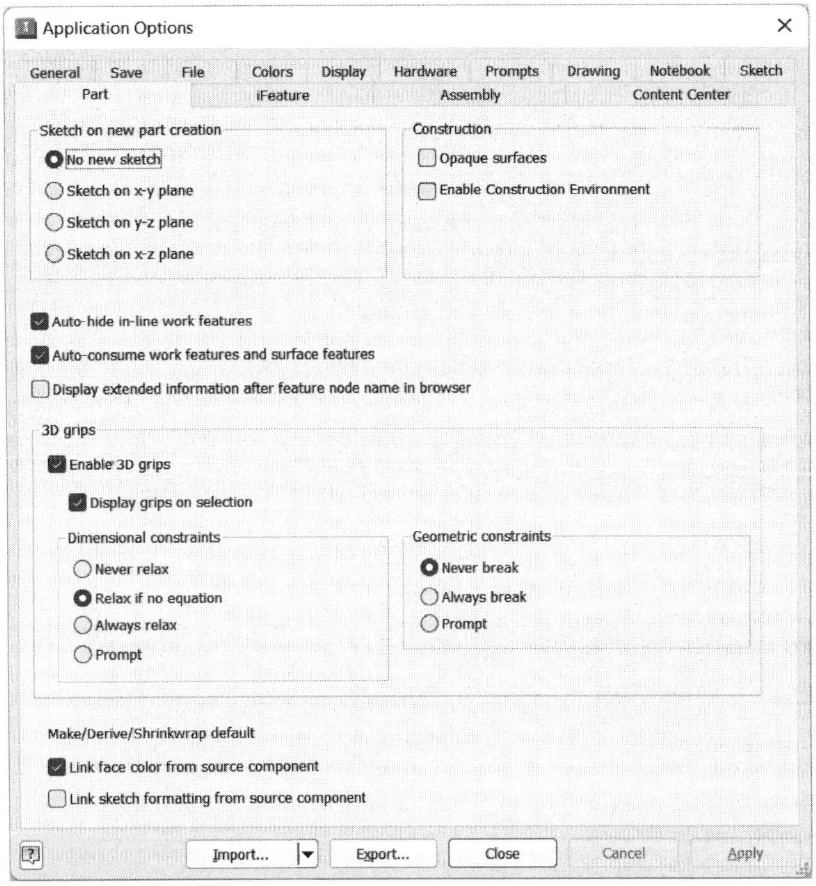

Figure 2.1

A common option that you may want to change is the first option: Sketch on New Part Creation. This option controls if and how a sketch is created when a part file is created.

No new sketch

When checked, Inventor does not set a sketch plane when you create a new part (this is the default setting).

Sketch on x-y plane

When checked, Inventor sets the x-y plane as the current sketch plane when you create a new part.

Sketch on y-z plane

When checked, Inventor sets the y-z plane as the current sketch plane when you create a new part.

Sketch on x-z plane

When checked, Inventor sets the x-z plane as the current sketch plane when you create a new part.

Sketch Options

Autodesk Inventor sketching options can be customized to your preferences. Click File tab > Options, and then click on the Sketch tab as shown in the following image. Descriptions of the most common Sketch options follow. For more information about the Application Options consult the help system. These settings are global and affect active and new Inventor documents.

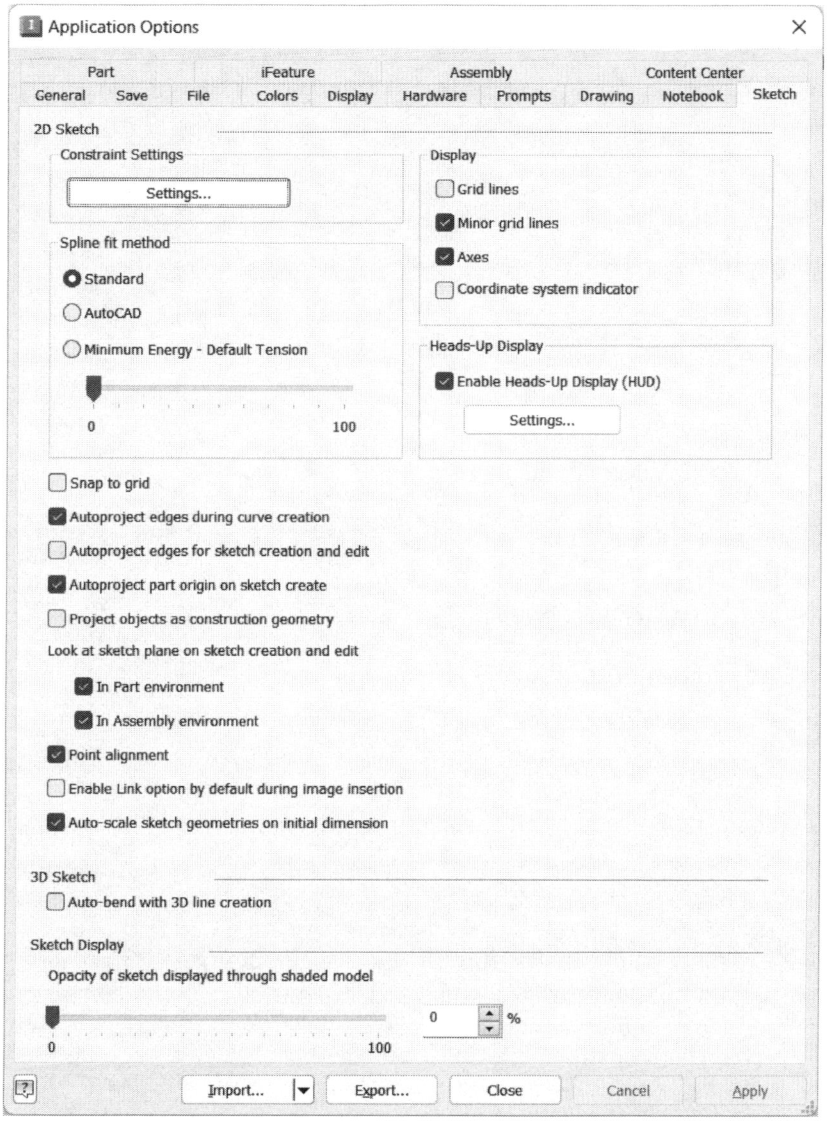

Figure 2.2

Following are descriptions of the common settings that you may want to change.

Constraint Settings

Click the Settings button to control how sketch constraints and dimensions behave.

Display

Grid lines

Toggles both minor and major grid lines on the screen on and off. To set the grid distance, open a part, or other file and click the Tools tab > Options panel > Document Settings command, and on the Sketch tab of the Document Settings dialog box, change the Snap Spacing and Grid Display.

Minor grid lines

Toggles the minor grid lines displayed on the screen on and off.

Axes

Toggles the lines that represent the X and Y-axis of the current sketch on and off.

Coordinate system indicator

Toggles the icon on and off that represents the X-, Y-, and Z-axes at the 0, 0, 0 coordinates of the current sketch.

Snap to Grid

When checked, endpoints of sketched objects snap to the intersections of the grid as the cursor moves over them.

Autoproject edges during curve creation

When checked and while sketching, place the cursor over an object and it will be projected onto the current sketch. You can also toggle Autoproject on and off while sketching by right-clicking and selecting Autoproject from the menu.

Autoproject edges for sketch creation and edit

When checked, and when a new sketch is created or edited, all the edges that define the plane are automatically projected as reference geometry.

Project objects as construction geometry

When checked, automatically projects selected geometry as construction geometry.

Look at sketch plane on sketch creation and edit

When checked, automatically changes the view orientation to look directly at the new or active sketch.

Autoproject part origin on sketch create

When checked, the parts origin point will automatically be projected when a new sketch is created. It is recommended to keep this setting on.

Point alignment

When checked, automatically infers alignment (horizontal and vertical) between endpoints of newly created geometry. No sketch constraint is applied. If this option is not checked, points can still be inferred; this technique is covered later in this chapter in the Inferred Points section.

Sketch Display

Change the opacity of a sketch that is displayed when the model is shaded.

UNITS

Autodesk Inventor uses a default unit of measurement for every part and assembly file. The default unit is set from the template file from which you created the part or assembly file. When specifying numbers in dialog boxes with no unit, the default unit will be used.

You can change the default unit in the active part or assembly document by clicking the Tools tab > Options panel > Document Settings button and click the Units tab as shown in Figure 2.3. The unit system values change for all the existing values in that file.

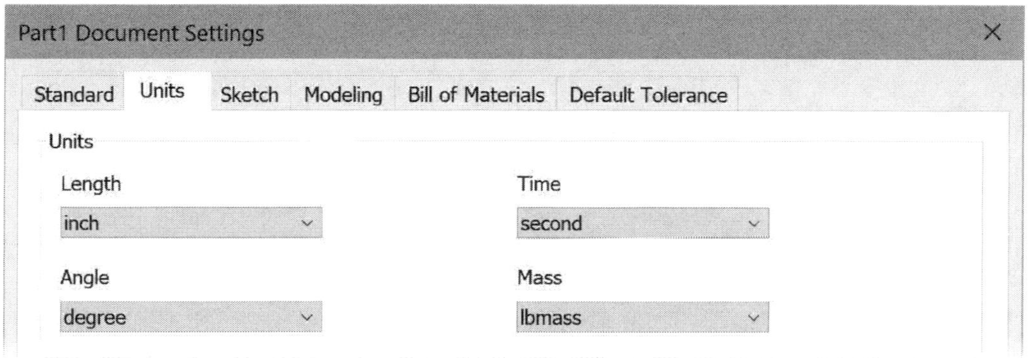

Figure 2.3

TIP: In a drawing file, the appearance of dimensions is controlled by dimension styles. Drawing settings are covered in Chapter 5.

You can override the default unit for any value by entering the desired unit. If you were working in a metric file whose unit is set to mm, for example, and you placed a 20 mm horizontal dimension as shown in the following image on the left, and you edited the dimension to 1 in (adding the unit) as shown in the middle image, the dimension would appear on the screen in the default units which would be 25.4 as shown in the right image.

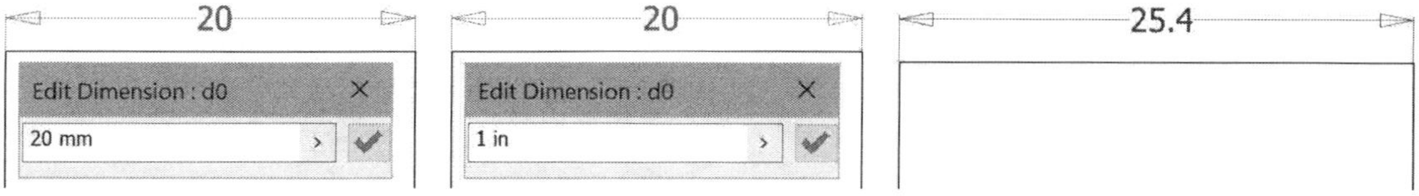

Figure 2.4

When you edit a dimension, the overridden unit appears in the Edit Dimension dialog box. For the previous example when the 25.4 mm dimension is edited, 1 in is displayed in the Edit Dimension dialog box as shown in Figure 2.5.

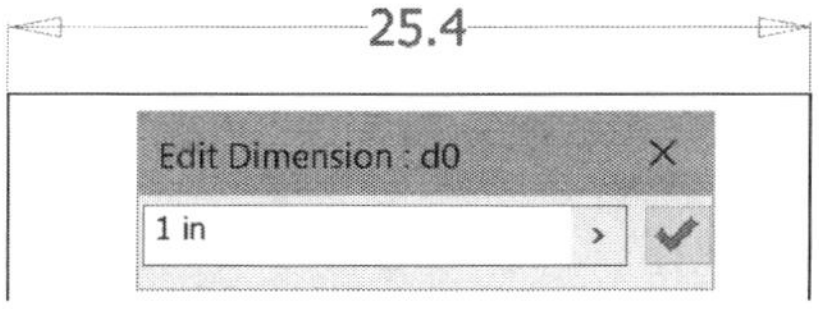

Figure 2.5

TEMPLATES

Each new file is created from a template. You can modify existing templates or add your own templates. As you work, make note of the changes that you make to each file. You then create a new template file or modify an existing file that contains all the changes and save that file to your template directory, which by default in Windows 10 is *C:\Users\Public\Public Documents\ Autodesk\Inventor 2027\Templates*. You can also create a new subdirectory under the templates folder, and place any Autodesk Inventor file in this new directory. After adding an Inventor file, the new tab will appear, and it will be available as a template.

You can use one of two methods to share template files among many users. You can modify the location of templates by clicking the File tab > Options button > File tab, and modifying the Templates location as shown in the following image. The Templates location will need to be modified for each user who needs access to templates that are not stored in the local location.

Figure 2.6

You can also change the unit of measurement (inches or millimeters) for the default part and assembly template files and set the default drawing standard (ANSI, DIN, ISO, etc.) for the default drawing template by clicking the Tools tab > Application Options > File > Configure Default Template button as shown in the previous image. Then make the changes in the Configure Default Template dialog box as shown in the image.

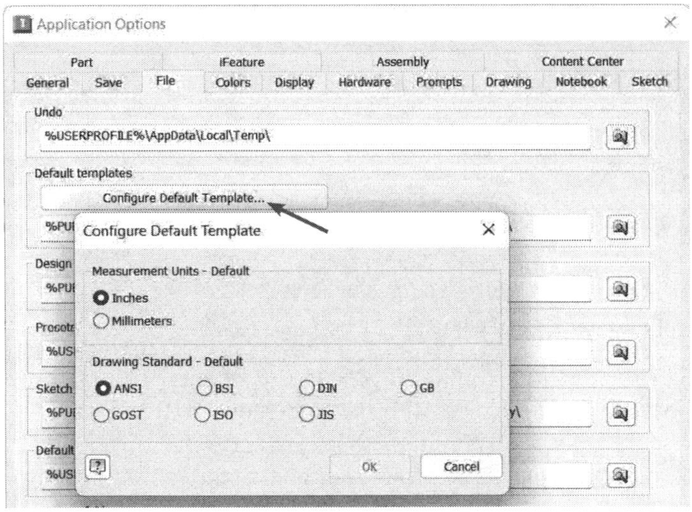

Figure 2.7

You can set the Templates location in each project file. This is useful if you need different templates for each project. While editing a project file, change the Templates location in the Folder Options area. The following image shows the default location in Windows 10. The Template location in the project file takes precedence over the Templates option in the Application Options, File tab.

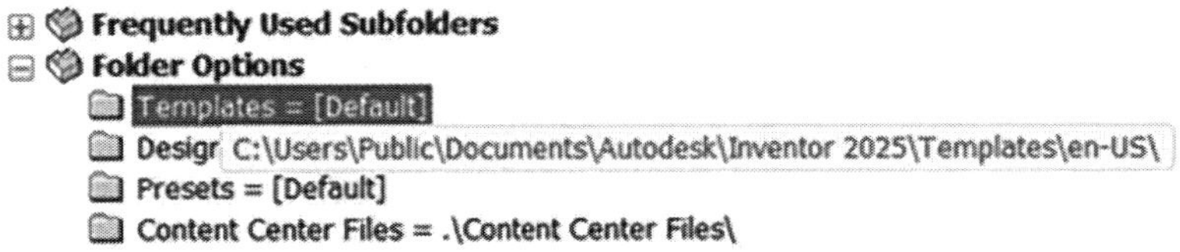

Figure 2.8

TIP: Template files have file extensions that are identical to other files of the same type, but they are in the template directory. Template files should not be used as production files.

CREATING A PART FILE

The first step in creating a part is to start or create a new part file in an assembly. You can use the following methods to create a new part file:

- In the Quick Access toolbar click the down arrow on the New icon, and click Part as shown in the following image on the left. This creates a new part file based on the default unit as was discussed in the previous Templates section.
- Click the down arrow by New on the Home page and choose part (bottom)
- From the File tab click New > Part as shown at right.

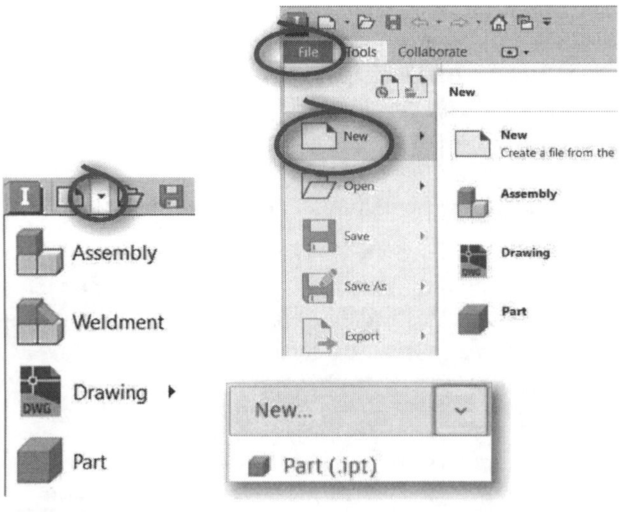

Figure 2.9

TIP: The default unit for the part and assembly templates and the standard for the drawing template is set in the Application Options dialog box > File tab > Configure Default Template.

Creating a Part File from a Specified Template

You can also create a part file from a template that is not the default location by clicking the New file command from one of these areas:

- Quick Access toolbar as shown in the following image on the left.
- File tab, as shown in the middle image.
- Or press CTRL + N.

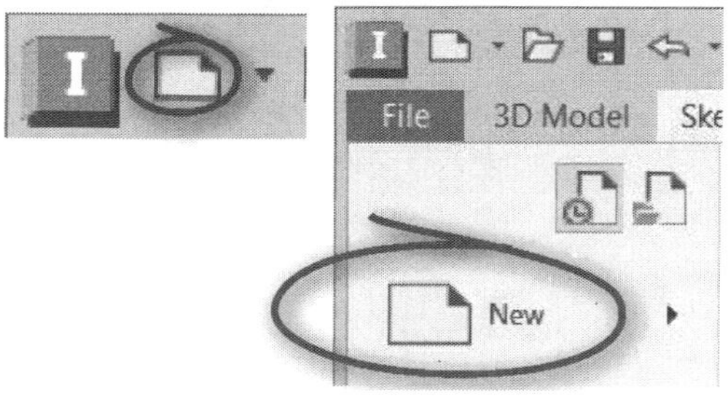

Figure 2.10

The Create New File dialog box appears. Then click the desired templates folder on the left side of the Create New File dialog box and then from the Part section on the right side of the dialog box click on the desired part template file, as shown in the following image.

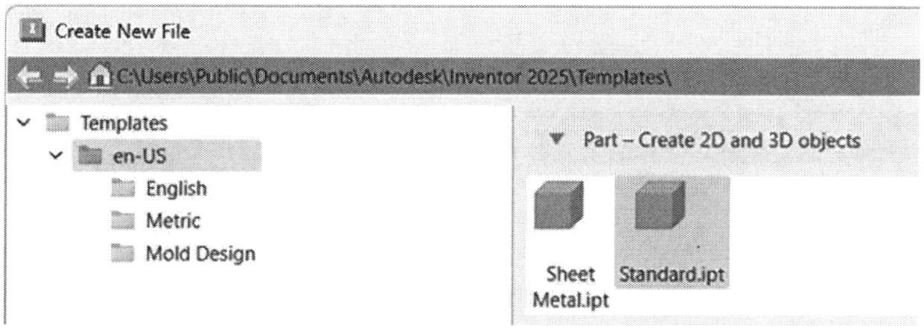

Figure 2.11

After starting a new part file using one of the previous methods, Autodesk Inventor's screen will change to reflect the part environment.

Sketches and Origin (Default) Planes

Before you start sketching, you select a plane on which to draw. A sketch is a plane on which 2D objects are sketched. You can use any planar part face or work plane to create a sketch. By default, when you create a new part file no sketch is created, and you will select an origin plane to sketch on. You can change the default plane on which you will create the sketch by selecting the File tab > Options and clicking on the Part tab. Select the sketch plane to which new parts should default.

Each time you create a new Autodesk Inventor part or assembly file, there are three planes (XY, YZ, and XZ), three axes (X, Y, and Z), and the center (origin) point at the intersection of the three planes and axes. You can use these default planes to create an active sketch. To see the planes, axes, or center point, expand the Origin entry in the browser by clicking on the left side of the text. You can then move the cursor over the names, and they will appear in the graphics window. The following image on the left and the middle image illustrate the default planes, axes, and center point with their visibility on. To leave the visibility of the planes or axes on, right-click in the browser while the cursor is over the name and click Visibility from the menu. When a plane is visible you can display the plane's label by moving the cursor over a plane in the browser or over a plane's edge in the graphics window as shown in the following image on the right.

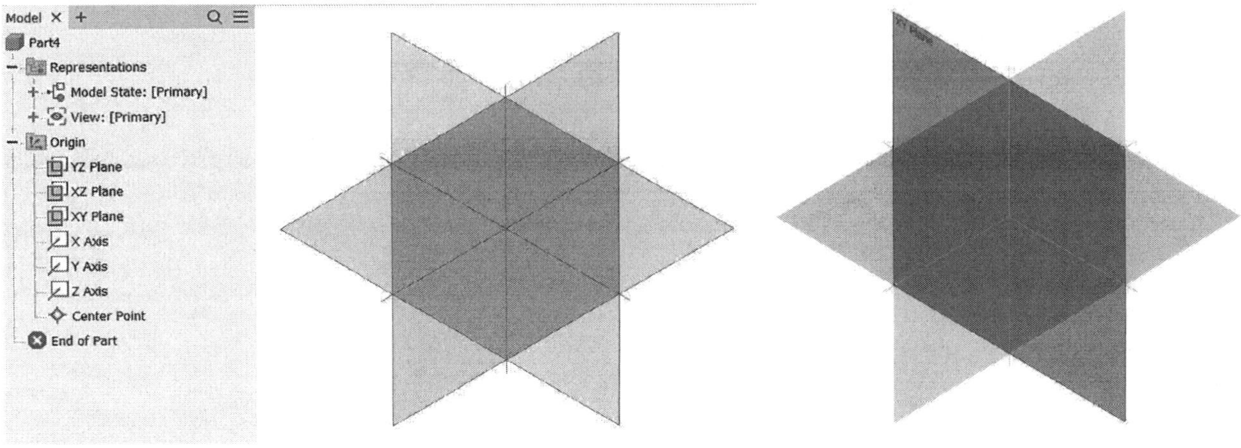

Figure 2.12

Origin 3D Indicator

When working in 3D, it is common to get your orientation turned around. By default, in the lower left corner of the graphics screen, there is an XYZ axis indicator that shows the default (world) coordinate system as shown in the following image on the left. The direction of these planes and axes cannot be changed. The arrows are color-coded:

- Red arrow = X axis
- Green arrow = Y axis
- Blue arrow = Z axis

In the Application Options dialog box > Display tab, you can turn the axis indicator and the axis labels on and off as shown in the following image on the right.

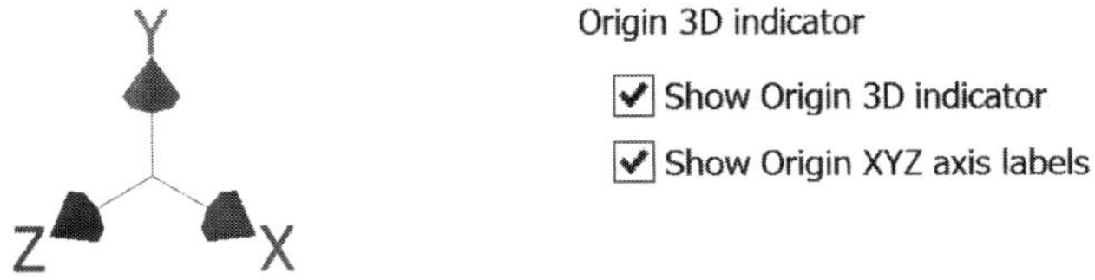

Figure 2.13

By default, Inventor will automatically project the origin point (0,0) when a new sketch is created in a part file. The origin point can be used to constrain a sketch to the 0, 0 point of the sketch. If desired, you can turn this option off by clicking the Tools tab > Application Options or from the File tab click Options > Sketch tab, and then uncheck Autoproject part origin on sketch create as displayed in the following image.

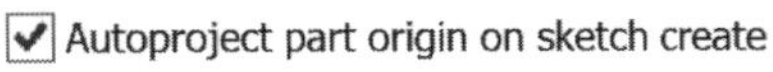

Figure 2.14

New Sketch

By default, when you create a new part file no sketch is active. You can define a plane from the origin folder to be the default by selecting a default plane from the File tab > Options > Part tab. Issue the 2D Sketch command to create a new sketch on a planar part face or a work plane or to activate a non-active sketch in the part. When you are in a part file that does not have a sketch defined and when you start the 2D Sketch command, the origin planes will be displayed in the graphics window, and you can select one of these planes to create the sketch on.

To create a new sketch or make an existing sketch active, use one of these methods:

- Click the 3D Model tab > Sketch panel > Start 2D Sketch as shown in the following image on the left or from the Sketch tab > Sketch panel > Start 2D Sketch. Then click a planar face, a work plane, or an existing sketch in the browser.
- Press the S key (a keyboard shortcut) and click a planar face of a part, a work plane, or an existing sketch in the browser.
- While not in the middle of an operation, right-click in the graphics window, and select New Sketch from the marking menu as shown in the middle image. Then click a planar face, a work plane, or an existing sketch in the browser.
- While not in the middle of an operation, click a planar face of a part, a work plane, or an existing sketch in the browser. Then right-click in the graphics window, and click 2D Sketch from the mini-toolbar. The following image on the right shows the mini-toolbar after selecting an origin plane.

TIP: You can start the command first, and then select a plane ***or*** you can select a plane first and then start the command.

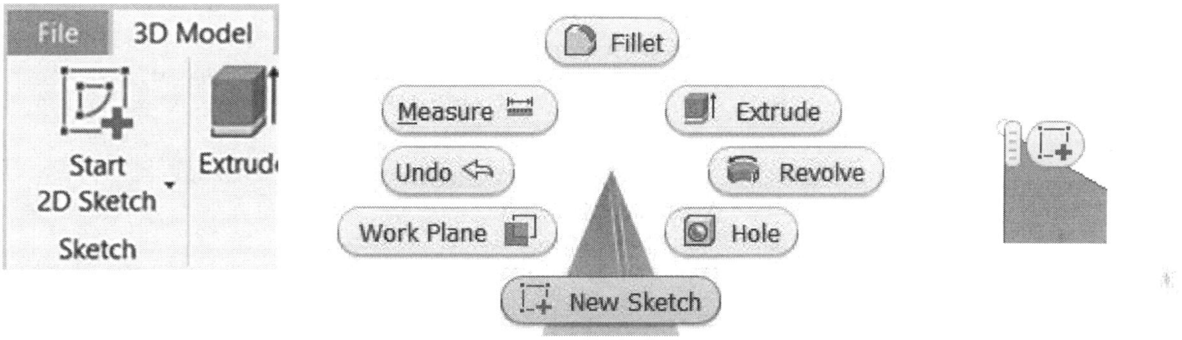

Figure 2.15

After creating a sketch, a Sketch entry will appear in the browser as shown in the following image, and a Sketch tab will appear in the ribbon. By default, after you have defined a sketch, the X and Y-axes will align automatically to this plane, and you can begin to sketch.

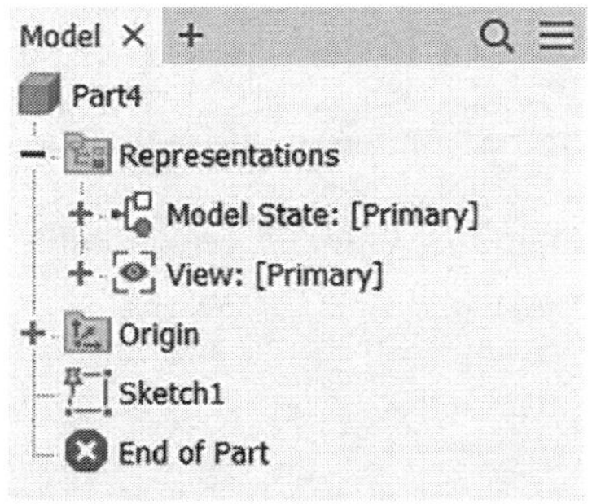

Figure 2.16

SKETCHING THE 2D OUTLINE OF THE PART

As stated at the beginning of this chapter, 3D parts usually start with a 2D sketch of the outline shape of the part. You can create a sketch with lines, arcs, circles, splines, or any combination of these elements. The next section will cover sketching strategies, commands, and techniques.

Sketching Overview

When deciding what outline to start with, analyze how the finished shape will look. Look for the 2-dimensional shape that best describes the part. When looking for this outline, try to look for a flat 2-dimensional shape that can be extruded or revolved to create a shape that other features can be added to, to create the finished part. It is usually easier to sketch 2-dimensional geometry than 3-dimensional geometry. As you gain modeling experience, you can reflect on how you created the model and think about other ways that you could have built it. There is usually more than one way to generate a given part.

When sketching, draw the geometry so that it is close to the desired shape and size— you do not need to be concerned about exact dimensional values. Even though Inventor allows islands in the sketch (closed objects that lie within another closed object) it is NOT recommended to sketch islands (when you extrude a sketch, island(s) may become voids in the solid). A better method is to place features, which make editing a part easier. For example, instead of sketching a circle inside a rectangle to represent a hole, extrude a rectangle and then place a hole feature.

The following guidelines will help you successfully generate sketches:

- Select a 2-dimensional outline that best represents the part. The 2D outline will be used to create the base feature. A base feature is the first feature. It is the feature other features will add material to or remove material from.
- Draw the geometry close to the finished size. If you want a 20-inch square, for example, do not draw a 200-inch square. Use dynamic input to define the size of the geometry. Dynamic input is covered in a later section in this chapter.
- Create the sketch proportional in size to the finished shape. When drawing the first object, verify its size in the lower-right corner of the status bar. Use this information as a guide.
- Draw the sketch geometry so it doesn't lie over other geometry, that is, a line on top of another line.
- Do not allow the sketch to have a gap; the geometry should start and end at a single point, just as the start and end points of a rectangle share the same point.
- Keep sketches simple. Leave out fillets and chamfers when possible. You can easily place them as features after making the sketch into a solid. The simpler the sketch, the fewer constraints and dimensions are required to constrain the model.

SKETCH COMMANDS

Before you start sketching the outline of the part, examine the 2D sketching commands that are available. After creating a sketch, the 2D sketch tab is current in the ribbon. The most frequently used commands will be explained throughout this chapter. Consult the help system for information about the remaining commands.

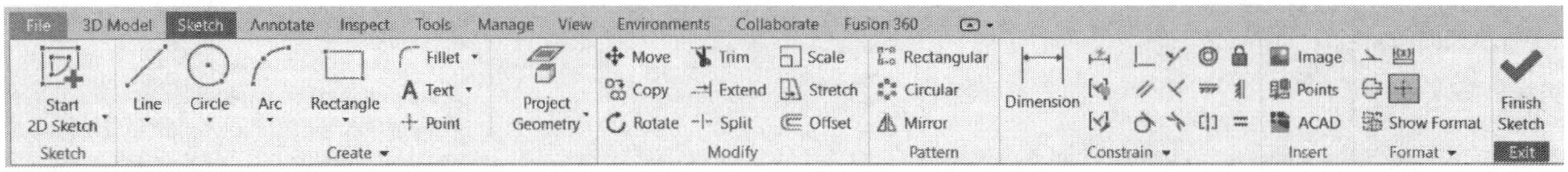

Figure 2.17

Using the Sketch Commands

After starting a new part, a sketch will automatically be active so that you can now use the sketch commands to draw the shape of the part. To start sketching, issue the sketch command that you need, click a point in the graphics window, and follow the prompt on the lower-left corner of the status bar. The sections that follow will introduce techniques that you can use to create a sketch.

Dynamic Input in the sketch environment makes a Heads-Up Display (HUD), which shows information near the cursor for many sketching commands that helps you keep your eyes on the screen. While using the Line, Circle, Arc, Rectangle, or Point commands, you can enter values in the input fields. You can toggle between the value input fields by pressing the Tab key. The following image shows examples of entering Cartesian coordinates and Polar coordinates.

TIP: If no data is entered in the input fields and you click in the graphics window to locate geometry, dimensions will NOT automatically be placed. You can manually place dimensions and constraints after the geometry is sketched.

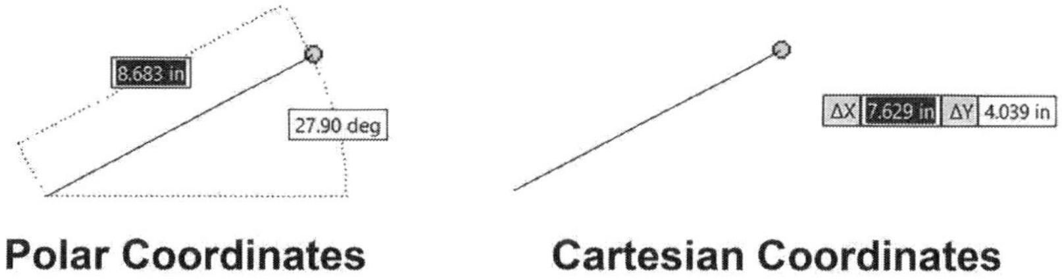

Figure 2.18

Dimension Input

When defining lengths and angles for a second point, the dimensional values change as you move the cursor. Press the Tab key to move to the next input field or click in another cell. After entering a value and pressing the Tab key, the value will be locked and a lock icon will appear to the right of the value as shown in the following image. After a dimension's value is locked, the parametric dimension will be created after clicking a point or pressing the Enter key. You can change the value in an input field by either clicking in the field or pressing the Tab key until the field is highlighted and then typing in a new value.

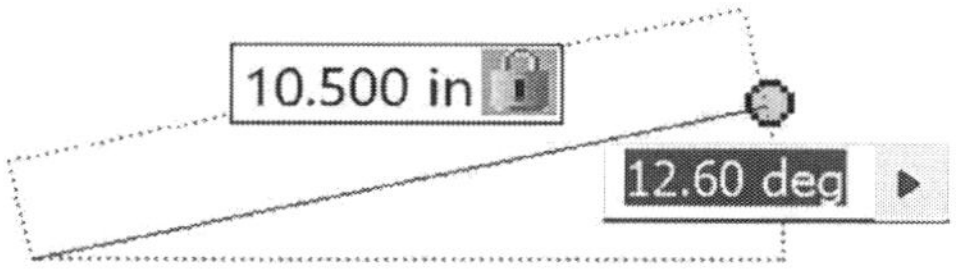

Figure 2.19

Line Command

The Line command is one of the most powerful commands that you will use to sketch. Not only can you draw lines with it, but you can also draw an arc from the endpoint of a line segment. To start sketching lines, click the Line command from the Sketch tab > Create panel as shown in the following image on the left, or right-click in a blank area in the graphics window and click Create Line from the marking menu as shown in the middle image, or press the L key on the keyboard. After starting the Line command, you will be prompted to click a first point, select a point in the graphics window, and then click a second point. The image on the right shows the line being created with the dynamic input as well as the horizontal constraint.

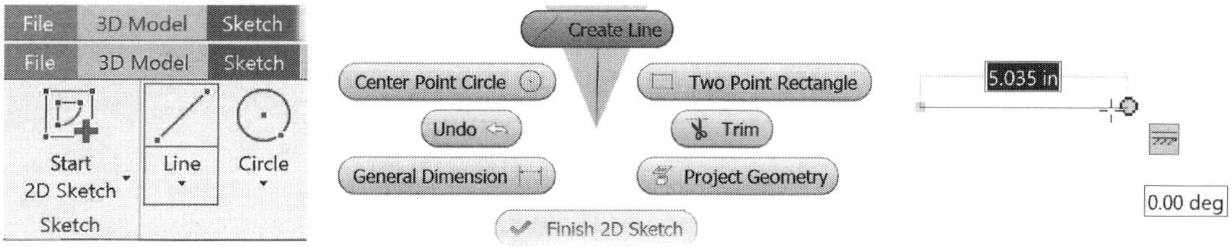

Figure 2.20

Drawing an Arc within the Line Command

You can continue drawing line segments, or you can sketch an arc from the endpoint. To make an arc from the line command, move the cursor over the endpoint of a line segment or arc until a small gray circle appears (Figure 2.21 left). Click the small circle and holding the left mouse button down, drag the cursor in the direction to define the arc.

Up to eight different arcs can be drawn, depending upon how you move the cursor. The arc will be tangent to the horizontal or vertical edges that are displayed from the selected endpoint.

Figure 2.21 right shows an arc that is tangent to the sketched line being drawn.

Figure 2.21

TIP: When sketching, look at the bottom-right of the status bar at the bottom of the screen to see the coordinates, length, and angle of the objects you are drawing.

The following image shows the status bar when a line is being drawn.

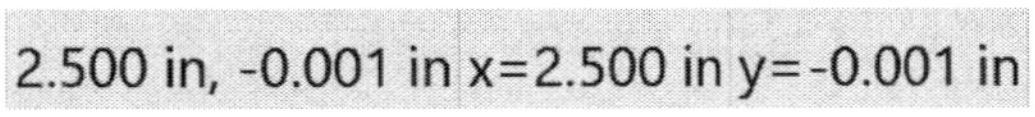

Figure 2.22

Object Tracking – Inferred Points

If the Point Alignment On option is checked from the Sketch tab of the Application Options, dashed lines will appear on the screen as you sketch. These dotted lines represent the endpoints; midpoints; and theoretical intersections of lines, arcs, and center points of arcs and circles that represent their horizontal, vertical, or perpendicular positions.

As the cursor gets close to these inferred points, it will snap to that location. If that is the point that you want, click that point; otherwise, continue to move the cursor until it reaches the desired location. When you select inferred points, no constraints (geometric rules such as horizontal, vertical, collinear, and so on) are applied from them. Using inferred points helps create more accurate sketches. The following image shows the inferred points from two midpoints that represent their horizontal and vertical position.

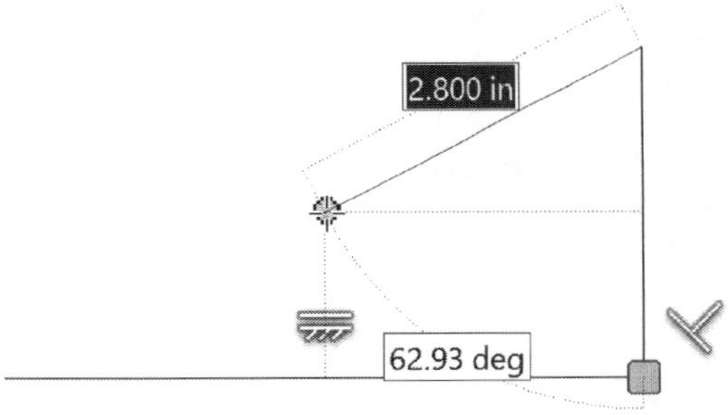

Figure 2.23

Automatic Constraints

As you sketch, a small constraint symbol appears representing geometric constraint(s) that will be applied to the object. If you do not want a constraint to be applied, hold down the CTRL key when you click to create the geometry.

Figure 2.24 shows a line being drawn from the arc, tangent to the arc, and parallel to the angled line, and the dynamic input is also displayed. The symbol appears near the object from which the constraint is coming. Constraints will be covered in the next section.

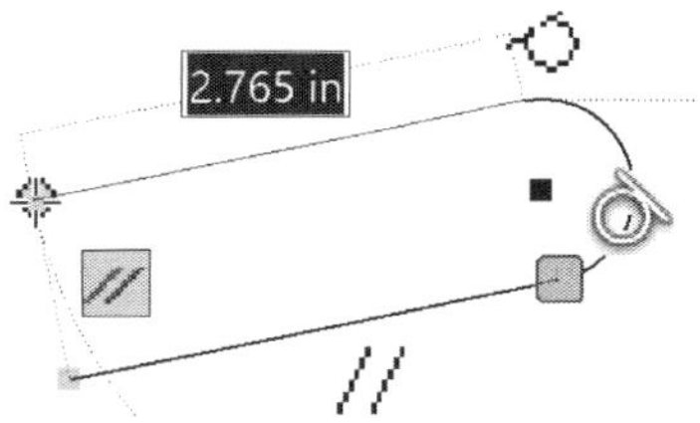

Figure 2.24

Scrubbing

As you sketch, you may prefer to apply a constraint different from the one that automatically appears on the screen. You may want a line to be perpendicular to a given line, for example, instead of being parallel to a different line. The technique to change the constraint is called scrubbing.

To place a different constraint while sketching, move the cursor so it touches (scrubs) the other object to which the constraint should be related. Move the cursor back to its original location, and the constraint symbol changes to reflect the new constraint. The same constraint symbol will also appear near the scrubbed object, representing that it is the object to which the constraint is matched. Continue sketching as normal.

The following image shows the top horizontal line being drawn with a parallel constraint that was scrubbed from the bottom horizontal line. Without scrubbing the bottom horizontal line, the applied constraint would have been perpendicular to the right vertical line.

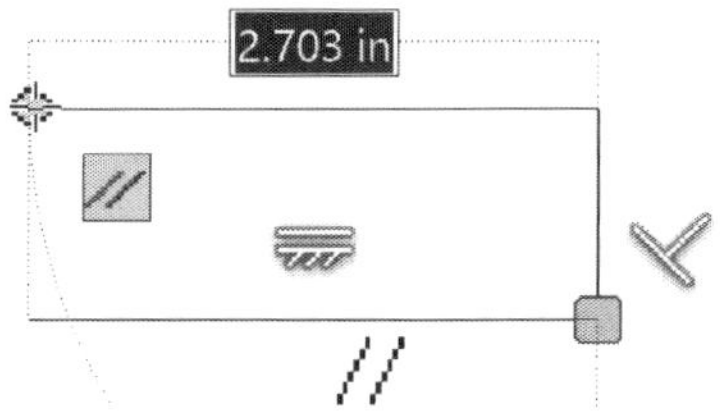

Figure 2.25

Deleting Objects

To delete objects first cancel the command that you are in by pressing the ESC key. Then select objects to delete, and either press the DELETE key or right-click and choose Delete as shown in Figure 2.26.

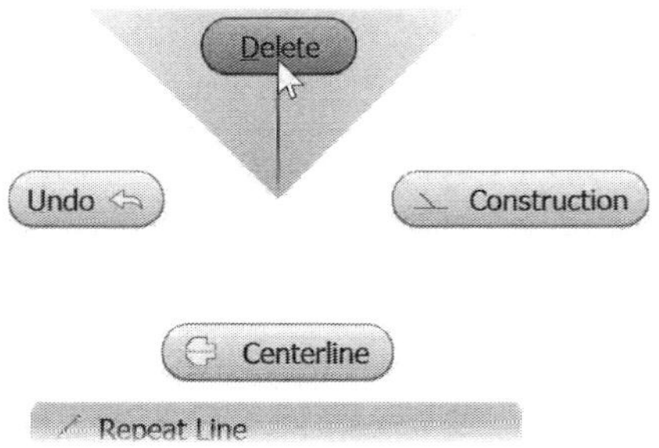

Figure 2.26

Common Sketch Commands

The following table lists common 2D sketch commands. Some commands are available by clicking the down arrow in the lower-right corner of the top command in the panel. Consult the help system for more information.

Command	Function
Center-point Circle	Creates a circle by clicking a center point for the circle and then a point on the circumference of the circle.
Tangent Circle	Creates a circle that will be tangent to three lines or edges by clicking the lines or edges.
Three-Point Arc	Creates an arc by clicking a start and endpoint and a point that will lie on the arc.
Tangent Arc	Creates an arc that is tangent to an existing line or arc by clicking the endpoint of a line or arc and then clicking a point for the other endpoint of the arc.
Center-Point Arc	Creates an arc by clicking a center point for the arc and then clicking a start and endpoint.
Two-Point Rectangle	Creates a rectangle by defining a point and then clicking another point to define the opposite side of the rectangle. The edges of the rectangle will be horizontal and vertical. If values were entered, dimensions will be placed on the rectangle.
Three-Point Rectangle	Creates a rectangle by clicking two points that will define an edge and then clicking a point to define the third corner. You can also type values to define the three points of the rectangle, and dimensions will be created that define the size of the rectangle.
Two-Point Center Rectangle	Creates a rectangle by defining a center point and another point to define the rectangle's size or type values for the center point and its X and Y values of the rectangle. The edges of the rectangle will be horizontal and vertical, and if values were entered, dimensions will be created.
Three-Point Center Rectangle	Creates a rectangle by defining a center point, a point to define the rectangle's starting point and its angle and another point size or type values for the center point, and size of the rectangle. The edges of the rectangle will be horizontal and vertical, and if values were entered, dimensions will be created.
Center to Center Slot	Creates a slot by defining the center-to-center distance, angle, and the diameter.
Overall Slot	Creates a slot by defining the overall distance, angle, and then the diameter.
Center Point Slot	Creates a slot by defining the center-to-center distance, angle, and the diameter.
Three Point Arc Slot	Creates an arc slot by defining a start point, end point, and an angle, a radius of the center of the slot and then the diameter of the slot.
Center Point Arc Slot	Creates an angled slot by defining a radius of the center of the slot and a starting angle, an ending angle and then the diameter of the slot.
Fillet	Creates a fillet between two nonparallel lines, two arcs, or a line and an arc at a specified radius. If you select two parallel lines, a fillet is created between them without specifying a radius. When the first fillet is created, a dimension will be created. If many fillets are placed in the same operation, you choose to either apply or not apply an equal constraint.
Chamfer	Creates a chamfer between lines. There are three options to create a chamfer: both sides equal distances, two defined distances, or a distance and an angle.
Polygon	Creates an inscribed or a circumscribed polygon with the number of faces that you specify. The polygon's shape is maintained as dimensions are added.
Mirror	Mirrors the selected objects about a centerline. A symmetry constraint will be applied to the mirrored objects.

Rectangular Pattern	Creates a rectangular array of a sketch with rows and columns that you specify.
Circular Pattern	Creates a circular array of a sketch with copies and spacing that you specify.
Offset	Creates a duplicate of the selected objects that are a given distance away. By default, an equal-distance constraint is applied to the offset objects.
Trim	Trims the selected object to the next object it finds. Click near the end of the object that you want trimmed. While using the Trim command, hold down the SHIFT key to extend objects. If desired, hold down the CTRL key to select boundary objects. While in the Trim command you can also hold down the left mouse button and move the cursor to dynamically trim geometry. While in the Dynamic mode you can hold down the Shift key to dynamically extend geometry.
Extend	Extends the selected object to the next object it finds. Click near the end of the object that you want extended. While using the Extend command, hold down the SHIFT key to trim objects. If desired, hold down the CTRL key to select boundary objects. While in the Extend command, you can also click and hold down the left mouse button and move the cursor to dynamically extend geometry. While in the Dynamic mode you can hold down the Shift key to dynamically trim geometry.

Selecting Objects

After sketching objects, you may need to move, rotate, or delete some or all the objects. To edit an object, it must be part of a selection set. There are multiple methods that you can use to place objects into a selection set.

- **CTRL or SHIFT Keys**. You can select objects individually by clicking on them. To manually select multiple individual objects, hold down the CTRL key or SHIFT key while clicking the objects. You can remove selected objects from a selection set by holding down the CTRL or SHIFT key and reselecting them. As you select objects, their color will change to show that they have been selected.
- **Window**. You can select multiple objects by defining a selection window. Not all commands allow you to use the selection window technique and only allow single selections. To define the window, click a starting point. With the left mouse button depressed, move the cursor to define the box. If you draw the selection window from left to right (solid lines), as shown in the following image on the left, only the objects that are fully enclosed in the window will be selected.
- **Crossing window**. If you draw the selection window from right to left (dashed lines), as shown in the following image on the right, a crossing window is used and all the objects that are fully enclosed in the selection window and the objects that are touched by the window will be selected.
- You can use a combination of the methods to create a selection set.

When you select an object, its color changes according to the color style that you are using. To remove all the objects from the selection set, click in a blank section of the graphics window.

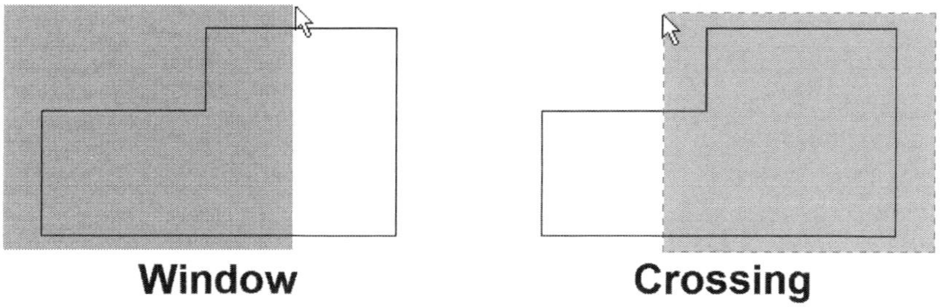

Figure 2.27

EXERCISE 2-1: CREATING A SKETCH WITH LINES

In this exercise, you create a new part file and 2D Sketch geometry using basic construction techniques. In this exercise, no dimensions will be created.

1. Click New from the Quick Access toolbar, click the English folder, and then double-click Standard (in).ipt or if inch is the default unit; from the left side of the Quick Access toolbar you can click the down arrow on the New icon and click Part.

2. Click Start 2D Sketch from the 3D Model tab > Sketch panel and then select the XY origin plane in the graphics window as shown in the following image.

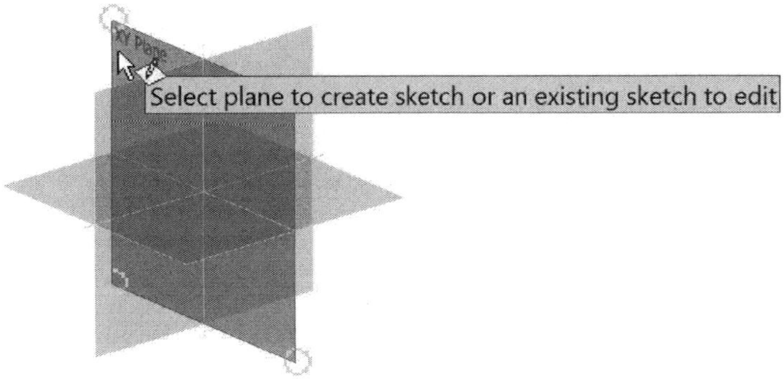

Figure 2.28

3. Start the Line command from the Sketch tab > Create panel.
4. Click on the origin point in the graphics window, move the cursor to the right about **4 inches**, and when the horizontal constraint symbol displays, click to specify a second point as shown in Figure 2.29. You may need to zoom back and pan the screen to see the entire line.

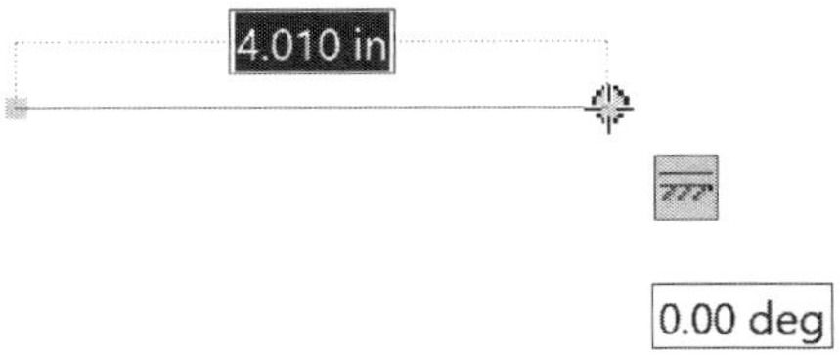

Figure 2.29

TIP: Symbols indicate the geometric constraints. In the image above, the symbol indicates that the line is horizontal. When you create the first entity in a sketch, make it close to final size.

5. Move the cursor up until the perpendicular constraint symbol displays beside the first line and then click to create a perpendicular line that is approximately **2 inches** as shown in Figure 2.30 left.

6. Move the cursor to the left and create a horizontal line perpendicular to the vertical line that is approximately **1 inch**. The perpendicular constraint symbol is shown in Figure 2.30 right.

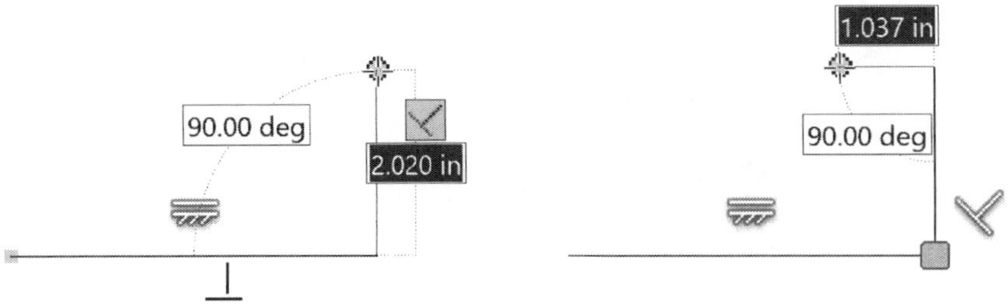

Figure 2.30

7 Move the cursor downward to create a line perpendicular to the top horizontal line and about **1 inch**.

8 Move the cursor left to create a line approximately **2 inches** long and perpendicular to the inside vertical line.

9 Move the cursor up and notice the perpendicular constraint symbol is displayed; to apply a parallel constraint instead, move (scrub) the cursor over the inside vertical line to create a relationship to it. Click when an *inferred line* (horizontal dotted line) appears from the top point as shown in Figure 2.31 left.

10 Move the cursor to the left until an inferred vertical line appears from the bottom-left point as shown in Figure 2.31 right and then click to locate the point.

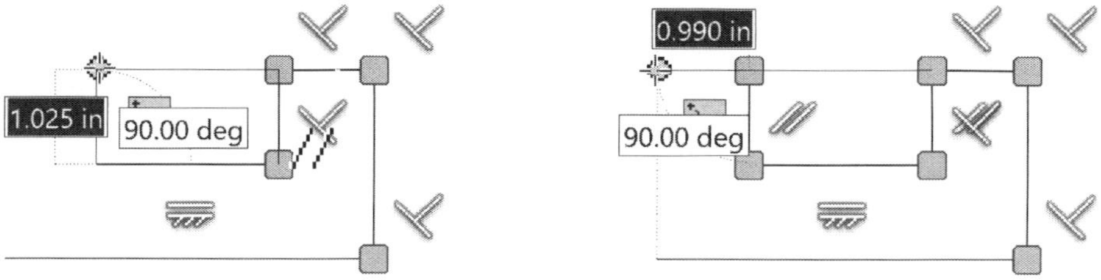

Figure 2.31

11 To close the profile right-click and choose Close from the menu.

12 Figure 2.32 left shows the sketch constraints and the image on the right without the constraints.

> **TIP:** To see the sketch constraints select the geometry using window or crossing window. Constraints will be covered in the next section in this chapter.

13 Right-click in the graphics screen, and choose Finish 2D Sketch.

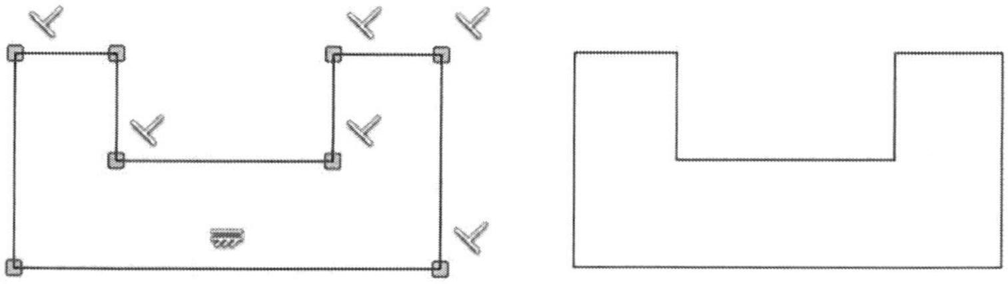

Figure 2.32

14 Close the file. Do not save changes.

End of exercise.

EXERCISE 2-2: CREATING A SKETCH WITH TANGENCIES

In this exercise, you create a new part file, and then you create a profile consisting of lines and tangent arcs.

1 Click the New command, and then double-click Standard (in).ipt; or if inch is the default unit, from the left side of the Quick Access toolbar you can click the down arrow of the New icon, and select Part.

2 Click the Start 2D Sketch command on the 3D Model tab > Sketch panel and then select the XY origin plane.

3 Start the Line command by right-clicking in a blank area in the graphics window and click Create Line from the marking menu.

4 Click on the projected origin point in the middle of the graphics window, and create a horizontal line to the right of the origin point and type **3** (inches will be assumed as the unit because the part file is based on the unit of inch) in the input field. Press the tab key and move the cursor until the horizontal constraint symbol appears and then click. If the second point of the line lies off the screen, roll the mouse wheel away from you to zoom out, hold down the mouse wheel, and drag to pan the view.

5 Create a perpendicular line, move the cursor up until the perpendicular constraint appears, type **1.5** in the input field as shown in the following image on the left, and then press enter.

6 In this step, you infer points, meaning that no sketch constraint is applied. Move the cursor to the intersection of the midpoints of the right-vertical line and bottom horizontal line. Dotted lines (inferred points) appear as shown in the image on the right, and then click to create the line. No dimension was created since a value was not entered.

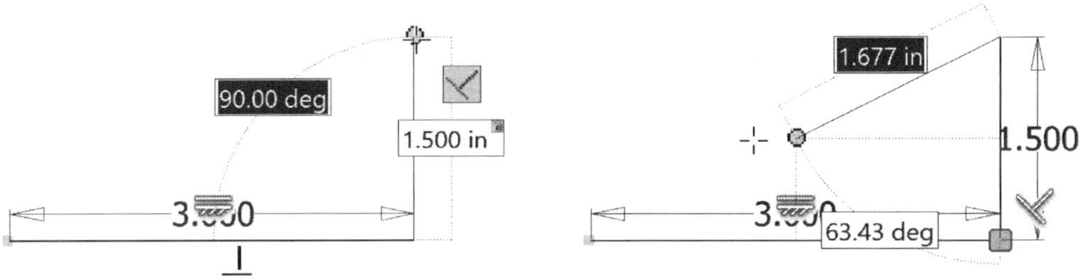

Figure 2.33

7 Next create a line parallel to the bottom line. If needed scrub the bottom line by moving the cursor over the bottom line (do NOT click), and then move the cursor up and to the left until the vertical inferred line and the constraints are displayed as shown in the following image on the left, and then click to create the line.

8 Next sketch an arc while in the line command. With the Line command active move the cursor over the left endpoint of the top horizontal line until the gray circle appears, click on the gray dot at the left end of the line, holding down the left mouse button drag the cursor to the left and then down to preview a tangent arc. Do not release the mouse button.

9 Move the cursor over the left endpoint of the first line segment until a coincident constraint (green circle) and the two tangent constraints at start and end points of the arc are displayed as shown in the following image on the right.

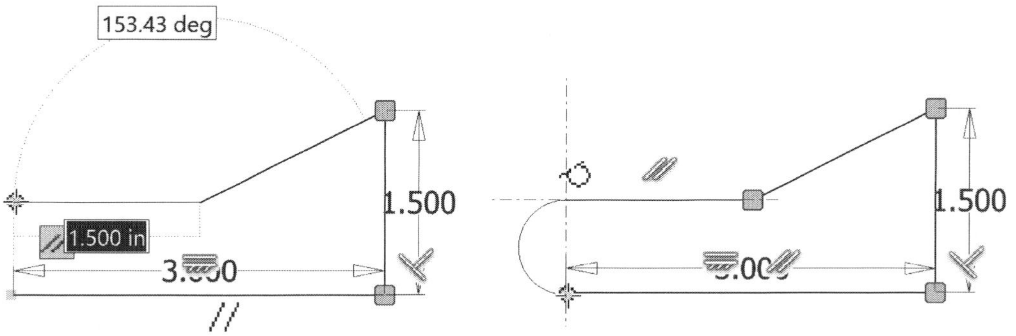

Figure 2.34

10 Release the mouse button to create the arc.

11 Right-click in the graphics window, and choose OK from the marking menu. Later in this chapter you will learn how to create dimensions.

12 Click Finish Sketch from the Sketch tab > Exit panel.

13 Close the file. Do not save changes.

End of exercise.

CONSTRAINING A SKETCH

After you draw a sketch, you may want to add geometric constraints to it to add design intent. Geometric constraints apply behavior to a specific object or create a relationship between two objects. An example of using a constraint is applying a vertical constraint to a line so that it will always be vertical. You can apply a parallel constraint between two lines to make them parallel to one another; then as you change the angle of one of the lines, the angle of the other line will change with it. You can apply a tangent constraint to a line and an arc or to two arcs.

When you add a constraint, the number of constraints or dimensions that are required to fully constrain the sketch will decrease. On the bottom-right corner of Autodesk Inventor, the number of constraints or dimensions will be displayed like what is shown in the following image. A fully constrained sketch is a sketch whose objects cannot move or stretch.

Figure 2.35

Constrain to the Origin Point

When sketching, it is recommended to constrain a point on the sketch to the origin point with a coincident constraint or dimension a point on the sketch to the origin point so it cannot move. You could apply a fix constraint instead of using the origin point, but it is not recommended. When a sketch is constrained to the origin point, Inventor will change the color of constrained objects. If the sketch is not constrained to the origin point, objects are free to move in the sketch and the color of the objects will not change.

TIP: Autodesk Inventor does not force you to fully constrain a sketch. However, it is recommended that you fully constrain a sketch, as this will allow you to better predict how future changes will affect the sketch and part.

Constraint Types

Autodesk Inventor has 12 geometric constraints that you can apply to sketch geometry. The following image shows the constraint types that can be applied from the Sketch tab > Constrain panel.

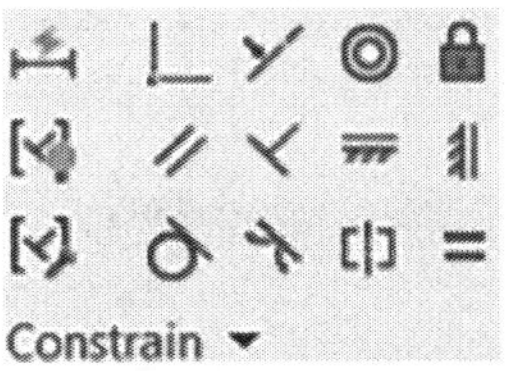

Figure 2.36

The following chart describes the geometric constraints.

Icon	Constraint	Function
	Coincident	A point is constrained to lie on another point or curve (line, arc, etc.).
	Collinear	Two selected lines will line up along a single line; if the first line moves, so will the second. The two lines do not have to be touching.
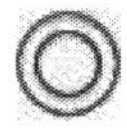	Concentric	Arcs and/or circles will share the same center point.
	Fix	Applying a fix constraint to a point prevents the constrained point from moving. Multiple points in a sketch can be fixed. If you select a line segment, the angle of the line will be fixed and only its length can change.
	Parallel	Two selected lines will remain parallel to one another.
	Perpendicular	Lines will remain at 90° angles to one another.
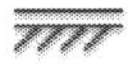	Horizontal	Line is positioned parallel to the X-axis, or a horizontal constraint can be applied between any two points in the sketch. The selected points will be aligned such that a line drawn between them will be parallel to the X-axis. Tip: Your keyboard's dash "-" key is a quick shortcut to start the Horizontal Constraint command.
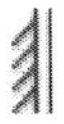	Vertical	Line is positioned parallel to the Y-axis, or a vertical constraint can be applied between any two points in the sketch. The selected points will be aligned such that a line drawn between them will be parallel to the Y-axis.
	Tangent	An arc, circle, or line will remain tangent to another arc or circle.
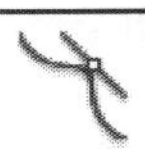	Smooth (G2)	A spline and another spline, line, or arc that connect at an endpoint with a coincident constraint will represent a smooth G2 (continuous curvature) condition.
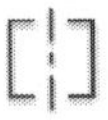	Symmetry	Selected points defining the geometry are made symmetric about a selected line.
	Equal	If two arcs or circles are selected, they will have the same radius or diameter. If two lines are selected, they will become the same length. If one of the objects changes, so will the constrained object. If the Equal constraint is applied after one of the arcs, circles, or lines has been dimensioned, the second arc, circle, or line will take on the size of the first one. If you select multiple similar objects (lines, arcs, etc.) before selecting the command, the constraint is applied to all of them.

Adding Constraints

As stated previously, you can apply constraints while you sketch objects. You can also apply additional constraints after the sketch is drawn. However, Inventor will not allow you to **over-constrain** the sketch or add duplicate constraints. If you add a constraint that conflicts with another, you will be warned with the message, "Adding this constraint will over-constrain the sketch." For example, if you try to add a vertical constraint to a line that already has a horizontal constraint, you will be alerted.

To add a constraint, follow these steps:

1 Click a constraint from the Constrain panel, or right-click in the graphics window and choose Create Constraint from the menu and select specific constraint from the menu (see Figure 2.36.)

2 Click the object or objects then apply the constraint.

Showing Constraints

To display the geometric constraints that are applied to a sketch, do one of the following:

- Select the geometry in the graphics window by selecting individual objects or using a window or crossing selection (see "Selecting Objects" on page 49 if needed.)
- Click the Show Constraints command from the Status Bar as shown in Figure 2.37 left or from the Constrain panel as shown in the middle image.
- Right-click in a blank area in the graphics window and click Show All Constraints from the menu.
- Press the F8 key.

The constraints on the selected geometry will be displayed. The yellow squares represent coincident constraints; move the cursor over a yellow square to display the two coincident constraints for the point. The image on the right shows all the constraints in a sketch.

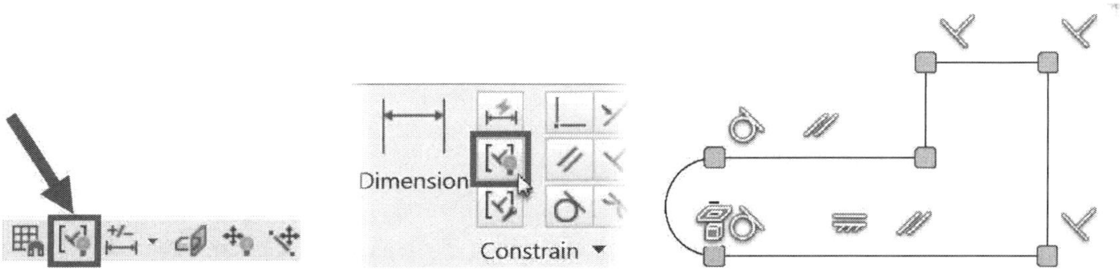

Figure 2.37

Modifying Constraint Size

You can set the Annotation Scale to modify the size of the constraint icons displayed on the screen. Use the Tools tab > Options panel > Application Options, and near the middle right of the General tab modify the Annotation Scale factor.

Figure 2.38 shows the Annotation Scale increased from 1.0 to 1.5. This setting also changes the size of the dimensions in a sketch. This change has no effect on the size of dimensions in a drawing.

Figure 2.38

Deleting Constraints

To delete the constraint(s), select a constraint or multiple constraints using one of the selection methods. Right-click and click Delete from the menu as shown in the following image on the left.

As an alternate method to deleting a constraint, you can press the Delete key once the constraint is selected.

- To delete all constraints except the coincident constraints, select or use the window or crossing selection technique, right-click and click Delete Constraints from the menu as shown in the image on the right.

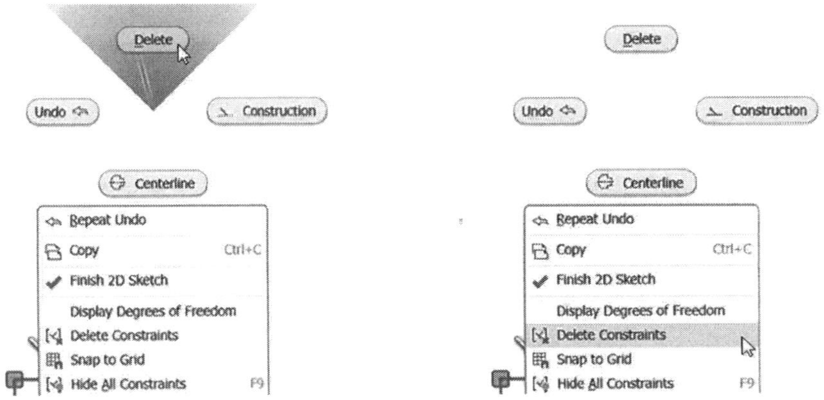

Figure 2.39

Hiding Constraints

You can hide the display symbol for individual or all geometric constraints. To perform this task, do one of the following:

To hide a constraint:

- Move the cursor over a constraint, right-click and click Hide from the marking menu as shown in the following image on the left.

To hide all constraints, do one of the following:

- Move the cursor over a constraint, right-click and click Hide All Constraints from the marking menu as shown in the image on the left.
- Click Hide All Constraints on the Status Bar as shown in the middle image. This is the same icon you selected to Show All Constraints.
- Right-click in a blank area in the graphics window and click Hide All Constraints on the menu as shown in the following image on the right.
- Press the F9 key.

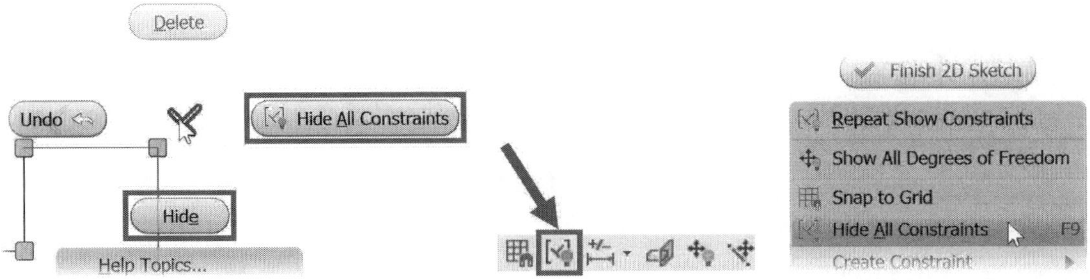

Figure 2.40

Construction Geometry

Construction geometry can help you create sketches that would otherwise be difficult to constrain. You can constrain and dimension construction geometry like normal geometry, but construction geometry will not be recognized as a profile edge in the part when you turn the sketch into a feature. Normal sketch geometry by default is visible in features created from the sketch.

Construction geometry can reduce the number of constraints and dimensions required to fully constrain a sketch, and can help define the sketch. For example, a construction circle that is tangent to the inside of a hexagon (drawn with individual lines and not the Polygon command) can drive the size of the hexagon. Without construction geometry, the hexagon would require six constraints and dimensions. With construction geometry, it would require only three constraints and dimensions; the circle would have tangent or coincident constraints applied to it and the hexagon.

To create construction geometry change the line style before or after you sketch the geometry using one of the following ways:

- After creating the sketch, select the geometry that you want to change and click the Construction icon on the Format panel as shown in Figure 2.41.
- Before sketching, click the Construction icon on the Format panel. All geometry created will be construction until the Construction command is deselected. Remember to click the Construction icon to turn it off afterward to avoid all you draw being construction geometry.

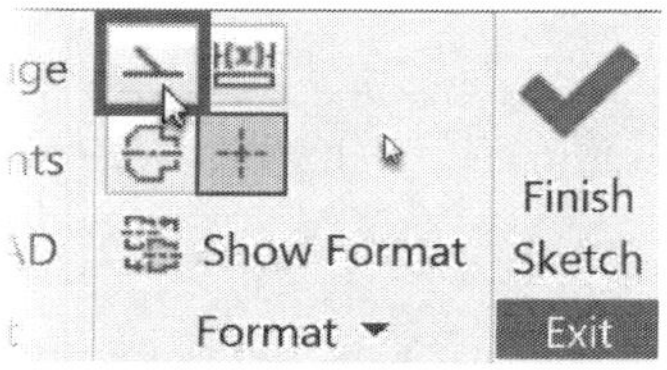

Figure 2.41

After turning the sketch into a feature, the construction geometry will be *consumed* with the sketch and is maintained in the sketch. When you edit a feature's sketch that has construction geometry, the construction geometry will reappear during editing and disappear when the part is updated.

You can add or delete construction geometry in a sketch just like normal sketch geometry can be. In the graphics window, construction geometry displays as a dashed line, lighter colored and thinner than normal geometry.

Figure 2.42 left shows a sketch with an angled construction line. The angled line has a coincident constraint applied to every endpoint that it touches. The image at right shows the sketch after it has been extruded. Notice that the construction line was not extruded.

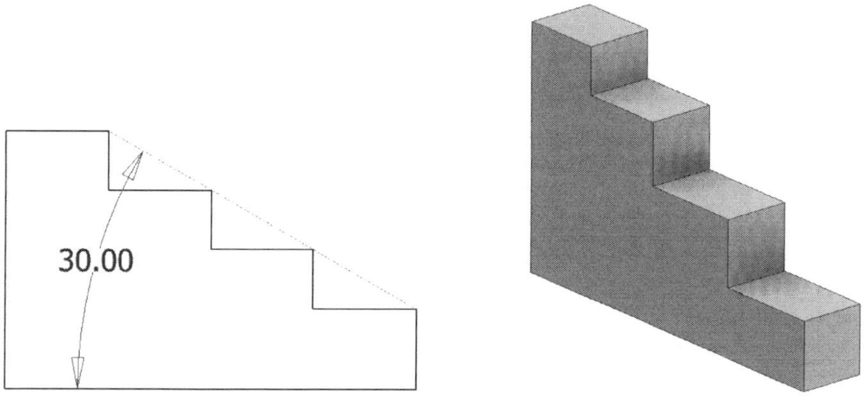

Figure 2.42

Number of Required Constraints or Dimensions

While constraining and dimensioning a sketch, there are multiple ways to determine the number of constraints or dimensions that are required to fully constrain the sketch. When you add a constraint or dimension, the number of constraints or dimensions needed to constrain the sketch decreases. A fully constrained sketch is a sketch whose geometry cannot move or stretch.

The bottom-right of the status bar shows the number of constraints or dimensions needed to fully constrain the sketch (Figure 2.43 left.) When no additional constraints or dimensions are needed to define the sketch, the message Fully Constrained appears at the bottom-right of the status bar (middle image), and in the browser, a pushpin icon appears to the left of the Sketch entry (right.)

Figure 2.43

Degrees of Freedom

To see the areas in the sketch that are NOT constrained, you can display the degrees of freedom. While a sketch is active, click Show Degree of Freedom on the status bar, as shown in Figure 2.44 left, or right-click in a blank area of the graphics window and choose Display Degrees of Freedom from the menu.

Lines and arcs with arrows will appear on your sketch as shown in the middle image. As you add constraints and dimensions to the sketch, degrees of freedom will disappear.

To remove the degree of freedom symbols from the screen, use the status bar to click Hide All Degrees of Freedom, as shown in Figure 2.44 right, or right-click in a blank area in the graphics window and uncheck Display Degrees of Freedom from the menu.

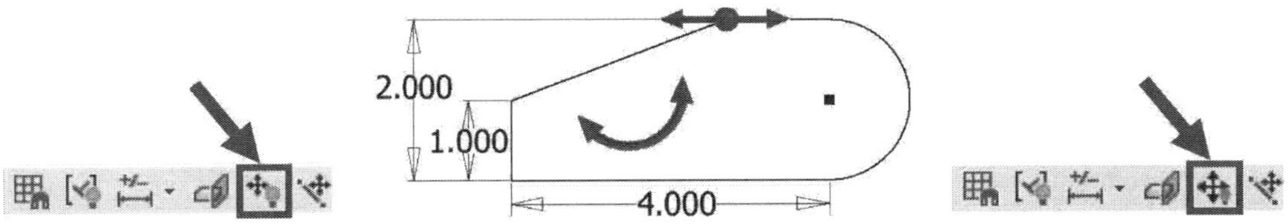

Figure 2.44

Dragging a Sketch

Another method to determine if an object is constrained is to try to drag it to a new location. While not in a command, click a point or an edge, or select multiple objects on the sketch. With the left mouse button depressed, drag it to a new location. If the geometry stretches, it is under constrained.

For example, if you draw a rectangle that has two horizontal and two vertical constraints applied to it and you drag a point on one of the corners, the size of the rectangle will change, but the lines will maintain their horizontal and vertical behaviors. If dimensions are set on the object, they will prevent the object from stretching.

EXERCISE 2-3: ADDING AND DISPLAYING CONSTRAINTS

In this exercise, you add geometric constraints to sketch geometry to control the shape of the sketch.

1 Click the New command, click the English folder, and double-click Standard (in).ipt.

2 Click the Start 2D Sketch command on the 3D Model tab > Sketch panel and then select the XY origin plane.

3 Sketch the geometry as shown in the following image, with an approximate size of **2 inches** in the X (horizontal) direction and **1 inch** in the Y (vertical) direction. Do not apply dimensions dynamically. Place the lower-left corner of the sketch on the origin point. Right-click in the graphics window, and then click OK. By starting the line at the origin point, that point is constrained to the origin with a coincident constraint.

4 Click Show All Constraints on the Status Bar, or press the F8 key. Your screen should resemble the following image.

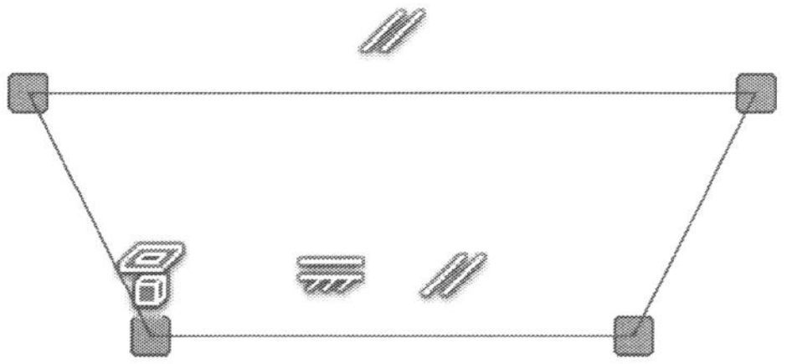

Figure 2.45

5 If another constraint appears, place the cursor over it, right-click, and then click Delete from the marking menu.

6 On the Constrain panel, click the Parallel constraint icon.

7 Select the two angled lines. Depending upon the order in which you sketched the lines, the angles may be opposite of the following image on the left.

8 Press the ESC key to stop adding constraints.

9 View the constraints (you may press the F8 key to refresh the visible constraints). Your screen should resemble the following image.

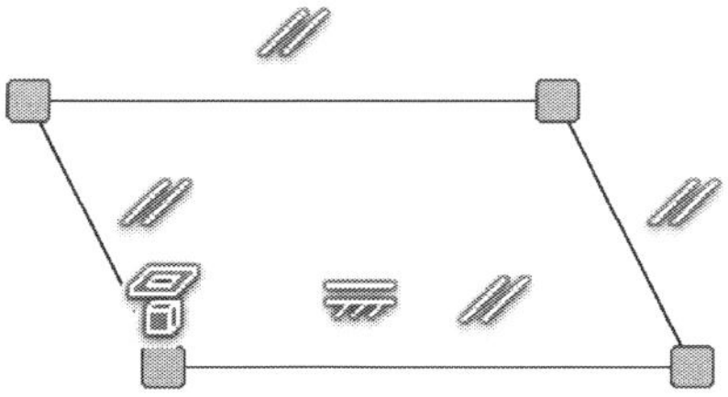

Figure 2.46

10 Select the top horizontal line in the sketch and drag the line. Notice how the sketch changes its size, but not its general shape. Try to drag the bottom horizontal line. The line cannot be dragged as it is constrained.

11 Select the endpoint on the bottom-right horizontal line, and drag the endpoint. The lines remain parallel due to the parallel constraints. Notice on the bottom-right of the Status Bar that 3 dimensions are needed to fully constrain the sketch.

12 On the Sketch tab > Constrain panel, click the Perpendicular constraint icon.

13 Select the bottom horizontal line and the angled line on the right side.

14 Even though it may appear that the rectangle is fully constrained, the height and length of the rectangle still need to be defined with dimensions. Notice on the bottom-right that the Status Bar is down to 2 dimensions to fully constrain the sketch.

15 Place the cursor over the icon for the parallel constraint on the right-vertical line, right-click, and click Delete from the marking menu as shown in the following image on the right. The parallel constraint that was applied to both angled lines is deleted and 3 dimensions would be needed to fully constrain the sketch.

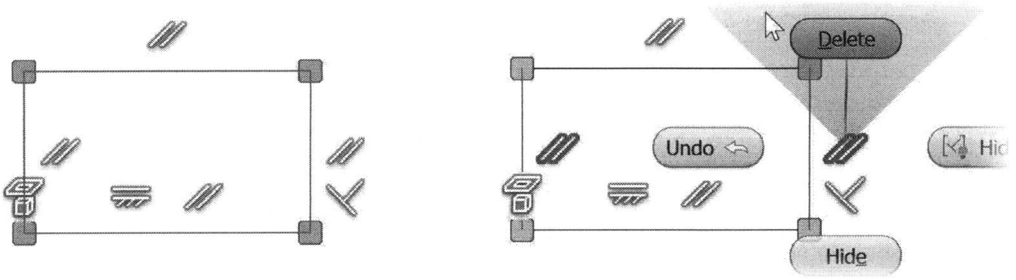

Figure 2.47

16 Right-click the constraint on the top-left corner of the rectangle and click Hide. Now you can click and drag that corner. Notice that the line on the right does not move because it is perpendicular to the bottom horizontal line.

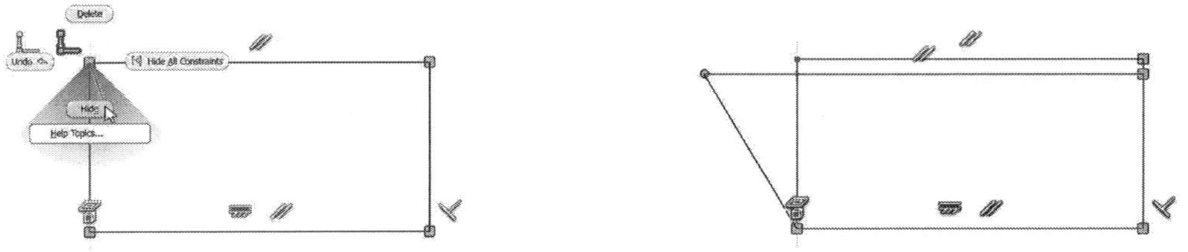

Figure 2.48

17 Click Hide All Constraints on the Status Bar, or press the F9 key.

18 Use a window (drawn left to right) to select the 4 lines and then right-click and choose Delete from the marking menu.

19 Use the Line command to sketch the geometry as shown in the following image with an approximate size of **3 inches** in the X direction and **1.5 inches** in the Y direction. Place the lower-left point of the sketch on the origin (projected center point). Do not apply dimensions dynamically. Right-click in the graphics window, and then click OK.

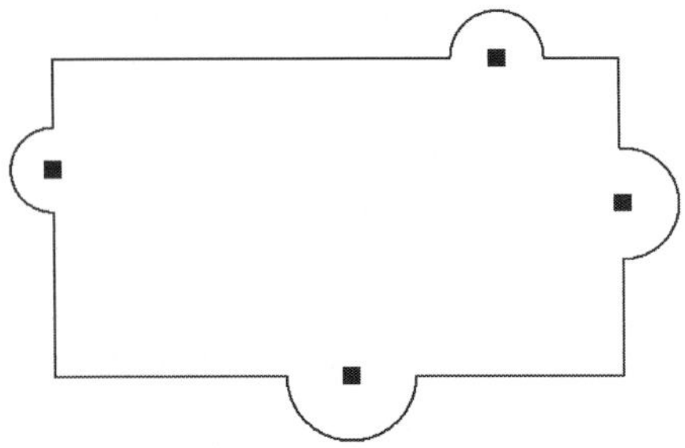

Figure 2.49

20 Inspect the constraints by dragging different points and edges.

21 Next make the arcs equal in size. From the Constrain panel, click the Equal constraint command or press the = key on the keyboard.

22 Select the arc on the left and then the bottom arc.

23 Select the arc on the left and then the arc on the right side.

24 Select the arc on the left and then the arc on the top.

25 Next align the line segments if necessary. From the Constrain panel, click the Collinear constraint command. **Note:** if the endpoints and center point of the arcs are aligned horizontally  or vertically  when they were sketched, you will receive a message "Adding this constraint will over-constrain the sketch." If you see this message click Cancel in the dialog box for steps 25 a. b. c. and d.

a. Select the two bottom-horizontal lines.

b. Select the two top-horizontal lines.

c. Select the two left-vertical lines.

d. Select the two right-vertical lines.

26 To stop applying the collinear constraint, either right-click and click Cancel (ESC) from the marking menu or press the ESC key.

27 Next you will align the top and bottom arcs vertically. On the Constrain panel, click the Vertical constraint command.

28 Select the center point of the bottom arc and then click the center point of the top arc.

29 Next you will align the left and right arcs horizontally. On the Constrain panel, click the Horizontal constraint command.

30 Click the center point of the left arc, and then click the center point of the right arc.

31 To stop applying the constraints, right-click and click Cancel (ESC) from the marking menu, or press the ESC key.

32 If desired, you can move the arcs by clicking and dragging on them.

33 Display all the constraints by pressing the F8 key. Your screen should resemble the following image.

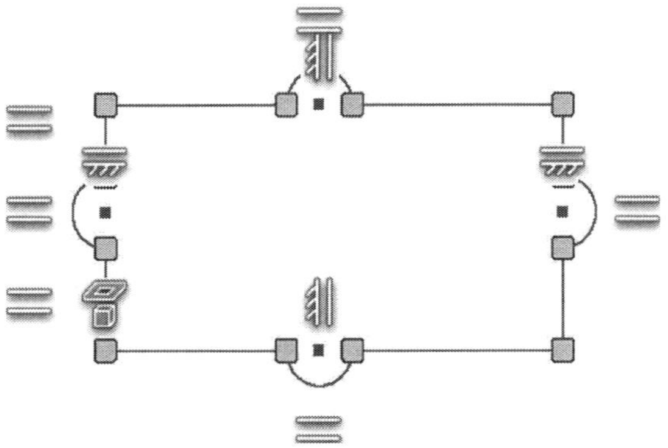

Figure 2.50

34 Hide all the constraints by pressing the F9 key.

35 Click on a line in the sketch and drag the line. Try dragging different lines, and notice how the sketch changes.

36 Next delete geometry. as directed:

a. Press the ESC key twice to cancel any command.

b. Click a point above and to the left of the top arc and drag a window down and to the right so it encompasses the arc on the right as shown in the following image on the left.

c. Release the mouse button and press the Delete key on the keyboard.

37 Close the open line segments. Drag the open endpoints onto each other until your sketch resembles the following image on the right. A green circle will appear when the two endpoints are near each other; this applies a coincident constraint.

38 Note, you could have connected the two open endpoints by using the coincident constraint.

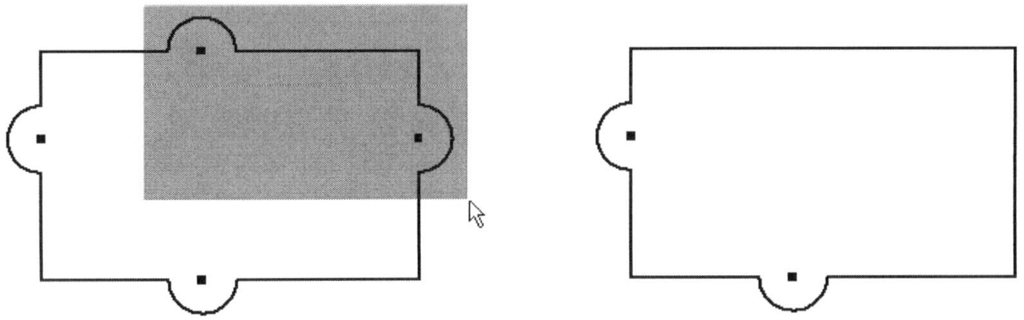

Figure 2.51

39 Next center the arcs in the middle of the sketch. On the Constrain panel, click the Vertical constraint command, click the center point on the bottom arc and then the midpoint of the top horizontal line as shown in the following image on the left.

On the Constrain panel, click the Horizontal constraint command, click the center point on the left arc and the midpoint of the right vertical line as shown in the following image on the right.

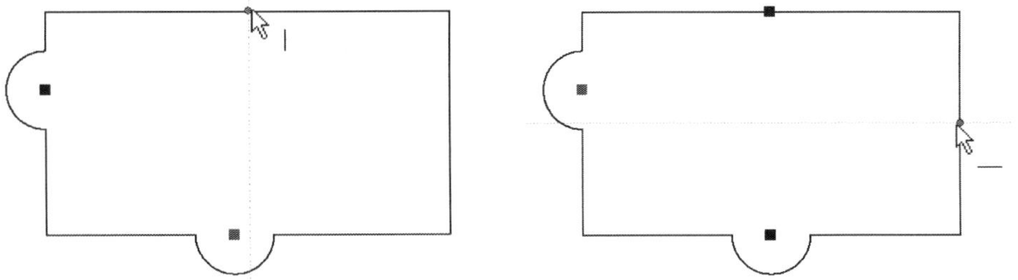

Figure 2.52

40 Click on different points and drag them, notice how the sketch changes shape, but the arcs are always centered as shown in the following image.

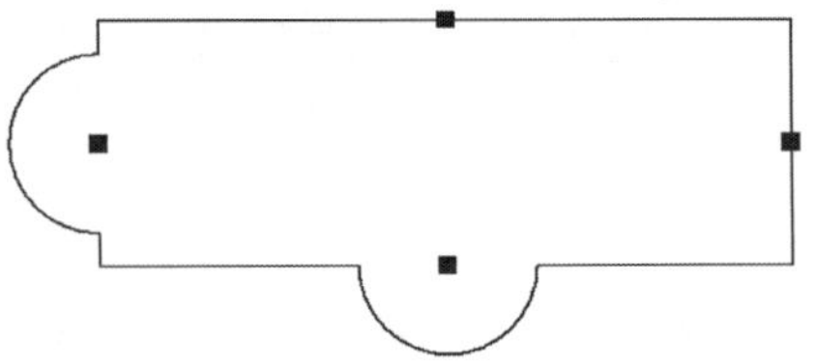

Figure 2.53

41 Close the file. Do not save changes.

Note: Dimensions are needed to fully constrain the sketch. Dimensions are covered in the next section.

End of exercise.

ADDING DIMENSIONS MANUALLY

The last step to constraining a sketch is to add dimensions that were not added dynamically. The dimensions you place will control the size of the sketch and can also appear in the part drawing views when they are generated. When placing dimensions, try to avoid having extension lines go through the sketch, as this will require more cleanup when drawing views are generated. Click near the side from which you anticipate the dimensions will originate in the drawing views.

All dimensions that you create are parametric as well as the dynamic dimensions that are placed automatically when sketching geometry. Parametric means that they will change the size of the geometry.

Scale Sketch

If the sketch is not constrained to the origin point and no dimension was dynamically added to the sketch when it was created, then the entire sketch will be uniformly scaled when the first dimension is added.

General Dimensioning

The General Dimension command can create linear, angle, radial, or diameter dimensions one at a time. The following image on the left shows an example of a dimensioned sketch. To start the General Dimension command, follow one of these techniques:

- Click the General Dimension command from the Sketch tab > Constrain panel as shown in the following image in the middle.
- Right-click in the graphics window and click General Dimension from the marking menu as shown in the image on the right.
- Press the shortcut key D.

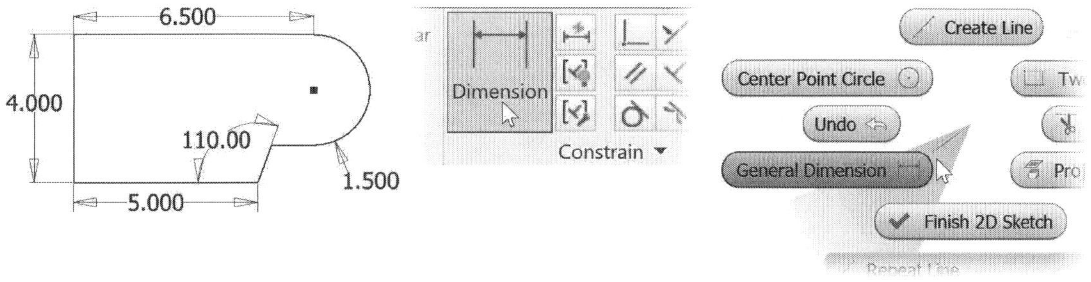

Figure 2.54

When you place a linear dimension, the extension line of the dimension will snap automatically to the nearest endpoint of a selected line; when an arc or circle is selected, it will snap to its center point. To dimension to a tangent point of an arc or circle, see "Dimensioning to a Tangent of an Arc or Circle" later in this chapter.

After you select the General Dimension command, follow these steps to place a dimension:

1. Select the geometry to be dimensioned.
2. After selecting the geometry, a preview image will appear attached to your cursor showing the type of dimension. If the dimension type is not what you want, right-click, and then select the correct style from the menu. After changing the dimension type, the dimension preview will change to reflect the new style.
3. Click to place the dimension.
4. Enter a value for the dimension.

The next sections cover how to dimension specific objects and how to create specific types of dimensioning with the Dimension command.

Dimensioning Lines

There are multiple techniques for dimensioning a line. Use the Dimension command and do one of the following:

- Click near two endpoints, move the cursor until the dimension is in the correct location, and click.
- To dimension the length of a line, click anywhere on the line; the two endpoints will be selected automatically. Move the cursor until the dimension is in the correct location and click.
- To dimension between two parallel lines, click one line and then the next, and then click a point to locate the dimension.
- To create a dimension whose extension lines are perpendicular to the line being dimensioned, click the line and then right-click. Click Aligned from the menu, and then click a point to place the dimension.

Dimensioning Angles

To create an angular dimension, use the General Dimension command, click two lines whose angle you want to define, move the cursor until the dimension is in the correct location. Place the dimension by clicking on a point.

Dimensioning Arcs and Circles

To dimension an arc or circle: start the General Dimension command, click on the circle's circumference, move the cursor until the dimension is in the correct location, and click. By default, when you dimension a circle, the default is a diameter dimension; when you dimension an arc, the result is a radius dimension. To change a radial dimension to diameter or a diameter to radial, right-click before you place the dimension and select the style from the Dimension Type menu.

To dimension the angle of the arc: start the Dimension command, click the arc's circumference, click the arc's center point, and then place the dimension. You can also click the arc's center point and then its circumference.

You can also add an arc length dimension this way. Start the dimension command, then click on the arc, right-click and choose Arc Length from the Dimension Type menu, then click a point to locate the dimension.

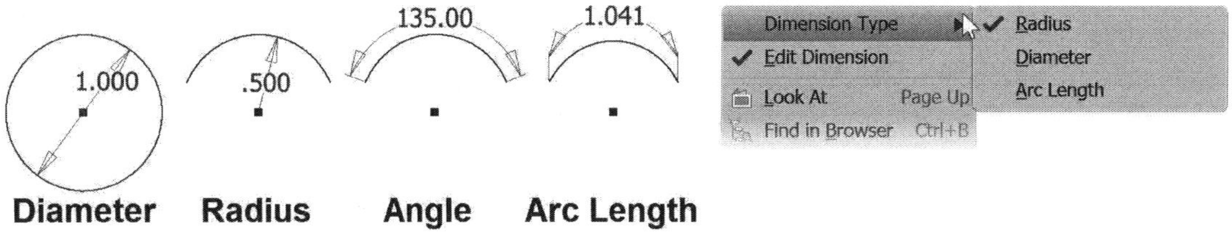

Figure 2.55

Dimensioning to a Tangent of an Arc or Circle

To dimension to a tangent of an arc or circle, follow these steps:

1. Start the General Dimension command.
2. Select a line that is parallel to the tangent arc or circle that will be dimensioned, labeled (A) in the following image.
3. Move the cursor over the arc or circle, and click when the midpoint is highlighted, labeled (B) in the following image.
4. Move the cursor until the dimension is in the desired location and click to create the dimension.

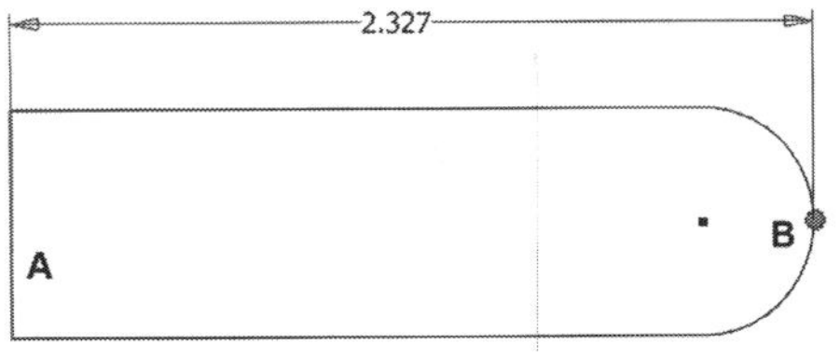

Figure 2.56

To dimension to two tangents, follow these steps:

1. Start the General Dimension command.
2. Select an arc or circle that includes one of the tangents to which it will be dimensioned. The following image illustrates an example of dimensioning a slot; the first selection is labeled (A).
3. Move the cursor over a second arc or circle and click when the midpoint is highlighted, as shown below, labeled (B).
4. Move the cursor until the dimension is in the desired location. Then click to create the dimension,

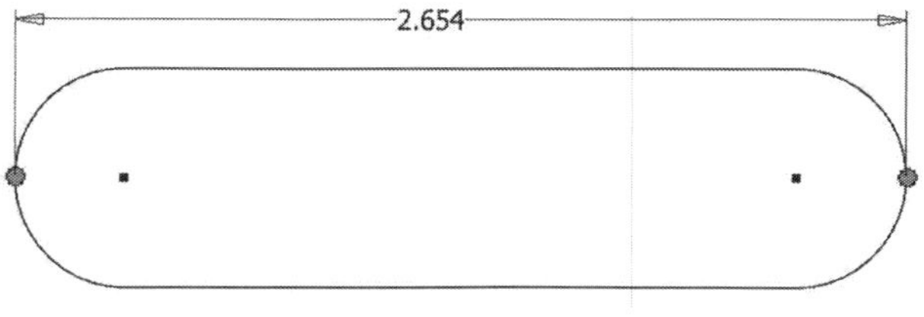

Figure 2.57

Entering and Editing a Dimension Value

After placing the dimension, you can change its value. Depending on your settings, the Edit Dimension dialog box may or may not appear automatically after you place the dimension. The Edit Dimension options can be set by any one of the following:

- Click the Tools tab > Options panel > Application Options. On the Sketch tab of the Application Options dialog box, from the Constrain Settings area, click the Settings button and then click the box next to "Edit dimension when created" as shown in Figure 2.58 left.
- Set the option by right-clicking in the graphics window while placing a dimension and click Edit Dimension from the menu as shown in Figure 2.58 right. This method will change the application option "Edit Dimension when created" as previously described.

If "Edit dimension when created" is checked, the Edit Dimension dialog box appears automatically after you place the dimension. Otherwise, the dimension will be placed with the default value.

Figure 2.58

To edit a dimension that has already been created, double-click on the dimension, and the Edit Dimension dialog box will appear, as shown in Figure 2.59.

Enter the new value and unit for the dimension; then either press ENTER or click the checkmark in the Edit Dimension dialog box. If no unit is entered, the units that the file was created with will be used. Enter the exact value — do not round up or down. The accuracy of the dimension that is displayed in a sketch is set in the Document Setting. For example, if you want to enter 4 1/16 as a decimal, enter 4.0625 in not 4.06 in. You can set how many decimal places to display on the drawing later, but create your model with as much accuracy as possible.

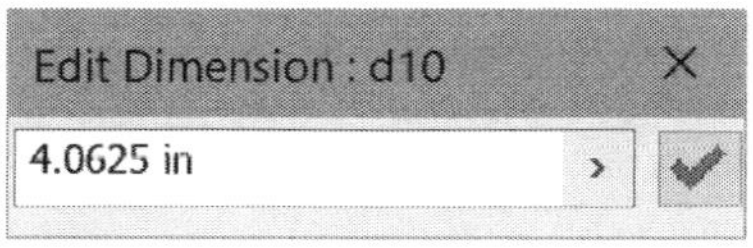

Figure 2.59

Fractions

Inventor allows you to enter a fraction when such a value is required. With the Length unit in the Tools tab > Options panel > Document Settings > Units tab set to any non-metric unit, as shown in Figure 2.60 left, when a fraction is entered, it will display as such in the graphics window and in the Edit Dimension dialog box. If the Length unit is set to a metric unit and a fraction is entered, the decimal equivalent will be displayed in the graphics window but the fraction will be shown in the Edit Dimension dialog box.

After entering a fraction, you can click on the right-faced arrow and set the type of dimension to display: Decimal, Fractional, or Architectural as shown in the middle of the following image.

When entering fractions do not use a dash to separate the fraction, just add a space. For example, enter 4 1/16, *not* 4-1/16, because Inventor interprets the — as part of an equation and would return the value 3.9375. Figure 2.60 right shows the fraction displayed in the graphics window.

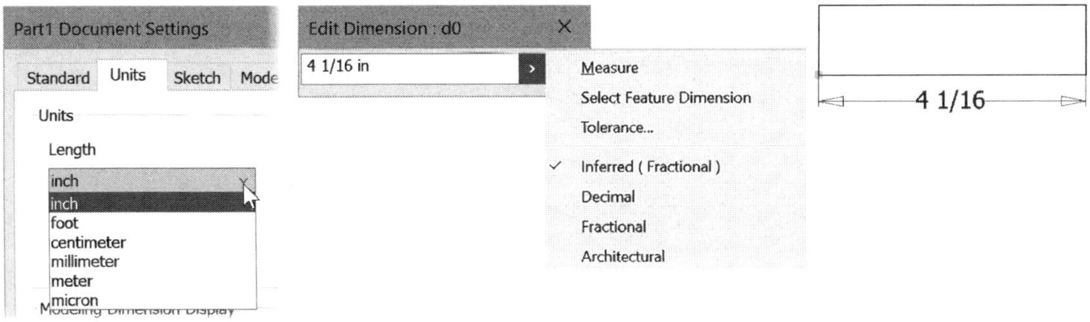

Figure 2.60

TIP: When placing dimensions, it is recommended that you place the smallest dimensions first. This will help prevent the geometry from flipping in the opposite direction.

Repositioning a Dimension

Once you place a dimension, you can reposition it, but the origin points cannot be moved. Follow these steps to reposition a dimension:

1. Exit the current operation either by pressing ESC twice or right-clicking and then clicking Cancel (ESC) from the marking menu.
2. Move the cursor over the dimension until the move symbol appears as shown in Figure 2.61.
3. With the left mouse button depressed, move the dimension to a new location and release the button.

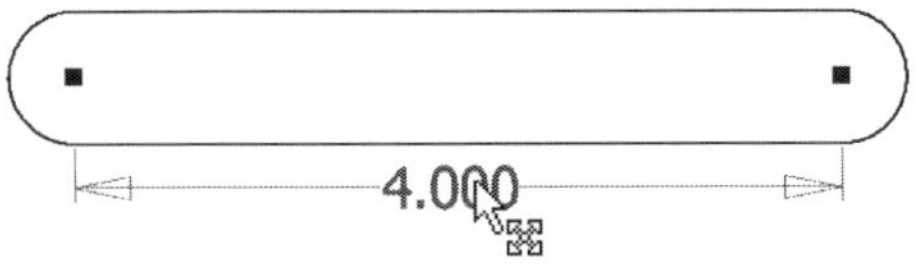

Figure 2.61

Fully Constrained Sketch

As described in the "Constraining the Sketch" section, as you add constraints and dimensions to a sketch, the number of required dimensions decreases. When no more constraints or dimensions are needed to constrain the sketch, instead of a number in "dimensions needed" on the bottom-right of the status bar Fully Constrained will show as in Figure 2.62 left. The icon to the left of the Sketch entry in the browser displays a push-pin when the sketch is fully constrained as shown in Figure 2.62 right.

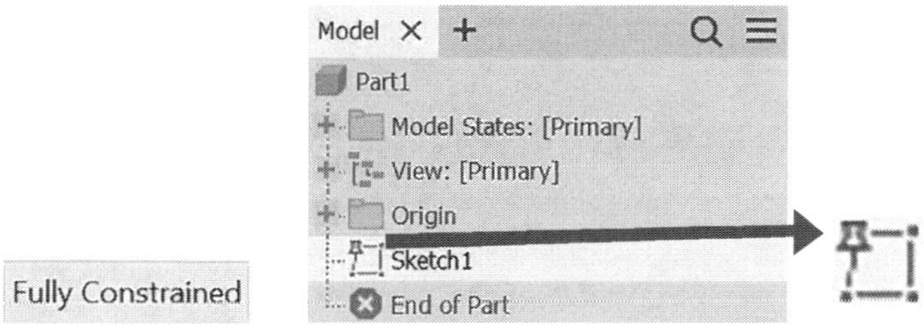

Figure 2.62

Over Constrained Sketch

As explained in the "Adding Constraints" section, Inventor does not allow over-constrained sketches or duplicate constraints. The same is true when adding dimensions. If you add a dimension that conflicts with another constraint or dimension, you will be warned that this dimension will over-constrain the sketch or that it already exists. You can either cancel the operation and no dimension will be placed, or accept the warning and a *driven* dimension will be created.

A *driven dimension* is a *reference* dimension. It is not a parametric dimension—it reflects the size of the points to which it is dimensioned. If the part changes, the driven dimension updates to show the new value. A driven dimension will appear with parentheses around the dimension's value—for example, (2.500). When you place a dimension that will over-constrain a sketch, a dialog box will appear like the following image.

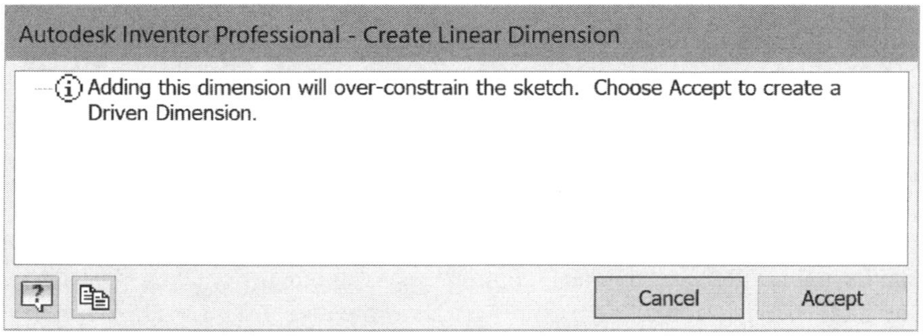

Figure 2.63

Relax Mode

When you try to place a constraint or add a dimension and receive the over-constrained message, if you want the constraint or dimension to take precedence, you can turn on relax mode. With relax mode on, when you reapply the constraint or add the dimension, the conflicting constraint will automatically be deleted, except for Coincident, Smooth, Tangent, Symmetry, Pattern, and Project constraints. A conflicting dimension will become a driven dimension. While in relax mode, if you are unable to add the new constraint or dimension, you may need to manually delete one of the Coincident, Smooth, Tangent, Symmetry, Pattern, and Project constraints.

Turn on relax mode by clicking the Relax Mode icon on the Status Bar as shown in Figure 2.64 left. You can also click Constrain Settings from the Sketch tab > Constrain panel, then use the Relax Mode tab to check Enable Relax Mode as shown in Figure 2.64 middle. When you apply a constraint or dimension that would over-constrain a sketch, a dialog box appears stating a constraint or dimension will be deleted to solve the conflict as shown in Figure 2.64 right.

Another method to remove conflicting constraints or dimensions is to drag a point or edge while in Relax Mode and conflicting constraints or dimensions will be deleted.

TIP: When done constraining a sketch, turn off Relax Mode to return to normal editing. If you don't turn it off, sketch points and objects can be changed by accidentally dragging them.

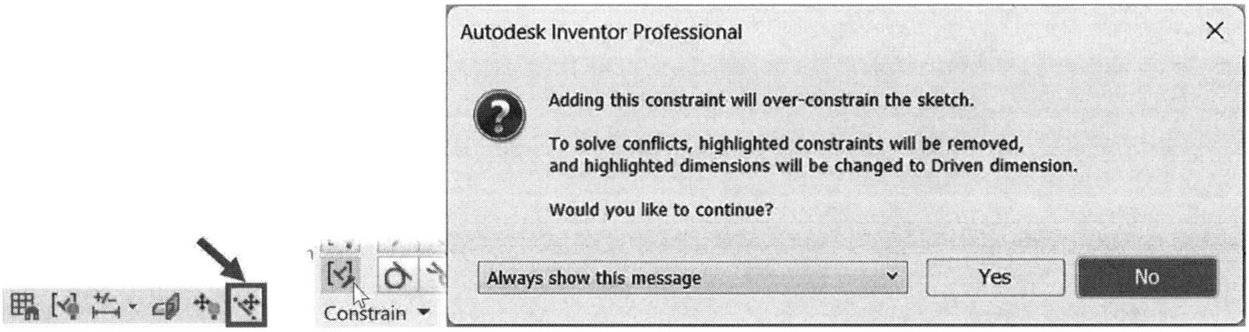

Figure 2.64

Figure 2.65 left shows a fully constrained rectangle with geometric constraints visible. With Relax Mode on a 120-degree angle dimension was placed. Clicking Yes in the warning dialog box caused the dimension to be created, and the perpendicular constraint in the lower-right corner deleted as shown in the middle image. The image at right shows the sketch after a perpendicular constraint was applied to the bottom-horizontal and left-vertical lines. Notice the 120-degree dimension was maintained.

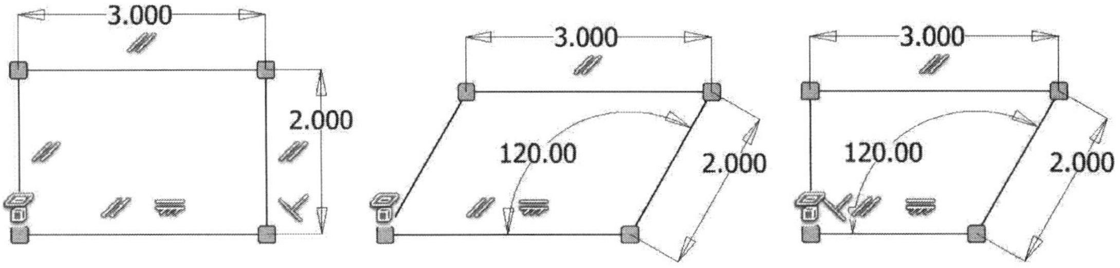

Figure 2.65

Driven Dimension

Driven dimensions can also be created using the Driven Dimension option. A driven dimension does NOT reduce the dimensions needed to constrain the sketch; it only reports the length of the object. You use this option to show reference dimensions.

Select Driven Dimension from the Sketch tab > Format panel, as shown in Figure 2.66 left. With the Driven Dimension icon highlighted, use the Dimension command as usual to create a driven dimension. (Don't forget to turn it off when you are finished.) Driven dimensions are represented in the sketch with parentheses around the value; parametric dimensions do not use parentheses. Without the Driven Dimension option active, regular parametric dimensions are created, which is the default.

The Driven Dimension option can also be used to change an existing dimension to either a driven or back to normal by selecting it and clicking Driven Dimension. Figure 2.66 right shows an example of a 5.250 driven dimension referencing the overall length of the sketch. The three parametric dimensions control the length of the sketch. If a parametric dimension value changes, the driven dimension updates to reflect the change in overall length.

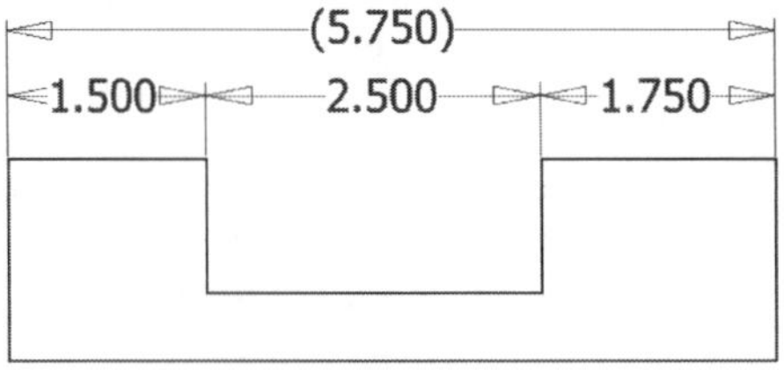

Figure 2.66

TIP: Avoid over using driven dimensions as they do not parametrically control the size of the sketch. They are only used for reference.

EXERCISE 2-4: CONSTRAINING AND DIMENSIONING A SKETCH

In this exercise, you add dimensional constraints to a sketch. Note: this exercise assumes that the "Edit dimension when created" and "Autoproject part origin on sketch create" options are checked in the Application Options dialog box under the Sketch tab. Experiment with Autodesk Inventor's color schemes to see how the sketch objects change color when they are constrained.

1 Click New, choose the English folder. Double-click Standard (in).ipt to select it.

2 Click Start 2D Sketch from the 3D Model tab > Sketch panel. Select the XY origin sketch plane.

3 Use the Line command from the Sketch tab > Create panel. Draw a line starting from the origin point, then move the cursor to the right, type **5** in the distance input. Press Tab and click a point when the horizontal constraint shows below the degrees input field as shown in Figure 2.67 left.

4 Next place an angled line and a dynamic dimension to define the angle. Press Tab and type **150** for the angle input, press the Tab key, and then click a point to the upper right as shown in Figure 2.67 right. The distance should be about 2 inches but the dimension is not needed to define this sketch.

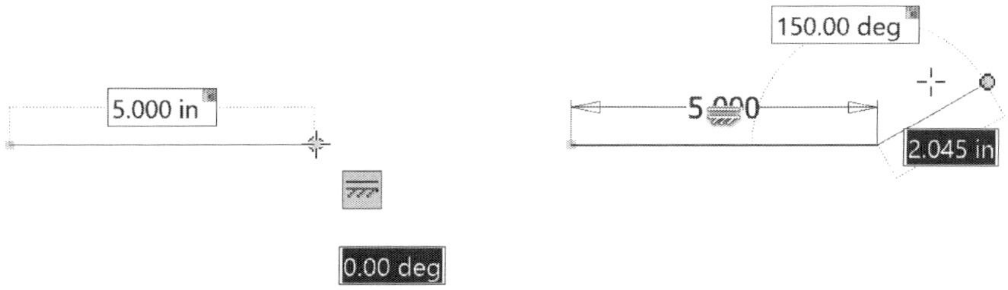

Figure 2.67

5 Sketch the geometry shown in the following image. When sketching, ensure that a perpendicular constraint is **not** applied between the two angled lines. If needed, hold down CTRL while sketching the top angled line to prevent the sketch constraint being applied. Make the arc tangent to both adjacent lines.

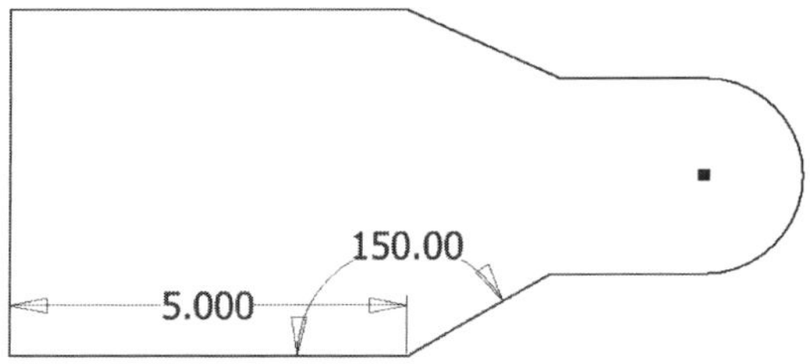

Figure 2.68

6 Horizontal Constraint command from the Sketch tab > Constrain panel to add a horizontal constraint between the midpoint of the left vertical line and the center of the arc as shown in Figure 2.69 left.

7 Use Vertical Constraint from the Sketch tab > Constrain panel and add a vertical constraint between the endpoints of the angled lines nearest to the right side of the sketch as shown in Figure 2.69 right.

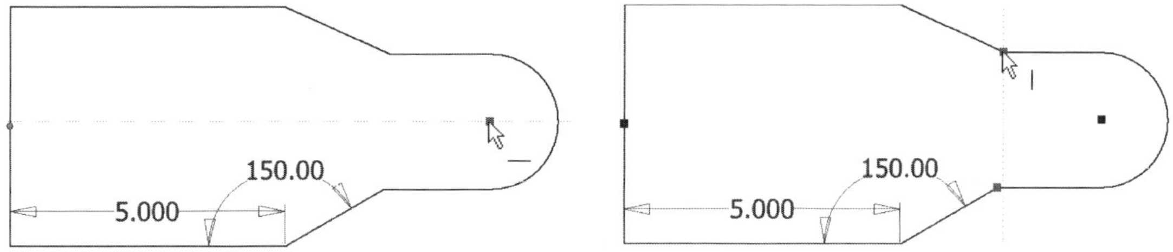

Figure 2.69

8 Press the = key and select the two angled lines to add an equal constraint to them.

9 Click General Dimension from the Sketch tab > Constrain panel and add a radial dimension to the arc. Move the cursor to position the dimension near the lower right corner of the sketch, then click a point and enter the value **1.5.** If the Edit Dimension dialog doesn't appear, double-click the dimension and change it to 1.5. Click the checkmark to finish.

> **TIP:** To show the Edit Dimension dialog box during dimension placement, right-click in the graphics window and click Edit Dimension from the menu.

10 While still in the General Dimension command, add a vertical dimension to the vertical line, click a point to locate the dimension to its left, enter **5**, and click the checkmark. When complete, your sketch should resemble the following image. Notice on the Status Bar bottom right, 1 dimension is required to fully constrain the sketch.

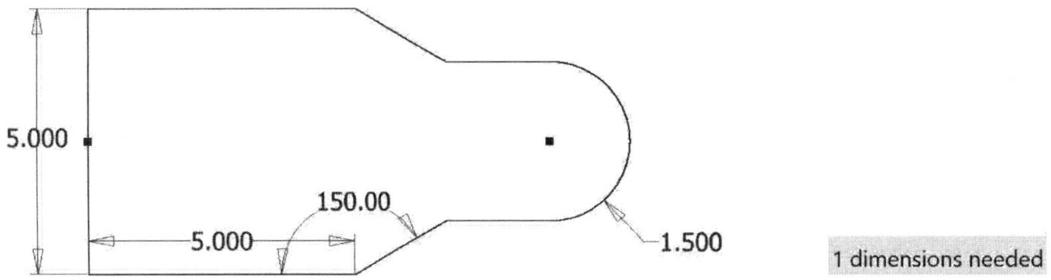

Figure 2.70

11 Add a horizontal dimension by selecting the left vertical line and the center point of the arc (or on top of the arc). Locate the dimension above the sketch. Enter **10** for the value as shown in Figure 2.71.

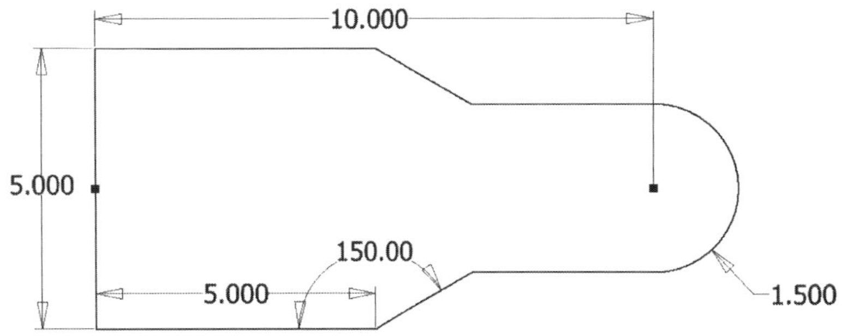

Figure 2.71

12 Press ESC twice to end the command. The Status Bar dimensions required section should display "Fully Constrained." In the browser, the icon to the left of the sketch should show a pushpin.

13 Try clicking and dragging different points of the sketch. The points do not change as it is fully constrained and dimensioned.

14 Click the Relax Mode icon in the Status Bar to turn it on as shown in Figure 2.72.

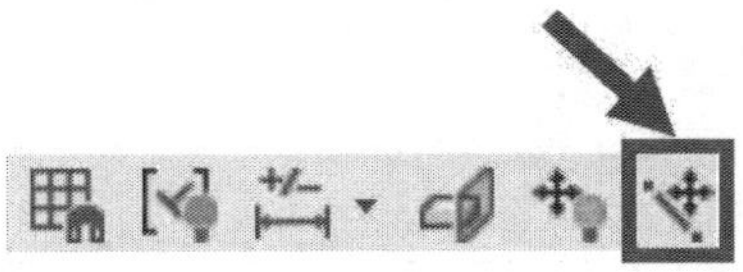

Figure 2.72

15 Click and drag different points on the sketch. The dimensions will change to reflect their new value.

16 Use Undo from the Quick Access toolbar to return the sketch to the values shown in Figure 2.71.

17 While still in Relax Mode, use General Dimension to add an overall horizontal dimension by clicking the vertical line (not an endpoint) and moving the cursor near the right tangent point of the arc until the middle point of the arc is highlighted and a glyph of a dimension appears, as shown in Figure 2.73 left.

- Locate the dimension above the 10.000 dimension.
- Click Yes to allow the highlighted dimension to be changed to a driven dimension.
- Accept the default length of **11.500** by clicking the green check mark in the Edit Dimension dialog.
- Notice that a perpendicular constraint was removed.

18 Select the 10.000 dimension and toggle it to a driven dimension by clicking Driven Dimension from the Sketch Tab > Format panel.

19 Add a perpendicular constraint between the left vertical line and the top horizontal line. Your sketch should resemble Figure 2.73 right.

Note that another option is to change the 10.000 dimension to a driven dimension and then add the 11.500 dimension.

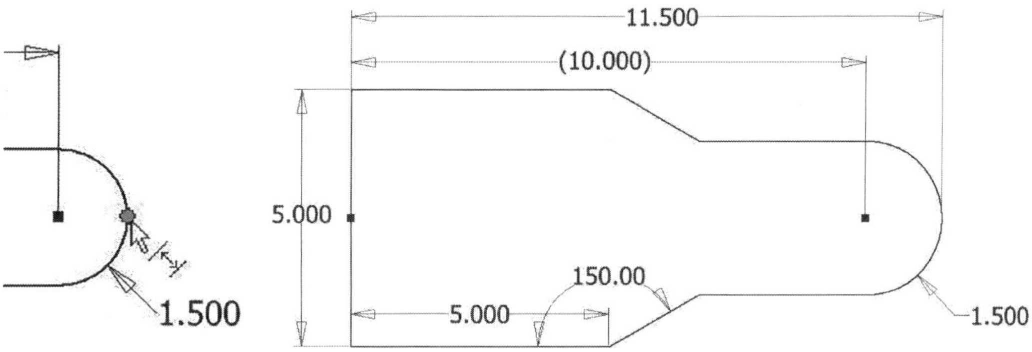

Figure 2.73

20 Click Relax Mode from the Status Bar to turn it off as shown in Figure 2.74.

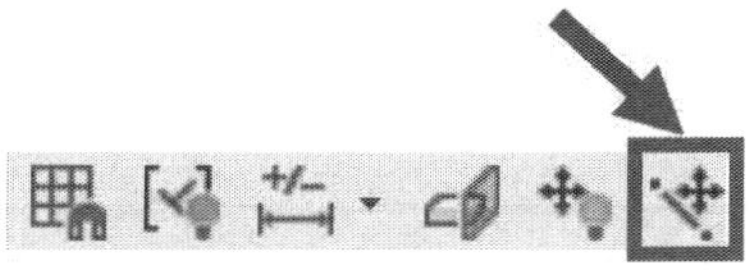

Figure 2.74

21 Edit some sketch dimensions by double-clicking on the dimension's value and typing in a new one. Press ENTER on the keyboard or click the green check mark in the Edit Dimension dialog box when done. Examine how the sketch changes. The arc should remain in the middle of the vertical line. Notice how the driven dimension changes when the horizontal, angle or radial dimension values change.

22 Delete the horizontal constraint between the center of the arc and the midpoint of the vertical line by selecting the center point of the arc, select the horizontal constraint above the center point, right-click and click Delete from the menu as shown in the following image.

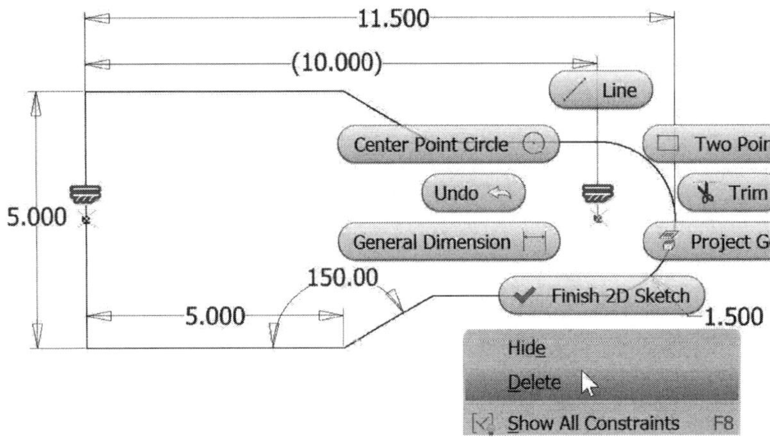

Figure 2.75

23 Prove the arc can move by clicking and dragging on the center point of the arc.

24 Display the visibility of all the constraints by using window to select all of the geometry.

25 Click in an empty area of the graphics window so the constraints disappear.

26 From the Status Bar, click Show All Constraints, as shown in Figure 2.76.

Figure 2.76

27 Practice deleting and adding other constraints.

28 Practice adding and deleting dimensions.

29 Close the file. Do not save changes.

End of exercise.

MEASURE COMMAND

The measure command assists in analyzing sketch, part, and assembly models by measuring position, length, angle, radius, diameter, perimeter, and area.

The Measure command is not a replacement for dimensions; it is an additional tool to give you more information. You can measure distances, angles, and loops, and you can perform area calculations. You can start the measure command first and then select the geometry, or select the geometry and then start a measure command.

The Measure command is located on the Inspect tab > Measure panel as shown in Figure 2.77 left. If desired, you can add it to the Quick Access toolbar by clicking the down arrow to the right of the Quick Access toolbar and choosing Measure from the menu as shown in Figure 2.77 middle image.

Once the Measure command is added you can access it as shown in Figure 2.77 right.

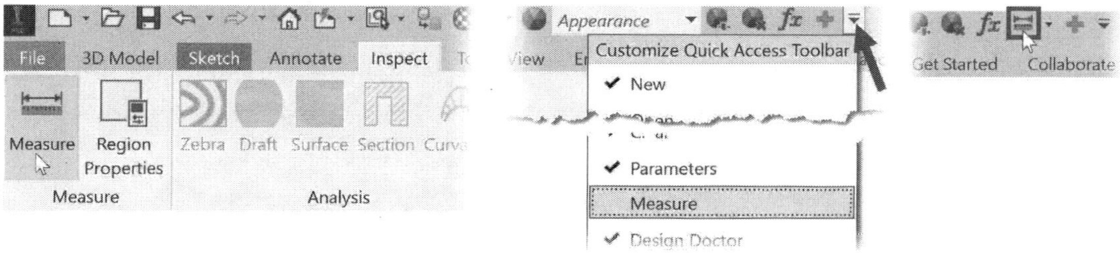

Figure 2.77

Position and Distance Between Two Points

Select a point and the X, Y, Z position data will be displayed in the Measure tool panel as shown in the following image on the left. Select two points and the position data of both points will be displayed as well as the distance between two points as shown in the following image on the right.

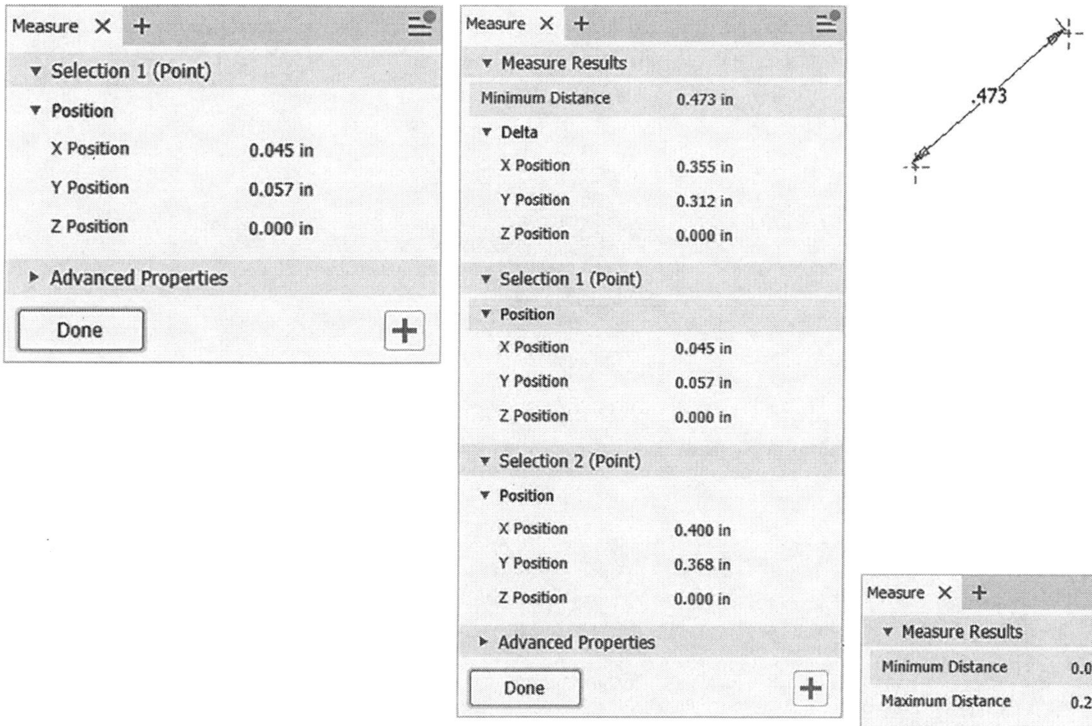

Figure 2.78

Length, Radius, Diameter

Measures the distance for an edge or line, diameter or radius of a cylindrical face or circle, distance between two points, and distance between two components in an assembly (covered in Chapter 6 Assemblies).

Angle

Measures the angle by selecting two edges, two lines, an arc, select two points and then hold down the Shift key and select a third point, or select a line or an edge and a planar face.

Figure 2.79 shows the results after measuring two edges.

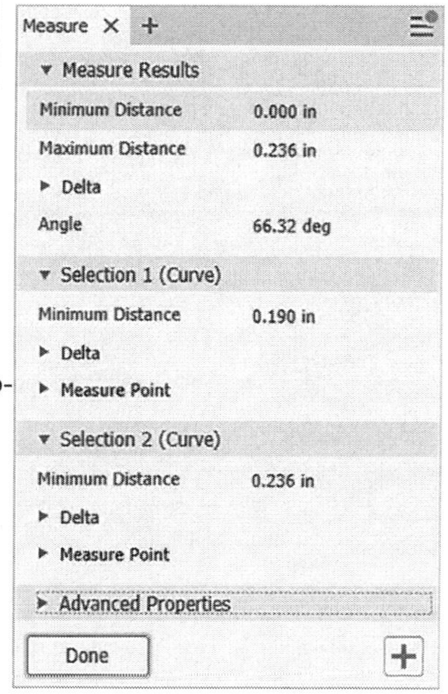

Figure 2.79

Loop

Start the Measure command and then move the cursor over an edge and let the cursor stay still for two seconds and then click Curve Loop from the toolbar as shown in the following image on the left. The loop distance will appear in the Measure property panel as shown in the following image on the right.

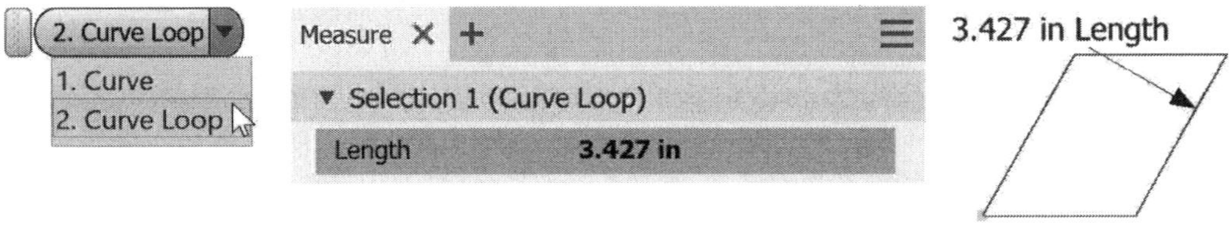

Figure 2.80

Measure Perimeter and Area

Measure the area by selecting inside an enclosed face of a 3D solid or surface. To get area data of a closed 2D Profile, you can also use the Region Properties command that is covered next.

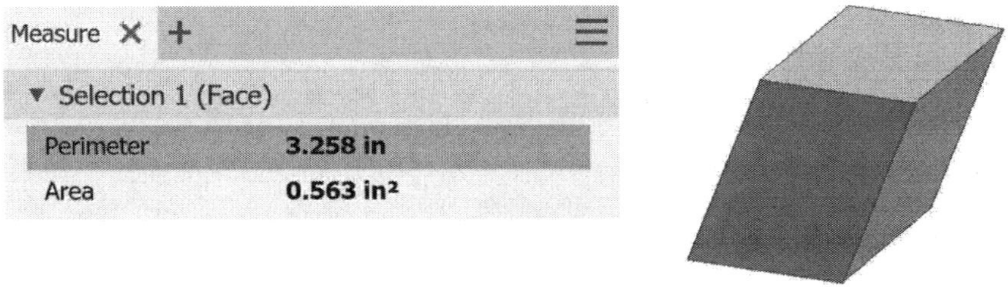

Figure 2.81

TIPS: Copy measured data to the clipboard by right-clicking in the cell containing the data and choosing Copy from the menu. ■ To restart measure selection, click anywhere in the graphics window.

Set Precision and Dual Units

To set the precision and display the results in dual units, click the right facing arrow to the left of the Advanced Properties area as shown in Figure 2.82.

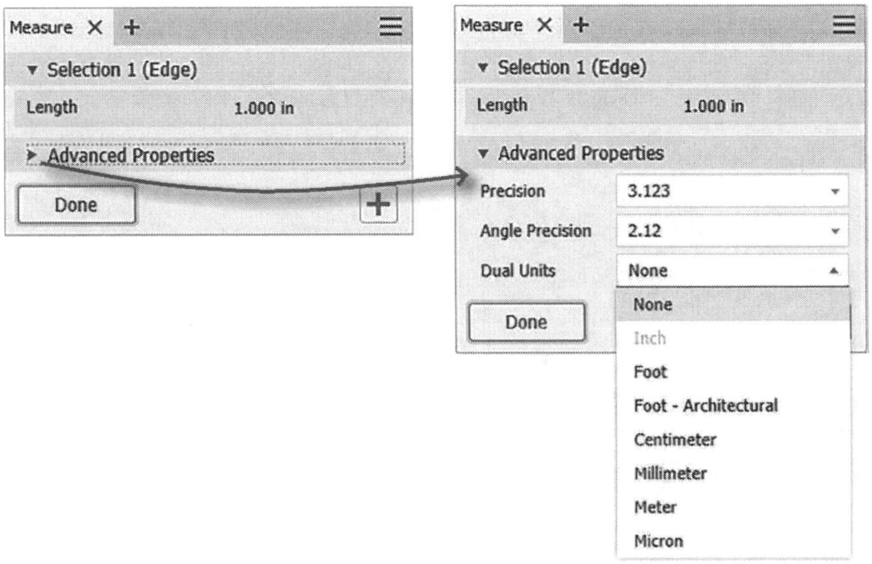

Figure 2.82

REGION PROPERTIES

The Region Properties command located on the Inspect tab > Measure panel can be used to determine the area, perimeter, and moments of inertia of a closed 2D sketch while in Sketch mode (Figure 2.83.)

Figure 2.83

Measurements are taken from the sketch coordinate system (0,0). After selecting inside a closed region, click the Calculate button. The properties can be displayed in the document's default units or dual units of your choice. The following image shows the region properties of a closed profile that has a void consisting of a slot.

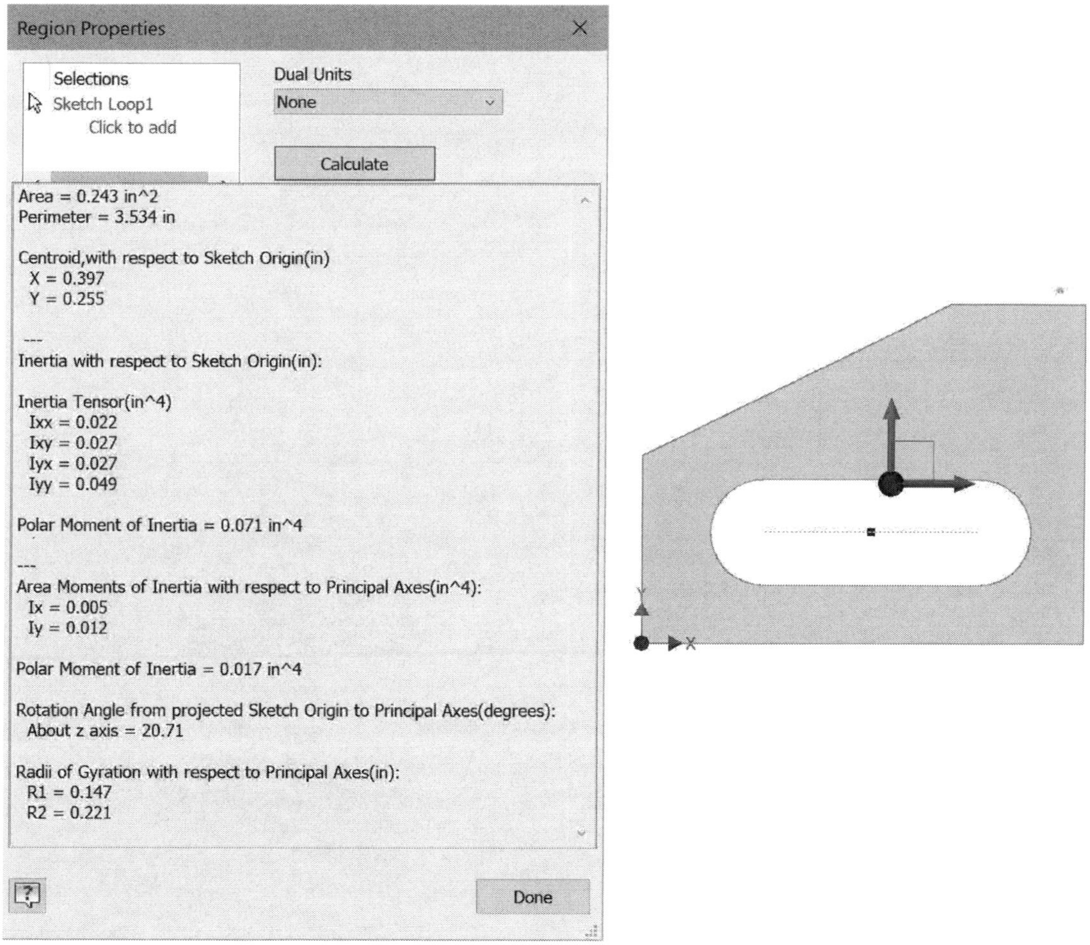

Figure 2.84

INSERTING AN AUTOCAD FILE

You may have legacy AutoCAD files or receive AutoCAD files that you need to convert into Inventor parts. In this section, you learn how to insert AutoCAD data into a sketch. When importing a 2D DWG file into Autodesk Inventor, you can either copy the contents from the DWG file to the clipboard via Autodesk Inventor or AutoCAD and paste the contents into Autodesk Inventor, or use an import wizard that guides you through the process. In this section, you learn how to insert AutoCAD 2D data into a sketch in an Inventor part file.

TIP: AutoCAD does not need to be installed to import AutoCAD geometry into Autodesk Inventor.

Inserting 2D AutoCAD Data into a Sketch

This section describes how to insert AutoCAD 2D data into the active sketch in a part or drawing. To insert AutoCAD data into the active sketch, follow these steps:

1. Start a new Inventor part file or open an existing part file.
2. Create a new sketch or make an existing sketch active.
3. Click the Insert AutoCAD File command on the Sketch tab > Insert panel as shown in the following image.

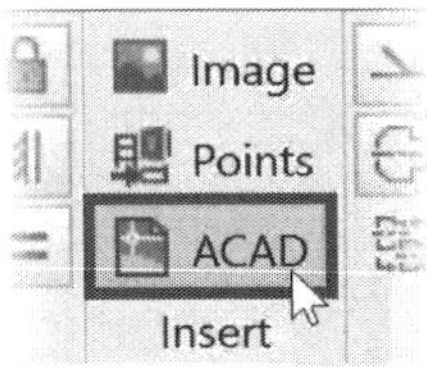

Figure 2.85

4. The Open dialog box will appear. Browse to and either double-click the desired DWG file or click on the DWG file and then click Open.

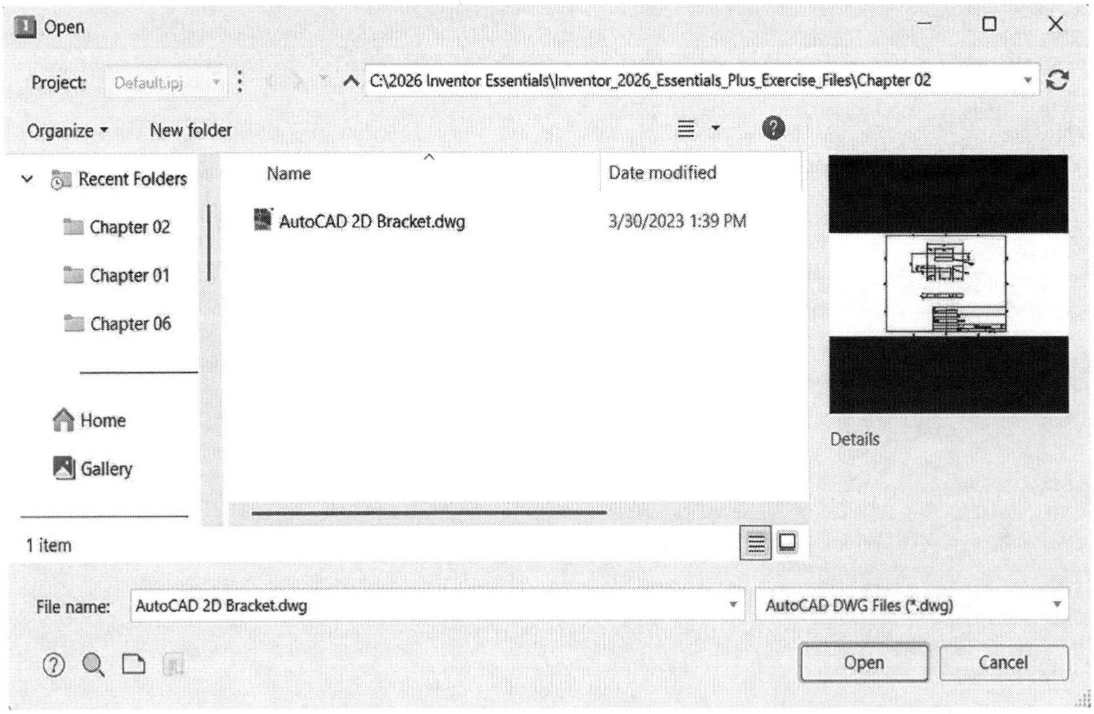

Figure 2.86

5. The Layers and Objects Import Options dialog box appears. In the Selective import section in the upper-left corner of the dialog box, uncheck the layer names you do not want data imported from as shown in the following image.

6. To select specific objects to insert, uncheck the All option, and then select the desired data in the preview window. In the preview window, you can zoom and pan as needed.

7. You can change the background color of the preview image by clicking the black or white icon at the top-right corner of the dialog box.

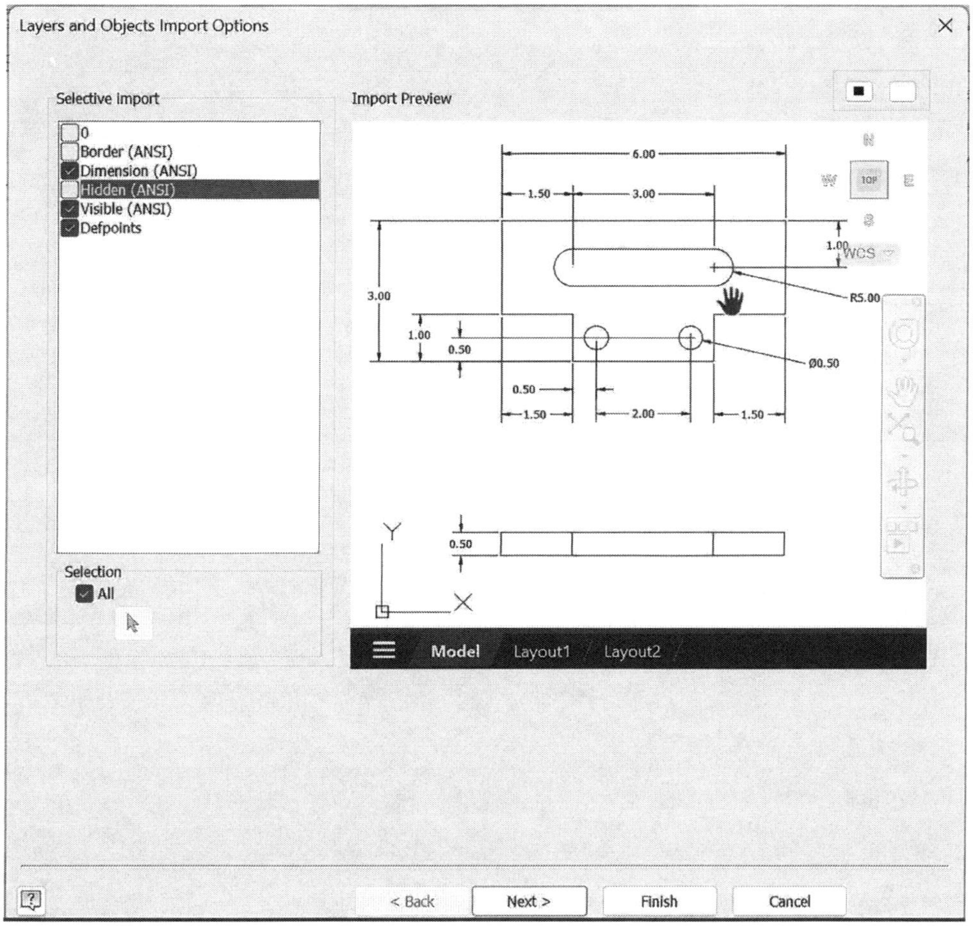

Figure 2.87

8. Click the Next button to go to the next step. In the Import Destination Options dialog box specify the units in which the data was created as shown in the following image.

9. Check the options to Constrain End Points and Apply geometric constraints as shown in Figure 2.88. The Apply geometric constraints option will add sketch constraints to geometry that is parallel, perpendicular and tangent.

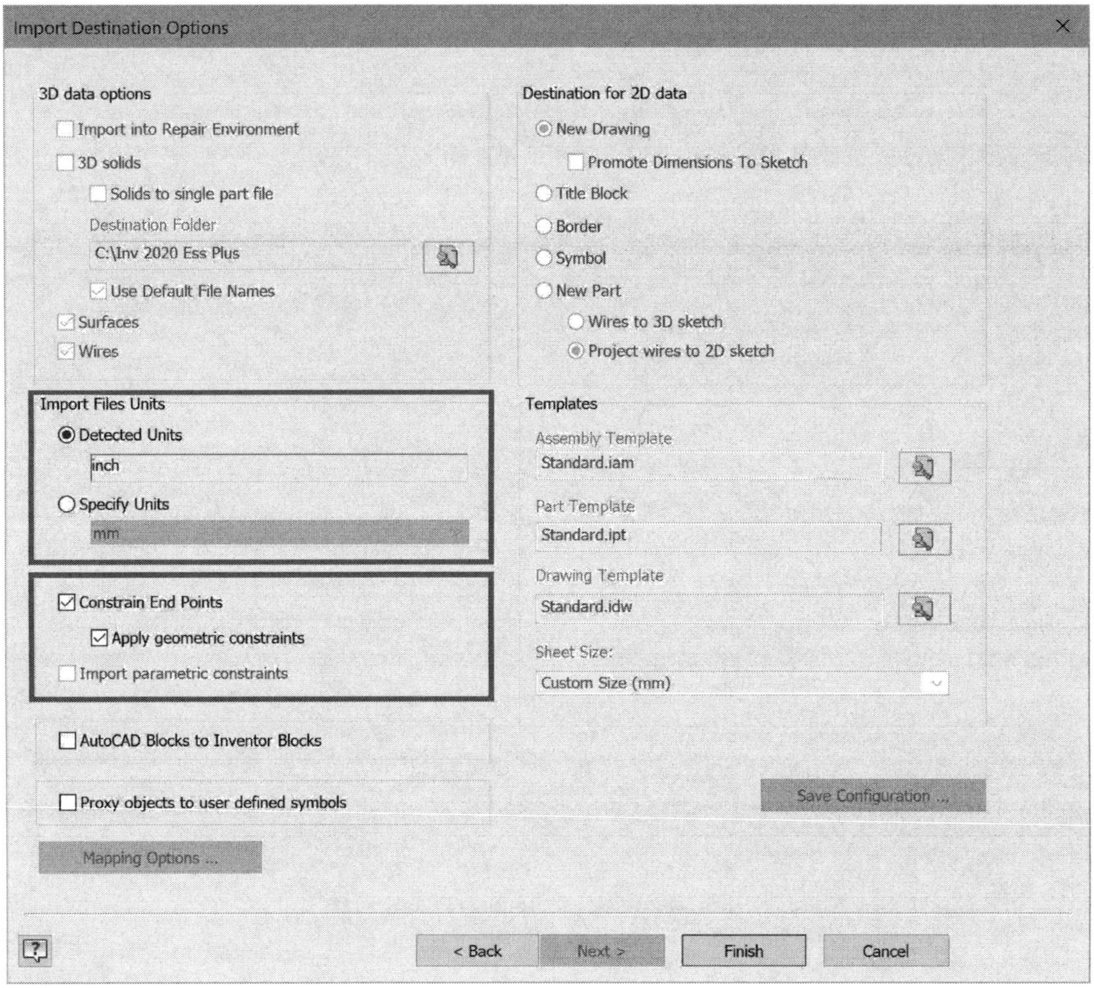

Figure 2.88

10. To import the data, click Finish.
11. Use the Zoom All command or double-click the wheel to see all the geometry.
12. Delete unnecessary geometry, constraints, and dimensions.
13. Add geometry (if needed), constraints, and dimensions to fully constrain the sketch.

Insert AutoCAD File with Associativity

If you want to import AutoCAD data as an underlay, which allows you to project the imported geometry onto a sketch, use the Import command on the Manage tab > Insert panel. If the imported AutoCAD data changes, you can update the geometry by clicking the Update command on the Quick Access toolbar.

OPEN OTHER FILE TYPES

Inventor can open parts and assemblies exported from other CAD systems. When files from other CAD systems are opened in Inventor, they will be imported as solids or surface models depending upon the original file and the components will NOT have feature history and an assembly will NOT have any assembly constraints.

You can add features to imported parts and edit the geometry by using Inventor's Direct Edit command. For files that are imported as an assembly, you can add assembly constraints.

To open file types such as DXF, Alias, Catia, IDF Board Files, IGES, JT, Parasolid, PRO/E, SAT, STEP, SolidWorks, and Unigraphics NX, click the File tab > Open or click Open on the Quick Access toolbar.

You can also select the Import CAD Files option using the Import DWG command from the File tab > Open > Import CAD Files; this command will import AutoCAD data into a new drawing, title block, border, symbol or part file without having to first create a new file.

In the Open dialog box, choose the desired file format from the Files of type list. See the help system for more information about the different file types.

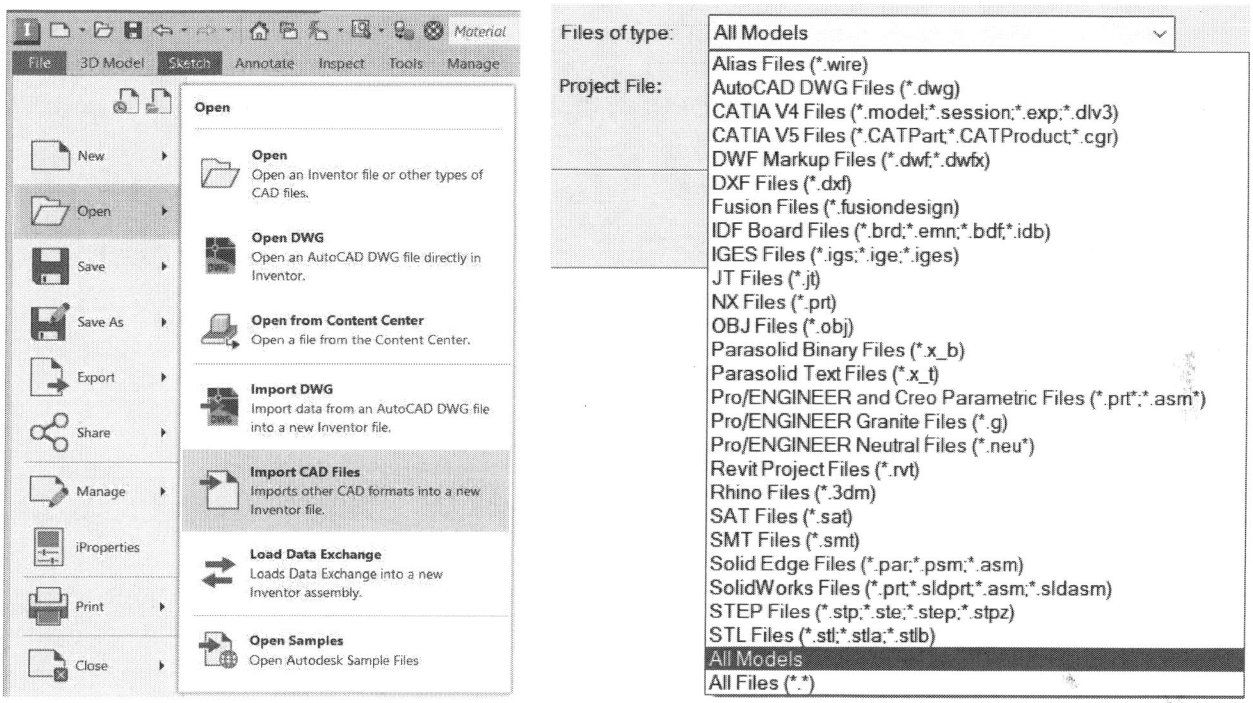

Figure 2.89

EXERCISE 2-5: INSERTING AUTOCAD DATA

In this exercise, you insert AutoCAD data into a sketch and add constraints to fully constrain the sketch.

1. Use the New command, English tab, and choose Standard (in).ipt, or if inch is already the default unit, from the left side of the Quick Access toolbar you can click the down arrow of the New icon, and select Part.

2. Click Start 2D Sketch from the 3D Model tab > Sketch panel. Select the XY origin sketch plane.

3. Click ACAD from the Sketch tab > Insert panel.

4. From the Frequently Used Subfolder area (dialog box upper-left) click the Chapter 02 subfolder and then in the file area double-click to choose the file *AutoCAD 2D Bracket.dwg*.

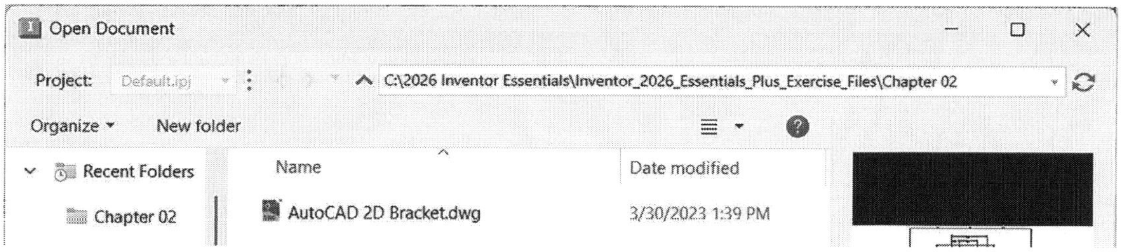

Figure 2.90

5 From the Layers and Objects Import Options dialog box *uncheck* the layers that are not needed. In the upper left corner of the dialog box *uncheck* layers 0, Border (ANSI) and Hidden (ANSI) as shown in Figure 2.91 (1).

6 In the Selection area near the bottom left corner of the dialog box uncheck All, labeled (2) in the following image.

7 Select the geometry and dimensions to insert, use the window selection–click a point above and to the left of the geometry labeled (3) and then click a point below and to the right of the geometry labeled (4). Note that if you click and drag you can draw an irregular shape around the geometry.

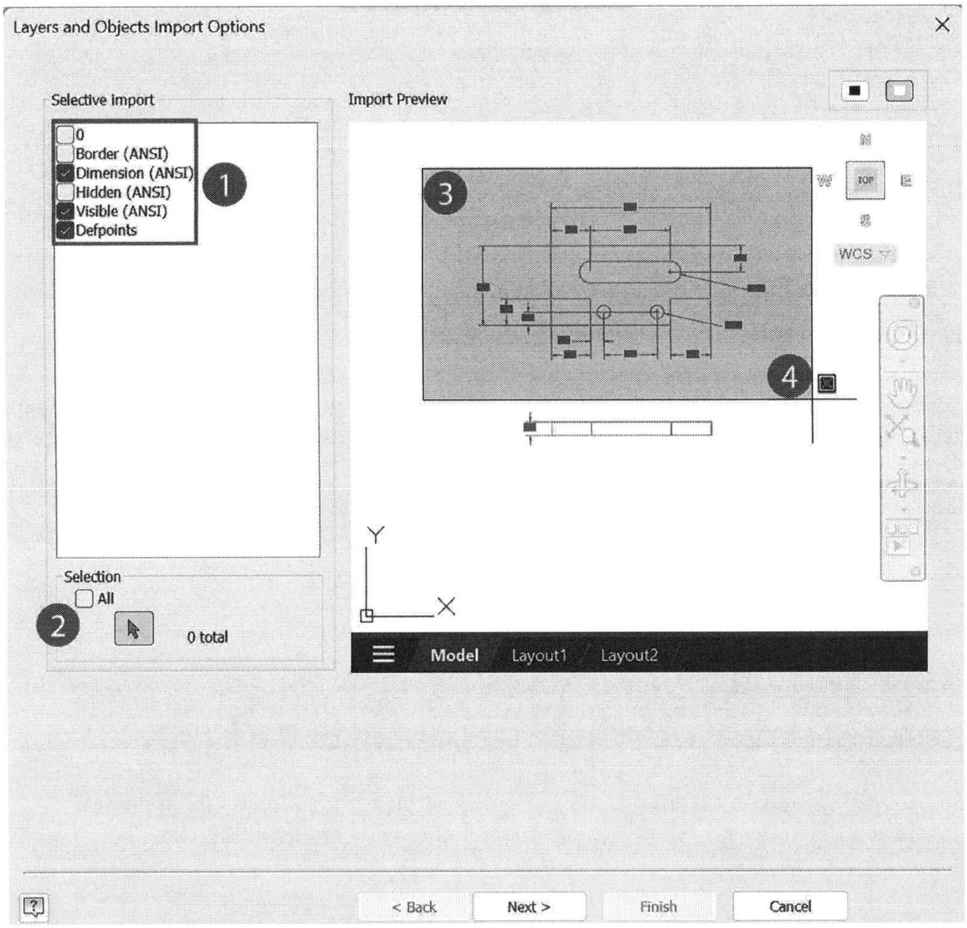

Figure 2.91

8 In the Selection area of the dialog box, verify that 27 total objects are selected as shown in Figure 2.92. If not, reselect all the data in the top view.

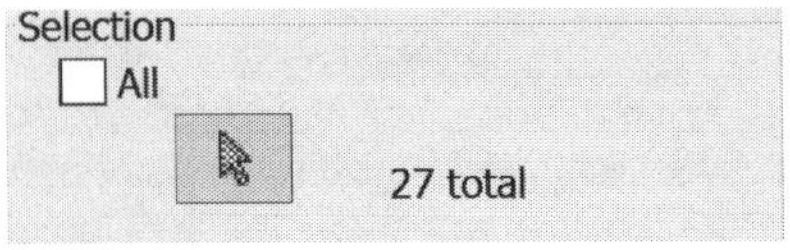

Figure 2.92

9 Click Next from the bottom of the dialog box.

10 In the Import Destination Options dialog ensure inch is set as the detected unit.

11 Check the Constrain End Points and the Apply geometric constraints options as shown in Figure 2.93.

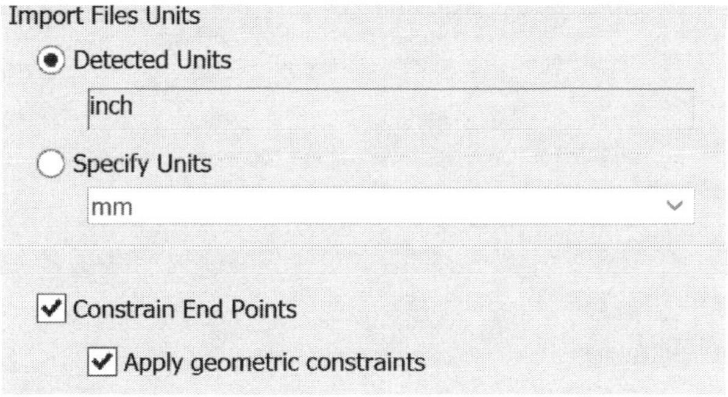

Figure 2.93

12 Click Finish from the bottom of the dialog box.

13 If needed, display all the geometry by double-clicking the mouse wheel.

14 Apply a horizontal constraint between the center points of the two circles labeled (1) in Figure 2.94.

15 Apply a collinear constraint between the two middle horizontal lines labeled (2).

16 Press the ESC key twice to cancel the command.

17 The sketch is free to move. To constrain the sketch to the origin, drag the lower-left corner of the sketch labeled (3) in the following image to the origin point of the sketch (0,0). Or you could add a coincident constraint between the origin point and the left point on the bottom line.

Note that you may need to zoom out to see the origin below and to the left of the inserted geometry.

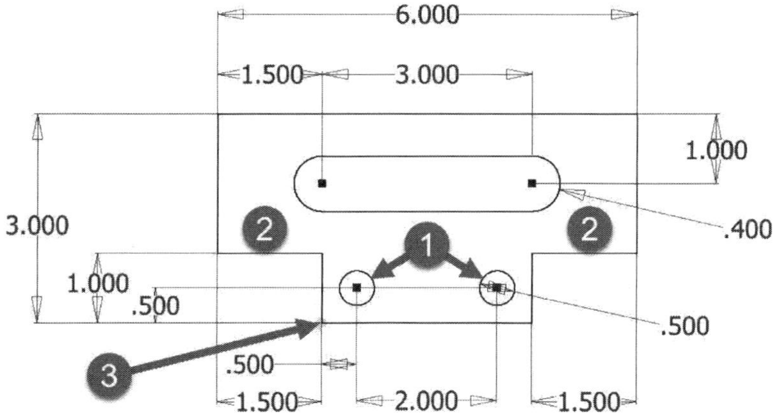

Figure 2.94

18 The lower-right of the Status Bar should show "1 dimensions needed" to constrain the sketch.

19 Drag the top-right endpoint of the upper horizontal line up. The sketch will be rotated slightly as shown in Figure 2.95 left.

20 Apply a horizontal constraint to the lower horizontal line, and this will fully constrain the sketch as shown in Figure 2.95 right. Reposition the dimensions as needed.

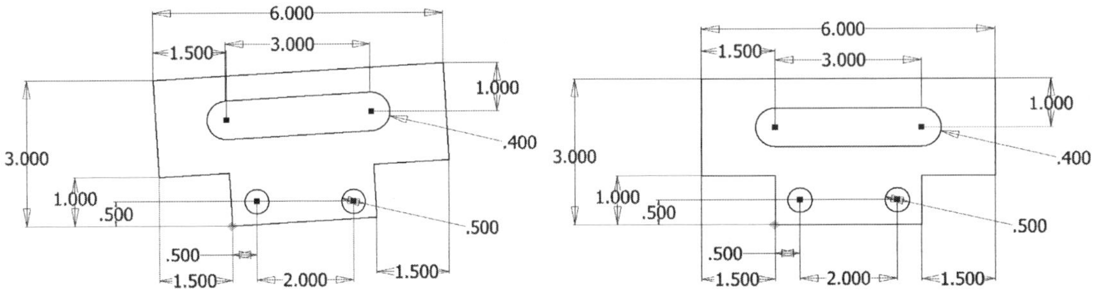

Figure 2.95

21 Press the F8 key to see all constraints.

22 Press the F9 key to hide all constraints.

23 The AutoCAD dimensions on the sketch are now parametric and can be edited. Practice editing the values of the dimensions by double-clicking on a dimension's value and enter a new value.

24 Close the file. Do not save changes.

End of exercise.

APPLYING YOUR SKILLS

Skills Exercise 2-1

Start a new part file based on the Standard (in).ipt, choose the XY sketch plane. Create the sketch and add geometric and dimensional constraints to fully control the size and shape for the part below.

- Assume that the top and bottom horizontal lines are collinear.
- The center points of the arcs are aligned vertically.
- The sketch is symmetric about the left and right sides.
- The bottom angled lines should be coincident with the center point of the lower arc (if the arc is drawn via the line command, the center point of the arc will automatically be coincident with the line it was drawn from).

When done, close the file and do not save the changes.

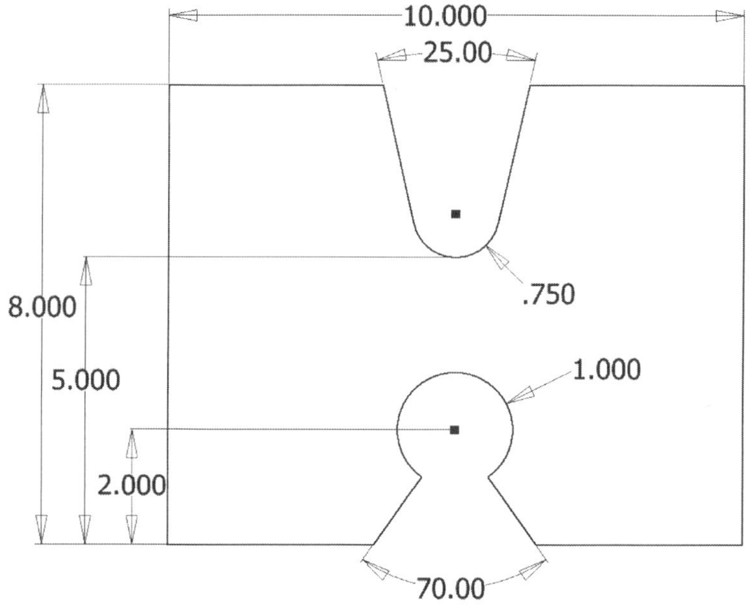

Figure 2.96

Skills Exercise 2-2

Start a new part file based on the Standard (in).ipt template. Create a fully constrained sketch on the XY plane using lines and arcs as shown in the following image. First sketch the two circles and align their center points horizontally. Then create the two lines and place a vertical constraint between the line endpoints on both ends. Add the dimensions. Check that your sketch is fully constrained. When done, close the file and do not save the changes.

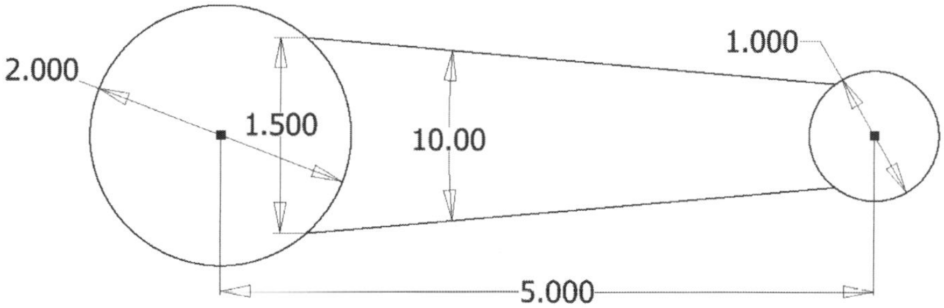

Figure 2.97

CHECKING YOUR SKILLS

Use these questions to test your knowledge of the material in this chapter.

1. True__ False__ While sketching, by default, geometric constraints are not applied to the sketch.
2. True__ False__ When you sketch and a point is inferred, a constraint is applied to represent that relationship.
3. True__ False__ It is recommended to never fully constrain a sketch.
4. True__ False__ When working on a millimeter part, you cannot input inch units.
5. True__ False__ After a sketch is fully constrained, you cannot change a dimension's value.
6. True__ False__ A driven dimension is another name for a parametric dimension.
7. True__ False__ Dimensions placed dynamically are not parametric.
8. True__ False__ You can only import 2D AutoCAD data into Autodesk Inventor.
9. Explain how to draw an arc while using the Line command.
10. Explain how to remove a geometric constraint from a sketch.
11. Explain how to change a vertical dimension to an aligned dimension while placing the dimension.
12. Explain how to create a dimension that is tangent to two arcs.
13. True__ False__ AutoCAD needs to be installed to insert AutoCAD geometry.
14. True__ False__ When a sketch is extruded that contains construction geometry, the construction geometry is deleted.
15. Explain how to change the unit type in a part file.
16. Explain where you would turn on Relax Mode.
17. True__ False__ When a pushpin appears in the Sketch entry in the browser, the sketch is fully constrained.
18. True__ False__ By default an arc length dimension can only be a driven dimension.
19. Explain how to draw a rectangle that is centered on the origin point.
20. True__ False__ When creating the first 2D sketch, you must select an origin plane to sketch on.

3 Creating & Editing Sketched Features

INTRODUCTION

After you have drawn, constrained, and dimensioned a sketch, your next step is to turn the sketch into a 3D part. This chapter takes you through the process to create and edit sketched features and to create features using primitive shapes.

OBJECTIVES

After completing this chapter, you will be able to perform the following:

- ☐ Describe what a feature is used for in the modeling process
- ☐ Describe the functions of Autodesk Inventor's browser
- ☐ Use direct manipulation techniques to create and edit a part
- ☐ Extrude a sketch into a part
- ☐ Revolve a sketch into a part
- ☐ Edit features of a part
- ☐ Edit the sketch of a feature
- ☐ Create a sketch on a plane
- ☐ Create sketched features using one of three operations: cut, join, or intersect
- ☐ Project edges of a part
- ☐ Create primitive features

FEATURES

After creating, constraining, and dimensioning a sketch, the next step is to turn the sketch into a 3D feature. The first sketch of a part used to create a 3D feature is referred to as the *base feature*. Features are the building blocks of part creation.

In addition to the base feature, you can create secondary *sketched features* on planar faces or work planes. Use the Extrude, Revolve, Sweep, or Loft commands to create sketched features in a part. Secondary features are covered later in this chapter.

You can also create *placed features* such as fillets, chamfers, and holes for features that have been created. Placed features are covered in Chapter 4. A plate with a hole in it, for example, would have a base feature representing the plate and a hole feature representing the hole. As features are added to the part, they appear in the browser and the history of the part or assembly, that is, in the order in which the features are created or the parts are assembled. Features can be edited, deleted from, or reordered in the part as required.

You can also add or subtract material to or from existing features in a part using modify commands like Split and Combine.

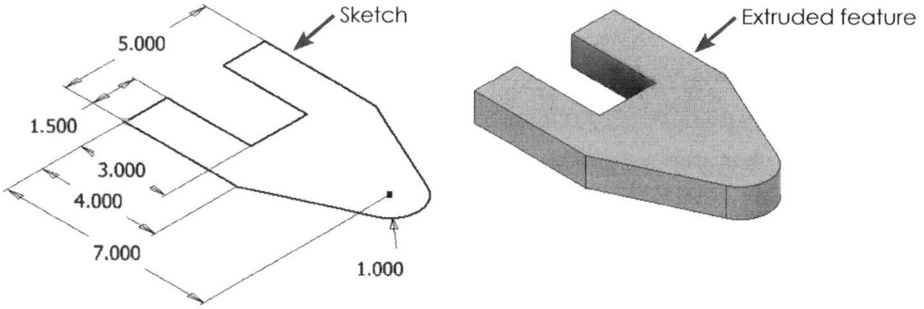

Figure 3.1

Consumed and Unconsumed Sketches

You can use any sketch as a profile in feature creation. A sketch that has not yet been used in a feature is called an *unconsumed sketch*. When you turn a 2D sketched profile into a 3D feature, the sketch becomes a subordinate part of the feature and is said to be "consumed." Figure 3.2 shows unconsumed Sketch2 in the browser. Sketch1 is consumed by the Extrusion1 feature. If desired, you can display more information in the browser about the features by clicking on the menu icon ≡ in the browser > Display Preferences and click Show Extended Names as shown at the right.

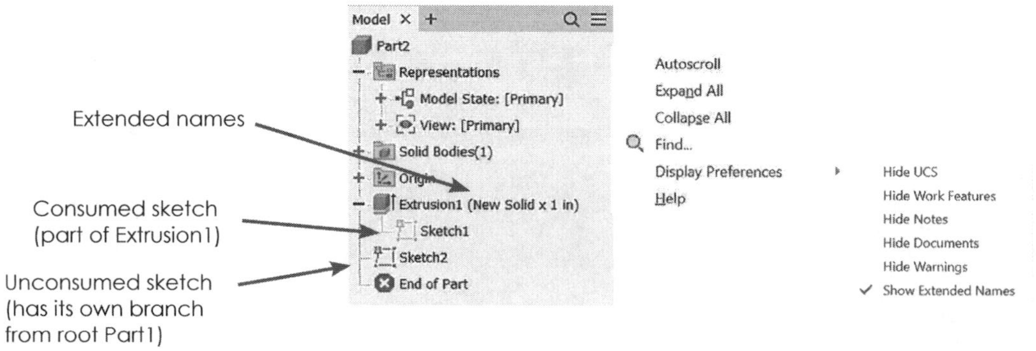

Figure 3.2

Although a consumed sketch is not visible as you view the 3D feature, you may need to access sketches and change their geometric or dimensional constraints to modify their associated features. A consumed sketch can be accessed from the browser by right-clicking and selecting Edit Sketch from the menu or by using the direct-manipulation mini-toolbar. The editing process will be covered later in this chapter. You can also edit the sketch by starting the Sketch command and selecting the sketch from the browser.

UNDERSTANDING THE BROWSER

The Inventor browser is an important part of managing and editing your parts, features, and assemblies. By default the browser is docked along the left side of the screen. It displays the history of the file.

In the browser you can create, edit, rename, copy, delete, and reorder features or parts. You can expand or collapse the feature's details by clicking the +/– symbols along the left of the items in the browser.

To expand all the features, move the cursor into a blank area in the browser, right-click in a blank area in the browser and choose Expand All from the menu. Figure 3.3 shows a browser with the features expanded.

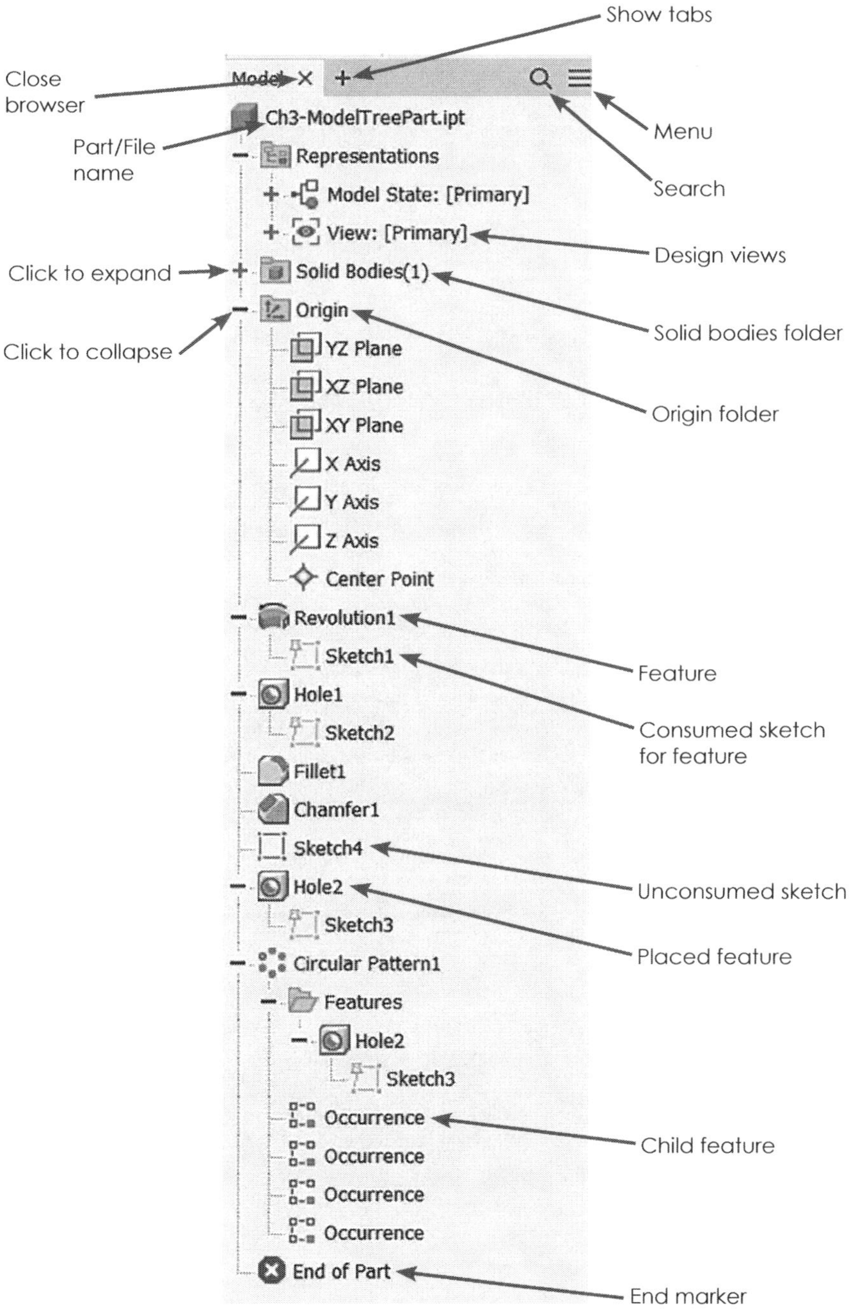

Figure 3.3

As you model a part it gets more complex and so does the information in the browser. Dependent features are indented to show that they relate to the item listed above it. This is referred to as a *parent-child relationship*. For example, if a hole is a child feature of an extruded rectangle and the extrusion is deleted, the hole will also be deleted.

Each feature in the browser is given a default name. The first extrusion, for example, will be named Extrusion1, the next Extrusion2 and so on. It is a good practice to rename the features to descriptions that are meaningful for your part. Click to select an item in the browser so it becomes highlighted, pause briefly and click it again. The typing cursor appears in the item's name so you can edit it.

The browser can help you to locate features in the graphics window. To highlight a feature in the graphics window, simply move your cursor over the feature name in the browser.

To zoom in on a selected feature, right-click on the feature's name in the browser and select Find in Window or press the END key on your keyboard.

Unlike toolbars, the browser can be resized while it is docked. To close the browser, click the X in its upper-left corner.

If the browser is not visible on the screen, you can show it by clicking the + (Show tabs) from the upper left corner of the graphics window, to show the menu, then click Model Browser (Figure 3.4).

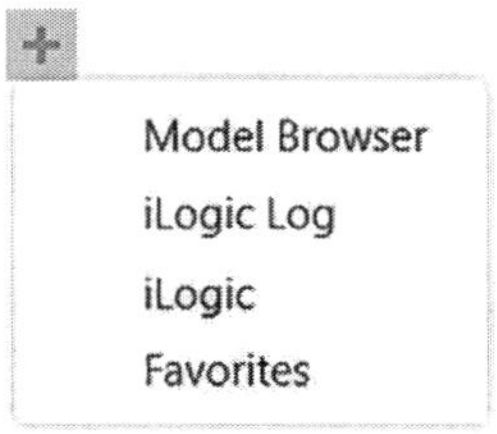

Figure 3.4

Specific functionality of the browser will be covered throughout this book in the pertinent sections.

SWITCHING ENVIRONMENTS

Up to this point, you have been working in the sketch environment where the work is done in 2D. The next step is to turn the sketch into a feature. To do so, you need to exit the sketch environment and enter the part environment. Several methods can be used to do this:

- Click the Finish Sketch command on the right side of the Sketch tab > Exit panel as shown in the following image on the left.
- In a blank area in the graphics window right-click and click Finish 2D Sketch from the marking menu as shown in the middle image.
- Press the S key on the keyboard.
- Click the 3D Model tab as shown in the image on the right and click on a command.
- Enter a shortcut key to initiate a feature creation command.

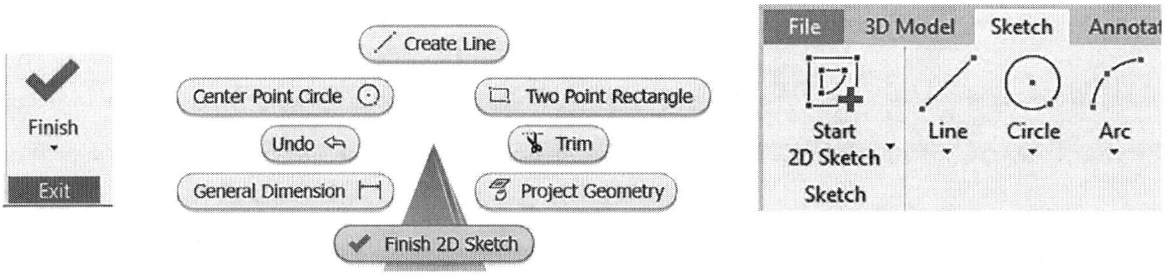

Figure 3.5

3D Modeling Commands

When you exit the sketch environment, the 3D Model tab is active. Figure 3.6 shows the 3D Model tab. Many of these commands will be covered throughout this book.

Figure 3.6

DIRECT MANIPULATION

Direct manipulation allows you to start common commands and operations by clicking directly on the geometry in the graphics window. The direct manipulation tools appear as mini-toolbars or in-canvas buttons. Each direct manipulation option is explained next.

Mini-Toolbars

Mini-toolbar options are different depending upon what geometry is selected. After selecting the geometry click on the desired option from the mini-toolbar. The options will be explained later in more detail.

Sketch

While not in the sketch environment and not in a command, select a sketch and a mini-toolbar appears and provides you the option to Extrude, Revolve, Edit Sketch, or Make Sketch Invisible.

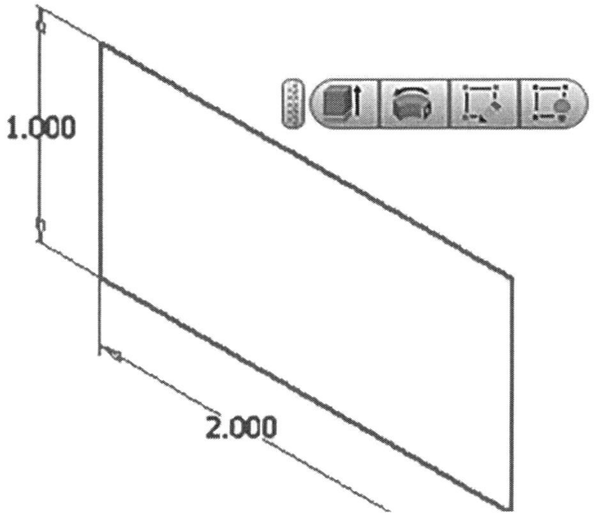

Figure 3.7

Face

While not in a command and you select a face, a mini-toolbar appears that allows you to quickly edit the extrude, edit the sketch, share the sketch, make the sketch visible, or create a new sketch.

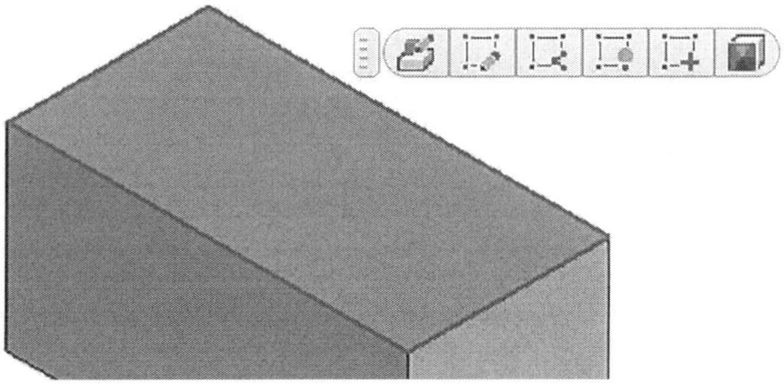

Figure 3.8

Edge

While not in a command and you select an edge, the mini-toolbar appears and provides you the option to Fillet or Chamfer the selected edge(s).

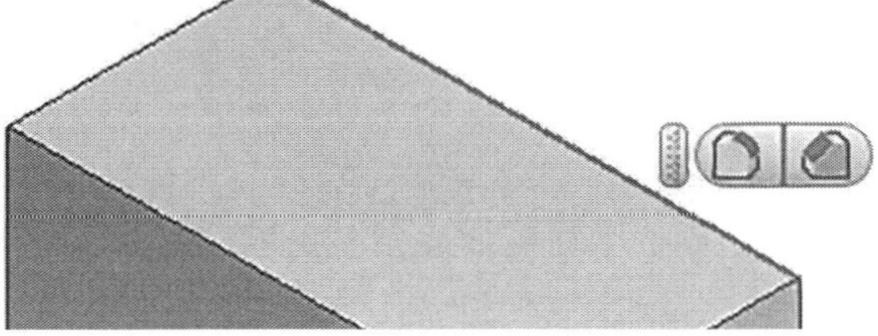

Figure 3.9

PROPERTY(IES) PANELS

Many of Inventor's modeling commands have a property panel that allows you to enter data to complete the command. The following image shows the Extrude command property panel.

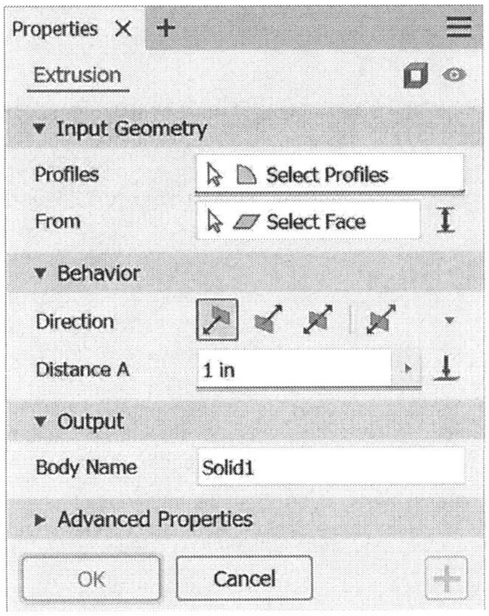

Figure 3.10

The main areas in a property panel and best practices are as follows.

- Workflow. Enter data from the top down. Entered data affects properties in sections below it, but a property does not affect the preceeding properties.
- Advanced menu where you can show and hide presets (reusable feature definitions), context entries and help.
- Surface mode. The default extrude type is a solid body. Click to turn ON Surface mode and instead of creating a solid, you will get just exterior surfaces. This setting is saved as part of a preset. For this text, create solids unless you are directed otherwise.
- Turns the preview on and off.
- Red text and = Input needed. Enter data or select required geometry.
- When you've entered the required data to create the feature the + icon in the lower-right corner of the property panel will change to green. Click this icon to create the feature and keep the property panel open where you can create a new feature with the same settings or modify the settings as required.
- **Docking.** The property panel can dock on an edge of the graphics window, within the browser, or can float.

EXTRUDE A SKETCH

The most common method for creating a feature is to extrude a sketch and give it depth along the Z-axis. Before extruding, it is helpful to view the part in an isometric view. When you exit the first sketch the viewpoint automatically changes to the home view. Inventor previews the extrusion depth and direction in the graphics window.

To extrude a sketch, follow one of these techniques:

- Finish the sketch, and click the Extrude command on the 3D Model tab > Create panel as shown in Figure 3.11 left.
- While not in a sketch, click on any geometry on the sketch and the mini-toolbar will appear, then click the Create Extrude button as shown in the middle image.

- Alternately, you can right-click in the graphics window and select Extrude on the marking menu as shown in Figure 3.11 at the right.
- Press the shortcut key E.

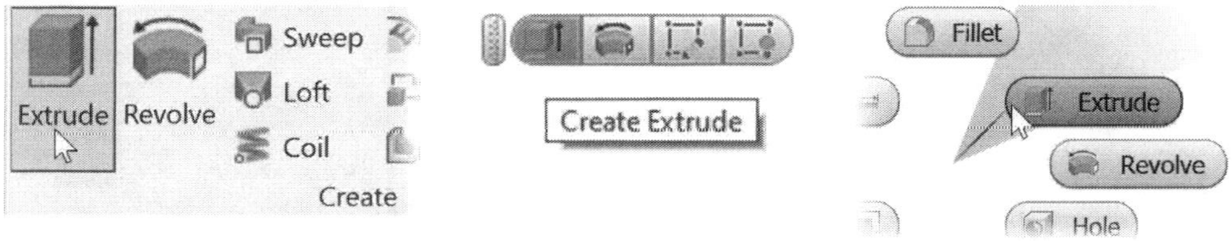

Figure 3.11

After starting the Extrude command, the Extrude Property panel appears. Figure 3.12 left shows the options available for the first sketch. The right image shows operations available when creating additional extrude features.

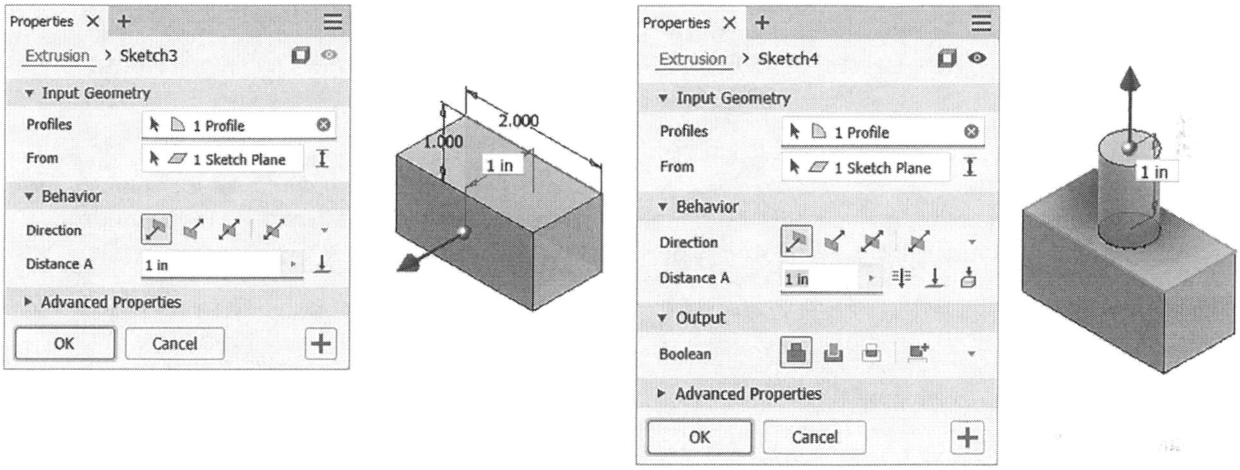

Figure 3.12

As you make changes in the property panel, the graphics window updates to represent the chosen values and options. When you have entered the values and the options you need, click OK to create the extruded feature. The common options to create an extrude feature are explained next.

TIP: Red text and/or the ⚠ icon appears in the property panel to alert you that input from you is needed to complete that function.

Input Geometry

Use this section to select the area to extrude and define where the extrude feature starts.

Profiles	Click this area to choose the sketch that you want to extrude. If there are multiple closed profiles, you can select on the sketch profile(s) or inside the profile(s) that will be extruded. If there is only one closed profile, Inventor will select it for you and you can skip this step. If you select the wrong profile or sketch area, click the ⊗icon and choose the desired profile or sketch area. To remove a selected profile, hold the CTRL key and click the area you wish to remove.
From	Select where the extrude feature will start.

Behavior Use this section to set the direction and distance of the extruded feature.

Direction Use this section to set the direction for the extruded feature.

	Default	Extrudes in the positive direction.
	Flipped	Flips the direction to the opposite side.
	Symmetric	The extrusion goes an equal distance in the negative and positive directions.
	Asymmetric	Define a different distance for the negative and positive direction.

Distance

In this area you define the length of the extruded feature. You can either enter a value in the property panel or in the cell in the graphics window where the profile will be extruded. You can also click and drag the arrow in the graphics window.

Another option is to click in the distance cell and then click the small arrow to the right of the cell and then click the Measure option as shown in Figure 3.13. From there you can select two points, two faces or an edge to set the value. You can also use the menu to select from the list recently used values. After a value is determined, a preview image of the extrusion appears in the graphics window.

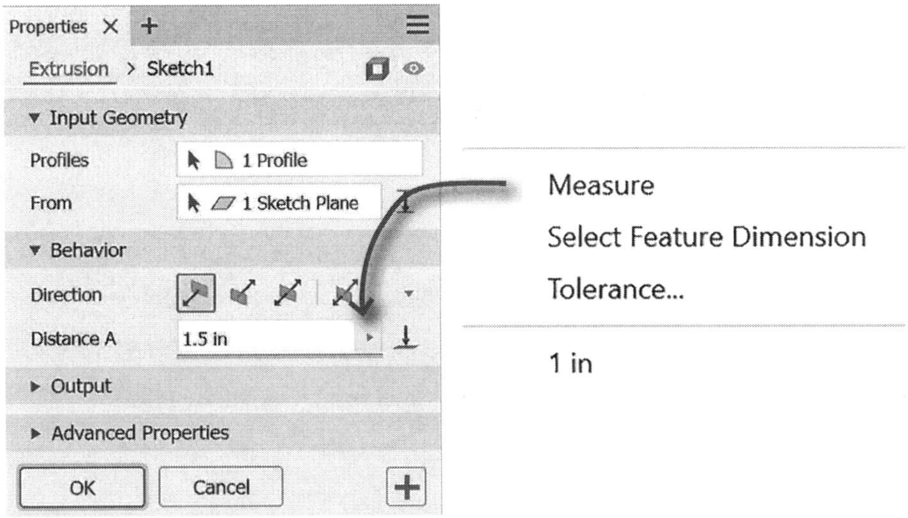

Figure 3.13

Extents

The Extent options to the right of the Distance cell determine how the extruded sketch terminates. Note that some options are not available until a base feature exists.

	Through All	Extrudes the profile all the way through the part in one or both directions.
	To Next	This option automatically extrudes the profile until it reaches a plane or face or lets you select a plane or face to terminate the extrusion. The sketch must be fully enclosed in the area to which it is terminating. If it is not fully enclosed, use the To termination type with the Extend to Surface option. Click the Direction button to determine the extrusion direction.

	To	Extrudes the profile until it reaches a selected face, plane or point. After selecting a face, plane or point you'll get two additional options: **Extend face to end feature**: When on, the terminating face is extended so a profile that extends beyond the selected face will calculate correctly. Figure 3.14 left shows a hexagon being extruded to the top rectangular face with the extend face option on. If the option were off, the feature would fail to compute because the terminating rectangular face does not fully encompass the extruded hexagon. **Minimum Solution:** When on, the feature terminates on the first possible solution for the selected surface. Figure 3.14 middle shows the outer circle extruded to the outside face of a circular tube with the Minimum Solution option on. Figure 3.14 right shows the results with Minimum Solution turned off.

Figure 3.14

Output

If this is the first sketch that you create a solid from, it is referred to as a base feature and the Output section is not available. After a base feature has been established and you start the extrude command, the Output options will be available and with these options you add or remove material from the model.

	Join	Adds material to the part.
	Cut	Removes material from the part.
	Intersect	Removes material, keeping what is common to the model and the new feature.
	New Solid	Creates a new solid body. The first solid feature created uses this option by default. Select this option to create a new body in a part file that has an existing solid body.

Advanced Properties

The Advanced Properties section, Figure 3.15, contains additional options to refine the feature being created:

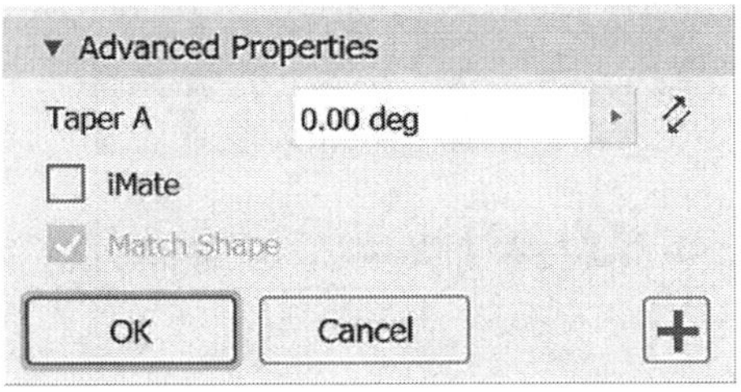

Figure 3.15

Taper

Taper extrudes the sketch and applies a taper angle to the feature. To extend the taper angle from the part, give the taper angle a positive or negative number. A positive taper increases the volume while a negative value decreases the volume of the resulting extruded feature. Enter a taper value in the property panel or click on the sphere in the graphics window and drag the arrow or enter a value as shown in the following image.

iMate

Places an iMate on a closed loop of the extrude feature. iMates are used to automate the assembly process.

Match Shape

Select an open profile in a part file and specify to Match Shape and select the side to keep.

EXERCISE 3-1: EXTRUDING A SKETCH

In this exercise, you create a base feature by extruding an existing profile. You will examine the direction options available in the Extrude dialog box.

1 Open *ESS_E03_01.ipt* from the Chapter 03 subfolder.

2 From the 3D Model tab > Create panel click the Extrude command or use the Create Extrude button from the mini-toolbar in the sketch. There is only one closed profile so it is automatically selected.

3 From the Extrude Properties panel, set Distance to **.5**, notice the extrusion preview in the graphics window.

4 In the graphics window click and drag the arrow on the extrusion until a distance of **1.000** shows in the Extrude dialog box and mini-toolbar as shown in Figure 3.16, and then release the mouse button.

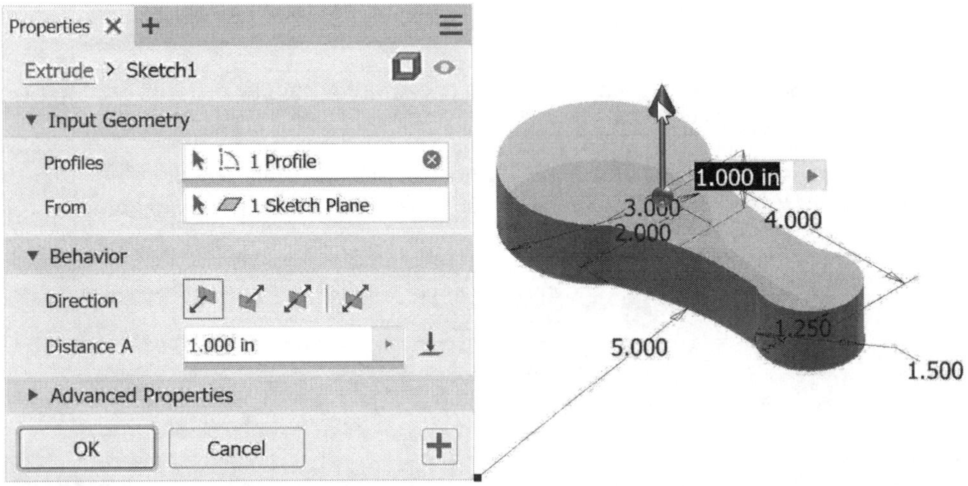

Figure 3.16

5 In the Extrude Properties panel expand the Advanced Properties section.

6 Change the taper by typing **20 deg** in the Taper area.

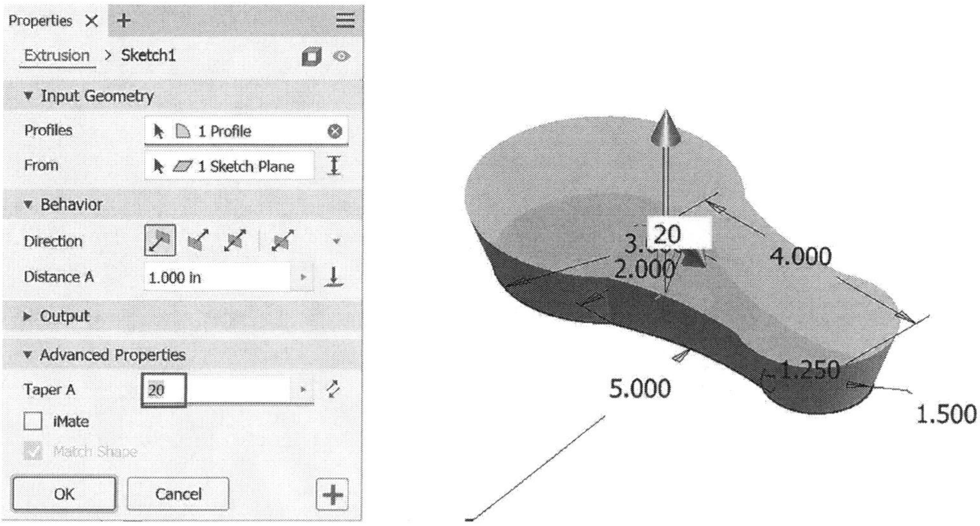

Figure 3.17

7 

In the graphics window, adjust the taper by clicking on the sphere and dragging the arrow to **10°** and then to **-10°.**

8 In the Extrude Properties panel change the direction of the extrusion to go in the opposite direction by clicking the Flipped icon, labeled (1) in the following image.

9 Change the direction to Asymmetric (extrude different values in each direction) labeled (2) in the following image and enter different values for each direction.

10 Change the extent to Symmetric labeled (3) in the following image.

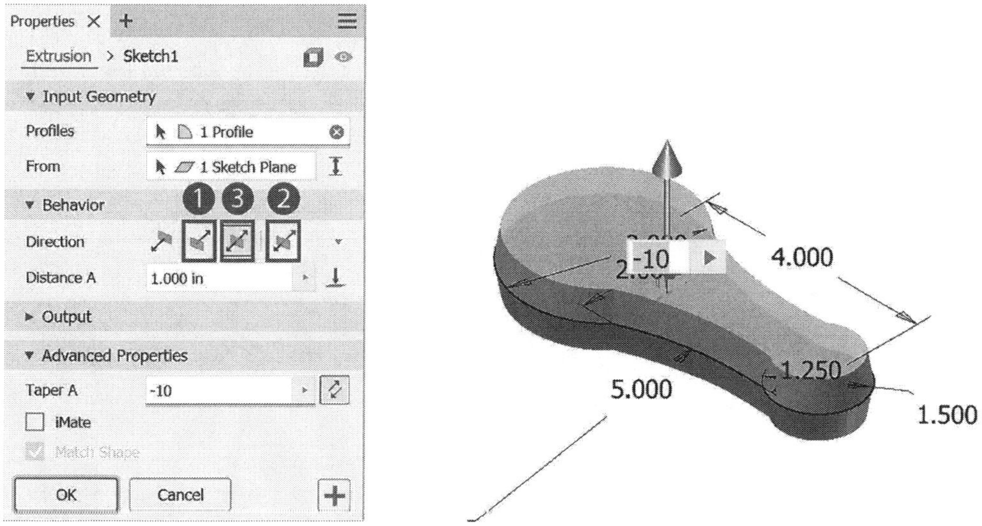

Figure 3.18

11 In the graphics window click and drag the arrow to different values.

12 Practice changing the values and directions to see the results in the Extrude Properties panel.

13 When done, click OK and the extrusion feature will be created. Later in this chapter you will learn how to edit features where you can change the values of a feature after it was created.

14 Close the file. Do not save changes.

End of exercise.

LINEAR DIAMETER DIMENSIONS

Another method for creating a part is to *revolve* a sketch to create cylindrical features. The steps to revolve are similar to those for extrude.

- You create the sketch, but the sketch is a *half-section* of the completed feature.
- You add constraints and dimensions and then revolve the sketch about a centerline, straight edge or axis.
- To define the sketch, you can use radial or diameter dimensions. To place diameter dimensions on a sketch you use a centerline or a normal line to define the linear diameter dimension.

The following sections explain how to create a diameter dimension with a centerline and a normal line.

Centerline

To create a centerline, activate the Centerline command on the Sketch tab > Format panel, as shown in Figure 3.19 left, and then sketch a line. Click on the centerline to deactivate it. If the command is not deactivated, all geometry that is sketched will be a centerline. You can also change an existing sketch entity to a centerline by selecting it and then select the Centerline command.

To create a linear diameter dimension by using a centerline, follow these steps:

1. Draw a sketch that represents a half-section of the finished feature or part.
2. If needed draw a line that will be used as the centerline to revolve about or use a line in the sketch that the sketch will be revolved around.
3. Change an existing line into a centerline that the sketch will be revolved around by clicking the Centerline command on the Sketch tab > Format panel. The centerline will be treated as a normal line and can be used to define a closed profile.
4. Start the Dimension command.
5. Click the centerline, and a point or edge to be dimensioned. It does not matter the order that the entities are selected; however, one of the selections needs to be the centerline, not just an endpoint of the centerline.
6. Move the cursor until the diameter dimension is in the correct location and click.
7. The following image on the right shows the linear diameter dimension created using the centerline to define the axis of revolution.

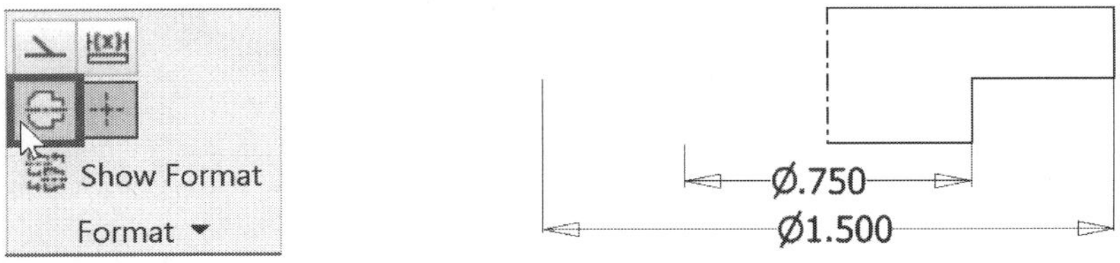

Figure 3.19

Normal Line

When a normal line is used to revolve the sketch around, you can create linear diameter (diametric) dimensions for sketches that represent a quarter outline of a revolved part. To create a linear diameter dimension, follow these steps:

1. Draw a sketch that represents a half-section of the finished feature or part. Remember a full section is as though you cut the part in half and looked at what remains. A half-section is half of that full section view.
2. Draw a line that will be used as the centerline to revolve about or use a line in the sketch that the sketch will be revolved around.
3. Start the Dimension command.
4. Click the line (not an endpoint) that the sketch will be revolved around (axis of rotation).
5. Click the other point or line to be dimensioned.
6. Right-click and select Linear Diameter from the menu as shown in Figure 3.20 left.
7. Move the cursor until the diameter dimension is in the correct location and click. Figure 3.20 right shows a sketch with a linear diameter dimension placed.

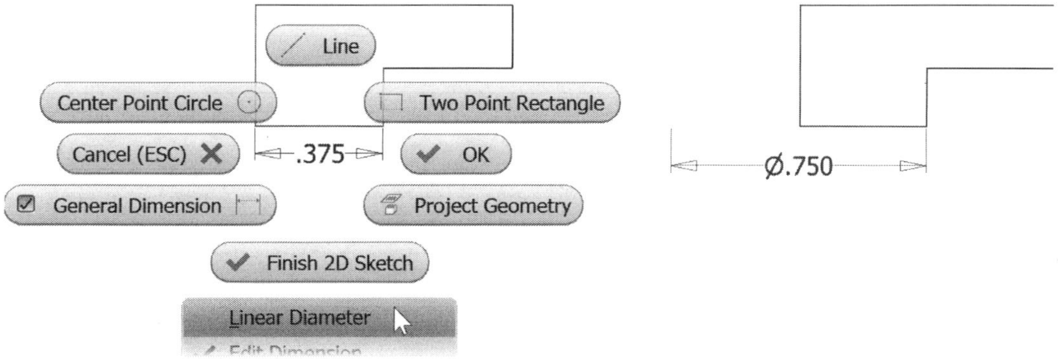

Figure 3.20

REVOLVE A SKETCH

After defining the sketch that will be revolved, use one of the following techniques to create a revolve feature:

- Click Revolve from the 3D Model tab > Create panel as shown in Figure 3.21 left.
- While not in sketch mode, click on the sketch geometry to show the mini-toolbar, then click Create Revolve button, as shown in the middle image.
- Right-click in the graphics window and choose Revolve from the marking menu (Figure 3.21 right).
- Press the R key.

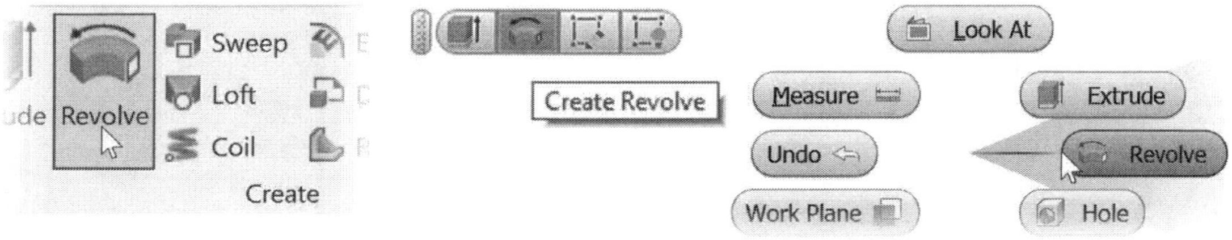

Figure 3.21

When the command starts, the Revolve Property panel appears. This is where you'll input data to create the revolve feature. The Revolve Property panel options will be discussed in the next section. Figure 3.22 left shows an example of a constrained sketch with a centerline. The image on the right shows the model after the sketch has been revolved.

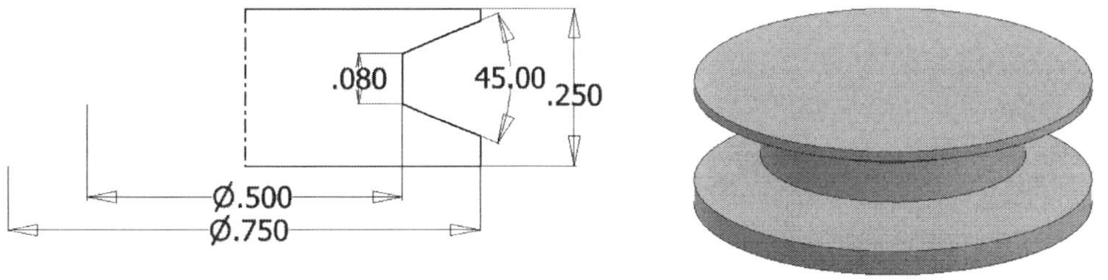

Figure 3.22

After starting the Revolve command, the Revolve Property panel appears. Figure 3.23 left shows options that are available when revolving the first sketch and the image on the right shows the operations that are available when creating additional revolve features.

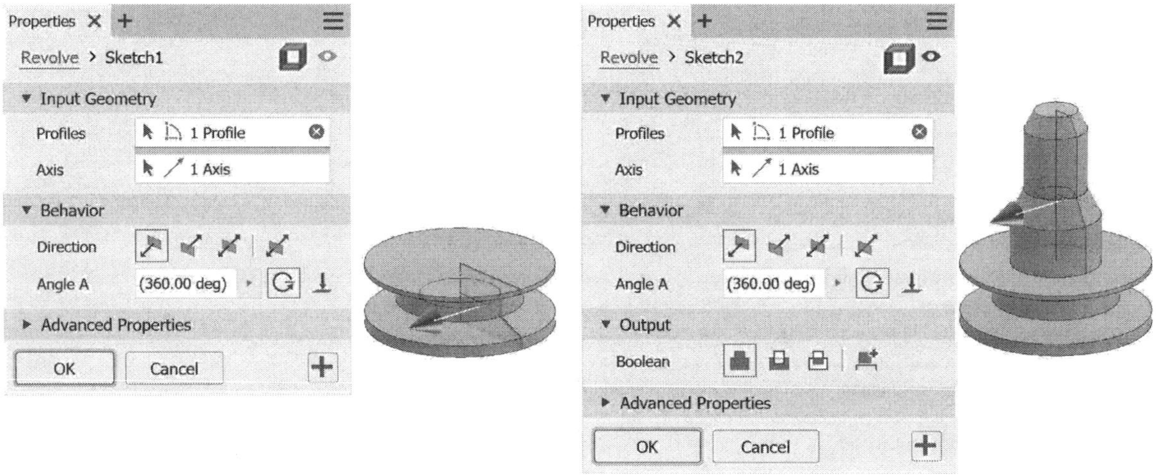

Figure 3.23

When you make changes in the property panel, the shape of the sketch changes in the graphics window to represent your choices. When done making selections, click the OK button to create the revolved feature. The common options to create a revolve feature are explained next.

Input Geometry

Use this area to select the area to revolve and define where the revolve feature should start.

Profiles	Click this area to choose the sketch to revolve. If there are multiple closed profiles, you can click on the sketch profile(s) or inside the profile(s) to revolve. If there is only one closed profile, Inventor selects it automatically. If you select the wrong profile or sketch area, click the ⊗icon and choose again. To remove a selected profile, hold the CTRL key and click the area you wish to remove.
Axis	Click a straight edge, centerline, work axis, or origin axis about which the profile(s) should be revolved. See the previous section on how to create a centerline and diametric dimensions.

Behavior

Use this area to set the direction and angle of the revolve feature.

Direction

Use this area to set the direction of the revolve feature.

	Default	Revolves in the positive direction.
	Flipped	Flips the direction to the opposite side.
	Symmetric	The revolved feature extends an equal angle in the negative and positive directions.
	Asymmetric	Define a different angle for the negative and positive direction.

Angle

This option revolves a profile a specified angle; the default is a full revolution of 360 degrees. You can enter a value in the Revolve Property panel or drag the arrow in the graphics window.

Extents

The Extent options are to the right of the Angle cell.These options determine how the revolved sketch terminates. Note that some options are not available until a base feature exists.

	Full	Revolves the profile 360 degrees.
	Angle	To revolve the profile less than 360 degrees click this icon or enter a value in the Angle cell.
	To	Revolves the profile until it reaches a selected face, plane or point. After selecting a face, plane or point you'll get two additional options: **From**: Select a plane, face, or point to start the revolve feature from. **To:** Select a plane, face, or point to terminate the revolve feature on.
	To Next	This option automatically revolves the profile until it reaches a plane or face or you can select a plane or face to terminate the revolution. The sketch must be fully enclosed in the area to which it is terminating.

Output

The first solid you create is referred to as a base feature and the Output section is not available. Once a base feature is established, the property panel Output options are available. You use these options to add or remove material from the model.

	Join	Adds material to the part.
	Cut	Removes material from the part.
	Intersect	Removes material, keeping what is common to the model and the new feature.
	New Solid	Creates a new solid body. The first solid feature created uses this option by default. Select this option to create a new body in a part file that has an existing solid body.

Advanced Properties

The Advanced Properties section (Figure 3.24) contains additional options to refine the feature being created:

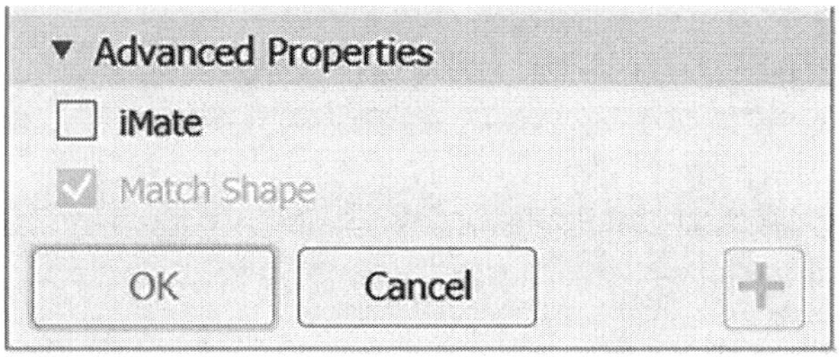

Figure 3.24

iMate

Places an iMate on a closed loop of the revolve feature. iMates are used to automate the assembly process.

Match Shape

Select an open profile in a part file, specify to Match Shape and select the side to keep.

EXERCISE 3-2: REVOLVING A SKETCH

In this exercise, you create a sketch and then create a revolved feature to complete the part. This exercise demonstrates how to revolve sketched geometry about an axis to create a revolved feature.

1 Use New from the Quick Access toolbar, choose the English tab, and then double-click Standard (in).ipt; or if inch is the default unit, from the left side of the Quick Access toolbar you can click the down arrow on the New icon, and click Part.

2 Click Start 2D Sketch from the 3D Model tab Sketch panel, then select the XY origin plane in the graphics window.

3 Create the sketch geometry as shown. Locate the lower endpoint of the geometry on the origin point. Select the left vertical line and then click Centerline from the Format panel to change the line to a centerline as shown in Figure 3.25 left.

4 Click Dimension from the Sketch tab > Constrain panel.

5 Add the linear diameter dimensions shown in the following image on the right by selecting the centerline and an endpoint on the sketch. Then place the dimension.

Note for clarity the remaining dimensions to fully constrain the sketch are not shown.

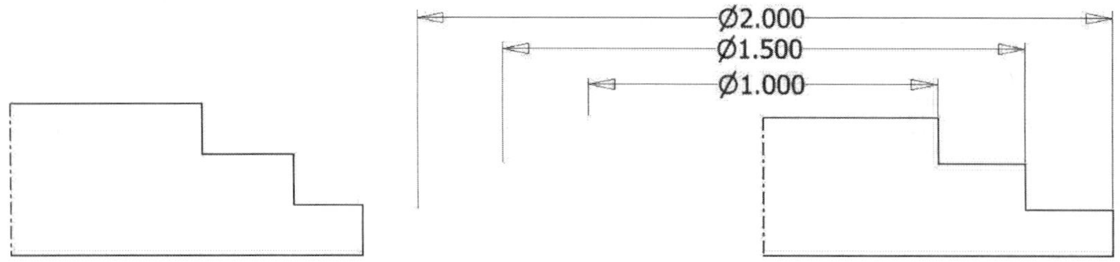

Figure 3.25

6 Right-click in the graphics window, and click Finish 2D Sketch from the marking menu.

7 Click Revolve from the 3D Model tab > Create panel, or click on the sketch and use the mini-toolbar to choose Create Revolve. As there is only one closed profile, it is automatically selected and as there is only one centerline, it is automatically selected as the axis.

8 From the Revolve Properties panel, change the Angle to 45 deg or drag the arrow on the part until 45.00 deg shows. The preview of the revolve updates to reflect the change as shown in Figure 3.26.

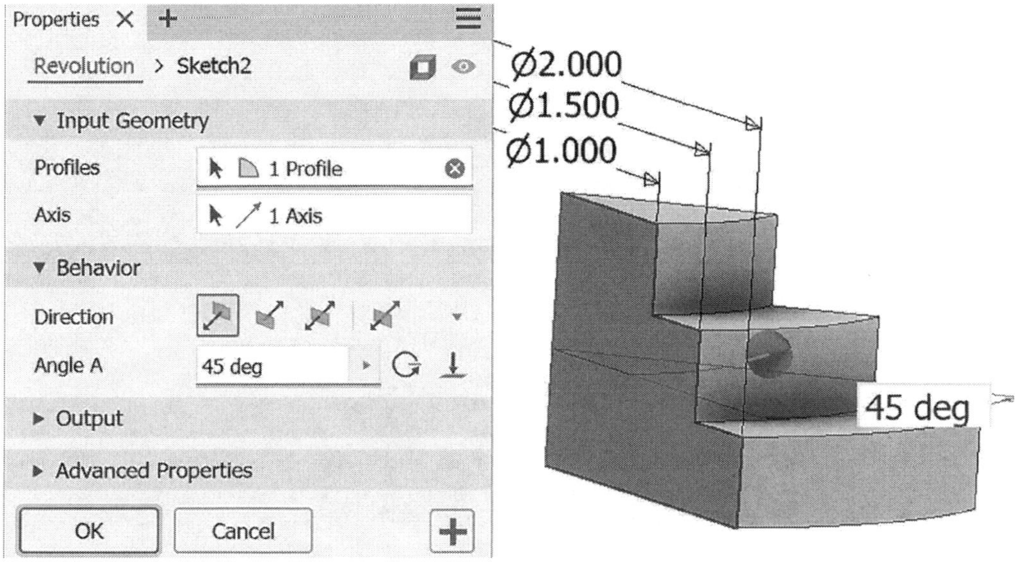

Figure 3.26

9 Change the direction of the revolve to counterclockwise by selecting the left Direction 2 button from the dialog box or click and drag the arrow in the graphics window to go 45.00 deg in the opposite direction. The preview image will reverse the direction counterclockwise.

10 From the Revolve Properties panel, choose Symmetric; it's the third icon from the left in the Direction area.

11 Enter 90 deg for the angle. The preview should resemble the following image.

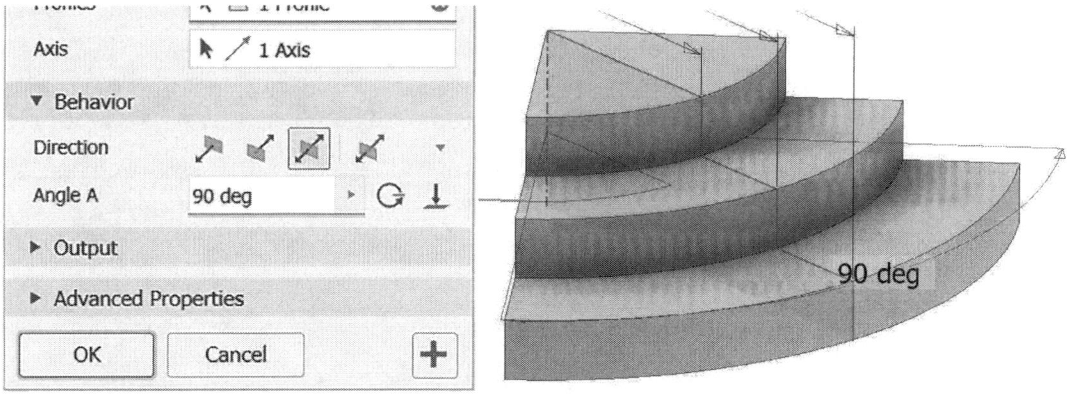

Figure 3.27

12 From the Revolve Properties panel change the Extents to Asymmetric; it's the fourth icon from the left in the Direction area.

13 In the graphics window click and drag the arrows to different values, similar to Figure 3.28.

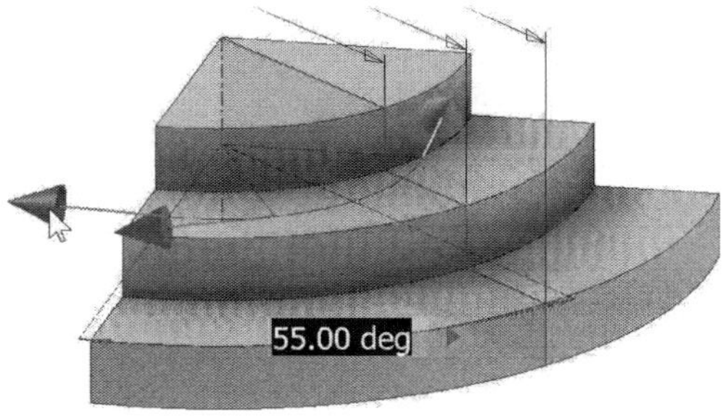

Figure 3.28

14 From the Revolve dialog box, change Extents to Full as shown in Figure 3.29 left.

15 Click the OK button in the Revolve dialog box to create the feature. Your part should resemble the image on the right.

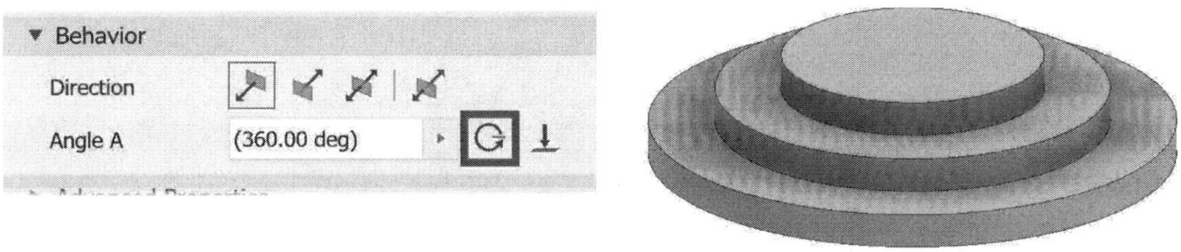

Figure 3.29

16 Close the file. Do not save changes.

End of exercise.

PRIMITIVE SHAPES/FEATURES

A method to quickly create basic shapes is to use the primitive commands: box, cylinder, sphere or torus as shown in the following image. Primitive shapes are created from a sketch and are extruded (for box and cylinder) or revolved (for sphere and torus). The advantage to using the primitive commands is they combine these operations to reduce clicks. To create a primitive shape/feature, follow these steps.

- Turn on the Primitive panel by selecting the Primitive panel visibility option from the drop-down arrow on the right side of the panels as shown in Figure 3.30 left.
- Then click one of the Primitive commands from the 3D Model tab > Primitives panel as shown at the right.

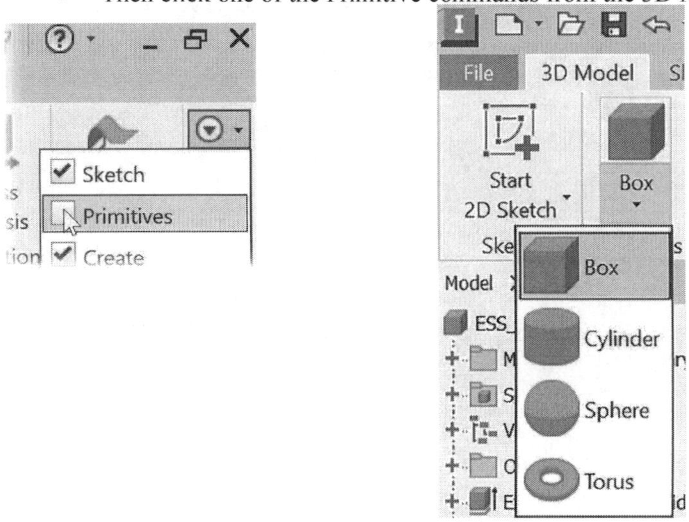

Figure 3.30

TIP: Features created using Primitive commands can be edited like any other feature. If a primitive is a secondary feature, you will need to edit its sketch to locate it. Editing a feature and a feature's sketch will be covered later in this chapter.

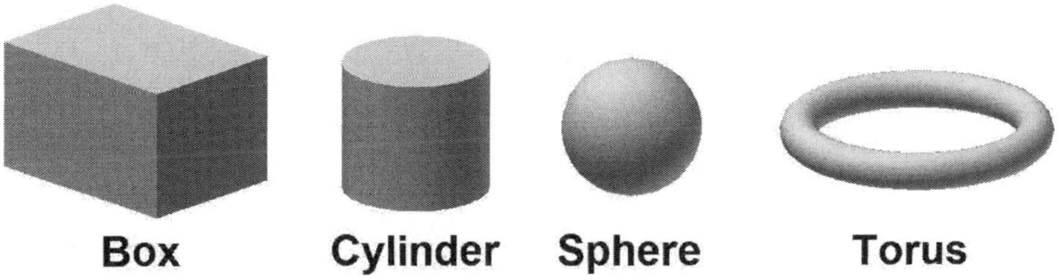

Figure 3.31

Select a plane to locate the primitive; the plane can be an origin plane, planar face on the part or a work plane. Depending on the primitive that you are creating you will do the following.

Box

1. Select a point to locate the center of the rectangle
2. Enter a value for the horizontal size
3. Press the Tab key
4. Enter a value for the vertical size
5. Press ENTER to extrude the rectangle
6. Enter a value for the extrusion distance
7. Click OK to create the feature

Cylinder

1. Select a point to locate the center of the circle
2. Enter a value for the diameter of the circle
3. Press ENTER to extrude the circle
4. Enter a value for the extrusion distance
5. Click OK to create the feature

Sphere

1. Select a point to locate the center of the circle
2. Enter a value for the diameter of the circle
3. Press ENTER to revolve the circle
4. Click OK to create the feature

Torus

1. Select a point to locate the center of the torus
2. Enter a value and then either press ENTER or click a point to define the radius of the torus
3. Enter a value for the diameter of the circle
4. Press ENTER to revolve the circle

5. Click OK to create the feature

SECONDARY 2D SKETCHED FEATURES

A secondary 2D sketched feature is created from a sketch on the planar face of a part or on a work plane. The sketch is then used to add or remove material from the part. The basic steps are as follows:

1. Create a new 2D sketch on a planar face on the part or on a work plane.
2. Draw geometry to define the outline of the profile.
3. Add constraints and dimensions.
4. Use Extrude, Revolve, Sweep, or Loft and perform a Boolean operation to add material, remove material, or keep what is common between the part and the feature that is being created. The following image shows a part (cylinder) and a sketch (circle) on the left and the affect the three Boolean operations would have on the part.

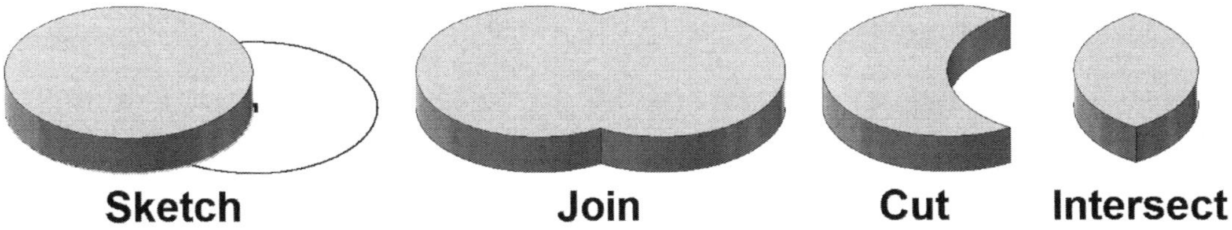

Figure 3.32

5. Use the Extents options in the Extrude or Revolve commands to control how the feature will terminate. For more information about Extents refer to the Extents section in the Extrude a Sketch and the Revolve a Sketch section covered earlier in this chapter.

There are no limits to the number of sketched features that can be added to a part. Each sketched feature is created on a plane, and multiple features can reference the same plane. You sketch geometry, apply constraints and dimensions exactly as you did with the first sketch. In addition to constraining and dimensioning the new sketch, you can also constrain and dimension the new sketch to the existing edges on the part. The edges you dimension to do not need to lie on the active sketch plane; the dimensions will be placed on the current sketch plane but reference the selected edge.

In the following section, you learn how to sketch on a plane and create a feature from the sketch.

Create a new 2D Sketch

As stated previously, each 2D sketch must lie on a plane and only one sketch can be active. A sketch must be created on a planar face, a work plane, or an origin plane. The planar face on the part does not need to have a straight edge. For example, a cylinder has two planar faces, one on the round-looking top and the other on the round-looking bottom of the part. Neither plane has a straight edge, but a sketch can be placed on either one.

To make a sketch active, use one of the following methods:

- Click the Start 2D Sketch command from the 3D Model tab > Sketch panel as shown in Figure 3.33 left, and then select the plane where you want to create the sketch.
- While not in the middle of an operation, right-click in the graphics window, and click New Sketch from the marking menu as shown in the middle image. Then click a planar face, or a work plane.
- While not in a sketch, click on the planar face or work plane that you want to create the active sketch on. Then click the Create Sketch button from the mini-toolbar as shown in Figure 3.33 right.
- Press the S key, and then click the plane where you want to place the sketch.

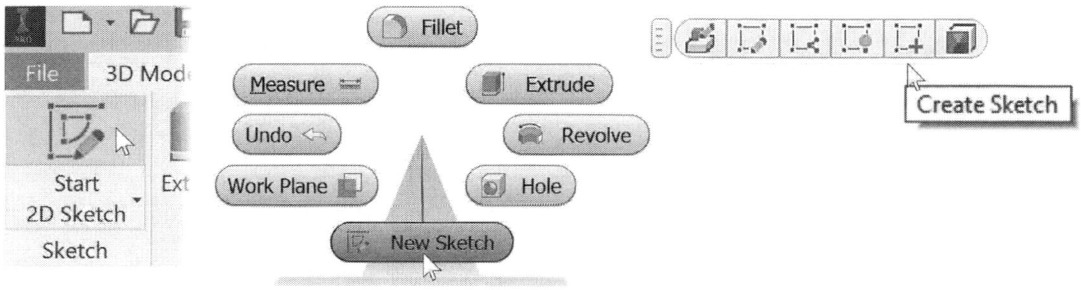

Figure 3.33

Sketches appear in the browser with the name Sketch followed by a number which increments for each new sketch that is created. When you create a new sketch, or make a sketch active, by default the view automatically changes so you are looking straight at the sketch. This is controlled by Application Option > Sketch tab > Look at sketch plane on sketch creation and edit. To manually change the view to look straight at a plane, click the Look At command on the Navigation Bar and then select the plane.

SELECT OTHER-FACE CYCLING

Inventor has dynamic face highlighting that helps you to select the correct face, edge, part etc. for your intended operation. As you move the cursor over them different edges and faces are highlighted.

To *cycle* to a face that is behind another one, start a command and hold the cursor over the face in front of the one that you want to select for two seconds to show the Select Other tool (Figure 3.34 left). Select the drop-down arrow and move the cursor over the objects in the list until the correct face is highlighted and then click. The number of objects that appear in the list depends upon the geometry and location of the cursor. You can also access Select Other by right-clicking on the desired item in the graphics window and choosing Select Other from the marking menu (Figure 3.34 middle).

TIP: To set the wait time before the Select Other option appears, click Tools tab > Options panel > Application Options > General tab and in the Selection area enter the "Select Other" delay (sec) as shown in Figure 3.34 right. If you do not want the Select Other command to open automatically, type OFF in the field. The default value is 2.0 seconds.

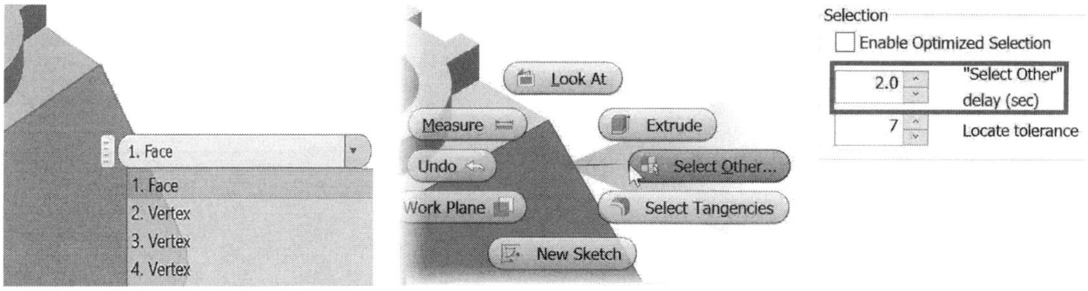

Figure 3.34

SLICE GRAPHICS

While creating parts, you may need to sketch on a plane that is difficult to see because features are obscuring the view. The Slice Graphics option temporarily removes the portion of the part that obscures the active sketch plane on which you want to sketch. Figure 3.35 left shows a revolved part with an origin plane visible and the Slice Graphics menu. The image on the right shows the graphics sliced and the origin plane visibility turned off.

To temporarily slice the graphics screen, follow these steps:

1. Create a sketch or make a sketch active where the graphics need to be sliced.
2. Rotate the model so the correct side will be sliced. The side that faces out will be sliced away.
3. With the sketch active, start the Slice Graphics command by performing one of the following:
    a. Right-click and select Slice Graphics from the marking menu as shown in the following image on the left.
    b. Click Slice Graphics from the bottom of the Status Bar as shown in the middle image.
    c. Press the F7 key.
    d. Click Slice Graphics on the View tab > Appearance panel.
4. The graphics will be sliced on the active sketch as shown in the following image on the right.
5. Use sketch commands from the Sketch tab to create geometry on the active sketch.
6. To exit the sliced graphics environment, right-click and select Slice Graphics from the marking menu, click Slice Graphics from the bottom of the Status Bar, press the F7 key or click the Finish Sketch button on the Sketch tab > Exit panel.

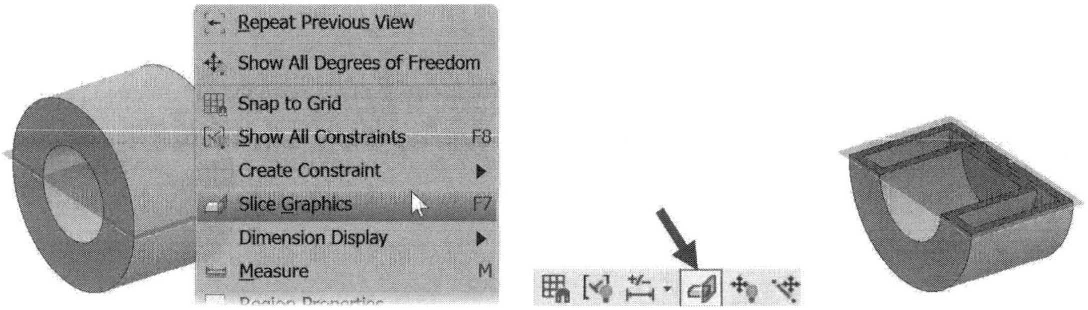

Figure 3.35

EXERCISE 3-3: SKETCH FEATURES

In this exercise, you sketch an oblong shape on an angled face.

1 Open *ESS_E03_03.ipt* from the Chapter 03 folder.

2 Select the top-inside angled face and choose Create Sketch from the mini-toolbar (Figure 3.36 left) to start a new sketch on that plane.

3 The view changes so you are looking straight at the sketch—if not, click the Look At command from the Navigation Bar, and then select the top angled plane.

4 Next create a center-to-center slot. Click Center to Center Slot from the Sketch tab > Create panel (it may be under the Two Point Rectangle command as shown in Figure 3.36 middle). Use the values of **1.500** for the center-to-center distance and **1.000** for the diameter as shown in Figure 3.36 right.

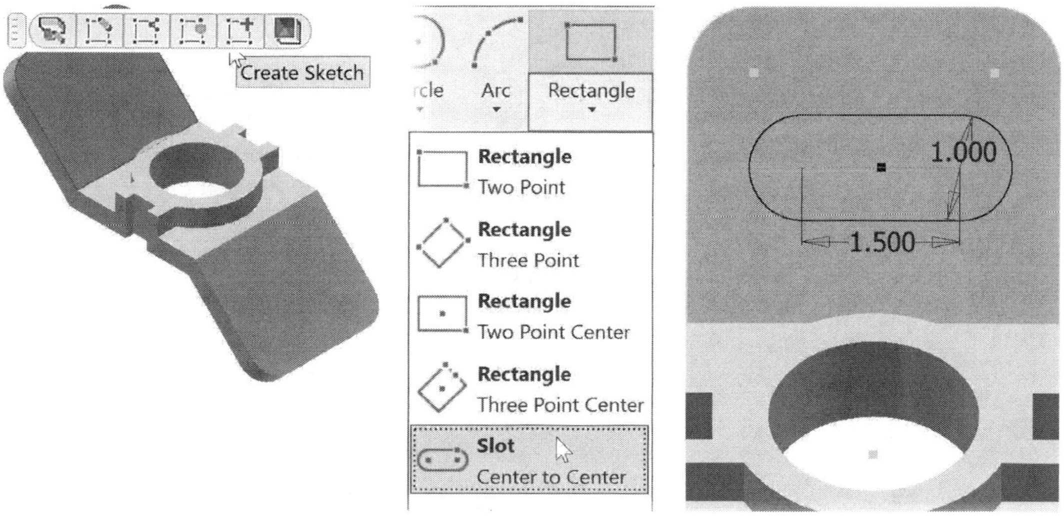

Figure 3.36

5 Next center the slot. Use Horizontal constraint from the Constrain panel (Figure 3.37 left) and select the center mid-point of the slot and the midpoint of the top horizontal edge of the part (Figure 3.37 right).

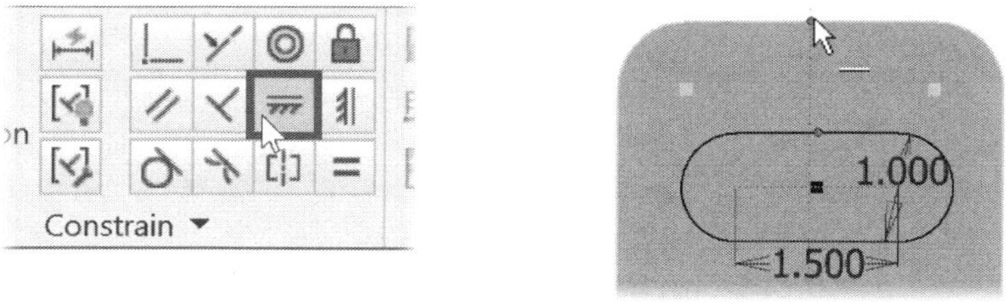

Figure 3.37

TIP: The horizontal and vertical direction is based on the sketch's coordinate system that is automatically aligned to the selected face when a sketch is created.

6 To fully constrain the slot, add a **1.500** vertical dimension as shown in Figure 3.38.

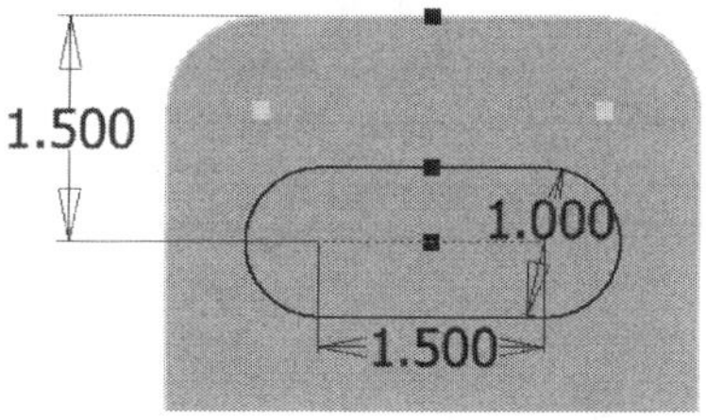

Figure 3.38

7 Click Finish Sketch from the Exit panel.

8 In the graphics window, click on the slot and choose Create Extrude from the mini-toolbar.

a. For the Extrude Properties (Figure 3.39 left):
   i. For the profile, click inside the slot.
   ii. Flip the Direction of the extrusion.
   iii. Change the Distance to Through All.
   iv. Change the Output to Cut (to remove the material from the slot).

b. Click OK. The completed part is shown in Figure 3.39 right.

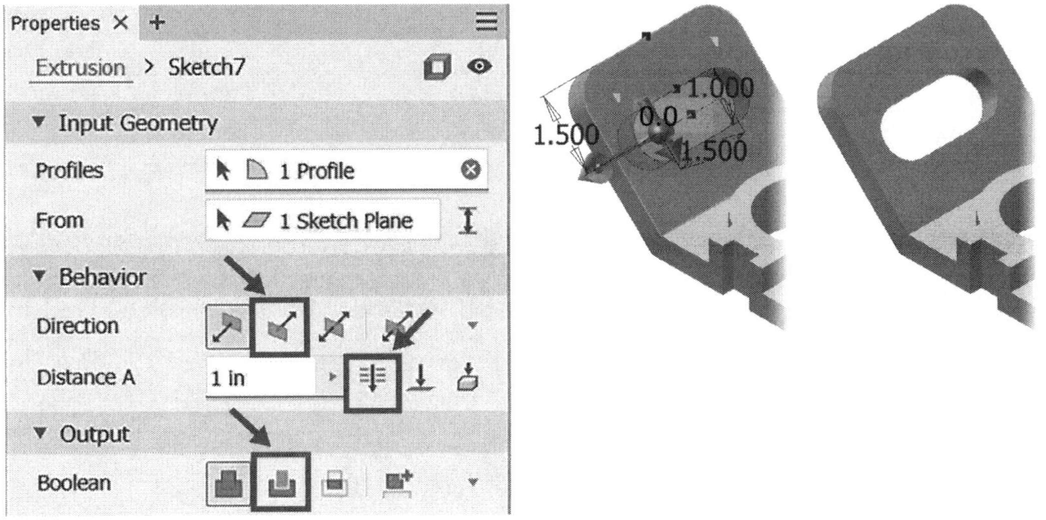

Figure 3.39

9 Practice creating sketched features and try different options.

10 Close the file. Do not save changes.

End of exercise.

EDITING A FEATURE AND SKETCH

You can edit features after they are created to change options and values such as operation, extents, distance, taper etc. After you create a feature, its sketch dimensions are consumed. They are not gone, but in the Browser, the sketch is listed subordinate to the feature. To change the dimensions or shape of the sketch, you find and edit it.

Edit Feature Command

There are multiple methods that you can use to edit the feature. There is no preferred method. Use the method that works best for your workflow.

- While not in a sketch, click the feature in the graphics window to select it. Click Edit *Feature type* from the mini-toolbar. Figure 3.40 left shows the mini-toolbar button for editing an extrusion.
- Double-click on the feature's icon in the browser.
- In the browser, right-click the feature's name and click Edit Feature from the menu (Figure 3.40 right).

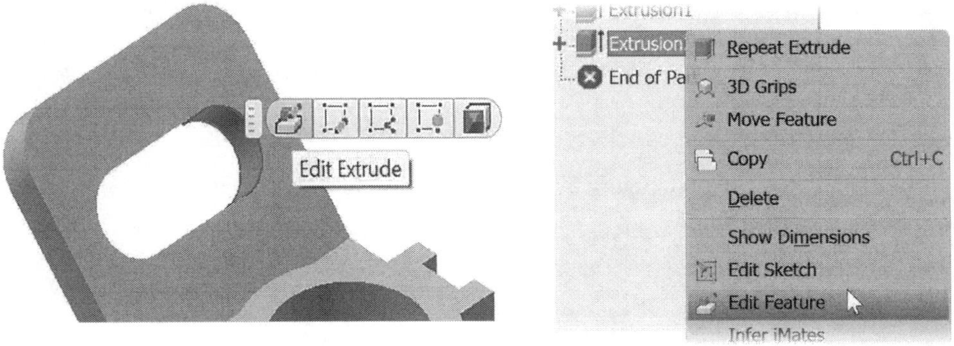

Figure 3.40

When editing a feature, the dialog box that was used to create the feature appears. You can use it to change the feature's values and options, except for the join operation on a base feature and the type of output (solid or surface).

Editing a Feature's Sketch

In this section, you learn how to add and delete constraints, dimensions, and geometry of a 2D sketch used to create a feature. To edit a feature's sketch, while not in sketch mode, do one of the following:

- From the graphics window, click on the feature whose sketch you want to edit. Choose Edit Sketch from the mini-toolbar as shown in Figure 3.41 left.
- From the browser, right-click on the feature's name and choose Edit Sketch from the menu Figure 3.41 middle.
- From the browser, expand the feature and below it right-click on the sketch and choose Edit Sketch from the menu (Figure 3.41 right).
- From the browser, expand the feature and double-click on the sketch icon.

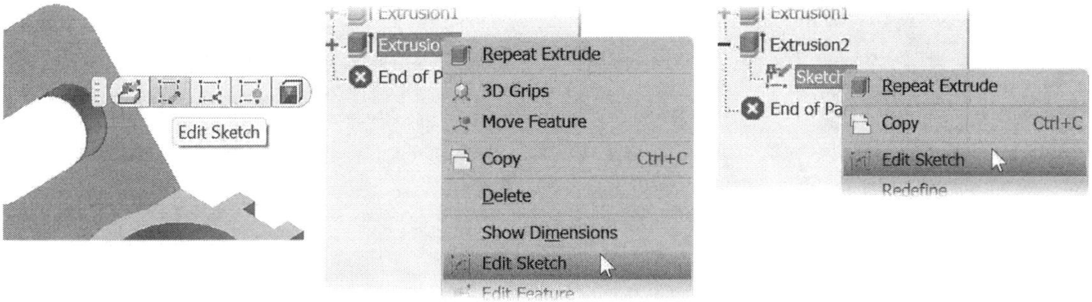

Figure 3.41

You return to sketch more where you can add or remove geometry lines, arcs, circles, and splines of the sketch. To delete an object, right-click it and select Delete from the menu, or press the DELETE key. If you delete an object from the sketch that has dimensions associated with it, the dimensions are no longer valid for the sketch, and they will also be deleted. You can also delete the entire sketch and replace it with an entirely new sketch. When replacing entire sketches, you should first delete other features that would be consumed by the new objects and re-create them.

You can also add or delete constraints and dimensions. When the dimensions are visible on the screen, double-click the dimension text that you want to edit. The Edit Dimension dialog box appears. Enter a new value, and then click the checkmark in the dialog box or press the ENTER key. Continue to edit the dimensions and when finished, click Finish Sketch from the ribbon or click the Local Update button on the Quick Access toolbar as shown in the following image. The new value(s) will be used to regenerate the sketch and feature.

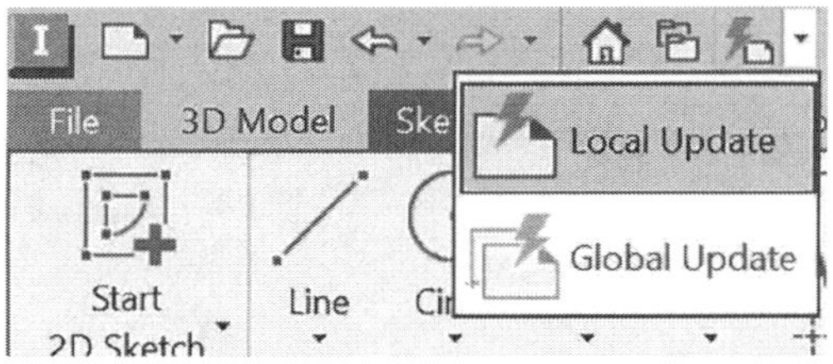

Figure 3.42

Renaming Features and Sketches

By default, each feature is given a name. These feature names may not help you when trying to locate a specific feature of a complex part, as they will not be descriptive to your design intent. The first extrusion, for example, is given the name Extrusion1 by default, whereas the design intent may be that the extrusion is the thickness of a plate. To rename a feature, slowly click the feature name twice and when the cursor appears, enter a new name. Spaces are allowed.

Deleting a Feature

You may delete a feature after it has been placed. To delete a feature, right-click its name in the browser, and select Delete from the menu, as shown in Figure 3.43. The Delete Features dialog box appears allowing you to select to delete consumed and/or dependent sketches and features. To delete multiple features, hold down the CTRL or SHIFT key and select them from the browser, then right-click one of the items and select Delete from the menu.

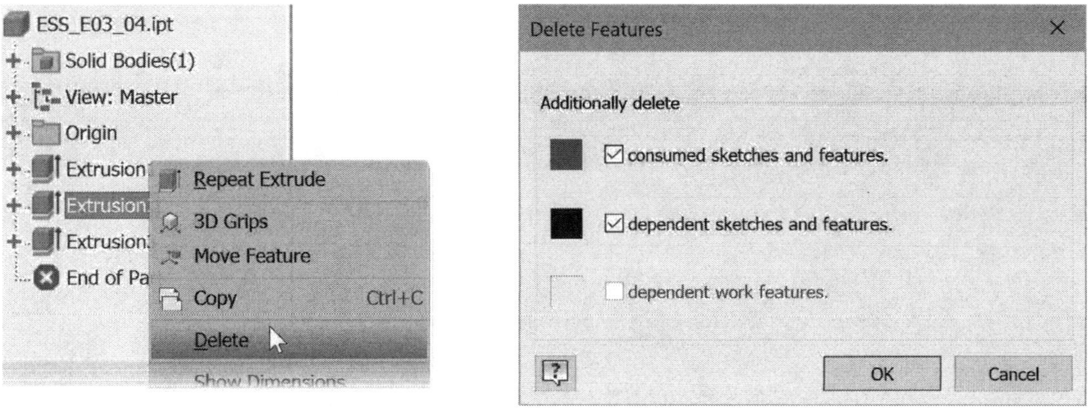

Figure 3.43

Failed Features

In the browser, failed features are shown in red with a yellow triangle and exclamation point symbol as shown in Figure 3.44 left. After updating the part, this alerts you that the values or settings for that feature did not regenerate successfully. To see the feature in its last successful state, move the cursor over its name in the browser to highlight it in the graphics window. Figure 3.44 right shows the last successful fillet feature. You can edit the values, enter new values, or select different settings to define a valid solution. Once you define a valid solution, the feature should regenerate without error.

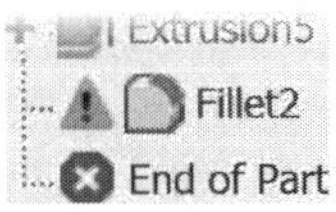

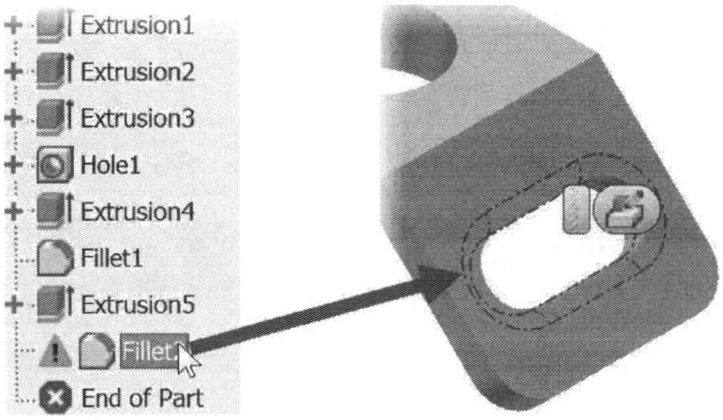

Figure 3.44

EXERCISE 3-4: EDITING FEATURES AND SKETCHES

In this exercise, you edit an extrusion's consumed sketch to update the part. In the next section, you learn how to create sketched features that were created in this model.

1. Open *ESS_E03_04.ipt* from the Chapter 03 folder.
2. Edit the sketch of Extrusion1 by selecting the top planar face and then click Edit Sketch from the mini-toolbar as shown in Figure 3.45. Right-clicking Extrusion1 in the browser and choosing Edit Feature from the menu also works.

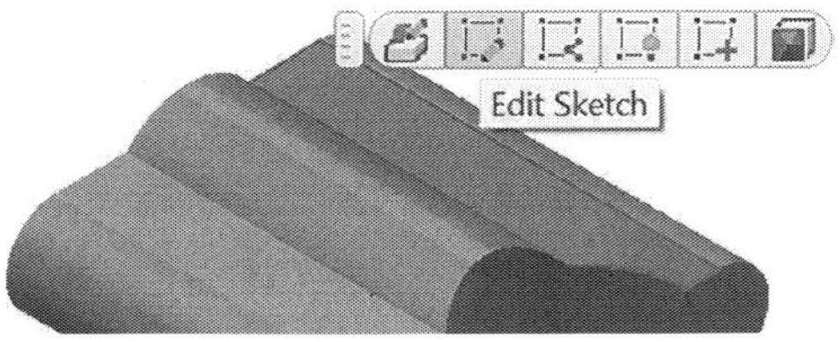

Figure 3.45

a. Double-click the .375 radius dimension. Use the Edit Dimension dialog box to change the value to **.25**, and then press Enter or click the checkmark in the dialog box.

b. Double-click the 2.000 dimension and change the value to **3** and then press Enter or click the checkmark in the dialog box. The sketch should resemble Figure 3.46 left.

c. Click Finish Sketch from the Sketch tab > Exit panel. The feature updates with the changed values as shown on the right.

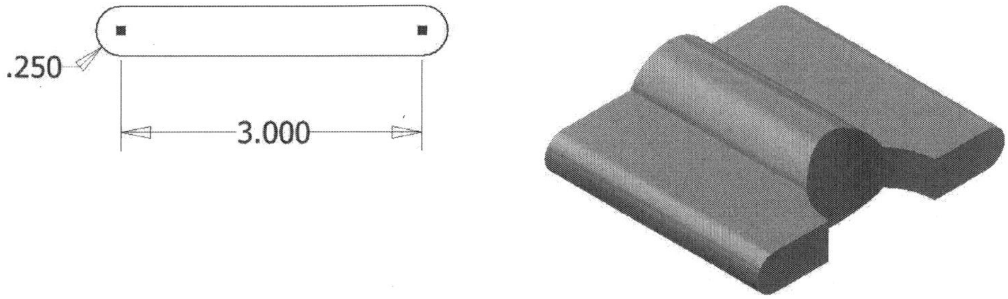

Figure 3.46

3 Next edit the extents used to create Extrusion2. From the graphics window, select the circular face and choose Edit Extrude from the mini-toolbar as shown in Figure 3.47 (1).

a. Use the Extrude dialog box to change the Boolean operation to Join as shown (2).

b. Change the Extents option to To labeled (3).

c. Select the bottom face labeled (4). The extrusion now stops at this face regardless of dimension changes in the part.

d. Click the OK button in the Extrude dialog box.

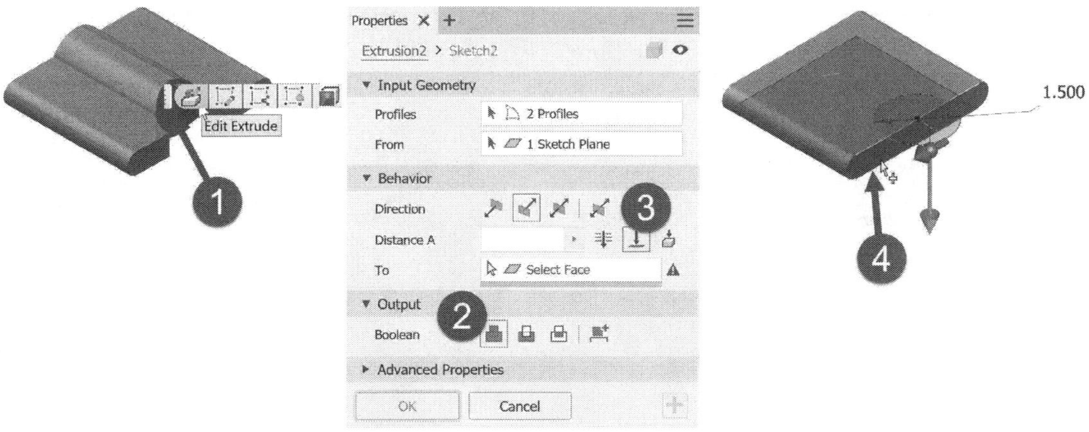

Figure 3.47

The feature will update as shown in Figure 3.48 on the left.

4 From the browser, right-click Extrusion3 and click Delete. Click OK to delete consumed sketches and features. When done, your part should resemble Figure 3.48 right.

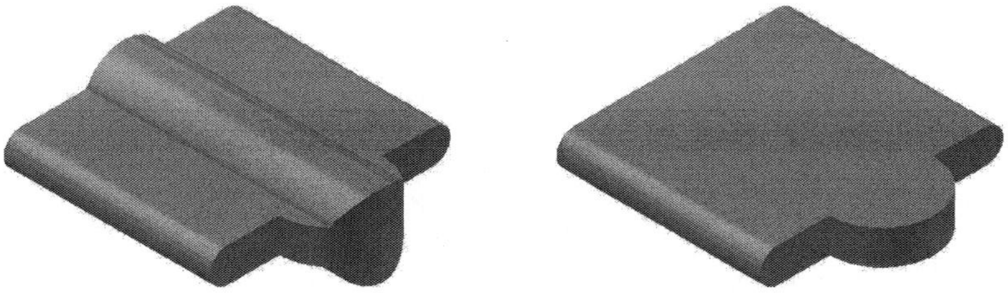

Figure 3.48

5 Practice editing the sketch dimensions and features.

6 Close the file. Do not save changes.

End of exercise.

PROJECTING GEOMETRY

You frequently need to reference existing faces, edges, or loops from features to create new features. The Project Geometry command is used to project an edge(s), face, point, or loop onto the active sketch. By default, projected geometry maintains an associative link to the original geometry. If you project the face of a feature onto another sketch and modify the original, the projected geometry will update to reflect that change.

Projecting Edges

In this section, you learn how to use the Project Geometry command that can project selected edges, vertices, work features, curves, the silhouette edges of another part in an assembly, or other features in the same part to the active sketch. There are four project commands available on the Sketch tab > Create panel: Project Geometry, Project Cut Edges, Project Flat Pattern and Project to 3D Sketch, as shown in the following image.

Project Geometry

Use to project geometry from a sketch or feature onto the active sketch.

Project Cut Edges

Use to project part edges that touch the active sketch. The geometry is only projected if the uncut part intersects the sketch plane. For example, a sphere with a sketch plane at its center projects as a circle onto the active sketch.

Project Flat Pattern

Use to project a selected face or faces of a sheet metal part flat pattern onto the active sheet metal part sketch plane.

Project to 3D Sketch

Use to project geometry from the active 2D sketch onto selected faces to create a 3D sketch.

Project DWG Geometry

Use to project geometry from AutoCAD geometry that is referenced from the Import command.

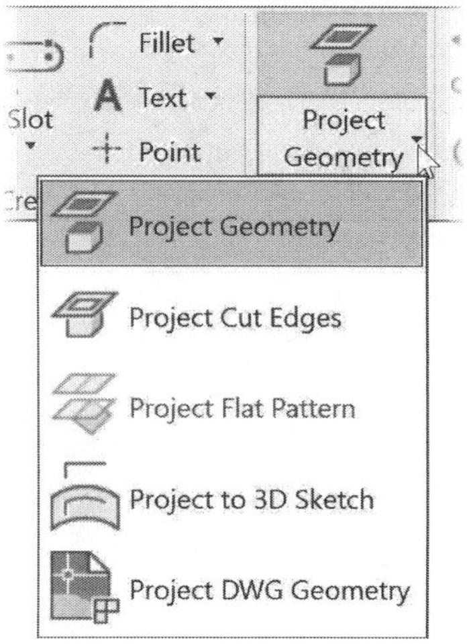

Figure 3.49

To use the Project Geometry command, follow these steps:

1. Create a 2D sketch or make an existing 2D sketch active onto which the geometry will be projected.
2. Click the Project Geometry command from the Sketch tab > Create panel.
3. Select the geometry to be projected onto the active sketch. Clicking a point in the middle of a face projects all edges of the face onto the active sketch. To project all edges defining a perimeter (a loop), right-click and then use the Select Other / Face Cycling command until they all appear highlighted, as shown in Figure 3.50 left. The image on the right shows the projected geometry.
4. To exit the operation, right-click and choose OK from the marking menu or press the ESC key or choose another command.

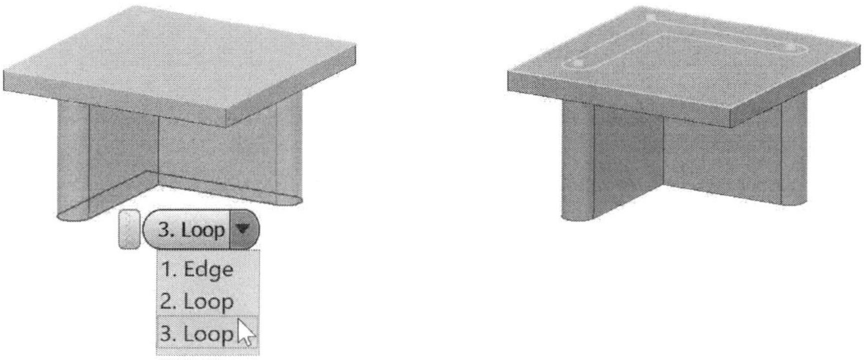

Figure 3.50

Internal islands defined on a face are projected and update accordingly. For example, the face shown on the left in Figure 3.51 projects all edges onto the active sketch like the following image on the right.

To disassociate projected geometry from the original geometry, select the profile or click the Show Constraints option on the Status Bar and then delete the reference constraints that were created for each projected edge/curve.

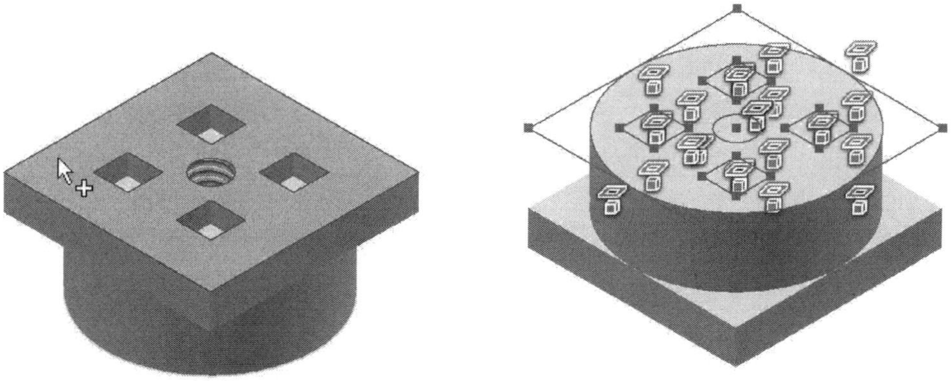

Figure 3.51

EXERCISE 3-5: PROJECTING GEOMETRY

In this exercise, you will project geometry to create a feature, edit the original sketch and update the linked feature and then delete the sketch constraints that were created from the projected geometry.

1 For now, when a new sketch is created you automatically project all the edges that lie on that plane as reference geometry. Turn on the Autoproject edges for sketch creation and edit option from the Application Options dialog box > Sketch tab as shown in the following image.

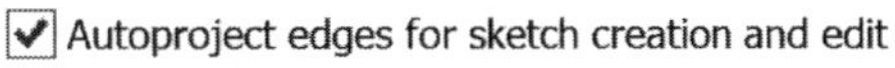

Figure 3.52

2 Open *ESS_E03_05.ipt* from the Chapter 03 folder.

3 Change your viewpoint so it resembles Figure 3.53 left.

4 Create a sketch on the back face of the cylinder as shown in Figure 3.53 left.

5 Rotate your view to show the back of the part as shown in Figure 3.53 right.

6 Click Project Geometry from the Sketch tab > Create panel. Select the front planar face of Extrusion2 (by selecting the inside of the planar face, all the edges that define the sketch will be projected) as shown in Figure 3.53 right.

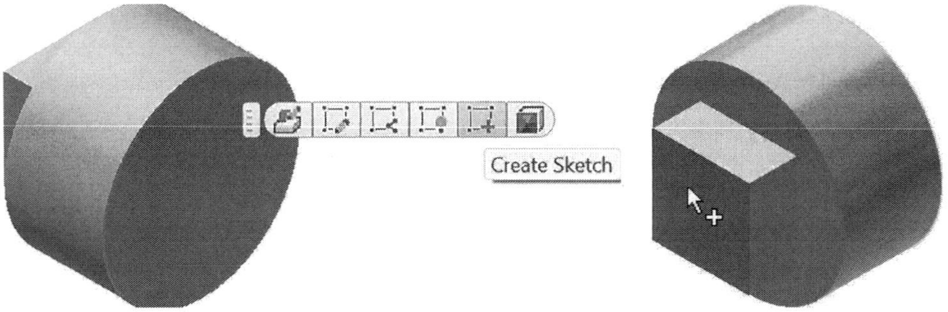

Figure 3.53

7 Rotate your view to show the back of the part again. Notice the edges of the front face are projected onto the active sketch and the sketch is fully constrained.

8 Right-click and choose Finish 2D Sketch from the marking menu.

9 Click on one of the projected lines and click Create Extrude from the mini-toolbar. Extrude the inside area of the projected geometry **.5 inches**, cutting material into the part as shown in Figure 3.54 left.

10 From the browser, right-click Extrusion2 (the original extruded square) and choose Edit Sketch from the menu.

11 Double-click to edit the 1.000 dimension and change it to **.5** and press enter. The sketch should resemble the image on the right. To better see the sketch, you can rotate your viewpoint.

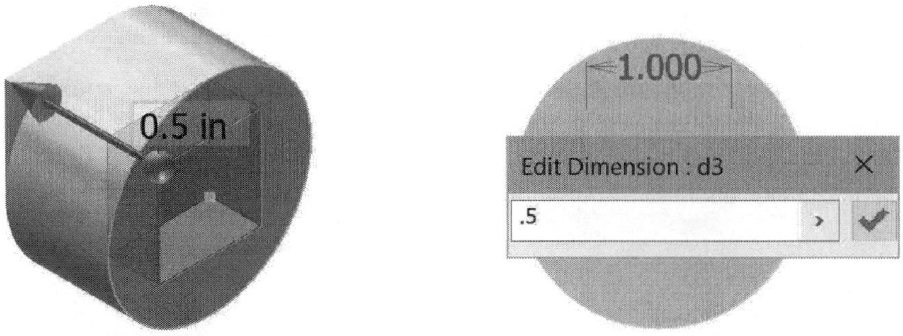

Figure 3.54

12 Click Finish Sketch from the Sketch tab > Exit panel. If needed rotate your view to show the back of the part. Notice Extrusion3 has updated to reflect the change to the projected geometry.

13 Edit the sketch of Extrusion 3 by clicking on one of the internal planar faces of the removed material and choose Edit Sketch from the mini-toolbar as shown in Figure 3.55.

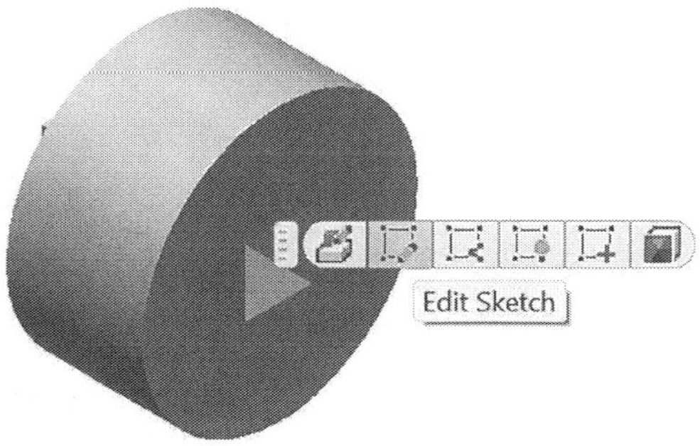

Figure 3.55

14 Next delete the reference constraints that were automatically created when the geometry was projected. To do this, drag a window around the square profile (left to right) to select it.

15 Move the icons that represent the constraints so they are not on top of each other.

Note you can select a line to see which constraint is associated with it.

16 Hold down the Ctrl key and select the four reference constraints related to the projected lines as shown in Figure 3.56 left. Right-click in a blank area in the graphics window and choose Delete from the marking menu.

17 Click and drag the endpoints of the lines and they will drag freely.

18 If desired, add constraints and dimensions or delete the four lines and sketch a new profile.

19 Finish the sketch and rotate the part so you see that Extrusion3 is a different size and shape than Extrusion2. The image on the right shows an example of the new shape.

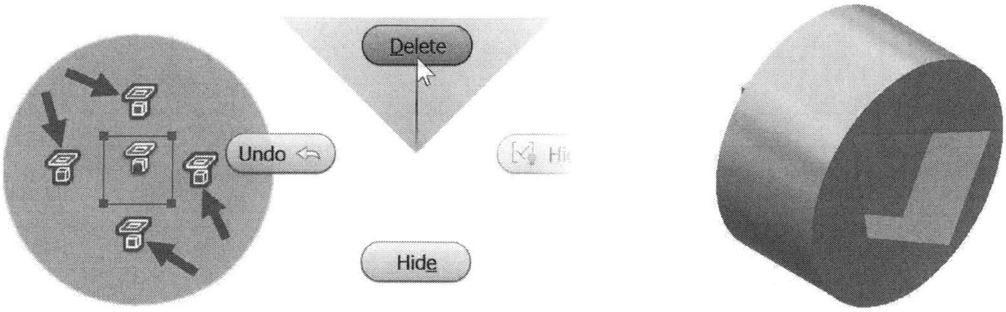

Figure 3.56

20 Close the file. Do not save changes.

End of exercise.

PART MATERIAL, PROPERTIES AND APPEARANCE

When modeling parts you can define the material and color so it correctly represents the manufactured part. The part materials have density values that affect a part's mass properties, see Figure 3.59.

Part Material

To specify a material to define a part's physical properties and appearance, use one of these techniques:

- Click the icon arrow in the Material area of the Quick Access toolbar and use the Materials Browser to select a material from the Inventor Materials Library as shown in Figure 3.57.
- Right-click the part's name from the browser and choose iProperties from the menu (Figure 3.58 left).
- Click the File tab > iProperties as shown in Figure 3.58 right.

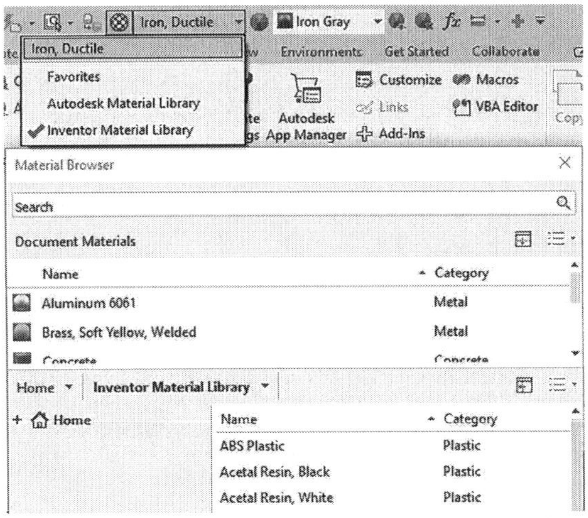

Figure 3.57

Figure 3.58

With the last two techniques listed above, the iProperties dialog box appears. From the dialog box click the Physical tab and select a material from the material drop-down list as shown in Figure 3.59 left.

To update the physical properties, click the Update button in the iProperties dialog box. Figure 3.59 right shows updated physical properties. Note that the mass, area, volume and center of gravity are also displayed.

Click the OK button in the dialog box to complete the operation and the appearance of the part in the graphics window changes to reflect the selected material.

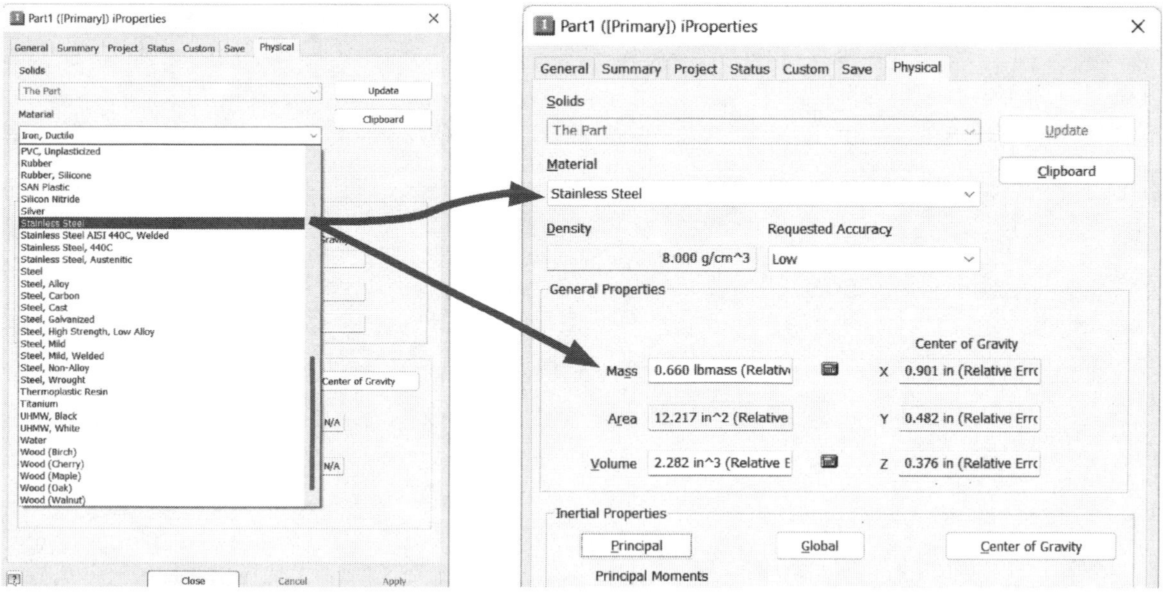

Figure 3.59

Appearance

After applying a material, you may override the color / appearance of the part, for example, if the part is to be painted red. You can override the appearance (color) of a part by clicking on the down arrow in the Appearance area of the Quick Access toolbar as shown in Figure 3.60. The appearance override only changes how the part appears and has no effect on the physical material or the mass of the part.

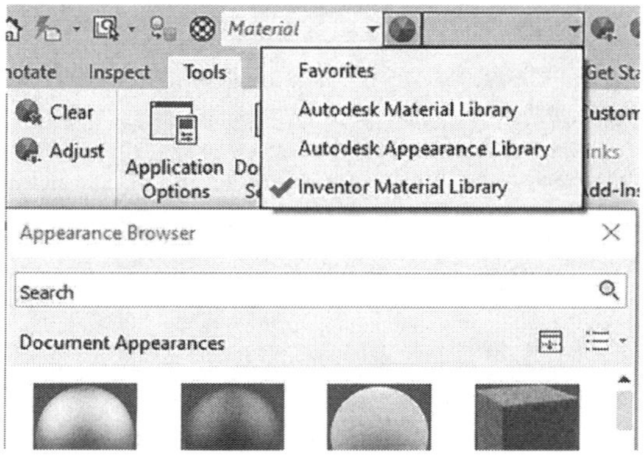

Figure 3.60

Additional Appearance Commands

You can adjust shadows, reflections, lighting style, perspective viewpoint, ground plane, and easily change the visual style using commands found on the View tab > Appearance panel (Figure 3.61). A brief description of each command follows.

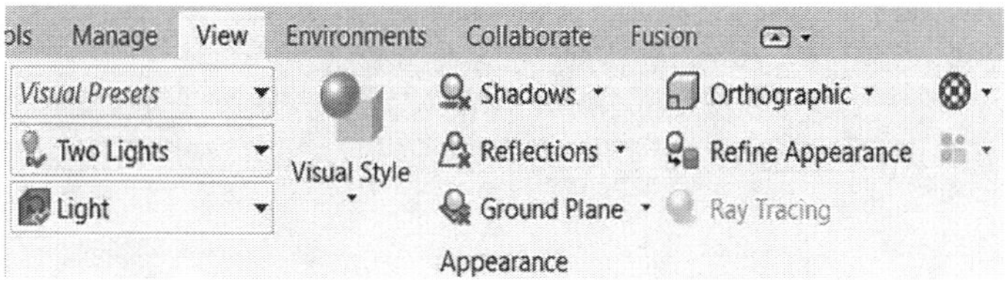

Figure 3.61

Visual Style

Use this to set the appearance of model faces and edges in the graphics window. There are eleven styles to choose from as shown in Figure 3.62 left. You can add Visual Styles to the Navigation Bar where it is easy to select from the listed styles as shown in Figure 3.62 right. Visual styles do not change the part's properties; they only change how it is displayed. You can combine visual styles with the other appearance options to show your part(s) clearly.

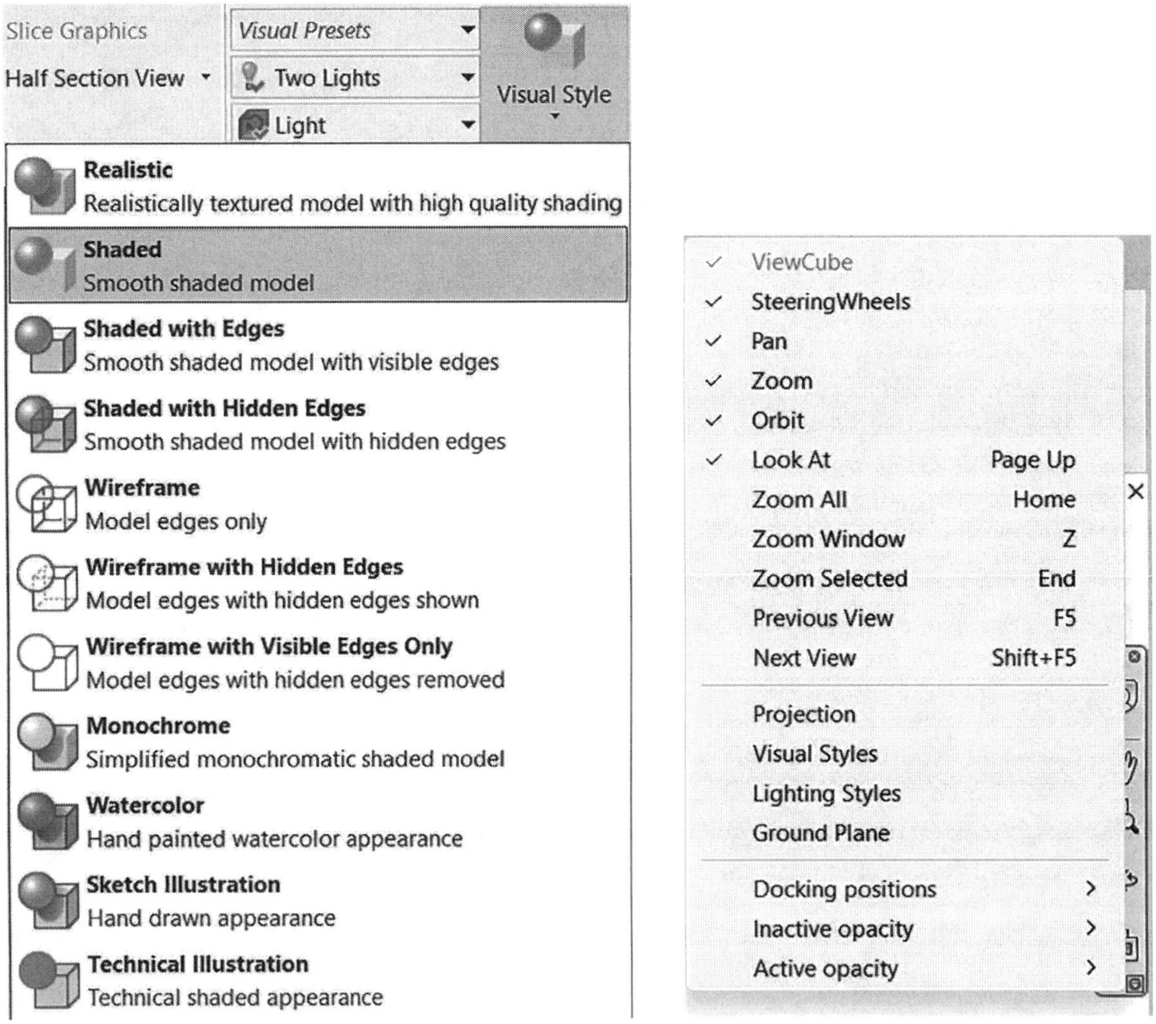

Figure 3.62

Visual Presets

Each preset option changes the quality of lighting and appearance. Choose from Render Quality, High Quality, or Performance.

Lighting Style

Use this to set lighting options. Under the Lighting Style dropdown, click the Settings option. Use the dialog box that appears to control the light's direction, color, add lights, and control brightness and ambience. From this dialog box, you can turn on Image Lighting, which allows you to set a background image and set the scale of the image.

With a scene image set, you can rotate your viewpoint 360 degrees. To move the image, select the Ground Plane Settings option, change the Position & Size option to Manual adjustment and a triad will appear. Select the arrows on the triad to move the image.

Color Scheme

Customize the background of your graphics window.

Shadows

To give your model a realistic look, you can choose different options for displaying shadows. By default, shadows are turned off.

Reflections

Click the Reflections command to toggle reflections on / off. The reflection shows on the ground plane parallel to the Bottom plane of the ViewCube. You can change which ViewCube plane is the top using the navigation commands to produce a view with the bottom plane oriented as you want. To do so, right-click on the ViewCube and choose Set Current View as Top.

Ground Plane

Use the Ground Plane command to toggle the ground plane on and off. You can change the ground plane orientation by adjusting the viewpoint with any of the navigation commands to produce a view parallel to the bottom surface of the part. Right-click on the ViewCube and choose Set Current View as Top. Adjust the ground plane's appearance by clicking the down arrow next to Ground Plane and choosing Settings.

Orthographic / Perspective

Use this to set the viewpoint to orthographic or perspective. When set to orthographic, the viewpoint is parallel, meaning that lines are projected perpendicular to the plane of projection. When set to perspective, the geometry on the screen converges to a vanishing point like the way the human eye works. You can adjust the perspective by holding down the CTRL and SHIFT keys and turning the mouse wheel.

Refine Appearance

This is used to smooth the appearance of the image, but is only active when coarse faceting is enabled.

Ray Tracing

Ray tracing produces images with excellent results, but the calculations required for it make it slow to process. It is enabled only for Realistic and Monochrome visual styles. With one of those active, you click to enable ray tracing and a toolbar appears in the lower right of the display with three modes of ray tracing from which to choose. The toolbar auto-hides after a couple of seconds, but you can move the cursor over it to cause it to redisplay.

Textures On / Textures Off

If you have a material or appearance with a texture that is applied to a part, you can toggle the texture on and off. This is helpful for improving graphics performance.

APPLYING YOUR SKILLS

Skills Exercise 3-1

In this exercise, you create a bracket from extruded features. Assume the part is symmetric about the center of the horizontal slot.

1. Start a new part based on the English Standard (in).ipt template.
2. Create a sketch on the XY origin plane.
3. Sketch the outline for the base feature as shown in the following image.
4. Add geometric constraints and dimensions. Make sure the sketch is fully constrained.
5. Extrude the base feature.
6. Create the three remaining features to complete the part.

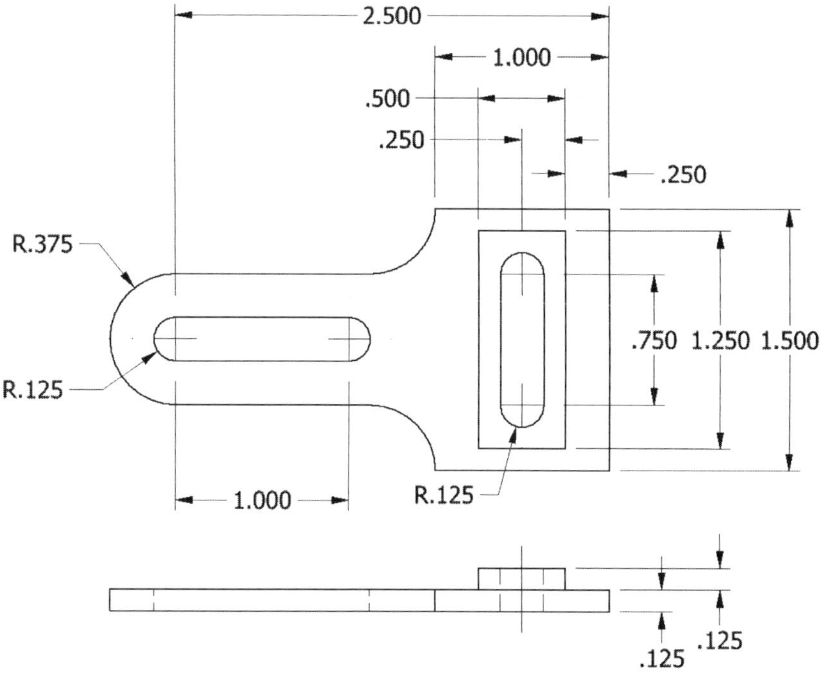

Figure 3.63

The completed part should resemble the following image. When done, close the file. Do not save changes. End of exercise.

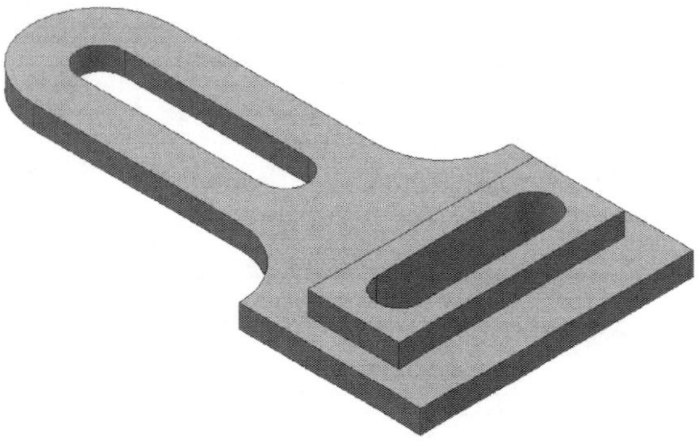

Figure 3.64

Skills Exercise 3-2

In this exercise, you create a connecting rod and add draft to extrusions during feature creation.

1. Start a new part based on the English Standard (in).ipt template.
2. Create a sketch on the XY origin plane.
3. Sketch the outside shape of the connecting rod as shown in the following image.
4. Add geometric constraints and dimensions to fully constrain the sketch.
5. Extrude the base feature using the Symmetric option, adding a -10° taper.
6. Create a separate feature for each pocket by projecting geometry. The sides of the pocket are parallel to the sides of the connecting rod.

Note you can create half of the part and mirror it to create the other half. If desired, complete the exercise as directed and repeat the exercise by modeling half the part and mirroring it. Consult Help for information on mirroring features.

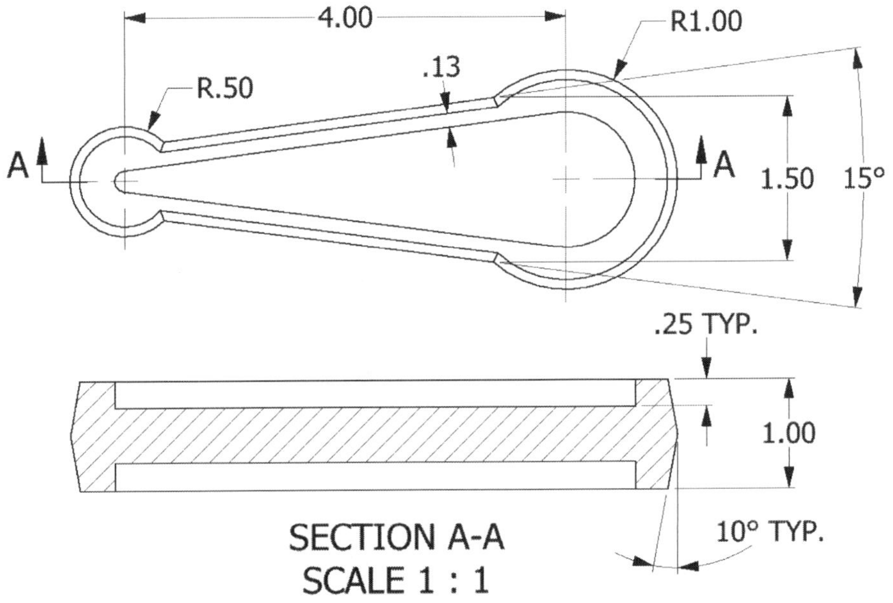

Figure 3.65

The completed part should resemble the following image. When done, close the file. Do not save changes. End of exercise.

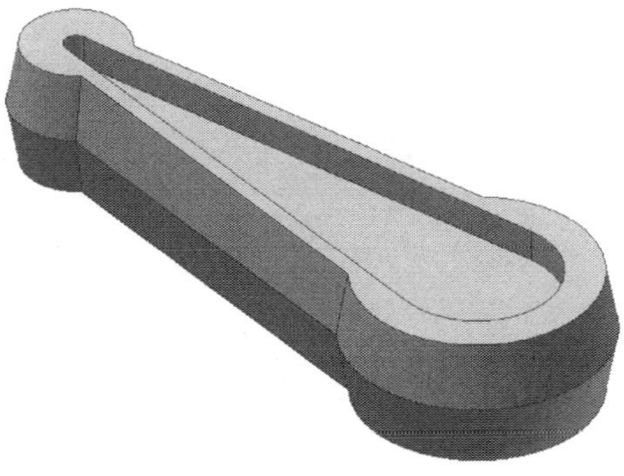

Figure 3.66

Skills Exercise 3-3

In this exercise, you create a pulley using a revolved feature. Assume that the part is symmetric about the middle of the part vertically.

1. Start a new part based on the English Standard (in).ipt template.
2. Create a sketch on the XY origin plane.
3. Sketch the cross-section of the pulley as shown in the following image.

TIP: To create a centerline, draw a line, select it, and then click the Centerline command on the Sketch tab > Format panel.

4. Apply appropriate sketch constraints.
5. Add dimensions.
6. Revolve the sketch using the Full Extents.

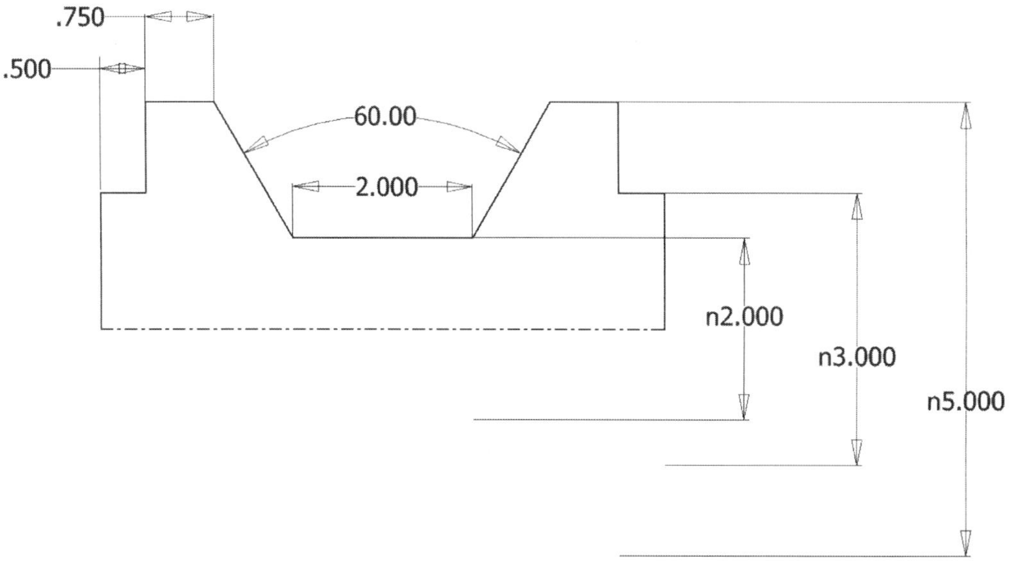

Figure 3.67

7. The completed part should resemble the following image. When done, close the file. Do not save changes. End of exercise.

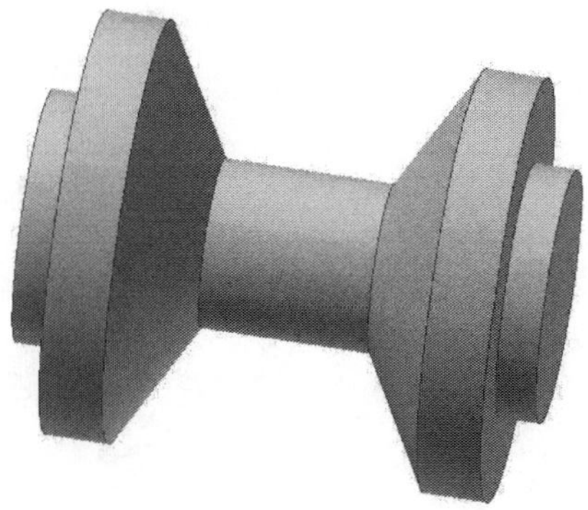

Figure 3.68

CHECKING YOUR SKILLS

Use these questions to test your knowledge of the material covered in this chapter.

1. What is a base feature?
2. True__ False__ When creating a base feature with the Extrude command, the only option you can define is the distance of the extrusion.
3. When creating a revolve feature, which objects can be used as an axis of revolution?
4. Explain how to create a linear diameter (diametric) dimension in a sketch.
5. Name two ways to edit a feature.
6. True__ False__ Once a sketch becomes a base feature, you cannot delete or add constraints, dimensions, or objects to the sketch.
7. Name three operation types used to create sketched features.
8. True__ False__ A direct manipulation technique can only be started by clicking on a face of the part.
9. True__ False__ Once a sketched feature exists, its extents type cannot be changed.
10. True__ False__ By default, geometry that is projected from one face to a sketch will update automatically based on changes to the original projected geometry.
11. Explain what the asymmetric option does for the extrude and revolve commands.
12. Where do you set the physical material property of a part?
13. True__ False__ After setting a part's material properties, the Appearance setting in the Quick Access toolbar must be set to Color Matching for the appearance to match the material of the part.
14. True__ False__ The Project Geometry command is used to copy objects onto any selected face.
15. True__ False__ By default, when a feature is deleted, the feature's sketch will be maintained.

4 Creating Placed Features

INTRODUCTION

In this chapter, you will learn how to create placed features. Many placed features are predefined, except for specific values, and only need to be located. You can edit placed features in the browser like you do for sketched features. When you edit a placed feature the dialog box that you used to create it will open.

When creating a part, it is usually better to use placed features instead of sketched features when possible. For example, to make a through hole as a sketched feature, you can draw a circle profile, dimension it, and then extrude it with the cut operation. By creating a hole as a placed feature, you can select the type of hole, size it, and then place it using a dialog box. When drawing views are generated, the type and size of the placed holes are easy to annotate, and they automatically update if the hole type or values change. If a circle was extruded to create the hole feature, the only information that you can retrieve in a drawing is the hole's diameter.

OBJECTIVES

After completing this chapter, you will be able to perform the following:

- ☐ Create fillets
- ☐ Create chamfers
- ☐ Create holes
- ☐ Shell a part
- ☐ Create work axes
- ☐ Create work points
- ☐ Create work planes
- ☐ Create a UCS
- ☐ Pattern features
- ☐ Export a file for 3D printing

FILLETS

Adding rounded corners is accomplished with the Fillet command. Fillet features consist of both fillets and rounds. Fillets add material to interior edges to create a smooth transition from one face to another. Rounds remove material from exterior edges. The following image shows a part without fillets on the left and the part with fillets and holes on the right.

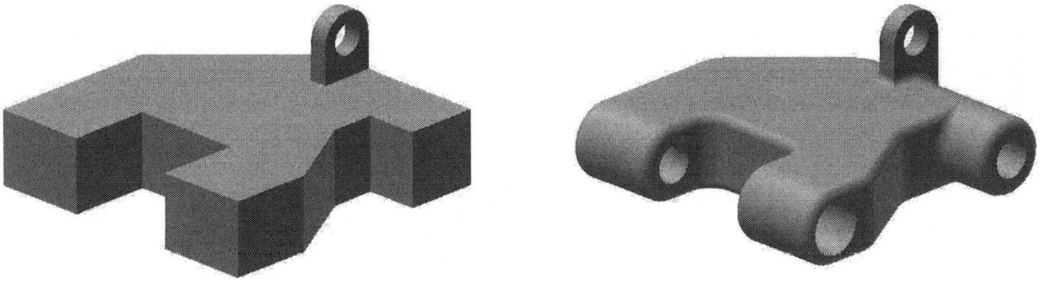

Figure 4.1

To create a fillet, you select the edge(s) to fillet. The fillet is created between the two faces that share the edge. You can select two or three faces to form a fillet between. When placing a fillet between two faces, the faces do not need to share a common edge. When creating a part, it is good practice to create fillets and chamfers as some of the last features in the part since fillets add complexity to the part, which in turn adds to the size of the file. Fillets also remove edges that you may need to place other features.

Select the Fillet command from the 3D Model tab > Modify panel, as shown in Figure 4.2 left. While not in a sketch, click on an edge of a part to show the mini-toolbar; choose Create Fillet as shown Figure 4.2 right or press the F key. After you start the command, a dialog box appears as shown in Figure 4.3.

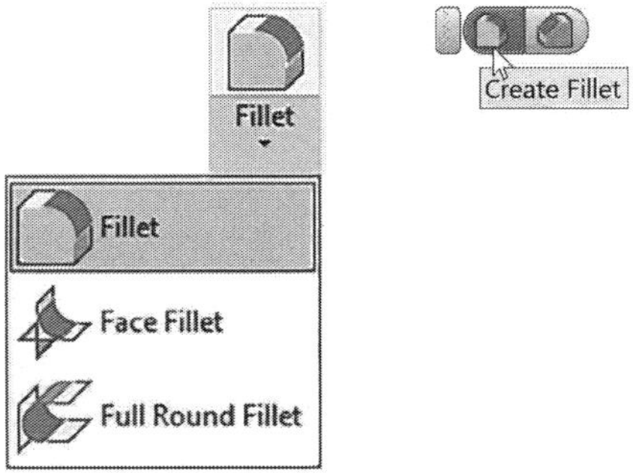

Figure 4.2

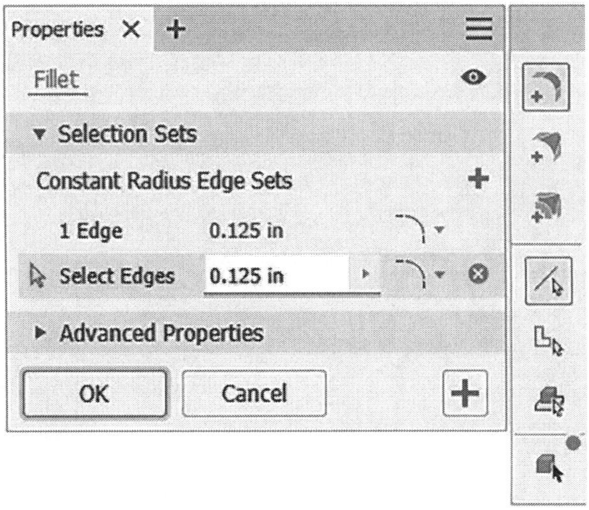

Figure 4.3

There are three types of fillets: Edge, Face, and Full Round. When the Edge option is selected, you see a toolbar to the right (although you can reposition it) as shown in Figure 4.3. The top 3 icons are for: Constant, Variable, and Setbacks. The lower icons set the selection priority, making it easier to choose the desired item.

The preview icon (looks like an eye) at the top right of the dialog box. It is used to preview/not preview the options for the fillet on the part. The preview only works when a valid fillet can be created from the values you entered.

Each fillet feature can contain multiple selection sets, each having its own unique fillet value. There is no limit to the number of selection sets that can exist in a single instance of the feature. An edge, however, can only exist in one selection set. To create another selection set, use the dialog box Click to add area and then select the edges to add and change the radius as needed. If you change the value, all the fillets in that selection group will change.

To remove an edge or face from your selection set, choose the selection set that includes the edge or face so that they are highlighted. Hold down the CTRL key, and click the edge(s) or face(s) to be removed from the selection set.

After the edges are selected, enter the desired values for the fillet. As changes are made in the dialog box, a representation of the fillet is previewed in the graphics window. Click OK or Apply to create the fillet.

To edit a fillet's type and radius, use one of the following methods:

- On the part, click on the face of the fillet and click Edit Fillet from the mini-toolbar.
- Right-click on the fillet's name in the browser and select Edit Feature from the menu.
- Double-click on the feature's name or icon in the browser.

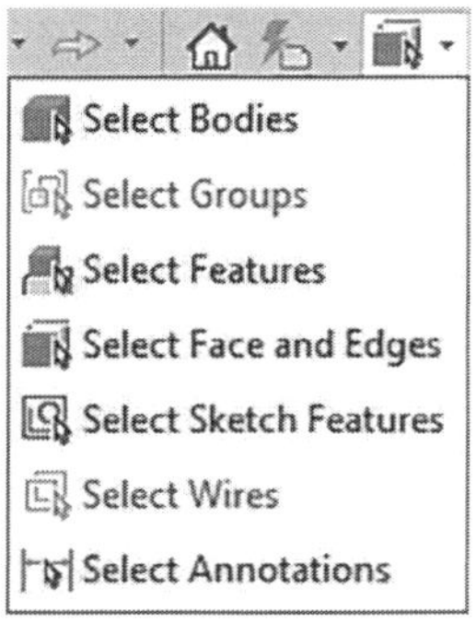

Figure 4.4

TIP: To use window or crossing to select multiple edges you must first set the *selection filter* to Faces and Edges. Use the Quick Access Toolbar as shown in Figure 4.4.

The options for each fillet type are described in the following sections.

Edge Fillet

An edge fillet is the default and has the same radius from beginning to end. There is no limit to the number of part edges that you can fillet with a constant radius fillet. You can select the edges as a single set or as multiple sets, and each set can have its own radius value. The following section describes the options for creating an edge fillet.

Edges

By default after starting the Fillet command, you can click edges and they appear in the first selection set. You can continue to select multiple edges. To remove an edge from the set, hold down the CTRL key and select it.

Radius

Enter a size for the fillet. The size of the fillet will be previewed on the selected edges.

Continuity

To adjust the continuity of the fillet, select a Tangent, Smooth G2 (continuous curvature), or Inverted fillet (Figure 4.5). Figure 4.6 shows examples of these fillets.

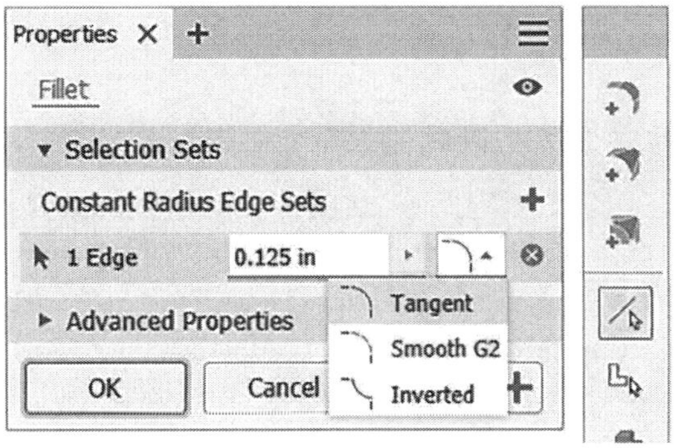

Figure 4.5

Figure 4.6

Select Mode

Edge: Edges and edges that are tangent to the selected edge will be filleted. This is the default option.

Loop: Click the Loop mode to have all the edges that form a closed loop with the selected edge filleted.

Feature: Click the Feature mode to select all the edges of a selected feature.

Solids

If multiple solid bodies exist, select the solid body option to apply All Fillets or All Rounds.

All Fillets

Click the All Fillets option to select all concave edges of a part that have not already been filleted. This operation adds material to the part. Figure 4.7 middle shows an example of All Fillets applied to a part.

All Rounds

Click the All Rounds option to select all convex edges of a part that have not already been filleted. This operation removes material from the part. Figure 4.7 right shows an example of All Rounds applied to a part.

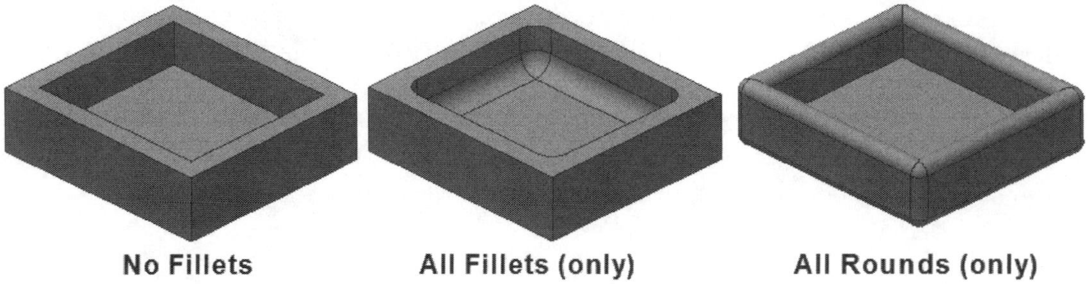

Figure 4.7

Click the arrow to expand the Advanced Properties to see more options as shown in Figure 4.8.

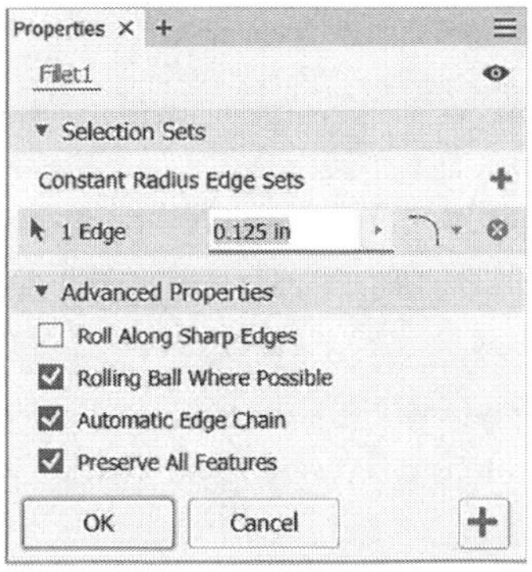

Figure 4.8

Roll along Sharp Edges

Choose this option to adjust the specified radius to preserve the edges of adjacent faces.

Rolling Ball Where Possible

Click this option to create a fillet around a corner that looks as though a ball has been rolled along the edges that define a corner, as shown in Figure 4.9 middle. When the Rolling ball where possible solution is possible, but it is not selected, a blended solution is used, as shown in the following image on the right.

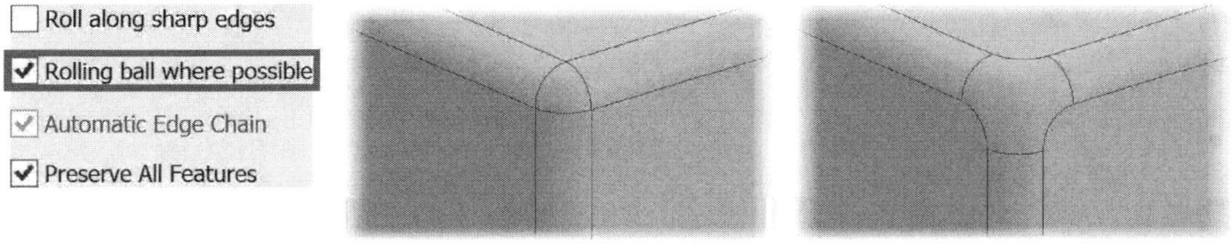

Figure 4.9

Automatic Edge Chain

Choose this option to automatically select edges that are tangent to the selected edge. If you don't want tangent edges filleted, deselect this option before selecting an edge.

Face Fillet

When the Face fillet type is selected the Properties dialog box is as shown in Figure 4.10. Create a face fillet by selecting two or more faces; the faces do not need to be adjacent. If a feature will be consumed by the fillet, the internal volume of the feature is filled in by the fillet.

Figure 4.10

To create a face fillet, follow these steps:

1. Start the command by selecting Face Fillet.
2. For Faces A, select one or more tangent contiguous faces on the part to which the fillet will be tangent.
3. Make Faces B button active, select one or more tangent contiguous faces on the part to which the fillet will be tangent.

Check Include Tangent Faces to automatically chain all faces that are tangent to faces in the selection set.

Figure 4.11 left shows a part that has a gap between the bottom and top extrusion. The middle image shows the preview of the face fillet. The image on the right shows the completed face fillet. Notice the rectangular extrusion on the top face is consumed by the fillet.

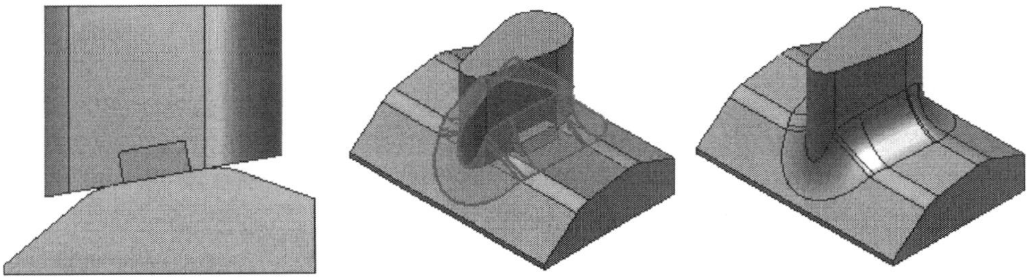

Figure 4.11

Full Round Fillet

Click the Full Round fillet type in the Fillet dialog box as shown in the following image. The full round fillet option creates a fillet that is tangent to three faces; the faces do not need to be adjacent.

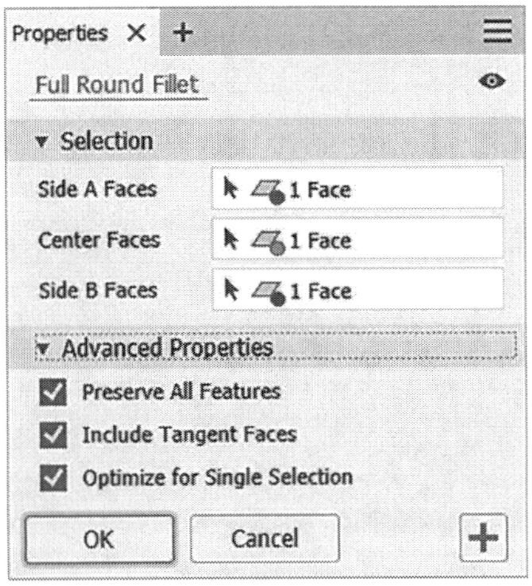

Figure 4.12

To create a Full Round fillet, follow these steps:

1. Start the fillet command and click the Full Round option.
2. With the Side Face Set 1 button active, select a face on the part where the fillet will start and to which it will be tangent.
3. With the Center Face Set button active, select a face on the part to which the middle of the fillet will be tangent.
4. With the Side Face Set 2 button active, select a face on the part that the fillet will end at and to which it will be tangent.

5. Check the Include Tangent Faces option to automatically chain all faces that are tangent to faces in the selection set.
6. Check the Optimize for Single Selection to automatically make the next selection set button active after selecting a face.

Figure 4.13 shows two examples of three faces selected on the left and the resulting full round fillet on the right.

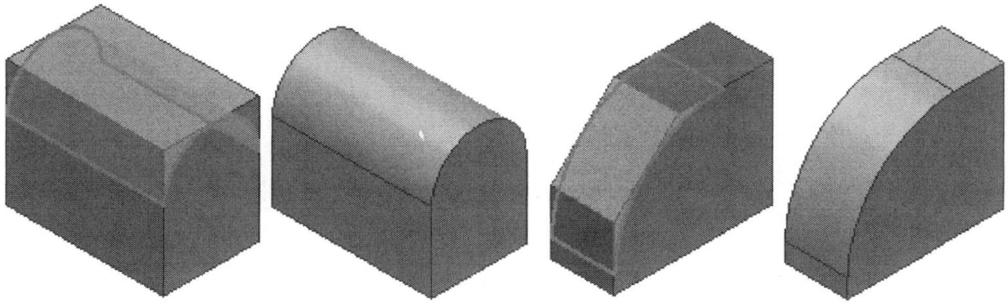

Figure 4.13

CHAMFERS

Chamfer features are used to bevel edges. When you create a chamfer on an interior edge, material is added to your model. When you create a chamfer on an exterior edge, material is removed from your model, as shown in Figure 4.14 right.

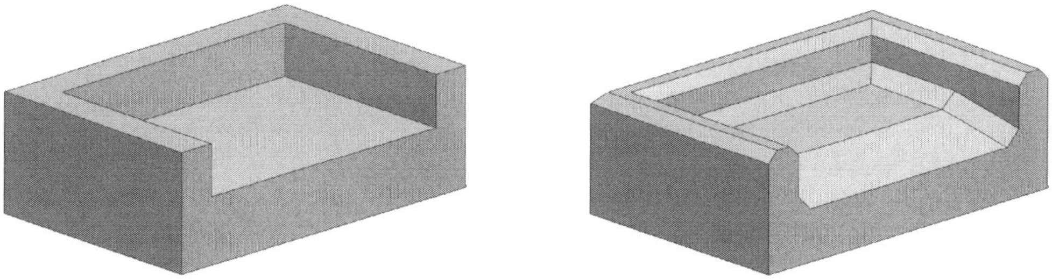

Figure 4.14

To create a chamfer feature, follow the same steps that you used to create fillet features. Click the common edge, and the chamfer is created between the two faces sharing the edge. To create a chamfer feature, click on the 3D Model tab > Modify panel > Chamfer as shown in Figure 4.15 left, or while not in a sketch, click on an edge and the mini–toolbar will appear, then click the Create Chamfer button in the mini-toolbar, as shown in the image on the right.

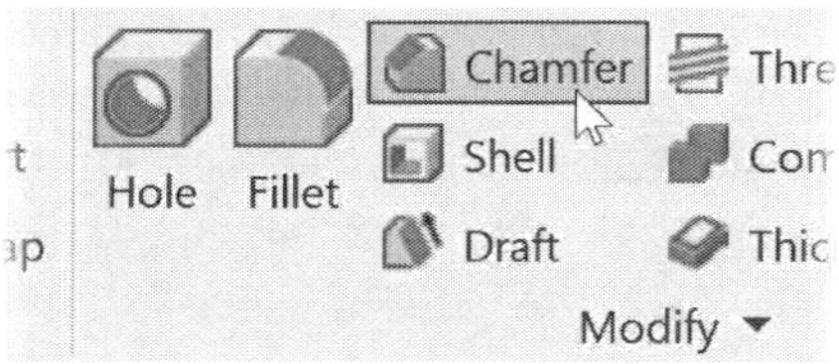

Figure 4.15

After you start the chamfer command, the Chamfer dialog box will appear. Select one of three methods to create chamfer features; the methods are described in the next section. As with fillet features, you can select multiple edges to be included in a single chamfer feature. From the dialog box, click a method, click the edge or edges to chamfer, enter a distance and/or angle, and then click OK.

Methods

There are three methods options found on the left column of the Chamfer dialog box used to create a chamfer:

- Distance
- Distance and Angle
- Two Distances.

Distance

After starting the Chamfer command, the Distance option is the default and creates a 45° chamfer on the selected edges. The Distance option in the Chamfer dialog box is shown in the following image. Figure 4.16 right illustrates the Distance option. You set the size of the chamfer by typing a distance in the dialog box. The value is the offset from the common edge of the two adjacent faces. You can select a single edge, multiple edges, or a chain of edges. A preview image of the chamfer appears on the part. If you select the wrong edge, hold down the CTRL key and select the edge to remove it.

Edges: Select an edge or edges to be chamfered.

Distance: Enter a distance to be used for the offset in both directions from the selected edge(s).

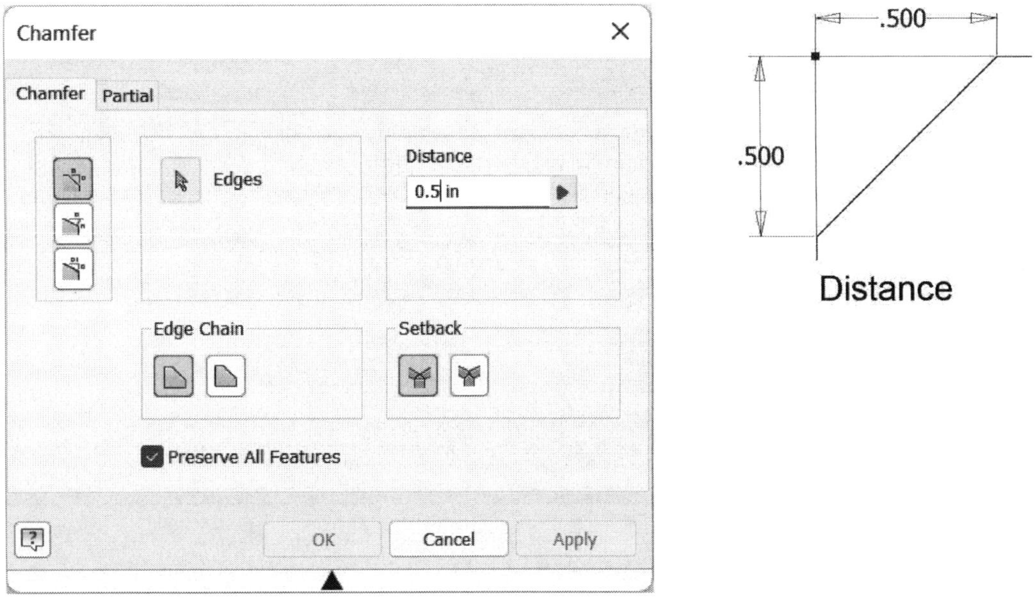

Figure 4.16

Distance and Angle

Click the Distance and Angle option to create a chamfer offset from a selected edge on a specified face, at the defined angle. In the Chamfer dialog box select the Distance and Angle option as shown in the following image. Enter an angle and distance for the chamfer, then click the face to which the angle is applied and specify an edge to be chamfered. You can select one edge or multiple edges. The edges must lie on the selected face. A preview image of the chamfer appears on the part. If you have selected the wrong face or edge, click on the Edge or Face button, and choose a new face or edge. Figure 4.17 right illustrates the use of the Distance and Angle option.

Edges: Select an edge or edges to be chamfered.

Face: Select a face on which the chamfer angle will be based.

Distance: Enter a distance to be used for the offset from the selected edge(s).

Angle: Enter a value that will be used for the chamfer.

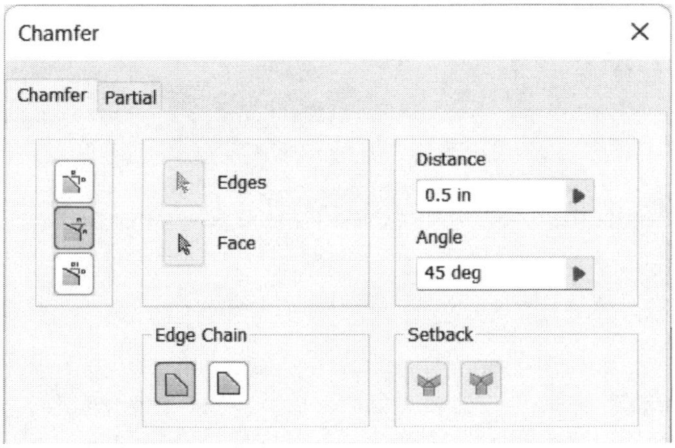

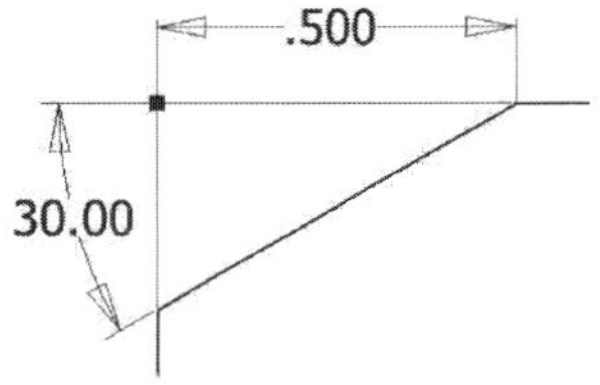

Distance and Angle

Figure 4.17

Two Distances

Click the Two Distances option to create a chamfer offset from two faces, each being the amount that you specify. In the Chamfer dialog box select the Distance and Angle option as shown in Figure 4.18. Click an edge first, and then enter values for Distance 1 and Distance 2. A preview image of the chamfer appears. To reverse the direction of the distances, click the Flip button. When the correct information about the chamfer is in the dialog box, click the OK button. You can only use a single edge or chained edges with the Two Distances option. The following image on the right illustrates the use of the Two Distances option.

Edge: Select an edge or edges to be chamfered.

Flip: Click the flip button to reverse the direction of the chamfer.

Distance1: Enter a distance to be used for the offset in one direction from the selected edge(s).

Distance2: Enter a distance to be used for the offset in one direction from the selected edge(s).

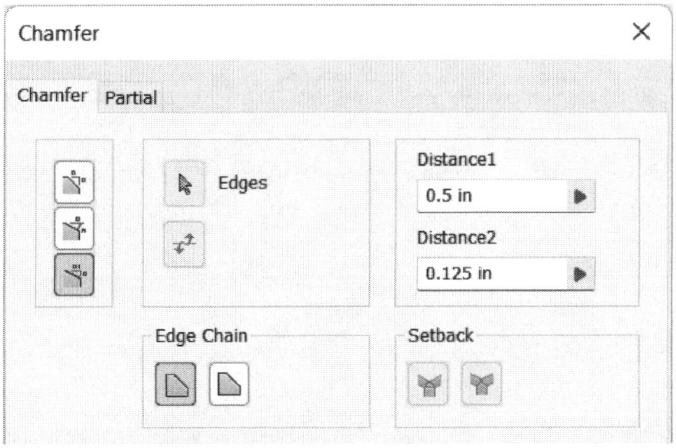

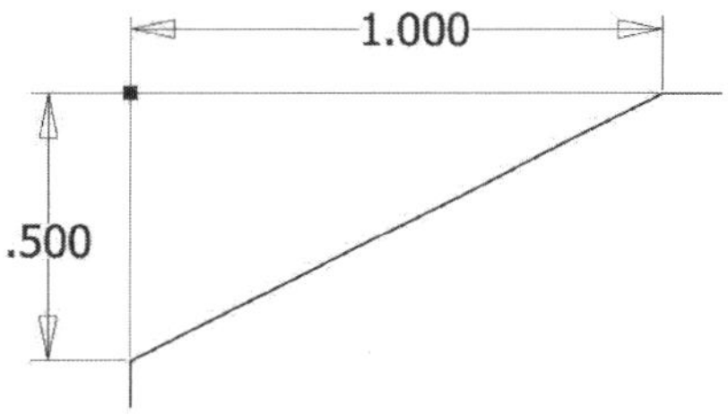

Two Distances

Figure 4.18

Near the bottom-half of the Chamfer dialog box there are more options as described next.

Edge Chain

Click this option to include tangent edges in the selection set automatically, as shown in Figure 4.19.

Setback

When the Distance method is used and three chamfers meet at a vertex, click this option to have the intersection of the three chamfers form a flat edge (left button) or to have the intersection meet at a point as though the edges were milled (right button), as shown in Figure 4.19.

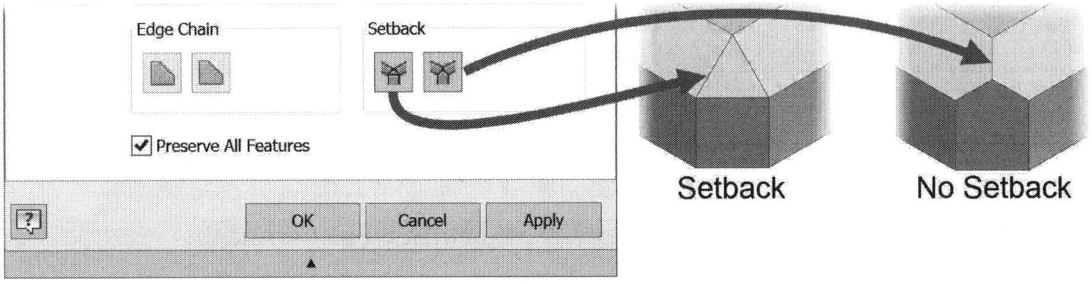

Figure 4.19

Preserve All Features

Click this option to verify all features that intersect with the chamfer and to calculate their intersections during the chamfer operation. If the checkbox is clear, only edges that are part of the chamfer operation are calculated.

Create Partial Chamfer

When creating a chamfer, you may only want it to span a portion of an edge and not the entire length of an edge. To chamfer a portion of an edge, follow these steps.

1. Start the Chamfer command on the 3D Model tab > Modify panel or click on an edge and click the Chamfer button on the mini–toolbar.
2. If needed select an edge to place a partial chamfer.
3. Click the Partial tab as shown in the following image on the right.
4. Click a point on the model to base an offset distance from; the following image on the left shows the endpoint on the right edge being selected.
5. In the Chamfer dialog box enter a Start distance or move the sphere closer to the start point on the edge to chamfer; this distance will be where the chamfer will begin. Note that the value can be zero.
6. In the Chamfer dialog box enter a value for where the chamfer will End or move the sphere furthest from the start point on the edge to chamfer. This value is calculated from the start point.
7. Change the size and type of chamfer as desired by clicking the Chamfer tab.
8. Near the bottom of the dialog box, select an option to create a Driven Dimension for the chamfer option: To Start, Chamfer, or To End.

Figure 4.20 left shows a partial chamfer being placed with a starting point on the right endpoint. The middle image and dialog box on the right side shows a partial chamfer being placed .5 inches in from the right, the length of the chamfer is 1.5 inches from the start point, the chamfer stops .375 inches from the endpoint, and the length of the original selected edge is 2.375 inches.

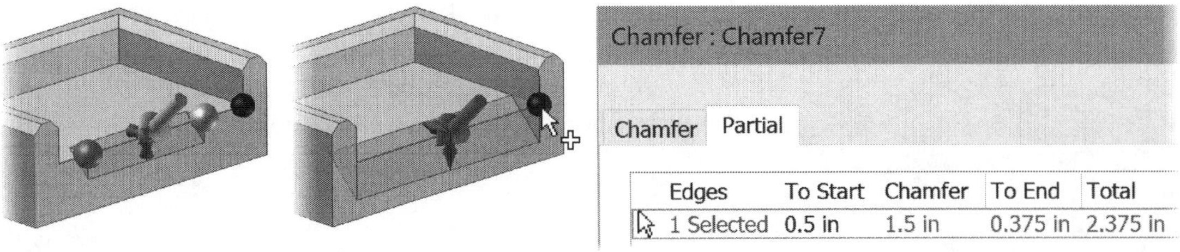

Figure 4.20

To edit a chamfer feature use one of the following methods:

- In the graphics window select on the chamfered face (not the edge) and choose Edit Chamfer from the mini-toolbar.
- Double-click the chamfer's name or icon in the browser.
- Right-click the chamfer's name in the browser, and select Edit Feature from the menu.

EXERCISE 4-1: CREATING FILLETS AND CHAMFERS

In this exercise, you create constant radius fillets, full round fillets, face fillet, and chamfers.

1 Open *ESS_E04_01.ipt* from the Chapter 04 folder.

2  Click Fillet from the 3D Model tab > Modify panel to create an edge fillet. Click the inside edge of the cutout as shown in Figure 4.21, or select the inside edge of the cutout first, and use Create Fillet from the mini-toolbar.

3 Click on the first entry in the Radius column, and enter **0.25 in** for the radius as shown in Figure 4.21.

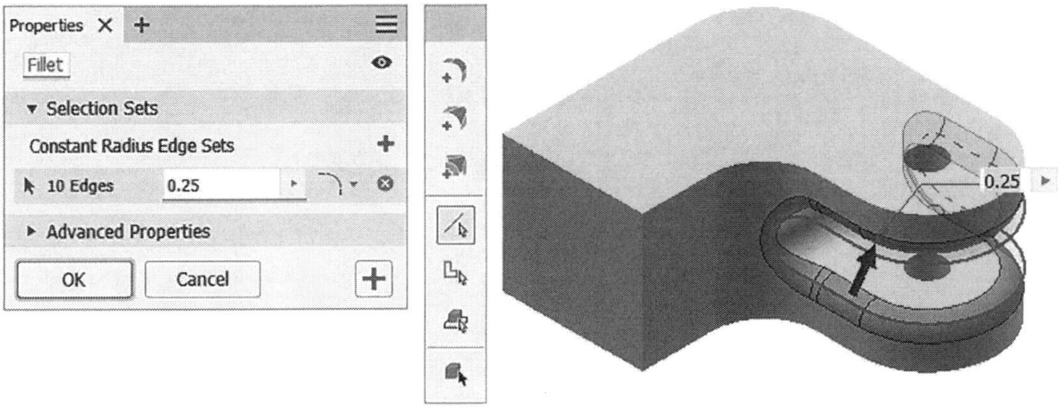

Figure 4.21

4 Click OK.

5  Next, create a Full Round Fillet tangent to the front three faces. Click the Full Round Fillet icon.

6 Click the front-inside face, then the left-front face, and then the back-vertical face of the part, as shown in Figure 4.22 labeled (1), (2), and (3).

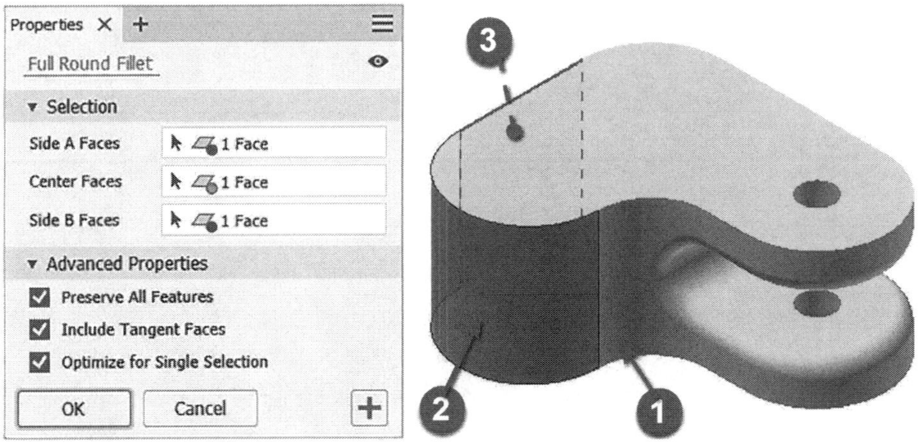

Figure 4.22

7 Click OK in the dialog box to create the fillet and close the dialog box.

8 In the browser, right-click on Extrusion4 (Suppressed), and from the menu, choose Unsuppress Features.

9  Next, create a Face Fillet by clicking the icon from the Modify panel. In the Face Fillet dialog box click the cylindrical face of Extrusion4 and then the full round fillet that you created in the last step, as shown in Figure 4.23 labeled (1) and (2).

10 In the Fillet dialog box, type **.375** in the Radius field.

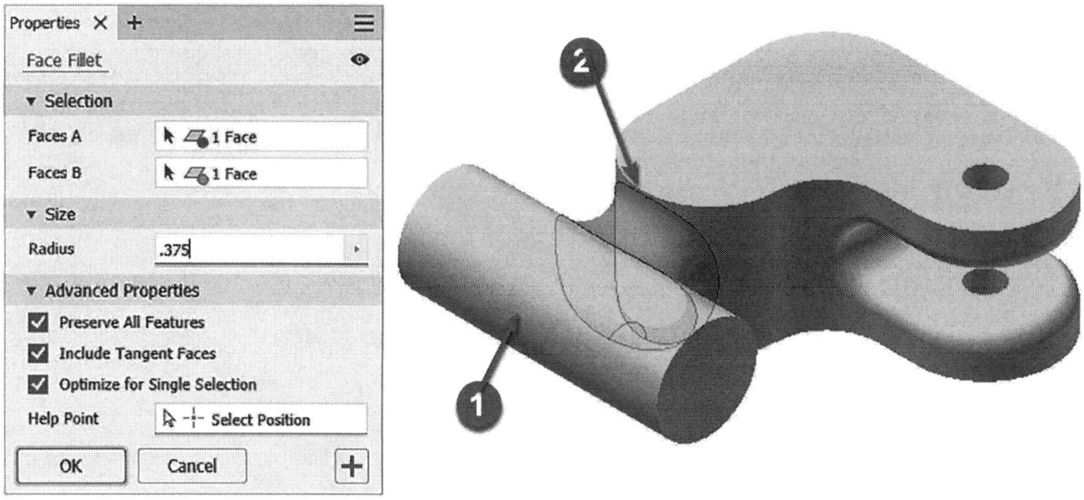

Figure 4.23

11 Click OK to create the fillet and rotate the viewpoint to examine the fillet features.

12 Move the cursor into a blank area in the graphics window, right-click, and click Fillet from the top of the marking menu. 

13 Set the Select Mode to Loop as shown in Figure 4.24.

14 Select the top and bottom edges of the model labeled (1) and (2).

15 In the Fillet dialog box change the Radius to **0.125 in**. To create the fillets and close the dialog box, click OK.

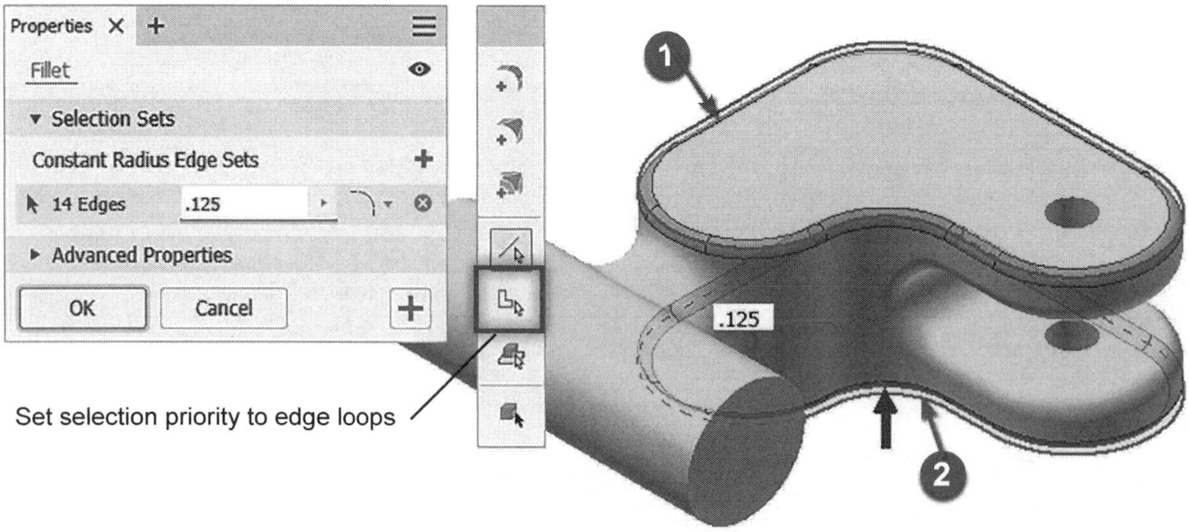

Figure 4.24

16 Click the back cylindrical edge on Extrusion4 and click Create Chamfer from the mini-toolbar as shown in Figure 4.25 left.

17 In the Chamfer dialog box, change the Distance field to **0.25**, as shown in Figure 4.25 right.

18 To create the chamfer and keep the dialog box open, click the Apply button in the Chamfer dialog box.

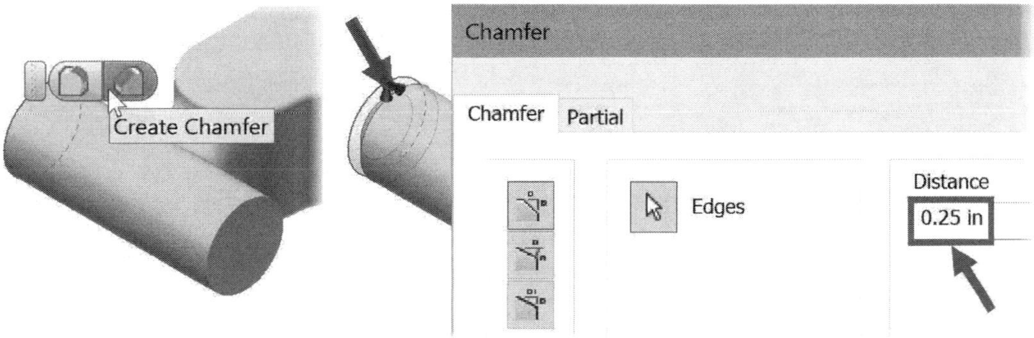

Figure 4.25

19 In the Chamfer dialog box, click the Distance and Angle option labeled (1) as shown in Figure 4.26. Enter **0.25 in** in the Distance field labeled (2) and **60 deg** in the Angle field labeled (3).

20 Click the front-circular face of Extrusion4 labeled (4) and click the front-circular edge of that face labeled (5).

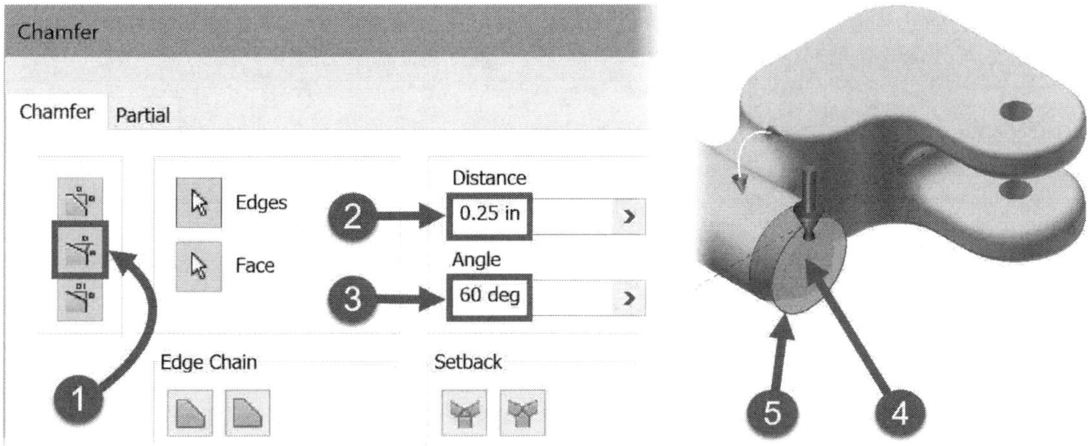

Figure 4.26

21 Click OK to create the chamfer. When done, your model should resemble Figure 4.27.

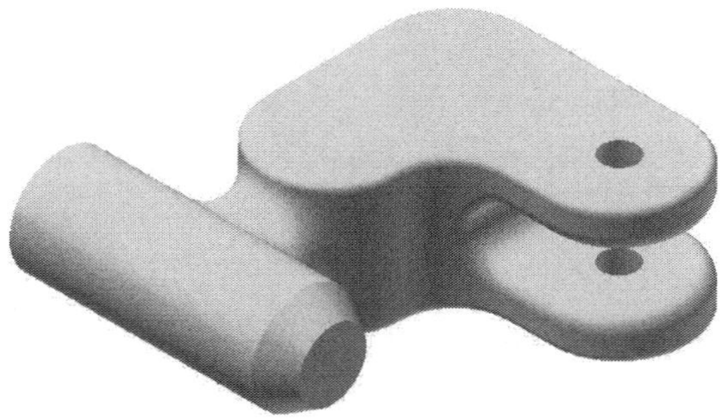

Figure 4.27

22 Practice editing and placing fillets and chamfers.

23 Close the file. Do not save changes.

End of exercise.

HOLES

The Hole command lets you create drilled, counterbore, spotface, countersink, clearance, tapped, and taper tapped holes, as shown in Figure 4.28. You can place holes using sketch geometry or existing points, or edges of a part, and specify the drill point and thread parameters.

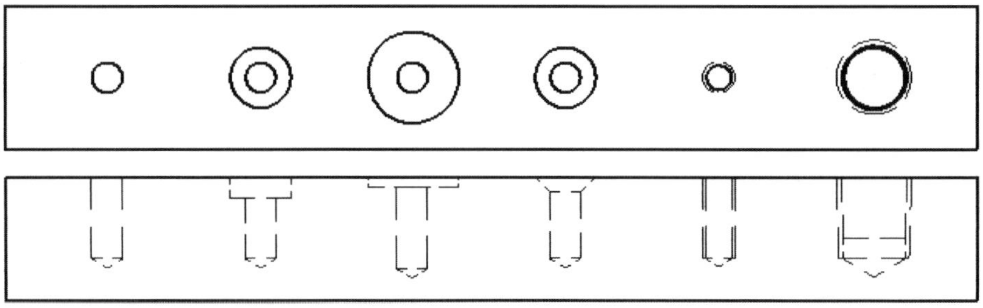

Figure 4.28

Creating Hole Features

To create a hole feature, follow these steps:

1. Create or open a part that you want to place a hole on.
2. Click the Hole command from the 3D Model tab > Modify panel, as shown in Figure 4.29 left, or press the shortcut key H. You could also right-click and click Hole from the marking menu.

The Hole Properties panel appears, as shown in Figure 4.29 right.

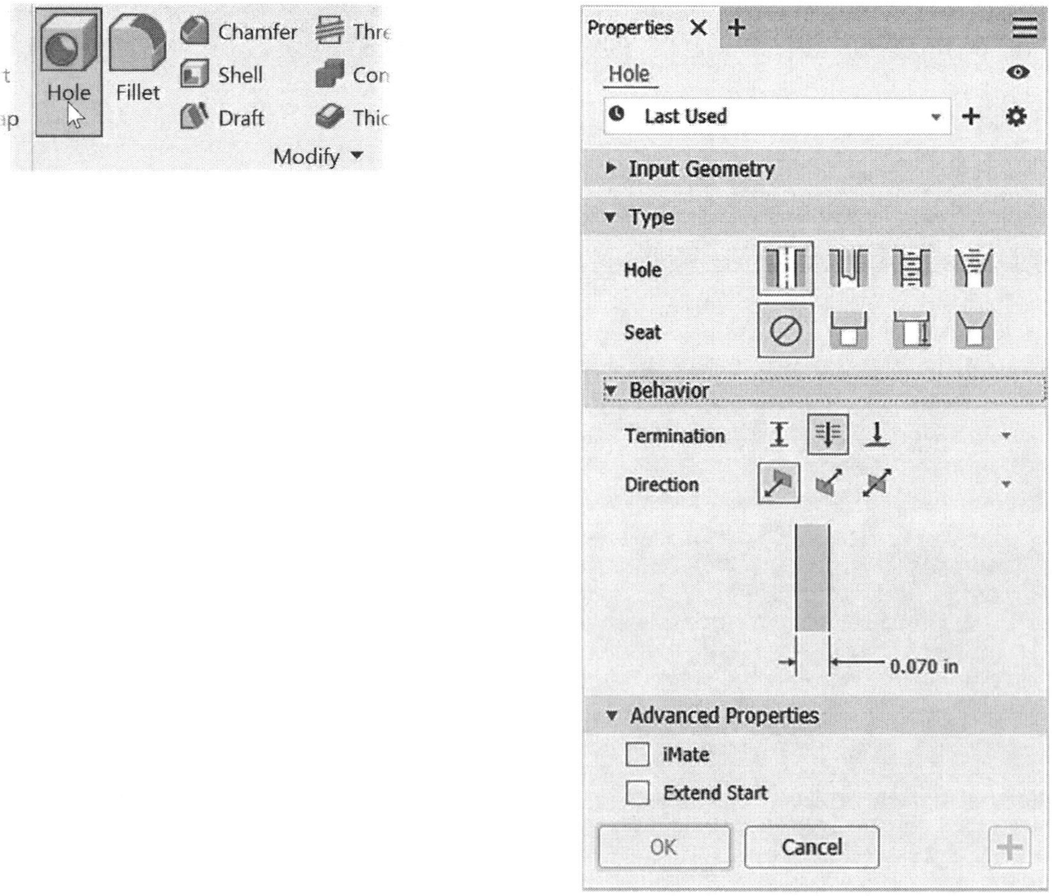

Figure 4.29

3. If available you can select a preset. If a preset is not available, you can add a preset by entering the data for a specific hole type and size, and then click the + icon in the Hole preset area in the dialog box.
4. Locate the hole's position by clicking on a work point, or on a planar face. The hole center is placed where you click. You can move unconstrained hole centers to a new location.
   a. Click a reference edge(s) to place dimensions. Linear placement dimensions can be selected, edited, or deleted while you are in the Hole command. Select the dimension and change the value or use the Delete key to remove the dimension.
   b. To create concentric holes, place the hole center and click the model edge or curved face the hole is to be concentric with.
   c. If a work point is selected to locate the hole, you need to set the direction by selecting a work axis, edge, or a face.
   d. If a sketch with more than one hole center is visible, all hole centers without an associated hole will be selected. If needed, you can edit the sketch by clicking the breadcrumb at the top of the property panel Hole# > Sketch#.
5. To remove a hole center from the selection and retain it in the sketch, press the CTRL key and select the hole center. To remove the hole and hole center, select and delete the hole center point.
6. Select the desired hole type, and then set the hole size, and termination as needed.

TIP If a sketch contains many hole centers for holes of various sizes, you can quickly deselect all selected hole centers using the Clear Selections button at the end of the Position(s) list in the Placement area of the dialog box as shown in Figure 4.30.

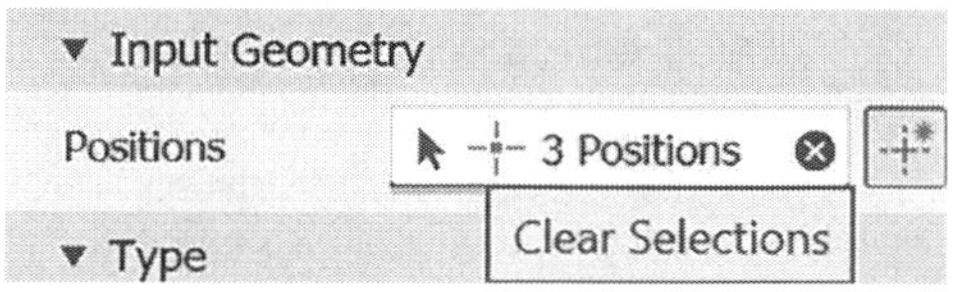

Figure 4.30

Sketch Points

When creating a hole(s) feature with the From Sketch option, all center points that reside in the sketch are automatically selected as centers for the hole feature. This can speed up the process of creating multiple holes, and the holes will reside as a single hole feature. You can also select endpoints of lines, arcs, splines, center points of arcs and circles, or spline control points as hole centers. Points can be deselected or selected by holding down the CTRL or SHIFT key and then select the points individually or use a window or crossing section method.

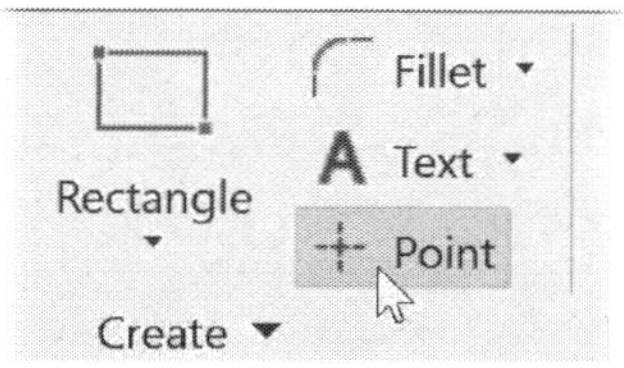

Figure 4.31

Holes Dialog Box

In the Holes dialog box, you define the placement method, type of hole, its termination, and options such as type of drill point, angle, and tapped properties.

Hole Types

Set the hole type that you want to create: simple (drilled), clearance, tapped, or taper tapped.

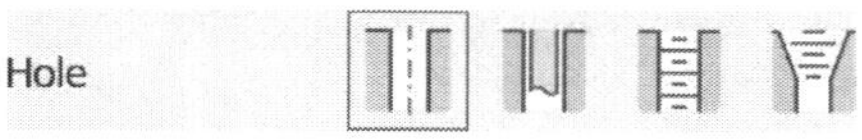

Figure 4.32

Hole Seat

Set the hole seat that you want to create: none, counterbore, spotface, or countersink.

Figure 4.33

Change Hole's Dimensions

To change the diameter, depth, countersink, counterbore diameter, countersink angle, or counterbore depth of the hole, click the dimension in the dialog box, and enter a desired value, or select a value in the drop down menu. The following image shows the options for a tapped hole.

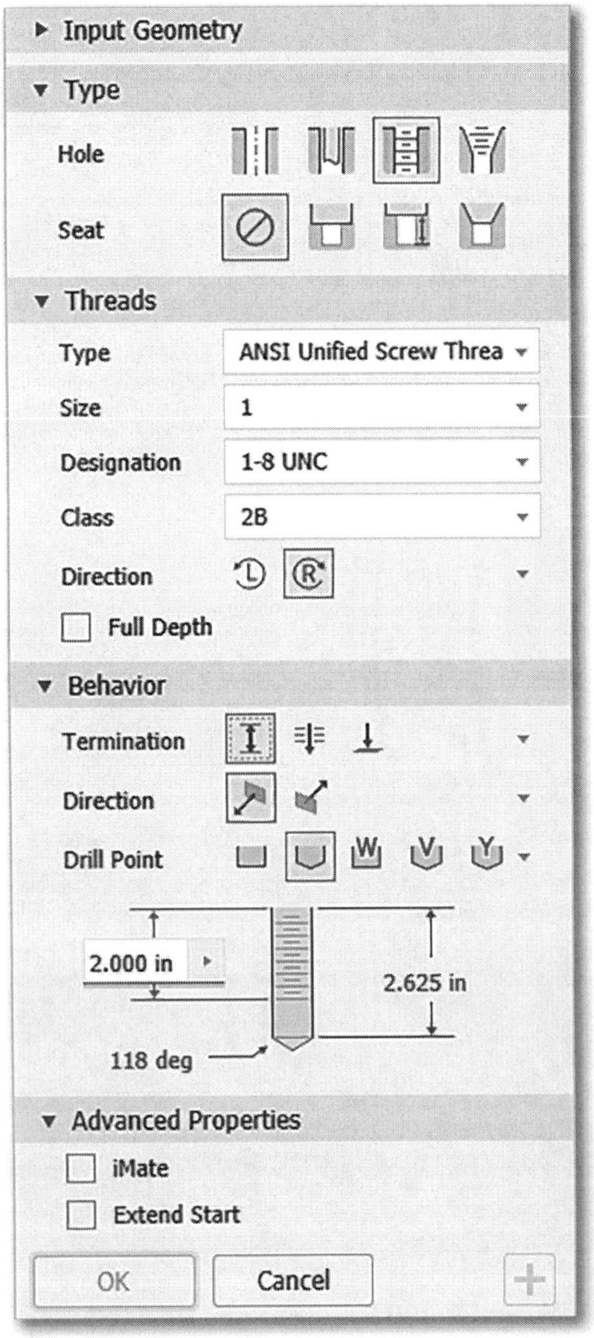

Figure 4.34

Termination

Select how the hole will terminate.

Figure 4.35

Distance

Specify a distance for the depth of the hole. Use the direction buttons to flip the direction in which the hole is drilled.

Through All

Choose to extend the hole through the entire part in one direction. Use the direction buttons to flip the direction that the hole is drilled, or select the symmetric option for the hole to go through to the part in both directions.

To

Select a plane at which the hole will terminate.

Drill Direction

Adjust the direction of the hole by selecting the desired flip option.

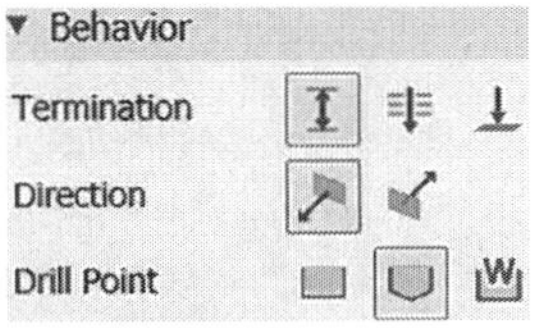

Figure 4.36

Drill Point

If the hole termination is set to Distance or To you can select a flat, angle, flat W, angle V, or angle Y drill point.

Figure 4.37

Advanced Settings

To see advanced settings, click the small arrow to the left of the Advanced Settings option in the dialog box.

iMate

When selected, an iMate is automatically created when the hole is being created.

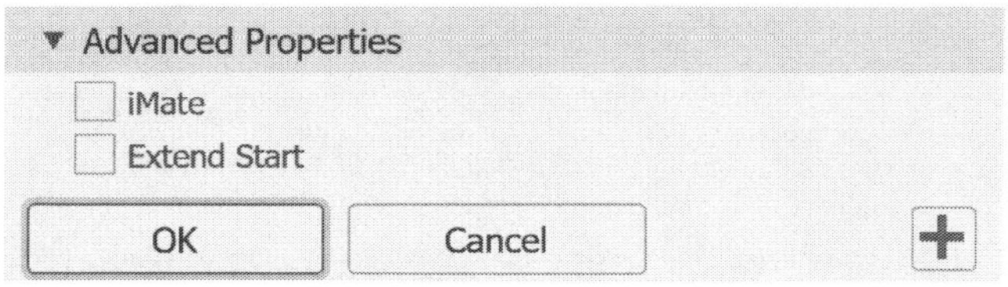

Figure 4.38

Extend Start

Click the Extend Start option to extend the start face of a hole to the first place where there is no intersection with the part and remove the section of material from the part. The following image on the left and in the middle shows a hole placed that interferes with a fillet that has material left above the hole. The image on the right shows the same hole with the Extend Start option checked, and you can see that the material around the fillet is removed.

Figure 4.39

Editing Hole Features

To edit the type of hole feature, use one of the following methods:

- Click on the circular face of the hole that you want to edit and click Edit Hole from the mini-toolbar.
- Double-click the feature's name or icon in the browser.
- Right-click the hole's name or icon in the browser, and select Edit Feature from the menu.

EXERCISE 4-2: CREATING HOLES

In this exercise, you add drilled, tapped, and counterbore holes to a part.

1 Open *ESS_E04_02.ipt* from the Chapter 04 folder.

2 First place a linear hole. Click Hole from the 3D Model tab > Modify panel.

a. Select the top planar face of the extruded rectangle labeled (1) in Figure 4.40.

b. For Reference 1, select the left top vertical edge labeled (2) and enter a value of **.375 inches**; do not press ENTER.

c. Reference 2 will be current, select the horizontal edge labeled (3) and enter a value of **.375 inches.**

d. Verify that the Hole Type is set to Simple labeled (4).

e. Verify that the Termination is set to Through All labeled (5).

f. Change the hole's diameter to **.25 inches** labeled (6).

g. Click OK to create the hole.

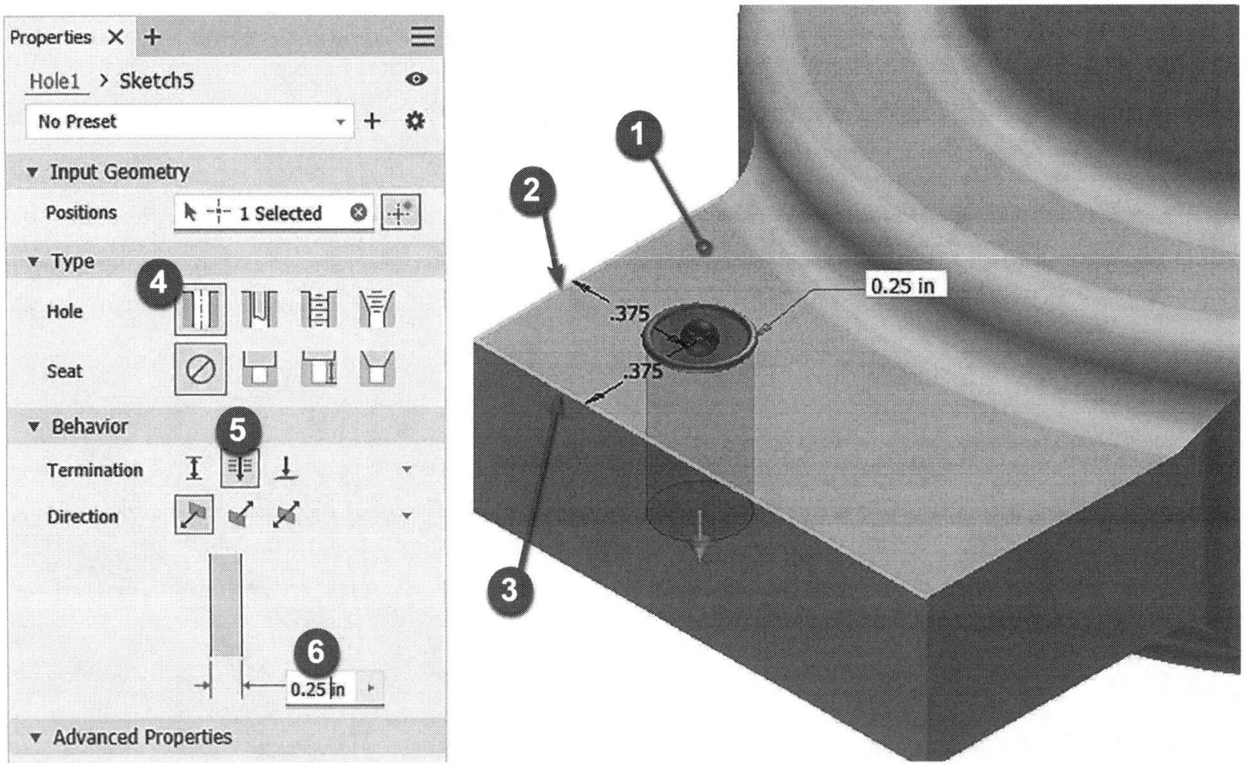

Figure 4.40

3 Next, you create multiple holes based on Sketch points and a center point. Click the top planar face of Extrusion1 and click Create Sketch from the mini-toolbar as shown in the following image on the left.

4 Click the Point, Center Point command in the Sketch tab > Create Panel, and place two points in the middle of the sketch labeled (1) and (2) in the following image on the right.

5 Add a horizontal constraint between the left point labeled (1) in the following image on the right and the center point that was automatically projected from the center of the right circular edges labeled (3) in the following image on the right. If the center point was not projected, use the Project Geometry tool to project the arc.

6 Add another horizontal constraint between the points labeled (2) and (3) in the following image on the right.

7 Add a **1.250 inch** and a **1.000 inch** horizontal dimension between the points as shown in Figure 4.41 right.

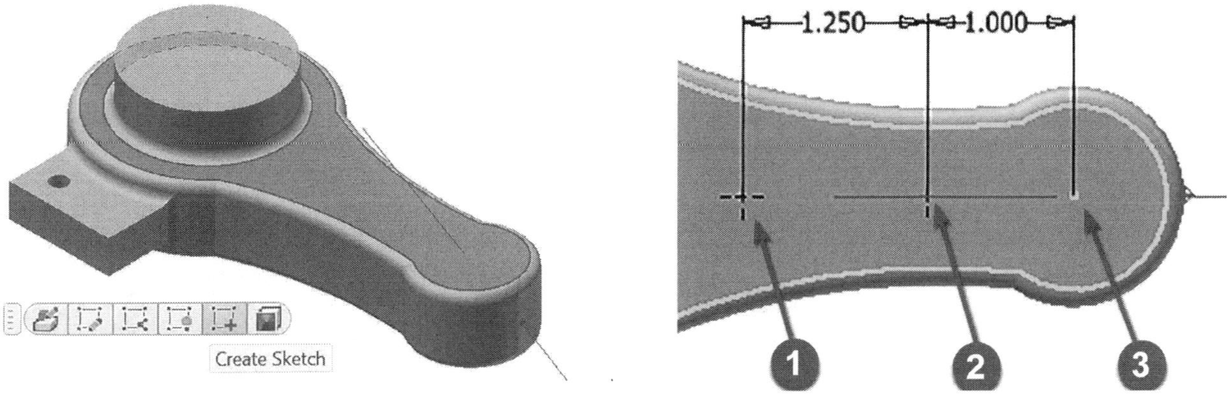

Figure 4.41

8 Right-click in the graphics window and click Finish 2D Sketch from the marking menu.

9 Start the Hole command by pressing the H key. Both Center Points should automatically be selected; if not, select the center-points.

10 Note the arc center point is automatically projected when the sketch is created, labeled (3) in Figure 4.41.

a. Ensure that the Hole Type is set to Simple Hole labeled (1) in the following image.
b. Change the Hole Seat to Countersink labeled (2) in the following image.
c. Ensure the Termination is set to Through All labeled (3).
d. Set the diameter of the countersink to **0.5 inches** labeled (4).
e. Set the diameter of the hole to **0.25 inches** labeled (5) and verify the angle is set to 82 degrees as shown in the following image.
f. Create the holes and keep the Hole Properties panel open by clicking the + icon in the lower-right corner of the Properties panel.

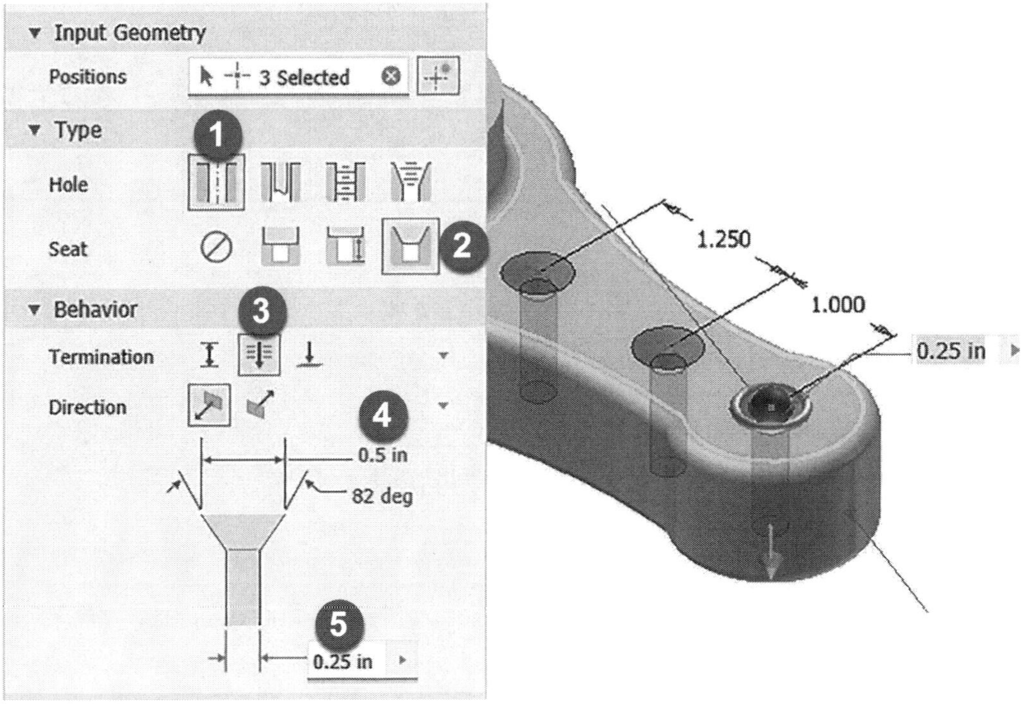

Figure 4.42

11 Next, you create a Taper Tapped hole that is concentric to the top of the left cylinder.

a. While still in the Hole command select the top face of the cylinder Extrusion3 labeled (1) in the following image.

b. For the concentric reference, select the top circular edge of the cylinder labeled (2).

c. Click the Hole Type to Taper Tapped labeled (3).

d. Change the Hole Seat to None labeled (4).

e. Change the Thread Type to NPT labeled (5).

f. Change the size to **1/2** labeled (6).

g. Ensure that the Termination is set to Through All labeled (7).

h. Create the holes and keep the Hole Properties panel open by clicking the + icon in the lower-right corner of the Properties panel.

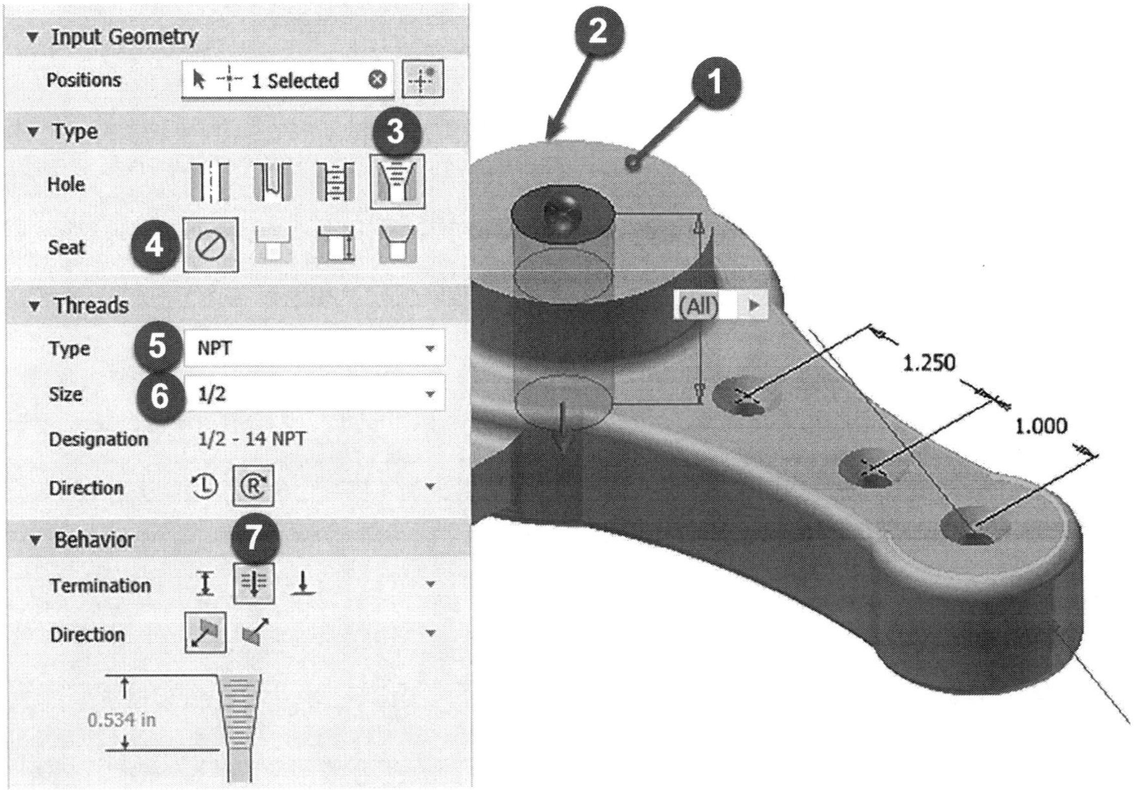

Figure 4.43

12 Lastly, you create a tapped hole through the end of the part at an angle using a work point and work axis. Work points and work axis will be covered later in this chapter.

i. For the Point, select the work point on the outside of the cylinder labeled (1).

j. For the Direction, select the work axis that goes through the cylinder labeled (2).

k. Change the Hole Type to Tapped hole labeled (3).

l. Ensure the Hole Seat is set to None labeled (4).

m. Verify that the Thread Type is set to ANSI Unified Screw Threads labeled (5); if not, make it current.

n. Change the size to **0.25 inches** labeled (6).

o. For the Thread, check the Full Depth option labeled (7).

p. Change the Termination to Distance labeled (8).

q. Change the distance value to **0.75 inches** labeled (9).

r. Zoom in on the start of the hole and notice that there is a small amount of material at the bottom of the hole that remains because the hole is positioned on the circular face and at an angle.

s. Remove this small amount of material by expanding the Advanced Properties option in the dialog box and checking the Extend Start option labeled (10).

t. If desired, you can add this tapped hole to the Last Used section by clicking the New Preset (+) icon near the top of the dialog box.

u. Click OK to create the hole.

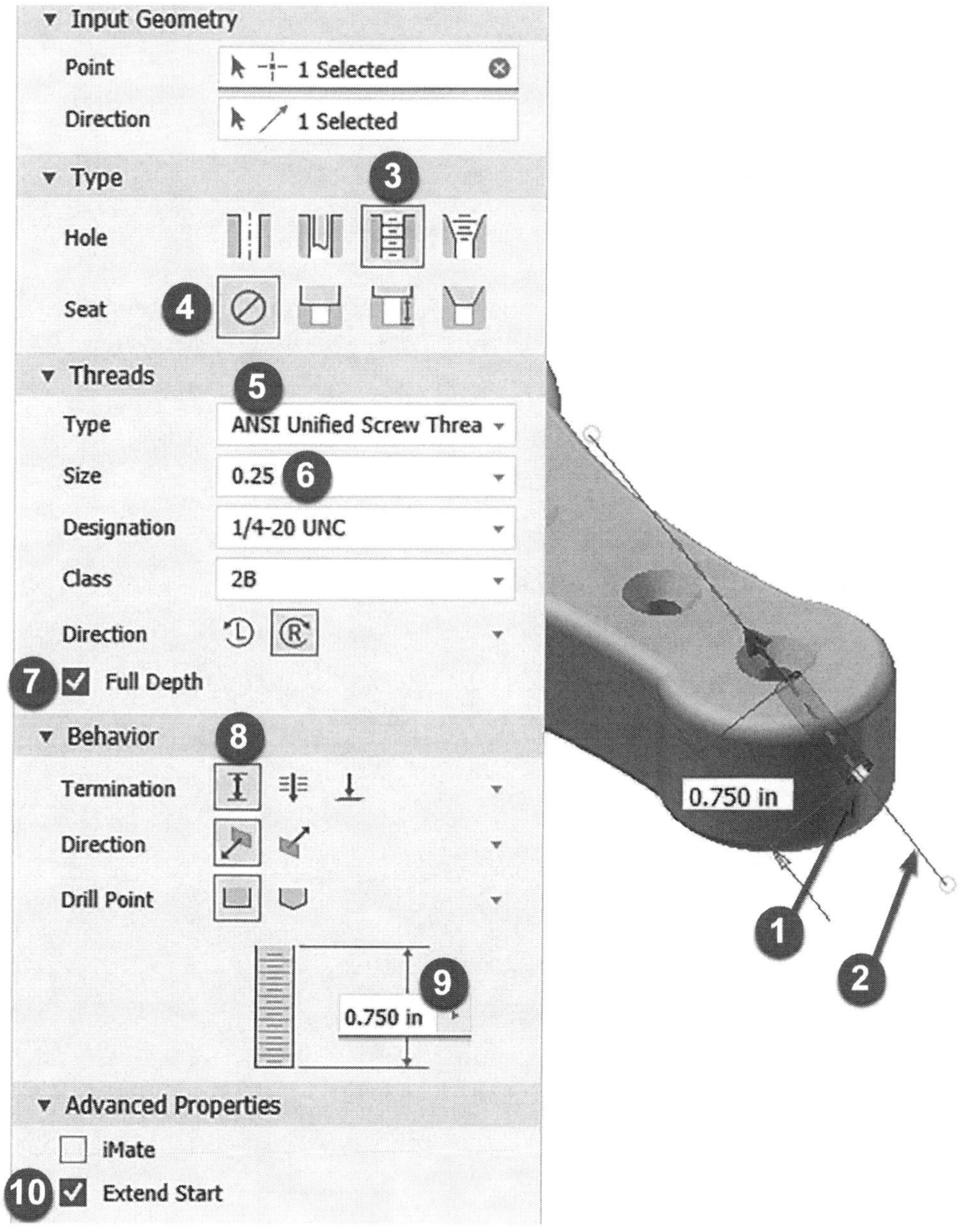

Figure 4.44

13 The completed part is shown in the following image. Rotate the viewpoint and examine the holes.

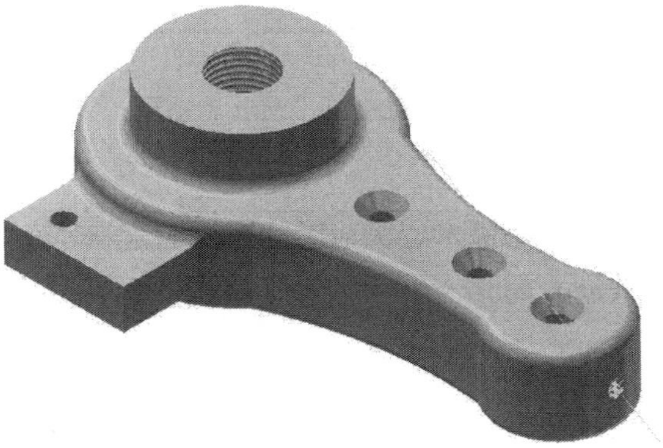

Figure 4.45

14 Practice editing the holes and placing new holes.

15 Close the file. Do not save changes.

End of exercise.

SLOT

Use the Slot command to create elongated holes for adjustment, alignment, or fastening. Slots can be placed directly on planar faces or workplanes, or by using existing sketch center points. Dimensions, orientation, and termination are controlled in the Property panel.

Figure 4.46

Creating a Slot

To create a slot:

1. Create or open a part.
2. Click 3D Model tab > Modify panel > Slot (Figure 4.47 left).
3. Select a planar face, workplane, or sketch point to place the slot center.
4. Define the slot size, orientation, and termination in the Property panel as shown in Figure 4.47 right.
5. Click OK (or press Ctrl+Enter).

When you finish a slot, a preset named Last Used is created automatically.

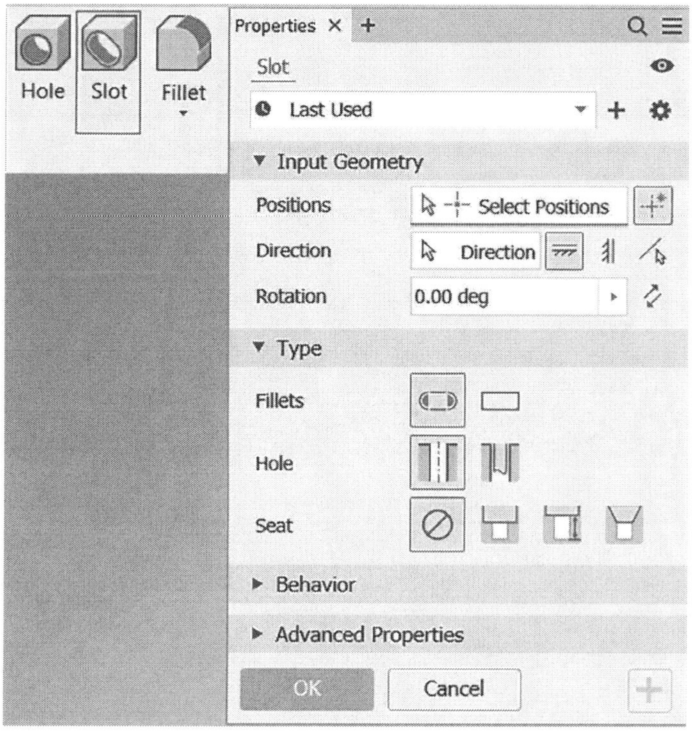

Figure 4.47

Input Geometry

You can place slots using the following options:

a. Positions— Select a planar face or workplane. A sketch is created automatically and a slot center is placed where you click. Use Allow Center Point Creation to control whether new center points can be placed on faces (ON by default). Click reference edges and enter distances to constrain location.

b. Direction — Choose a horizontal, vertical, or custom direction.

c. Rotation—Use Rotation to enter an angle relative to the selected direction. Click Flip Direction to reverse the angle.

Slot Type

Define the end condition and fastening intent.

- Fillets— Choose between rounded ends (radius = half slot width) or sharp corners with square ends.
- Hole Type—Select Simple to manually define the slot width or Clearance for a slot width driven by fastener size (Standard, Type, Size, Fit).
- Seat Options—Define the seat as either None, Counterbore, Spotface, or Countersink. Additional behavior parameters appear for counterbore, spotface, or countersink options.

Behavior

In the Behavior tab, common parameters such as slot width and depth can be set. Values can be entered directly or adjusted using manipulators in the graphics window. Termination options include Distance for specifying depth, Through All for cutting through the entire part, or To, which terminates at a selected planar face.When using Distance or Through All, choose the cut direction (Default or Flipped). When Distance is selected, End Mill provides the option to specify the bottom shape.

Editing Slots

To edit a slot, double-click the feature in the browser, or right-click and choose Edit Feature.

EXERCISE 4-3: CREATING A SLOT

In this exercise, you will recreate the exercise from Exercise 3-3 using the Slot command.

1. Open *ESS_E04_03.ipt* from the Chapter 04 folder.
2. Click the Look At command from the Navigation Bar and select the top angled plane so that you are looking straight at the surface.
3. Select the Point command from the Sketch tab > Create panel.
4. Create a point anywhere on the top plane of the part.
5. Use the Horizontal constraint from the Constrain panel to center the point horizontally by selecting the point and the midpoint of the top horizontal edge as shown in Figure 4.48 left.
6. Center the point vertically by adding a **1.500** vertical dimension with the top edge (Figure 4.48 right).

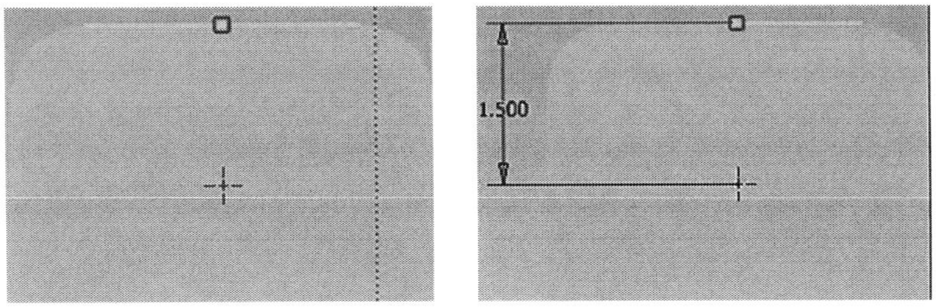

Figure 4.48

7. Click Finish Sketch in the Exit panel.
8. Next, select the Slot command on the 3D Model tab > Modify panel. The slot is placed automatically on the centerpoint that you created.
9. In the Slot Property Panel, change the direction to Vertical. Your screen should resemble the following image.

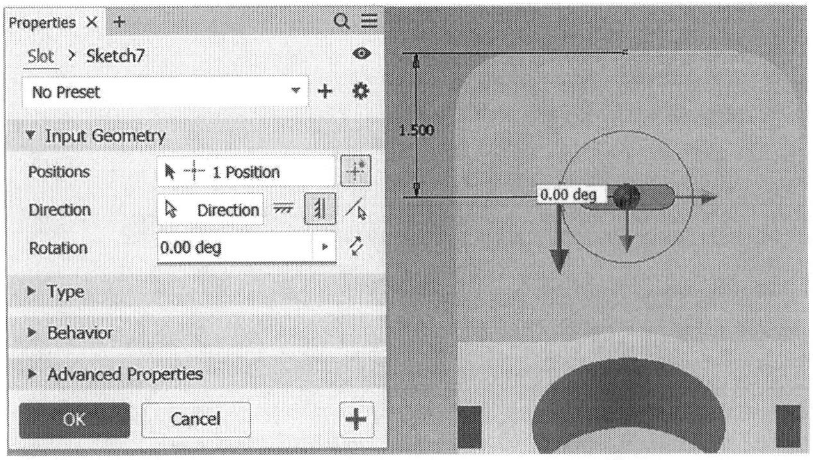

Figure 4.49

10 From the Behavior tab, change the Center Marks to Arc Centers Mark, and make the length of the slot **1.500 in** and the height to **1.000 in**. Check that Termination is set to Through All and that the cut direction is set to Default as shown in Figure 4.50.

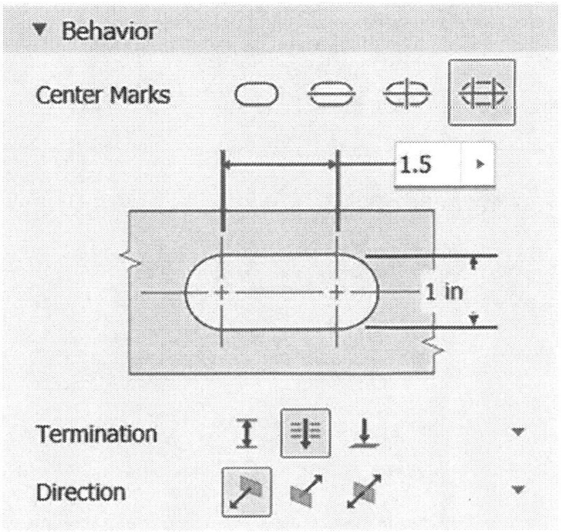

Figure 4.50

11 Click OK to make the slot. The completed slot is shown in Figure 4.51.

Figure 4.51

12 Practice editing the slot by changing the Fillets, Seats, and Termination settings.

13 Close the file and do not save changes.

End of exercise.

SHELLING

As you design parts, you may need to create a model that is made with thin walls but is not a sheet metal part. The easiest way to create a thin-walled part is to create the main shape and then use the Shell command to remove material.

The term shell refers to giving a wall thickness to the outside shape of a part and removing the remaining material. Essentially, you are scooping out the inside of a part and leaving the walls a specified thickness, as shown in the following image. You can offset the wall thickness in, out, or evenly in both directions. If the part you shell contains a void, such as a hole, the feature will have the thickness built around it.

A part may contain more than one shell feature, and individual faces of the part can have different thicknesses. If a wall has a different thickness than the shell thickness, it is referred to as a unique face thickness. If a face that you select for a unique face thickness has faces that are tangent to it, those faces will also have the same thickness. You can remove faces from being shelled, and these faces remain open. If no face is removed, the part is hollow on the inside.

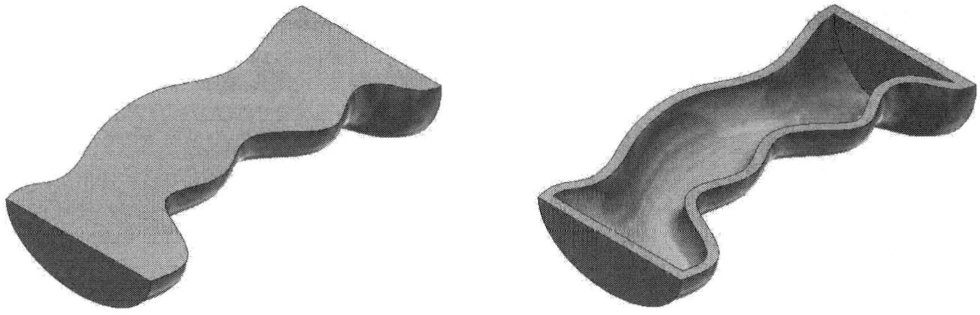

Figure 4.52

To create a shell feature, follow these steps:

1. Create a part that will be shelled.
2. Start the Shell command from the 3D Model tab > Modify panel, as shown in Figure 4.53 left.
3. The Shell dialog box will appear, as shown in Figure 4.53 right.

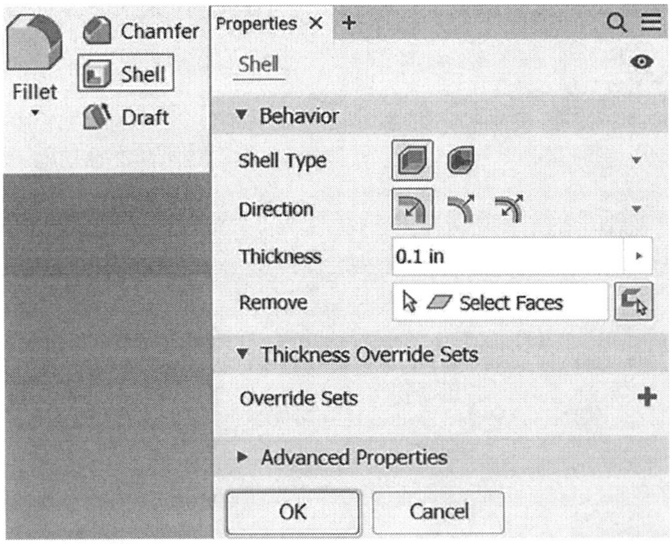

Figure 4.53

4. In the dialog box, select the type of shell, the direction for the shell, the faces that you want to remove, and enter a thickness value and then click OK. The part is now shelled.

To edit a shell feature, right-click the name of the shell feature in the browser, and select Edit Feature from the menu. Alternately, you can double-click the feature's name or icon in the browser.

The following explains the Shell command options.

Shell Type

Choose between a shell with sharp corners or a shell with rounded corners.

Direction

In this area, you set the direction that the shell thickness will go.

Inside

Click this button to offset the wall thickness into the part by the given value.

Outside

Click this button to offset the wall thickness out of the part by the given value.

Both

Click this button to offset the wall thickness evenly into and out of the part by the given value.

Remove Faces

Click the Remove Faces button, and then click the face or faces to be left open. To deselect a face, click the Remove Faces button, and hold down the CTRL key while you click the face.

Automatic Face Chain

When you are removing faces and this option is checked, faces that are tangent to the selected face are automatically selected. Uncheck this option to select only the selected face.

Solids

If multiple solid bodies exist, select the solid body to shell.

Thickness

Enter a value or select a previously used value from the drop-down list to be used for the shell thickness.

Thickness Override Sets

Thickness override is available by clicking the plus icon that is located on the right side of the dialog box, as shown in the following image on the left.

To create a shell set with a specific thickness, click on Select Faces, select the face, and enter a value. A part may contain multiple faces that have different unique thicknesses. The image on the right shows a unique thickness applied to the left face.

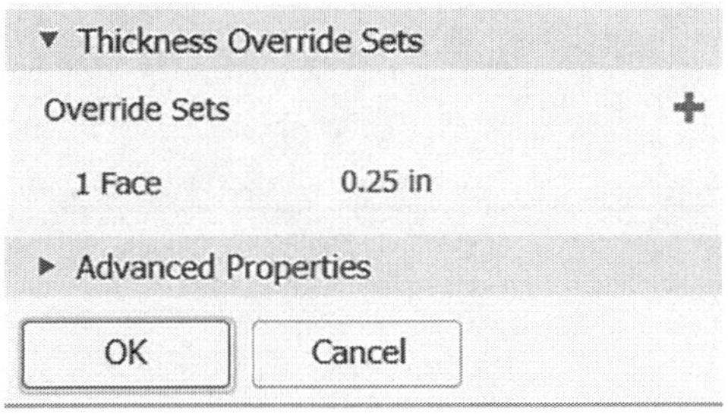

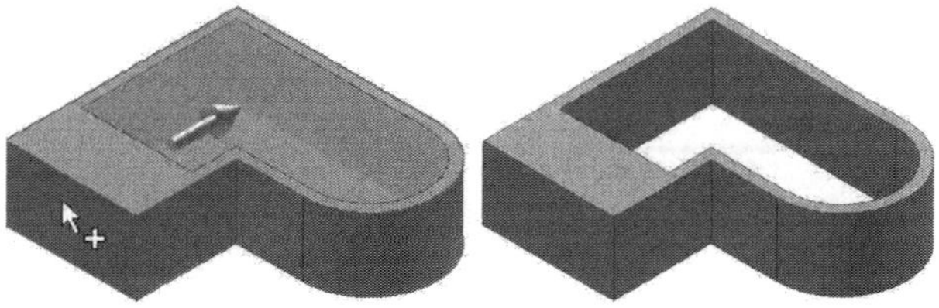

Figure 4.54

Advanced Properties

In the Advanced Properties section, configure approximation behavior by enabling Allow Approximation to permit deviations when a precise offset is impossible, and specify the method of your choosing.

EXERCISE 4-4: SHELLING A PART

In this exercise, you will use the Shell command to create a shell on a part.

1 Open *ESS_E04_04.ipt* in the Chapter 04 folder.

2 Click the Shell command on the 3D Model tab > Modify panel.

3 Remove the top face by selecting the top face of the part as shown in the following image.

4 Type a thickness of **0.0625** inches.

5 Click OK to create the shell feature.

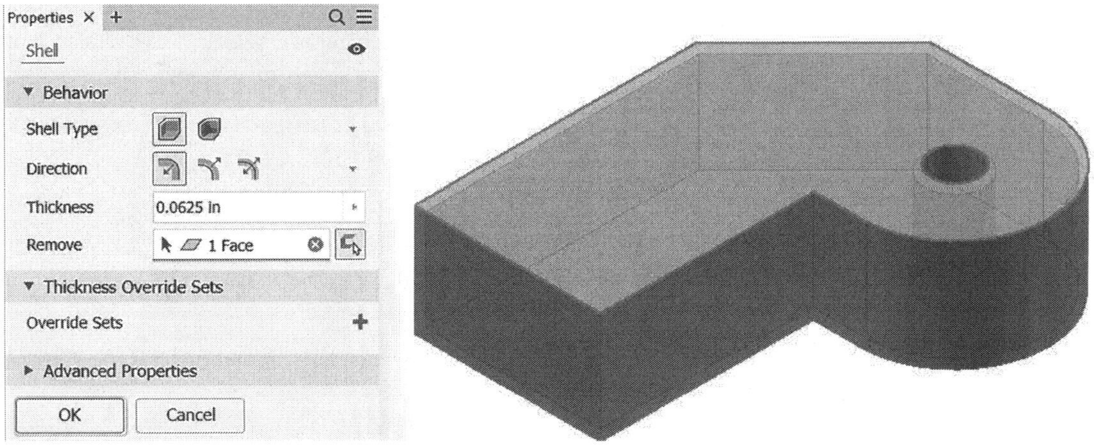

Figure 4.55

6 Rotate the viewpoint and examine the shell. Only the top face should have been removed.

7 Change to the Home View by clicking the Home icon above the ViewCube.

8 Edit the Shell feature that you just created. Right-click on Shell1 from the browser and select click Edit Feature as shown in the following image.

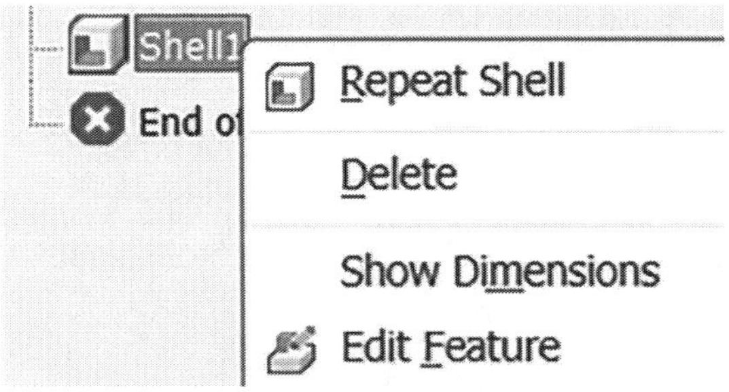

Figure 4.56

9 Click the Thickness Override Sets dropdown arrow and click the plus icon on the right-hand side of the area, and select the left outside-vertical face as shown in the following image.

10 Enter a value of **.5**, as shown in the following image.

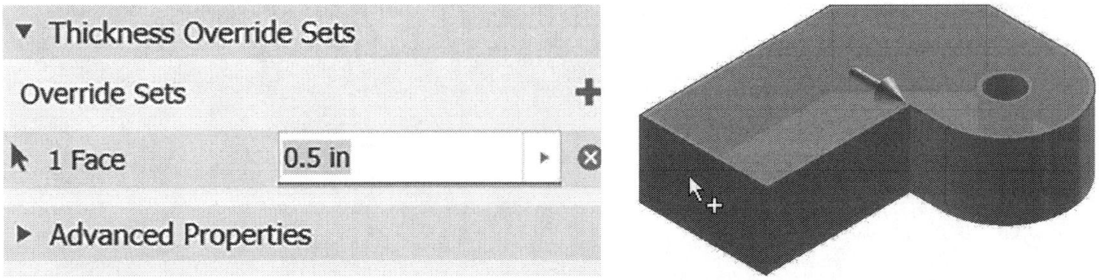

Figure 4.57

11 Click OK to update the shell. Your part should resemble the following image.

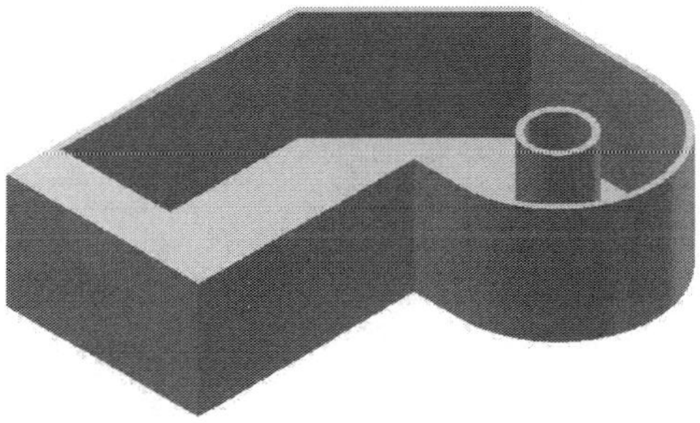

Figure 4.58

12 Practice editing the shell feature. Change the thickness direction, thickness, and delete the unique thickness by selecting the unique thickness entry in the dialog box and then press the Delete key.

13 Close the file without saving changes.

End of exercise.

WORK FEATURES

When you create a parametric part, you define how the features of the part relate to one another. A change in one feature results in appropriate changes in all related features. Work features are special construction features that are attached parametrically to part geometry or other work features. You typically use work features to help position and define new features in your model. There are three types of work features: work planes, work axes, and work points. Use work features in the following situations:

- To position a sketch for new features when a planar part face is not available.
- To establish an intermediate position that is required to define other work features. You can create a work plane at an angle to an existing face, for example, and then create another work plane at an offset value from that plane.
- To establish a plane or edge from which you can place parametric dimensions and constraints.
- To provide an axis or point of rotation for revolved features and patterns.
- To provide an external feature termination plane off the part, such as a beveled extrusion edge, or an internal feature termination plane in cases where there are no existing surfaces.

Creating a Work Axis

A work axis is a feature that acts like a construction line. You can use a work axis to create work planes, work points, and subsequent part features. You can also use work axes as an Axis of rotation for the revolve command, axes of rotation for polar arrays, or to constrain parts in an assembly using assembly constraints. Their length always extends beyond the part—as the part changes size, the work axis also changes size. A work axis is tied parametrically to the part. As changes occur to the part, the work axis will maintain its relationship to the points, edge, or cylindrical face from which you created it. To create a work axis, use the Work Axis command on the 3D Model tab > Work Features panel or press the key / (forward slash). The Axis command will allow you to select all geometry in the graphics window to create an axis. Another option is to click the down arrow in the lower right corner of the Axis button and select an option for creating an axis using only certain types of geometry. These options only allow you to select geometry that is allowed for the active command; other geometry is filtered out and cannot be selected. For example, if the Through Two Points option is selected, only points can be selected in the graphics window.

You can also create a work axis when using the Work Plane or Work Point commands; right-click and select Create Axis when one of these work feature commands is active. Then, after the work axis is created, the command you were running will be active. The created axis will be indented as a child of the work axis or the work plane in the browser.

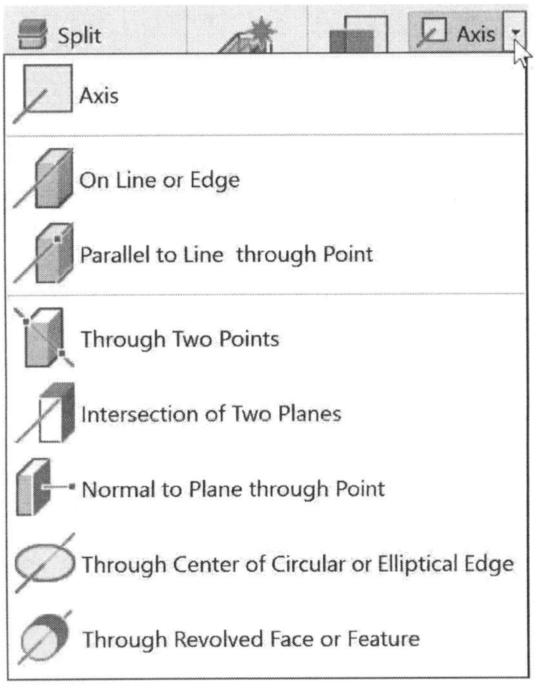

Figure 4.59

Use one of the following methods to create a work axis:

Axis Type	Process
Axis	Select geometry to create a work axis that is dependent upon the selected geometry; no geometry is filtered.
On Line or Edge	Select a line or a linear edge and a work axis is created that is collinear to the selected line or linear edge.
Parallel to Line through Point	Select a linear edge or sketch line and an endpoint, midpoint, sketch point, or work point and a work axis is created that is parallel to the selected line or linear edge and through the selected point.
Through Two Points	Select two endpoints, intersections, midpoints, sketch points, or work points and a work axis is created that goes through the two points.
Intersection of Two Planes	Select two nonparallel work planes or planar faces and the work axis is created coincident at the intersection of two planes.
Normal to Plane through Point	Select a planar face or work plane and an endpoint, midpoint, sketch point, or work point and a work axis is created that is perpendicular to the plane and point.
Through Center of Circular or Elliptical Edge	Select a circular edge, elliptical edge or an edge of a fillet and a work axis is created through the center point and perpendicular to the edge.
Through Revolved Face or Feature	Select a circular face and a work axis is created that goes through the center of the circular face.

EXERCISE 4-5: CREATING WORK AXES

In this exercise, you create a work axis to position a circular pattern. Circular patterns are covered later in this chapter.

1 Open *ESS_E04_05.ipt* in the Chapter 04 folder.

2 Select the inside angled planar face and choose Create Sketch from the mini-toolbar as shown in Figure 4.60 left.

3 If the view did not change and you are looking directly at the face, click the Look At command on the Navigation bar and then in the browser select Sketch5, the sketch that you just created.

4 Click the Point, Center Point command in the Sketch tab > Create panel, and place a point near the middle of the sketch.

5 Add a vertical constraint between the center point you just created and the midpoint on the top line.

6 Apply a horizontal constraint between the point and the midpoint of the edge on the right, as shown in Figure 4.60 right.

Figure 4.60

7 Finish the sketch by right-clicking in the graphics window and click Finish 2D Sketch on the marking-menu.

8 From the 3D Model tab > Work Features panel click the down arrow on the right of the Work Axis command and from the list click Normal to Plane through Point.

9 Select the angled planar face, and then select the center point. The work axis is created through this point and normal to the plane, as shown in Figure 4.61 left.

TIP: One use of a work axis is to use it as an axis of rotation when creating a circular pattern. The following image on the right shows a hole that was patterned around the work axis. Patterns are covered later in this chapter.

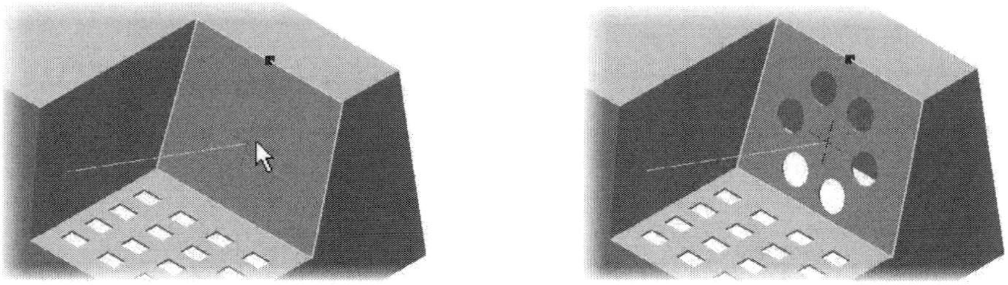

Figure 4.61

10 From the 3D Model tab > Work Features panel click the down arrow on the right of the Work Axis command and from the list click Through Center of Circular or Elliptical Edge.

11 Select the top circular edge on the back of the part as shown in Figure 4.62 left. The axis is created through the center of the circular edge.

12 Undo the last work axis you just created. Click the Undo command from the Quick Access toolbar and the work axis will be removed.

13 From the 3D Model tab > Work Features panel click the down arrow on the right of the Work Axis

command and from the list click Through Revolved Face or Feature.

14 Select the back-circular face on the part as shown in Figure 4.62 right. The axis is created through the center of the circular face.

TIP: The results for the work axis in step 11 and 14 are the same. The Through Center of Circular or Elliptical Edge or the option Through Revolved Face or Feature create an identical work axis since both sets of geometry share the same centers.

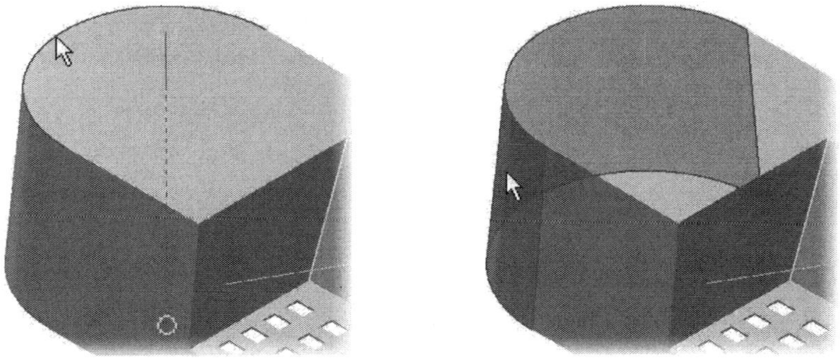

Figure 4.62

15 Practice creating work axes using different options.

16 Close the file. Do not save changes.

End of exercise.

Creating Work Points

A work point is a feature that you can create on the active part or in 3D space. Work points are created relative to selected geometry; if the selected geometry changes location, the work point will move to keep the relationship to the new location of the geometry. Work points can be used to create other work features such as the following:

- work axes
- work planes
- placed holes using the On Point option
- placed 3D lines on the work points.

To create a work point, use the Point command on the 3D Model tab > Work Features panel or press the key (.), a period. The point command will allow you to select all geometry in the graphics window to create a work point. Another option is to click the down arrow in the lower right corner of the Point button and select an option for creating a work point using only filtered geometry type(s) as shown in Figure 4.63. These options allow you to select geometry that is filtered for the active command; other geometry is filtered out and cannot be selected. For example, if the Intersection of Three Planes option is selected, only planes can be selected in the graphics window.

You can also create a work point when using the Work Plane or Work Axis commands; right-click and select Create Point when one of these work feature commands is active. Then, after the work point is created, the command you were running will be active. The created point will be indented as a child of the work axis or the work plane in the browser.

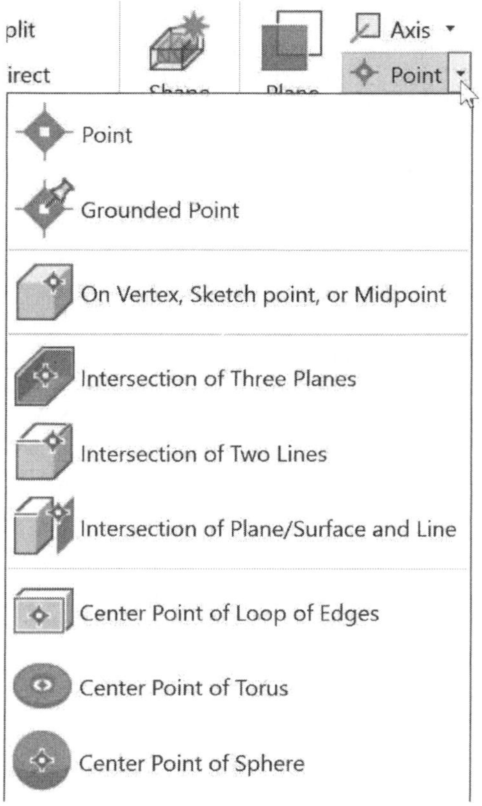

Figure 4.63

Use one of the following methods to create a work point:

Work Point Type	Process
Point	Select the desired geometry. No filters are applied.
Grounded Point	Select a work point, midpoint, or vertex. Grounded work points will be covered in the following section.
On Vertex, Sketch point, or Midpoint	Select a 2D or 3D sketch point, vertex, or the endpoint or midpoint of a line or linear edge.
Intersection of Three Planes	Select three work planes or planar faces or combination of work planes or planar faces.
Intersection of Two Lines	Select any two lines including linear edges, 2D or 3D sketch lines, or work axes.
Intersection of Plane/ Surface and Line	Select a planar face, work plane, or surface and then select a work axis, sketch line, straight edge, or work axis. A work point will be created where the two selections intersect.
Center Point of Loop of Edges	Select an edge of a closed loop of edges and the work point will be created at the center of the loop or edge.
Center Point of Torus	Select a torus and a work point will be created at the center of the torus.
Center Point of Sphere	Select a sphere and a work point will be created at the center of the sphere.

Grounded Work Points

You can create grounded work points that are positioned in 3D space. Grounded work points are not associated with the part or any other work features, including the original locating geometry. When you modify surrounding geometry, the grounded work point remains in the specified location. To create a grounded work point, use the Grounded Work Point command on the 3D Model tab > Work Features panel and click the arrow next to the Point command, as shown in the following image, or press the hot key (;), a semicolon.

After starting the Grounded Point command, select a vertex, midpoint, sketch point, or work point on the model. When you have selected the vertex or point, a triad and a mini-toolbar will appear, as shown in the following image. The initial orientation of the triad matches the principal axes of the part. These colors represent the three axes: red X, green Y, and blue Z.

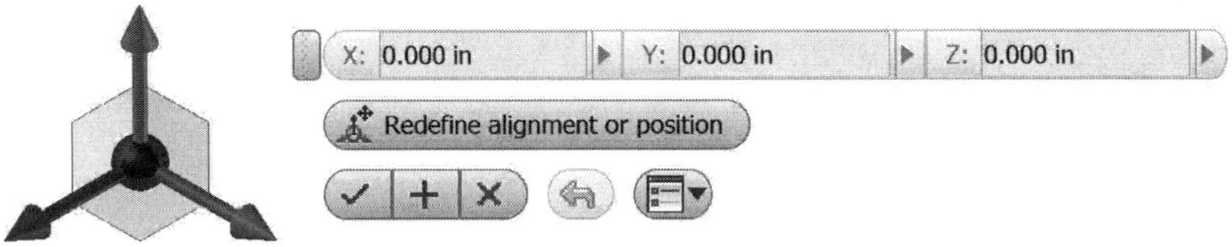

Figure 4.64

To precisely position the grounded work point relative to the selected point, enter values in the mini-toolbar. You can also select areas of the triad to move the triad and locate the grounded work point in the desired direction, as shown in the following image and as described in the following sections.

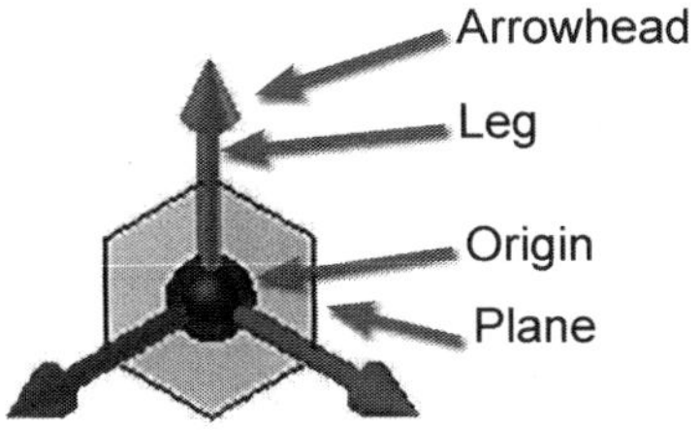

Figure 4.65

Arrowheads
Select an arrowhead to specify a position along an axis, and move the cursor or enter a value.

Legs
Select a leg to rotate about that axis, and move the cursor or enter an angle.

Origin
Click to move the triad freely in 3D space, move it to a selected vertex or point, or to enter X, Y, or Z coordinates.

Planes
Select a plane to restrict movement to the selected plane.

Once you position the triad, click Apply or OK in the mini-toolbar to create the grounded work point. By clicking Apply the mini-toolbar will remain open, allowing you to create another grounded work point. You can identify a grounded work point in the browser by the thumbtack icon that is placed on the work point. The following image shows a regular work point that is placed based on geometry (Work Point1) and a grounded work point based on values you enter (Work Point2) represented in the browser.

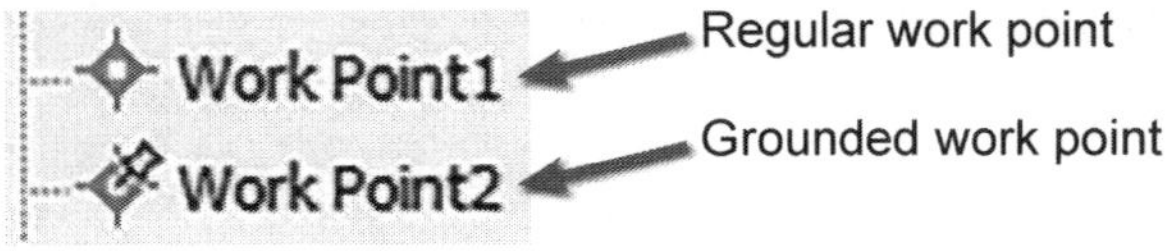

Figure 4.66

Creating Work Planes

Before introducing work planes, it is important that you understand when you need to create a work plane. You can use a work plane when you need to create a sketch and no planar face exists at the desired location, or if you want a feature to terminate at a plane and no face exists to select. If you want to apply an assembly constraint to a plane on a part and no part face exists, you will need to create a work plane. If a face exists in any of these scenarios, you should use it and not create a work plane. A new sketch can be created on a work plane.

A work plane looks like a rectangular plane. It is tied parametrically to the part. Though extents of the plane will appear slightly larger than the part, the plane is in fact infinite. If the part or related feature moves or resizes, the work plane will also move or resize. For example, if a work plane is tangent to the outside face of a 2-inch diameter cylinder and the cylinder diameter changes to 3 inches, the work plane moves with the outside face of the cylinder. You can create as many work planes on a part as needed, and you can use any work plane to create a new sketch. A work plane is a feature and is modified like any other feature.

Before creating a work plane, ask yourself where this work plane needs to exist and what you know about its location. You might want a plane to be tangent to a given face and parallel to another plane, for example, or to go through the center of two arcs. Once you know what you want, select the appropriate options and create a work plane. There are times when you may need to create an intermediate work plane before creating the final work plane. You may need to create a work plane, for example, that is at 30° and tangent to a cylindrical face. You should first create a work plane that is at a 30° angle and located at the center of the cylinder; then create a work plane parallel to the angled work plane that is also tangent to the cylinder.

You can also create a work plane while using the Work axis or Work Point commands; right-click and select Create Plane when one of these work feature commands is active. Then, after the work plane is created, the command you were running will be active. The created plane will be indented as a child of the work axis or the work point in the browser.

Work planes can also be associated to the Origin planes, axis and points that exist in every part. These origin entities initially have their visibility turned off, but you can make them visible by expanding the Origin folder in the browser, right-clicking on a plane or planes, and selecting Visibility from the menu. You can use the origin planes to create a new sketch or to create other work planes.

To create a work plane, click the Work Plane command on the 3D Model tab > Work Features panel, or press the hot key (]), an end bracket. The Work Plane command allows you to select all geometry types in the graphics window to create a work plane. Another option is to click the down arrow in the bottom of the Plane button and select an option for creating a work plane by only allowing you to select filtered geometry types as shown in the following image. These options will select predetermined geometry and will filter out all other types of geometry. For example, if the Three Points option is selected, only points can be selected in the graphics window.

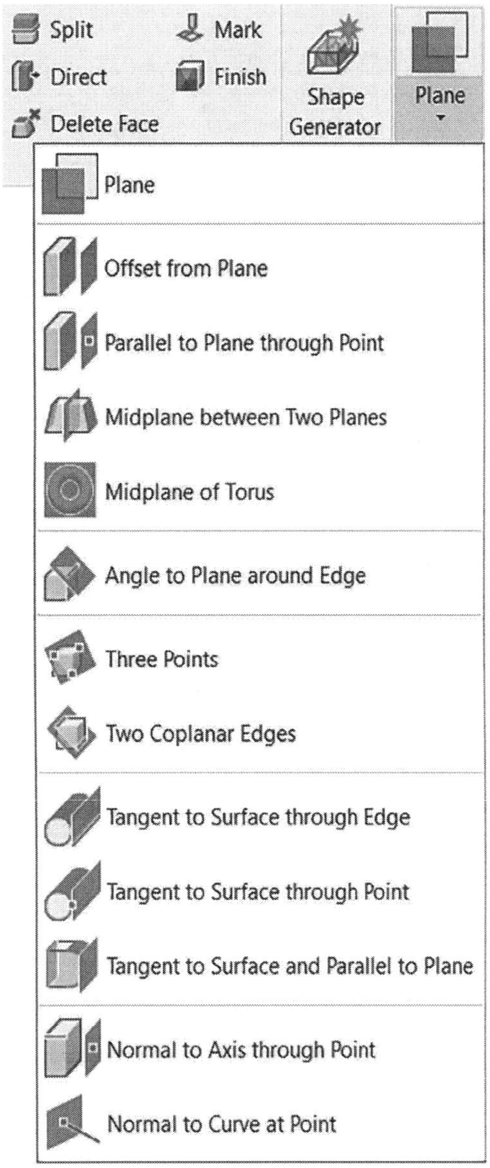

Figure 4.67

Use one of the following methods to create a work plane:

Work Plane Type	Process
Plane	Use this option to create a work plane that is dependent upon the selected geometry; no geometry is filtered.
Offset from Plane	Click a planar face and then drag the new work plane to a selected location. To specify the offset distance, enter a value in the input field.
Parallel to Plane through Point	Select a planar face or work plane and a point; order of selection does not matter. The work plane will be created that is parallel to the selected plane and at the point.
Midplane between Two Planes	Select two planar faces or work planes. A work plane is created at the midplane of the two planes.
Midplane of Torus	Select a torus and a work plane is created that goes through the center / midplane of the torus.

Angle to Plane around Edge	Select a planar face or a work plane and then select an edge or line that is parallel to the plane and enter the desired angle in the input field.
Three Points	Select a combination of three points. The points can be endpoints, intersections, midpoints, or work points. Once selected, a work plane is created that goes through the three points.
Two Coplanar Edges	Select two coplanar work axes, edges, or lines and a work plane is created that goes through the axis, edges or lines.
Tangent to Surface through Edge	Select a circular face and a linear edge and a work plane is created that passes through the linear edge and is tangent to the circular face.
Tangent to Surface through Point	Select a circular face and an endpoint, midpoint, or work point and a work plane is created that passes through a point and is tangent to the circular face.
Tangent to Surface and Parallel to Plane	Select a circular face and a planar face or work plane, and a work plane is created that is parallel to a plane and tangent to the circular face.
Normal to Axis through Point	Select a linear edge or axis and a point, and a work plane is created that is perpendicular to the edge or axis and passes through a point.
Normal to Curve at Point	Select a nonlinear edge or sketch arc, circle, ellipse, or spline and a vertex, edge midpoint, sketch point, or work point on the curve, and a work plane is created that is normal to a circular edge and passes through a point.

When you are creating a work plane, and more than one solution is possible, the Select Other-Face Cycling command appears. Click the forward or reverse arrows from the Select Other command until you see the desired solution displayed. Click the checkmark in the selection box. If you clicked a midpoint on an edge, the resulting work plane links to the midpoint. If the selected edge's length changes, the location of the work plane will adjust to the new midpoint.

TIP: The order in which points or planes are selected is irrelevant.

UCS—User Coordinate System

A User Coordinate System (UCS) is like the data in the origin folder; it contains three work planes, three axes, and a center point. Unlike the data in the origin folder you can create as many UCSs as required and position them as needed. The following list shows the common uses for a UCS.

- Create a UCS where it would be difficult to create a work plane, for example, a compound angle.
- Locate a sketch on a UCS plane.
- Start and terminate features on UCS planes.
- A UCS axis can be used as a rotation axis to pattern features or parts.
- A UCS axis can be used to rotate parts.
- In an assembly, you can constrain UCSs of two parts.
- In a part or an assembly, you can measure to the planes, axis or point in a UCS.
- Measure to the planes, axis or origin.

To start the UCS command, click 3D Model tab > Work Features panel UCS as shown in the following image.

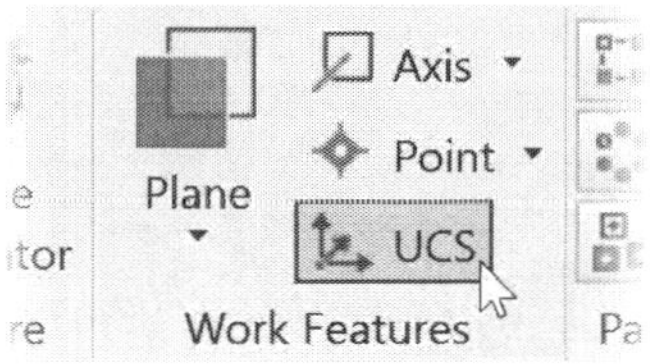

Figure 4.68

There are multiple methods to place a UCS:

- In a part file, a UCS can be placed on existing geometry.
- A UCS can be positioned using absolute coordinates in a part or an assembly file.

The first method to place a UCS is by selecting existing geometry:

1. Start the UCS command, click 3D Model tab > Work Features panel > UCS.
2. Select a point to locate the origin.
3. Select a point to define the direction of the X axis.
4. Select a point to define the direction of the Y axis.

When selecting points to locate or position a UCS, valid inputs are vertex of an edge, midpoint of an edge, sketch, work point origin, solid circular edge, or solid elliptical edge.

The following images show a UCS being placed at the top left vertex in the left image. The X axis is positioned by selecting a vertex as shown in the middle image, and the Y axis is positioned by selecting a vertex as shown in the image on the right. The midpoints of the edges could have also been used to align the UCS.

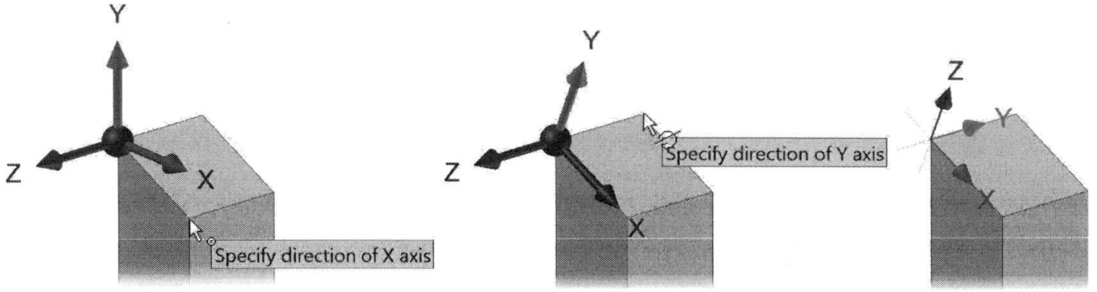

Figure 4.69

The second method to place a UCS is to enter absolute coordinates:

1. Start the UCS command, click 3D Model tab > Work Features panel > UCS.
2. Enter a value for the X, Y, and Z location. Press the Tab key to switch between the cells as shown in the following image.
3. After the three points are defined, press Enter or click in the graphics window.
4. To define the direction of the UCS, click on an arrow and enter a value.
5. To rotate the UCS about an axis on the UCS, click on the shaft of an arrow and enter a value.

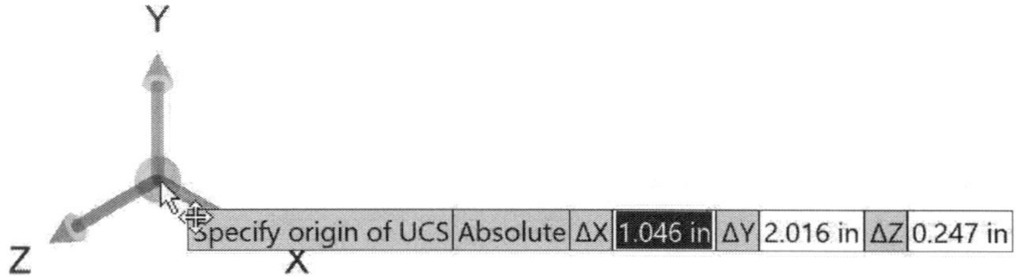

Figure 4.70

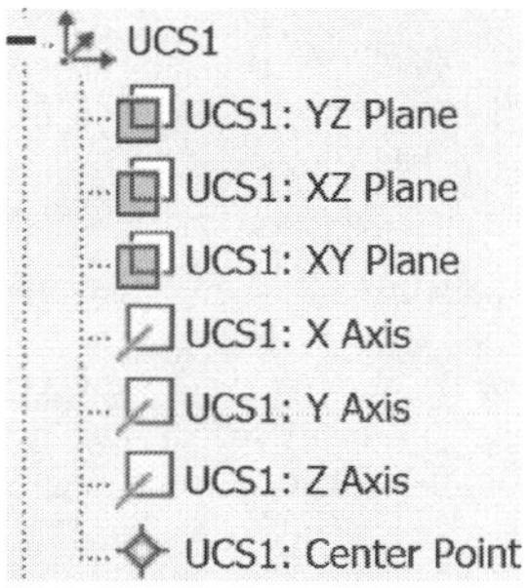

Figure 4.71

After creating a UCS, it will appear in the browser as shown in Figure 4.71. The UCS in the browser can be treated like the origin folder; turn the visibility on and off and measure to the planes, axis and center point.

Figure 4.72

Feature Visibility

You can control the visibility of the origin planes, origin axes, origin point, user work planes, user work axes, user work points, 3D annotations (covered in chapter 5), and sketches by either right-clicking on them in the graphics window or on their name in the browser and selecting Visibility from the menu. You can also control the visibility for all origin planes, origin axes, origin point, user work planes, user work axes, user work points, 3D Annotations, sketches, solids, and UCS triad, planes, axis, and points from the View tab > Visibility panel > Object Visibility. Visibility can be checked to turn visibility on or cleared to turn it off, as shown in Figure 4.72.

Editing the UCS

To edit a UCS, move the cursor over the UCS in the graphics window or in the browser and double-click, or right-click and choose Redefine Feature as shown in Figure 4.73 left. Select on the desired UCS segment: arrow, leg, or origin, and enter a new value or drag to a new location. When done relocating the UCS, right-click and click Finish from the menu. Do NOT click Done [ESC] as this will cancel the operation. Figure 4.73 right shows the prompt to select a segment to edit.

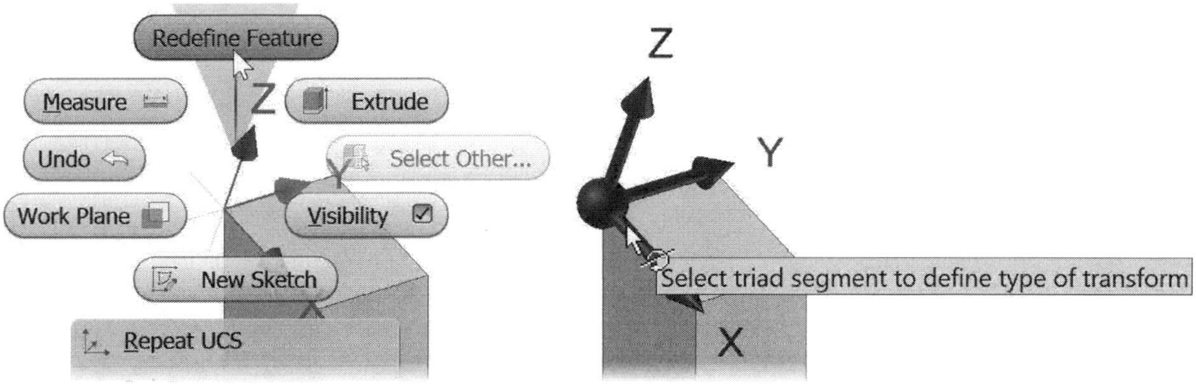

Figure 4.73

EXERCISE 4-6: CREATING WORK PLANES AND A UCS

In this exercise, you create work planes that are used to create features.

1 Open *ESS_E04_06-1.ipt* from the Chapter 04 folder.

2 Use these steps to create an angled work plane.

a. From the 3D Model tab Work > Features panel click the down arrow on the lower-right corner of the Work Plane command and from the list click Angle to Plane around Edge.

b. Click the top face of the rectangle and then select the top-back left edge, and then enter a value of **-30** as shown in the following image on the left. Click the green checkmark in the mini-toolbar to create the work plane.

3 Create a sketch on the top face of the extrusion; do not select the work plane. Click on the top face of Extrusion1 and click the Create Sketch command on the mini-toolbar as shown in Figure 4.74 right.

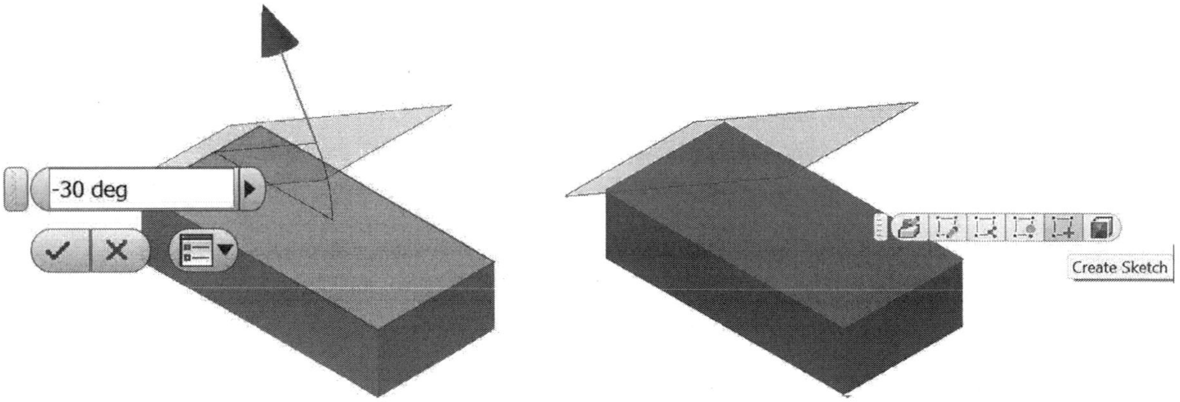

Figure 4.74

4 Create and dimension a **.5 inch** diameter circle and apply a horizontal constraint between the center of the circle and the midpoint of the right vertical edge as shown in Figure 4.75.

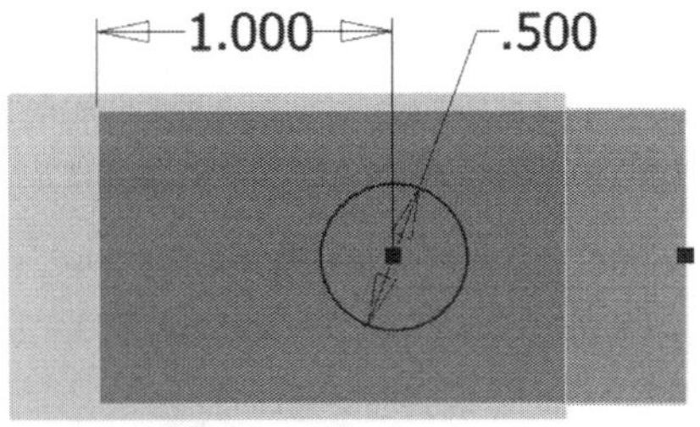

Figure 4.75

5 Finish the sketch.

6 Extrude the circle with the Extents set to To. Select an edge of the work plane as shown in Figure 4.76.

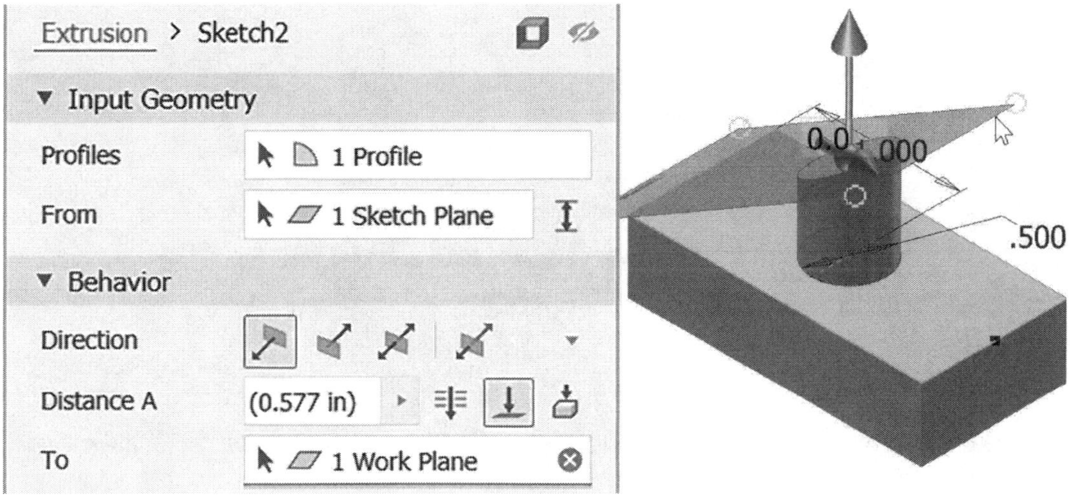

Figure 4.76

7 Edit the angle of the work plane. In the graphics window click on an edge of the work plane you created in step 3, then click Edit Dimension from the mini-toolbar as shown in Figure 4.77, or double-click on the work plane. Enter **-40 degrees** for the angle and then click the green check mark from the mini-toolbar; if the part does not automatically update, click the Update command from the Quick Access toolbar.

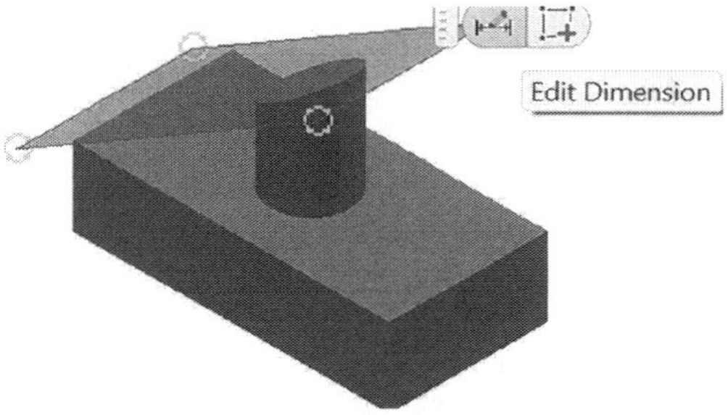

Figure 4.77

8 Edit the angle of the work plane and change the angle back to **-30 degrees**.

9 Turn off the visibility of the work plane by moving the cursor over its edge in the graphics window, right-click and click Visibility from the marking menu.

10 Use the next steps to create a UCS. You will use it later to define a plane.

a. Click the UCS command from the 3D Model tab > Work Features panel.

b. Align the UCS on the top face of Extrusion1 by selecting the three vertices in the order as shown in Figure 4.78 left.

c. Edit the UCS by double-clicking on one of the arrowheads on the UCS in the graphics window.

d. Click the X leg of the UCS and enter a value of **-40** as shown in Figure 4.78 right. Press ENTER on the keyboard to see the change.

e. To complete the edit, right-click and click Finish from the marking menu.

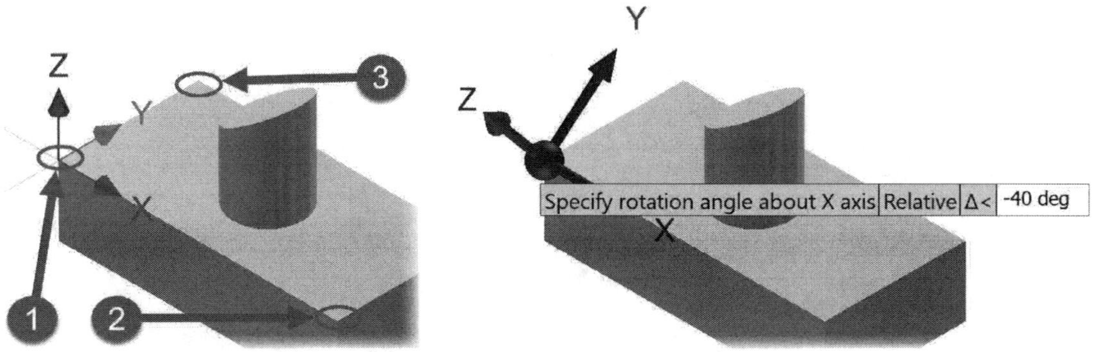

Figure 4.78

11 In the next step you edit Extrusion2 and terminate it on one of the UCS planes, and the UCS must exist in the browser before Extrusion2. Reorder the UCS by dragging the UCS entry in the browser so it is above Extrusion2 as shown in the following image.

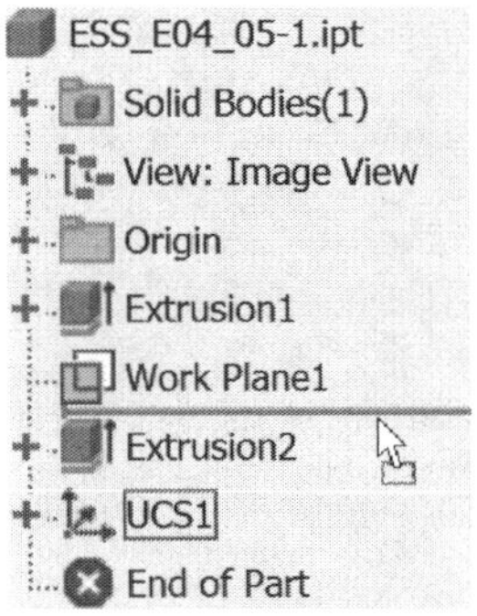

Figure 4.79

12 Edit Extrusion2 by clicking on the circular face of the cylinder and click Edit Extrude from the mini-toolbar.

a. To select a different plane to terminate the extrusion, clear the current selection by clicking the X in the To section in the Extrusion Properties panel 

.

b. In the browser expand the UCS1 entry and then click UCS1: XY Plane as shown in the following image on the left.

c. To complete the edit, click OK in the Extrusion Properties panel.

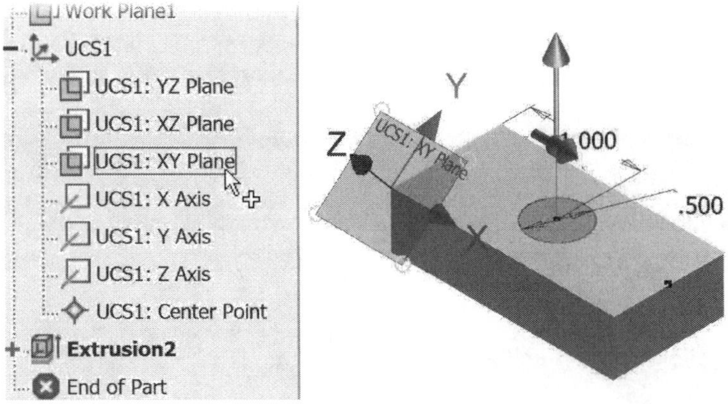

Figure 4.80

13 Right-click on the UCS in the graphics window. Click Visibility from the marking menu to turn it off.

14 Next create a work plane centered between two parallel planes.

a. From the 3D Model tab > Work Features panel click the down arrow below the Plane command, and from the list click Midplane between Two Planes.

b. Use the Select Other operation to select the back-right planar face and then select the front-left face as shown in the following image on the left.

Note: If two adjacent planes are selected, an angular work plane will be created that bisects the two planes.

15 Click one of the edges of the work plane you just created. Choose Create Sketch from the mini-toolbar.

16 Slice the graphics by pressing the F7 key or click Slice Graphics from the Status Bar.

17 To project the edges of the part, click the Project Cut Edges command from the Sketch tab > Create panel (the Project Cut Edges command may be under the Project Geometry command).

18 Create and dimension a **.375 diameter** circle on the horizontal edge of the projected geometry as shown in Figure 4.81 right.

TIP: If the circle is placed at the midpoint of the projected edge, the constraint must be deleted before placing the horizontal dimension.

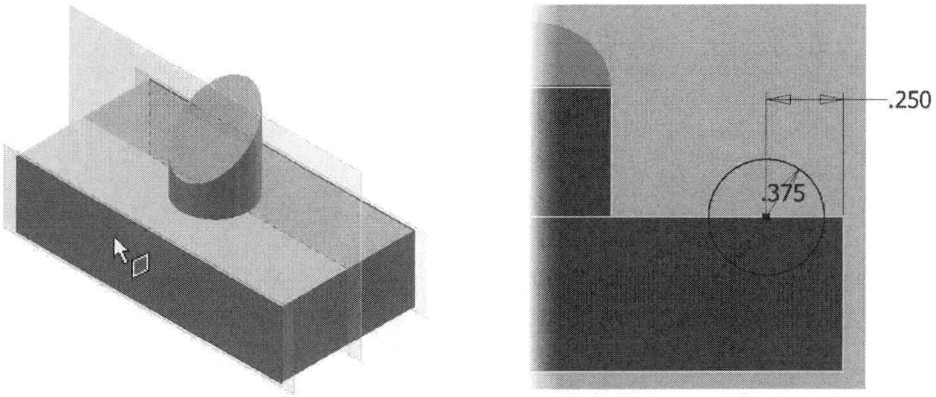

Figure 4.81

19 Finish the sketch by clicking Finish Sketch from the Sketch tab > Exit panel.

20 Extrude the circle **.5 inches** with the Symmetric option, as shown in Figure 4.82.

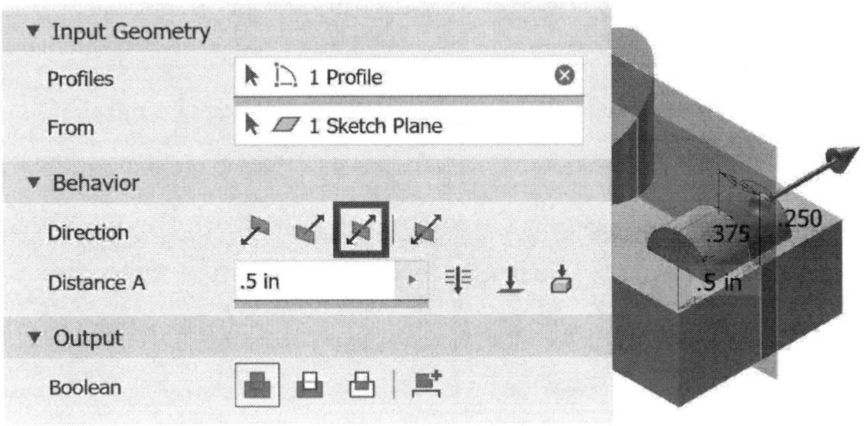

Figure 4.82

21 Turn off the visibility of the work plane by moving the cursor over the work plane in the graphics window or in the browser and right-click and click Visibility from the marking menu. Your screen should resemble the following image on the left.

22 To verify that the symmetric extrusion will be maintained in the center of the part if the width changes, edit Sketch1 of Extrusion1, and change the 1 inch vertical dimension to **1.5 inches.**

23 If needed, click the Local Update command on the Quick Access toolbar. Your screen should resemble Figure 4.83 right.

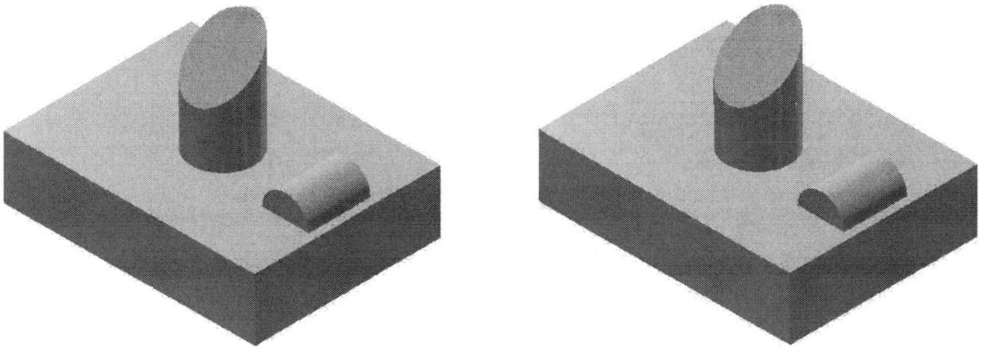

Figure 4.83

24 Use the Free Orbit command to view the part from different perspectives. Notice how Extrusion2 still terminates at the UCS XY plane and Extrusion3 is still in the middle of the part.

25 Close the file. Do not save changes.

End of exercise.

EXERCISE 4-7: CREATING HOLES THROUGH CYLINDERS

In this exercise, you use work planes to place two holes on a cylinder.

1 Open *ESS_E04_06-2.ipt* from the Chapter 04 folder.

2 Create a work plane that is parallel to an origin plane and tangent to the cylinder.

a. In the browser, expand the Origin folder.

b. Create a work plane on the outside of the cylinder. From the 3D Model tab > Work Features panel click the down arrow below the Work Plane command and from the list click Tangent to Surface and Parallel to Plane.

c. In the browser click the YZ plane under the Origin folder.

d. Click near the front-right of the cylinder, as shown in Figure 4.84 left.

3 Create a sketch on the work plane. In the graphics window select an edge of the work plane, and click Create Sketch on the mini-toolbar.

4 Use the Project Geometry command from the Create panel and project the Z axis of the Origin folder onto the sketch.

5 Next you place a Point, Center Point that will be used to locate a hole.

a. Place a Point, Center Point on the projected axis—this will center the point in the center of the cylinder horizontally as shown in the middle image.

b. Add a **1.000 inch** vertical dimension, as shown in the middle image.

6 Finish the sketch.

7 Press H on the keyboard to start the Hole command and place a **.5 inch** diameter Through All simple hole located at the center point.

8 Turn off the visibility of the work plane and your screen should resemble Figure 4.84 right.

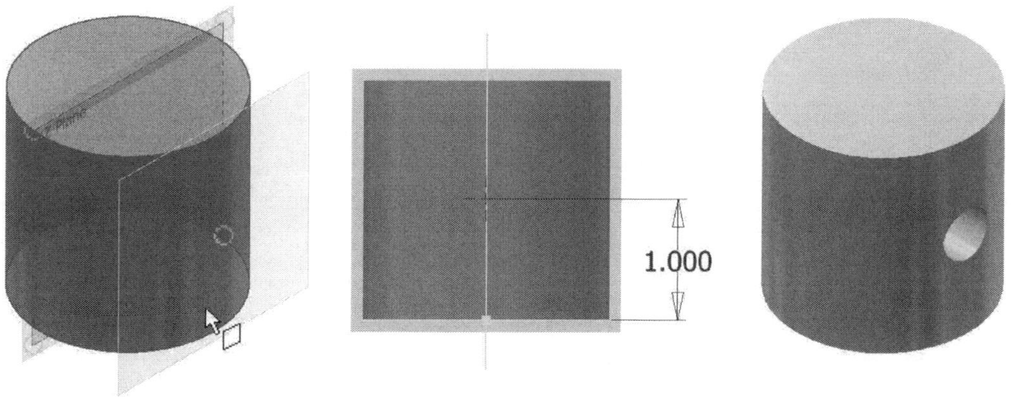

Figure 4.84

9 Next, you place a hole on the cylinder at an angle. First you create an angled work plane in the middle of the cylinder.

a. From the 3D Model tab > Work Features panel click the down arrow below the Work Plane command and from the list click Angle to Plane around Edge.

b. Select the plane to base the angle on; in the browser click the YZ plane under the Origin folder.

c. Select the axis to rotate the plane about; in the browser click the Z Axis under the Origin folder.

d. In the Angle dialog box, enter **315 degrees** as shown in the following image on the left.

e. Create the work plane by clicking the green check mark in the mini-toolbar.

10 Next, create a work plane that is parallel to an angled plane and tangent to the cylinder.

a. From the 3D Model tab > Work Features panel click the arrow below the Plane command and from the list click Tangent to Surface and Parallel to Plane.

b. In the graphics window click the angle work plane that you just created.

c. Click a point near the front face of the cylinder, as shown in Figure 4.79 right.

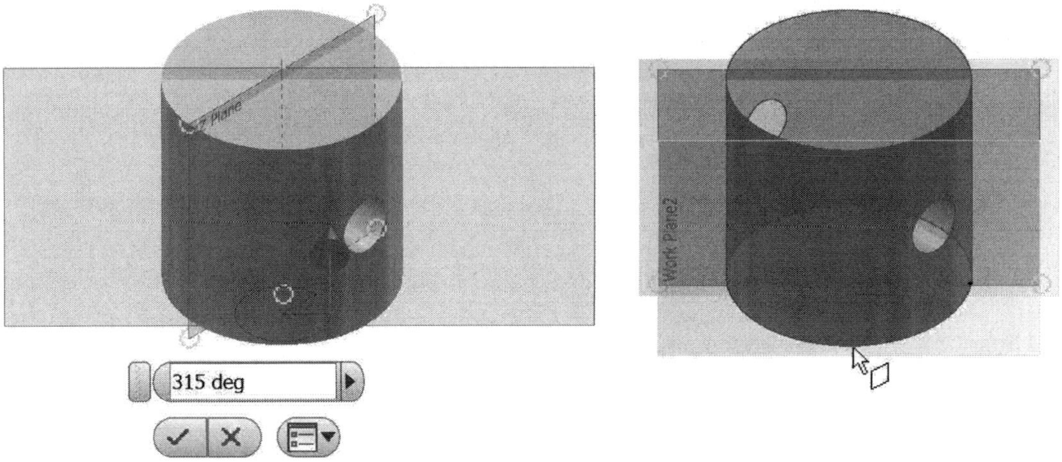

Figure 4.85

11 Turn off the visibility of the inside angled work planes. Move the cursor over the work plane. Right-click, and click Visibility from the marking menu.

12 Create a new sketch on the last work plane you created.

13 Next you place another Point, Center Point that will be used to locate a hole.

a. Place a Point, Center Point in the middle of the sketch.

b. Apply a vertical constraint between the origin point that was automatically projected onto the sketch and the Point, Center Point labeled (1) as shown in Figure 4.80 left.

c. Apply a **1.5 inch** vertical dimension as shown in the image on the left.

14 Finish the sketch.

15 Start the Hole command, and place a counterbore hole with the size of your choice, and locate it on the center point you just created.

16 Turn off the visibility of the work plane and your screen should resemble the image on the right.

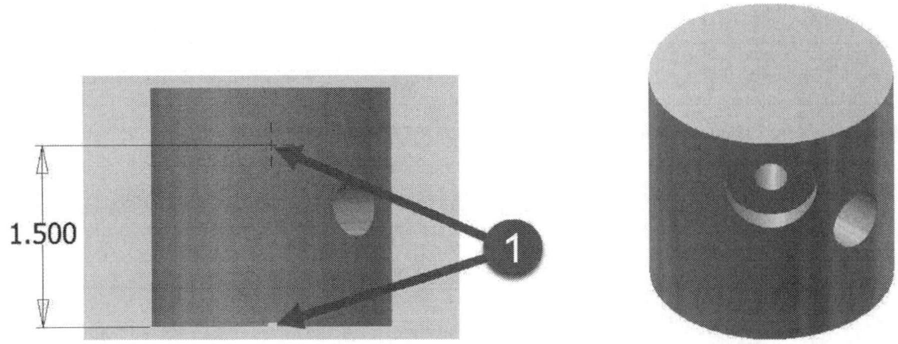

Figure 4.86

17 Close the file. Do not save changes.

End of exercise.

PATTERNS

There are two types of feature patterns that you can create: rectangular and circular. The pattern is represented as a single feature in the browser, and the original feature and individual feature occurrences are listed under the pattern feature. You can suppress the entire pattern or individual occurrences except for the first occurrence.

Both rectangular and circular patterns have a child relationship to the parent feature(s) that you patterned. If the size of the parent feature changes, all the child features will also change. If you patterned a hole, and the parent hole type changes, the child holes also change.

Because a pattern is a feature, you can edit it like any other feature. You can pattern the base part or feature, as well as patterns. A rectangular pattern repeats the selected feature(s) along the direction set by one or two edges on the part or lines that reside in a sketch. These edges do not need to be horizontal or vertical, as shown in the following image. A circular pattern repeats the feature(s) around an axis, a cylindrical or conical face, or an edge.

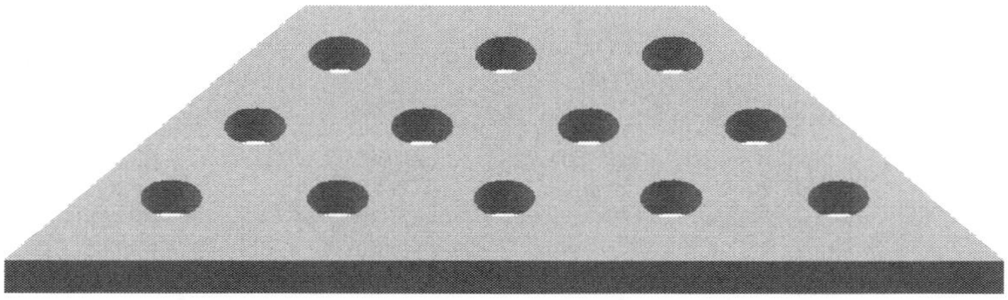

Figure 4.87

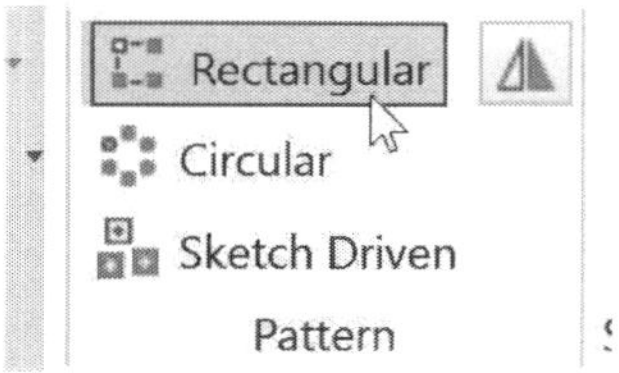

Figure 4.88

Rectangular Patterns

To create a rectangular pattern, follow these steps.

1. Click Rectangular Pattern from the 3D Model tab > Pattern panel, as shown in Figure 4.82.
2. The Rectangular Pattern dialog box appears, as shown in Figure 4.83.
3. Click one or more features to pattern.
4. Click the Select Direction button in the Direction A area to define the direction that the feature will travel.
5. Click an edge or line on a sketch to define the direction.
6. Define the count and the spacing or distance by entering values in the Number and Distance input fields in the dialog box.
7. Choose how spacing is measured with Distribution.

8. If needed, expand the Direction B area in the dialog box to define the direction that the feature will travel that is not parallel to Direction A and enter values for the count and spacing.
9. As you enter values the pattern is previewed in the graphics window.
10. Click OK to create the pattern.

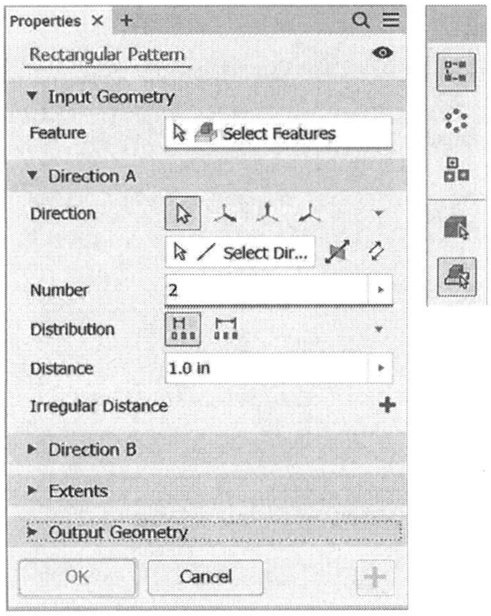

Figure 4.89

Rectangular Pattern Dialog Box

The following options are available in the Rectangular Pattern dialog box.

Pattern Individual Features

Click this button in the upper right tool palette to pattern a feature or features. When you select this option, you activate a features button, as described below.

Pattern the Entire Solid

Click this button in the upper right tool palette to pattern a solid body. When you select this option, you select the entire part as the item to pattern. You also have the Include Work Features option when patterning an entire solid.

Features

Select and then click the feature(s) to be patterned from either the graphics window or the browser. You can add or remove features to or from the selection set by holding down the CTRL key and clicking them.

Solid

If multiple solid bodies exist, select the solid body that you want the feature(s) patterned on.

Direction A

In Direction A, you define the first direction for the alignment of the pattern. It can be an edge, an axis, or a path.

Select Direction. Click this arrow button, and then click an edge or sketch that defines the alignment along which you will pattern the feature.

Flip. If the pattern in the preview image shows the wrong direction, click this button to reverse it.

Midplane. Check this option to have the occurrences patterned on both sides of the selected feature. The midplane option is independent for both Direction 1 and Direction 2.

Number. Enter a value or click the arrow to choose a previously used value that represents the number of feature(s) you will include in the pattern along the selected direction or path.

Distribution. Click the Incremental icon for specified spacing between features, or the Fitted icon to evenly divide a set length.

Distance. Enter a value or click the arrow to choose a previously used value that represents the distance between the patterned features.

Irregular Distance. Click the plus icon to add a pattern member with a non-uniform spacing.

Direction B

In Direction B, you can define a second direction for the alignment of the pattern. It can be an edge, an axis, or a path, but it cannot be parallel to Direction A. The same options are available for Direction B that are available for Direction A.

Boundary

Available in part modeling only, select a sketch, face, or multiple faces to pattern features within a boundary. You may offest a boundary using the Offset button. There are 3 patterning options to choose from: Include Geometry, Include Centroids, and Base Points.

Creation Method

Optimized. Click this option to pattern the feature's faces instead of the feature(s) to calculate all the occurrences in the pattern. This option is ideal when the occurrences you are creating do not intersect and are identical. It can improve the performance of pattern creation.

Identical. Click this option to use the same termination as that of the parent feature(s) for all the occurrences in the pattern. This is the default option.

Adjust. Click this option to calculate the termination of each occurrence individually. Since each occurrence is calculated separately, the processing time can increase. You must use this option if a parent feature terminates to a face or plane.

EXERCISE 4-8: CREATING A RECTANGULAR PATTERN

In this exercise, you create a rectangular pattern of a hole in a cover plate.

1 Open *ESS_E04_07.ipt* in the Chapter 04 folder.

2 Click the Rectangular Pattern command in the 3D Model tab > Pattern panel.

a. Click the small hole feature on the lower-left corner of the part as the feature to be patterned; it is labeled (1) in the following image.

b. In the Direction A area of the dialog box, click the Direction button labeled (2), and then select the bottom horizontal edge of the part labeled (3).

c. If the preview of the pattern shows the pattern is going in the negative direction, change the pattern's direction by clicking the Direction A Flip button labeled (4).

d. Enter **5** for the Number labeled (5) and **.625 inches** for the Distance labeled (6).

e. Click the Direction B Path button labeled (7), and select the vertical edge on the left side of the part labeled (8).

f. Enter **4** for the Number labeled (9) and **.625 inches** for the Distance labeled (10).

g. Click OK to create the pattern.

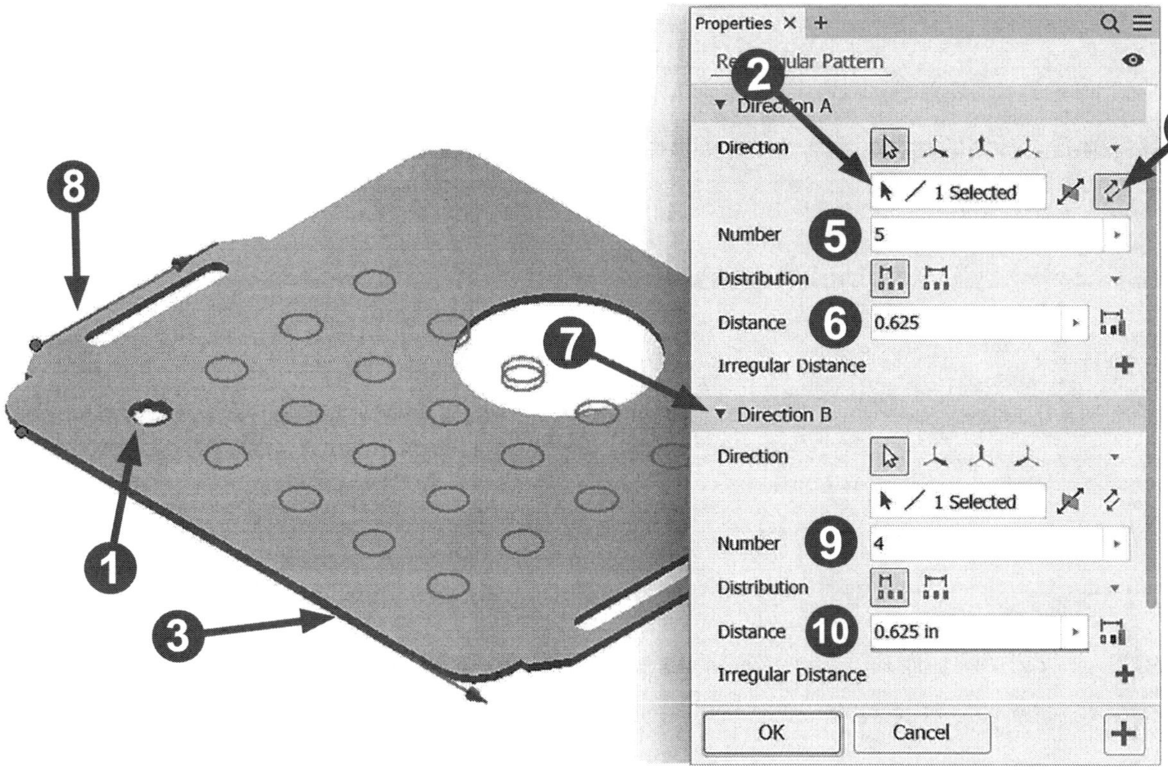

Figure 4.90

3 Next you suppress three of the holes that are not required in the design. Expand the Rectangular Pattern feature in the browser to display the occurrences.

4 In the browser, move the cursor over an occurrence and it highlights in the graphics window. Hold down the CTRL key and click to select the three occurrences of the holes as shown in the following image.

5 Right-click on any one of the highlighted occurrences in the browser, and click Suppress, as shown in the following image.

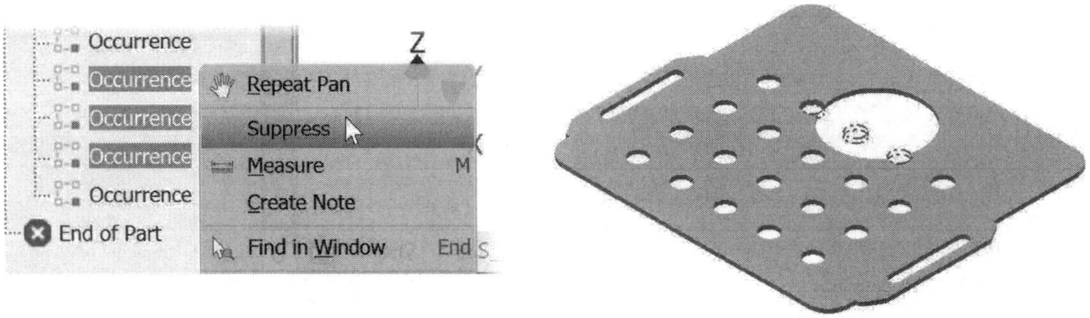

Figure 4.91

6 When done, your model should resemble the following image.

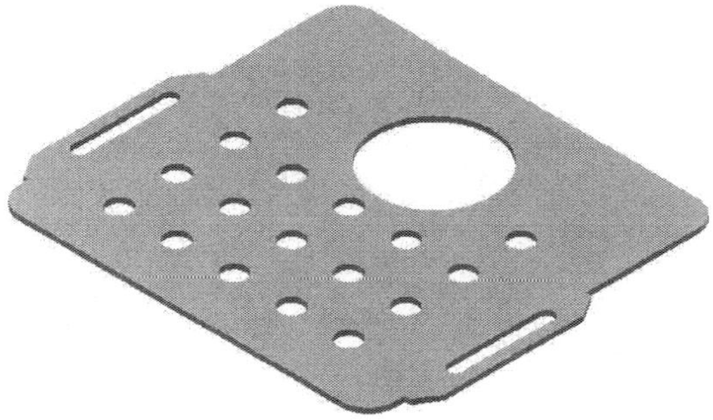

Figure 4.92

7 Practice editing the feature pattern, and change the count and spacing for each direction. The suppressed occurrences remain suppressed.

8 Close the file. Do not save changes.

End of exercise.

Circular Patterns

When creating a circular pattern, you must have a work axis, a part edge, or a cylindrical face around which the features will be patterned. To create a circular pattern, follow these steps:

1. Click Circular Pattern from the 3D Model tab > Pattern panel, as shown in Figure 4.93 left.
2. The Circular Pattern dialog box appears, as shown in Figure 4.93 right.
3. Click one or more features to pattern.
4. Click the Select Direction button to define the axis of rotation the feature will be patterned around.
5. Click a work axis, edge, or cylindrical face (the centerline of the cylinder will be used) to define the axis.
6. Define the Occurrence Count and Occurrence Angle by entering values in the Occurrence Count and Occurrence Angle cells in the dialog box.
7. As you enter values the pattern preview shows in the graphics window.
8. Click OK to create the pattern.

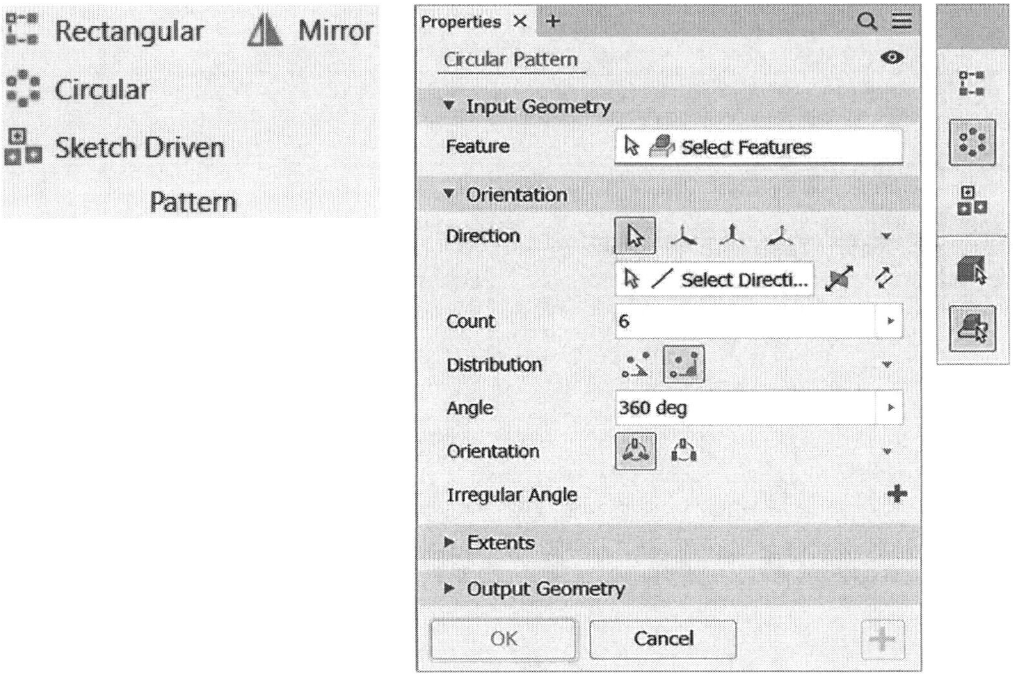

Figure 4.93

Circular Pattern Dialog Box

The following options are available in the Circular Pattern dialog box.

Pattern Individual Features

Click this button in the upper right tool palette to pattern a feature or features. When you select this option, the features button is available and operates as described below.

Pattern the Entire Solid

Click this button in the upper right tool palette to pattern a solid body. When you select this option, you select the entire part as the item to pattern. You also have the Include Work Features option when patterning an entire solid.

Features

Click this button, and then click a feature or features to be patterned. You can add or remove features to or from the selection set by holding down the CTRL key and clicking them.

Select Direction

Click the button and select an edge, axis, or cylindrical face (center) to define the rotation axis for the feature(s).

Midplane

Check this option to have the occurrences patterned evenly on both sides of the selected feature.

Flip

If the preview image shows the pattern going in the wrong direction, click this button to reverse its direction.

Occurrence Count

Enter a value or click the arrow to choose a previously used value that represents the number of feature(s) that you will include in the pattern. A positive number will pattern the feature(s) in the clockwise direction; a negative number will pattern the feature in the counterclockwise direction.

Distribution

Click the Incremental icon for specified spacing between features, or the Fitted icon to evenly divide a set length.

Occurrence Angle

Enter a value or click the arrow to choose a previously used value that represents the angle that you will use to calculate the spacing of the patterned features.

Orientation

Rotational. The default option, and rotates the geometry around the axis of rotation.

Fixed. Click the button to maintain the orientation of the selected geometry. The default uses the center of the profile for the base point.

Base Point. Click this option and select a vertex or point to redefine the pattern's base point.

Irregular Angle

Click the plus icon to add a pattern member with a non-uniform angle.

Boundary

Available in part modeling only, select a sketch, face, or multiple faces to pattern features within a boundary. You may offest a boundary using the Offset button. There are 3 patterning options to choose from: Include Geometry, Include Centroids, and Base Points.

Creation Method

Optimized. Click this option to pattern the feature's faces instead of the feature(s) to calculate all the occurrences in the pattern. This option is ideal when the occurrences you are creating do not intersect and are identical. It can improve the performance of pattern creation. This method is recommended when there are 50 or more occurrences.

Identical. Click this option to use the same termination as that of the parent feature(s) for all the occurrences in the pattern. This is the default option.

Adjust. Click this option to calculate each occurrence termination individually. Because each occurrence is calculated separately, the processing time can increase. You must use this option if a parent feature terminates to a face or plane.

TIP: For the axis of rotation you can use a work axis, an edge, or a cylindrical face (the centerline of the circular face will be used).

EXERCISE 4-9: CREATING CIRCULAR PATTERN

In this exercise, you create a circular pattern of a rectangular cut.

1. Open *ESS_E04_08.ipt* in the Chapter 04 folder.
2. Edit the Extrusion2 feature to verify that the extents for the extrusion is Through All.
3. Click the Cancel button to close the Extrude dialog box.
4. Click the Circular Pattern command in the 3D Model tab > Pattern panel.
    a. Click the extrude cut feature as the feature to pattern.
    b. In the Circular Pattern dialog box click the Select Direction button.

c. Specify the rotation axis, and in the graphics window, click the work axis. A preview of the pattern is displayed in the graphics window.

d. Type **8** in the Occurrence Count field as shown in the following image.

e. Click the OK button to create the pattern.

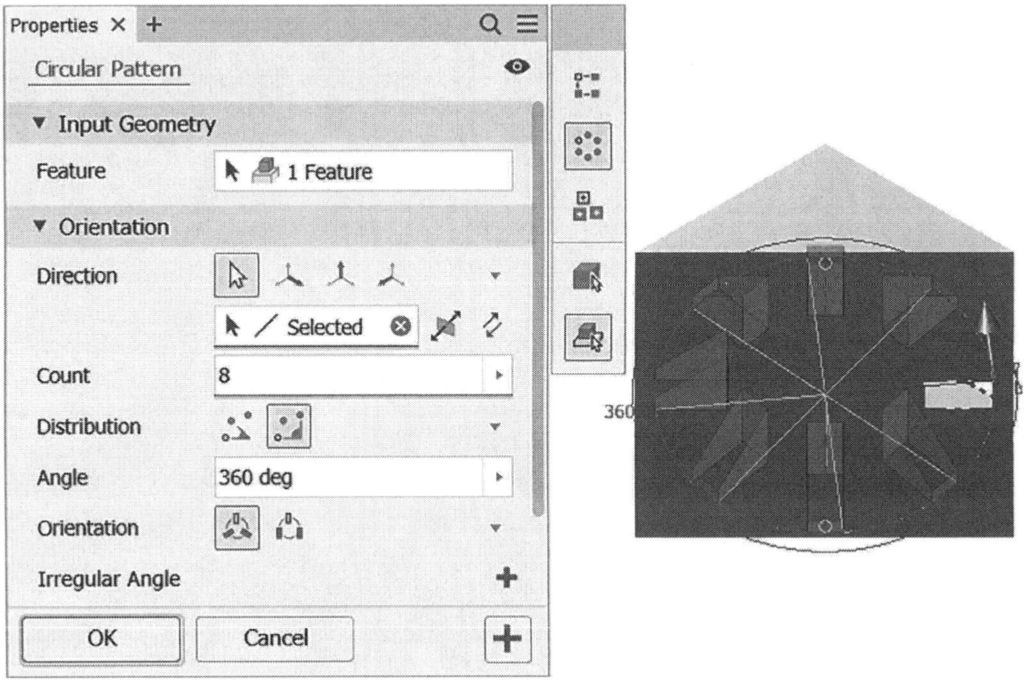

Figure 4.94

5 Rotate your viewpoint to show the back of the part, as shown in Figure 4.95. Notice that six holes do not go through because the Identical Creation Method was used to create the pattern. The patterned holes are identical to the original hole. If desired, change the visual style to Wireframe to verify that all the holes are identical.

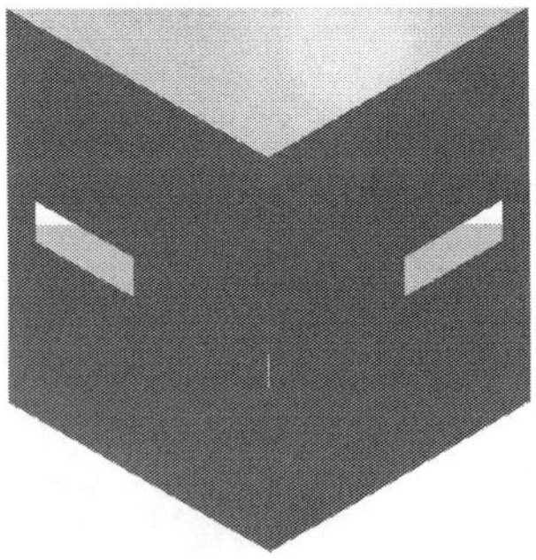

Figure 4.95

6 Next edit the pattern so all the extrusions go through the model. In the browser, right-click on Circular Pattern1 and click Edit Feature from the menu or click on one of the faces of a patterned extrusion (not the original extrusion) and click Edit Circular Pattern from the mini-toolbar.

7 In the Circular Pattern dialog box, click the Output Geometry arrow.

a. Under Creation Method, select the Adjust option, as shown in the following image on the left.

b. Click OK to create the circular pattern. When done, your model should resemble the Figure 4.96 right.

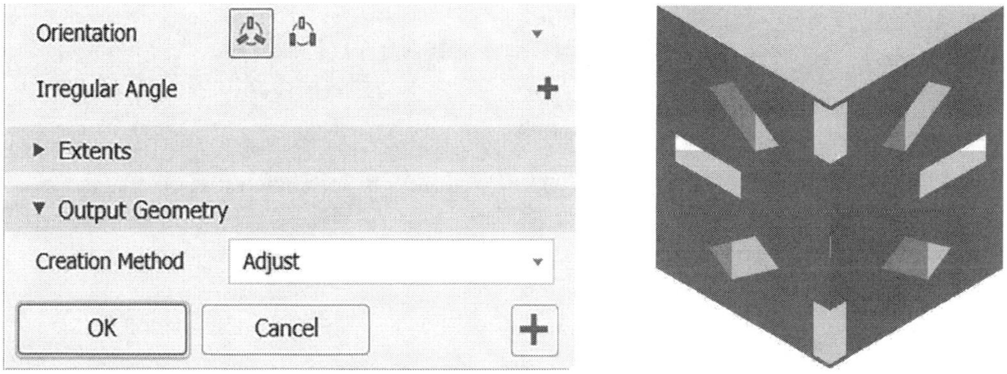

Figure 4.96

8 Change to the home view.

9 Edit the circular pattern, and change the Orientation to Fixed as shown in Figure 4.97.

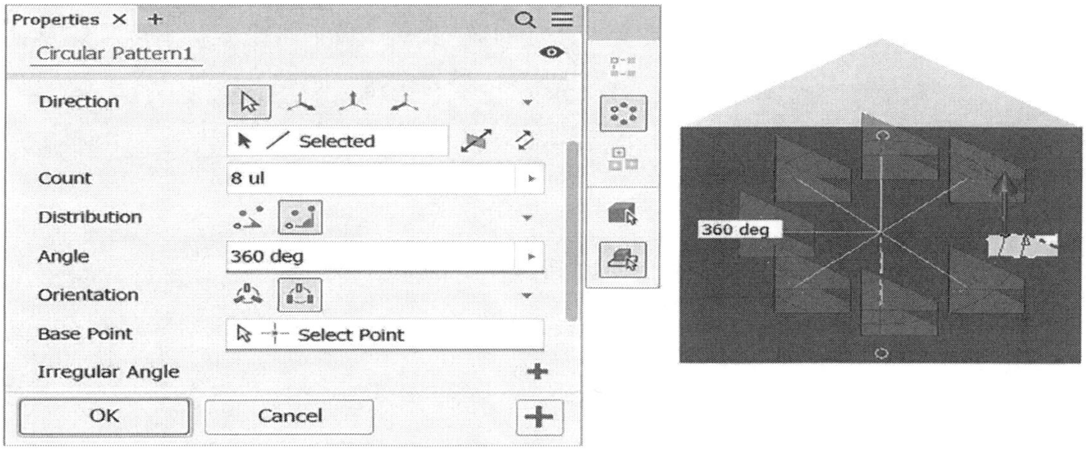

Figure 4.97

10 Next experiment with different options.

a. Edit the circular pattern and change the Base Point that the orientation is based on. Click the Base Point button in the dialog box for the Fixed orientation. Select points on the rectangle that represent the feature that is patterned, and then select the work axis.

b. Also try different combinations of count, angle, creation method, and distribution.

11 Close the file. Do not save changes.

End of exercise.

Linear Patterns—Pattern along a Path

You can also use the Rectangular Pattern command to pattern a feature about a path. You can define a path by a complete or partial ellipse, an open or closed spline, or a series of curves (lines, arcs, splines, etc.).

To pattern along a path, use the options described previously for rectangular patterns. The path can be either 2D or 3D geometry.

EXERCISE 4-10: CREATING A PATTERN ALONG A NONLINEAR PATH

In this exercise, you pattern a boss and hole along a nonlinear path.

1 Open *ESS_E04_09.ipt* in the Chapter 04 folder. Note that the visibility of the sketch that the pattern will follow is on.

2 Click the Rectangular Pattern command on the 3D Model tab > Pattern panel.

3 For the features to pattern, in the browser or in the graphics window, click both the Extrusion2 and Hole2 features as shown in the following image on the left.

4 In the Rectangular Pattern dialog box, in the Direction A area click the Select Direction button and select a line or an arc in the visible sketch. A preview of the pattern is displayed, as shown in Figure 4.98 right.

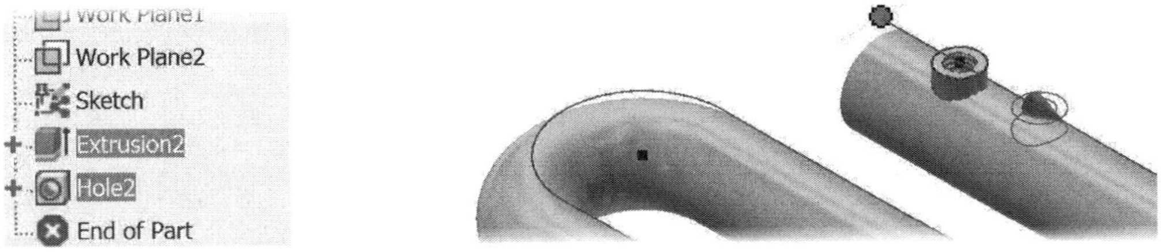

Figure 4.98

5 In the Number cell enter **40 ul**, and in the Distnce cell enter **1.25 inches** for the spacing. The preview shows 1.25 inches between each occurrence as shown in Figure 4.99.

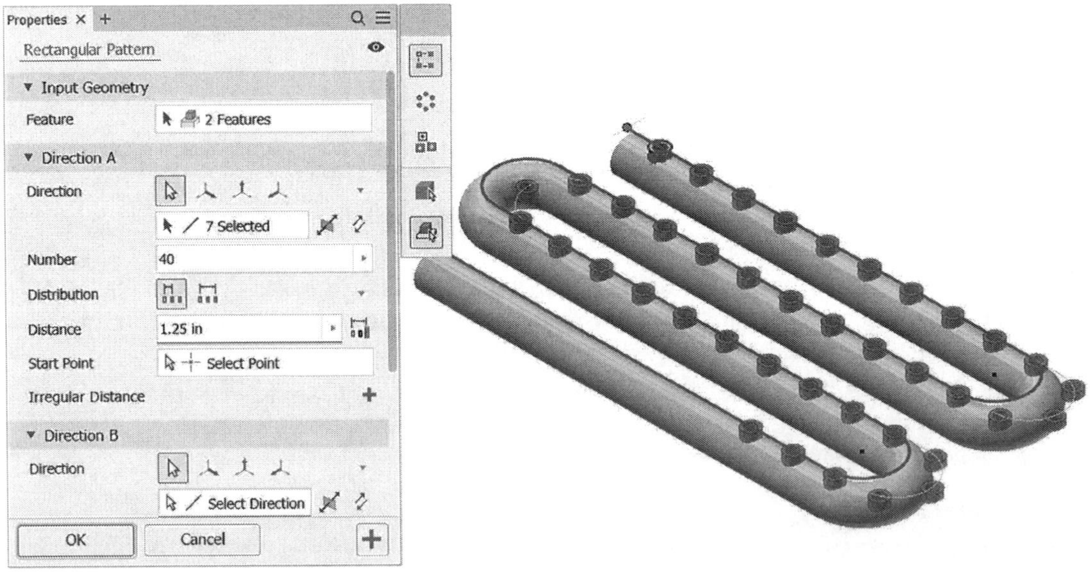

Figure 4.99

6 From the Distribution options, select the Fitted icon as shown in Figure 4.100. The preview shows 40 occurrences fit within 1.25 inches. Do NOT click OK.

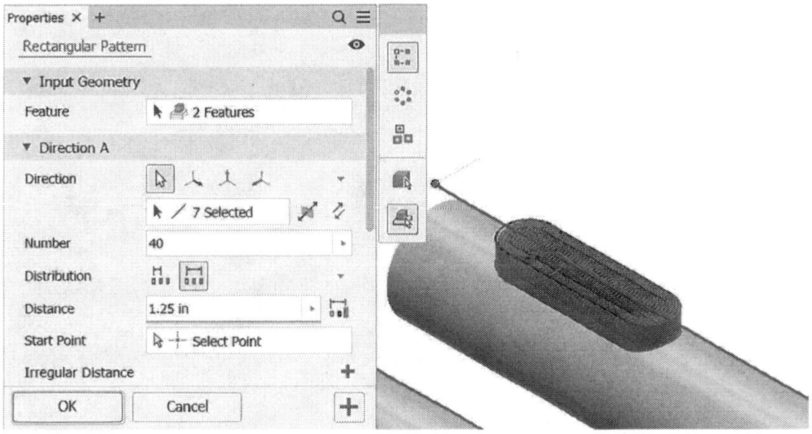

Figure 4.100

7 Next to the Distance cell, select Curve Length. The preview shows the 40 occurrences fitting within the entire length of the path.

8 Rotate the model to verify that the occurrences on the right side are hanging off the model; this is because the first occurrence is spaced 1 inch away from the start of the path.

9 To solve the starting point issue, click the More >> button.

10 In the Direction A area in the bottom area of the dialog box, click the Start Point button and then click the center point of the top of the first hole, as shown in Figure 4.101 left. The preview updates to show that all occurrences are now located on the part, as shown in the image on the right. However, the last occurrence is on the back edge of the part.

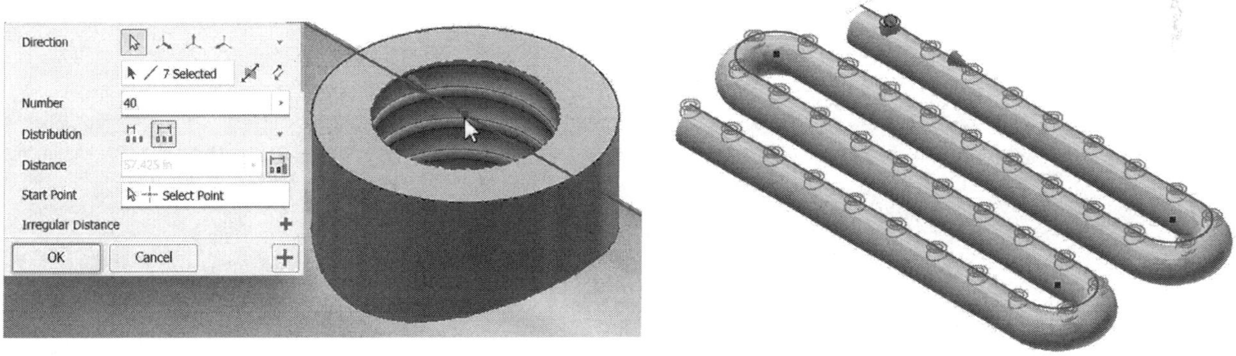

Figure 4.101

11 Next to the Distance cell, select Curve Length. The curve length remains in the distance area but the value can be edited.

12 Click in the distance area and enter an equation that uses the current distance (57.425) and subtracts 1 inch. Enter **57.425 in – 2 in**, as shown in the following image on the left, or enter a value of **55.425**.

13 Click OK to create the pattern.

14 In the browser, right-click on the entry Sketch, and uncheck Visibility from the menu. When done, your model should resemble Figure 4.102 right.

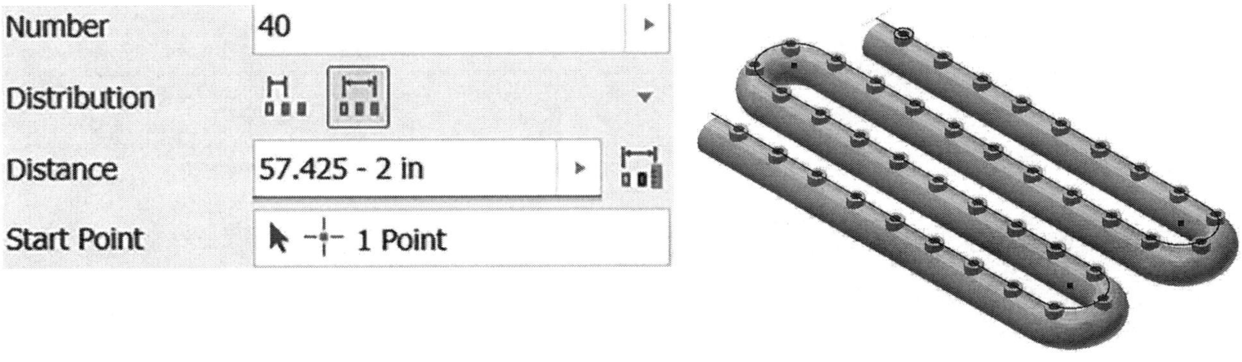

Figure 4.102

15 Edit the pattern trying different combinations.

16 Close the file.

17 Do not save changes.

End of exercise.

3D PRINTING – ADDITIVE MANUFACTURING

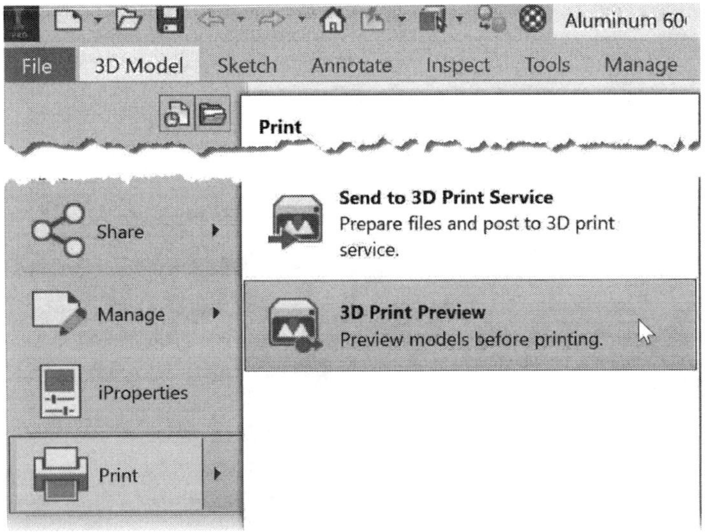

Figure 4.103

You can send a part or an assembly file to a 3D printer via an STL file. To generate an STL file from a part or assembly follow these steps.

1. Start the 3D Print Preview command from the File tab > Print as shown in Figure 4.103.
2. In the graphics window examine the preview of the STL file.
3. Use Options from the toolbar, shown in Figure 4.104, to change the format and resolution of the STL file.
4. To create an STL file, click the Save Copy As command on the left side of this toolbar.

Figure 4.104

When generating an STL file of a part or an assembly file, another option previews the model with the selected 3D printer. You can also reposition the model within the 3D printer space and adjust the STL file.

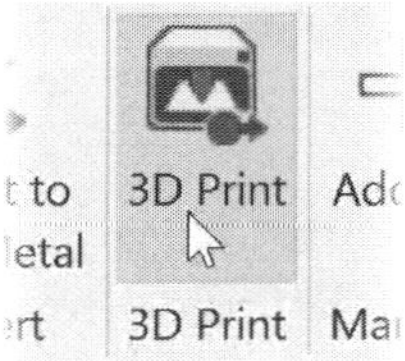

Figure 4.105

To use this method, follow these steps.

1. Start the 3D Print command on the Environment tab > 3D Print panel as shown in Figure 4.105.
2. Select your 3D printer from the list.
3. Adjust the orientation, position, add a partition, and mesh display using the commands in the 3D Print tab.

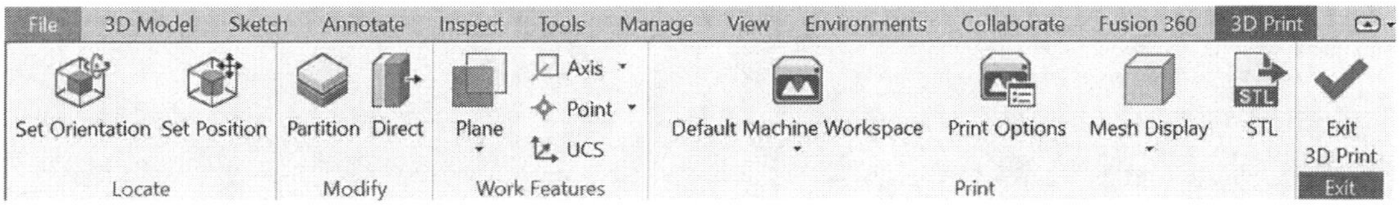

Figure 4.106

4. To create an STL file, click Save Copy As STL from the 3D Print tab > Print panel.
5. To see how the file will print in a 3D printer, select a 3D printer from the list. Figure 4.107 shows a part in an Object Connex 500 printer.
6. When done, click the Save Copy As STL command in the Print panel.

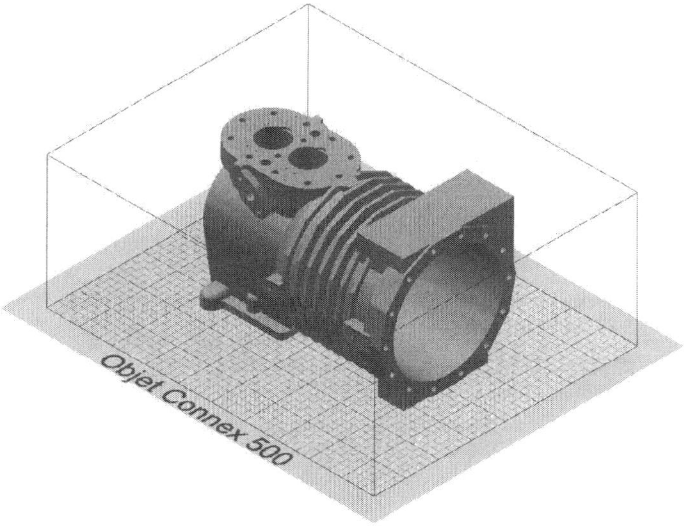

Figure 4.107

APPLYING YOUR SKILLS

Skills Exercise 4-1

In this exercise, you create a drain plate cover.

1. Start a new part based on the English Standard (in).ipt template.
2. Use the extrude, shell, hole, rectangular pattern, and the Fillet commands to create the part. Only fillet the inside vertical edges.

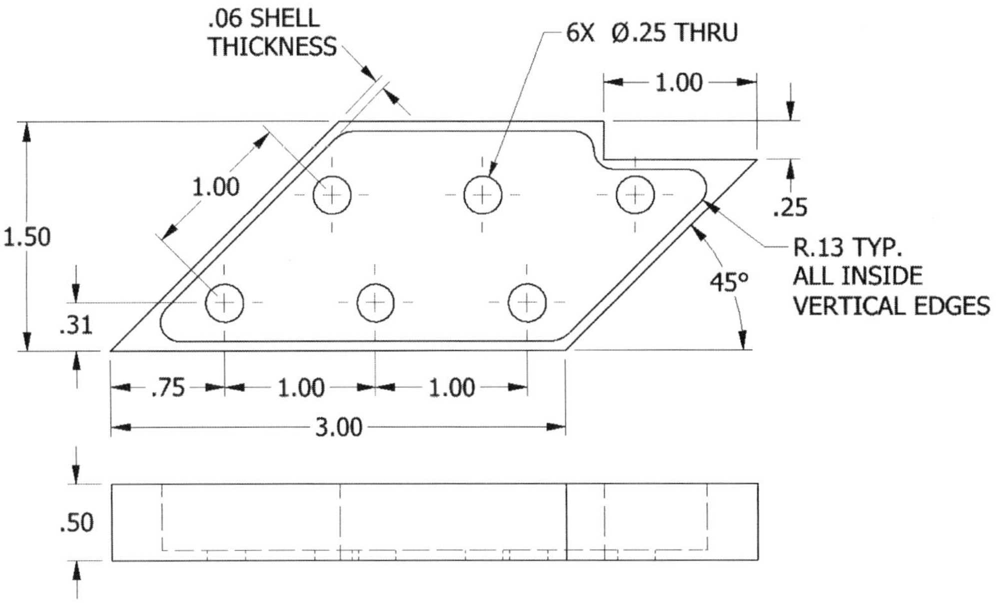

Figure 4.108

Skills Exercise 4-2

In this exercise, you create a connector.

1. Start a new part based on the English Standard (in).ipt template.
2. Use the revolve, work plane, hole, chamfer, fillet, and circular pattern commands to complete the part.

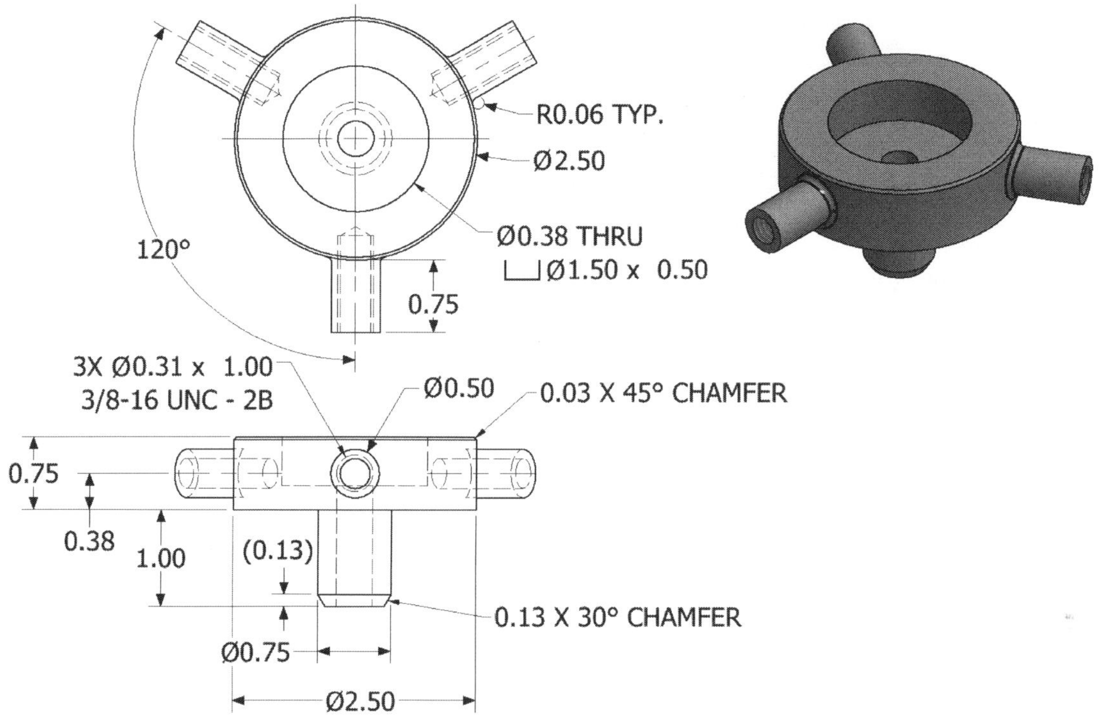

Figure 4.109

CHECKING YOUR SKILLS

Use these questions to test your knowledge of the material covered in this chapter.

1. True__ False__ When creating a fillet feature that has more than one selection set, each selection set appears as an individual feature in the browser.
2. When creating a fillet feature, what is a smooth radius transition?
3. True__ False__ When creating a fillet feature with the All Fillets option, material is removed from all concave edges.
4. True__ False__ When creating a chamfer feature with the Distance and Angle option, you can only chamfer one edge at a time.
5. True__ False__ When you are creating a hole feature, you must place points on a sketch.
6. What is a Point, Center Point used for?
7. True__ False__ When shelling a part, you can only have one unique face thickness in the part.
8. True__ False__ The only method to create a work axis is by clicking a cylindrical face.
9. True__ False__ Every new sketch must be placed on a work plane.
10. Explain the steps to create an offset work plane.
11. True__ False__ A work plane is only used to create a sketch on.
12. True__ False__ You cannot create work planes from the origin planes.
13. True__ False__ A UCS can only be placed on existing geometry.
14. True__ False__ When you are creating a rectangular pattern, the directions along which the features are duplicated must be horizontal or vertical.
15. True__ False__ When creating a circular pattern, you can only use a work axis as the axis of rotation.
16. True__ False__ To start the Fillet command from a mini-toolbar, select two planes that define the location of the fillet.
17. True__ False__ Use the Fillet command with the Face option to create a fillet that is tangent to three faces.
18. While in the shell command, explain how to create a unique face thickness.
19. Explain three reasons why you would create a UCS.
20. True__ False__ Use the Linear Pattern command to pattern a feature along a selected path consisting of multiple lines and arcs.

5 Creating & Editing Drawing Views

INTRODUCTION

After creating a part or assembly, the next step is to create 2D drawing views that represent it. You can create drawing views at any point after a part or assembly exists. The part or assembly does not need to be complete because the part and drawing views are associative in both directions (bidirectional). This means that if the part or assembly changes, the drawing views will automatically be updated. If a part's parametric dimension changes in a drawing view, the part will be updated before the drawing views are updated. This chapter guides you through setting up styles, creating drawing views of a single part, editing dimensions, and adding annotations.

When creating drawings, adhere to specific drawing standards that communicate information about designs in a consistent manner. There are many different drawing standards, such as ANSI and ISO. Companies usually adopt a drawing standard and modify it to meet their requirements. This book follows the ANSI drawing standard.

OBJECTIVES

After completing this chapter, you will be able to perform the following:

- Create base and projected drawing views from a part
- Create auxiliary, section, detail, and broken views
- Edit the properties and location of drawing views
- Retrieve and arrange model dimensions for use in drawing views
- Edit, move, and hide dimensions
- Add automated centerlines
- Add general, baseline, chain, and ordinate dimensions
- Add annotations such as text, leaders, Geometric Dimensioning & Tolerancing (GD&T), surface finish symbols, weld symbols, and datum identifiers
- Create hole notes
- Create a hole table
- Open a model from a drawing
- Open a drawing from a model
- Create a 3D PDF
- Add 3D annotations to a part or an assembly file

DRAWING OPTIONS

Before you create drawing views, set the drawing options to your preferences using the Application Options from the Tools panel to show the Application Options dialog box. Click the Drawing tab. Your screen should resemble Figure 5.1. Make changes to the options before creating the drawing views, otherwise the changes may not affect drawing views that you have already created. A description of the common options follows. For more information about the Drawing Application Options consult the help system.

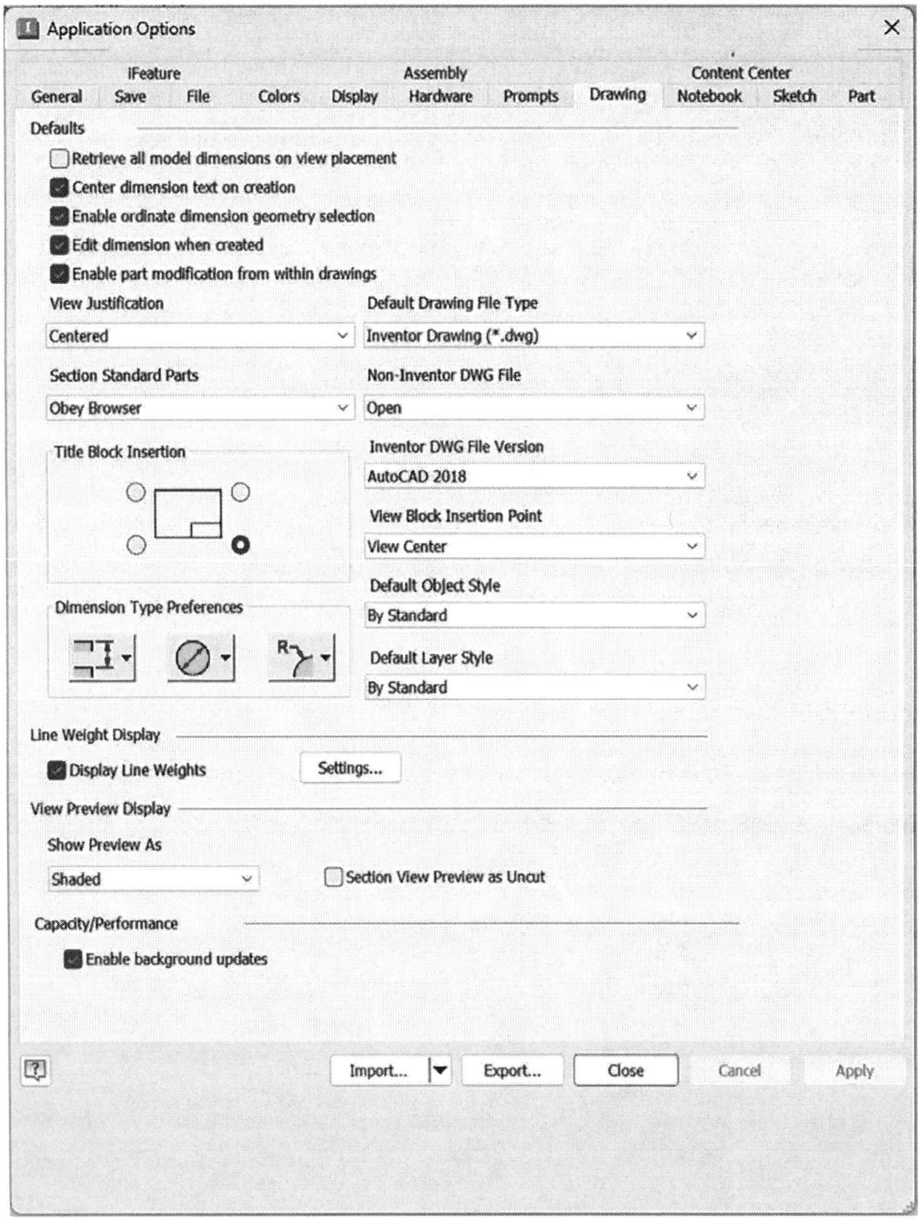

Figure 5.1

Retrieve All Model Dimensions on View Placement

Click this box to add applicable model dimensions to drawing views when they are placed. If the box is clear, model dimensions are not placed automatically. You can override this setting by manually selecting the All Model Dimensions option in the Drawing View dialog box when creating base views. Note when creating a drawing view of a complex part or large assembly, when this option is on, you may retrieve many more dimensions than what is required.

Center Dimension Text on Creation

Click this option to have dimension text centered as you create the dimension.

Edit Dimension When Created

Click this option to see the Edit Dimension dialog box every time a dimension is placed in a drawing.

Default Drawing File Type

In this section, select either Inventor drawing type, IDW or DWG, as the default file format. The IDW is a file format that can only be opened with Autodesk Inventor. When a drawing file is created from a DWG template, an Inventor DWG file is created and AutoCAD can open this file without translation. In AutoCAD, you can view, plot, and add annotations to the Inventor DWG file. You cannot edit the drawing views in AutoCAD as these are controlled by Inventor.

CREATING A DRAWING

To start a drawing from an existing part or assembly (IDW or DWG) use one of the following methods:

- From the Quick Access toolbar click the down arrow on the right side of the New icon and click Drawing from the drop list as shown in Figure 5.2 left. This creates a new drawing file based on the default drawing file type either DWG or IDW depending on what is set in the Application Options > Drawing tab.
- Click New > Drawing on the Home page as shown in the middle image.
- From the File tab click New > Drawing as shown in Figure 5.2 right and then choose from the recent templates.

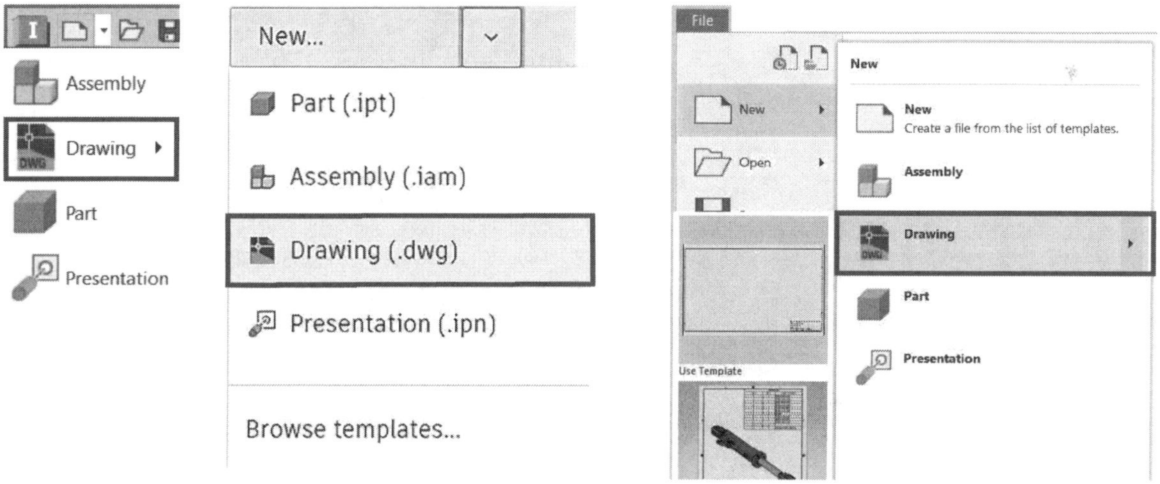

Figure 5.2

While working in a part or assembly file, you can also create a new drawing based on the default template by right-clicking on the file name in the browser, and clicking Create Drawing View from the menu as shown in Figure 5.3.

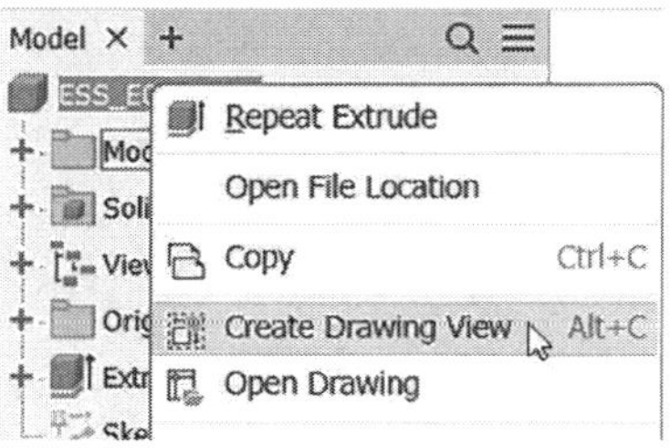

Figure 5.3

You can create a drawing file from a template that is not the default by clicking the New file command from one of these areas:

- Quick Access toolbar as shown in the following image on the left.
- File tab as shown in the left image.
- From the File tab click New > Drawing and then choose from the recent templates.
- Or press CTRL + N.

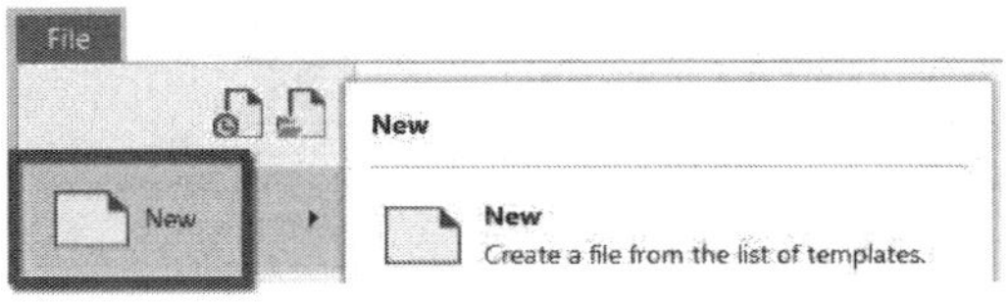

Figure 5.4

The Create New File dialog box appears. Then click the desired template folder on the left side of the dialog box, and in the Drawing - Create an annotated document area, click the desired template file as shown in the following image.

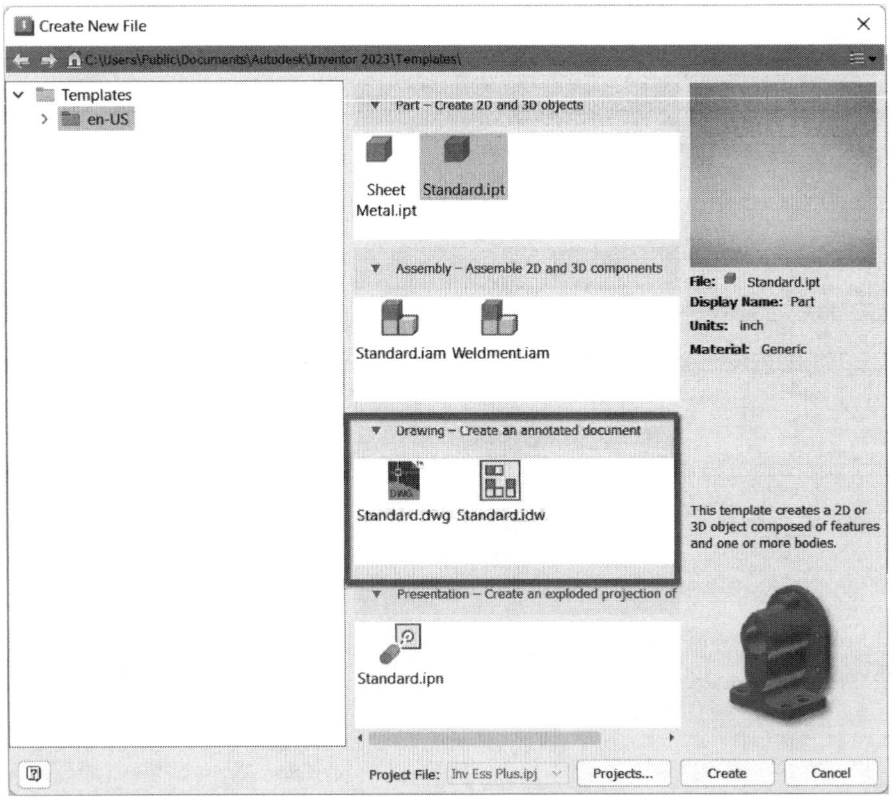

Figure 5.5

DRAWING SHEET PREPARATION

When you create a new drawing file using one of the provided template files, a drawing sheet with a default title block and border will appear. The template selected determines the default drawing sheet, title block, and border. The drawing sheet represents a blank piece of paper (sheet) on which you will add drawing views.

To create views that do not fit on a single sheet you can add sheets by clicking New Sheet from the Place Views tab > Sheets panel as shown in Figure 5.6 left. Alternately, you can right-click in the browser and choose New Sheet from the menu as shown in the middle image, or on the current sheet in the graphics window right-click and click New Sheet from the menu as shown in Figure 5.6 right.

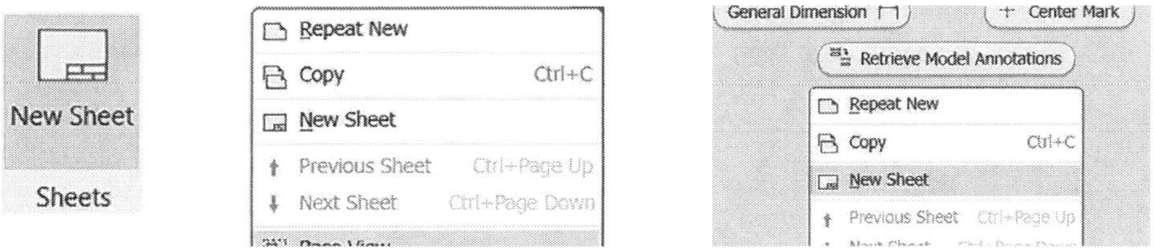

Figure 5.6

A new sheet appears in the browser and the graphics window with the same size, border and title block of the active sheet. To edit the sheet's size right-click on the sheet name in the browser, and select Edit Sheet from the menu, as shown in the following image on the left. Then select a size from the list, as shown in the following image on the right. To use your own values, select Custom Size from the list, and enter values for the height and width. From the dialog box, you can also change the sheet's name or slowly double-click on its name in the browser and then enter a new name.

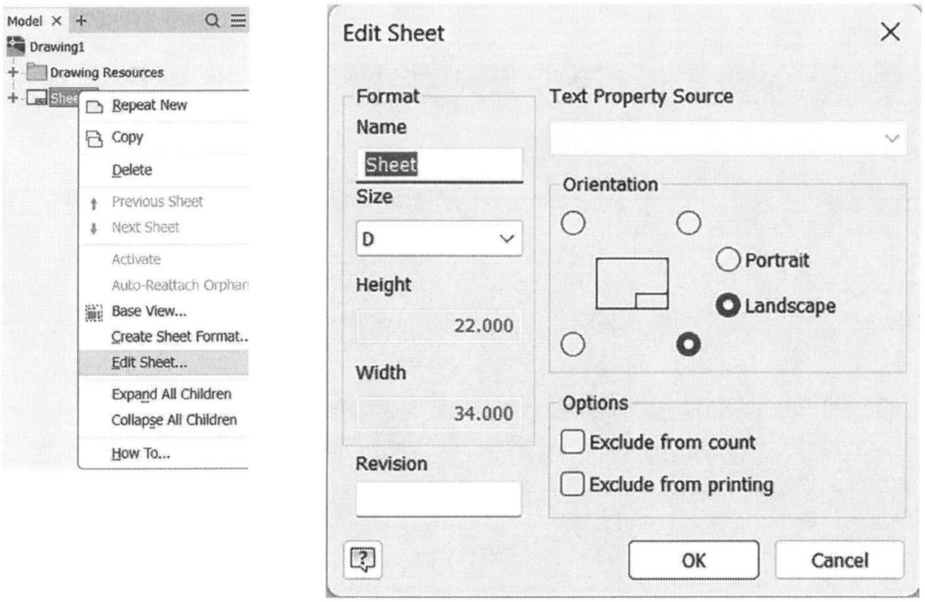

Figure 5.7

TIP: The sheet size is inserted full scale (1:1) and should be plotted at 1:1. The drawing views will be scaled to fit the sheet size.

Border

To change a border on a drawing sheet, you can either insert a default border or construct a customized border and insert it into a drawing sheet.

Changing a Border

To insert a default border, follow these steps:

1. If a border exists in the sheet, it must be deleted before a new border can be inserted. Make the sheet active: from the browser, right-click on the Default Border entry and click Delete from the menu as shown in Figure 5.8 left.

2. Insert a border by expanding Drawing Resources > Borders in the browser.
   a. To insert a generic border, double-click on the border's name and a generic border will be inserted.
   b. To control the border's appearance right-click on Borders > Default Border and click Insert Drawing Border, as shown in the following image in the middle. The Default Drawing Border Parameters dialog box will appear; modify the options as needed, click the More >> button to control text and layer data as well as the sheet's margin as shown in the following image on the right.
   c. Consult the Help system to learn how to create a new border.

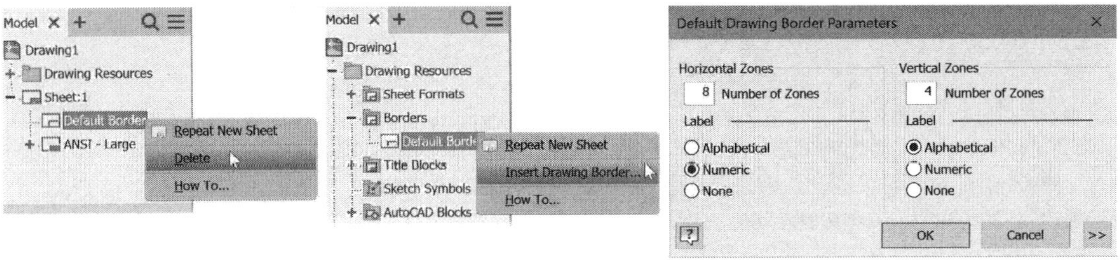

Figure 5.8

Title Block

To change a title block on a drawing sheet, you can either insert a default title block or construct a customized title block and insert it into a drawing sheet.

Inserting a Default Title Block

To insert a default title block, follow these steps:

1. If a title block exists in the sheet, it must be deleted before a new title block can be inserted.
2. Make the sheet active and in the browser right-click on the title block entry and click Delete from the menu as shown in Figure 5.9 left.
3. Insert a title block by expanding Drawing Resources > Title Blocks in the browser.
4. Either double-click on the title block's name or right-click on the title block's name and choose Insert from the menu, as shown in Figure 5.9 right.

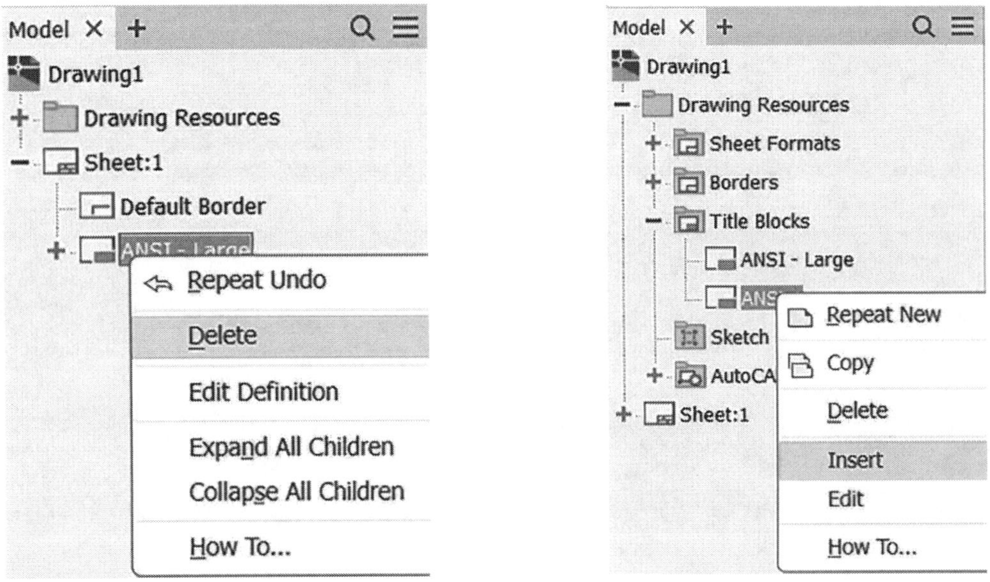

Figure 5.9

Edit Property Fields Dialog Box

To fill in title block information, expand the title block in the browser and double-click Field Text entry or right-click on the Field Text entry and choose Edit Field Text from the menu, as shown in Figure 5.10 left. This displays the Edit Property Fields dialog box. By default, the following information will already be filled in: Sheet Number, Number of Sheets, Author, Creation Date, and Sheet Size. To fill in other title block information such as Part Number, Company Name, Checked By, and so on, select the iProperties command button from the upper right corner of the dialog box, shown in the middle image. The Drawing iProperties dialog box appears, as shown in Figure 5.10 right, and you can fill in the information as needed. You can find most title block information under the Summary, Project, and Status tabs.

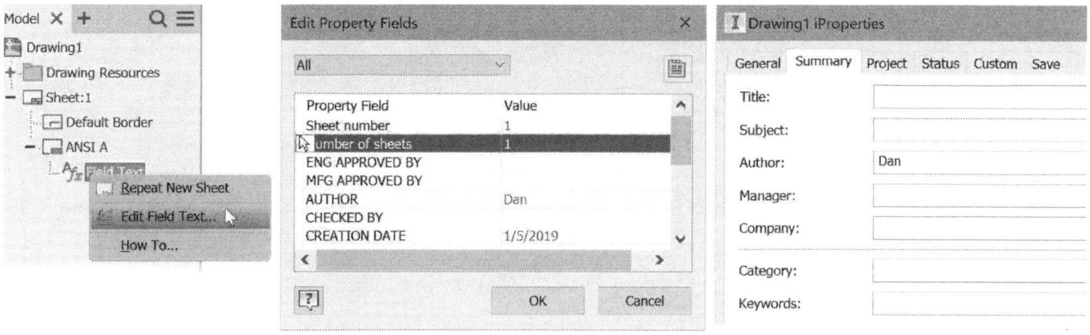

Figure 5.10

Save Drawing to a Template

After setting up a drawing, you can save it to the template folder to use to start new drawings of that style. From the File tab click Save As > Save Copy As Template as shown in Figure 5.11. Save the file to the default location: Templates (), English, Metric, or create a new folder. The location of the template files is set in the Application Options > File tab > Default templates field or may be overridden in the active project's Folder Options Templates.

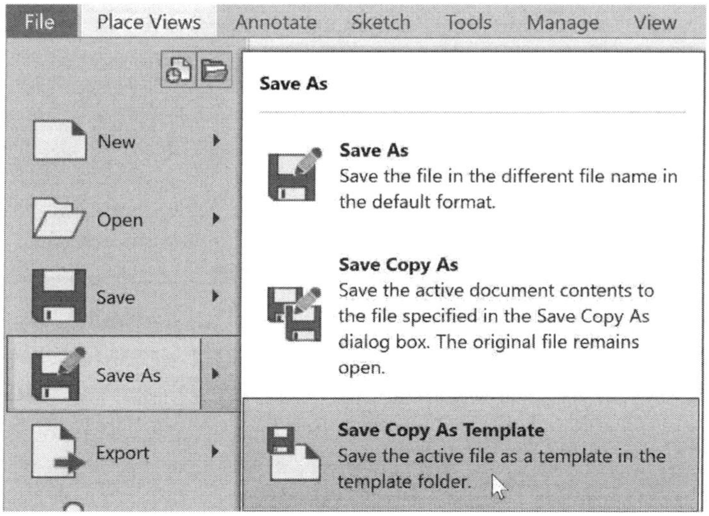

Figure 5.11

CREATING DRAWING VIEWS

After you set the drawing sheet format, border, and title block you can create drawing views from any existing part, assembly, or presentation file. The file from which you will create the views does not need to be open when a drawing view is created. It is suggested that both the file and the associated drawing file be stored in the same directory that is referenced in the project file. There are many different types of drawing views you can create. The following sections describe the view types.

Base View. This is the first drawing view of an existing part, assembly, or presentation file. It is typically used as a basis for generating the following dependent view types. You can create many base views in each drawing file.

Projected View. A dependent orthographic or isometric view generated from an existing drawing view.

- Orthographic (Ortho View): A drawing view projected horizontally or vertically from another view.
- Isometric View: A drawing view projected at a 30° angle from a given view. An isometric view can be projected to any of four quadrants.

Auxiliary View. A dependent drawing view perpendicular to a selected edge of another view.

Section View. A dependent drawing view that represents the area defined by a slicing plane or planes through a part or assembly.

Detail View. A dependent drawing view in which a selected area of an existing view is generated at a specified scale.

Broken View. A dependent drawing view showing a section of the part removed while the ends remain. Any dimension that spans over the break shows the actual object length.

Break Out View. A drawing view with an area of material removed to expose internal parts or features.

Crop View. This drawing view allows a view to be clipped based on a defined boundary.

Overlay View. This drawing view uses positional representations to show an assembly in multiple positions in a single view.

TIP: Inventor places a temporary raster image for drawing views while the precise view is being calculated. Small horizontal and vertical lines appear in the corners of the view and an angled line shows on the view's browser icon to let you know this is the case. You can continue to work on the drawing while the views are calculated.

Creating a Base View

A base view is the first view you create from the selected part, assembly, or presentation file. When you create a base view, the scale is set in the dialog box. From the base view, you can project other drawing views.

There is no limit to the number of base views you can create in a drawing based on different parts, assemblies, or presentation files. As you create a base view, you select the orientation of that view from the Orientation list on the Component tab of the Drawing View dialog box. By default, there is an option to create projected views immediately after placing a base view.

To create a base view, follow these steps:

1. Click the Base View command on the Place Views tab > Create panel as shown in the following image on the left, or right-click in the graphics window and click Base View from the marking menu as shown in Figure 5.12 right.

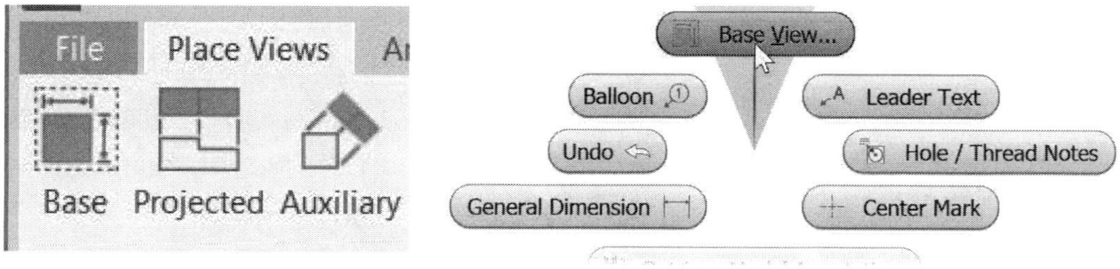

Figure 5.12

2. The Drawing View dialog box appears as shown in Figure 5.13 left. From the Component tab, select a document from the Select Document list labeled (1), and click the Open an existing file icon labeled (2) to navigate to and select the part, assembly, or presentation file from which to create the base drawing view.
3. After making the selection, a preview image appears attached to your cursor in the graphics window. Do not place the view until the desired view options have been set. You can change the viewpoint so it looks directly at a plane by clicking a plane on the ViewCube which is above and to the right of the view labeled (3) in Figure 5.13. You can change to an isometric view by clicking a corner on the ViewCube.
4. Select the style for the view labeled (4).
5. If desired, enter a name for the view labeled (5) and turn on/off the visibility of the label by clicking the light bulb.
6. Select the scale from the list or enter a scale for the view, labeled (6).
7. You can also cycle through predetermined view scales by clicking and dragging a corner of the view in the graphics window labeled (7). To freely scale the view in the graphics window, hold down the CTRL key while dragging.
8. Locate the base view by selecting a point in the graphics window. Move the cursor in the direction to place a projected orthographic or isometric view and click. Continue locating views and when done, click OK in the dialog box or right-click and choose OK from the marking menu.

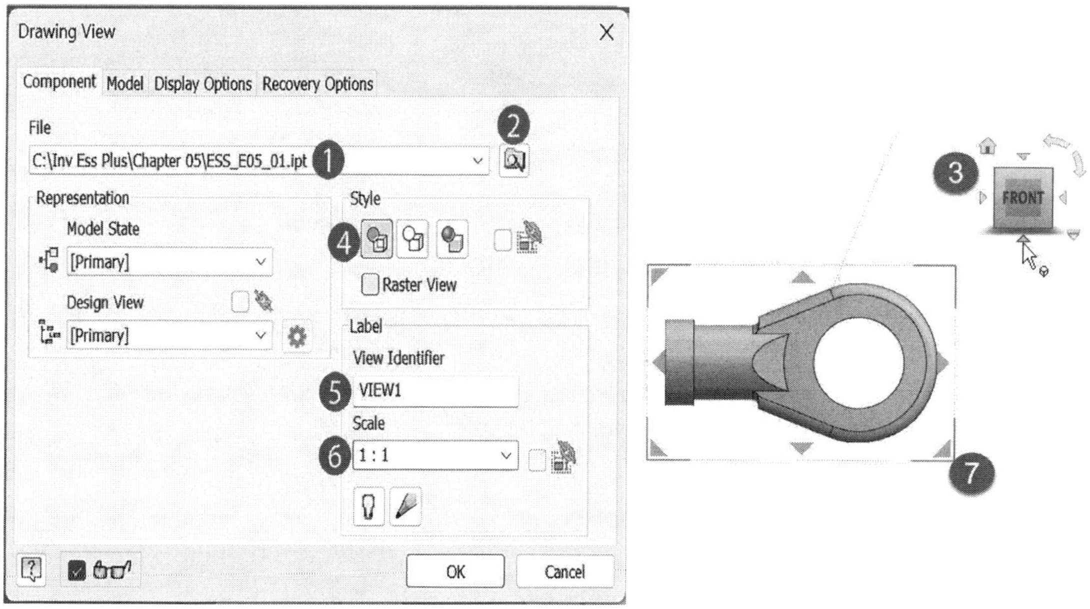

Figure 5.13

The Drawing View dialog box has four tabs: Component, Model State, Display Options, and Recovery Options.

The Component Tab

Use the options on this tab to select the file from which the views are created and how the views will appear. Consult the Help system for more information on these options.

The Model State Tab

Use the options on this tab to specify the state of the weldment or member state of an iAssembly or iPart to use in a drawing view. These options change based on the file that views are being created from.

The Display Options Tab

Use the options on this tab (Figure 5.14) to control how tangent edges and features like threads and work features are shown.

The Recovery Options Tab

Use the choices on this tab to include surface bodies, mesh bodies, or work features.

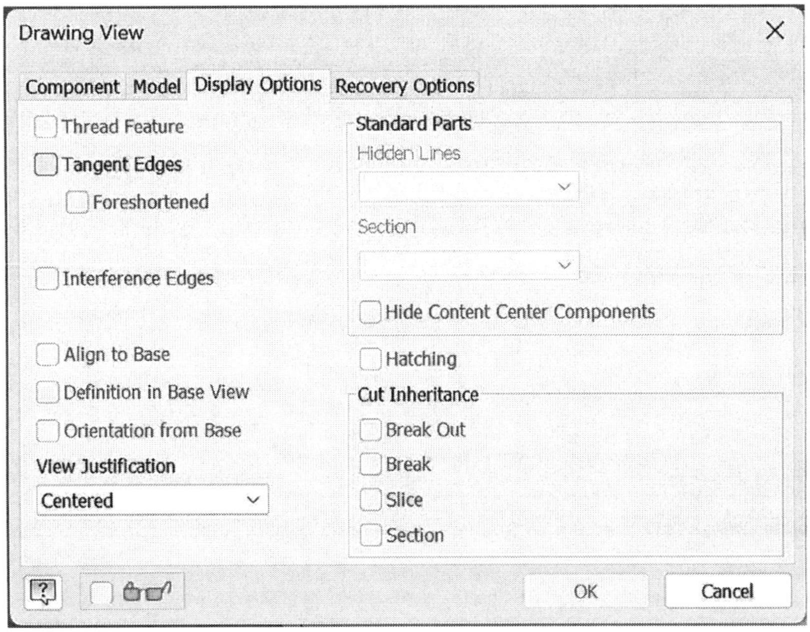

Figure 5.14

Creating Projected Views

An orthographic or isometric view can be projected from a base view or any other existing view. When you create a projected view, a preview image shows the orientation of the view you will create in the drawing. There is no limit to the number of projected views you can create.

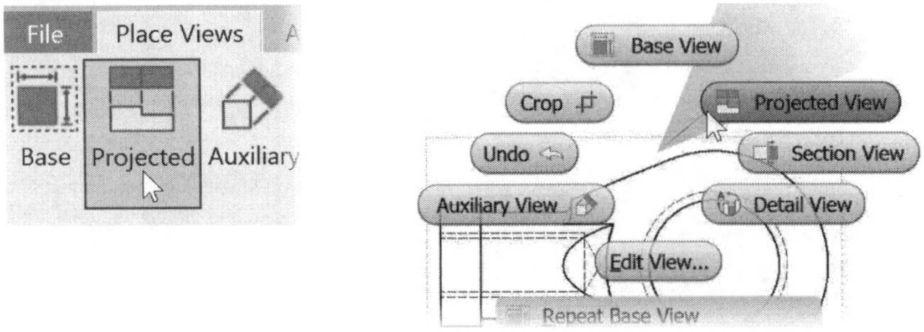

Figure 5.15

To create a projected drawing view, follow these steps:

1. Click Projected View from the Place Views tab > Create panel as shown in Figure 5.15 left, or right-click inside the bounding area of an existing view (displayed as dashed lines when the cursor moves into the view) and choose Projected View from the marking menu as shown in Figure 5.15 right.
2. If you selected the Projected View command from the Create panel, click inside the desired view to project a view from.
3. Move the cursor horizontally, vertically, or at an angle (to create an isometric view) to get a preview image of the view you will generate. Keep moving the cursor until the preview matches the view that you want to create, and then press the left mouse button. Continue placing projected views.
4. When finished, right-click, and select Create from the marking menu.

Moving Drawing Views

To move a drawing view, hover the cursor over the view until a bounding box consisting of dotted lines appears, as shown in Figure 5.16. Press and hold the left mouse button to drag the view to its new location. Release the mouse button when you are finished. As you move the view, a rectangle appears representing the bounding box of the view. If you move a base view, any projected children or dependent views also move with it to maintain view alignment. If you move an orthographic or auxiliary view, you are only able to move it along the axis in which it was projected from the part edge or face. You can move detail and isometric views anywhere in the drawing sheet.

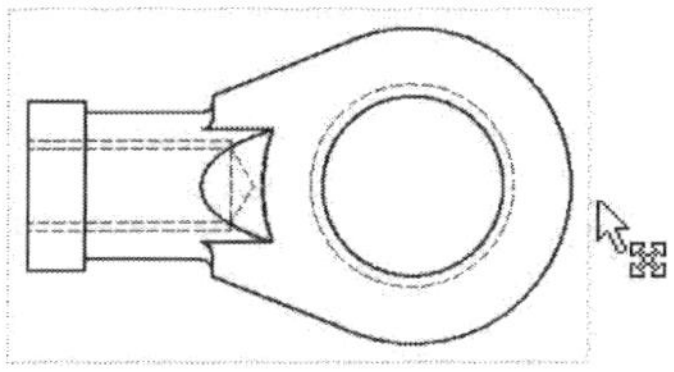

Figure 5.16

Drawing Zone Identifiers

In addition to the title block information, notice the numbers and letters outside of the border. Use these as reference points to locate a specific detail, view, or dimension on a large sheet. Numbers are located horizontally along the top and bottom of the border. Letters are located vertically along the left and right border edges. These reference numbers make it easy to describe where to find an item in a complex drawing. For example, you could say, "Look in C2." This identifies C along the vertical portion of the border and the number 2 along the horizontal portion of the border. Where the two references intersect, your search item is easily found.

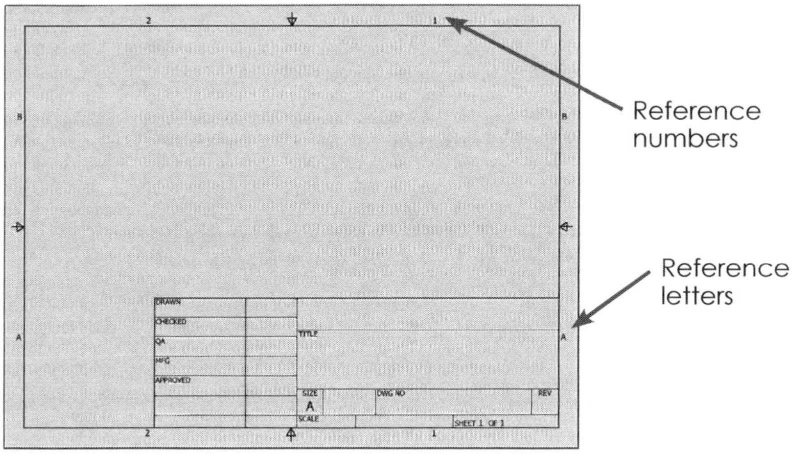

Figure 5.17

EXERCISE 5-1: CREATING A MULTIVIEW DRAWING

In this exercise, you create a base view, and then add projected views to create a multi-view orthographic drawing. Finally, you add an isometric view to the drawing.

1 Open the file *ESS_E05_01.idw* from the Chapter 05 folder. This drawing file contains a single sheet with a border and title block.

2 Click the Base View command from the Place Views tab > Create panel or right-click in the graphics window and choose Base View from the marking menu to show the Drawing View dialog box.

a. Select the source file. Use the File area from the Component tab to click the Open an existing file button labeled (1) in Figure 5.18. Browse to file *ESS_E05_01.ipt* from the Chapter 05 folder and double-click to select it.

b. From the graphics window, change to various viewpoints by clicking on different planes of the ViewCube above and to the right of the view. You can rotate the viewpoint by 90 degrees by clicking on the arrows on the ViewCube. Change to an isometric view by clicking on a corner on the ViewCube. When done experimenting, click the FRONT plane of the ViewCube (2).

c. In Style, verify that the Hidden Line button in the Drawing View dialog box is selected (blue background), labeled (3) and the Shaded option is not active, labeled (4).

d. In the Scale list, change the scale to 1:1, labeled (5).

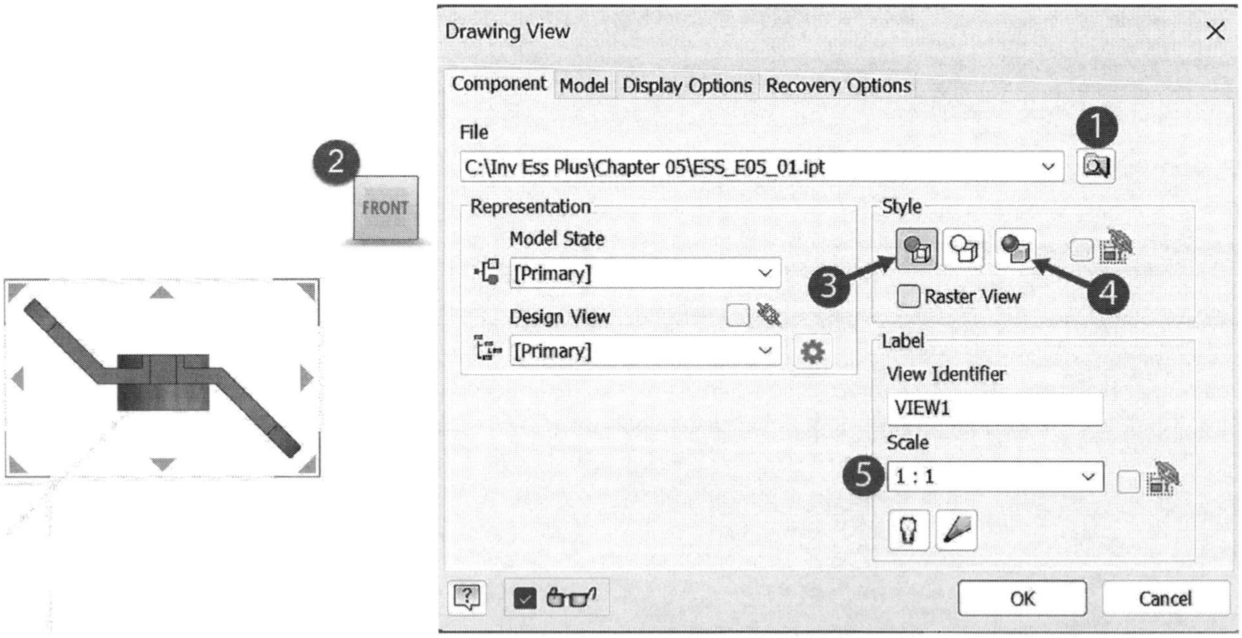

Figure 5.18

3 The view preview will be positioned near the lower-left corner of the sheet; move the cursor over the border of the previewed view, the move symbol will appear, click and drag the view until it is positioned in zone C6 as shown in the following image. To place the view, release the mouse button. To review drawing zones see "Drawing Zone Identifiers" on page 197.

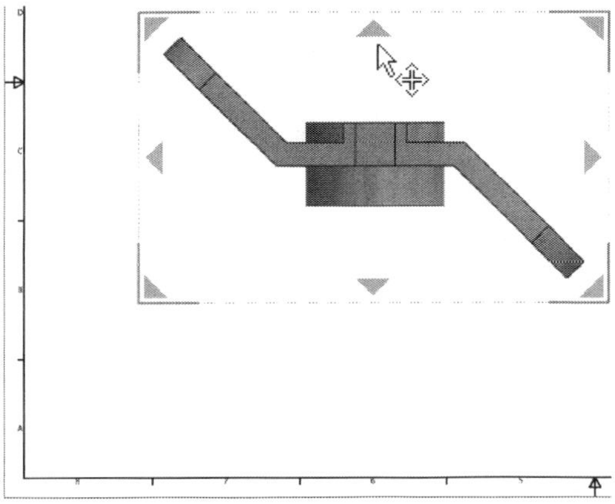

Figure 5.19

4 Next place orthographic projected views. While still in the Base View command move the cursor to the right of the base view. Click in zone C2 to place the right-side view.

a. Move the cursor to the top of the base view. Click in zone E6 to place the top view.

b. Right-click, and click OK from the marking menu to create the views. When done, your views should resemble the following image.

Note: You can create the right side and top view using the Projected View command.

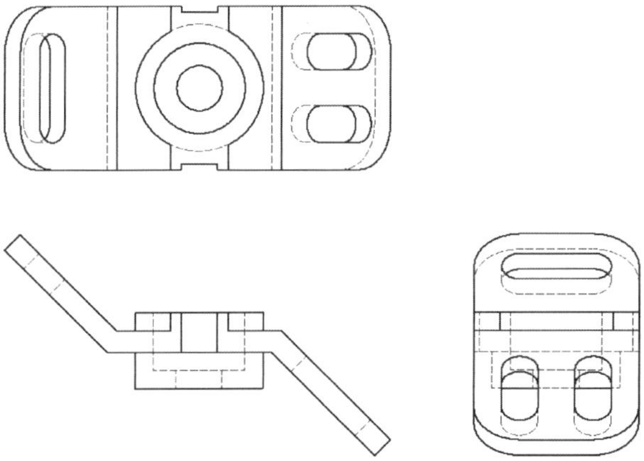

Figure 5.20

5 Zoom into the title block and verify that the scale is displayed as SCALE 1 : 1.

6 Zoom out so you see the entire drawing.

The views are crowded at scale 1:1. In the steps that follow, you reduce the size of all drawing views by editing the base view scale to 1:2. The dependent views will update automatically.

7 Edit the base view (front view) by either double-clicking on the base view or right-click the base view and choose Edit View from the marking menu.

TIP: To edit a view, right-click on the view border or inside the view. Do not right-click on the geometry.

8 From the Drawing View dialog box, select 1/2 from the Scale list, and click OK.

9 The scale of all views update, as shown in Figure 5.21.

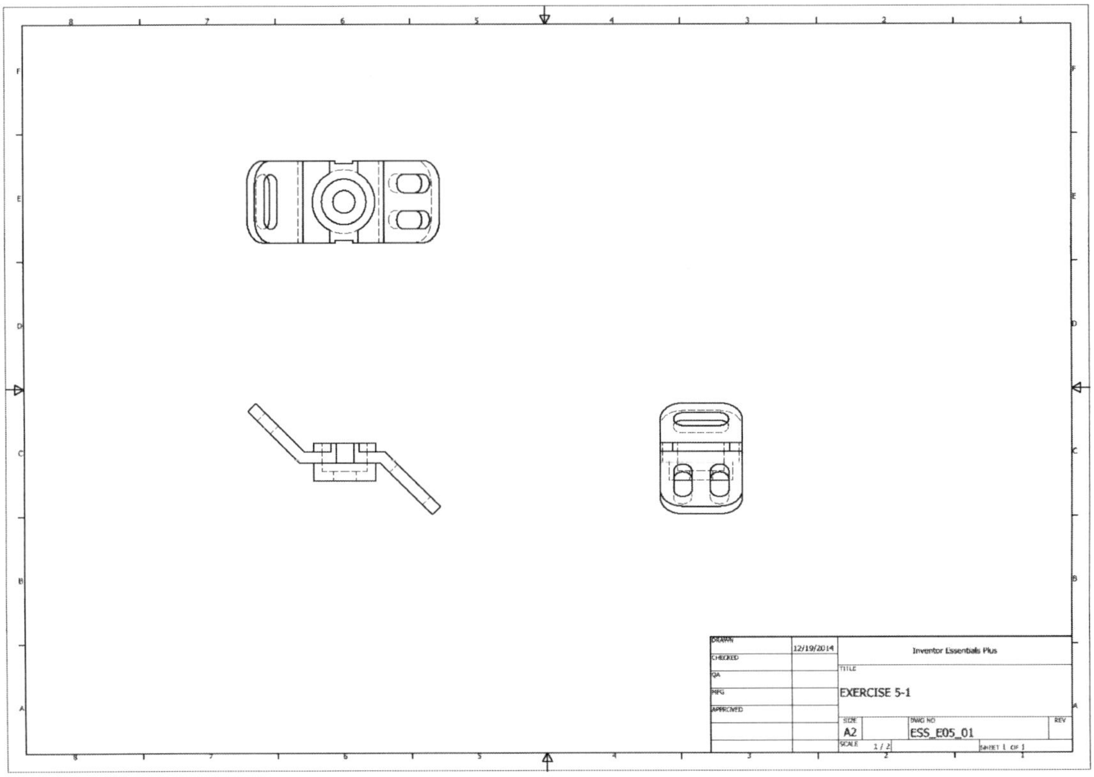

Figure 5.21

10 Zoom into the title block and verify that the scale was automatically updated to reflect the new scale.

SCALE 1 / 2

11 Next create an isometric view. Begin by clicking the Projected View from the Place Views tab > Create panel.

12 Click inside the base view (front view) and move your cursor in zone E3 and click to locate the isometric view, then right-click and choose create from the marking menu. Your drawing should resemble Figure 5.22.

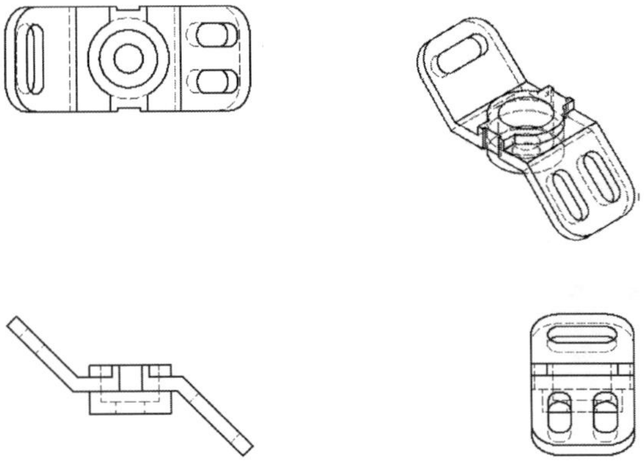

Figure 5.22

13 To move the views, hover the cursor near a border of a view until the move symbol appears, then click and drag the view until it is positioned. Reposition your drawing views as needed.

14 Edit the view by double-clicking in the isometric view.

15 In the Style area in the dialog box click the Shaded option as shown in Figure 5.23 left.

16 Click the Display Options tab and clear the check box for Tangent Edges as shown in the middle image.

17 Complete the edit by clicking OK.

18 The isometric view should resemble the image on the right.

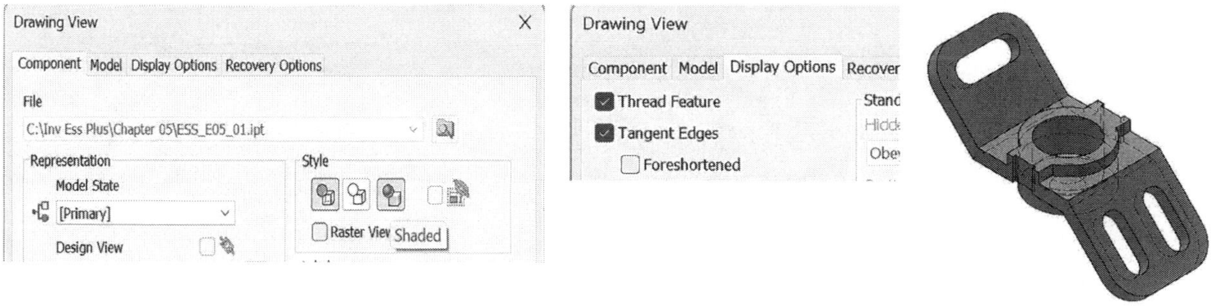

Figure 5.23

19 If desired practice editing and moving the drawing views.

20 Close all open files. Do not save changes.

End of exercise.

CREATING AUXILIARY VIEWS

An auxiliary view is a view that is projected perpendicular to a selected edge or line in a base view. It is designed primarily to view the true size and shape of a surface that appears foreshortened in other views.

To create an auxiliary drawing view, follow these steps:

1. Click the Auxiliary View command on the Place Views tab, as shown in Figure 5.24 left. You can also right-click inside the bounding area of an existing view, shown as a dotted box when the cursor moves into the view, and choose Auxiliary View from the marking menu as at right.

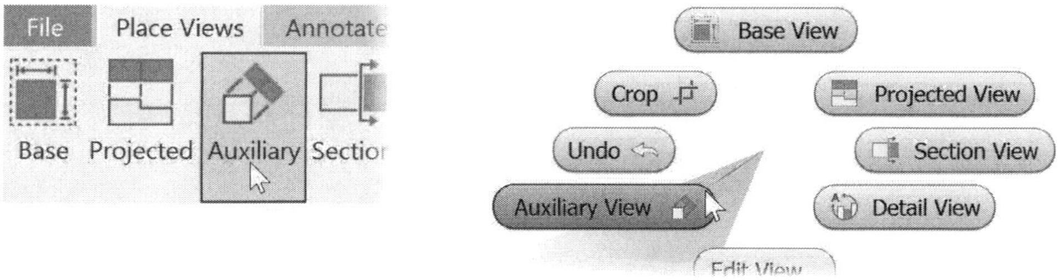

Figure 5.24

2. After using the Auxiliary View command from the Create panel, click inside the view from which the auxiliary view will be projected. The Auxiliary View dialog box will appear, as shown in Figure 5.25 left. Type in a name for View Identifier and a value for the Scale of the view, and then select one of the Style options: Hidden, Hidden Line Removed, or Shaded.
3. In the drawing view, select a linear edge or line from which the auxiliary view will be projected perpendicularly, as shown in Figure 5.25 right.

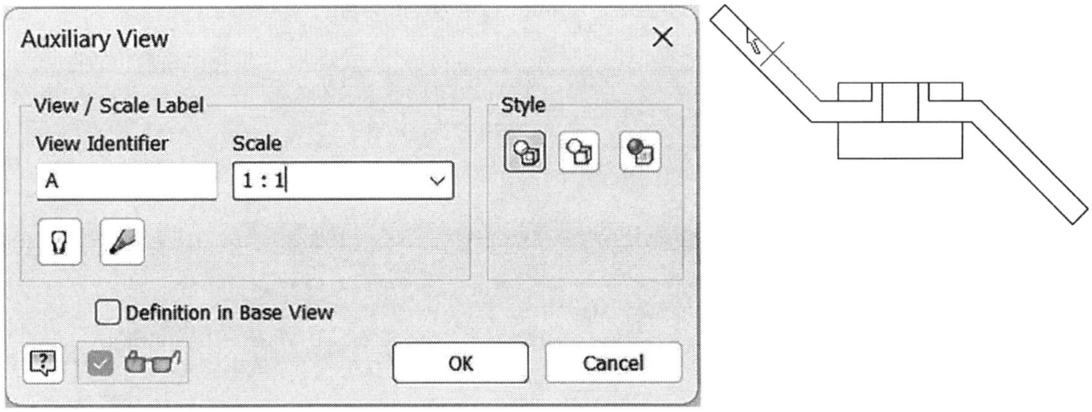

Figure 5.25

4. Move the cursor to position the auxiliary view, as shown in Figure 5.26 left.
5. Click a point on the drawing sheet to create the auxiliary view. The completed auxiliary view layout is shown in the image on the right.

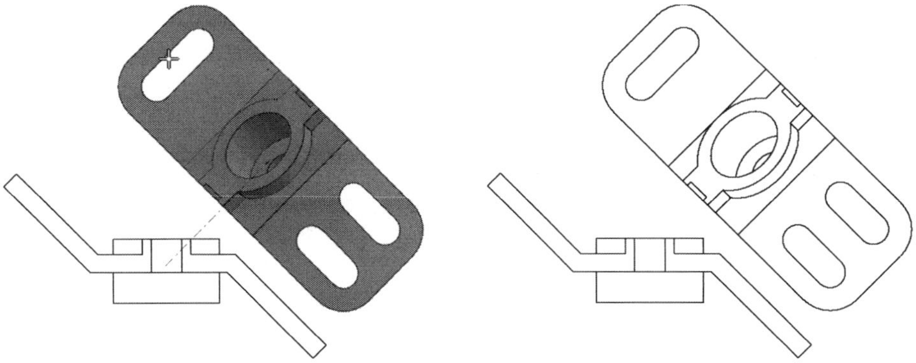

Figure 5.26

CREATING SECTION VIEWS

A section view is created by defining a section line or multiple section lines to represent the plane(s) cut through a part or assembly. The view shows the surface of the cut area and any geometry visible behind the cut face from the direction being viewed.

When defining a section, you sketch line segments that are horizontal, vertical, or at an angle. You cannot use arcs, splines, or circles to define section lines. When you sketch the section line(s), geometric constraints will automatically be applied between the line being sketched and the geometry in the drawing view. You can also infer points by moving the cursor over (or scrubbing) certain geometry locations, such as centers of arcs, endpoints of lines, and so on, and then moving the cursor away to display a dotted line showing that you are inferring or tracking that point.

To place a geometric constraint between the drawing view geometry and the section line, click in the drawing when a green circle appears; the glyph for the constraint will appear. If you do not want the section lines to have constraint(s) applied to them automatically when they are created, hold down the CTRL key when sketching the line(s). Because the area in the section view that is solid material appears with a hatch pattern by default, you may want to set the hatching style before creating a section view.

To create a section drawing view, follow these steps:

1. Click the Section View command on the Place Views tab > Create panel, as shown in Figure 5.27 left. You can also right-click inside the bounding area of an existing view, displayed as a dotted box when the cursor moves into the view. You can then click Section View from the marking menu as shown in the middle image.

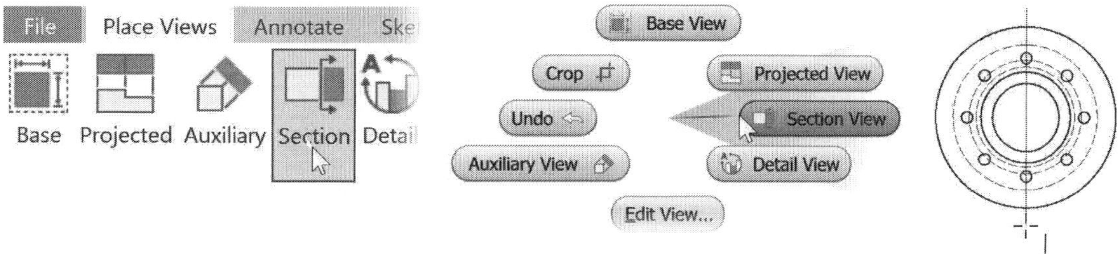

Figure 5.27

2. Click inside the view from which to create the section view to activate it. Sketch a line or lines that define where and how you want the view to be cut. In Figure 5.27 right, a vertical line is sketched through the center of the object. When sketching and a green dot appears, a coincident constraint will be applied.
3. When finished sketching the section line(s), right-click and select Continue from the marking menu as shown in Figure 5.28 left.
4. The Section View dialog box appears, as shown in Figure 5.28 right. Fill in the information for the label, scale, and style to appear in the drawing.

   When the Include Slice option is checked, a section view will be created with some components sliced and some components sectioned, depending on their browser attribute settings.

   Placing a check in the box next to Slice The Whole Part will override any browser component settings and will slice all parts in the view according to the Section line geometry. Components that are not crossed by the Section Line will not be included in the section operation.

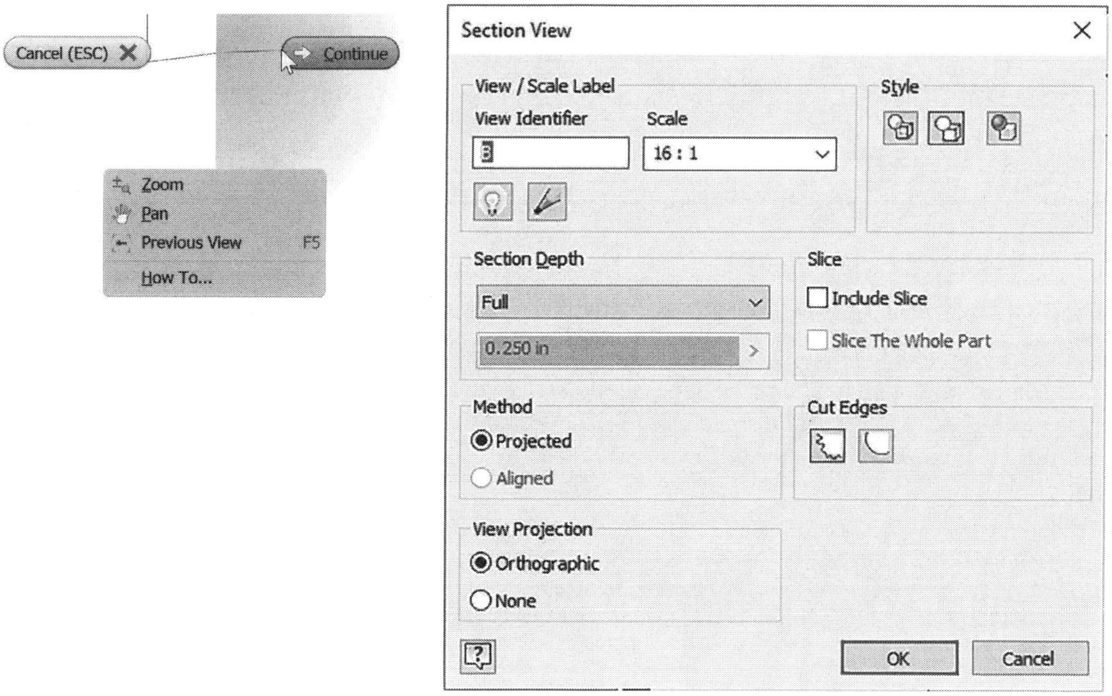

Figure 5.28

5. If the section line does not go entirely through the component(s) you can choose to have the section line be smooth (the default) or jagged by changing the Cut Edges option to Jagged. Figure 5.29 shows a section line that does not go through the part with examples of jagged on the left and smooth on the right.

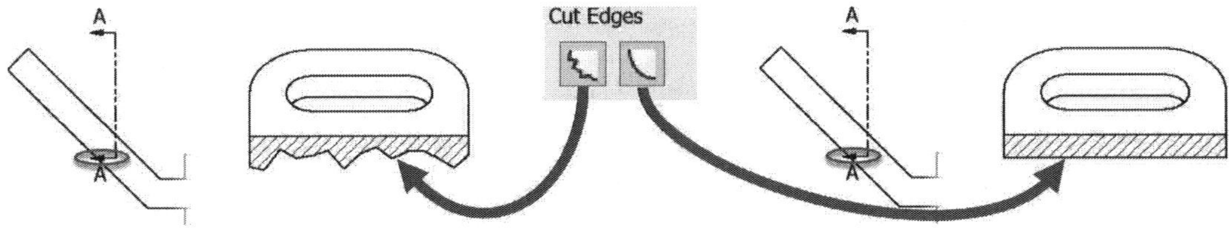

Figure 5.29

Modifying Section View and a Hatch Pattern

You edit the section view by right-clicking on a section line and click Edit Section Properties as shown in the following image on the left. In the Edit Section dialog box as shown in the image on the right, you can change the section depth, to include a slice and change the method to Projected or Aligned if the view supports it.

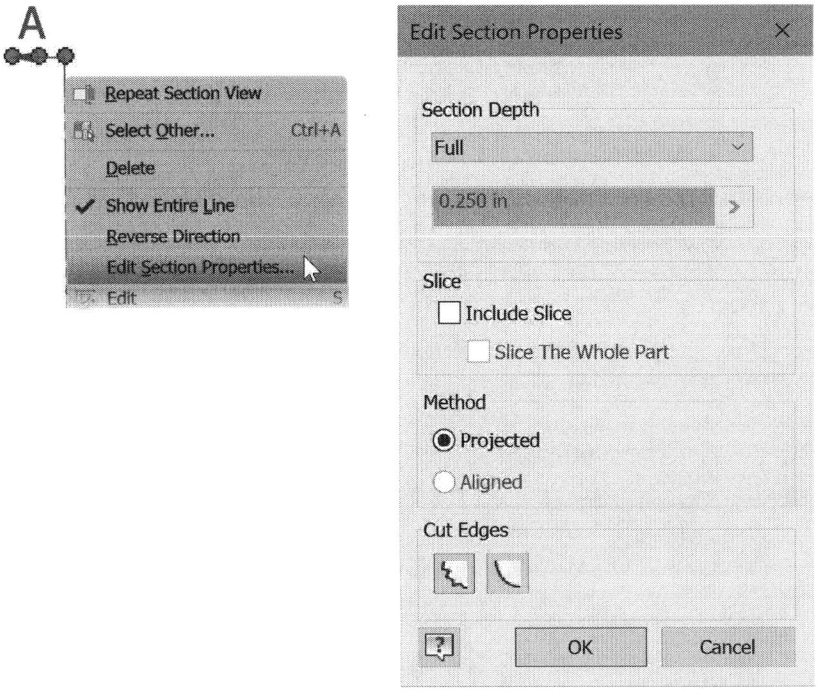

Figure 5.30

You can edit the hatch pattern by right-clicking on the hatch pattern in the section view and selecting Edit from the menu as shown in Figure 5.31 left. The Edit Hatch Pattern dialog box appears as shown in Figure 5.31 right.

Use the Edit Hatch Pattern dialog box to make the desired changes in the Pattern, Angle, Scale, Line Weight, Shift (shifts the hatch pattern a specified distance, but still stays within the section boundary), Double, and Color areas. Click the OK button to update the changes in the drawing.

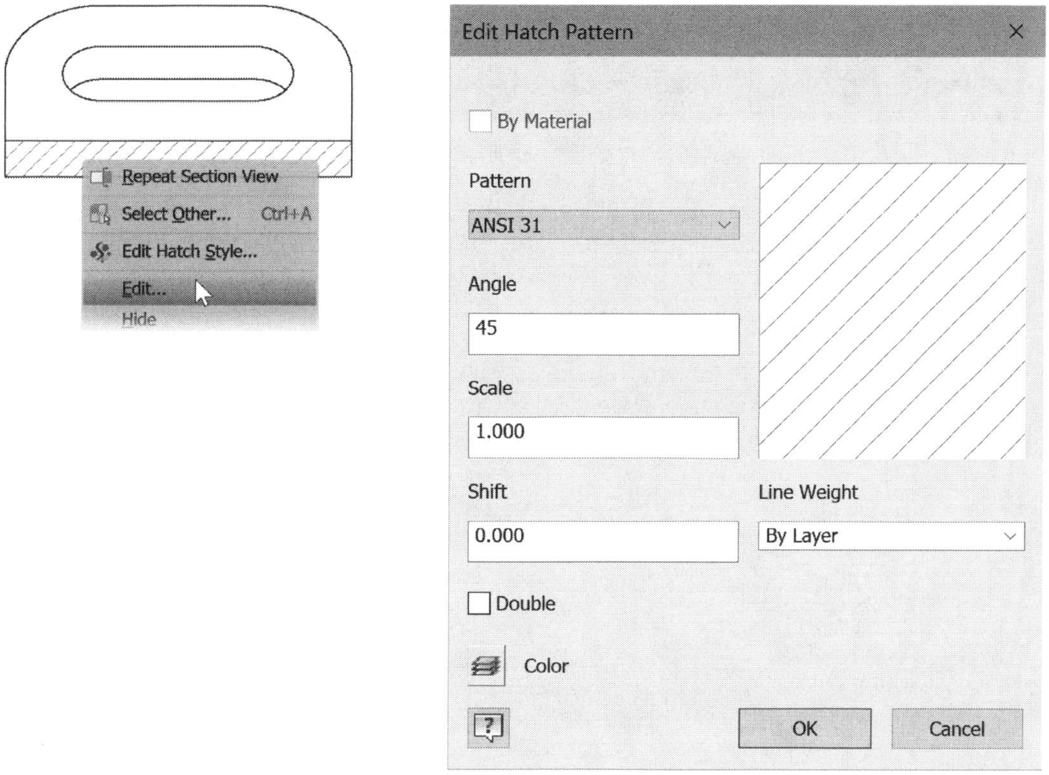

Figure 5.31

Hatching in Drawing Views

When an isometric view is projected from a section view, the hatch pattern displays. You can control the visibility of the hatch pattern in the section view or the projected isometric view by double-clicking in the view or right-clicking inside of the bounding box of the view, and choosing Edit View from the marking menu as shown in Figure 5.32.

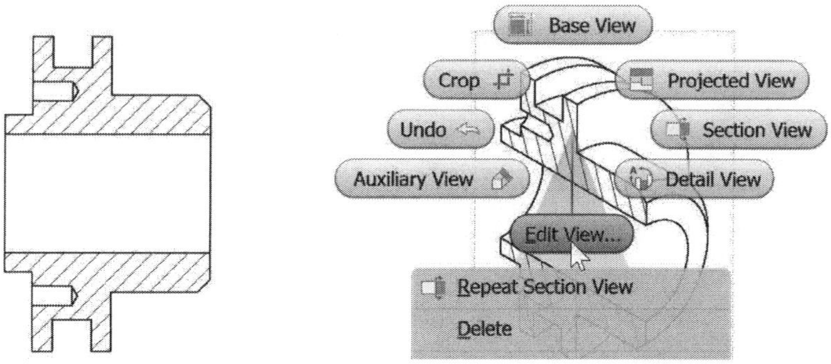

Figure 5.32

When the Drawing View dialog box appears, click on the Display Options tab. To see the hatch pattern, place a check in the box next to Hatching, as shown in the following image or uncheck it to turn off the hatch pattern.

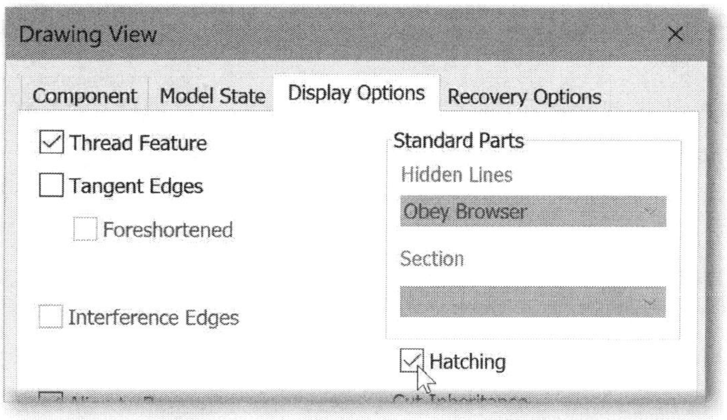

Figure 5.33

CREATING A DETAIL VIEW

A detail view isolates an area of an existing drawing view at a specified scale. You define a detail area by a circle or rectangle. Detail views can be placed anywhere on the sheet.

To create a detail drawing view, follow these steps:

1. Click Detail View from the Place Views tab > Create panel as shown in Figure 5.34 left. You can also right-click inside the bounding area of a view (shown as dashed lines when the cursor moves over it) then right-click and choose Detail View from the marking menu shown in the middle image.
2. With the Detail View command started, click inside the view from which to create a detail view.
3. The Detail View dialog box appears as shown in the image on the right. Fill in information for the label, scale, and style, and pick the desired view fence shape to define the view boundary.
   Do not click the OK button yet as you need to define the area the detail view is based on.

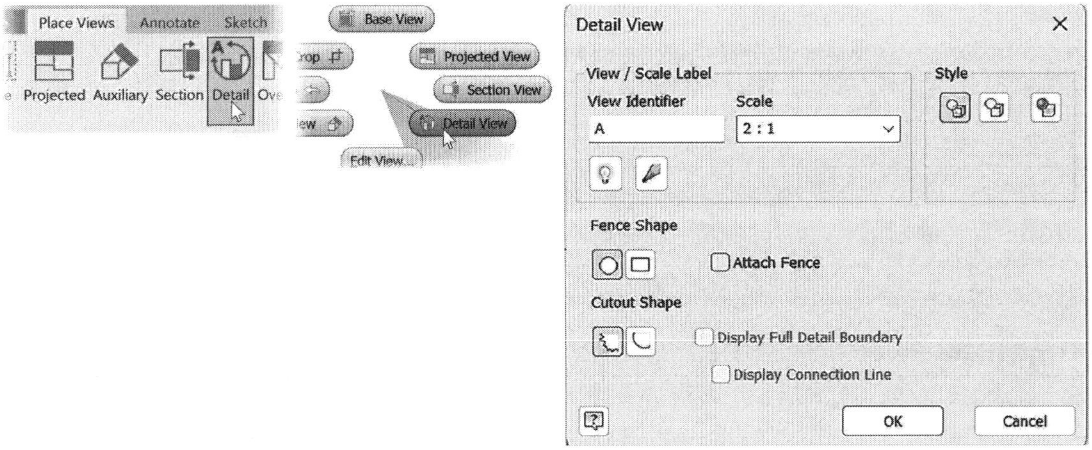

Figure 5.34

4. In the selected view, select a point to use as the center of the fence shape of the detail area. Figure 5.35 left shows the icon that appears when creating a circular fence. The image on the right shows the icon for creating a rectangular fence.

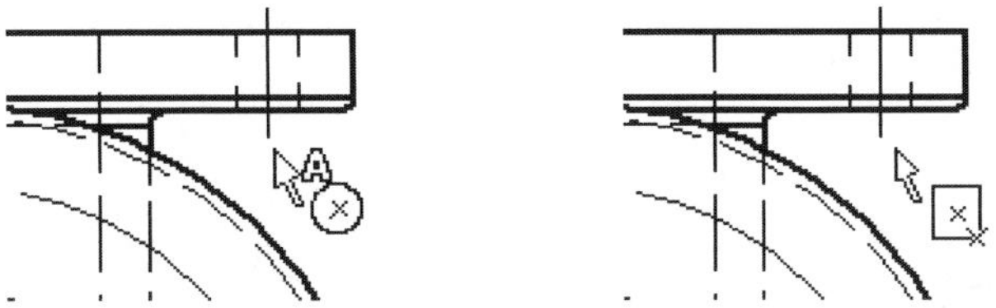

Figure 5.35

5. Click another point to define the radius of the detail circle, as shown in Figure 5.36 left, or the corner of the rectangle, as shown at right. As you move the cursor, a preview of the boundary will appear.

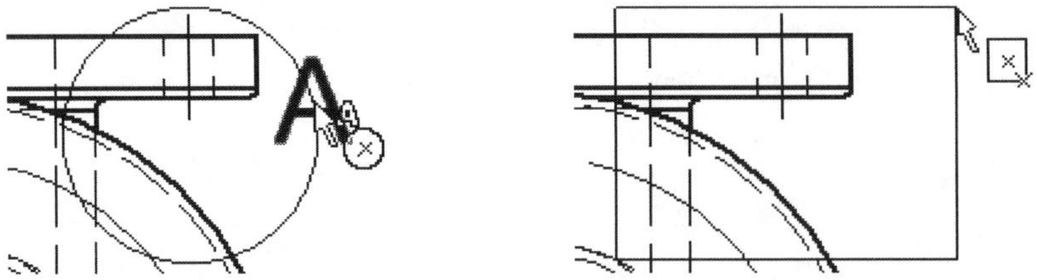

Figure 5.36

6. Select a point on the sheet to place the view. Figure 5.37 left shows the completed detail view based on a circular fence; a similar detail based on a rectangular fence is shown on the right.

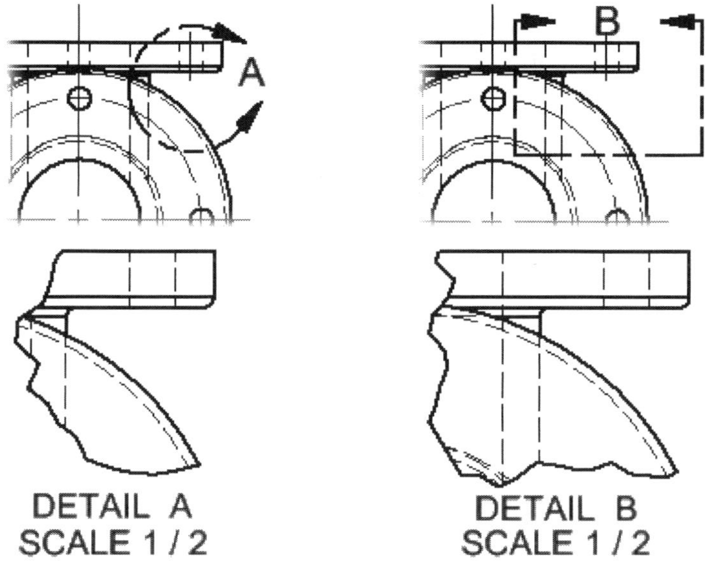

Figure 5.37

Modifying a Detail View

After a detail view is created, options are available to fine-tune the view. To access these options, right-click on the edge of the detail circle or rectangle and choose Edit Detail Properties, as shown in Figure 5.38. The detail view dialog box options are explained below.

- **Smooth Cutout Shape**. Placing a check in this box will affect actual detail view. The following image shows a smooth cutout shape instead of ragged.
- **Full Detail Boundary**. This option displays a full detail boundary in the detail view when checked, as shown in the following image.
- **Connection Line**. This option will create a connection line between the main drawing and the detail view, as shown in the following image.

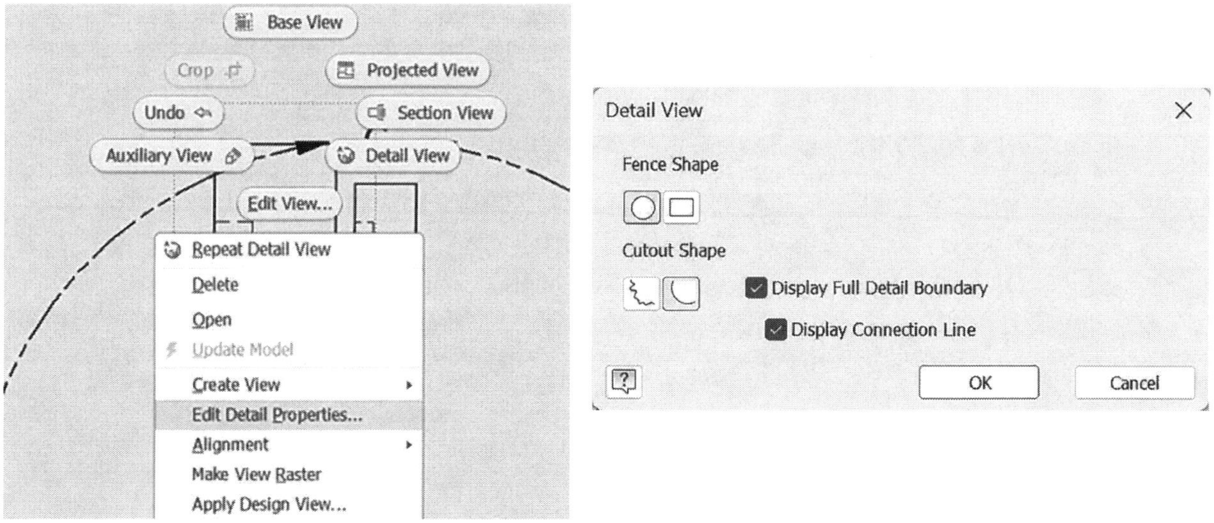

Figure 5.38

Once a connection line is created in a detail view, you can add a new vertex by right-clicking the detail boundary or connection line and clicking Add Vertex from the menu. Picking a point on the connection line allows you to add the vertex and then move it to a new position. You can remove a vertex by right-clicking it and choosing Delete Vertex from the menu.

EXERCISE 5-2: CREATING AUXILIARY, SECTION, AND DETAIL VIEWS

In this exercise, you create a variety of drawing views from a model of a cover.

1. Open the file *ESS_E05_02.idw* from the Chapter 05 folder. Zoom in on the top view.
2. First create an auxiliary view. Click Auxiliary View from the Place Views tab > Create panel.
    a. In the graphics window, click in the top view. Next, select the left-outside angled line, as shown in Figure 5.39 left.
    b. Click in zone B3 to define the projection direction and place the view to the upper-left of the front view as shown at right.

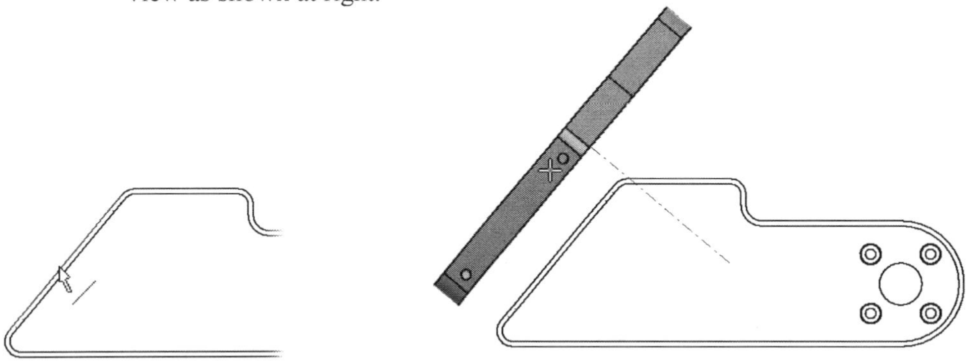

Figure 5.39

3. In the graphics window, select and drag on the border of the views to better fit them in the drawing.
4. Next create a section view. Click the Section View command on the Place Views tab > Create panel.
    a. In the graphics window, click in the top view.
    b. To define the first point of the section line, hover the cursor over the center point of the center hole until a green dot appears (do not click), move the cursor up and click a point directly above the hole (the dotted lines represent the inferred vertical point), as shown in Figure 5.40 left.
    c. Click a point below the geometry to create a vertical section line.
    d. Right-click and click Continue from the marking menu.
    e. In the Section View dialog box, change the View Identifier to A.
    f. Place the section view by clicking a point to the right of the top view, as shown in the middle image.
    g. If the section view does not show the top portion of the part, drag the endpoints of the section line in the top view so it is above the view; your view should resemble the image at the right.
    h. If the section view label (description) does not appear below the section view, edit the section view and in the Style area click on the .

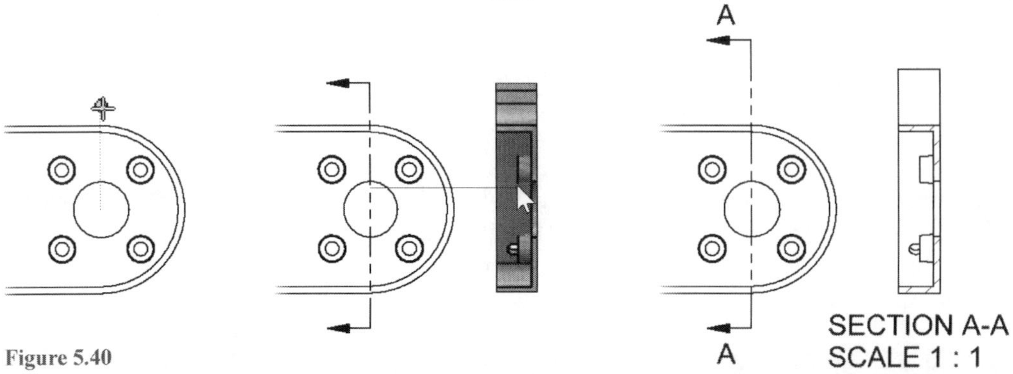

Figure 5.40

5 Drag the section line in the top view from left to right. If the section line can move, follow these steps.

a. Right-click on the section line and click Edit from the menu.

b. Use Project Geometry to project the large circle that is concentric to the right side of the part.

c. Apply a coincident constraint between the section line and the center point of the projected circle.

d. Click the Finish Sketch command from the Exit panel.

TIP: If the section sketch line cannot move and you need to reposition it, edit the section line as described in the previous step 5, but delete the constraint holding the line in place.

6 If desired, project an isometric view from the section view.

7 Finally, create a detail view. Click in the top view and click the Detail View from the marking menu.

a. Select a point near the lower-left of the top view to define the center of the detail view's circular boundary.

b. If needed, in the Drawing View dialog box change the Style Label to B.

c. Drag the circular boundary to include the entire lower-left corner of the view geometry.

d. Click to position the detail view below the break view. The image in the middle shows the completed detail in the top view and the image on the right shows the completed detail view.

TIP: Your view identifier (letter) may be different than what is shown.

8 Reposition the drawing views so they resemble the following image.

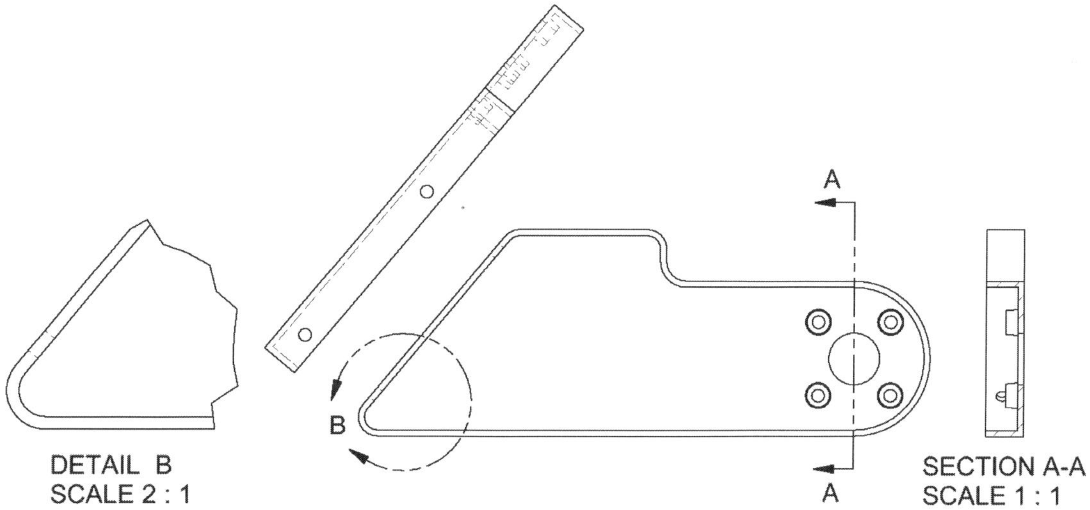

Figure 5.41

9 Close all open files. Do not save changes.

End of exercise.

CREATING BREAK VIEWS

When creating drawing views of long parts, you may want to remove a section or multiple sections from the middle of the part in a drawing view and show just the ends. This is referred to as a "break view." For example, to create a drawing view of a 2" x 2" x ¼" angle, that is 48 inches long and has chamfered ends, the end detail is small and relative to the part length (Figure 5.42). This makes it hard to see and to place dimensions.

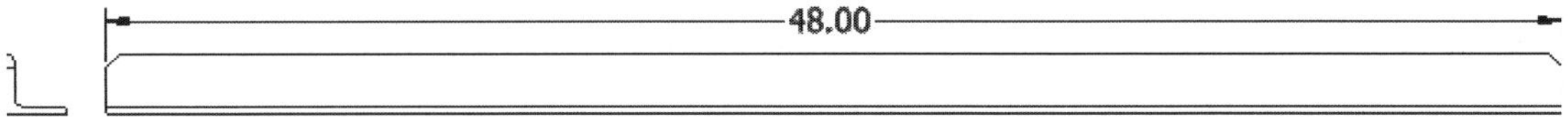

Figure 5.42

You can create a break view that removes the middle of section and leaves each end. When you place an overall length dimension that spans the break, it appears as the correct 48 inches. The dimension line shows a break symbol to note that it is based on a break view, as shown in Figure 5.43.

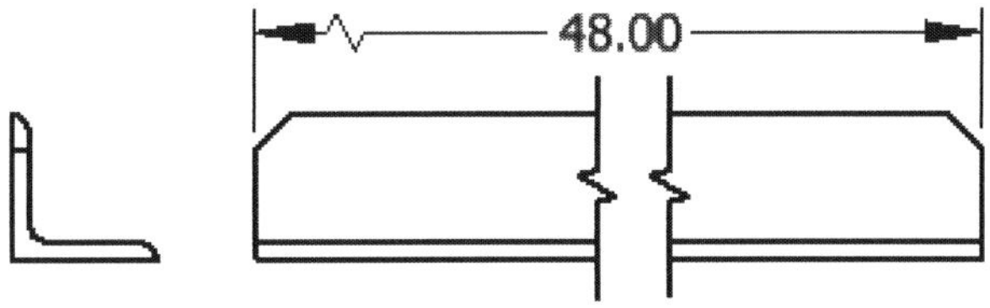

Figure 5.43

You can add as many break views to an existing drawing view as needed. The following view types can be changed into break views: part, assembly, projected, isometric, auxiliary, section, break out, and detail views. After creating a break view, you can move the breaks dynamically to change what shows in the broken view.

To create a break view, follow these steps:

1. Create a base or projected view or one that will eventually be shown as broken.
2. Click the Break command on the Place Views tab > Modify panel, as shown in the following image on the left.
3. If you selected the Break command on the Place Views tab > Modify panel, click inside the view that you want to break.
4. The Break dialog box will appear, as shown in the following image on the right. Do not click the OK button now, as this will end the command.

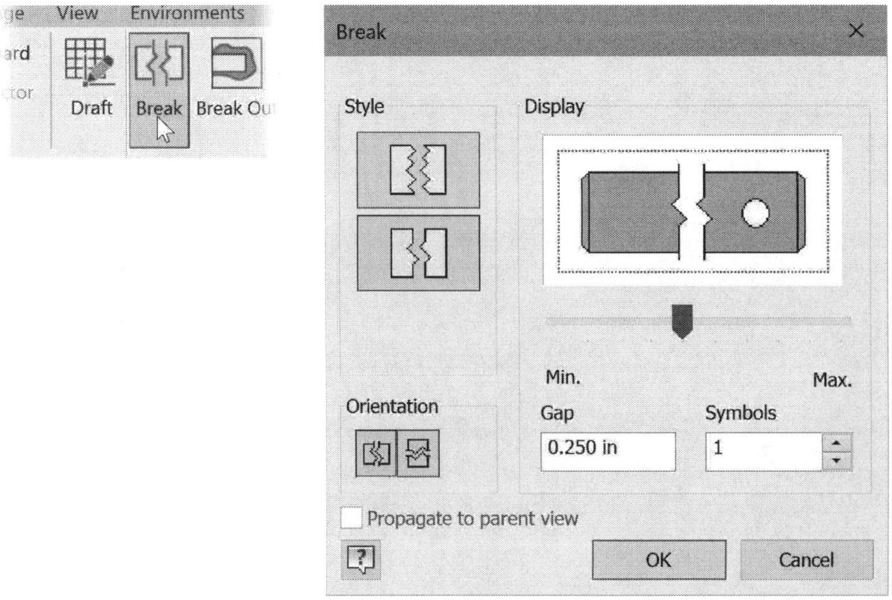

Figure 5.44

5. From the drawing view that will be broken, select a point where the break will begin, as shown in Figure 5.45.

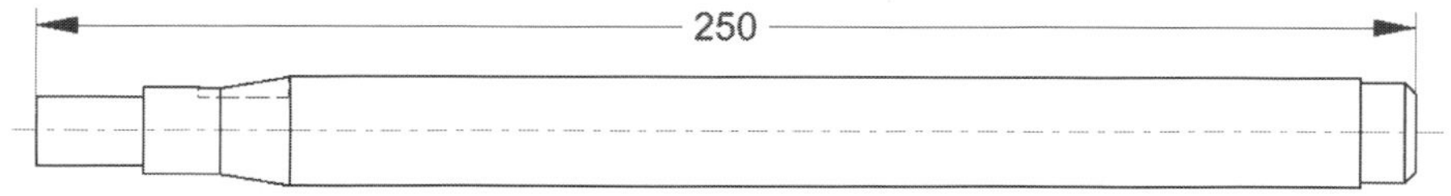

Figure 5.45

6. Select a second point to locate the second break, as shown in Figure 5.46. As you move the cursor, a preview image will appear to show the placement of the second break line.

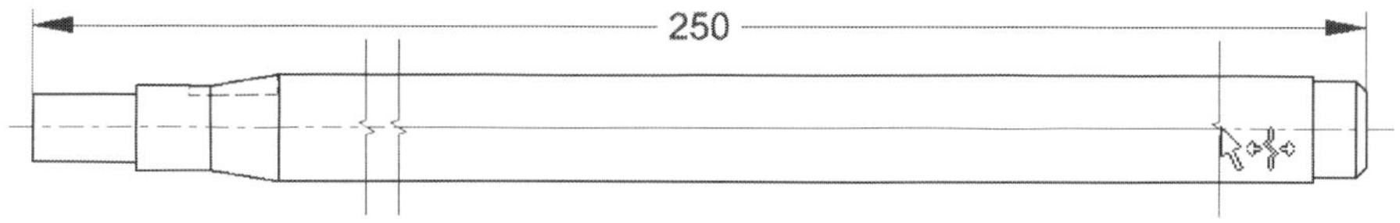

Figure 5.46

7. Figure 5.47 illustrates the result of creating a broken view with a dimension added. Notice that the dimension line appears with a break symbol to signify that the view is broken.

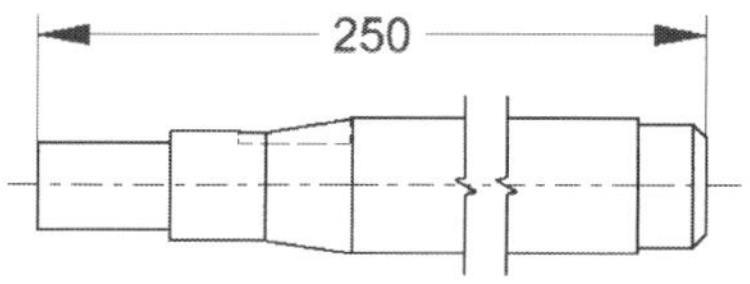

Figure 5.47

8. To edit the properties of the break view, move the cursor over the break lines, and a green circle will appear in the middle of the break. Right-click and select Edit Break from the menu. The same Break dialog box will appear. Edit the data as needed, and click the OK button to complete the edit.

9. To move the break lines, click on one of the break lines and, with the left mouse button pressed down, drag the break line to a new location, as shown in the following image. Drag the cursor away from the other break line to reduce the amount of geometry that is displayed, and drag the cursor into the other break line to increase the amount of geometry that is displayed. In either case the other break line will follow to maintain the gap size.

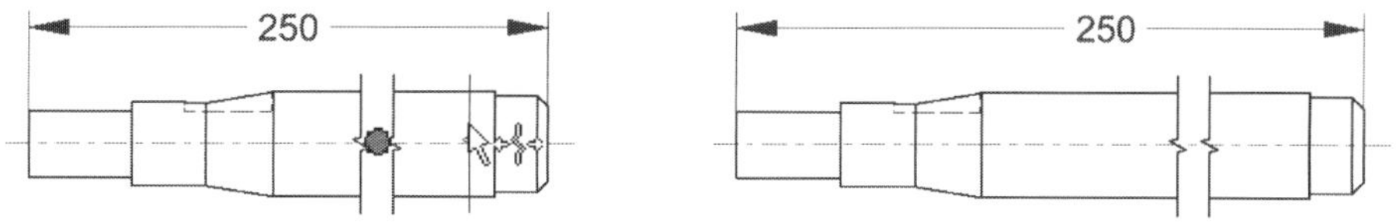

Figure 5.48

EXERCISE 5-3: CREATING BREAK VIEW

In this exercise, you create a break view and a break out view.

1 Open the file *ESS_E05_03.idw* in the Chapter 05 folder. This file has a base view and a side view that is too long to fit in the drawing.

2 Start the Break command from the Place Views tab > Modify panel.

3 In the graphics window, click in the right-side view to specify that it is the view that you want to break.

4 With the Break dialog box open, you define the area to remove. Move the cursor to the right of the left hole and click, and move the cursor to the left of the hole on the right side of the part and click. The following image shows the selections.

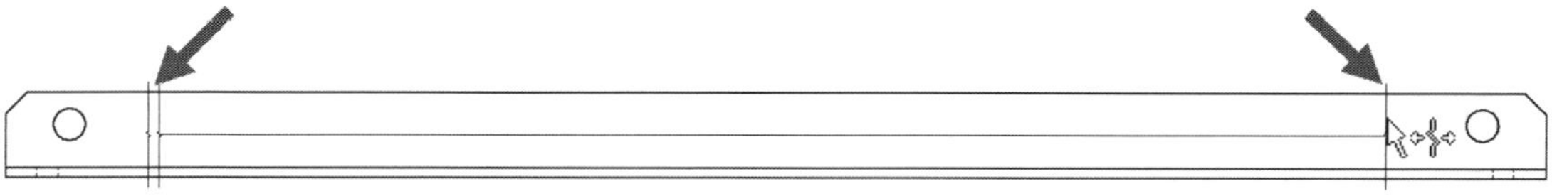

Figure 5.49

5 Move the right-side view so it is next to the front view as shown in the following image.

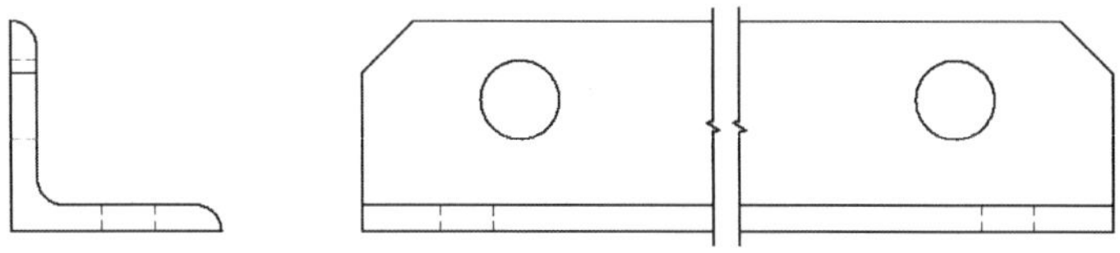

Figure 5.50

6 To edit the properties of the break view, move the cursor between the break lines, and a green circle will appear in the middle of the break. Right-click and select Edit Break from the menu as shown in the following image on the left.

7 Next you increase the size of the break line. In the Break dialog box move the Display slider to about 3/4, labeled (1) in the middle image.

8 Change the Gap between the break lines. In the Break dialog box change the gap to **0.50 inches** labeled (2) in the middle image. Click the OK button to complete the edit.

9 To move the break lines, click on one of the break lines and with the left mouse button pressed down, drag the break line to a new location. Drag the cursor away from the other break line to reduce the amount of geometry that is displayed, and drag the cursor into the other break line to increase the amount of geometry that is displayed. In either case the other break line will follow to maintain the gap size.

10 Next you will create a dimension. Press the D key on the keyboard and place the dimension on the two outside points. Click OK in the Edit Dimension dialog box. Notice the break symbol in the dimension as shown in the image on the right. Dimensioning a drawing will be covered later in this chapter.

11 When done, your screen should resemble the image on the right.

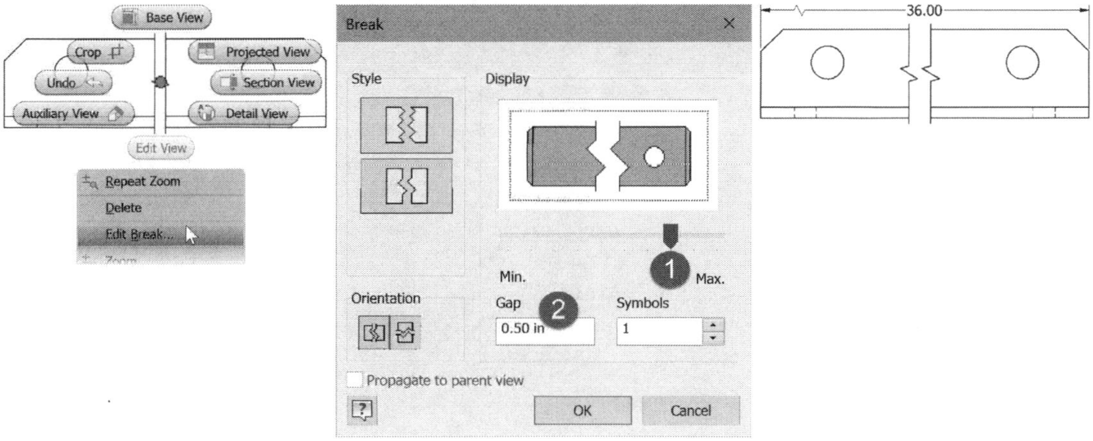

Figure 5.51

12 Close all open files. Do not save changes.

End of exercise.

EDITING DRAWING VIEWS

You may need to edit the properties of, delete a drawing view, or replace the component that a drawing references. The following sections discuss these options.

Editing Drawing View Properties

After creating a drawing view, you may need to change the label, scale, style, or hatching visibility, break the alignment constraint to its base view, or control the visibility of the view projection lines for a section or auxiliary view. To edit a drawing view, follow one of these steps:

- Double-click on or inside the bounding area of the view.
- Double-click on the view you want to edit in the browser.
- Right-click in the drawing view's bounding area or on its name, and select Edit View from the marking menu, as shown in the following image.

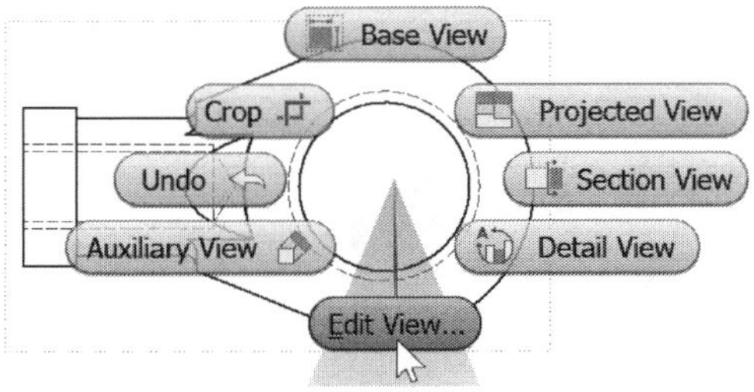

Figure 5.52

When you perform the operations listed above, the Drawing View dialog box and ViewCube appear as shown in Figure 5.53. The dialog box and ViewCube are the same as when the view was created. Make the necessary changes, and click the OK button to complete the edit. When you change the scale in the base view, all the dependent views will be scaled as well. You can also change the orientation of the base view; dependent views will then be updated to reflect the new orientation.

Just like when you created the base view, while editing it you can also create additional views by clicking and dragging on one of the arrows ▲ that the new view will be based on. See the example shown in Figure 5.53.

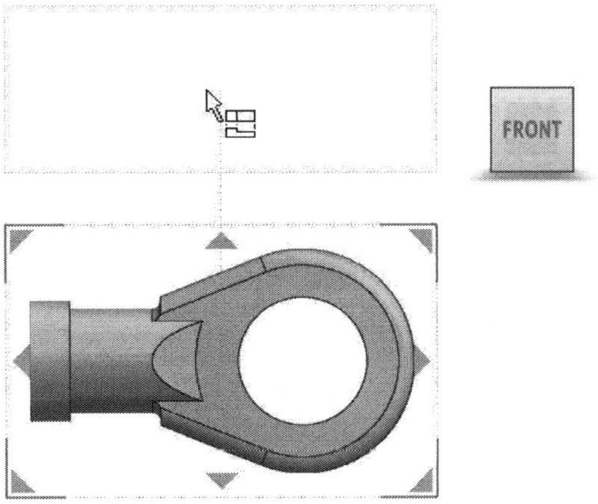

Figure 5.53

You can also change the scale of a base or isometric view while editing by clicking and dragging a corner of the view in the graphics window. 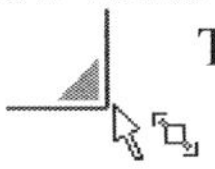 To freely scale the view, hold down the CTRL key while dragging.

Deleting Drawing Views

To delete a drawing view, either right-click in the bounding area of the drawing view or on its name in the browser and select Delete from the menu. You can also click in the bounding area of the drawing view, and press the Delete key on the keyboard. A dialog box appears, prompting to confirm the deletion of the view. If the selected view has a view that is dependent on it, you will be asked if the dependent views should also be deleted. By default, the dependent view(s) will be deleted. To exclude a dependent view from the delete operation, expand the dialog box by clicking on the >> button, and click on the Delete cell on the row of the view to change the option to No as shown in Figure 5.54.

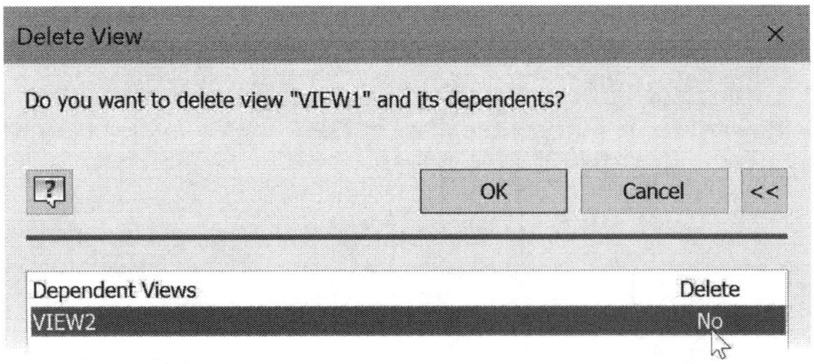

Figure 5.54

Break and Align Views

After creating or deleting drawing views you may need to move a view to a different location on the drawing, but the view can only move orthographically to its parent view. To break this relationship, click the Break Alignment command on the Place Views tab > Modify panel as shown in Figure 5.55.

To align a view to another view, follow the same process but from the drop-down list under Break Alignment click Horizontal to align a view horizontally to a selected view, Vertical to align a view vertically to a selected view and In Position to keep a view near its current position in relation to a selected view.

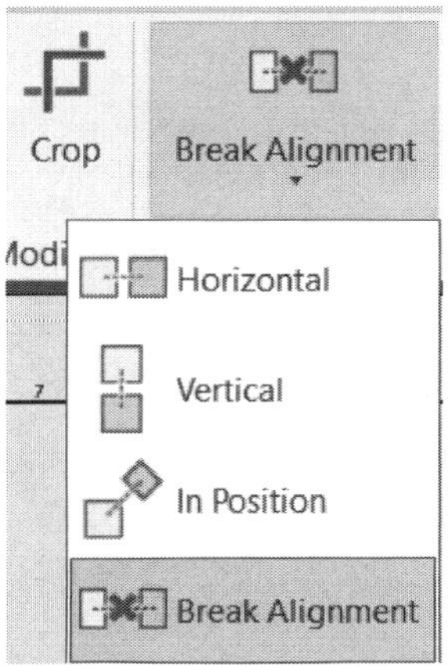

Figure 5.55

Replace Model Reference

While annotating a drawing you may want to change the component that the drawing references for the geometry. Parts, assemblies, and presentation files can be replaced, but they can only be replaced with the same type of file. The file replacing the existing file should be similar in shape and size, otherwise dimensions may be orphaned. To replace a model that a drawing view references, follow these steps:

1. Click the Replace Model Reference command on the Manage tab > Modify panel as shown in the following image on the left.
2. The Replace Model Reference dialog box will appear as shown in the following image on the right. All the files that are referenced in the drawing will appear in the list. In the dialog box click on the file to replace.
3. In the dialog box click the Select new model button.
4. Browse to the new file and select it.
5. In the Open dialog box click Open.
6. Click Yes in the warning dialog box to replace the model.
7. In the Replace Model Reference dialog box click OK to complete the command.
8. Clean up the drawing annotations and the view orientations as needed.

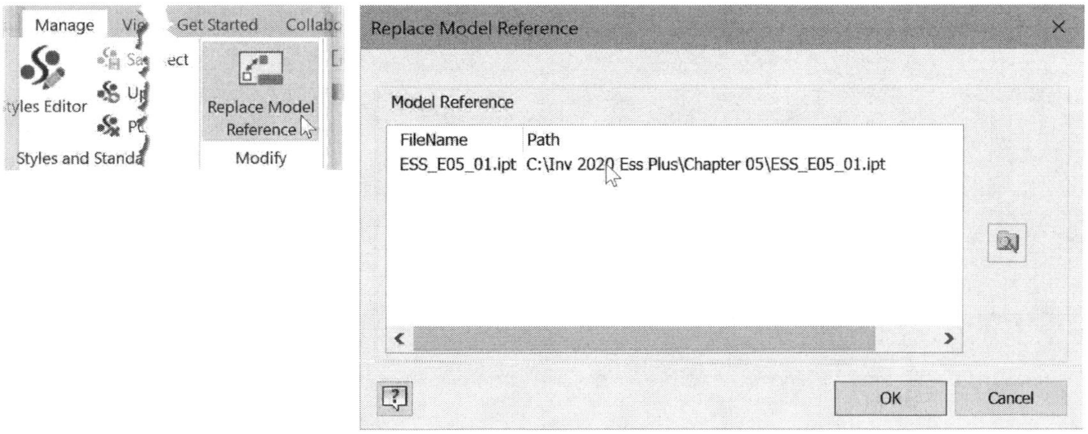

Figure 5.56

EXERCISE 5-4: EDITING DRAWING VIEWS

In this exercise, you will delete the base view while retaining its dependent views. The base view is not required to document the part, but its dependent views are. Next, you align the section view with the right-side view to maintain the proper orthographic relationship between the views. You also modify the hatch pattern of the section to better represent the material.

1 Open the drawing file *ESS_E05_04.idw* in the Chapter 05 folder. This drawing contains three orthographic views, an isometric view, and a section view.

2 First you delete a view. Right-click on the front view (the middle view on the left side of the drawing; this is also the base view as shown in the following image), and click Delete from the menu as shown Figure 5.57 left.

a. In the Delete View dialog box, notice that the projected views are highlighted on the drawing sheet. Select the More >> button in the dialog box.

b. Click on the word Yes in the Delete column for each dependent view to toggle each to No as shown in the following image on the right. This will delete the base view but keep the dependent views present on the drawing sheet.

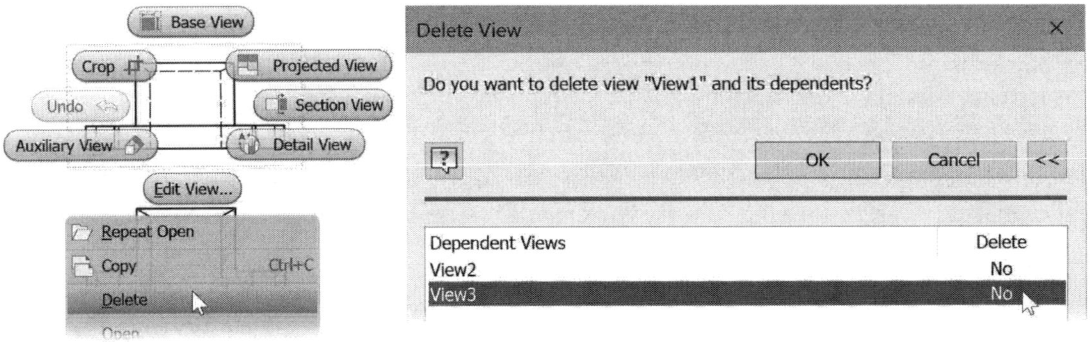

Figure 5.57

c. In the dialog box click OK to delete the base view and retain the two dependent views. Your drawing views should appear like the following image.

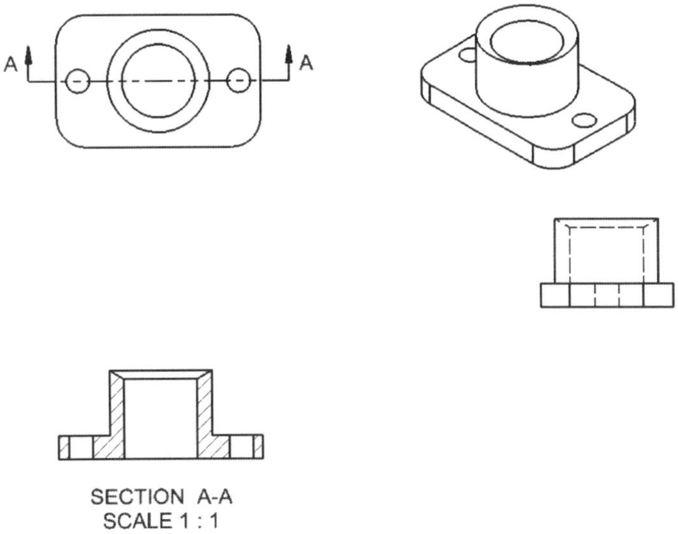

Figure 5.58

3 With the front view deleted, the section view needs to be aligned with the right-side view.

a. From the Place Views tab > Modify panel, click the down arrow under the Break Alignment command and click Horizontal from the drop list.

b. Select the right-side view as the view to align and then click the section view to align it to.

4 Move the new aligned views. Select on the border of the section view, and drag the section view up to the area previously occupied by the front view. Notice how the right-side view automatically aligns horizontally to the section view.

5 Move the isometric view by clicking and dragging the isometric view and the section view and notice the isometric view moves independently of the base view.

6 Next, align the section and isometric view.

a. From the Place Views tab > Modify panel, click the down arrow under the Break Alignment command and click In Position.

b. In the graphics window, select the isometric view as the view to reposition and then click the section view as the base view.

7 Move the section view, and notice that the isometric view now moves with the section view. Move the isometric view and notice that the view can only move along the same angle. Your drawing should resemble the following image.

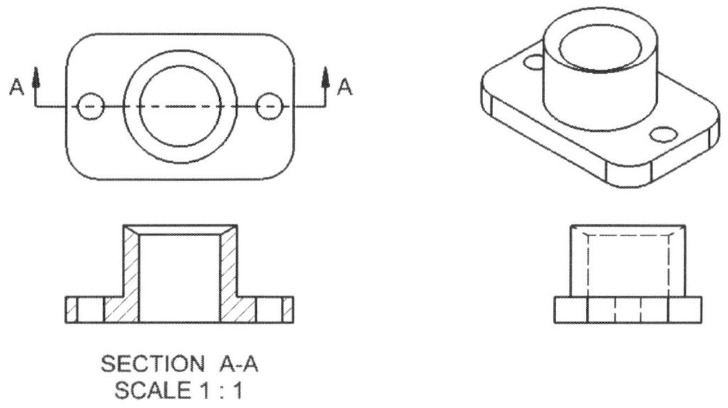

Figure 5.59

8 Next edit the section view hatch pattern to be bronze using the ANSI 33 hatch pattern.

a. Right-click on the hatch pattern in the section view, and click Edit from the menu as shown in the following image on the left.

b. The Edit Hatch Pattern dialog box is displayed. Select ANSI 33 from the Pattern list, and click OK in the Edit Hatch Pattern dialog box. The hatch pattern changes, as shown in Figure 5.60 right.

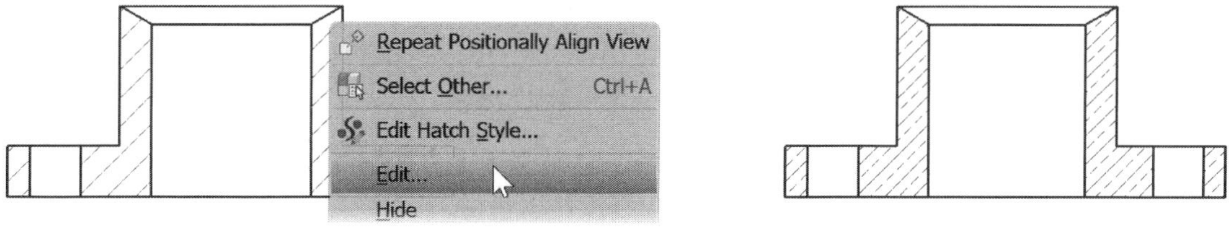

Figure 5.60

9 Lastly, you replace the part with a similar part but the holes are in different locations.

a. Click the Replace Model Reference command on the Manage tab > Modify panel.

b. In the Replace Model Reference dialog box click on the only file in the list.

c. In the Replace Model Reference dialog box click the Select new model button, and browse to the Chapter 05 folder and select the file *ESS_E05_04-Replace.ipt* and then click Open in the dialog box.

d. In the warning dialog box click Yes to replace the model.

e. In the Replace Model Reference dialog box click OK to complete the command.

10 Notice that the views were updated to reflect the new part that has four holes centered on the fillets and notice the hatch pattern and view alignment was maintained as shown in the following image.

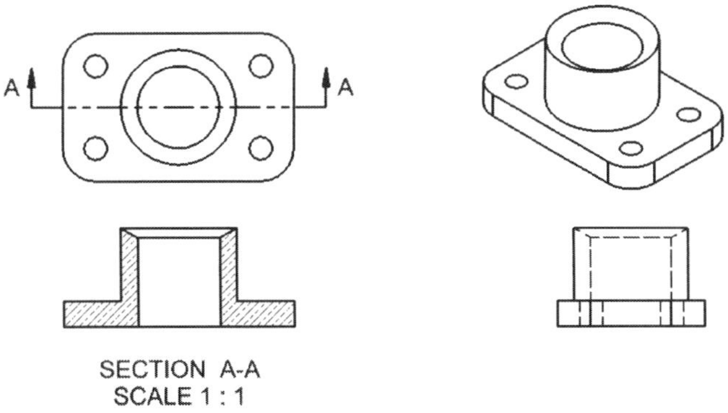

Figure 5.61

11 Close all open files. Do not save changes.

End of exercise.

ANNOTATIONS

To complete an engineering drawing, you must add annotations such as centerlines, dimensions, surface texture symbols, welding symbols, geometric tolerance symbols, text, bills of materials, and balloons. How these objects appear in the drawing is defined by styles. Styles are covered next.

Drawing Standards and Styles

Drawing standards control overall appearance and styles available in a drawing. Only one standard can be active at a time. Inventor uses styles to control how objects such as dimensions, centerlines, hole tables, and text appear. Styles are saved within an Inventor file or to a project library location. They can also be saved to a network location so that many users can access the same styles.

Once a style is used in a document, it is stored in the document. Changes made to the style are only saved in that document. The updated style library can be saved to the main style library but requires permission set in the local project file and rights of the folder that the styles are saved in. To help maintain standards, it is recommended that only a limited number of people have access to write to the style library. For more information on how to save styles to a style library consult the help system topic "Save styles to a style library."

Style Name/Value

Inventor uses the style name as its unique identifier. Only one name for the same style type can exist. For example, in a drawing, only one-dimension style with the name "Default ANSI" can exist. An object can only be associated to one style.

Editing a Style

After dimensions have been placed, you may want to change how all the dimensions appear. Instead of changing each dimension, you can alter the dimension style or create a new style. The easiest way to edit a dimension style is to move the cursor over a dimension whose style you want to change and right-click to choose Edit Dimension Style from the menu as shown in Figure 5.62.

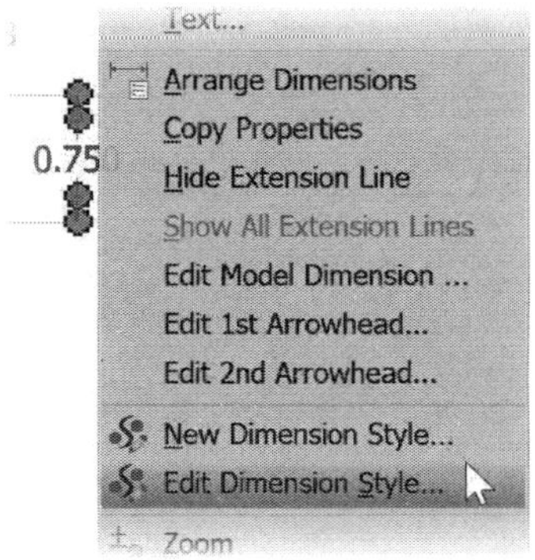

Figure 5.62

The Style and Standard Editor dialog box appears as shown in Figure 5.63. Make changes to the style as needed and then click the Save button on the top of the dialog box.

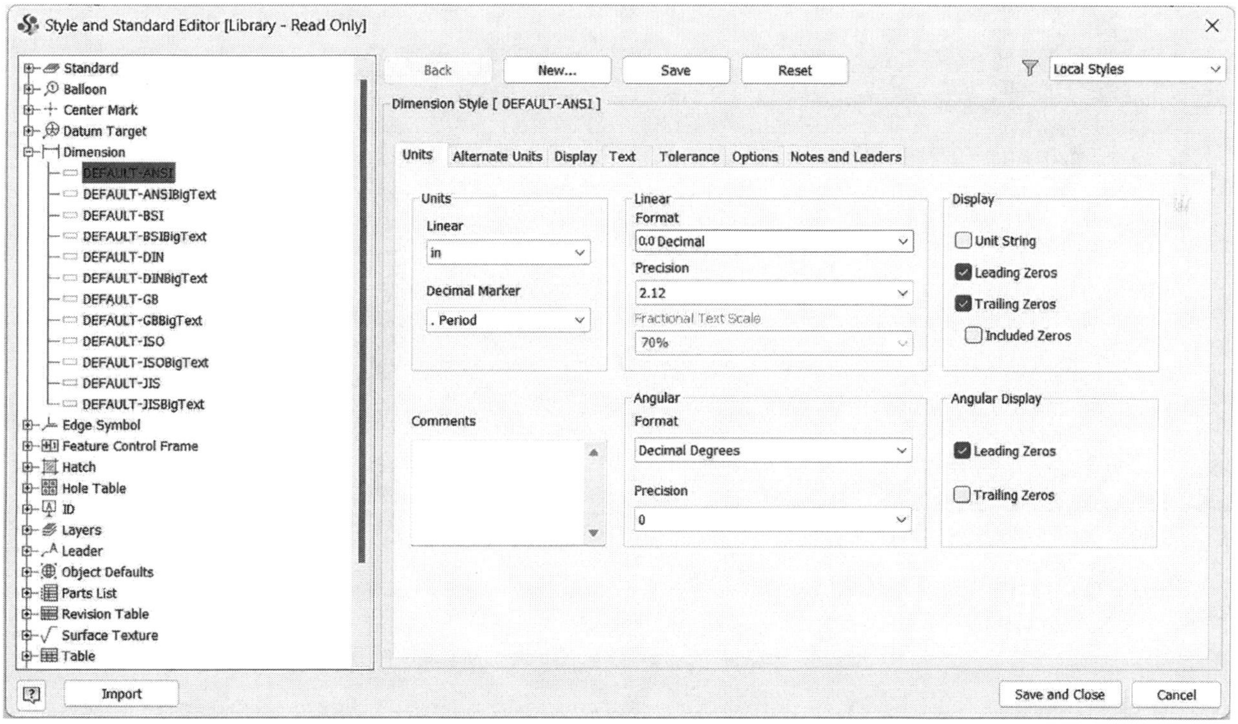

Figure 5.63

Creating a Style

Instead of editing a current dimension style, you can create a new style based on an existing style. To create a new style based on an existing style, follow these steps:

1. Click Styles Editor from the Manage tab > Styles and Standards panel.
2. Click and expand the style section for which you want to create a new style, for example Dimension.
3. Right-click the style on which the new style will be based, and select New Style, as shown in the following image on the left. This image shows a new style being created from the Default (ANSI) dimension style.
4. The New Local Style dialog box appears. Enter a style name, as shown in Figure 5.64 right. The Add to Standard option adds the new style to the available styles list in the active standard style.

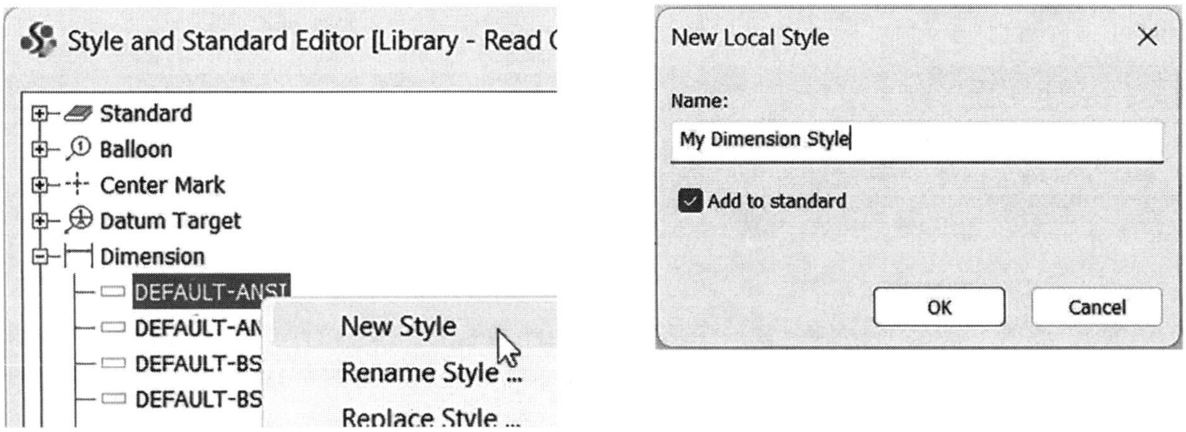

Figure 5.64

5. Make changes to the style, and save the changes.
6. To change existing objects to a different style, select the object or objects whose style you want to change (selected objects need to be the same object type, i.e., all dimensions), and then select a style from the Style area on the Annotate tab > Format panel as shown in the following image.

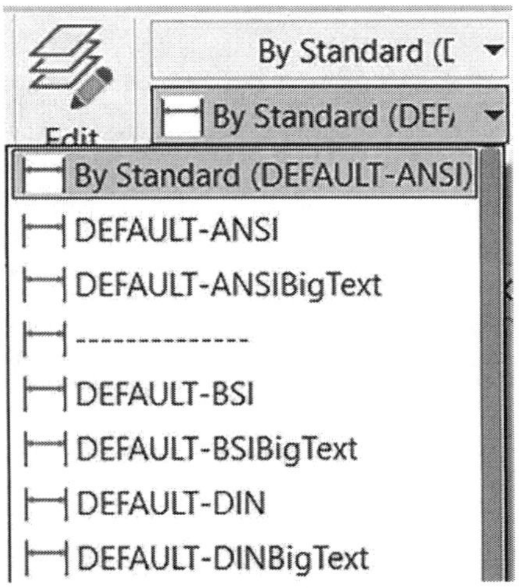

Figure 5.65

TIP: Styles will be covered through the remaining exercises in this chapter.

Centerlines and Center Marks

Four methods allow you to construct centerlines noting the centers of holes, circular edges, or the middle (center axis) of two lines. Use the Center Mark, Centerline, Centerline Bisector, and Centered Pattern commands from the Annotate tab, as shown in Figure 5.66. The centerlines are associated to the geometry that you select when you create them. If the geometry changes or moves, the centerlines update automatically to reflect the change.

Figure 5.66

Centerline

To add a centerline, follow these steps and refer to the following image:

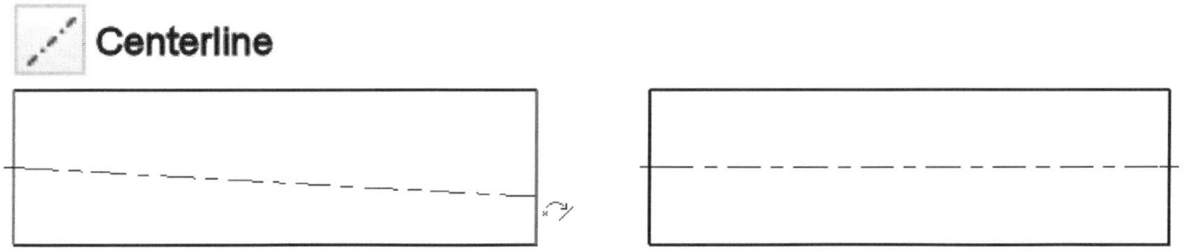

Figure 5.67

1. Choose the Centerline command from the Annotate tab > Symbols panel.
2. In the graphics window, select a geometric object for the start of the centerline.
3. Click a second geometric object for the ending location. You can also select multiple circles or points. If multiple points or circles are selected whose center points fall on a circle, a circle with the centerline line type (often referred to as a bolt circle) will be created.
4. Continue placing centerlines by selecting geometry.
5. Right-click and click Create from the menu to create the centerline. The centerline will be attached to the midpoints of selected edges or the center point of selected arcs and circles.

Centerline Bisector

To add a centerline bisector, follow the steps and refer to the following image:

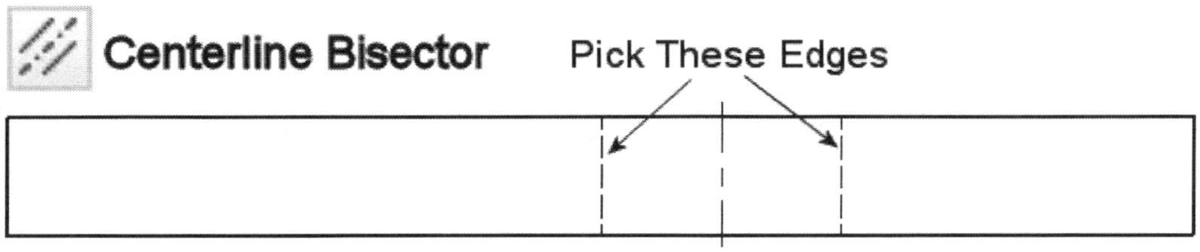

Figure 5.68

1. Click the Centerline Bisector command on the Annotate tab > Symbols panel.
2. In the graphics window, select two lines between which you want to place the centerline bisector. The lines do not need to be parallel to each other.
3. Continue placing centerline bisectors by selecting geometry.

Center Mark

To add a center mark, follow the steps and refer to the following image:

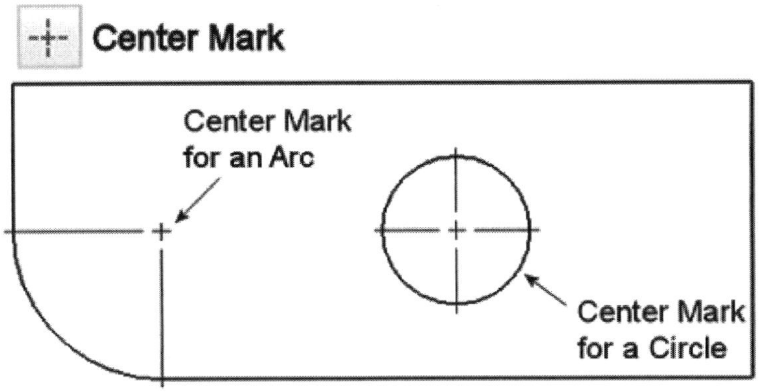

Figure 5.69

1. Click the Center Mark command on the Annotate tab > Symbols panel.
2. In the graphics window, select a center point of a circle or arc or select a circle or arc geometry in which you want to place a center mark.
3. Continue placing center marks by selecting arcs and circles.

Centered Pattern

To add a centered pattern, follow the steps and refer to the following image:

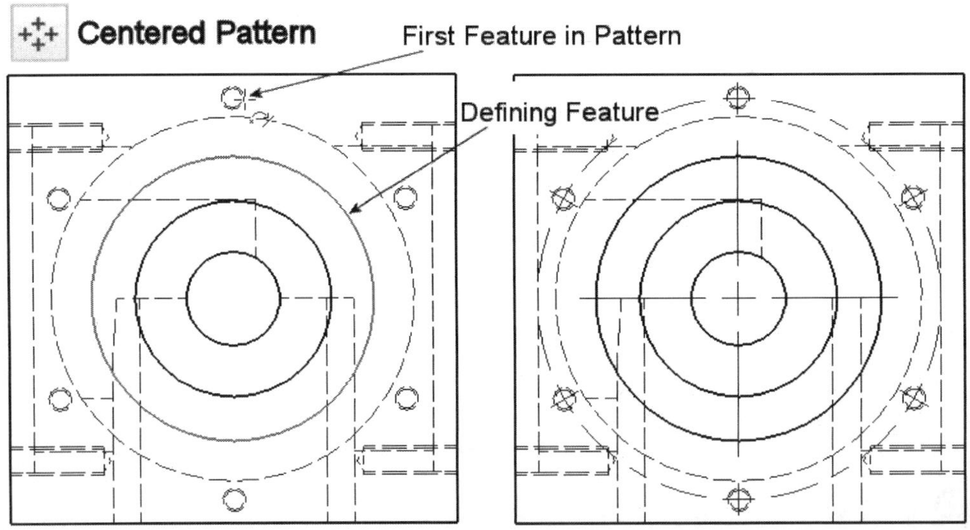

Figure 5.70

1. Click the Centered Pattern command on the Annotate tab > Symbols panel.
2. In the graphics window, select the circular face whose center point will be used as the center of the circle that a set of center marks will fall on, commonly referred to as a bolt circle.
3. Click the first feature of the pattern.
4. Continue selecting features in a clockwise or counterclockwise direction until all the features are added to the selection set.
5. Right-click, and click Create from the menu to create the centered pattern.

Automated Centerlines for Drawing Views

Creating centerlines automatically can eliminate a considerable amount of work. You can control which features automatically get centerlines and marks and in which views these occur.

You can create automated centerlines for drawing views by right-clicking on the view boundary, or hold down the Ctrl key and select multiple view boundaries and right-click to display the menu and click Automated Centerlines from the menu as shown in the following image on the left. You can also choose Automated Centerlines from the Annotate tab> Symbols panel. 

The Automated Centerlines dialog box is shown in Figure 5.71 right. Use this dialog box to set the type of feature(s) to which you will apply automated centerlines, such as holes, fillets, cylindrical features, etc., and to choose the projection type (plan or profile). The following sections discuss these actions in greater detail.

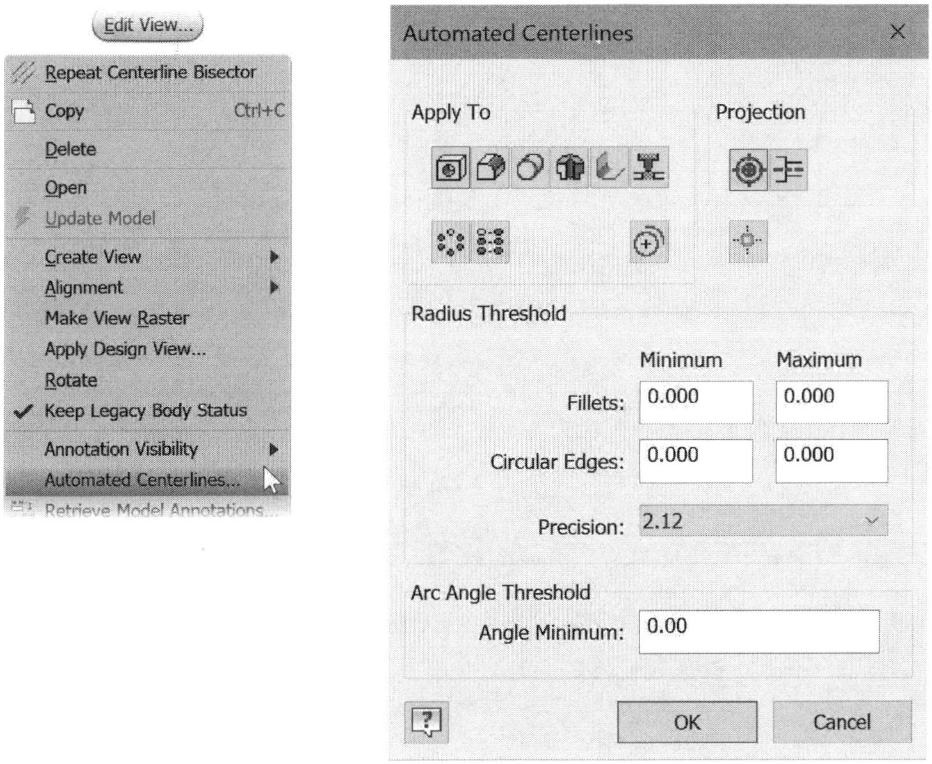

Figure 5.71

Apply To

This area controls the feature types for which to apply automated centerlines. Feature types include holes, fillets, cylindrical features, revolved features, circular patterns, rectangular patterns, sheet metal bends, punches, and circular sketched geometry. Click on the appropriate button to activate it, and automated centerlines are applied to all features of that type in the drawing. You can click on multiple buttons to apply automated centerlines to multiple features. To disable centerlines in a feature, click on the feature button a second time to deselect it.

Projection

Click on the projection buttons to apply automated centerlines to plan (axis normal) and/or profile (axis parallel) views.

Radius Threshold

Thresholds are minimum and maximum value settings and are provided for fillet features, arcs, and circles. Any object residing within a range should get the appropriate center mark. The values are based upon the model values, not the drawing values. This allows you to know what will or will not receive a centerline regardless of the view scale in the document. For example, if you set a minimum value of 0.50 for the fillet feature, a fillet that has a radius of 0.495 will not receive a center mark. A zero value on both threshold settings (min/max) denotes no restriction. This means that center marks will be placed on all fillets regardless of size.

Arc Angle Threshold

This option sets the minimum angle value for creating a center mark or centerline on circles, arcs, or ellipses.

TIP: Centerlines and center marks can be created in multiple views in one operation by holding down the CTRL key and selecting in the views and then starting the Automated Centerline command. The following image shows centerlines and center marks added to two views.

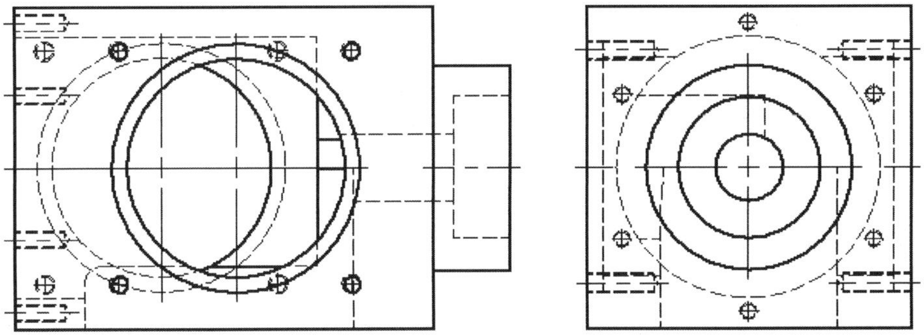

Figure 5.72

EXERCISE 5-5: ADDING CENTERLINES

In this exercise, you add center marks and centerlines, and use the Automated Centerlines command to annotate drawing views.

Both model dimensions and drawing dimensions are used to document feature size.

1 Open the drawing *ESS_E05_05.dwg* in the Chapter 05 folder. The drawing contains four views.

2 Next you add centerlines and center marks to the top view.

a. Zoom in on the top view.

b. Click Center Mark from the Annotate tab > Symbols panel.

c. Select the left outside arc as shown in Figure 5.73 left.

d. Right-click and click OK from the marking menu.

e. Notice the centerline does not extend beyond the center of the circle on the right side; this is because the arc is only on the left side. Click and drag on the right side of the centerline so it extends beyond the hidden line as shown in the middle image.

3 Next add a centerline.

a. Click Centerline from the Annotate tab > Symbols panel.

b. Select the two arcs that define the center slot in the middle of the part as shown in Figure 5.73 right.

c. Right-click and choose Create from the marking menu.

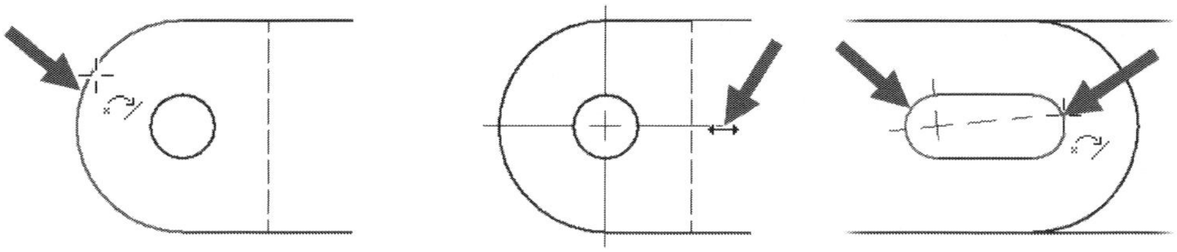

Figure 5.73

4 Next create a centered pattern (bolt circle) on the right-side of the top view.

a. Click Centered Pattern from the Annotate tab > Symbols panel.

b. First select the larger center circle labeled (1) in Figure 5.74 left. The center of this large circle will be the center of the centered pattern.

c. Select the six holes surrounding the circle (click on the circles or on the circles' center points).

d. Right-click and choose Create from the marking menu. The image on the right shows all the centerlines and center marks in the top view.

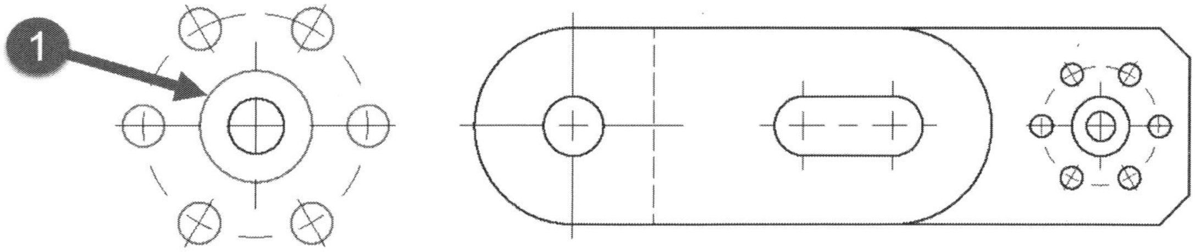

Figure 5.74

5 Pan your screen to show the front view.

6 Next create two centerlines that bisect two lines.

a. Click the Centerline Bisector command on the Annotate tab > Symbols panel.

b. In the front view, select the two hidden lines that represent the drilled hole on the left side of the part as shown in the following image on the left with the centerlines created.

c. Select the hidden lines that represent the center hole on the right side of the part, as shown in the enlarged image on the right.

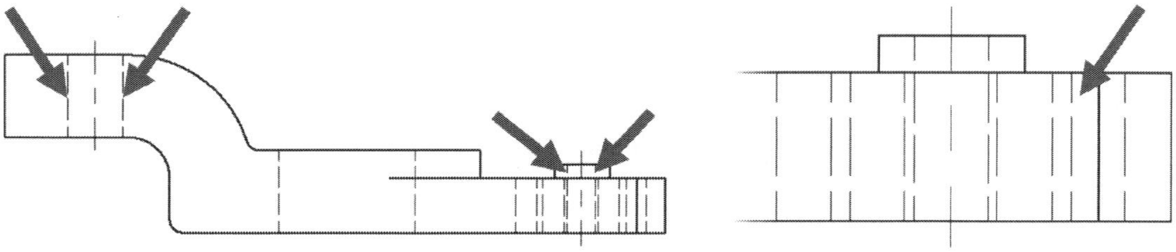

Figure 5.75

Next create centerlines and centermarks using the Automated Centerlines command.

7 Close the drawing without saving the changes and reopen it, or undo or delete all the centerlines and center marks in the drawing.

8 Hold down the Ctrl key and click in the top, front, and side views and then right-click and click Automated Centerlines from the menu.

a. Use the Automated Centerlines dialog box to make the selections as shown in Figure 5.76 left.

b. In the Apply To area select Hole Features, Fillet Features, Cylindrical Features, and Circular Patterned Features.

c. In the Projection area select Objects in View, Axis Normal and Objects in View, Axis Parallel.

d. Click OK to create the centerlines. Figure 5.76 right shows the views with the automatic centerlines.

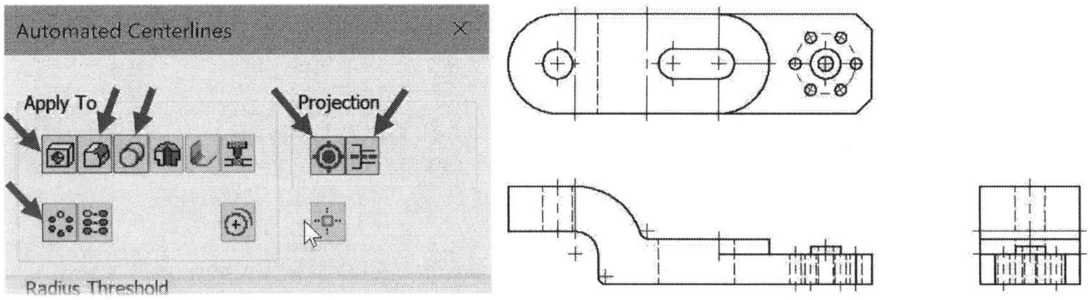

Figure 5.76

9 Zoom in and examine the centerlines that were created. If desired, resize the centerlines by dragging their endpoints.

10 If the centerline and hidden lines do not display correctly, use the next steps to adjust the global line scale.

a. To adjust the global line scale, click the Styles Editor command from the Manage tab > Styles and Standards panel.

b. In the Styles and Standard Editor on the left side expand the Standard entry and click Default Standard (ANSI) (the bold text denotes that this is the active standard).

c. On the right side of the dialog box, enter a new value in the Global Line Scale area.

d. To see the change in the centerlines and hidden lines in the graphics window, press Enter on the keyboard and then click the Save button on the top of the dialog box.

e. Try different values; when finished, click Done on the lower right corner of the dialog box. Styles will be covered later in this chapter.

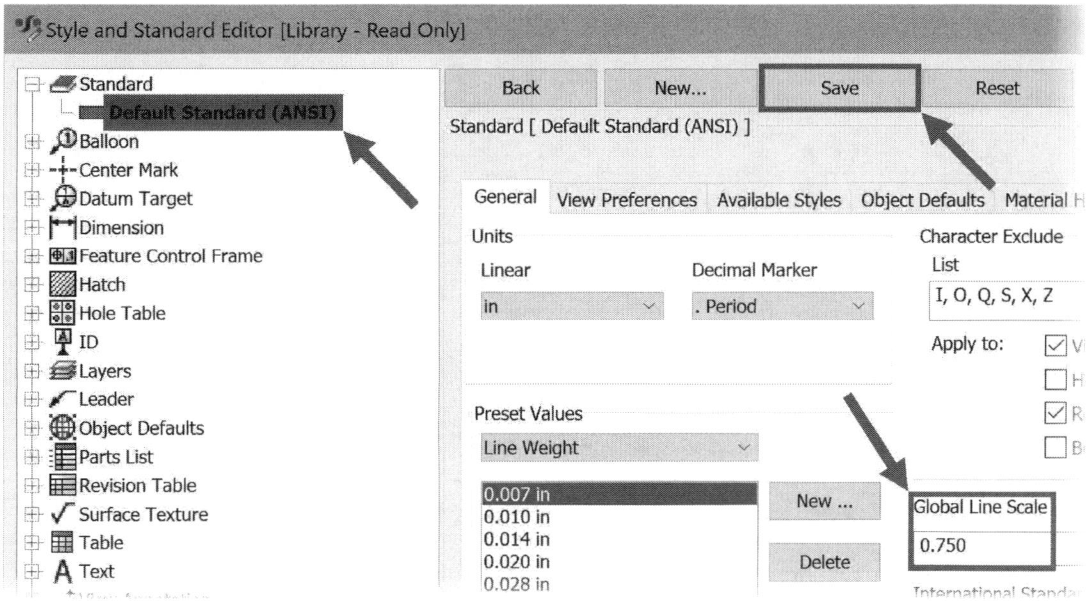

Figure 5.77

11 Close all open files. Do not save changes.

End of exercise.

ADDING DIMENSIONS TO A DRAWING VIEW

Once you have created the drawing view(s) and centerlines, the next step is to add dimensions to the drawing. You may want to alter the model dimensions, hide certain dimensions, add drawing (general) dimensions, or move dimensions to a new location. The following sections describe these operations.

Retrieving Model Dimensions

Model dimensions may not appear automatically when you create a drawing view. You can use the Retrieve Model Annotations command to select valid model dimensions for display in a drawing view. Only those dimensions that you placed in the model on a plane parallel to the view will appear.

To retrieve model dimensions, follow these steps:

1. Click Retrieve Dimension from the Annotate tab > Retrieve panel or right-click a blank area in the graphics window (not in a drawing view) and click Retrieve Model Annotations as shown in Figure 5.78.

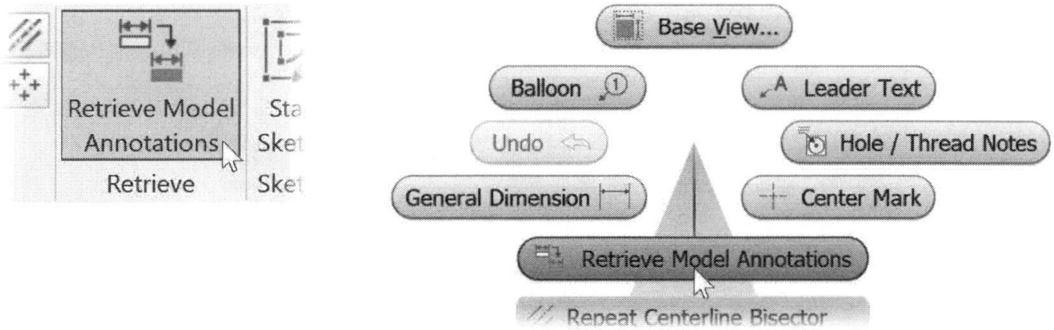

Figure 5.78

2. Select the drawing view in which to retrieve model dimensions.
3. To retrieve model dimensions based on one or more features, click the Select Features radio button. Then click the Select Dimension Source button in the dialog box and then select the desired features.
4. If you want to retrieve model dimensions based on the entire part in a drawing view, click the Select Parts radio button or click the Select Features radio button, then select the desired part or parts. You can select multiple objects by selecting them simultaneously. It is not necessary to hold the SHIFT or CTRL keys. You can also select objects with a selection window or crossing selection box.
5. To see all the dimensions that are available in the view, click the Select Dimension Source button. If the dimensions were selected with the Feature or the model, dimensions appear in the drawing view in preview mode.
6. Click the Apply button to retrieve the model dimensions, and leave the dialog box active for further retrieval operations.
7. Click OK to retrieve the selected dimensions, and dismiss the Retrieve Dimensions dialog box.

The following image shows the effects of retrieving model dimensions using the Select Features mode. In this example, the large rectangle of the block was selected. This resulted in the horizontal and vertical model dimensions being retrieved.

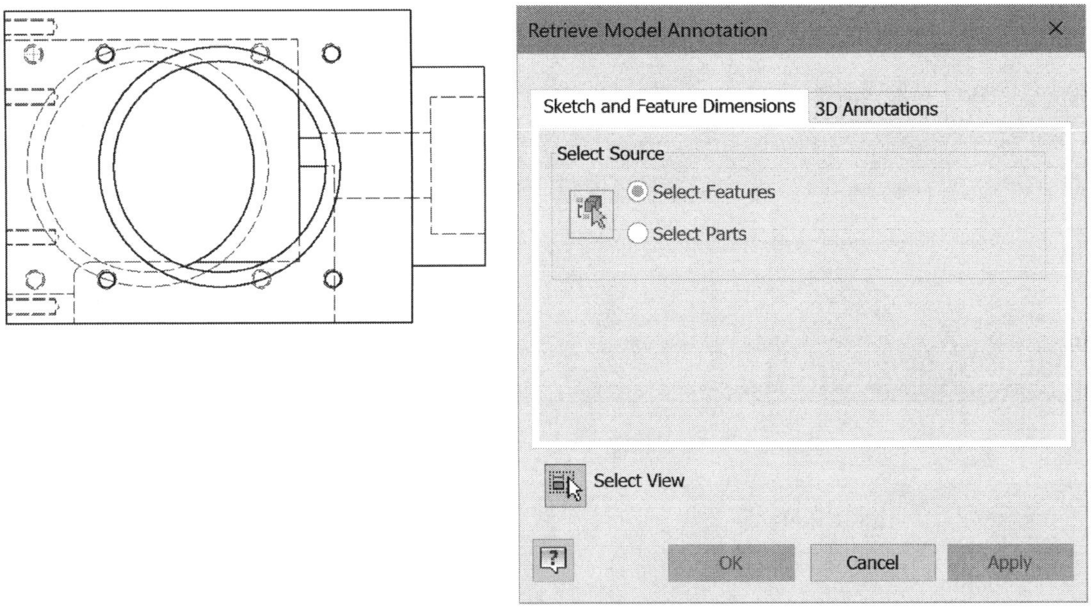

Figure 5.79

Figure 5.80 shows the result of retrieving model dimensions using the Select Parts option. In this example, the entire engine block was selected. This resulted in all model dimensions parallel to this drawing view being retrieved.

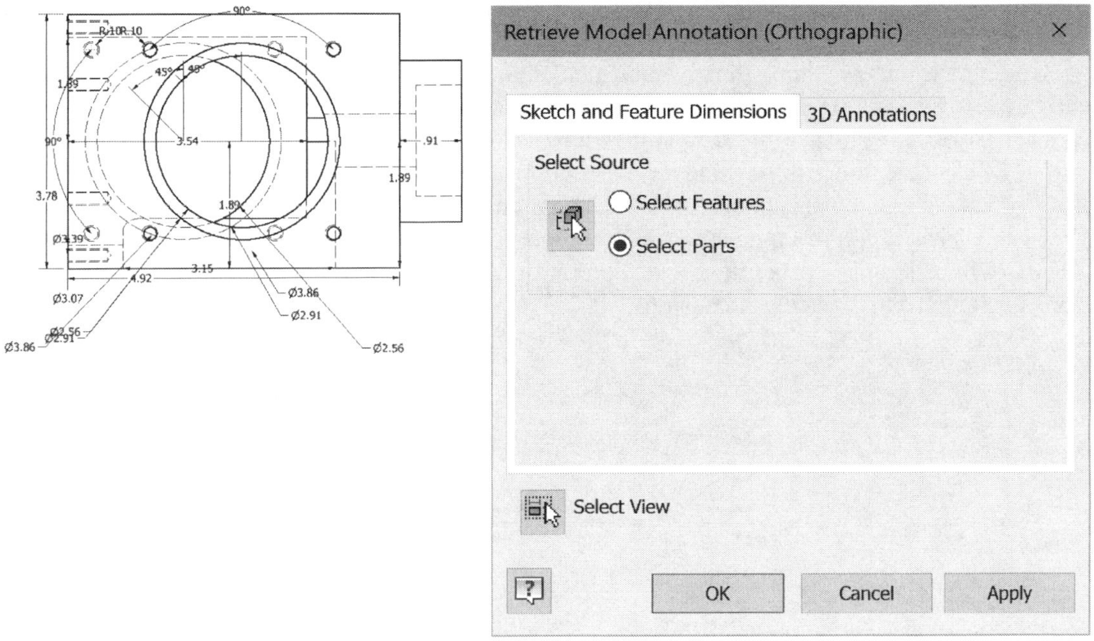

Figure 5.80

Selecting Dimensions – Additional Options

As you annotate a drawing you need to select multiple dimensions. For example, you would select multiple dimensions to arrange them, or change their dimension styles. To select multiple dimensions, you can use the same selection techniques, window, crossing, etc., that you learned about in the Selecting Objects section in Chapter 2. There are additional selection options available in a drawing. To use one of the selection options, click the down arrow next to the current selection command on the Quick Access toolbar, as shown in the following image on the left, or press down the Shift key and right-click and click the desired selection option from the menu as shown in the image on the right.

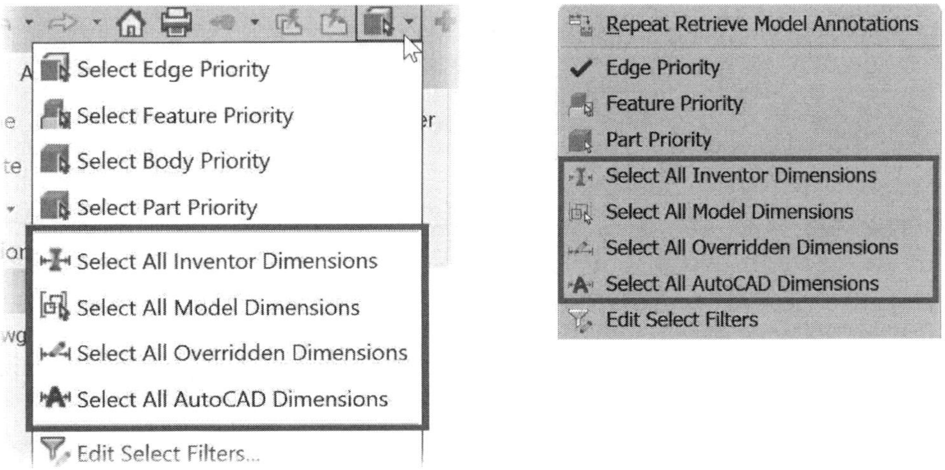

Figure 5.81

Following are explanations for the dimension selection options.

Select All Inventor Dimensions. This option selects all dimensions that are in the current sheet.

Select All Model Dimensions. This option selects all model dimensions that were retrieved in the current sheet.

Select All Overridden Dimensions. This option selects all dimensions whose values have been changed or are hidden.

Select All AutoCAD Dimensions. This option is only available when working in a DWG file and selects all AutoCAD dimensions in the current sheet.

Auto Arrange Dimensions

After retrieving model dimensions or adding dimensions manually (covered in the next section), the resulting dimensions usually need to be rearranged to conform to standard dimensioning practices. You can manually reposition the dimensions, or better is to start by having Inventor perform an automatic arrangement of the linear dimensions. To do this, select the dimensions using any selection technique or selection option. With all the dimensions selected, click the Arrange Dimensions command from the Annotate tab > Dimension panel as shown in Figure 5.82. You can also start the Arrange Dimensions command and then select the dimensions to arrange. The linear dimensions will be arranged according to the current dimension style. If needed you can manually arrange dimensions like radial and diameter dimensions.

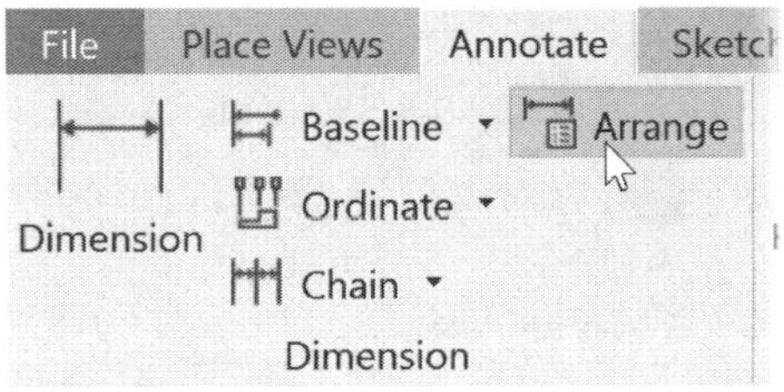

Figure 5.82

Moving and Centering Dimension Text

To move a dimension by either lengthening or shortening the extension lines or moving the text, position the cursor over the dimension until it becomes highlighted, as shown in the following image on the left. An icon consisting of four diagonal arrows will appear attached to your cursor. Notice the appearance of a centerline. This represents the center of the dimension text. You can drag the dimension text up, down, right, or left to better position it. To re-center the dimension text, drag the text toward the centerline. The dimension will snap to the centerline. In the image on the right, the centerline changes to a dotted line to signify that the dimension text is centered. This centerline action will work on linear dimensions including horizontal, vertical, and aligned but will not be displayed when repositioning the text of radius, diameter, or hole note dimensions.

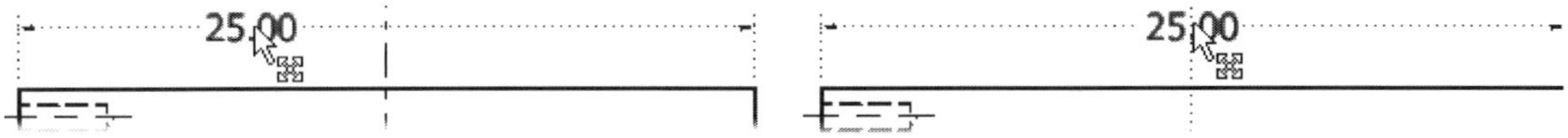

Figure 5.83

Editing a Model Dimension's Value in a Drawing

While working in a drawing view, you may find it necessary to change the model (parametric) dimensions of a part. You can open the part file, change a dimension's value, and save the part, and the change will then be reflected in the drawing views. For dimensions that were added to the drawing with the Retrieve Model Annotations command you can change a model dimension's value in the drawing by right-clicking on the dimension and selecting Edit Model Dimension from the menu, as shown in the following image on the left.

The Edit Dimension text box will appear. Enter a new value and click the checkmark, as shown in the image on the right, or press ENTER. The associated part will be updated and saved, and the associated drawing view(s) will be updated automatically to reflect the new value.

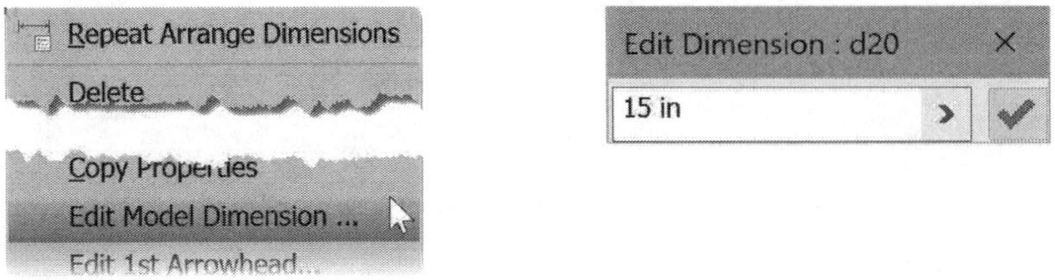

Figure 5.84

TIP: To reduce the possibility of modeling issues, it is recommended that you verify the edits in the part model.

Creating General Dimensions

After laying out the drawing views, you may find that a dimension other than the retrieved model dimension is required to define the part. You can add a general dimension to the drawing view. A general dimension is not a parametric dimension; it is associative to the geometry to which it is referenced. A general dimension reflects the size of the geometry being dimensioned. After you create a general dimension, and the value of the geometry that you dimensioned changes, the general dimension will be updated to reflect the change. You add a general dimension by using the General Dimension command on the Annotate tab > Dimension panel as shown in the following image on the left; right-click in a blank area in the graphics screen and click the General Dimension command from the marking as shown in the image on the right, or press the D key.

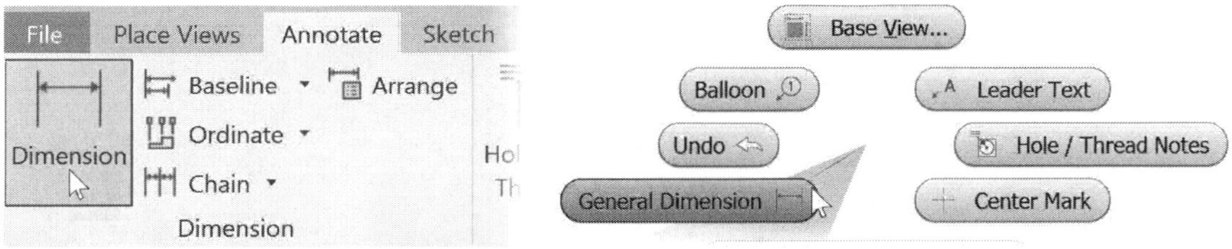

Figure 5.85

The general dimensions you create in a drawing follow a similar process that you did when creating a dimension in a part's sketch, except when a drawing dimension is placed, the Edit Dimension dialog box will appear as shown in Figure 5.86. In the Edit Dimension dialog box, the <<>> symbol represents that value of the geometry being dimensioned. You can add text or symbols (select a symbol from the right side of the dialog box) before or after the <<>>. You can also add a tolerance or add an inspection symbol to the dimension from the two other tabs. By default, when a drawing dimension is placed, the Edit Dimension dialog box will appear. You can turn off this behavior by unchecking the **Edit dimension when created** option at the bottom of the dialog box or from the Application Options > Drawing tab. You can edit an existing drawing dimension by double-clicking on a dimension or right-clicking on a dimension and click Edit from the menu. The Edit Dimension dialog box will appear. Edit the dimension as needed.

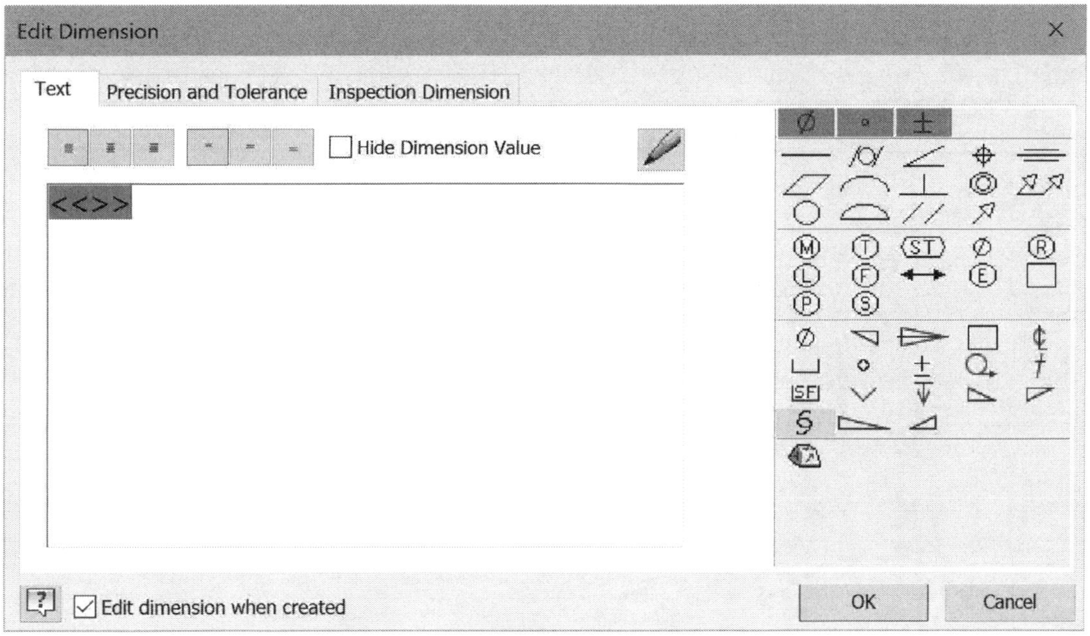

Figure 5.86

Adding General Dimensions to a Drawing View

The following image illustrates a simple object with various dimension types. General dimensions can take the form of linear dimensions labeled (A) and (B), diameter dimensions labeled (C), radius dimensions labeled (D), and angular dimensions labeled (E). When using the General Dimension command, Inventor automatically chooses the dimension type depending on the object you chose. In this example, the counterbore hole is dimensioned using the hole note command which will be covered later in this chapter.

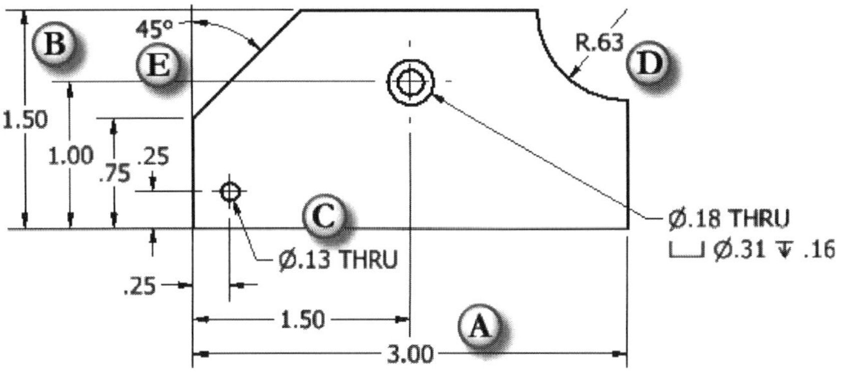

Figure 5.87

Adding Dimensions to an Isometric Drawing View

You can retrieve model dimensions or use the General Dimension command to add dimensions to isometric drawing views as well as to standard orthographic views. When placing a dimension on an isometric view, the dimension text, dimension lines, extension lines, and arrows are oblique and aligned to the geometry being dimensioned. Figure 5.88 shows an isometric view complete with oblique dimensions.

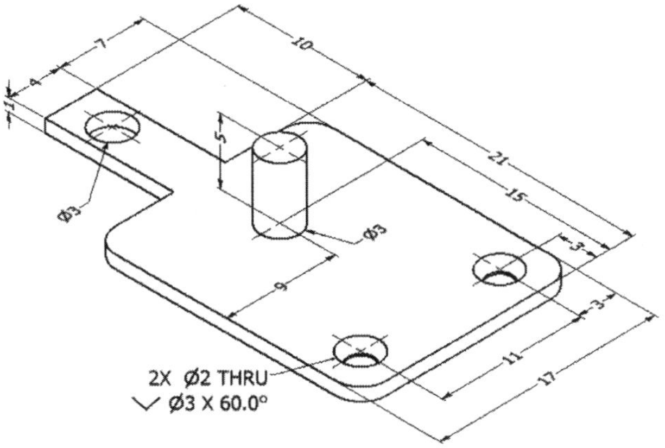

Figure 5.88

Depending on the object being dimensioned, additional controls are available to manipulate the isometric dimension being created. Once edges or points in a drawing are selected, the dimension previews on an annotation plane, as shown in Figure 5.89 left. If more than one annotation plane is possible, you can toggle between these planes by pressing the spacebar before locating the dimension. The image on the right shows the results after pressing the spacebar. Once the dimension is previewed correctly, locate the dimension by clicking.

Figure 5.89

EXERCISE 5-6: ADDING DIMENSIONS

In this exercise, you add dimensions to a drawing of a clamp, and edit and create a dimension style.

1 Open the drawing *ESS_E05_06.dwg* in the Chapter 05 folder. The drawing file contains four drawing views with centerlines.

2 In this step, you retrieve model dimensions for use in the front view.

a. Begin by zooming in on the front view.

b. Move the cursor in the front view, right-click and click Retrieve Model Annotations from the menu. The model dimensions appear on the view and the Retrieve Model Annotations dialog box opens.

c. From the Retrieve Model Annotations dialog box, click Select Features.

d. Select the dimensions to retrieve by dragging a selection window around all dimensions.

e. When finished, click the OK button. The model dimensions planar to the view are displayed, as shown in Figure 5.90.

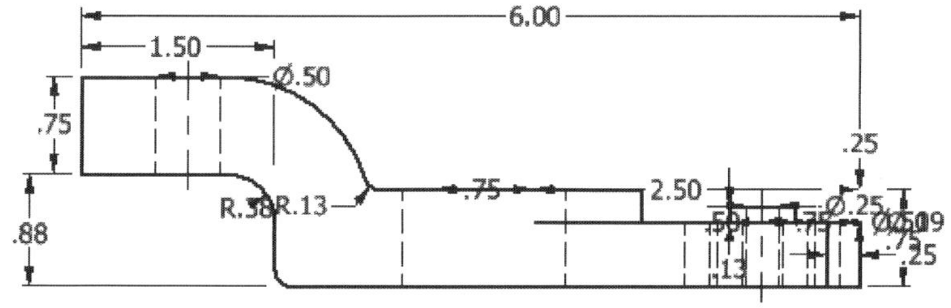

Figure 5.90

3 Instead of manually dragging dimensions to new locations, click the Arrange Dimensions command on the Annotate tab > Dimension panel.

a. Drag a selection window around all dimensions. This will select the dimensions to arrange.

b. When finished, right-click and click OK from the menu. The dimensions are arranged according to the settings in the current dimension style.

4 There are nine dimensions that were retrieved that are not needed. Hold down the Ctrl key and select the nine dimensions that are highlighted in the following image and press the Delete key on the keyboard or right-click and click Delete from the menu. The dimensions that were deleted can be retrieved again if needed.

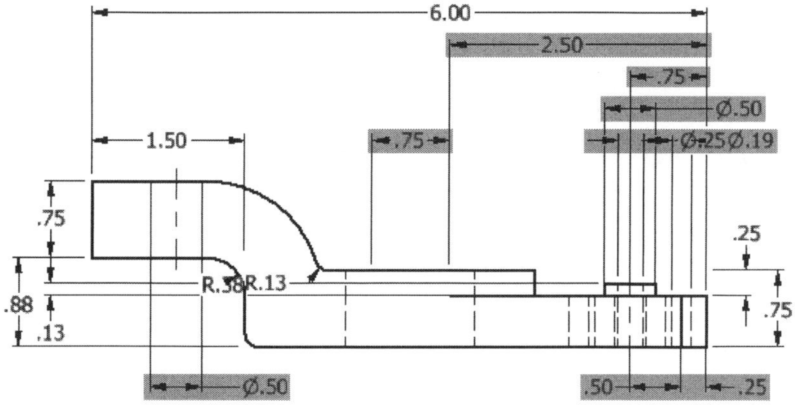

Figure 5.91

5 Use the Arrange Dimensions command again to arrange the dimensions. When done, your view should resemble the following image.

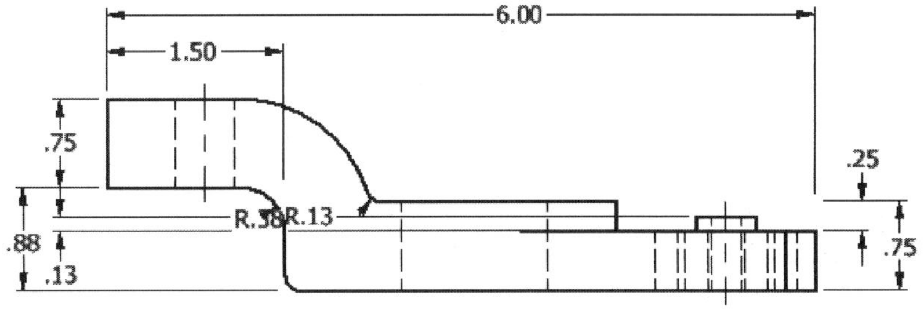

Figure 5.92

6 A few of the dimensions need to be repositioned. Click and drag the highlighted dimensions to locations like those in Figure 5.93. You can move the text by dragging a dimension's text; move the dimension by dragging a dimension's extension line. If needed delete dimensions that are not needed.

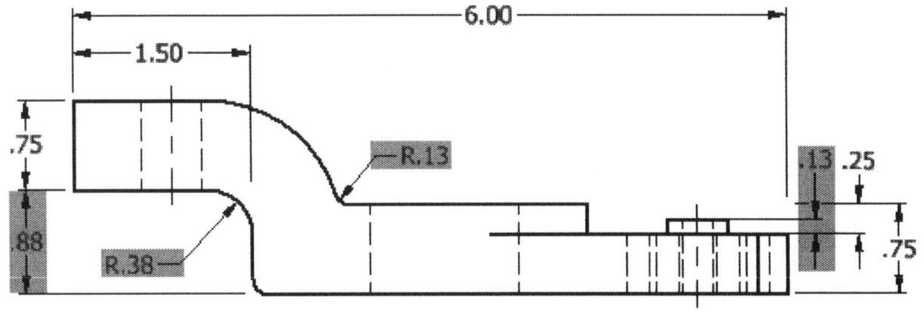

Figure 5.93

TIP: Before you reposition a dimension text, exit the active command.

7 In the top view, you will manually add dimensions. Pan to display the top view.

c. Click the General Dimension command on the Annotate tab > Dimension panel.

d. Place the nine dimensions as shown in Figure 5.94. To change a radius dimension to a diameter dimension or vice versa you must right-click and select the dimension type before the dimension is placed.

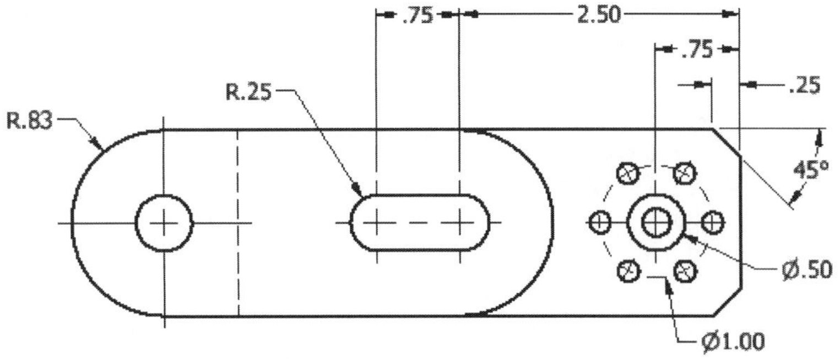

Figure 5.94

8 Next you add text to a dimension. Double-click on the .25 horizontal dimension and then type in **TYP.** after the <<>> as shown in Figure 5.95.

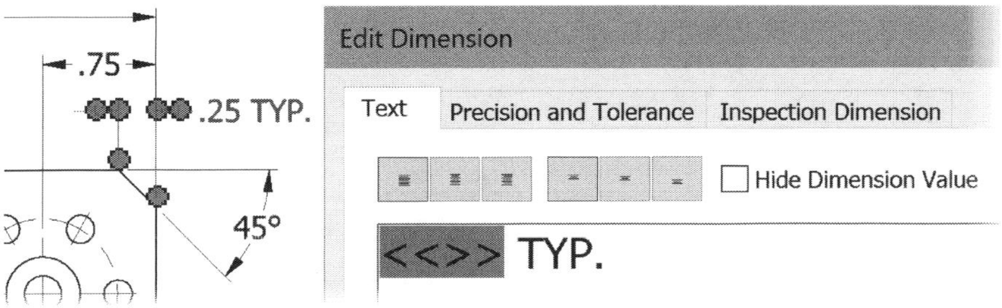

Figure 5.95

9 Next you change a model dimension's value via the drawing. In the front view, right-click on the 6.00 dimension and click Edit Model Dimension from the menu. Enter a value of **7 inches** and then click on the green check mark in the Edit Dimension dialog box.

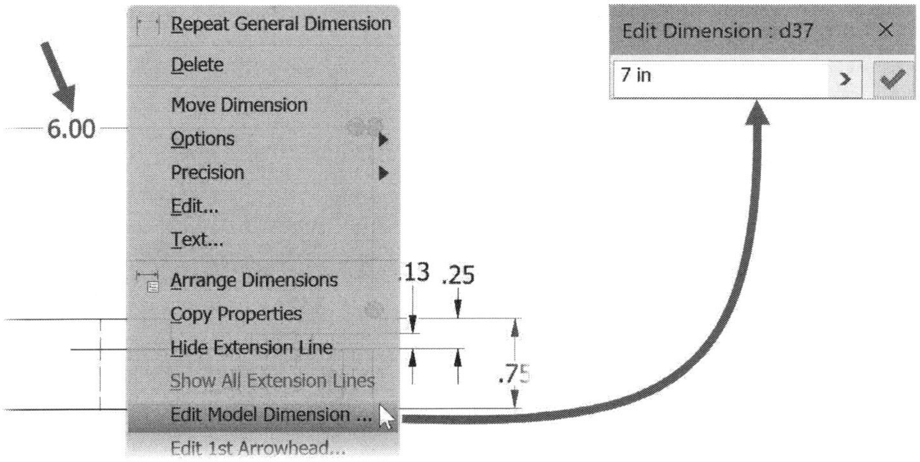

Figure 5.96

10 Notice how the drawing views and dimensions update to reflect the new value.

11 Next you will open the part file and change the horizontal dimension back to its original value. Move the cursor into a blank area in one of the drawing views and click Open from the menu as shown in Figure 5.97 left.

a. Edit Sketch1 under Extrusion1 as shown in the image on the right.

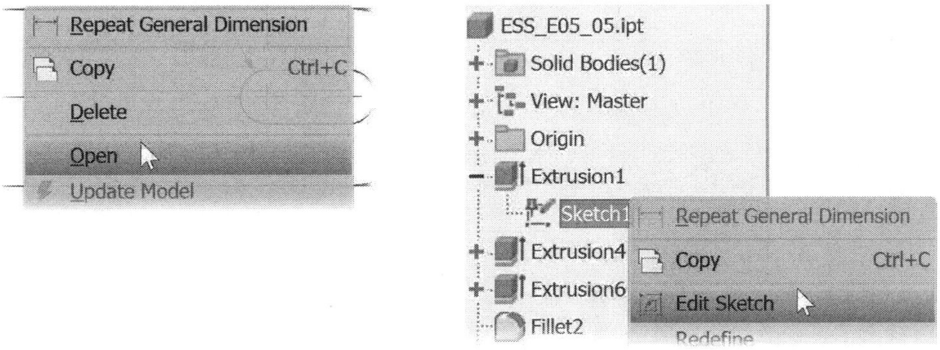

Figure 5.97

b. Double-click the **7** horizontal dimension and enter **6** and then press ENTER on the keyboard.

c. Finish the sketch and then save and close the part file.

12 The drawing should be the current file; if not, make it the active file by clicking on the *ESS_E05_06.dwg* tab at the bottom of the screen.

13 Notice how the drawing views automatically updated to reflect the updated model.

14 Next change the dimension style. Move the cursor over any dimension, right-click and click Edit Dimension Style from the menu. In the Style and Standard Editor dialog box make the following edits.

a. Ensure Default (ANSI) dimension style is current, labeled (1) in the following image.

b. From the Units tab, change the Precision to 1.1 (represents one decimal place), labeled (2).

c. In the Display area, check the Leading Zeros option, labeled (3).

d. Click Save on the top of the dialog box, labeled (4).

e. Close the dialog box by clicking Save and Close on the bottom-right corner of the dialog box.

f. All the dimensions should update to a single decimal place and a leading zero will be displayed on the dimensions smaller than one inch in length or radius.

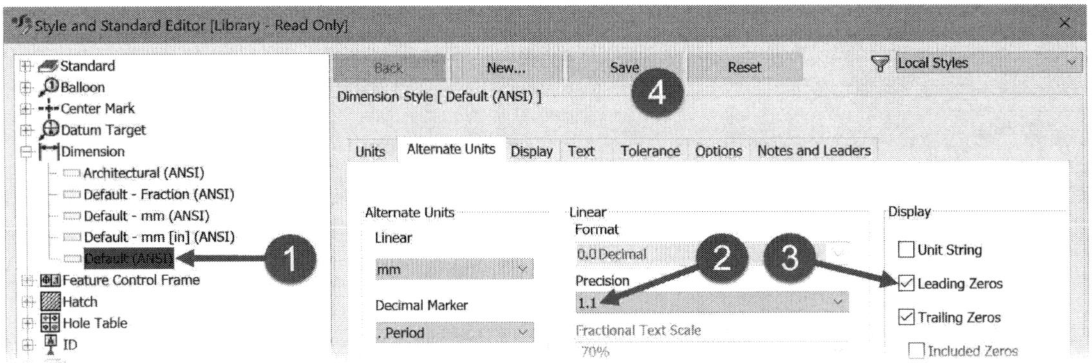

Figure 5.98

15 Next you create a new dimension style. Right-click on the 6.0 horizontal dimension and click New Dimension Style from the menu. The new dimension style will take on the properties of the dimension style of the selected dimension. In the New Dimension Style dialog box make the following edits.

a. In the upper left corner, enter a name for the new style as **3 Decimal (ANSI)**, labeled (1) in Figure 5.99.

b. From the Units tab change the Precision to 3.123, labeled (2).

c. Click OK in the dialog box to create the dimension style and close the dialog box.

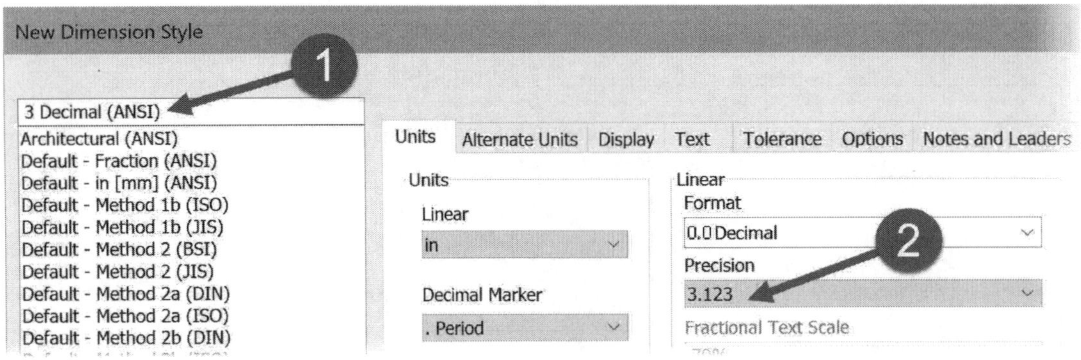

Figure 5.99

16 The 6.0 dimension should now be 6.000 but the other dimensions still display a single decimal place. Only the 6.000 dimension is using the new 3 Decimal place dimension style.

17 Next you change a few dimensions to the new style. In the drawing, hold down the CTRL key and select a few dimensions. From the Annotate tab > Format panel select the 3 Decimal (ANSI) dimension style from the list as shown in Figure 5.100 left.

TIP: Another option to change dimension properties is to replicate the properties of another dimension. To copy properties, right-click on the dimension whose properties you want to copy and click Copy Properties from the menu as shown in the following image on the right. Then select the dimension(s) you want the properties copied to. You can also right-click and click Settings from the menu and from the Copy Dimension Properties dialog box you can uncheck the properties you do not want copied.

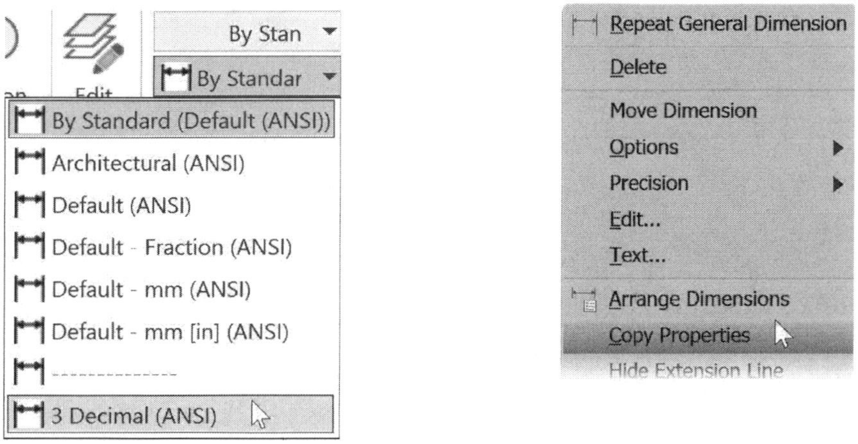

Figure 5.100

18 Practice editing and creating new dimension styles.

19 In Exercise 5-8, you will continue to work on a similar drawing where you will add hole notes, a texture symbol, and a text note.

20 Close all open files. Do not save changes.

End of exercise.

Creating Baseline Dimensions

To add multiple general dimensions to a drawing view in a single operation, use the Baseline Dimension command, which will place a baseline dimension about the selected geometry. The dimensions can be either horizontal or vertical. Two commands are available to assist with the creation of these dimensions: Baseline Dimension and Baseline Dimension Set.

Baseline Dimensions

The Baseline Dimension command allows you to create a group of horizontal or vertical dimensions from a common origin. However, the group of dimensions are not considered one group of objects; rather, each dimension that makes up a Baseline Dimension is considered a single dimension.

To create and edit baseline dimensions, follow these steps:

1. Select the Baseline Dimension command from the Annotate tab > Dimension panel as shown in the following image.

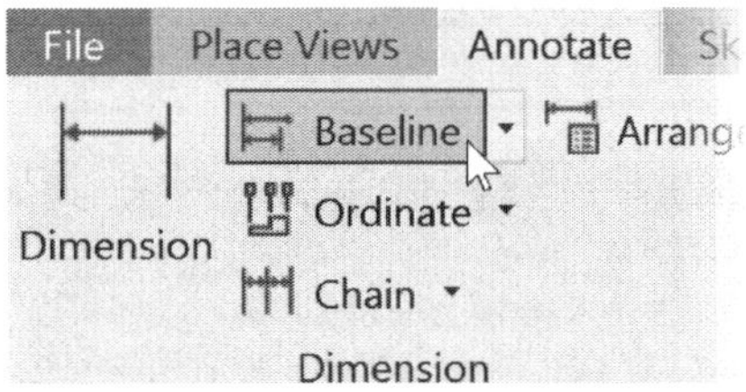

Figure 5.101

2. Individually select or drag a selection window around the geometry that you want to dimension. To window select, move the cursor into position at the first point of the box. Press and hold down the left mouse button, and move the cursor so the preview box encompasses the geometry that you want to dimension, as shown in Figure 5.102 left, and then release the mouse button.
3. When you are finished selecting geometry, right-click, and click Continue from the marking menu.
4. Move the cursor to a position where the dimensions will be placed. When the dimensions are in the correct place, click the point with the left mouse button to anchor the dimensions.
5. While still in the command you can change the origin of the dimensions by moving the cursor over the extension line that will be the origin, right-click and click Make origin from the menu.
6. To complete the operation, right-click and click Create from the marking menu.
7. After creating dimensions using the Baseline Dimension command, the same editing operations can be performed on these dimensions as general dimensions. To perform these edits, move the cursor over the baseline dimension, and small green circles will appear on the dimension. Right-clicking will display the menu, as shown in the image on the right.

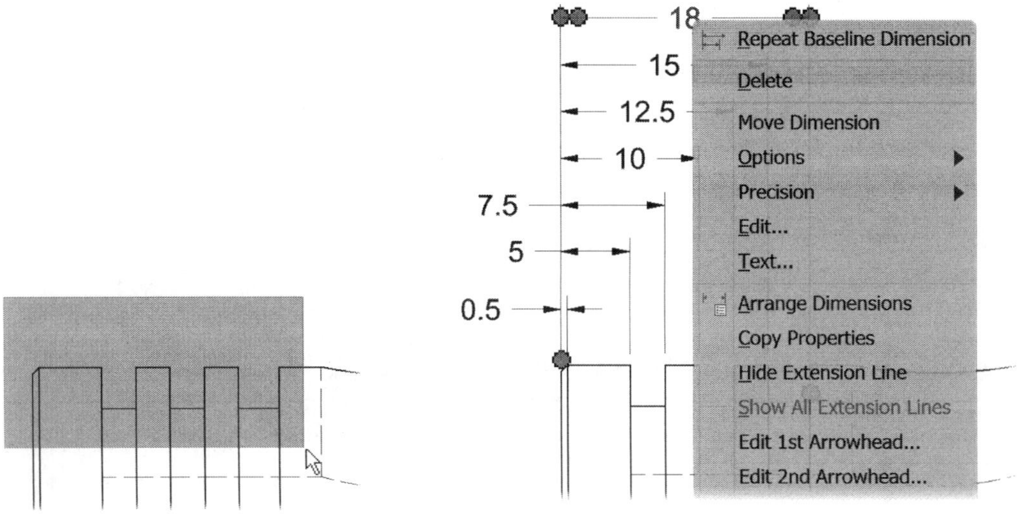

Figure 5.102

Creating Baseline Set Dimensions

Creating Baseline Set dimensions is like the Baseline Dimensions as explained in the previous section, except with the Baseline Set dimension command, the dimensions are grouped together into a set. Select the Baseline Set command from the Annotate tab > Dimension panel by clicking the down arrow next to the Baseline Dimension command, as shown in Figure 5.103.

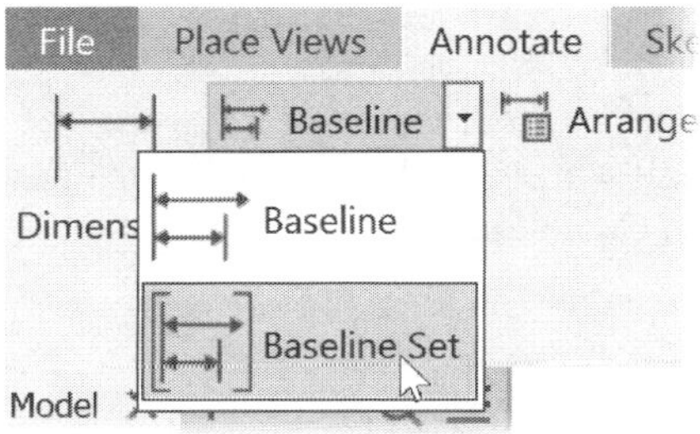

Figure 5.103

Members can be added and deleted from the set, a new origin can be placed, and the dimensions can be automatically rearranged to reflect the change. Right-clicking on any dimension in the set will highlight all dimensions and display the menu, as shown in Figure 5.104.

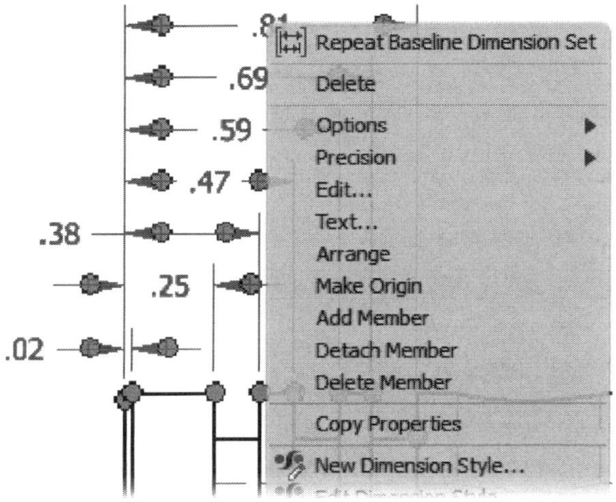

Figure 5.104

The main options for editing a baseline dimension set are described as follows:

Delete. This option will delete all baseline dimensions in the set. Move your cursor over any dimension in the set. When the green circles appear on each dimension, right-click, and select Delete from the menu.

Arrange. This command will rearrange the dimensions to the spacing defined in the active Dimension Style.

Make Origin. This option will change the origin of the baseline dimensions. Move the cursor over an extension line of the baseline dimension set that you want to define as the new baseline, right-click, and click Make Origin from the menu.

Add Member. This option will add a drawing dimension to the baseline dimensions. Move the cursor over the baseline dimension set, and when the green circles appear on the dimensions, right-click and click Add Member from the menu and then click a point that you want to add a dimension to. The dimension will be added to the set of baseline dimensions and rearranged in the proper order.

Detach Member. This option will remove a dimension from the set of baseline dimensions. Move your cursor over any dimension in the set. When the green circles appear on each dimension, right-click on the green circle on the dimension that you want to detach, right-click, and select Detach Member from the menu. The detached dimension can be moved and edited like a normal drawing dimension.

Delete Member. This option will erase a dimension from the set of baseline dimensions. Move your cursor over the dimension in the set that you want to erase, and then select Delete Member from the menu. The dimension will be deleted.

Creating Chain Dimensions

Another option for placing dimensions is to create chain dimensions, dimensions whose extension lines are shared between two dimensions. Chain dimensions can be arranged horizontally or vertically. Like the baseline dimension commands, there are two Chain dimension commands: Chain and Chain Set.

Chain Dimensions

The Chain Dimension command allows you to create a group of horizontal or vertical dimensions. However, the dimensions are not considered one group of objects; rather, each dimension that makes up the chain dimension is considered a single dimension.

To create and chain dimensions, follow these steps:

1. Select the Chain dimension command from the Annotate tab > Dimension panel as shown in the following image.

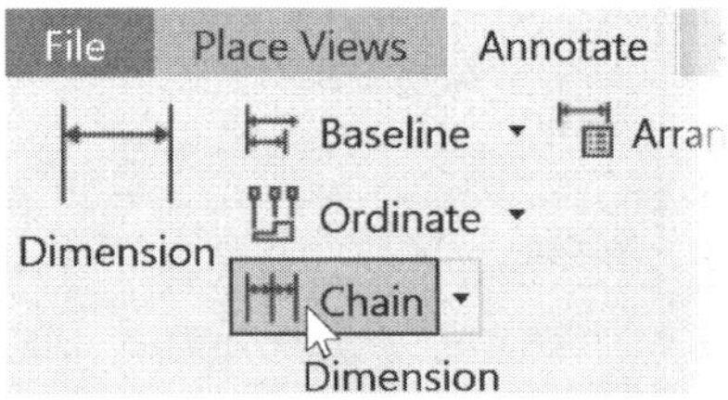

Figure 5.105

2. Individually select or drag a selection window around the geometry that you want to dimension. To window select, move the cursor into position where the first point of the box will be. Press and hold down the left mouse button, and move the cursor so the preview box encompasses the geometry that you want to dimension. Figure 5.106 left shows four objects that were selected.
3. When you are finished selecting geometry, right-click, and click Continue from the marking menu.
4. Move the cursor to a position where the dimensions will be placed and click to anchor the dimensions. The image on the right shows the placed dimensions.

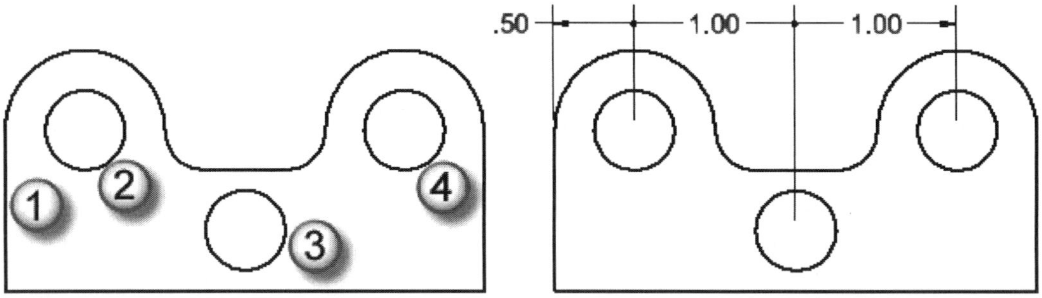

Figure 5.106

5. To complete the operation, right-click and click Create from the marking menu.
6. After creating dimensions using the Chain dimension command, the same editing operations can be performed on these dimensions as general dimensions.

Chain Set Dimensions

Creating a Chain Set dimensions is like Chain dimensions as explained in the previous section except with the Chain Set command; however, dimensions are grouped together into a set. Select the Chain Set command from the Annotate tab > Dimension panel by clicking the down arrow next to the Chain dimension command, as shown in Figure 5.107.

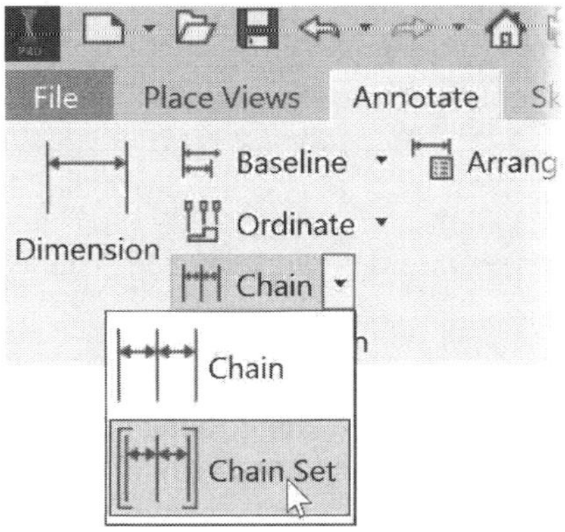

Figure 5.107

Members can be added and deleted from the set. Right-clicking on any dimension in the set will highlight all dimensions and display the menu, as shown in Figure 5.108.

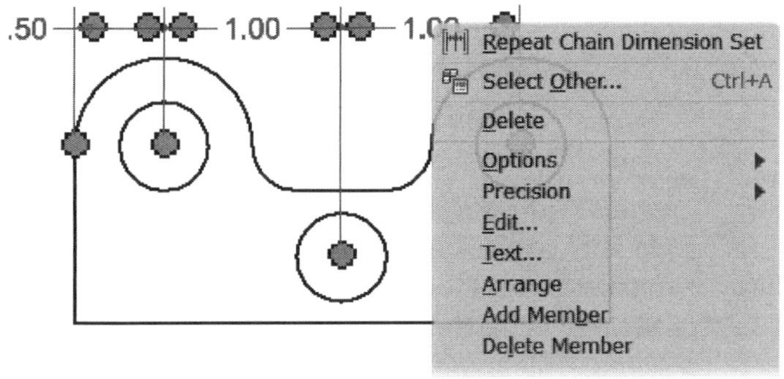

Figure 5.108

The main options for editing a chain dimension set are described as follows:

Delete. This option will delete all chain dimensions in the set. Move your cursor over any dimension in the set. When the green circles appear on each dimension, right-click, and select Delete from the menu.

Options. From the Options area, you can lock the chain, control the arrowheads and add a leader to a dimension that the cursor was over when you right-clicked.

Add Member. This option will add a drawing dimension to the chain dimensions. Move the cursor over the chain dimension set, and when the green circles appear on the dimensions, right-click and click Add Member from the menu and then click a point that you want to add a dimension to. The dimension will be added to the set of chain dimensions and rearranged in the proper order.

Delete Member. This option will erase a dimension from the set of chain dimensions. Move your cursor over the dimension in the set that you want to delete. When the green circles appear on each dimension, right-click, and then select Delete Member from the menu. The dimension will be deleted.

EXERCISE 5-7: CREATING BASELINE & CHAIN DIMENSIONS

In this exercise, you will add annotations using baseline and chain dimensions.

1 Open *ESS_E05_07.dwg* in the Chapter 05 folder.

2 In this step, you add baseline set dimensions. Click the Baseline Set command from the Annotate tab > Dimension panel by clicking the down arrow next to the Baseline Dimension command.

a. Select the three edges as shown in the following image on the left.

b. Right-click, and click Continue from the marking menu.

c. Click a point above the geometry to position the dimension set as shown in the middle image. The baseline dimension set is created

3 Next you change the origin for the dimensions to reference the opposite edge of the part. To perform this task, move the cursor over the right most vertical extension line, right-click, and click Make Origin from the menu. Notice how the dimensions are regenerated from the new origin, as shown in Figure 5.109 right.

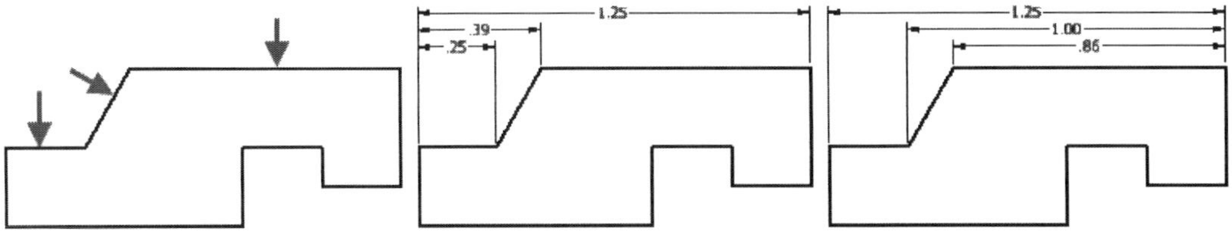

Figure 5.109

4 Next you add a dimension to the set. Move the cursor over the baseline dimension set; when the green circles appear on the dimensions, right-click and click Add Member from the menu and then click the point as shown in the following image on the left. Right-click and click Done from the menu and the dimension will be added to the set and the dimensions will be rearranged as shown in Figure 5.110 right. If the dimensions are too close to the geometry, click and drag the .50 dimension up.

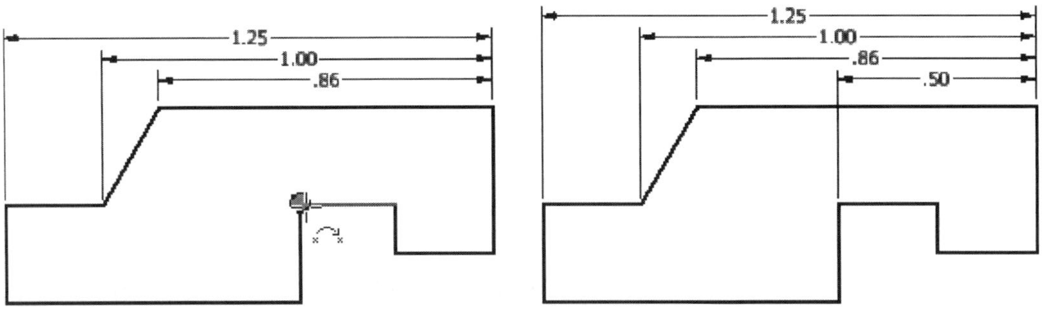

Figure 5.110

5 Next delete a dimension from the set, move the cursor over the .86 dimension, right-click and click Delete Member from the menu. The dimension will be deleted as shown in Figure 5.111 left. Notice the dimensions were not automatically arranged.

6 Next arrange the dimensions, move the cursor over the dimension set, and when the green circles appear on the dimension right-click and click Arrange from the menu. The dimensions will be rearranged as shown in Figure 5.111 right.

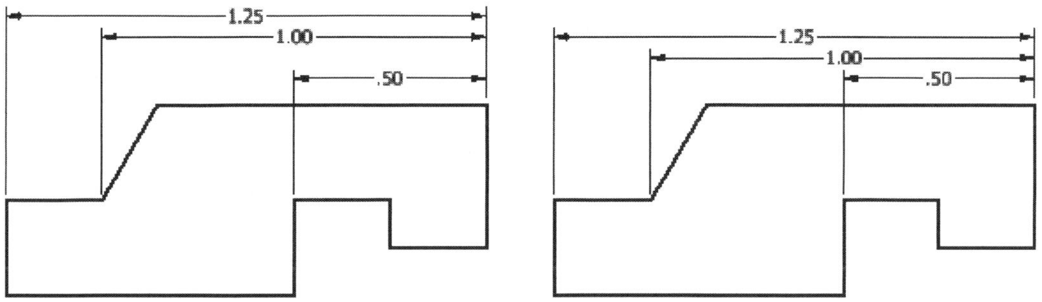

Figure 5.111

7 Next delete the baseline dimension set. Move the cursor over the dimension set; when the green circles appear on the dimensions, right-click and click Delete from the menu. The dimension set will be deleted.

8 Next you create chain set dimensions. Select the Chain Set command from the Annotate tab > Dimension panel by clicking the down arrow next to the Chain dimension command.

a. Window select the edges as shown in the following image on the left.

b. Right-click, and click Continue from the marking menu.

c. Click a point below the geometry to position the chain set as shown in Figure 5.112 right. The chain set dimensions are created.

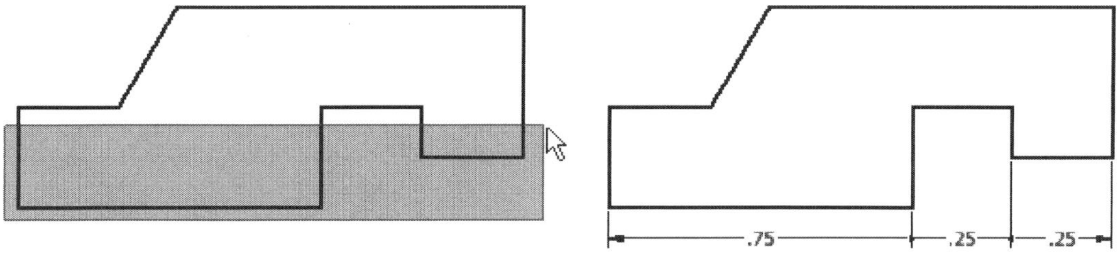

Figure 5.112

9 To add a dimension to the set, move the cursor over the chain set dimensions, and when the green circles appear on the dimensions, right-click and click Add Member from the menu and then click the top-left point as shown in Figure 5.113 left.

10 Right-click and click Done from the menu. The dimension will be added to the set and the dimensions will be rearranged as shown in the image on the right.

TIP: If a dimension does not fit within its extension lines, you can right-click on the dimension and from the menu click Options > Leader and then drag the dimension and leader as needed.

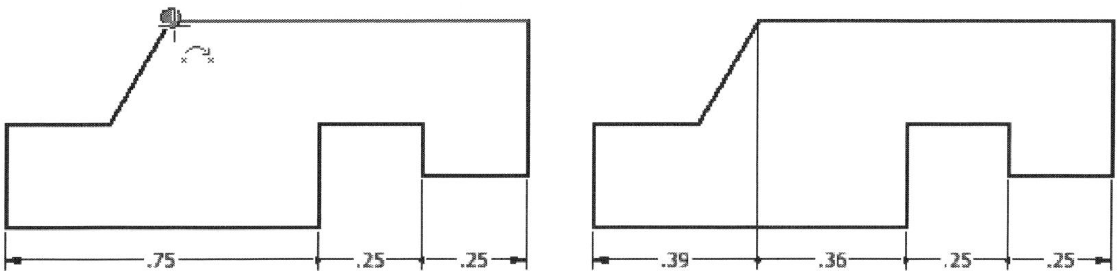

Figure 5.113

11 Delete a dimension, move the cursor over the .39 dimension, right-click, and click Delete Member from the menu.

12 To center the text, right-click on the dimension set and click Arrange from the menu. Notice the dimensions were automatically arranged as shown in the following image.

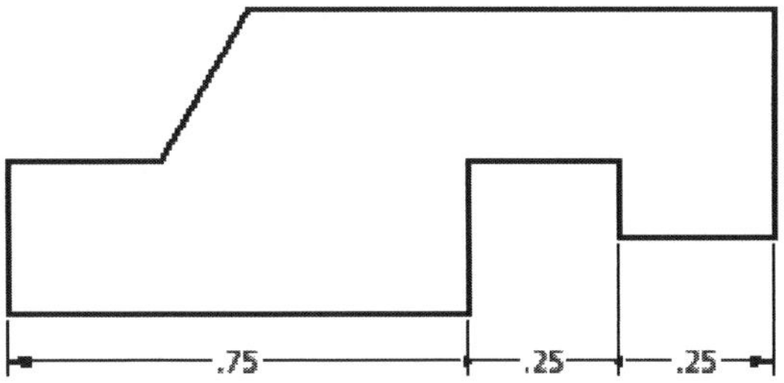

Figure 5.114

13 Practice adding and editing baseline and chain dimensions.

14 Close all open files. Do not save changes.

End of exercise.

Creating Ordinate Dimensions

Ordinate dimensions are used to indicate the location of a point along the X or Y axis from a common origin point. This type of dimensioning is especially suited for describing part geometry for numerical control tooling operations. Two commands are available for creating ordinate dimensions: Ordinate Dimension and Ordinate Dimension Set. Both commands will create drawing dimensions that reference the geometry and will be updated to reflect any changes in the geometry to which they are dimensioned. Ordinate dimensions can be placed on circular or straight edges.

Ordinate Dimensions

Ordinate dimensions created with the Ordinate Dimension command are recognized as individual objects, and an origin indicator will be created as part of the operation. If the origin indicator location is moved, the other ordinate dimensions will be updated to reflect the change.

To create an ordinate dimension using the Ordinate Dimension command, follow these steps:

1. Create a drawing view.
2. Click the Ordinate Dimension command on the Annotate tab > Dimension panel, as shown in the following image on the left.
3. In the graphics window, select in the view to dimension.
4. Select a point to set the origin indicator (the zero) for the dimensions as shown in Figure 5.115 right.

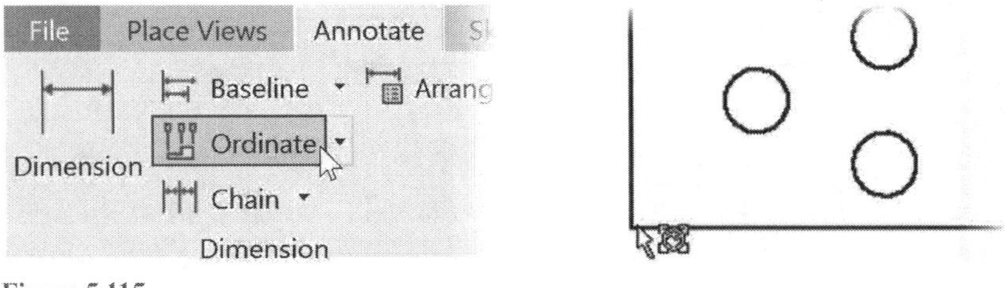

Figure 5.115

5. Select geometry that the ordinate dimensions will be applied to. When done selecting geometry, right-click and choose Continue from the marking menu.
6. Locate the dimensions horizontally or vertically by moving the cursor and then click a point when the dimensions are previewed in the correct orientation, as shown in the following image on the left. The dimensions are created.
7. To edit a dimension, right-click on a dimension and click the desired option from the menu, as shown in the following image on the right.

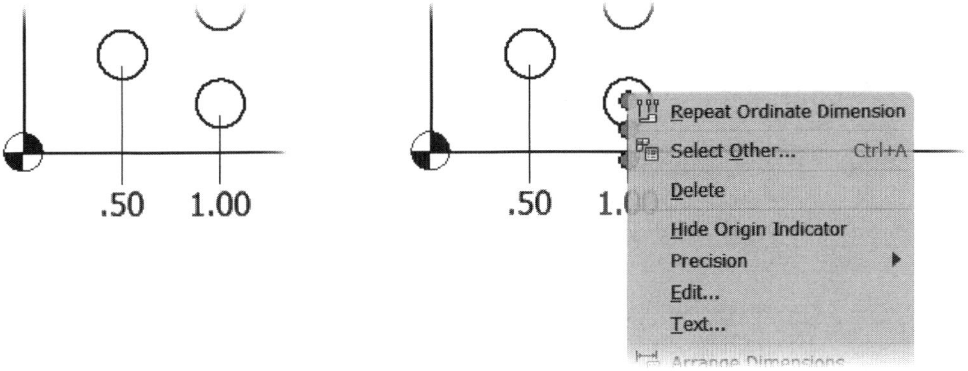

Figure 5.116

8. To move an ordinate dimension, click on the dimension, and then click and drag on an anchor point (a green circle), and drag it to the desired location.
9. To edit the origin indicator, do one of the following:
   - Move the cursor over the origin indicator, and drag it to the desired location.
   - Double-click the origin indicator, and enter the precise location in the Origin Indicator dialog box.

Ordinate Dimension Set

All the ordinate dimensions that are created in a single operation will be grouped together and can be edited individually or as a set. When creating an ordinate dimension set, the first dimension created will be used as the origin. The origin dimension needs to be a member of the set and can later be changed to a different dimension. If the location of the origin or origin member changes, the other members update to reflect the new location.

When you place ordinate dimensions, they automatically aligned to avoid interfering with other ordinate dimensions.

To create an ordinate dimension set, follow these steps:

1. In a drawing view, click the Ordinate Dimension Set command on the Annotate tab > Dimension panel by clicking the down arrow next to the Ordinate Dimension command, as shown in the following image on the left.
2. Select a point on the geometry to set the origin for the dimensions, as shown in Figure 5.117 right.

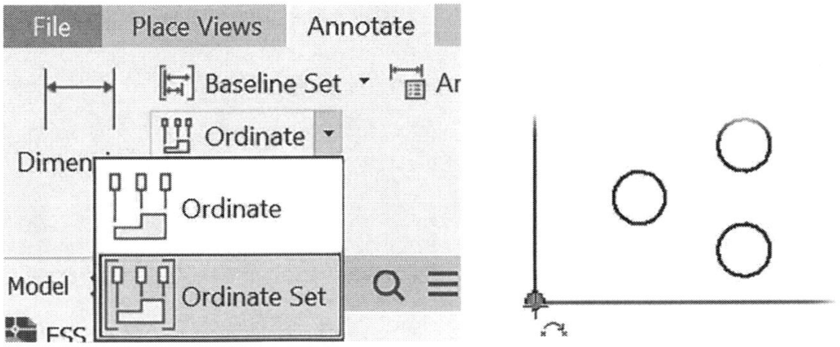

Figure 5.117

3. Select geometry that the ordinate dimensions will be applied to. When done selecting geometry, right-click and click Continue from the marking menu.

4. Locate the dimensions horizontally or vertically by moving the cursor and then click a point when the dimensions are previewed in the correct orientation, as shown in Figure 5.118 left. The image on the right shows the placed dimensions.

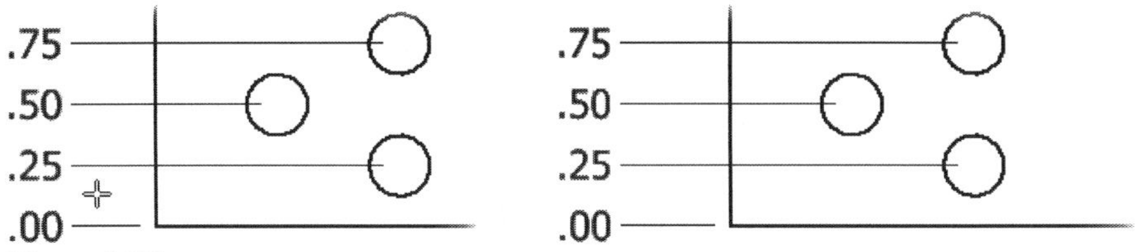

Figure 5.118

5. To change options for the dimension set, right-click, and then click Options, as shown in Figure 5.119 left. This can also be done after creating the dimensions by moving the cursor over the dimension set, right-click and click Options from the marking menu.

6. To create the dimensions set and end the command, press the ESC key.

7. To edit the origin, right-click on the dimension that will be the origin and click Make Origin from the menu as shown in the image on the right. The other dimension values will be updated to reflect the new origin.

8. To edit a dimension set, right-click on a dimension in the set and click the desired option from the menu as shown in the image on the right.

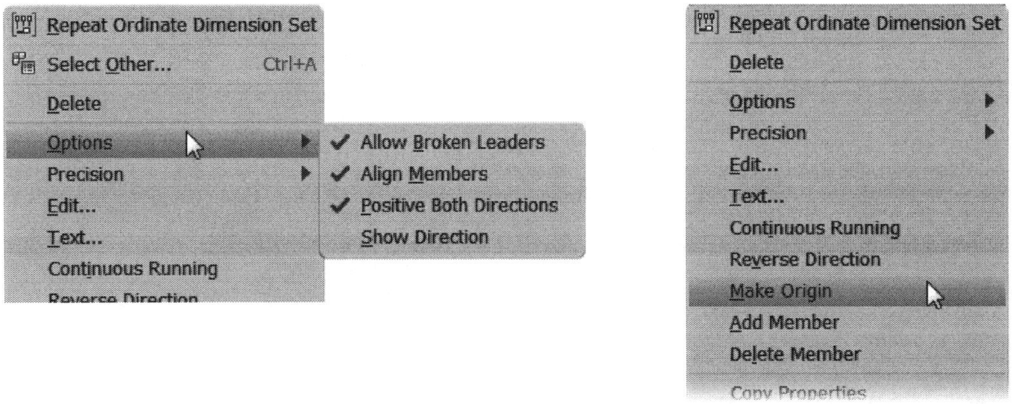

Figure 5.119

ADDING TEXT AND ADDITIONAL SYMBOLS

When documenting a drawing you will need to add additional annotations to describe the design. In this section, you learn how to add text and annotation symbols.

Text and Leader Text

To add text to a drawing, click either the Text or the Leader Text command on the Annotate tab > Text panel, as shown in Figure 5.120.

Figure 5.120

The Text command only adds text. The Leader Text command adds a leader with the text. Select the desired text command, and define the leader points and/or text location. Once you have chosen the location in the graphics window, the Format Text dialog box appears. When placing text using the Format Text dialog box, select the orientation and text style as needed and type in the text, as shown in Figure 5.121. Click the OK button to place the text in the drawing. To edit the text or text leader position, move your cursor over it, click one of the green circles that appear, and drag to the desired location.

To edit the text or text leader content, right-click on the text, and select Edit Leader Text or Edit Text from the menu.

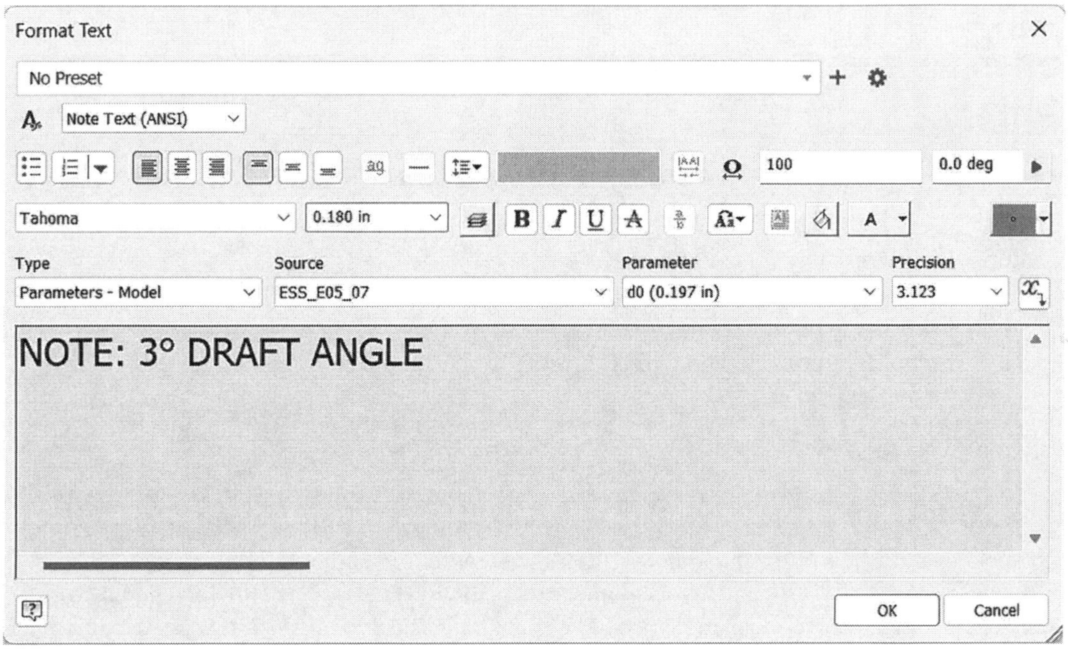

Figure 5.121

Annotation Symbols

To add more detail to your drawing, you can add symbols: surface texture symbols (GD& T), welding symbols, feature control frames, feature identifier symbols, datum identifier symbols, and datum targets by clicking the corresponding command on the Annotate tab > Symbols panel, as shown in the following image. Note that the Import AutoCAD Block command is only available when you are in an Inventor DWG file.

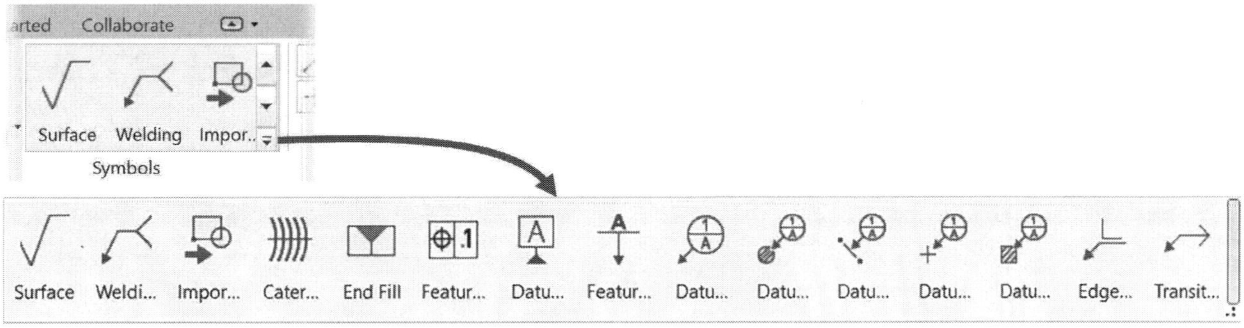

Figure 5.122

Follow these steps for placing symbols:

1. Click the appropriate symbol command on the Annotate tab > Symbols panel.
2. Select a point at which the leader will start. If you don't want a leader, click a point that the symbol will be placed and proceed to step 4.
3. Continue selecting points to position the leader lines.
4. If you only want to place symbols with one segment, before locating the symbol, right-click and click Single-segment leader from the menu. This setting is used for how the selected symbol is placed in the future.

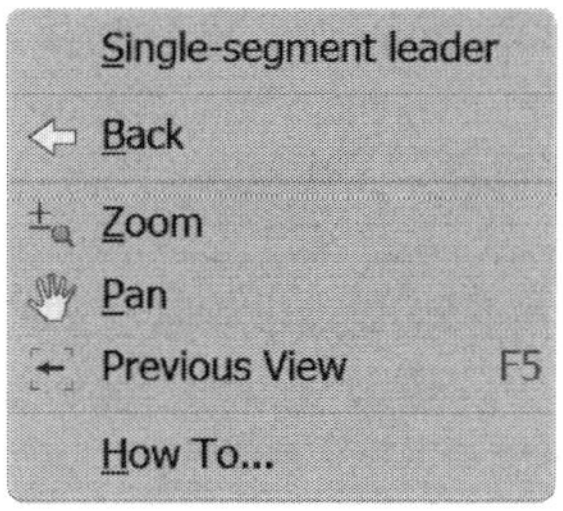

Figure 5.123

5. Right-click and click Continue from the marking menu. If the Single-segment leader option is selected, after clicking a second point, the symbol dialog box will automatically appear.
6. Fill in the information as needed in the dialog box.
7. When done, click the OK button in the dialog box.
8. To complete the operation, right-click, and click Cancel [ESC] from the marking menu.
9. To edit a symbol, move the cursor over it. When the green circles appear, right-click, and click the corresponding Edit option from the menu.

Hole and Thread Notes

Another annotation you can add is a hole or thread note. Before you can place a hole or thread note in a drawing, a hole or thread feature must exist. You can also annotate extruded circles using the Hole or Thread Note command. If the hole or thread feature changes in the part file, the note in the drawing will be updated automatically to reflect the change. The following image shows examples of counterbore, countersink, and threaded/tapped hole notes.

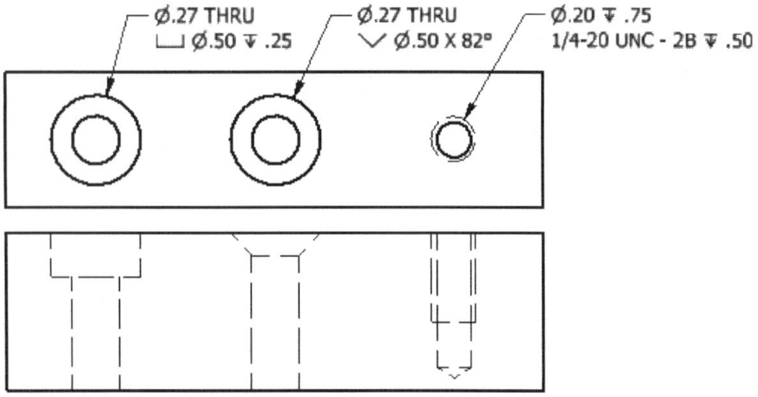

Figure 5.124

To create a hole or thread note, follow these steps:

1. Click the Hole and Thread Notes command on the Annotate tab > Feature Notes panel, as shown in the following image.

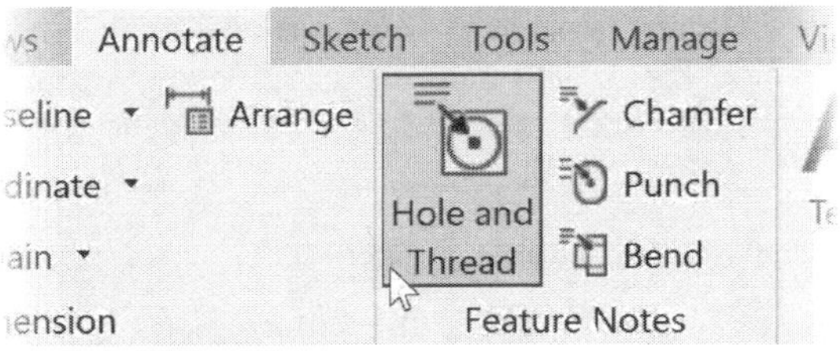

Figure 5.125

2. In a drawing view select a hole or thread feature to annotate.
3. Click a second point to locate the leader and the note.
4. To complete the operation, right-click, and select OK from the marking menu.

Editing Hole Notes

To edit a hole note, either double-click on an existing hole, or right-click on a hole note and click Edit Hole Note from the menu as shown in the following image.

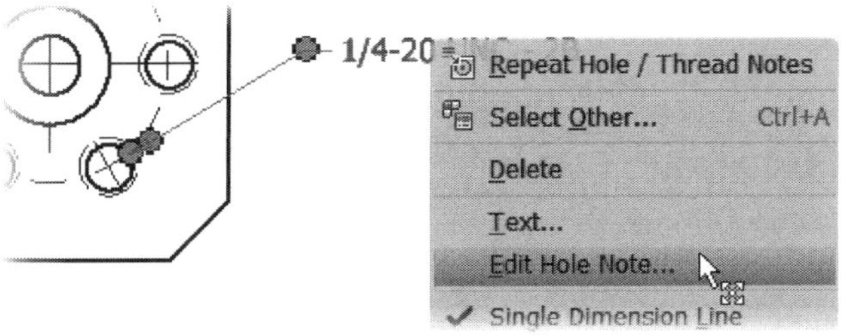

Figure 5.126

The Edit Hole Note dialog box appears; in the dialog box, you can add options to the hole note from the Values and Symbols area. For tapped holes, you can add information about the tap drill by checking the Tap Drill option on the right side of the dialog box. To add a quantity of the number of holes to the hole note, move the cursor to the beginning or the end of the hole note and then click the # (Quantity Note) in the Values and Symbols area. You can modify how the quantity note is calculated by clicking on Edit Quantity Note in the Options area of the Edit Hole Note dialog box. The following image on the left shows the Tap Drill and Quantity information added in the dialog box and the updated hole note in the image on the right.

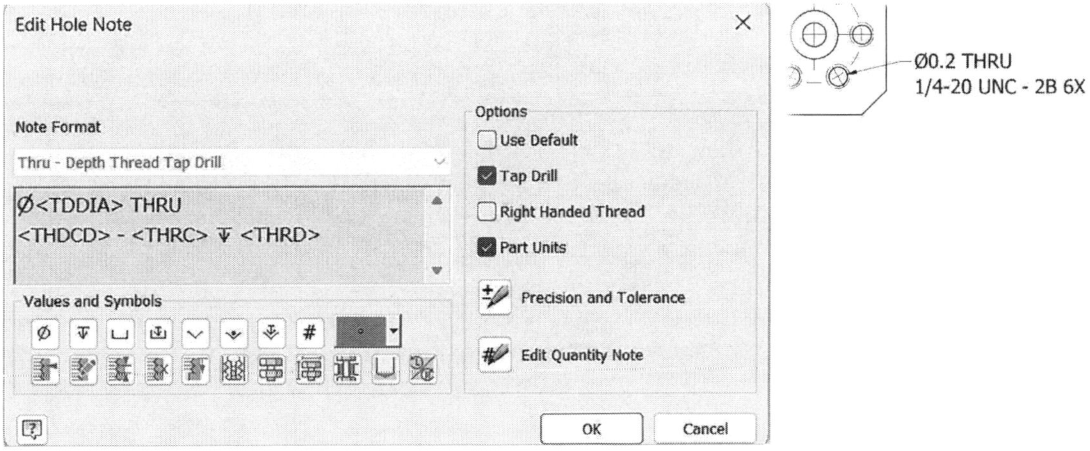

Figure 5.127

EXERCISE 5-8: ADDING ANNOTATIONS

In this exercise, you add hole notes, a surface texture symbol, and a text note.

1 Open the drawing *ESS_E05_08.dwg* in the Chapter 05 folder. The drawing file contains four drawing views with centerlines and dimensions.

2 First you add three hole notes. Begin by zooming in on the top view.

a. Click the Hole and Thread Notes command on the Annotate tab > Feature Notes panel.

b. Select the tapped hole on the left side of the top drawing view and click a point to place the hole note, as shown in the following image on the left, and complete the command by right-clicking and click OK in the marking menu.

c. Notice in the hole note that there is no information about the tap drill. To add tap drill information, double-click on the hole note and in the Options area, in the Edit Hole Note dialog box, check the Tap Drill option as shown in Figure 5.128 right.

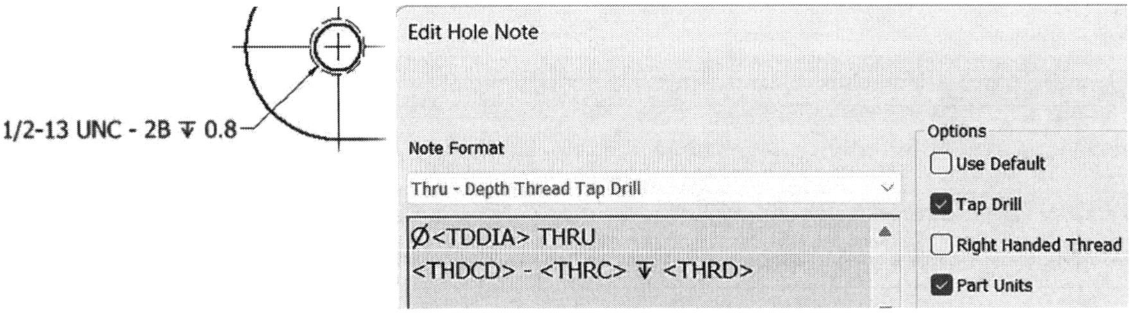

Figure 5.128

d. Click OK to finish the edit. Figure 5.129 left shows the updated hole note.

3 On the right side of the top view, add two hole notes as shown in Figure 5.129 right.

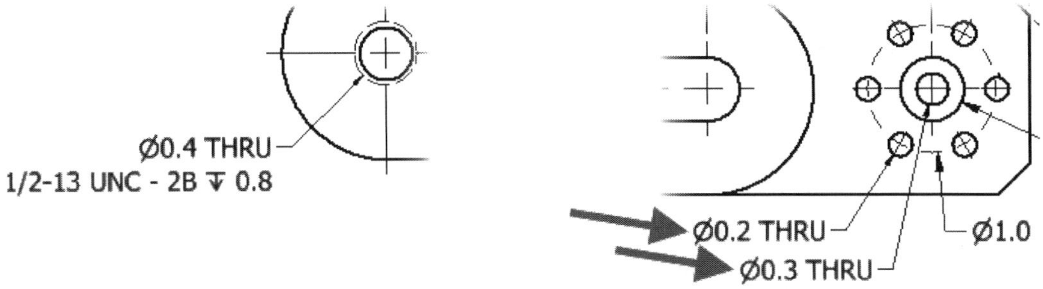

Figure 5.129

4 The Ø0.2 THRU hole note defines the diameter of the holes that fall on a bolt circle but the note does not define the number of holes. To add a quantity note, double-click on the Ø0.2 THRU hole note, and in the Note Format area, add a space after the word THRU and then click on #, which is the Quantity Note symbol, as shown in Figure 5.130 left.

5 Click OK to finish the edit. Figure 5.130 right shows the updated hole note.

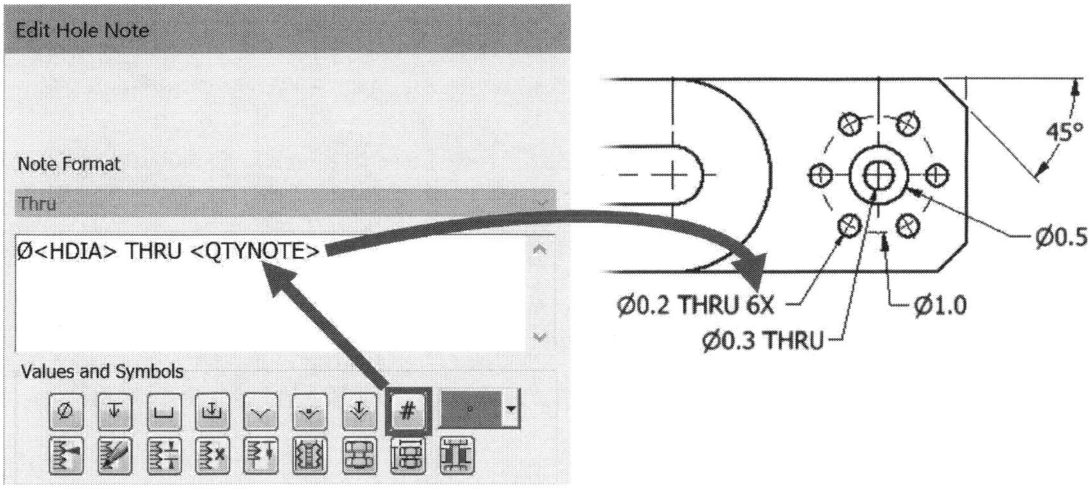

Figure 5.130

6 Next you add a surface finish note. Pan the screen so you can see the front view.

a. Click the Surface Texture Symbol command on the Annotate tab > Symbols panel.

b. Click on the middle horizontal edge of the part as shown in Figure 5.131 and then right-click and click Continue from the marking menu. Since the symbol will sit on the edge you don't need to click again to create an arrow.

c. In the Surface Texture dialog box enter **32** in the B Production method area as shown in the following image.

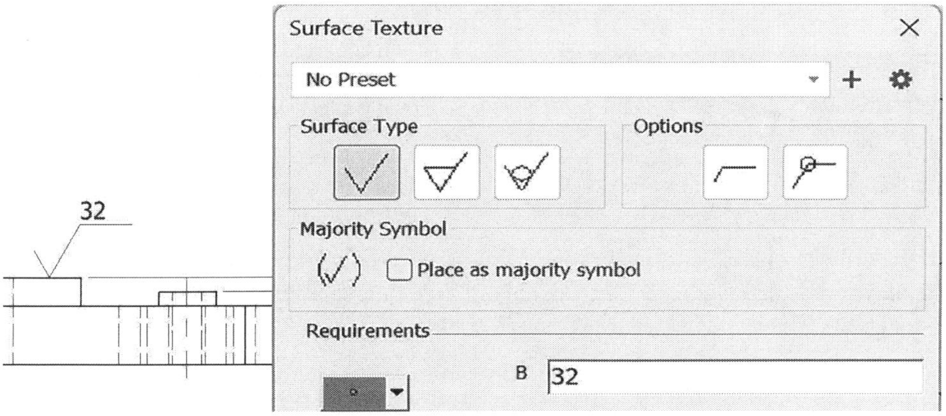

Figure 5.131

d. Click OK to create the surface texture symbol.

7 Next you add a text note. Click the Text command on the Annotate tab > Text panel.

a. To locate the position where the text will start, click in a blank area below the front view.

b. In the Format Text dialog box enter the text as shown in Figure 5.132. To add the ± symbol, select the ± symbol from the Insert symbol area in the dialog box.

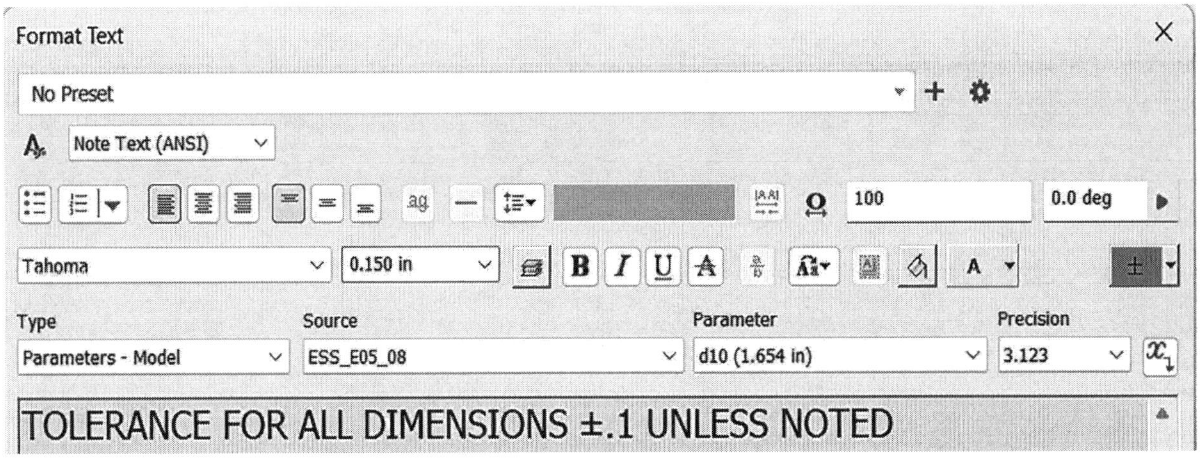

Figure 5.132

c. To create the text, click OK in the dialog box, then complete the command by right-clicking and click OK on the marking menu.

8 The text is displayed on a single line. To change the text so it appears on multiple lines, click on the text in the graphics window and then click and drag the lower-right green circle of the text's bounding area and drag it to the left as shown in the following image on the left. Release the mouse button and the text will be fitted with the new defined area as shown in the following image on the right.

Figure 5.133

9 Lastly, you arrange all the dimensions. Click the Arrange command in the Annotate tab > Dimension panel and drag a selection window around all the dimensions in the top and front views, right-click and click OK on the marking menu. The dimensions are arranged according to the settings in the current dimension style.

10 On the right side of the top view move the 45-degree dimension closer to its geometry. When done, your screen should resemble Figure 5.134.

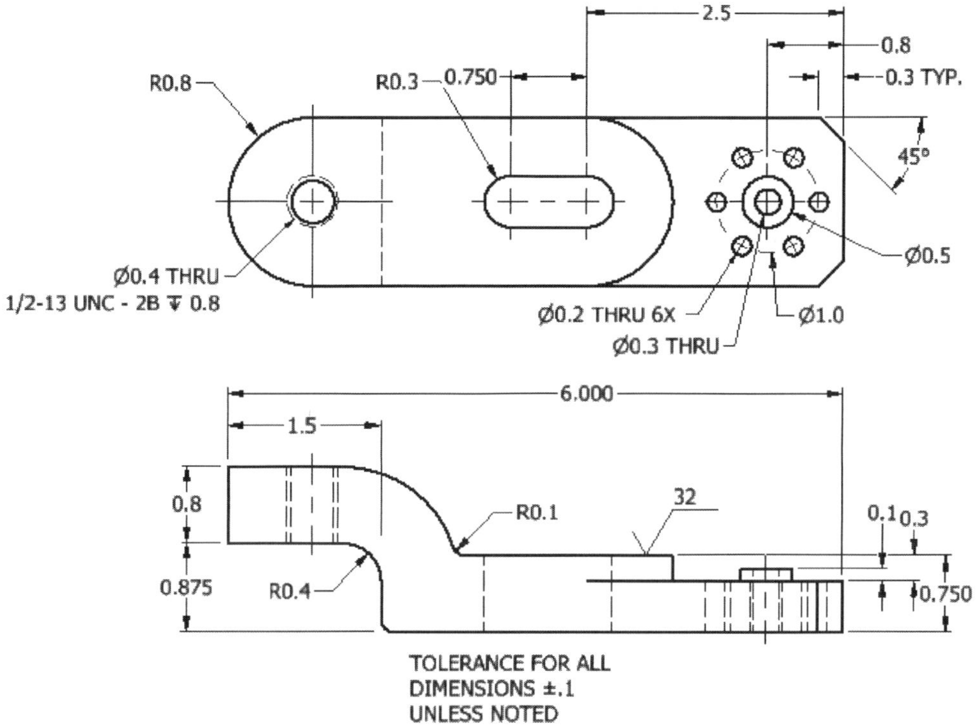

Figure 5.134

11 Close all open files. Do not save changes.

End of exercise.

Creating a Hole Table

If the drawing view that you are dimensioning contains holes, you can locate them by placing individual dimensions or by creating a hole table that will list the location and size of all the holes, or just the selected holes, in a view. The hole locations will be listed in both X and Y axis coordinates with respect to a hole datum that will be placed before creating the hole table.

After placing a hole table, if you add, delete, or move a hole, the hole table will automatically be updated to reflect the change after the part is saved and the drawing is the active document. Each hole in the table is automatically given an alphanumeric tag as its name. It can be edited by double-clicking on the tag in the drawing view or hole table. Alternately, you can right-click on the tag in the drawing view or hole table, and select Edit Tag from the menu. Type in a new tag name, and the change will appear in the drawing view and hole table. There are three commands for creating a hole table: Selection, View and Features, and they are located on the Annotate tab > Table panel, as shown in Figure 5.135.

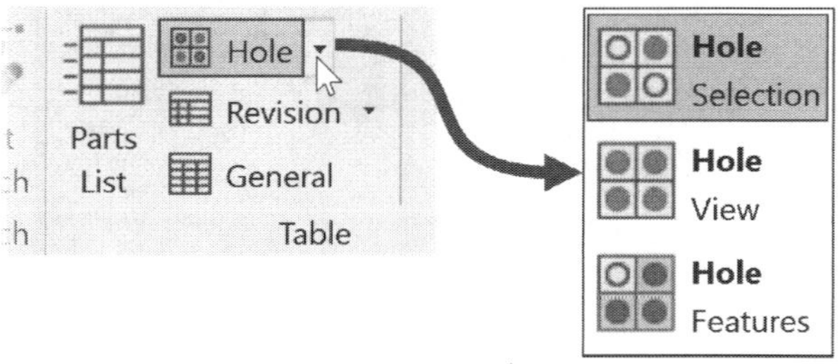

Figure 5.135

Following are descriptions of the hole table commands.

Hole Table Selection

When using the Hole Table Selection command, a hole table is created only from selected holes, as shown in Figure 5.136.

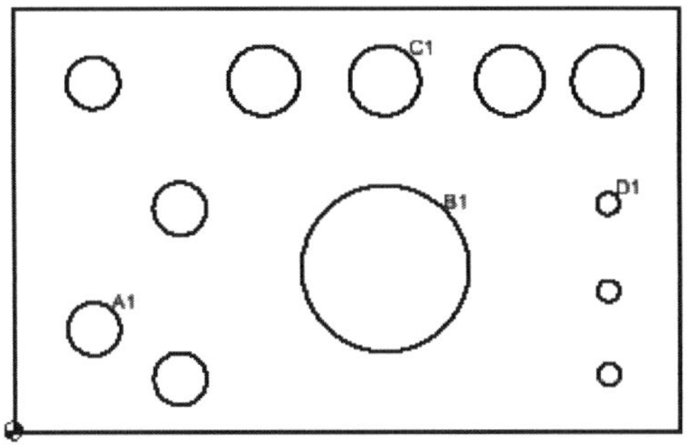

Hole Table			
HOLE	XDIM	YDIM	DESCRIPTION
A1	14.28	18.40	Ø0.38 THRU
B1	66.53	29.23	Ø1.18 THRU
C1	66.53	63.09	Ø0.50 THRU
D1	106.89	40.89	Ø0.16 THRU

Figure 5.136

Hole Table View

When using the Hole Table View command, a hole table is created based on all the holes in a selected view, as shown in the following image. Notice that all holes are given an alphanumeric identifier, which locates each hole by X and Y axis coordinates.

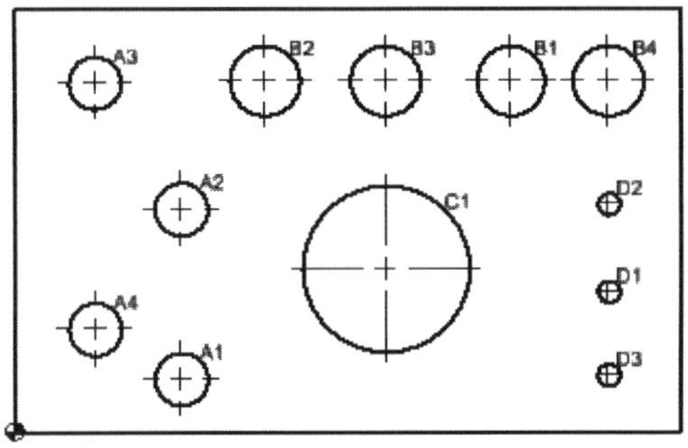

Hole Table			
LOC	XDIM	YDIM	SIZE
A1	29.77	9.39	Ø0.38 THRU
A2	29.77	40.02	Ø0.38 THRU
A3	14.28	62.73	Ø0.38 THRU
A4	14.28	18.40	Ø0.38 THRU
B1	89.23	63.09	Ø0.50 THRU
B2	44.91	63.09	Ø0.50 THRU
B3	66.53	63.09	Ø0.50 THRU
B4	106.89	63.09	Ø0.50 THRU
C1	66.53	29.23	Ø1.18 THRU
D1	106.89	25.12	Ø0.16 THRU
D2	106.89	40.89	Ø0.16 THRU
D3	106.89	10.11	Ø0.16 THRU

Figure 5.137

Hole Table – Selected Features

When using the Hole Table – Selected Features command, a hole table is created based on holes that are identical to the selected hole. The following image shows one of the larger holes, located in the upper part of the object (A4), selected using the Hole Features command.

When the hole table is created, all holes that share the same type and size are added to the hole table, as shown in Figure 5.138.

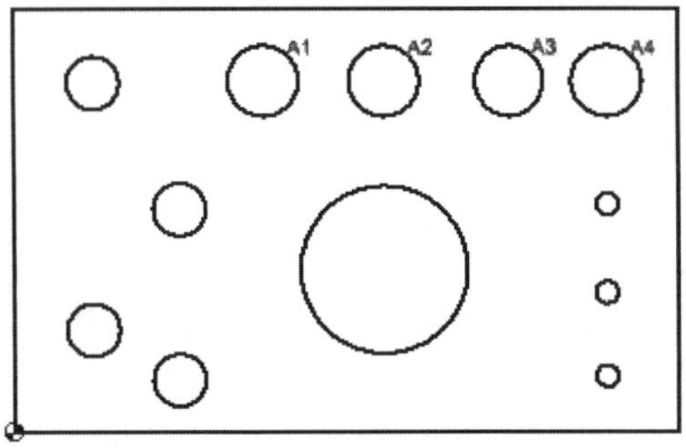

Hole Table			
HOLE	XDIM	YDIM	DESCRIPTION
A1	44.91	63.09	Ø0.50 THRU
A2	66.53	63.09	Ø0.50 THRU
A3	89.23	63.09	Ø0.50 THRU
A4	106.89	63.09	Ø0.50 THRU

Figure 5.138

To create a hole table based on an existing drawing view that contains holes in a plan view, follow these steps:

1. Select the desired Hole Table command from the Hole drop list on the Annotate tab > Table panel.
2. Select the view on which the hole table will be based.
3. Select a point to locate the origin (0,0). Typical origins include the corners of rectangular objects and the centers of drill holes used for data.
4. For the Hole Selection and the Hole Features commands select the hole(s) on which to base the hole table and then right-click and click Create from the marking menu.
5. Click a point in the drawing to locate the table.
6. The contents of the hole table can be edited by either double-clicking or right-clicking on the hole table and click Edit Hole Table from the menu.

EXERCISE 5-9: CREATING HOLE TABLES

In this exercise, you create a hole table for a shim plate. After creating the hole table, you will modify the plate model to add new holes and update the hole table accordingly.

1 Open the drawing file *ESS_E05_09.idw* in the Chapter 05 folder.

2 First create a hole table; click the Hole Table – View command from the Annotate tab > Table panel by clicking the down arrow next to Hole and click the Hole Table View command.

a. Click the front view of the plate as the view to which the hole table will be associated.

b. To place the hole table datum (0,0), acquire the theoretical intersection of the lower left corner of the front view by moving the cursor over the bottom horizontal and left vertical line until the intersection is displayed and then click, as shown in the following image on the left.

c. Then place the hole table on the right side of the sheet as shown in Figure 5.139 right.

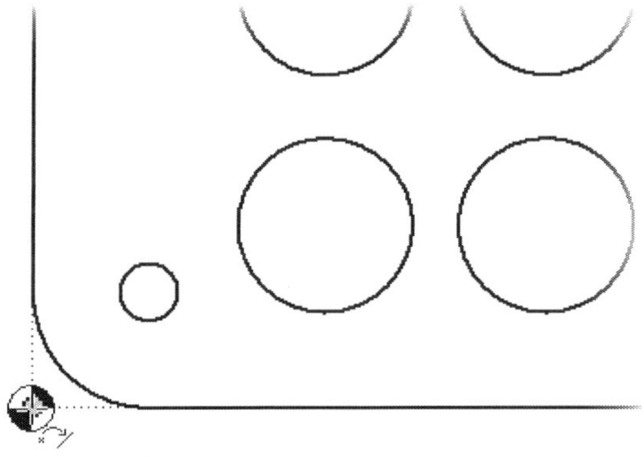

HOLE TABLE			
HOLE	XDIM	YDIM	DESCRIPTION
A1	.25	.25	∅0.13 THRU
B1	.63	.39	∅0.38 THRU
B2	1.10	.39	∅0.38 THRU
B3	1.57	.39	∅0.38 THRU
B4	2.05	.39	∅0.38 THRU
B5	2.52	.39	∅0.38 THRU
B6	.63	.91	∅0.38 THRU
B7	1.10	.91	∅0.38 THRU
B8	1.57	.91	∅0.38 THRU
B9	2.05	.91	∅0.38 THRU
B10	2.52	.91	∅0.38 THRU
C1	.38	2.00	∅0.25 THRU
C2	2.75	2.00	∅0.25 THRU

Figure 5.139

3 The top two holes in the plate need to be changed to Spotface holes. Open the part file by right-clicking in the front view and click Open from the menu as shown in Figure 5.140.

Figure 5.140

4 In the browser double-click on the Hole1 feature.

5 In the Hole Properties panel, change the hole type to Spotface and change the hole specifications, as shown in the following image on the left. To update the hole features, click OK in the dialog box.

6 Save the part file.

7 Make the drawing file current by clicking the tab labeled *ESS_E05_09.idw* near the bottom of the screen (do not close the part file). Note the updated hole table; the holes labeled with the letter C reflect the change to the holes, as shown in Figure 5.141 right.

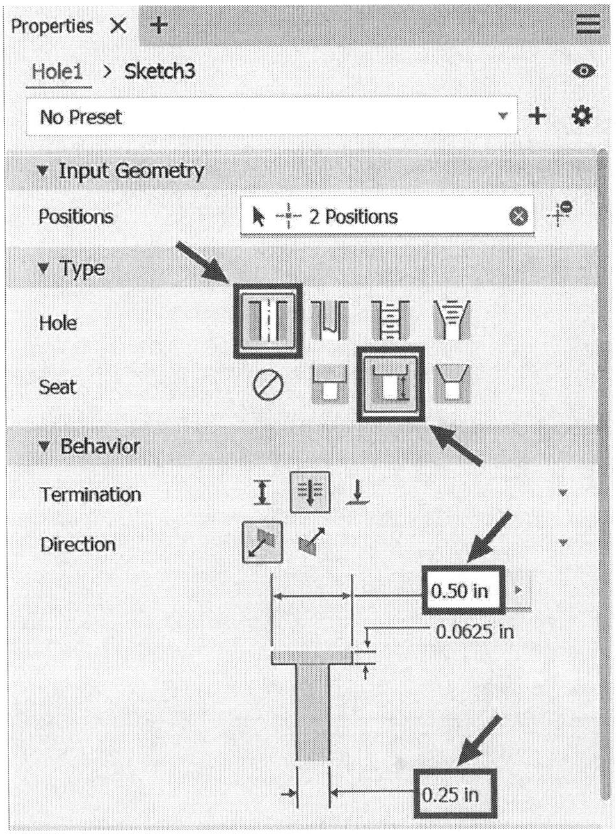

HOLE TABLE			
HOLE	XDIM	YDIM	DESCRIPTION
A1	.25	.25	Ø0.13 THRU
B1	.63	.39	Ø0.38 THRU
B2	1.10	.39	Ø0.38 THRU
B3	1.57	.39	Ø0.38 THRU
B4	2.05	.39	Ø0.38 THRU
B5	2.52	.39	Ø0.38 THRU
B6	.63	.91	Ø0.38 THRU
B7	1.10	.91	Ø0.38 THRU
B8	1.57	.91	Ø0.38 THRU
B9	2.05	.91	Ø0.38 THRU
B10	2.52	.91	Ø0.38 THRU
C1	.38	2.00	Ø0.25 THRU ⌴ Ø0.50
C2	2.75	2.00	Ø0.25 THRU ⌴ Ø0.50

Figure 5.141

8 Two countersink mounting holes need to be created in the part file. Make the part file *ESS_E05_09.ipt* current by clicking on its tab.

a. Add two center points by creating a sketch, placing two center points, adding a horizontal constraint between the points, and adding the dimensions as shown in Figure 5.142 left.

b. Create two countersink, through all holes with the values shown in the image on the right.

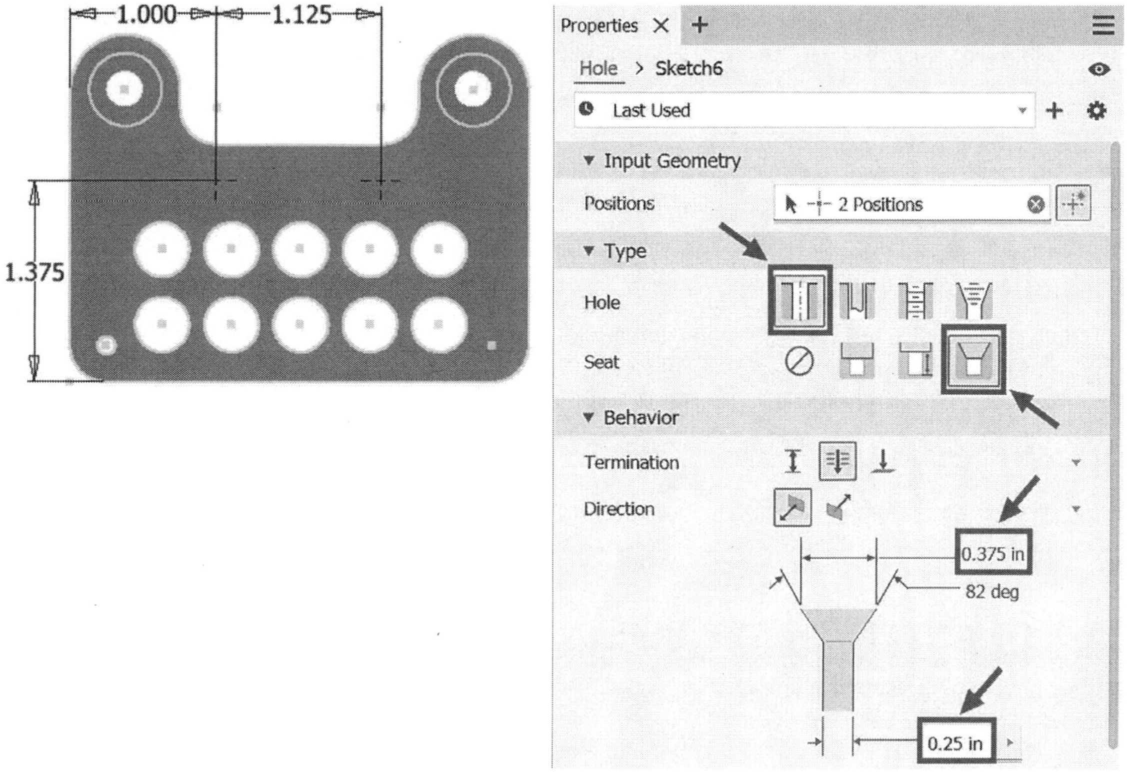

Figure 5.142

9 Save the part file.

10 Make the file *ESS_E05_09.idw* current. Notice that the new holes have automatically been added to the hole table as shown in Figure 5.143.

C2	2.75	2.00	⌴ Ø0.50
D1	1.00	1.38	Ø0.25 THRU ⌵ Ø0.38 X 82°
D2	2.13	1.38	Ø0.25 THRU ⌵ Ø0.38 X 82°

Figure 5.143

11 A hole tag will now be moved. Position the cursor over the tag A1 on the drawing view, then click on the green circle (location where the tag is attached to the geometry) and drag and position the A1 hole tag, as shown in the following image on the left.

12 Next double-click on the A1 tag to open the Format Text dialog box, add the text "**TOOLING HOLE**" after A1 to the hole tag, and click OK. The tag updates as shown in Figure 5.144 right.

Figure 5.144

13 The hole table updates to include – TOOLING HOLE after A1. You can also resize the hole table columns by clicking and dragging on a table border, as shown in Figure 5.145.

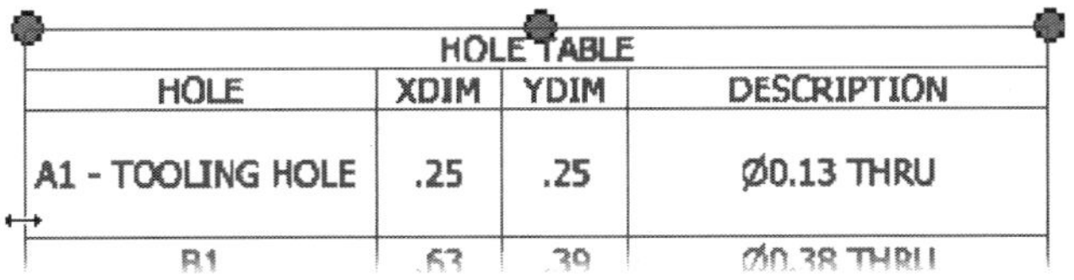

Figure 5.145

14 If desired, you can combine the descriptions that are identical in the hole table. Combine the descriptions by right-clicking on the hole table and click Edit Hole Table from the menu. In the Edit Hole Table dialog box, on the Options tab, select the Combine Notes option and click OK in the dialog box.

15 Close all open files. Do not save changes.

End of exercise.

SHORTCUT FOR OPENING REFERENCED FILES

To prevent laborious searches to open referenced part, assembly, presentation or drawing files, you can open the file directly from within the file they are referenced.

Opening a Model File from a Drawing

While working inside of a complex drawing view, you may need to make changes to a part, assembly, or presentation file. Rather than starting the open command and searching for the part, assembly, or presentation file, you can open them directly from the drawing in which they are referenced. The part, assembly or presentation file must have the same name as the drawing file. There are multiple techniques to do this.

- One technique is to right-click in a drawing view to display the menu as shown in the following image on the left, and then click Open to open the part file. In this example, an assembly file was opened from the drawing view.
- Another technique involves expanding the drawing view in the browser to expand the assembly and continue expanding the browser until the file you want to open is visible in the list. Right-click the desired file in the browser and select Open from the menu to display the part model as shown in the following image on the right.

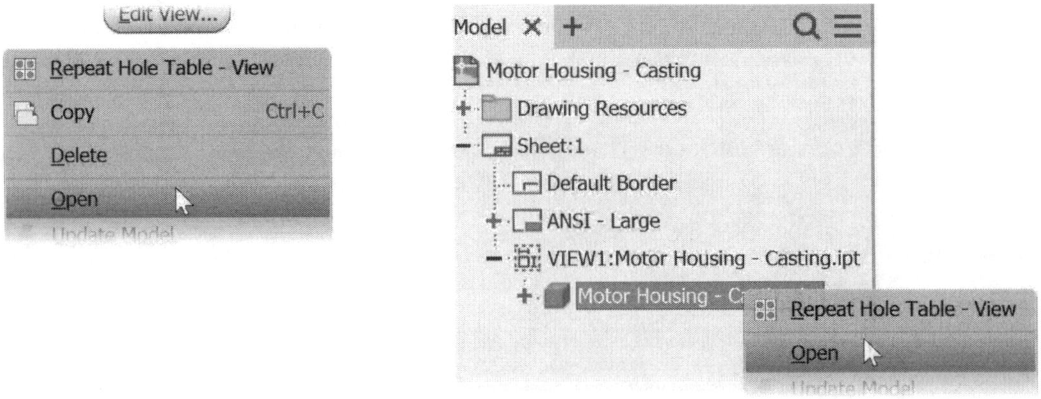

Figure 5.146

Opening a Drawing from a Model File

You can open a drawing from a part, assembly, or presentation file by right-clicking on the part, assembly presentation file name and clicking Open Drawing from the menu as shown in the following image. This will open the drawing file associated with the file. The drawing file must have the same name as the part, assembly or presentation file.

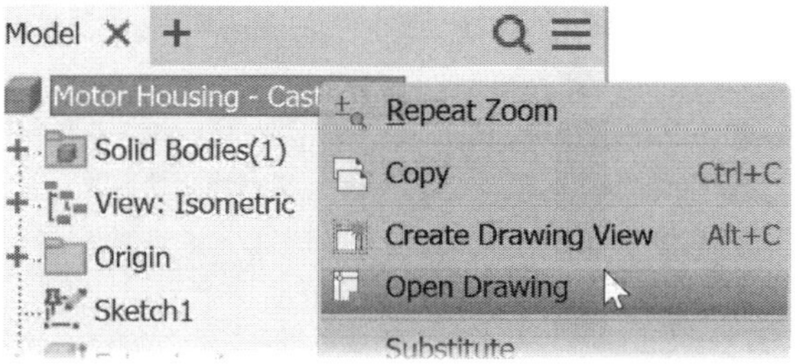

Figure 5.147

CREATE A 3D PDF

Another way to share your design is to output a part file or an assembly file to a 3D PDF; the 3D PDF can be viewed in a free Adobe Acrobat Reader. To create a 3D PDF, follow these steps.

1. Open a part or an assembly file.
2. Click the File tab > Export 3D PDF as shown in the following image.

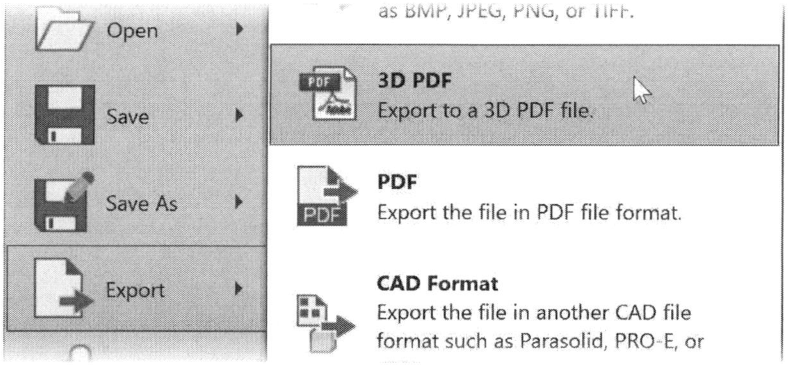

Figure 5.148

3. Make desired changes in the Publish 3D PDF dialog box as shown in the following image.

   A. Select the model properties to include in the 3D PDF.

   B. Select the design view representations to include in the 3D PDF.

   C. Set Visualization Quality and Export Scope.

   D. Set the location of the Template file.

   E. Set the location to output 3D PDF file.

   F. Select the "View PDF when finished" option to open the output file after the export finishes.

   G. If desired, attach files to the 3D PDF file.

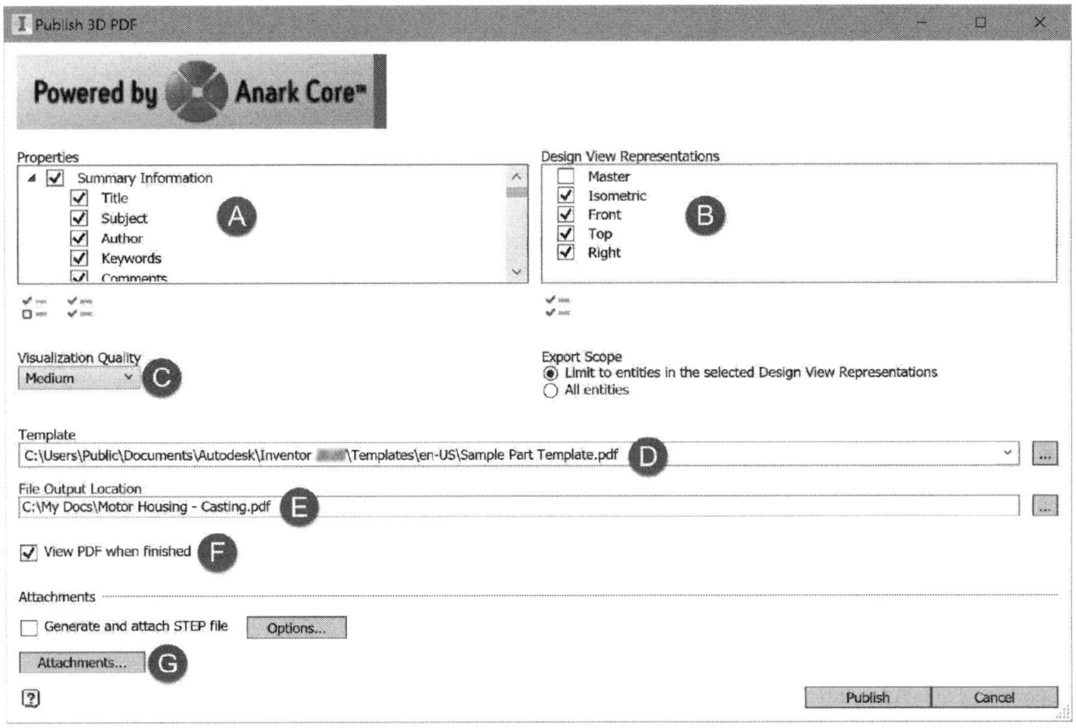

Figure 5.149

4. Click Publish.
5. View the 3D PDF in Adobe Acrobat Reader. The following image shows an example of a published 3D PDF.

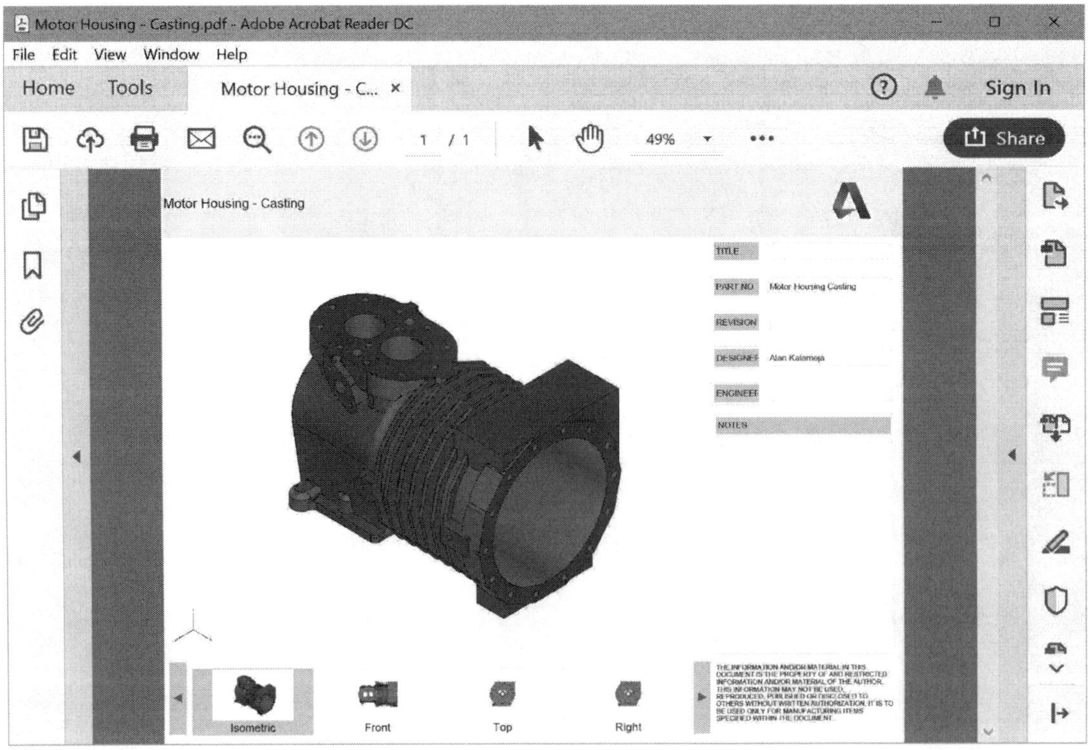

Figure 5.150

3D ANNOTATIONS / MODEL-BASED DEFINITION (MBD)

Another option to document the manufacturing of a part is to add 3D annotations or model-based definition such as Surface Finish, GD&T, and dimension data directly to the 3D part. This information can also be retrieved in a drawing just like dimensions can be retrieved.

Creating a 3D Annotation

To add a 3D annotation to a part, follow these steps.
Note that you can also add 3D annotations to an assembly file, except geometric annotations cannot be added to an assembly file.

1. Open or make active a part file that the 3D annotations will be added to.
2. Click the Annotate tab and the tools on the ribbon will change as shown in the following image.

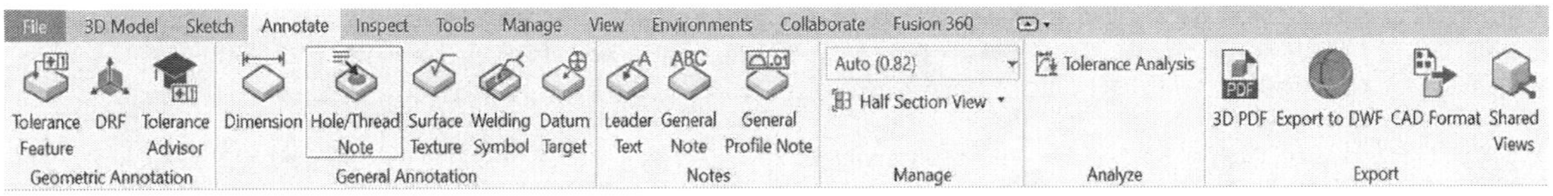

Figure 5.151

3. Click the desired command.
4. Select the geometry to associate the annotation to and then click the green check mark in the mini-toolbar. The following image shows a Tolerance Feature being attached to a hole.

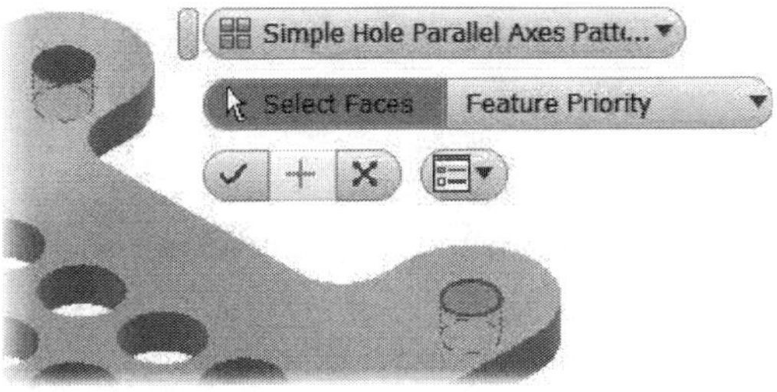

Figure 5.152

5. To place the annotation, click the green check mark and click a point. Make changes as needed by selecting options in the drop-down menus or the mini-toolbar. The following image shows the options for the Tolerance Feature.

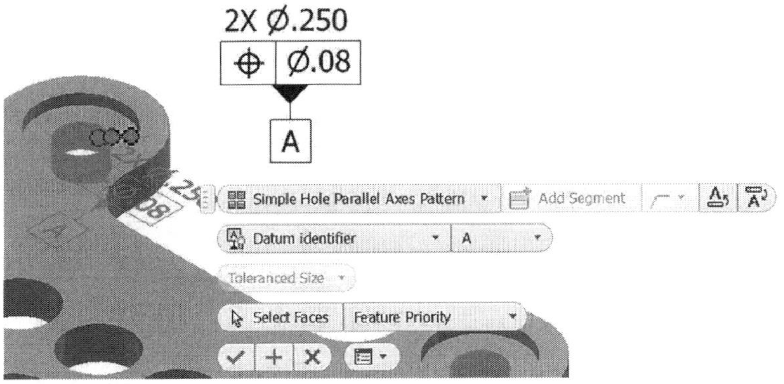

Figure 5.153

6. When done, click the green check mark in the mini-toolbar.
7. To edit a 3D Annotation, in the graphics window either double-click on it or right-click on it and click Edit from the menu.

8. Use the General Annotation and Text commands to document your part.

TIP When placing a dimension press the Spacebar to cycle through the planes where the dimension could be placed (this needs to be done while placing the dimension).

Tolerance Advisor

Use the Tolerance Advisor command to check the health of the tolerance you added to the part; it will check for the following:

- Potential problematic datums and Datum Reference Frames (DRFs)
- Information about which features are not fully controlled
- Errors in a tolerance feature

The following image on the left shows an example of errors in the Tolerance Advisor browser. To learn more about the error message, double-click on an error. The following image on the right shows the help that is displayed after double-clicking on the error "Zero-value (or inconsistent) tolerance not allowed."

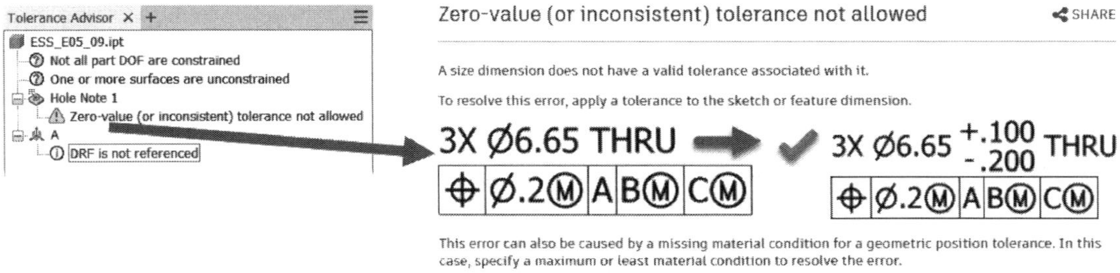

Figure 5.154

Controlling Visibility of Annotations

You can control the visibility of 3D Annotations by right-clicking on the annotation in the graphics window or in the browser and click Visibility as shown in the previous Figure.

You can also control the visibility of all 3D annotations with a single click with the Object Visibility command. Click the Object Visibility command on the View tab > Visibility panel and then check or uncheck 3D Annotations as shown in the following image. Check the option to turn the visibility on, and clear it to turn off the visibility.

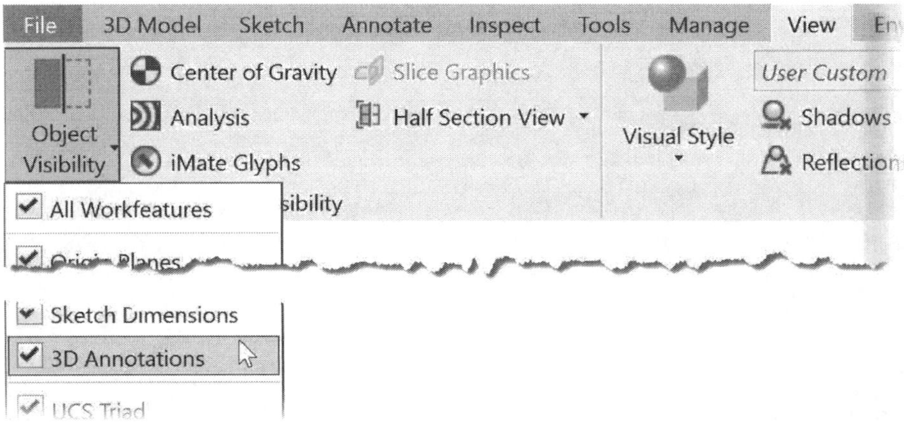

Figure 5.155

Retrieving 3D Annotations in a Drawing

To Retrieve 3D Annotations in a drawing use the Retrieve Model Annotations command to select 3D Annotations to display in a drawing view. Only those 3D annotations that you placed in the model will appear.

To retrieve 3D Annotations, follow these steps:

1. Create or open a drawing that contains a view(s) of a part or assembly that uses 3D annotations.
2. Start the Retrieve Model Annotations command on the Annotate tab > Retrieve panel as shown in Figure 5.156 left, or right-click in a blank area in the graphics window (not in a drawing view), and click Retrieve Model Annotations from the marking menu as shown in the image on the right.

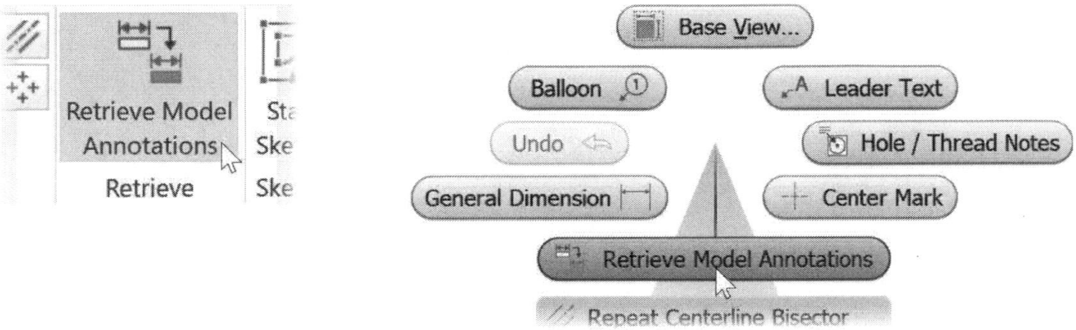

Figure 5.156

3. Select the drawing view in which to retrieve the 3D Annotations.
4. If needed select the 3D Annotations tab in the Retrieve Model Annotations dialog box, as shown in Figure 5.157.

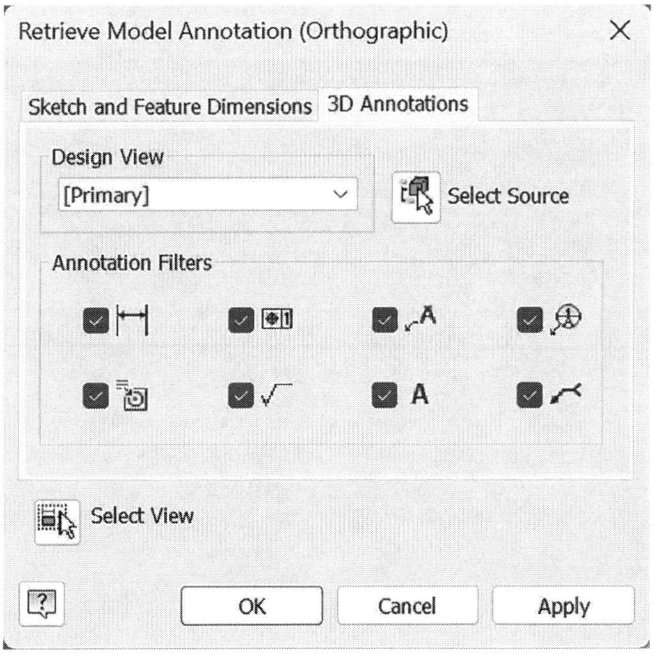

Figure 5.157

5. Uncheck the annotation types you do NOT want included in the drawing.
6. Click the OK button to retrieve the selected annotations.
   Note: The only dimensions that will be retrieved are the dimensions that are parallel to the selected view.
7. Reposition the annotations as needed.

APPLYING YOUR SKILLS

Skills Exercise 5-1

In this exercise, you create a drawing of a part.

1. Create a new drawing based on the English ANSI (in).idw template.
2. Change the sheet size to A.
3. Create drawing views from the file ESS_Skills_5-1.ipt from the Chapter 05 folder. Use a scale of 1:1 for all views.
4. Add centerlines, dimensions, and hole notes.
5. When done, your drawing should resemble the following image.

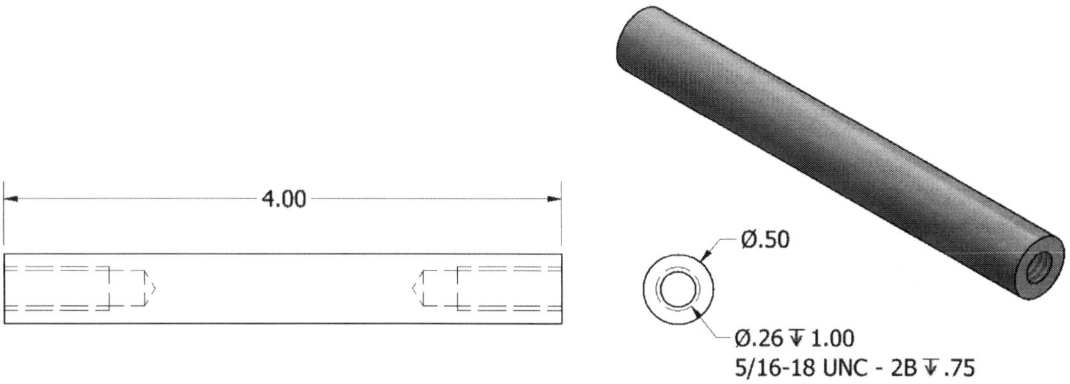

Figure 5.158

Skills Exercise 5-2

In this exercise, you create a drawing for a drain plate cover.

1. Create a new drawing using an English ANSI (in).dwg template.
2. Change the sheet size to A.
3. Create three views of the part ESS_Skills_5-2.ipt located in the Chapter 05 folder. Use a scale of 1:1 for all views.
4. Add center marks to the views.
5. Change the Default (ANSI) dimension style as follows:
    a. Set the length that the extension line extends beyond the dimension leader line to **.0625 inches**.
    b. Set the precision to **3.123** (three decimal places).
    c. Set text height to **.100 inches**.
6. Add dimensions and annotations.
7. When done, your drawing views should resemble the following image.

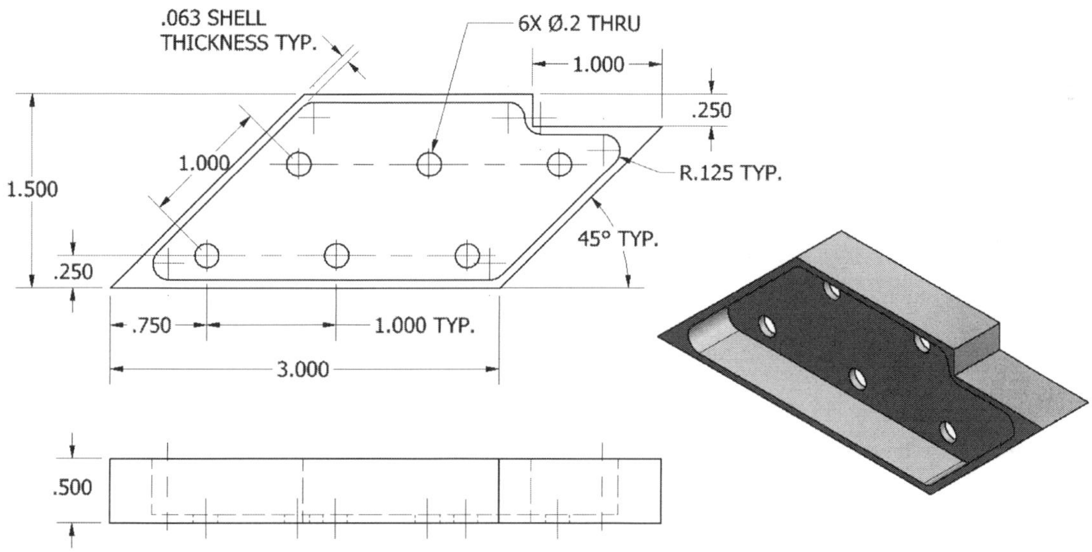

Figure 5.159

CHECKING YOUR SKILLS

Use these questions to test your knowledge of the material covered in this chapter.

1. True__ False__ A drawing can have an unlimited number of sheets.
2. Explain how to change a sheet's size.
3. True__ False__ There can only be one base view per sheet.
4. True__ False__ An inclined view is a view that is projected perpendicular to a selected edge or line in a view.
5. True__ False__ An isometric view can only be projected from a base view.
6. True__ False__ A section view is a view created by sketching a line or multiple lines that will define the plane(s) that will cut through a part or assembly.
7. True__ False__ Drawing dimensions can drive dimensional changes parametrically back to the part.
8. Explain how to shade an isometric drawing view.
9. True__ False__ When creating a hole note using the Hole and Thread Notes command, circles that are extruded to create a hole can be annotated.
10. While in a drawing file, explain how to open the part file that is referenced in the drawing.
11. True__ False__ When you create a title block in a drawing, the new title block will automatically be written back to the template file.
12. True__ False__ When creating a section view that goes through a slot at an angle, you use the projected option to prevent the slot from being distorted.
13. True__ False__ A hatch pattern in a section view is changed via the Document Settings.
14. True__ False__ Before creating a section view, you must first create a work plane in the part file that the section will be based on.
15. True__ False__ When creating a Detail View, you can use a circular or rectangular fence.
16. True__ False__ Use the Break Out command to remove a middle section from a drawing view of long part.
17. True__ False__ The Styles and Standards Editor only controls dimension styles.
18. True__ False__ The Centerline Bisector command can only be used to create a centerline between two parallel edges.
19. True__ False__ The Automatic Centerline command can only place centerlines and center marks in one view at a time.
20. True__ False__ 3D Annotations can be placed on both part and assembly files.

6 Creating & Documenting Assemblies

INTRODUCTION

In this chapter, you will learn how to place individual component files into an assembly file. You will also learn to create components in the context of the assembly file. After creating components, you will learn how to constrain the components to one another using assembly constraints, edit the assembly constraints, check for interference, and create presentation files that show how the components are assembled or disassembled. Manipulating and editing a Bill of Materials (BOM) is also discussed, including the placement of a Parts List and identifying balloons.

OBJECTIVES

After completing this chapter, you will be able to:

- ☐ Understand the assembly options
- ☐ Place components into an assembly
- ☐ Create components and assemblies
- ☐ Constrain components together using assembly constraints
- ☐ Edit assembly constraints
- ☐ Use assembly joints to control the location and motion of components
- ☐ Pattern components in an assembly
- ☐ Check parts in an assembly for interference
- ☐ Drive constraints
- ☐ Create a presentation file
- ☐ Create drawing views from an assembly or presentation file
- ☐ Manipulate and edit the Bill of Materials (BOM)
- ☐ Create individual and automatic balloons
- ☐ Create and perform edits on a parts list in a drawing

ASSEMBLY OPTIONS

Before creating an assembly, review the assembly option settings. From the Tools tab > Options panel click Application Options to show the Application Options dialog box. Click on the Assembly tab, as shown in the Figure 6.1. Consult the help system for information about the Assembly Application Options. These settings are global and will affect how new components are created, referenced, analyzed, or placed in the assembly.

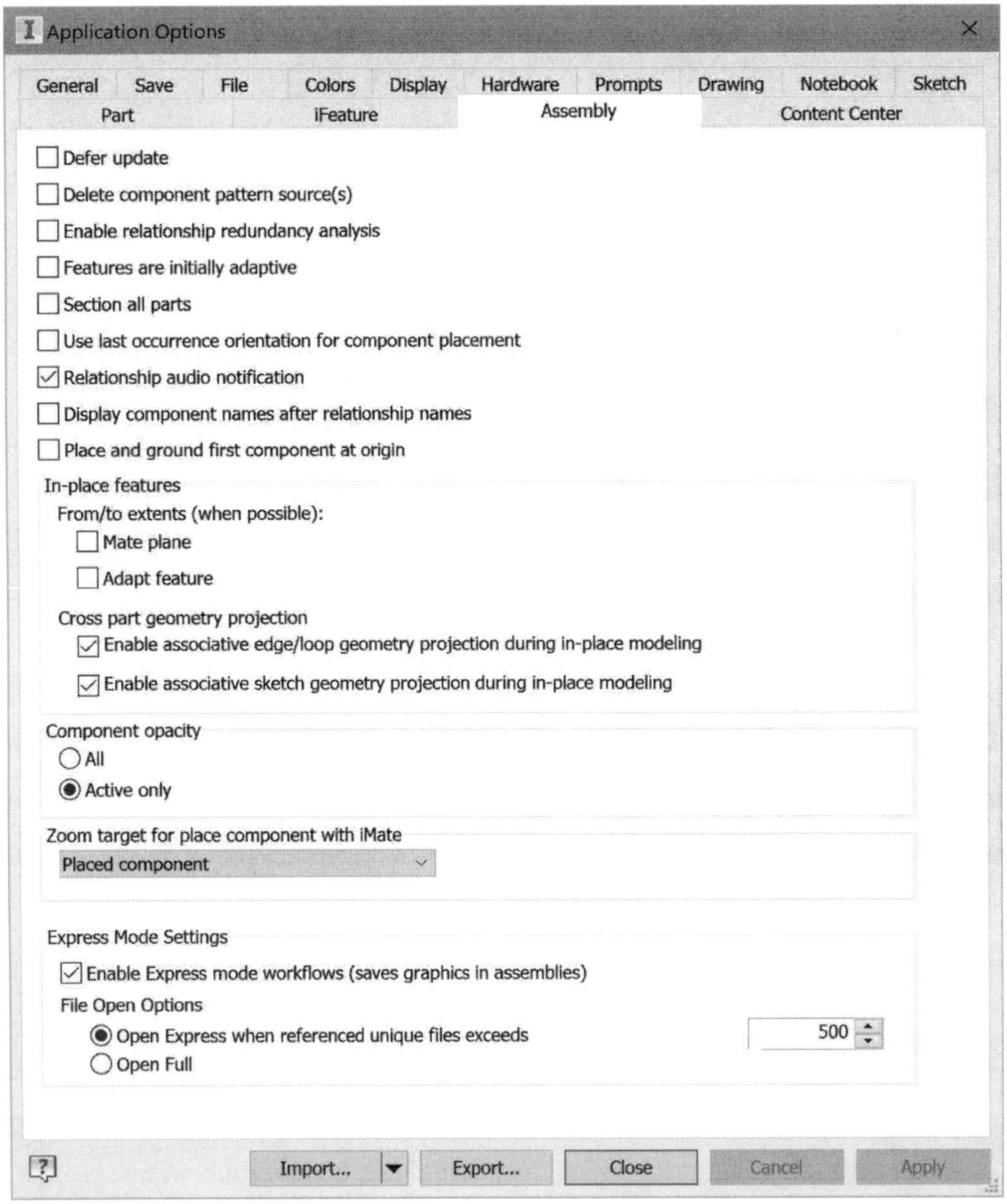

Figure 6.1

CREATING ASSEMBLIES

As you have already learned, part files have the .ipt extension, and consist of a single component. Assembly files (.iam file extension) bring multiple part files together. All the components in an assembly are referenced in, meaning that each component exists in its own component .ipt file, and its definition is linked into the assembly. You can edit the components while in the assembly, or you can open the component file and edit it. When you have made changes to a component and saved it, the changes are reflected in the assembly when you open or update it. There are three methods for creating assemblies: bottom-up, in-place, and a combination of both.

Bottom-up refers to an assembly in which all the components were created in individual component files and are referenced into the assembly. The in-place approach refers to an assembly in which the components are created from within the context of the assembly. In other words, the user creates each component from within the top-level assembly. Each component in the assembly is saved to its own .ipt file. The following sections describe the bottom-up and in-place assembly techniques.

Note that complex assemblies are typically made up of subassemblies, to let you better manage the large amount of data that is created when building assemblies.

To create a new assembly file use one of the following methods:

- From the Quick Access toolbar click the down arrow on the right side of the New icon and click Assembly from the drop list as shown in Figure 6.2 left.
- Click Assembly on the Home page as shown in the middle image.
- From the File tab click New > Assembly as shown at right.

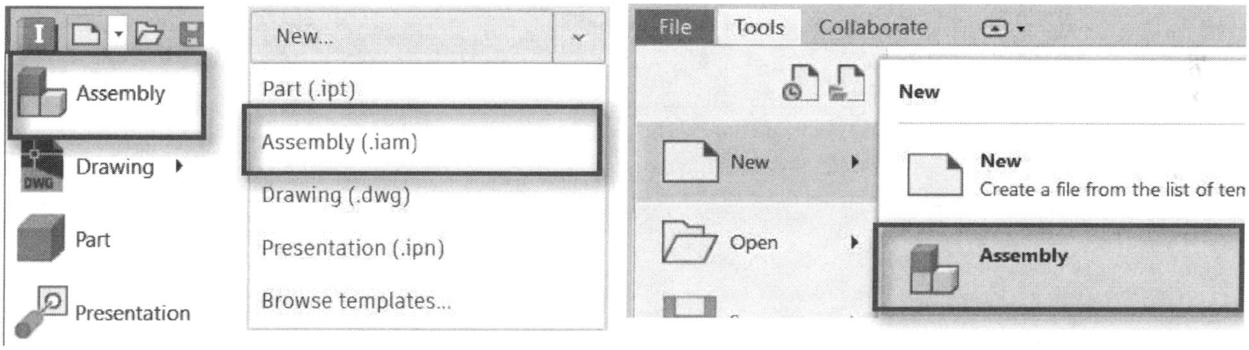

Figure 6.2

You can also create an assembly file from a template that is not the default by clicking the New file command from one of these areas:

- Quick Access toolbar as shown in Figure 6.3 left.
- File tab, as shown in the middle image.
- Or press CTRL + N.

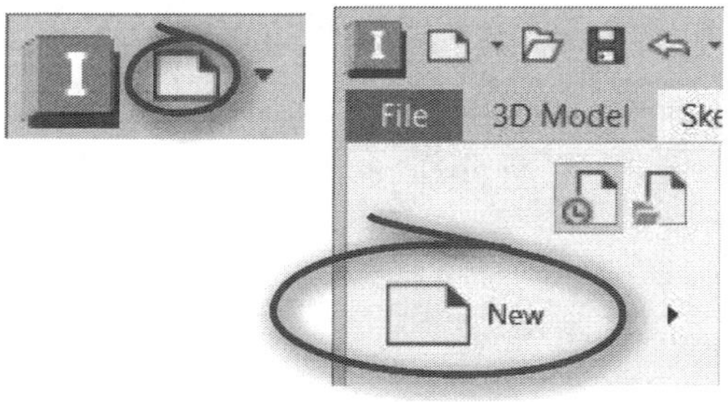

Figure 6.3

The Create New File dialog box appears; click the desired template folder on the left side of the dialog box, and in the Assembly – Assemble 2D and 3D components area, click the desired template file as shown in Figure 6.4.

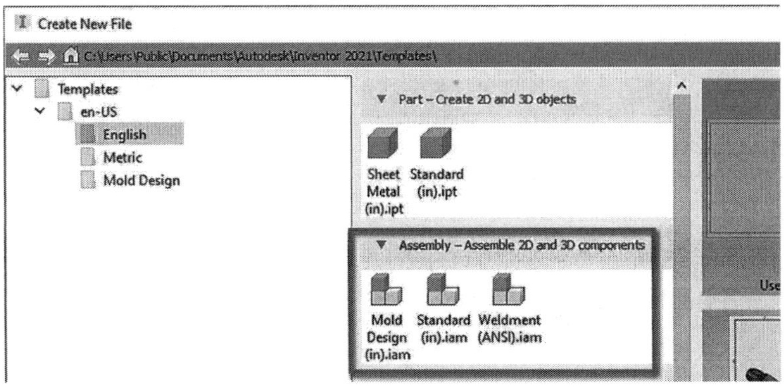

Figure 6.4

After creating a new assembly file, Inventor changes the Assembly environment as shown in Figure 6.5. The major assembly commands are covered throughout this chapter.

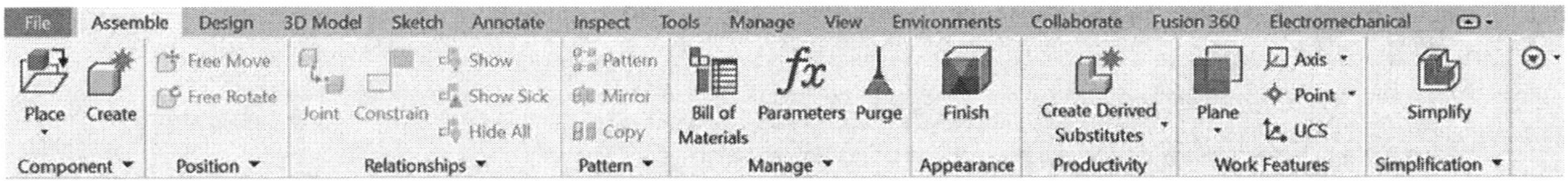

Figure 6.5

In the next sections, you learn approaches to creating an assembly. You can place existing components into an assembly, create a component in the context of the assembly, or use a combination of both techniques. With experience, you will determine which method works best for the assembly that you are creating. Whether you place components into an assembly or create components in the assembly, all the components are saved to their own individual .ipt files. The assembly is saved as an .iam file.

Placing Components

To place a component(s) first create the components in their individual files and then place them into an assembly. If you place an assembly file into another assembly, it will be brought in as a subassembly. Be sure to include the path(s) for the file location(s) of the placed components in the project file; otherwise, Inventor may not be able to locate the referenced component when you reopen the assembly.

To insert a component into the current assembly, use one of these ways to start the Place Component command:

- Click the Place Component command on the Assemble tab > Component panel as shown in the following image on the left.
  Note: The Place Component command may be under the Place from Content Center command.
- Right-click in the graphics window and click Place Component from the marking menu as shown in the following image on the right.
- Press the shortcut key P.

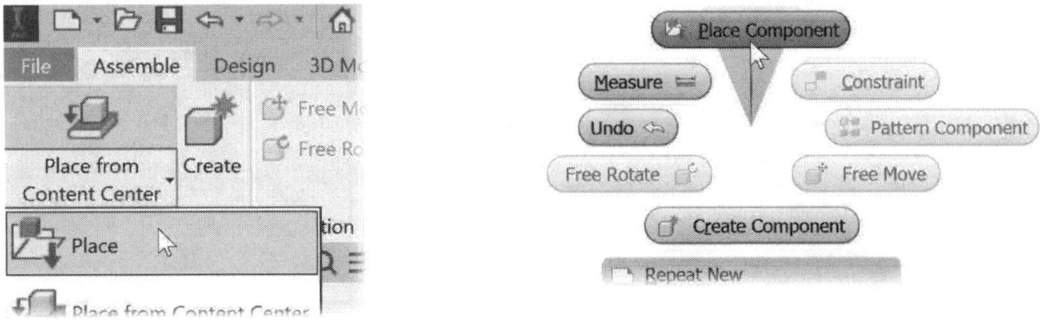

Figure 6.6

After starting the Place Component command, the Place Component dialog box appears as shown in Figure 6.7. Select the component to place and then either double-click or click the Open button in the dialog box. You can also select multiple components by holding down the CTRL or SHIFT key and select the components from the list and then click the Open button.

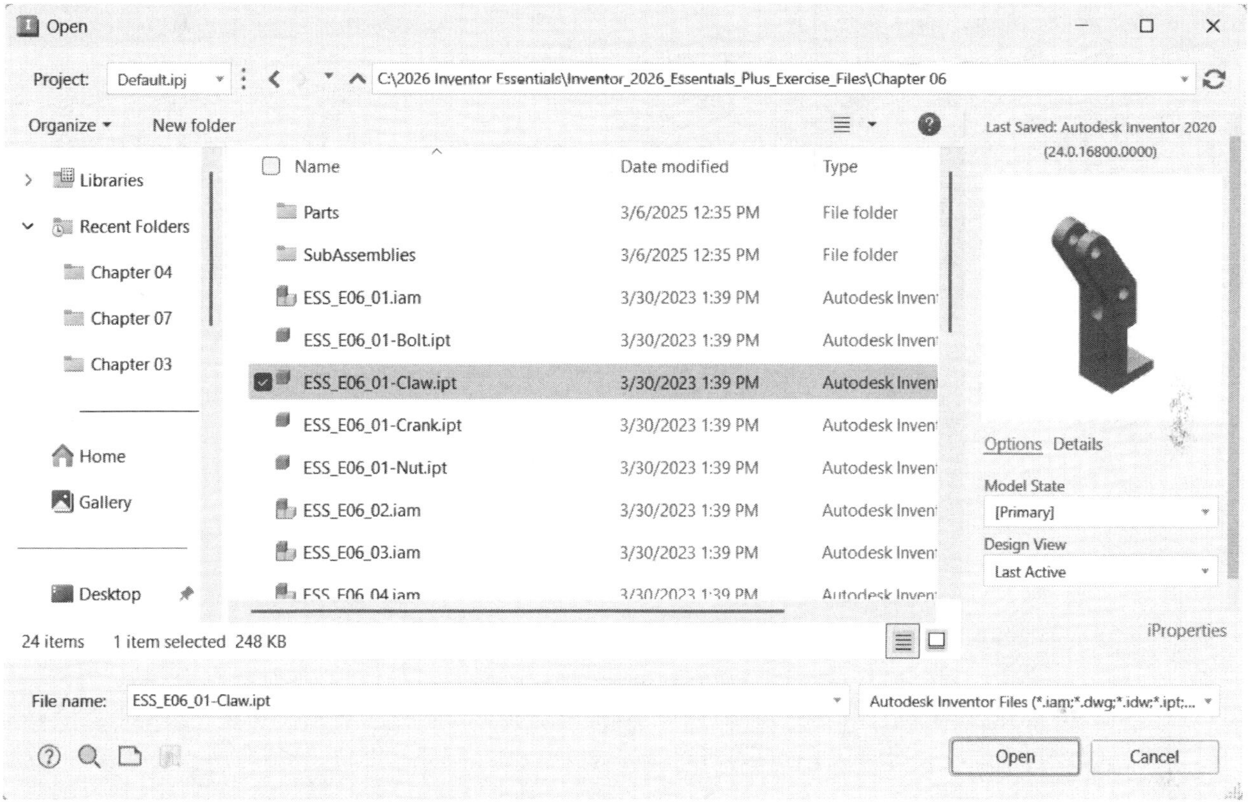

Figure 6.7

After selecting a component(s) they will be previewed in the graphics window. Before placing the component(s) you can rotate the component about the X, Y or Z axis (based on the coordinates of the component, NOT the assembly's coordinates) or ground the component at the origin (lock it from moving) by right-clicking and choosing that option from the marking menu as shown in Figure 6.8.

If the Place Grounded at Origin option is not selected, select placement points. If you need multiple occurrences of the component in the assembly, continue selecting placement points. When done, press the ESC key or right-click and click OK from the menu. To ground a component after it is placed, right-click on the component and choose Grounded from the menu.

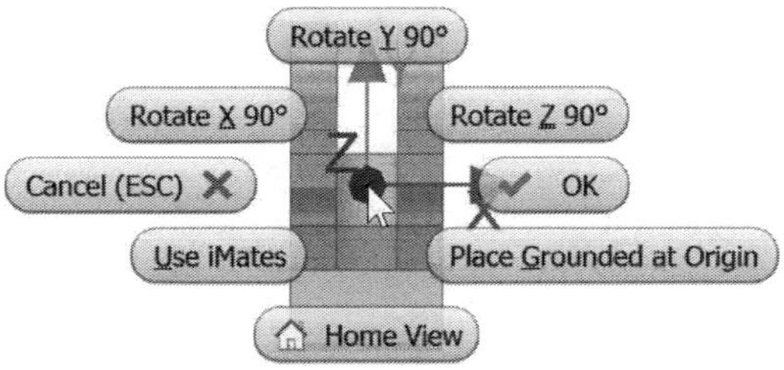

Figure 6.8

TIP: When placing or creating components in an assembly, it is recommended to list them in the order in which they are assembled. To reorder a component, in the browser click and drag on a component to move it up or down.

Placing Components from Another CAD System

You can also place components into an assembly that were modeled in a CAD system other than Autodesk Inventor without having to convert them to an Inventor file and maintain associativity. If the imported file changes, the updates will appear in Inventor. This is very handy as suppliers often provide models you can download. But the files are not always available as Inventor files. Figure 6.9 shows an example part from 3DContentCentral. You save lots of time by downloading stock parts and just inserting them into the assembly.

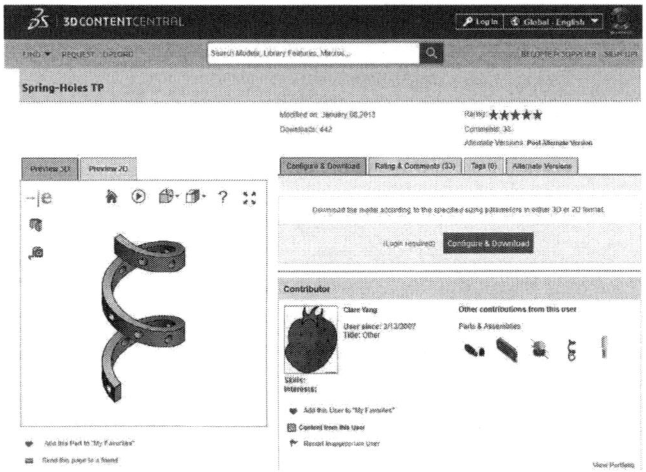

Figure 6.9

To place a component from another CAD system into an Inventor assembly follow these steps:

1. Click the Place Imported CAD Files command on the Assemble tab > Component panel as shown in the following image on the left. The Place Component command can also be used to import a file from another CAD system by changing the file filter option as shown in Figure 6.10 right.
2. Navigate to and select the desired file. Figure 6.10 right shows the type of files that can be placed.

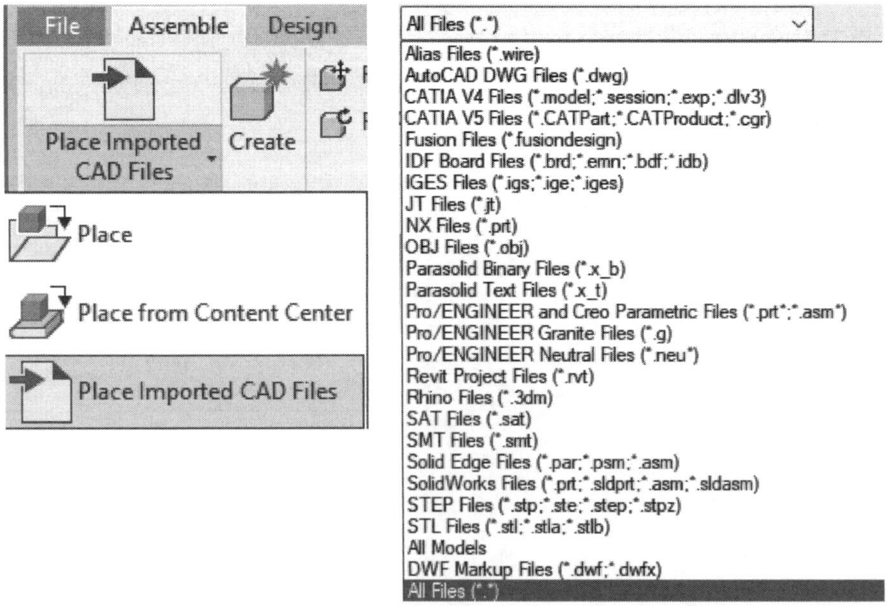

Figure 6.10

TIP: To reference Fusion Files (*.fusiondesign), you must have the Fusion 360 files saved to a Fusion Team Hub that you have access to, and you must install the Autodesk Desktop Connector.

3. In the Import options dialog box, select Reference Model or Convert Model. A referenced model maintains a link to the original file; if the original file changes the model will update in Inventor. The Reference Model option and its options are shown in the following image on the left. When the converted option is selected, the model is converted to an Inventor file(s). The Convert Model option and its options are shown in the following image on the right.

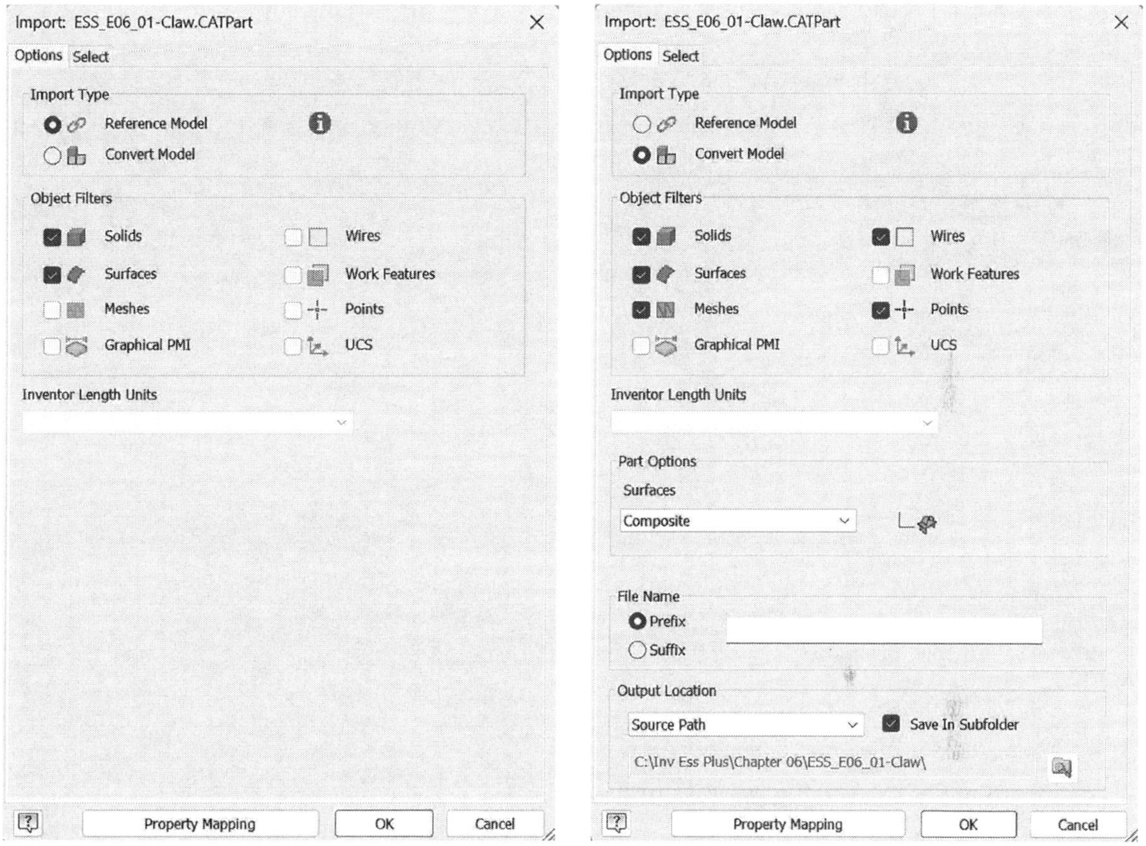

Figure 6.11

4. In the Import dialog box click the Select tab to choose which files will be imported. The following image shows the Select tab.

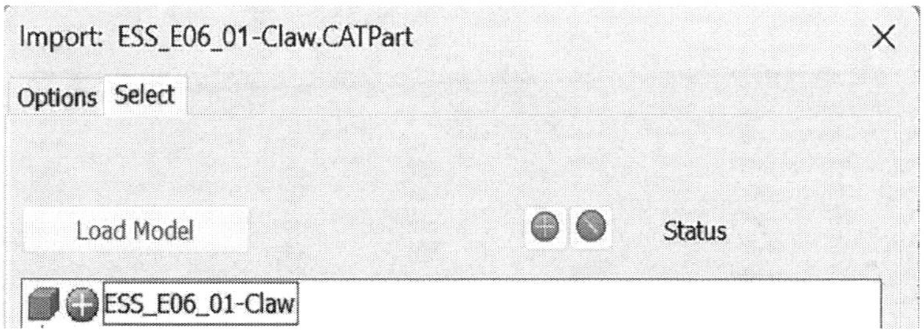

Figure 6.12

TIP: You can drag and drop to import part and assembly files. Select the file in the Windows Explorer and drag it onto the Inventor title bar and use the dialog box to translate it into a new Inventor file. You can do this with or without an Inventor file open.

Creating a Component in an Assembly

Most components in the assembly environment are created in relation to existing components in the assembly. When creating an in-place component, you can sketch on the face of an existing assembly component or a work plane and if desired, project geometry from another component to help create the new component.

While working in an assembly, to create a component in the context of the assembly, you can start the Create Component command one of these ways:

- Click Create from the Assemble tab > Component panel as shown in Figure 6.13 left.
- Right-click in the graphics window and click Create Component from the marking menu shown in the image on the right.
- Press the shortcut key N.

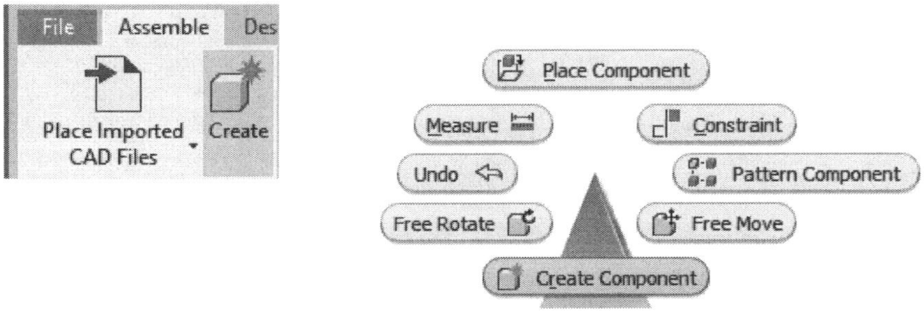

Figure 6.13

After starting the Create Component command, the Create In-Place Component dialog box appears as shown in the following image. Provide the following information in the dialog box.

- Enter a name for the file.
- Select the desired template.
- Select the location where the file will be saved.
- Select the BOM (bill of material) structure.
- To sketch on an existing plane of a part in the assembly keep the Constrain sketch plane to selected face or plane option checked. This option is not displayed if no component exists in the assembly.

Click OK, then select on a planar face, origin plane, or work plane to constrain the part to and the new part is created. The browser and ribbon switch to the part environment even though you are in an assembly.

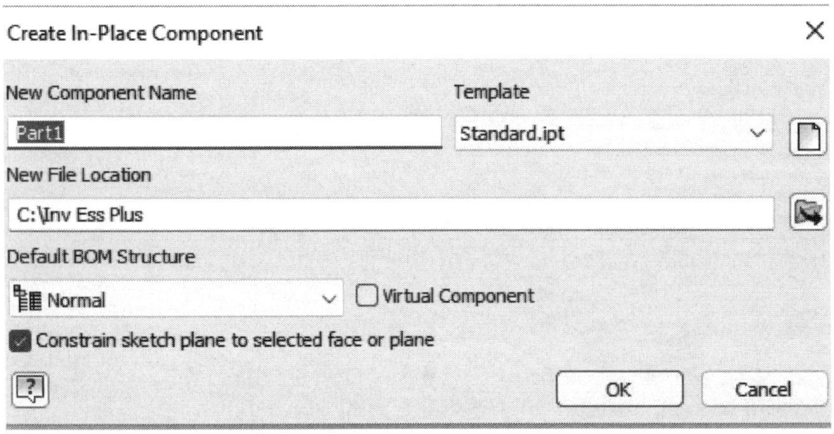

Figure 6.14

Notice that the 3D Model tab is current, as shown in Figure 6.15, and the component is nested in the browser. You can create this part using the same methods previously described in the book including the ability to project edges from other components.

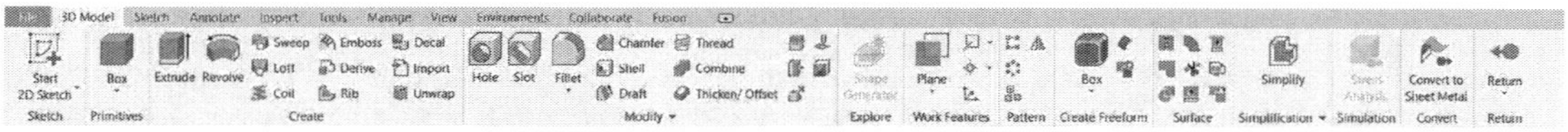

Figure 6.15

When finished modeling the part, return to the top-level assembly by double-clicking the assembly name in the browser or clicking the Return command on the 3D Model tab > Return panel. The Return command will be described in the upcoming Editing a Component in Place section. The assembly environment once again becomes active. Here you can assemble the components and analyze the assembly.

The Assembly Browser

While in an assembly, the browser displays the hierarchy of the part occurrences, subassembly occurrences, and assembly constraints that make up the assembly, as shown in the following image. Each occurrence of a component is represented by a unique name. In the browser, you can select a component for editing, move components between assembly levels, reorder assembly components, control component status, rename components, edit assembly constraints, and manage design views and representations.

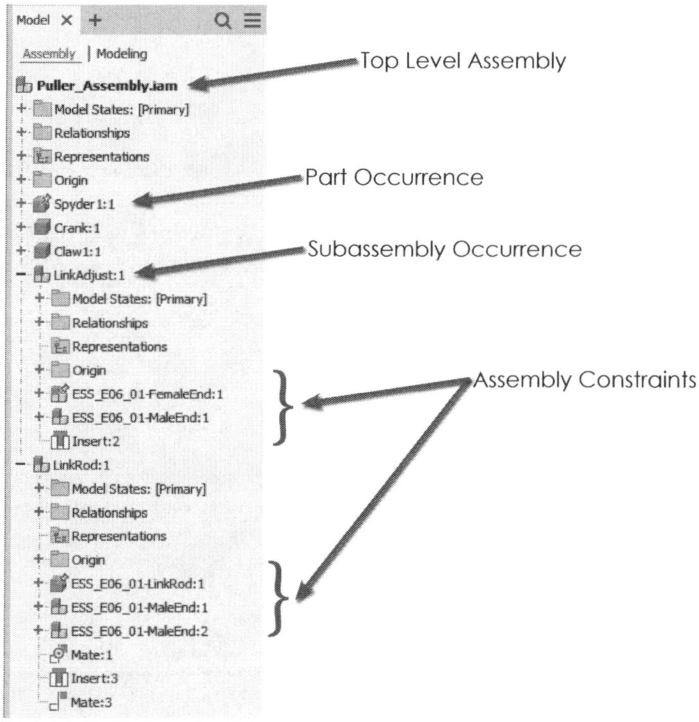

Figure 6.16

Occurrences

An occurrence is a copy of an existing component that has the same name as the original component with a colon and a sequenced number. For example, after placing a component named Bracket it will be the first occurrence, and in the browser it will be labeled Bracket:1 and a subsequent occurrence will be Bracket:2, as shown in the following image. If the original component changes, all the component occurrences update to reflect the change.

To create an occurrence, place the component. If the component already exists in the assembly, you can click the component's icon in the browser, and drag an additional occurrence into the assembly. You can also use the copy-and-paste method to place additional components by right-clicking on the component name in the browser, or on the component in the graphics window, and then click Copy from the menu. Then right-click and click Paste from the menu. You can also use the Windows shortcuts CTRL-C and CTRL-V to copy and paste the selected component.

If you want an occurrence of the original component to have no relationship with its source component, make the original component active, use the Save Copy As command by clicking the File tab > Save As > Save Copy As and enter a new name. The new component will have no relationship to the original, and then place it in the assembly using the Place Component command.

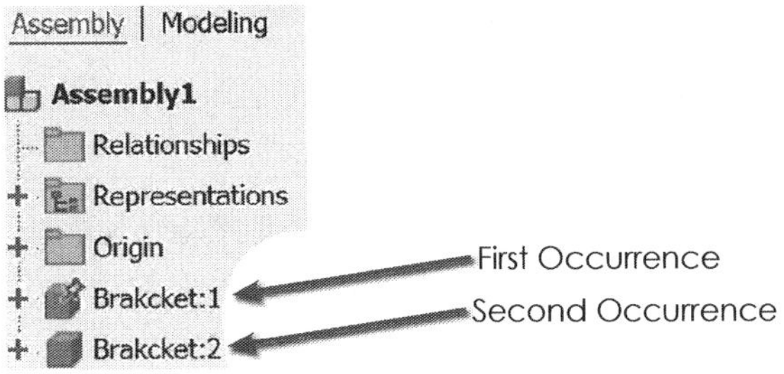

Figure 6.17

The Assembly Capacity Meter

To provide feedback on the resources used by an assembly model, use the Assembly Capacity Meter. This meter in the lower-right corner of the screen is only present in the assembly environment. Two types of information are displayed in the status bar of this meter. The first set of numbers reflects the number of occurrences in the active assembly and the second set of numbers displays the total number of files (also referred to as Documents) open in Inventor.

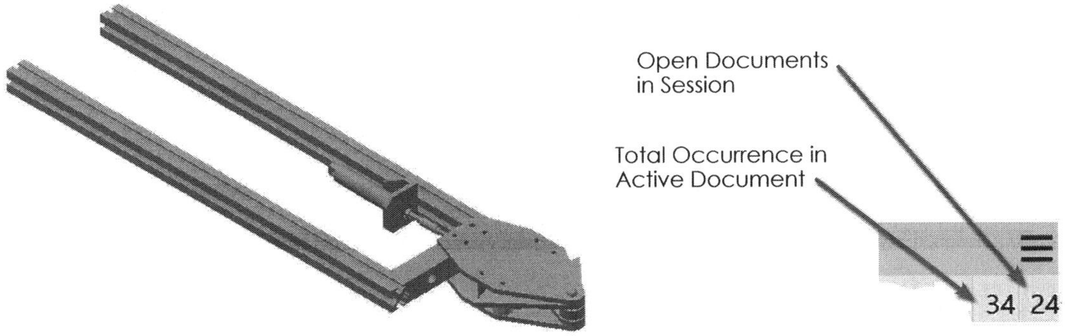

Figure 6.18

Editing a Component in Place

To edit a component while in an assembly, activate the component. Only one component in the assembly can be active at a time. To make a component active, double-click on the component in the graphics window, or double-click on the file name or icon in the browser. Alternately, right-click on the component name in the browser or graphics window and click Edit from the menu.

Once the component is active, the other components in the browser will appear shaded, as shown in Figure 6.19. In the graphics window, the non-active components will appear faded if the Component Opacity option in Application Options > Assembly tab is set to Active Only.

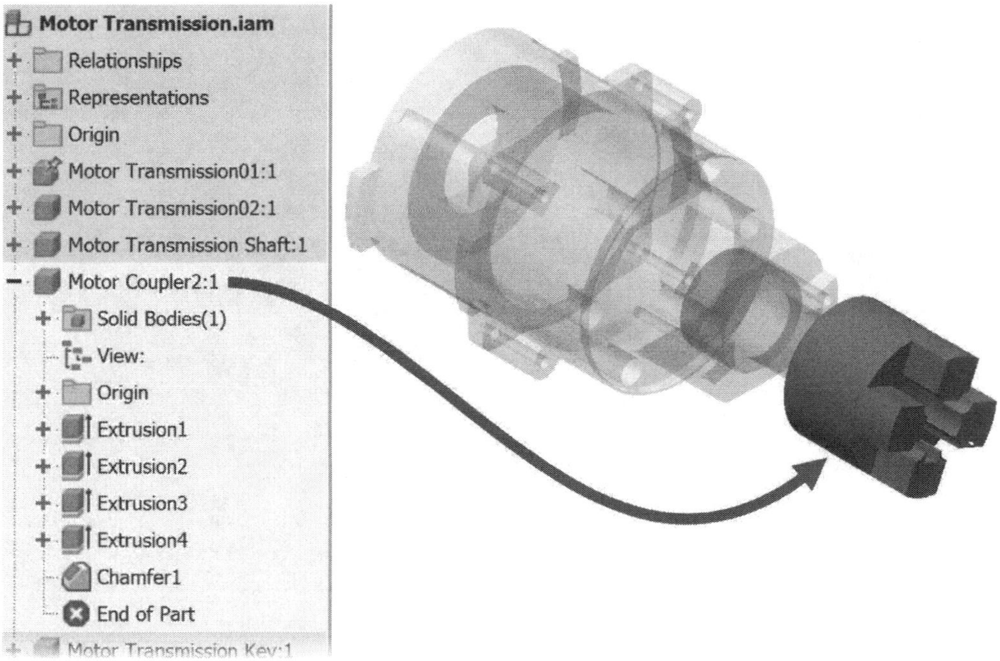

Figure 6.19

When the component is active you can edit the component as normal. If you save the file while in the editing mode, only the active component will be saved. To return to the assembly environment, click the Return button on the 3D Model panel > Return panel as shown in Figure 6.20 left. You can also double-click on the assembly name in the browser or right-click in the graphics window and select Finish Edit from the marking menu as shown in the following image on the right.

Under the Return button there are two other commands, as shown in the middle image; Return to Parent will return you up one level (from the part level to the part subassembly) and Return to Top will return you to the top level in the browser no matter how deep you are in the browser tree.

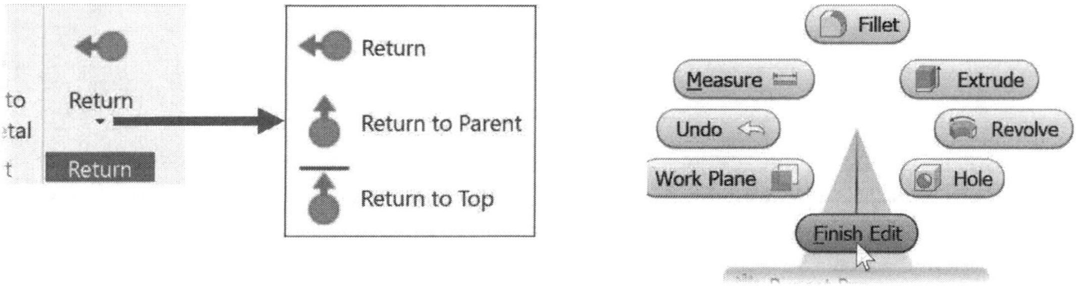

Figure 6.20

Opening a Component in an Assembly

You can also open a component in its own window by right-clicking on the component in the assembly's graphics window and choosing Open from the marking menu, as shown in Figure 6.21 left. In addition you can right-click on the component's name in the browser and click Open from the menu as shown in the image on the right. The component will open in a new window. Edit the component as needed, save the changes, activate the assembly file, and the changes will appear to the component(s) in the assembly.

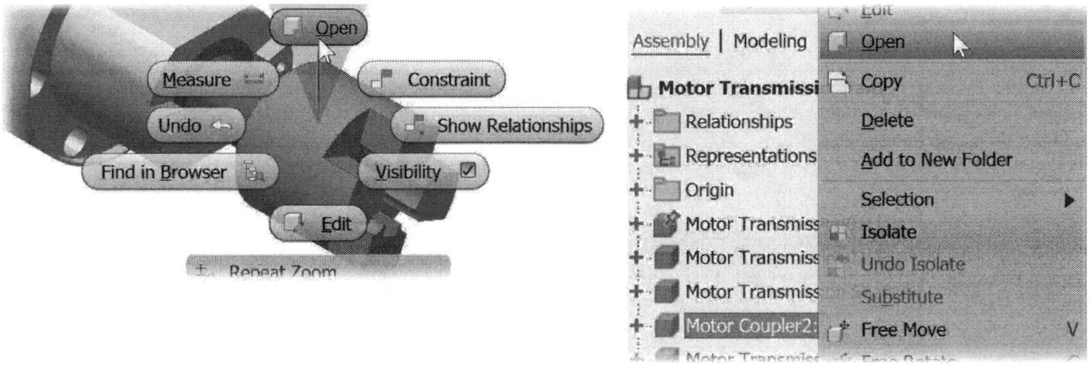

Figure 6.21

Grounded Component

When assembling components, it is recommended to have at least one component or subassembly grounded so it will not move. When applying assembly constraints, ungrounded and unconstrained components will move to the grounded component. There is no limit to how many components can be grounded but usually only one component in the assembly is grounded; otherwise, the entire assembly can move. To ground or unground a component, right-click on the component's name in the browser, and select or deselect Grounded from the menu (near the bottom third of the menu), as shown in Figure 6.22 left. A grounded component is represented with a pushpin superimposed on its icon in the browser, as shown on the right.

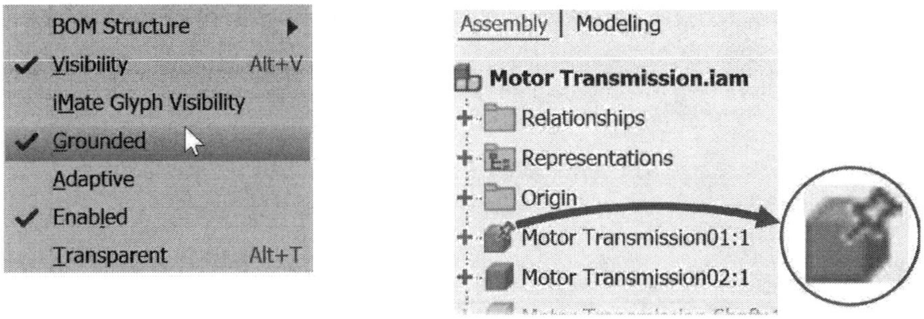

Figure 6.22

Degrees of Freedom (DOF)

You have learned how to create an assembly, but the components have no relationship to one another, unless you created a component in the context of an assembly and referenced a face on another component. For example, if you placed a bolt in a hole and move the hole, the bolt would not move to the new hole position. You apply assembly constraints to create relationships between components. With the correct constraint(s) applied, components will move as the parts change. From the previous example, if an insert constraint is applied between the hole and the bolt and the hole is changed to a new location, the bolt also moves to the new hole location.

In Chapter 2, you learned to apply geometric constraints to sketches. As constraints were applied, the number of dimensions or constraints required to fully constrain a profile was reduced. Similarly, when you apply assembly constraints, they reduce the degrees of freedom (DOF) that allow the components to move freely in space.

There are six degrees of freedom: three are translational and three are rotational. Translational means a component can move along an X, Y, or Z axis. Rotational means that a component can rotate about an X, Y, or Z axis. As you apply assembly constraints, the number of degrees of freedom decreases.

To see a graphical display of the degrees of freedom remaining on all the components in an assembly, select the Degrees of Freedom command on the View tab > Visibility panel as shown in Figure 6.23 left. The image on the right shows the resulting symbols visible on the components in the assembly.

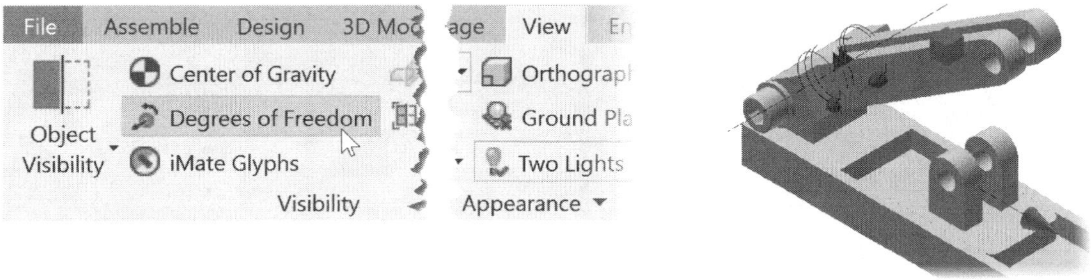

Figure 6.23

An icon appears in the center of the component showing the remaining degrees of freedom. The line and arrows represent translational freedom, and the arc and arrows represent rotational freedom. To toggle off the DOF icons, click Degrees of Freedom on the View menu a second time.

ADDING ASSEMBLY CONSTRAINTS

The following special terminology is used with assembly constraints:

Line. This can be the centerline of an arc, a circular edge, a cylindrical surface, a selected edge, a work axis, or a sketched line.

Normal. This is a vector perpendicular to a planar face.

Plane. A plane or face that includes the following:

- two non-collinear but coplanar lines or axes,
- three points,
- one line or axis and a point that does not lie on the line or axis.

When you use edges and points to select a plane, this creates a work plane referred to as a construction plane.

Point. This can be an endpoint or midpoint of a line, the center or end of an arc or circular edge, or a vertex created by the intersection of an axis and a plane or face.

Offset. This is the distance between two selected lines, planes, or points, or any combination of the three.

Inventor does not require components to be fully constrained. As discussed earlier, it is recommended to ground a component; then other components will move in relation to the grounded component.

Assembly Constraint Command

Inventor uses these five types of assembly constraints:

- mate
- angle
- tangent
- insert
- symmetry

It uses two types of motion constraints:

- rotation
- rotation-translation

and a transitional constraint, and constraint set.

You can access the constraints through the Constrain command found on the Assemble tab > Relationships panel as shown in Figure 6.24 left, by right-clicking and clicking Constraint from the marking menu as shown in the middle image, or by using the C key.

The Place Constraint dialog box appears, as shown in Figure 6.24 right. The dialog box is divided into four areas, which are described in the following sections. Depending upon the constraint type, the option titles may change.

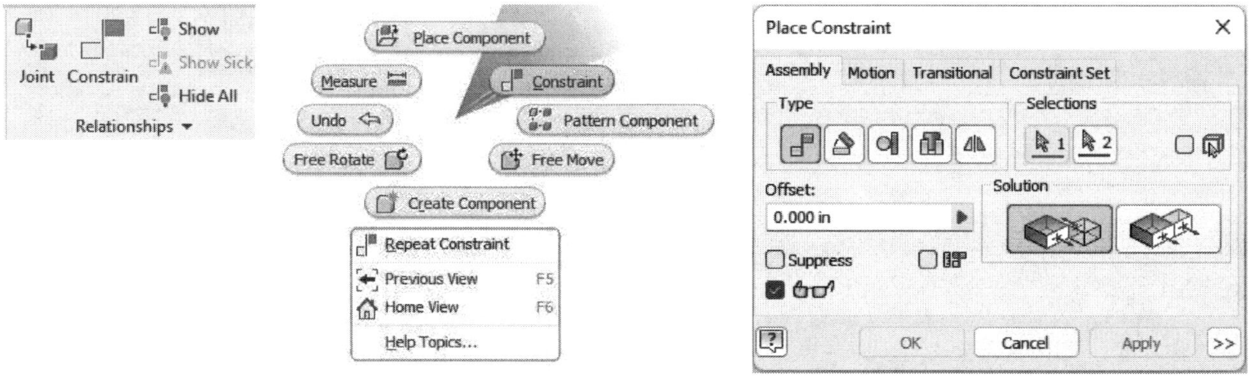

Figure 6.24

Assembly Tab Options

Type. Select the type of assembly constraint to apply: mate, angle, tangent, insert or symmetry.

Selections. Click the button with the number 1, and select a component's edge, face, point, and so on, on which to base the constraint type. Then click the button with the number 2, and select a component's edge, face, point, and so on, on which to base the constraint type. By default, the second arrow will become active after you have selected the first input.

Color-coding is also available to assist with the assembly process. For example, when picking a face with the number 1 button, the color blue is associated with this selection. In the same way, the color green is associated with the number 2 button selection. This schema allows you to better recognize the selections, especially if they need to be edited.

You can edit an edge, face, point, and so on, of an assembly constraint that has already been applied by clicking the number button that corresponds to the constraint and then selecting a new edge, face, point, and so on. While working on complex assemblies, you can click the box on the right side of the Selections section called Pick part first. If the box has a check, select the component before selecting a component's edge, face, point, and so on.

Offset/Angle. Enter or select a value for the offset or angle. The Offset option changes to Angle when the Angle constraint is being applied.

Solution. Select how the constraint will be applied; when two planar faces are selected, the surface normals will be pointing in the same or opposing direction.
Note: The Solution icons will change depending upon the current constraint type and the geometry that is selected.

Suppress. Use the check box to suppress the constraint being edited.

Show Preview. Click, and when constraints are applied to two components, you will see the under-constrained components previewed in their constrained positions. If you leave the box clear, you will not see the components assembled until you click the Apply button.

Predict Offset and Orientation. Click to display the existing offset distance between two components. This allows you to accept this offset distance or enter a new offset distance in the edit box.

Motion Tab

From the motion tab, you can apply a rotation or rotation translation constraint, which allows two components to rotate in relation to one another.

Transitional Tab

A transitional constraint will maintain contact between the two selected faces. You can use a transitional constraint between a cylindrical face and a set of tangent faces on another part.

Constraint Set Tab

This option will constrain two UCSs together.

Constraint Limits

While adding assembly constraints you may want to allow the component to move or rotate a specific value; or when constraining geometry that does not fit perfectly with the constraint value but may fit within a tolerance, you can apply a limit. By setting constraint limits you specify the distance or rotation that the geometry can deviate from an exact location without having the constraint fail. To set a limit for a constraint, click the More button >> on the lower-right corner of the Place Constraint dialog box and the Limits section will appear as shown in the following image on the left. An assembly constraint that has limits will appear in the browser with a +/symbol appended to its name as shown in the following image on the right.

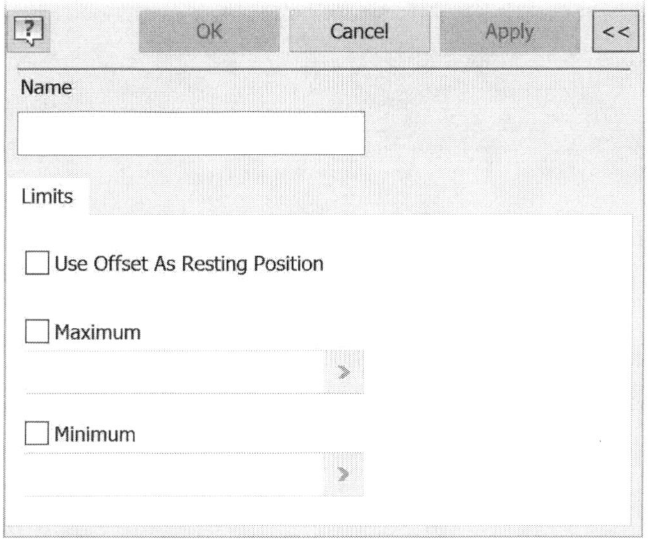

Figure 6.25

Assembly Constraint Examples

This section shows examples of common assembly constraints being applied.

Mate Constraint

When the Mate constraint, with the mate solution, is selected you can apply a constraint between planes, edges/lines, and points or a combination of the three. The following are a few examples of the mate constraint with different geometry selected.

Mate: Plane/Plane

When two planar faces are selected, the planes can be aligned so the surface normals are pointing towards each other (mate solution), as shown in the following image on the right, or the normals can be pointing in the same direction (flush). The flush solution is explained in a following section.

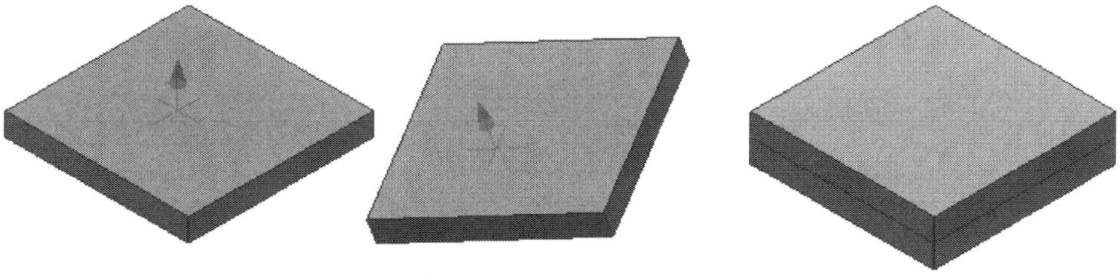

Figure 6.26

Mate: Edge/Edge

When two edges are selected, the edges will be collinear as shown in the following image.

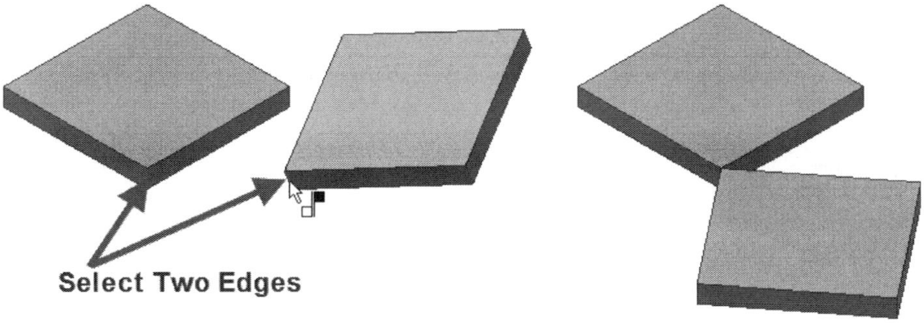

Figure 6.27

You can also align circular faces by selecting two circular faces or centerlines as shown in the following image.

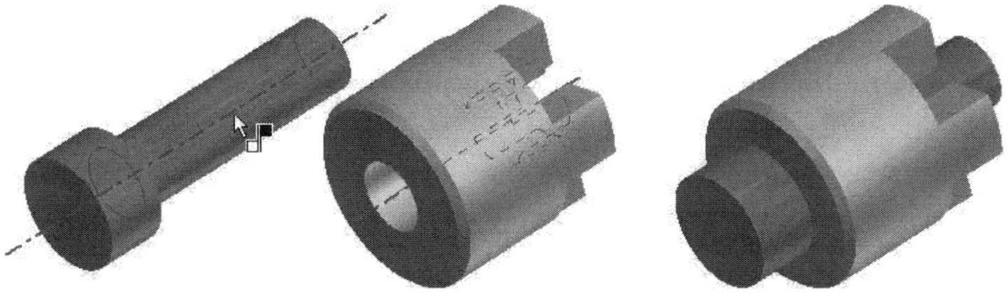

Figure 6.28

Mate: Point/Point

When two points, such as centers of arcs and circular edges, endpoints or the midpoint of an edge are selected, they will be coincident as shown in the following image.

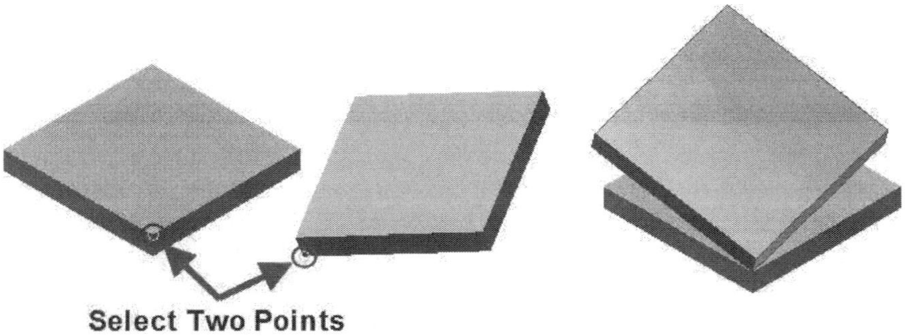

Figure 6.29

Mate with Flush Solution

When applying a mate constraint, you can use the flush solution that aligns two components so that the surface normal on the selected planar faces or work planes are pointing in the same direction, as shown in Figure 6.30.

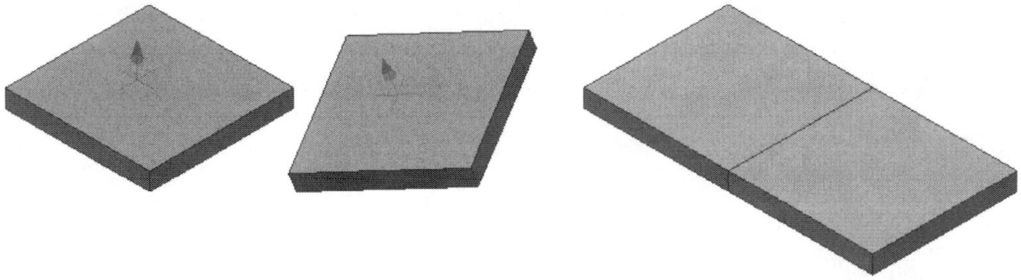

Figure 6.30

Angle Constraint

The angle constraint specifies the degrees between selected planes or faces or axes. The following image shows the angle constraint applied between two planes selected and a 30° angle applied.

Three solutions are available when placing an angle constraint: directed angle, undirected angle, and Explicit Reference Vector. The Explicit Reference Vector option requires a third selection that defines the Z axis.

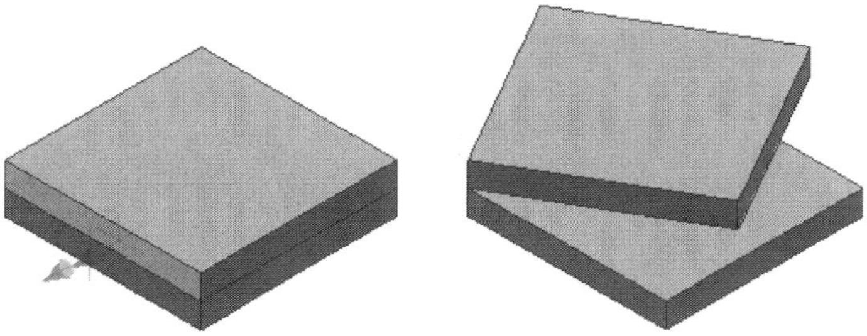

Figure 6.31

Tangent Constraint

The tangent constraint defines a tangent relationship between planes, cylinders, spheres, cones, and ruled splines. At least one of the faces selected needs to be a curve, and you can apply the tangency to the inside or outside of the curve. The following image shows the tangent constraint applied to one outside curved face and a selected planar face, as well as the piece with the outer and inner solutions applied.

Figure 6.32

Insert Constraint

The insert constraint is used to align circular edges removing five degrees of freedom (DOF) with one operation; the open degree of freedom allows the component to rotate. The insert constraint can only be applied between two circular edges. Select circular edges on two components and the centerlines of the selected circles or arcs will be aligned and the selected edges will be mated and centered together. This single constraint aligns and mates the selected geometry in one operation because a circular edge has a defined centerline/axis and a plane. The following image shows the insert constraint applied between two circular edges and the opposed solution applied.

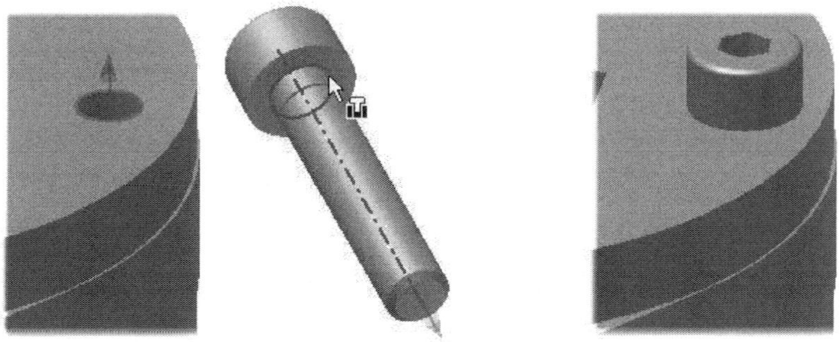

Figure 6.33

By default, the component that moved will be able to rotate. If you want to prevent the rotation when the Insert constraint is active, click the Lock Rotation option in the dialog box as shown in Figure 6.34.

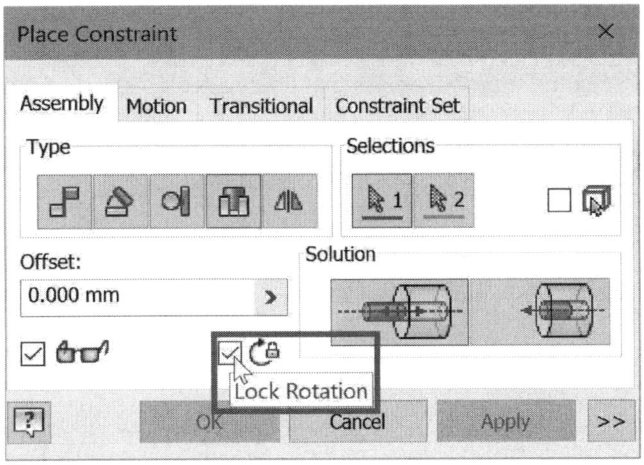

Figure 6.34

Symmetry Constraint

The Symmetry constraint allows you to make two parts or two subassemblies symmetric about a plane. If needed add a work plane that the parts will be symmetric about. Select two planar faces that will be symmetric labeled (1) and (2) in the following image on the left. Select a work plane or a planar face that the components will be symmetric about, labeled (3) in Figure 6.35 left. The following image on the right shows the symmetry constraint applied.

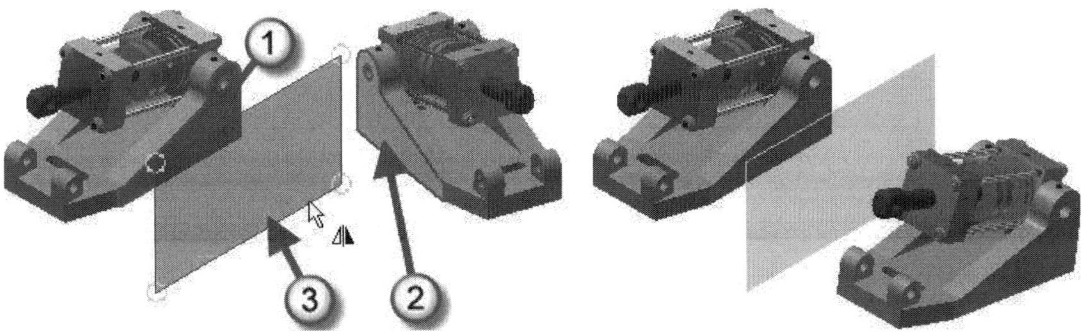

Figure 6.35

Motion Constraints

There are two types of motion constraints: rotation and rotation-translation, as shown in the Type section of the following image. Motion constraints allow you to simulate the motion relationships of gears, pulleys, rack and pinions, and other devices. By applying motion constraints between two or more components, you can drive one component and cause the others to move accordingly.

Both types of motion constraints are secondary constraints, which means that they define motion, but do not maintain positional relationships between components. Constrain your components fully before you apply motion constraints. You can later suppress constraints that restrict the motion of the components you want to animate.

Rotation Constraint Type

The rotation constraint defines a component that will rotate in relation to another component by specifying a ratio for the rotation between them. Use this constraint for showing the relationship between gears and pulleys. Selecting the tops of the gear faces displays the rotation glyph, as shown in Figure 6.36. You may also have to change the solution type from Forward to Backward, depending on the desired results.

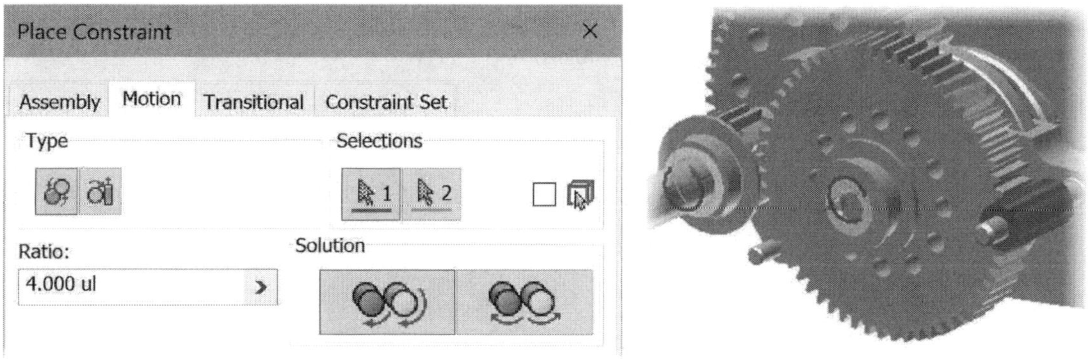

Figure 6.36

Rotation-Translation Constraint Type

The rotation-translation constraint defines the rotation relative to translation between components. This type of constraint is well suited for showing the relationship between rack and pinion gear assemblies. In a rack and pinion assembly, as shown in the following image, the top face of the pinion and one of the front faces of the rack are selected. You supply a distance the rack will travel based on the pitch diameter of the pinion gear, and then you can drive the constraints and test the travel distance of the mechanism.

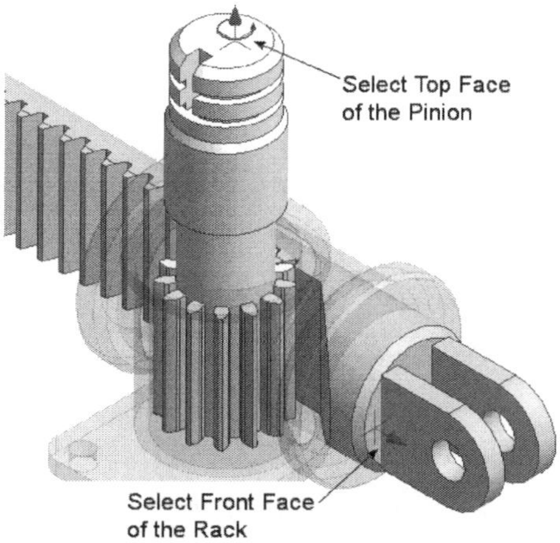

Figure 6.37

Transitional Constraint

The transitional constraint specifies the intended relationship between, typically, a cylindrical part face and a contiguous set of faces on another part, such as a cam follower in a cam slot. The transitional constraint maintains contact between the faces as you slide the component along open degrees of freedom. Access this constraint type through the Transitional tab of the Place Constraint dialog box, as shown in Figure 6.38.

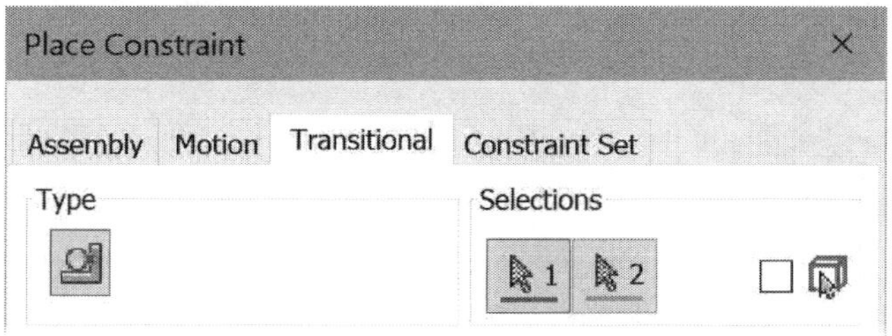

Figure 6.38

Select the moving face first on the cam, as shown in the following image on the left. Next, select the transition face, as shown in the middle of the following image. The transitional face will now contact and follow the cam rotation, as shown in the following image on the right.

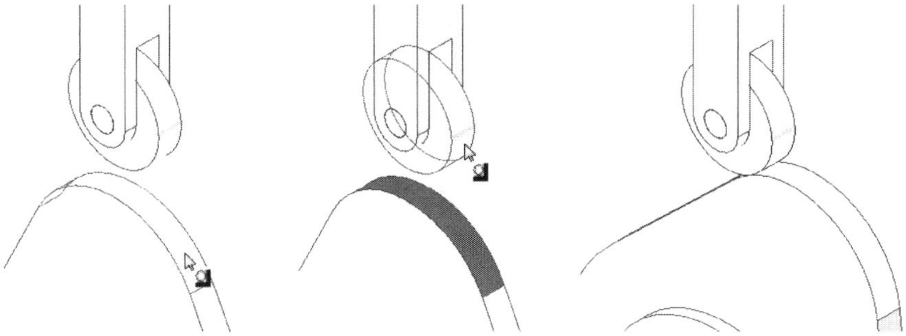

Figure 6.39

Constraint Set Constraint

If User Coordinate Systems are defined in individual part or assembly files, these UCSs can be constrained together. The buttons found under the Type and Selections areas of the dialog box allow for the selecting of individual UCSs and having the constraints be applied to the selections.

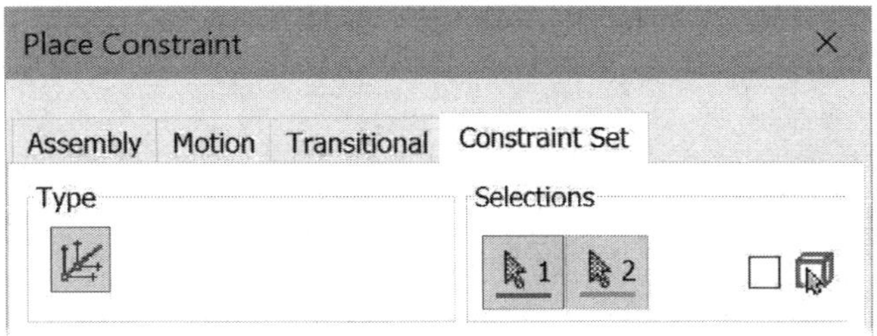

Figure 6.40

Selecting Geometry and the Select Other Command

After selecting the type of assembly constraint to apply, Selections button number 1 becomes active. If it does not automatically activate, click the button. Position the cursor over the face, edge, point, and so on, to apply the first assembly constraint. Use the Select Other command that you learned in Chapter 3 to select geometry that is behind other geometry. Previously you used Select Other to select faces in a part file, now you can cycle through geometry on different parts. To do this, move the cursor over the face, edge or point in front of the one you want to select and hold the cursor still for two seconds. When the Select Other tool appears, as shown in the Figure 6.41 left, click the drop-down arrow and choose from the objects in the list. The number of objects that appear in the list depends upon the geometry and the location of the cursor. You can also access the Select Other command by right-clicking while on the desired location in the graphics window and click Select Other from the menu as shown in the image on the right.

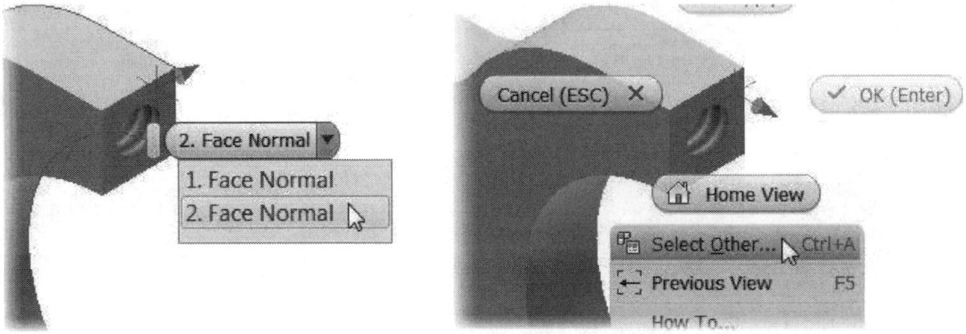

Figure 6.41

TIP: A work plane can also be used as a plane with assembly constraints, a work axis can be used to define a line, and a work point can be used to define a point.

Applying an Assembly Constraint

After selecting the first geometry, select the second set of geometry. You may need to cycle through the selection set until the correct feature is highlighted. With Show Preview option on in the Place Constraint dialog box, you hear a snapping sound and the components move to show the result of the assembly constraint. To change either selection, click on the button 1 or 2, and select the new input. Enter a value as needed for the offset or angle, and select the correct Solution option so that the desired outcome appears. Click the Apply button to complete the operation. Leave the dialog box open to define subsequent constraint relationships.

ALT + Drag Constraining Technique

Another way to apply an assembly constraint is to hold down the ALT key while dragging a part's edge or face to another part's edge or face. No dialog box appears. The key to dragging to apply a constraint is to select the correct area on the part. Selecting an edge will create a different type of constraint than if a face is selected. If you select a circular edge, for example, an insert constraint will be applied. To apply a constraint while dragging a part, you cannot have another command active.

To apply an assembly constraint using the ALT + drag technique, follow these steps:

1. Hold down the ALT key. Select the face, edge, or other location/entity on the part to constrain.
2. Select a planar face, linear edge, or axis to place a mate or flush constraint. Select a cylindrical face to place a tangent constraint. Select a circular edge to place an insert constraint.
3. Drag the part into position. As you drag the part over features on other parts, the preview shows the constraint type. If the face to constrain to is behind another face, hover the cursor until the Select Other command appears. Cycle through the possible selection options, and click to accept the selection.

MOVING AND ROTATING COMPONENTS

Temporarily moving or rotating a component makes it easier to see hidden geometry and to visualize the assembly and apply constraints. To do this, use the Free Move and Free Rotate commands on the Assemble tab > Position panel, as shown in the following image.

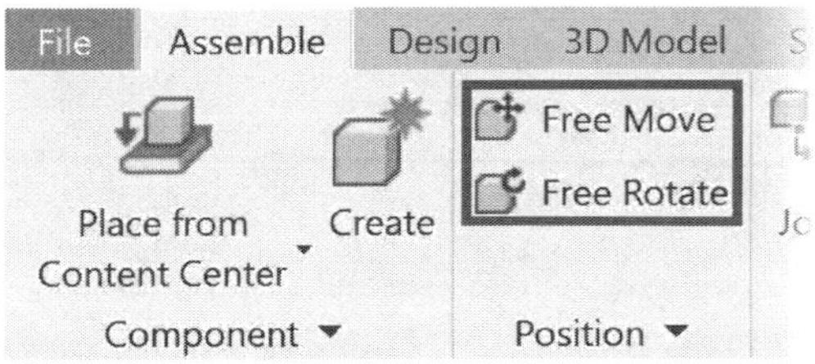

Figure 6.42

Free Move

Use Free Move to temporarily relocate a component in an assembly. Follow these steps:

1. Start the Free Move command.
2. Click and drag the component to a new location. If the component is constrained to other components, an elastic band will display these relationships as shown in Figure 6.43 right.

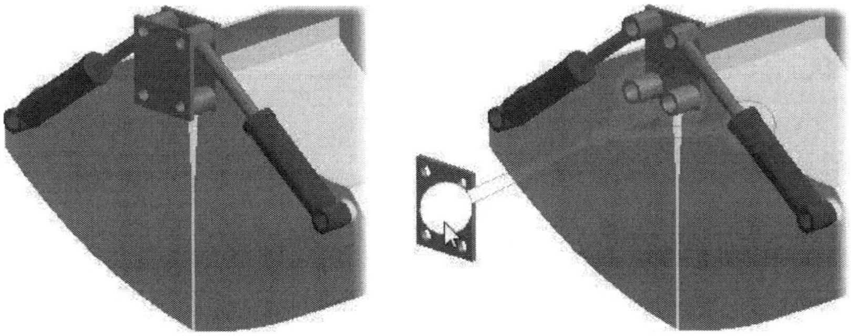

Figure 6.43

3. When the component is at its correct location, release the mouse button.
4. To return constrained components to their constrained position, click the Update command from the Quick Access toolbar.
5. An unconstrained component remains in its new location until you constrain it or move it.

When a constrained component is moved the following will appear:

- A glyph will appear for each assembly constraint or joint that was applied to the moved component. To see more information about the constraint or joint, move the cursor over the glyph as shown in the following image on the left.
- To alter a constraint or a joint, right-click on its glyph and click an option from the menu as shown in the image on the right. Editing constraints and creating joints will be covered in the next sections.

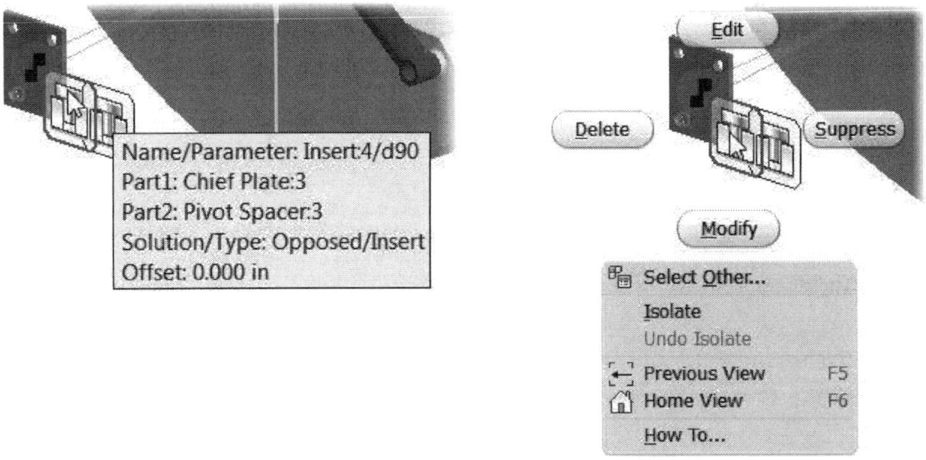

Figure 6.44

Free Rotate

Use the Free Rotate command to rotate a component in an assembly. This is very useful when constraining faces that are hidden from view. You can rotate the component and not have to change the viewpoint of the entire assembly.

Follow these steps to rotate a component in an assembly:

1. Activate the Free Rotate command, and select the component to rotate. Notice the appearance of the Rotate symbol on the selected component in the following image.
2. Drag your cursor until you see the desired view of the component.
3. Release the mouse button to complete the rotation.
4. To return constrained components to their constrained position, click the Update command on the Quick Access toolbar.

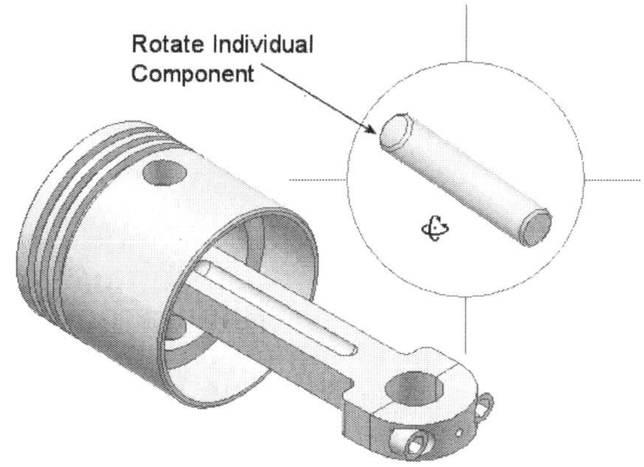

Figure 6.45

EDITING ASSEMBLY CONSTRAINTS

After you have added an assembly constraint, you may want to edit, suppress, or delete it to reposition the component. There are multiple ways to edit assembly constraints.

The first method is to activate the assembly or subassembly that contains the constraints you want to edit in the browser and display the constraints by expanding the component name, right-click on the constraint and click Edit from the menu as shown in Figure 6.46 left. The Edit Constraint dialog box appears; modify the constraint option as needed and when done, click OK. To edit the offset value, click on the constraint to change or right-click on the constraint and choose Modify from the menu as shown in the middle image. The Edit Dimension dialog box appears as shown in the image on the right; enter a new value for the offset distance.

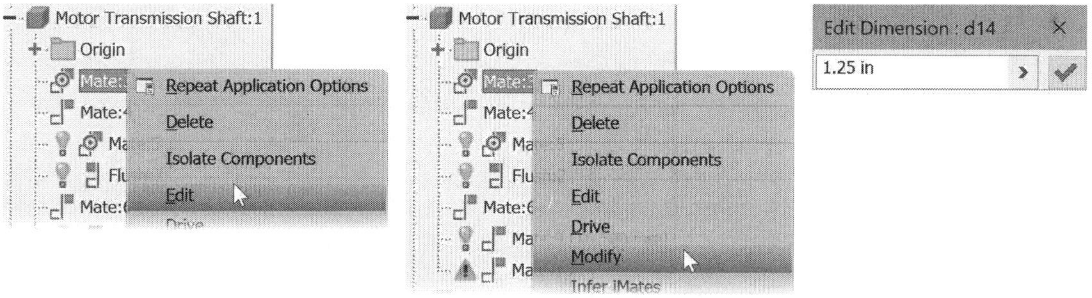

Figure 6.46

Another method is to display the assembly constraints on selected components using the Show Relationships command from the Assemble tab > Relationships panel as shown in Figure 6.47 left. After selecting a component, glyphs will appear for all assembly constraints and joints that have been applied to the component. To edit the constraint or joint, right-click and click Edit from the menu, as shown in the following image on the right. The Edit Constraint dialog box appears; modify the constraint option as needed and when done, click OK. After right-clicking on the glyph you can also use the other options in the menu to modify the constraint. To remove the glyphs from the screen, click Hide All Relationships from the Assemble tab > Relationships panel.

You can edit the constraint's options by right-clicking on the assembly constraint's name in the browser and clicking Edit from the menu. The Edit Constraint dialog box will appear. Modify the constraint as needed.

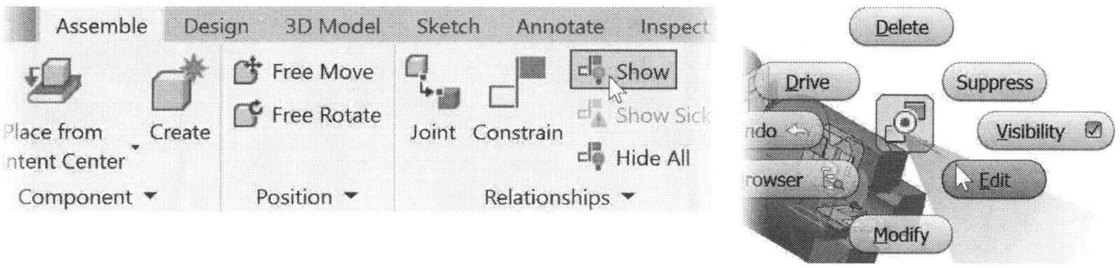

Figure 6.47

IDENTIFY UNDER-CONSTRAINED COMPONENTS

It is helpful when creating assemblies to be able to identify under-constrained parts. This is particularly so if you are going to do motion studies, or other analysis. An under-constrained assembly will not behave as you intend your design to function.

You can identify the constraint status of the parts in your assembly model from information in the assembly browser. There are three possible states that a part in the assembly may show.

- Fully constrained: [•] a solid dot in front of the component name indicates it is fully constrained.
- Under constrained: [○] an unfilled dot in front of the component name indicates it is under constrained. You should add additional constraints or determine the cause of this issue. A good start is to open the Edit Constraint dialog box to identify possible issues.
- Unknown: [-] a dash indicates the constraint status of the component is not known. You may need to use commands to recalculate the assembly information, in other words rebuild the assembly. You can find the rebuild command on the ribbon. From the Manage tab, Update panel click Rebuild All. If this does not resolve the issue, try the Edit Constraints dialog box.

Figure 6.48 shows the assembly browser with the constraint status identifiers. You may have noticed these as you worked through the chapter. They can be turned on or off by setting, Show Constraint State in Browser from the ribbon from the Tools tab, Options panel, choose Document Settings (Figure 6.49) to show the dialog box and click to make the Modeling tab active as shown in Figure 6.50. By default, the Show Constraint State in Browser setting is enabled. If it is not on your system, click so a check mark shows for this option.

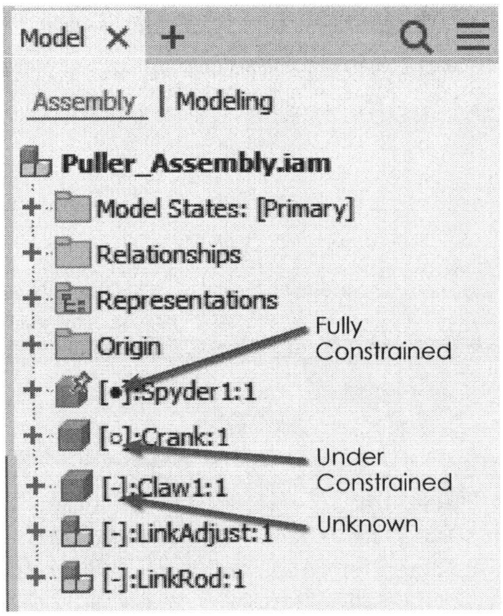

Figure 6.48

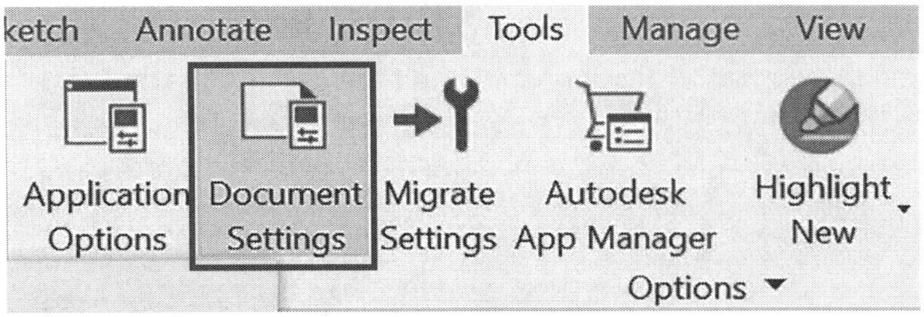

Figure 6.49

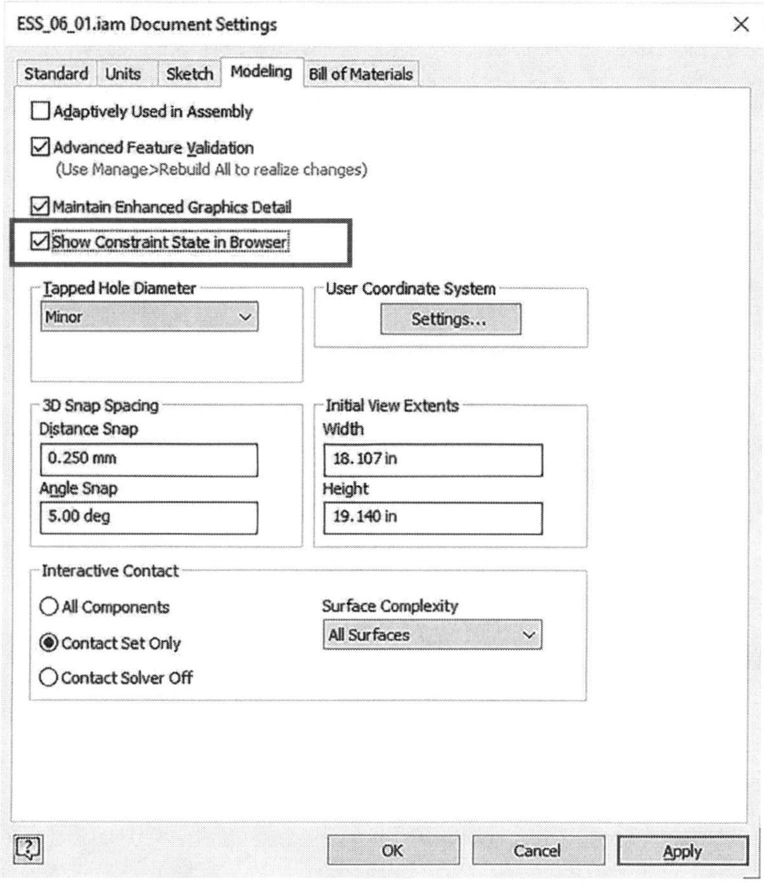

Figure 6.50

You can also use the Advanced Settings Menu from the Browser and choose Display Preferences to show a pop-up menu where you can choose to Hide Fully Constrained components (Figure 6.51). Notice that you can also turn on and off Show Constraint State using this menu.

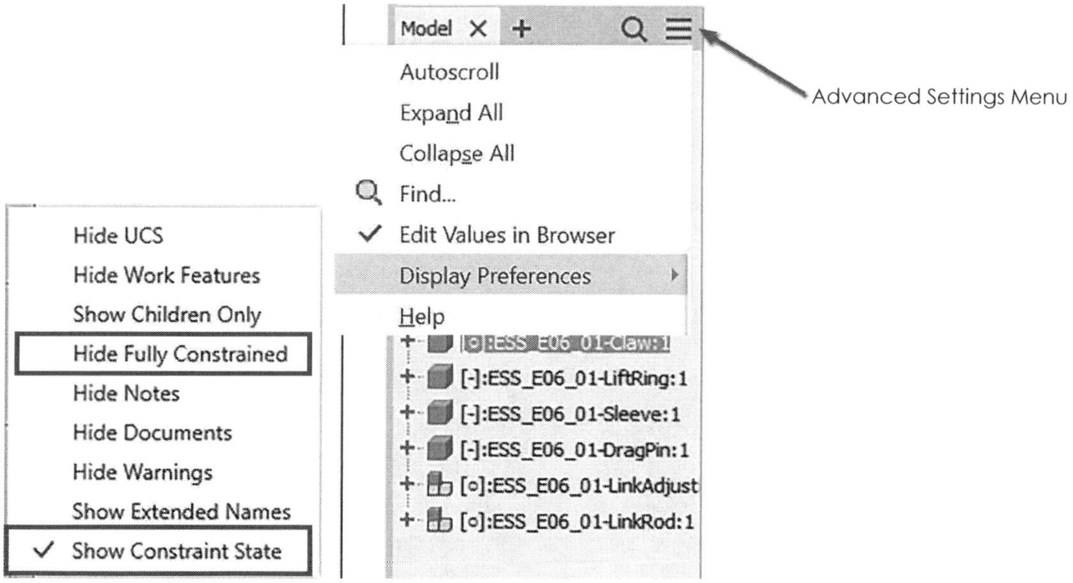

Figure 6.51

EXERCISE 6-1: ASSEMBLING PARTS

In this exercise, you open an assembly of a lifting mechanism, place additional components, and constrain them.

1. Open *ESS_E06_01.iam* in the Chapter 06 folder.
2. Begin assembling the connector and sleeve:
   a. Zoom in on the small connector and sleeve.
   b. Drag the connector so that the small end (blue component) is near the sleeve, as shown in the following image.

Figure 6.52

3. Next you add a mate constraint between the centerlines of both components.
   a. Click the Constrain command from the Assemble tab > Relationships panel. Mate is the default constraint.
   b. Move the cursor over the hole in the arm on the sleeve. Click when the centerline displays, as shown in the Figure 6.53 left.
   c. Move the cursor over the hole in the link. Click when the centerline displays, as shown in Figure 6.53 middle.
   d. If the green dot displays, move the cursor until the centerline displays, or use the Select Other tool to cycle through the available choices to select the centerline.
   e. Click Apply to accept this constraint.
4. Now add a mate constraint between planar faces of both parts.
   a. With the Constrain command still active, select the small planar face on the ball in the link end, as shown in Figure 6.53 right.
   b. Click the nearest inner planar face of the slot on the sleeve, as shown in Figure 6.53 right.
   c. Click OK to create this constraint and close the dialog box.

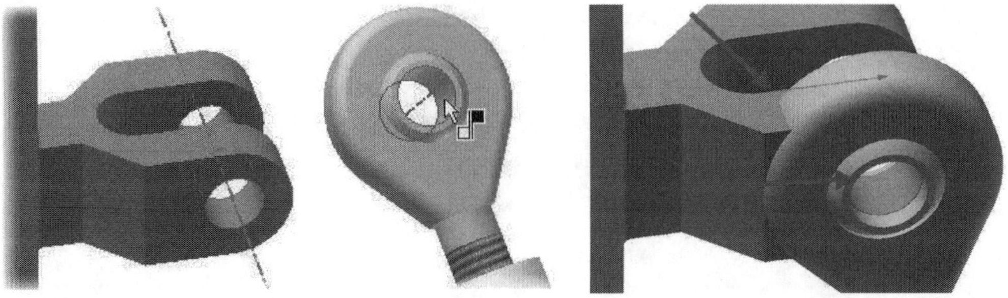

Figure 6.53

5 Drag the small link arm and the crank to see the effect of the constraints.

6 Next you place a crank into the assembly.

a. Click the Place Component command on the Assemble tab > Component panel and double-click on the file *ESS_E06_01-Crank.ipt* from the Chapter 06 folder.

b. Before placing the component, right-click and click Rotate Z 90° as shown in Figure 6.54 left.

c. Move the crank near the end of the middle of the right arm of the spyder component (yellow part) and click to place the crank as shown in the image on the right.

d. End the command by right-clicking and click OK on the marking menu.

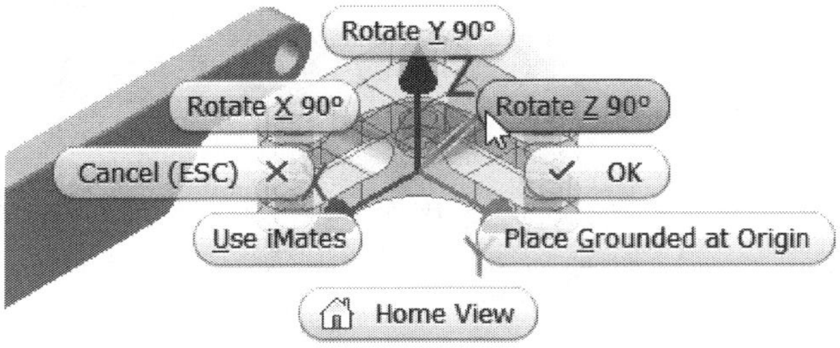

Figure 6.54

7 To assemble the crank to the spyder arm you will again use the Constrain Command, but simplify the process by applying an Insert constraint that will align the centerlines and edges of two holes in a single operation.

a. Click the Constrain command from the Assemble tab > Relationships panel.

b. In the Place Constraint dialog box, change the constraint type to Insert .

c. Select the inside edge of the hole on the crank and right edge of the hole in the middle of the spyder arm as shown in Figure 6.55 left.

d. Click OK to complete the command.

e. Drag the crank to see the effect of the constraint. Position it approximately as shown in Figure 6.55 right.

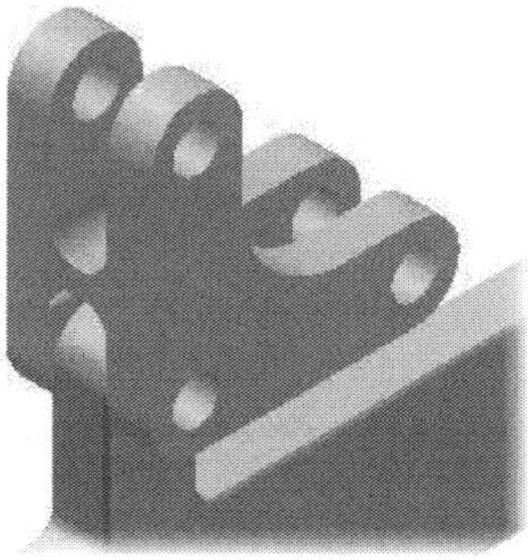

Figure 6.55

8 Next you constrain the crank and the small link. Use the Constrain command to place a mate constraint between the centerlines of the two holes, as shown in Figure 6.56 left.

9 Next you place a claw in the assembly.

a. Click the Place Component command from the Assemble tab > Component panel and double-click on the file *ESS_E06_01-Claw.ipt*.

b. Position the claw near the end of the right side of the spyder arm and click to place the component, as shown in Figure 6.56 right.

c. Complete the command by right-clicking and click OK on the marking menu.

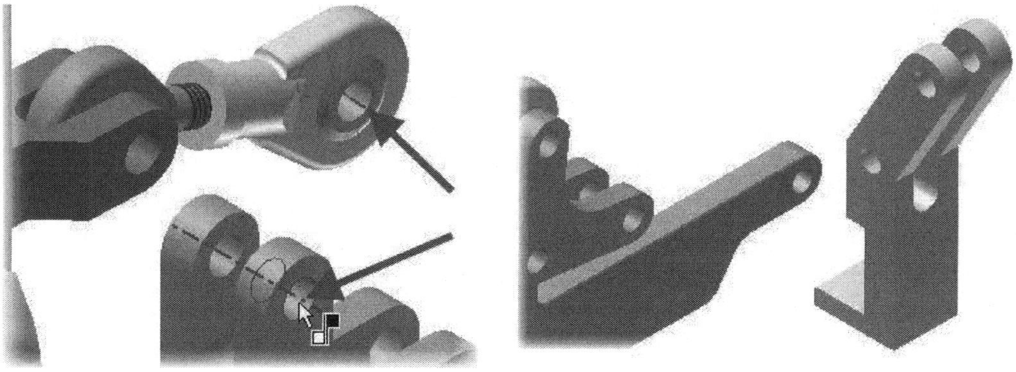

Figure 6.56

10 Next you rotate the claw so it will be easier to apply an assembly constraint to.

a. Start the Free Rotate command from the Assembly tab > Position panel.

b. In the graphics window, select the claw.

c. Then move the cursor in the rotate symbol (circle with lines at the quadrants) and click and drag, rotating the claw until you can see the inside face, like what is shown in Figure 6.57 left.

d. Right-click and click OK from the marking menu to exit the command.

11 Constrain the claw to the spyder.

a. Apply an Insert constraint to the outside edge of the hole on the end of the spyder arm and the inside edge of the hole on the claw as shown in Figure 6.57 right.

b. Click OK to place the constraint and close the Place Constraint dialog box.

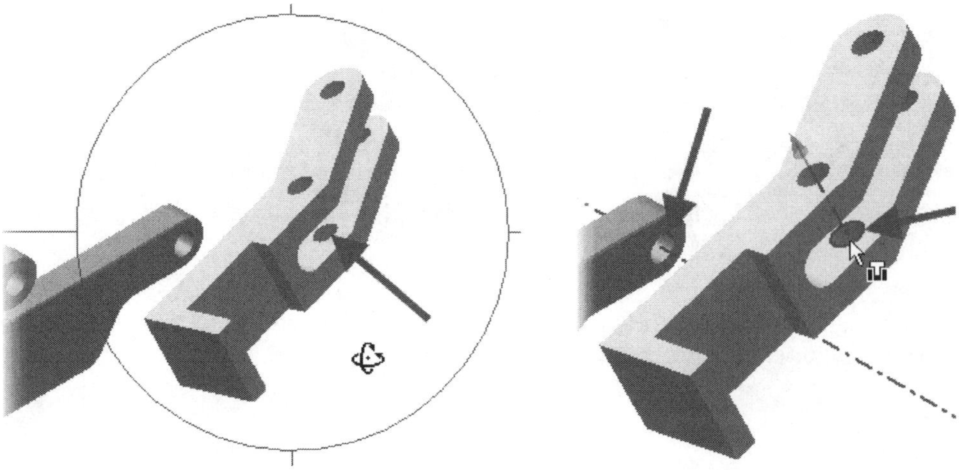

Figure 6.57

12 Next you assemble the link rod to the crank and claw.

a. Drag the long link rod down so it is closer to the crank.

b. Apply an Insert constraint to the left inside edge of the hole on the crank and the small planar face on the left ball in the link rod as shown in the following image on the left.

c. Click Apply to create the constraint and keep the Place Constraint dialog box open.

d. In the Place Constraint dialog box change the constraint type to Mate .

e. Move the cursor over a top hole on the claw and click when the centerline displays, as shown in the following image on the right.

f. Move the cursor over the right hole in the link rod and click when the centerline displays, as shown in the following image on the right.

g. Click OK to create the constraint.

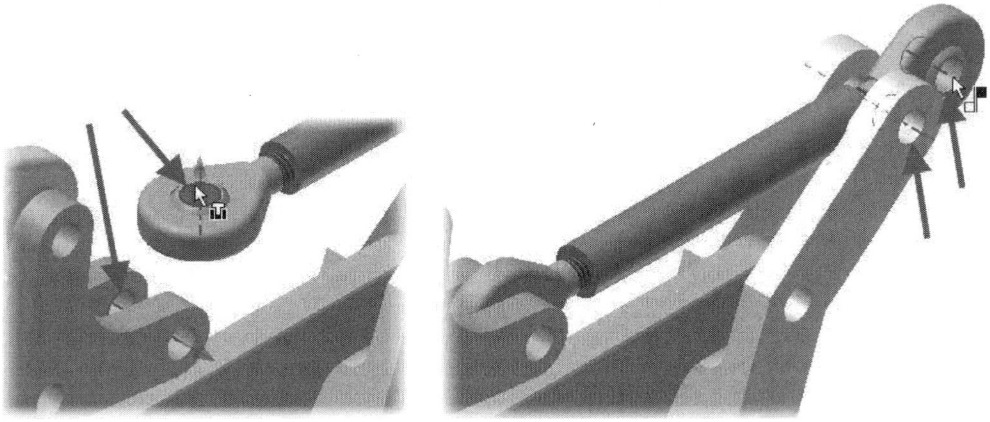

Figure 6.58

13 Drag the claw to see the effect of the assembly constraints, as shown in the following image on the left.

14 Next you add bolts and nuts to the spyder assembly.

a. Begin by clicking the Place Component command on the Assemble tab > Component panel. Hold down the CTRL key and select *ESS_E06_01-Bolt.ipt* and *ESS_E06_01-Nut.ipt* and click Open.

b. Place six nuts and bolts near their final position in the assembly, as shown in Figure 6.59 right.

c. Complete the operation by right-clicking and click OK on the menu.

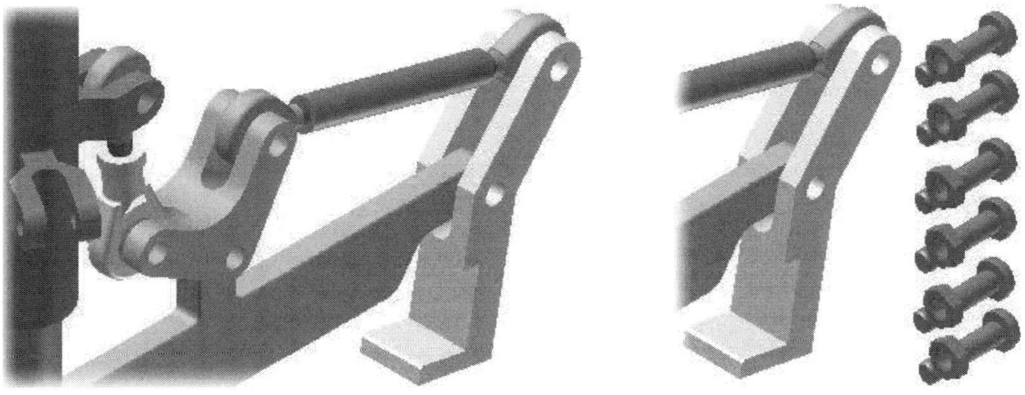

Figure 6.59

15 Place an Insert constraint with the ALT drag technique.

a. Press the ESC key twice to exit the active command.

b. Hold down the ALT key and then Click and drag on the outside circular edge of the bolt and an Insert glyph will appear as shown in Figure 6.60 left.

c. Release the ALT key but keep the left mouse button pressed.

d. Drag the cursor over the circular edge of the hole on the claw until it is previewed in the correct location as shown in the image on the right.

e. Then release the mouse button to create the constraint.

Figure 6.60

TIP: When applying an Insert constraint via the Place Constraint dialog box or with the ALT drag technique, ensure you select the edges that will be planar to one another.

16 Practice constraining the bolts and nuts into the holes in the spyder assembly. When done, your assembly should resemble the following image.

Figure 6.61

17 Try editing a few assembly constraints by using the Show Relationships command on the Assemble tab > Relationships panel.

18 Also edit an assembly constraint by right-clicking on the constraint in the browser and click Edit from the menu.

19 Close all open files. Do not save changes.

End of exercise.

ASSEMBLY JOINTS

Another option to position assembly components is to apply joints. Assembly joints are like constraints that fully define a component's movement and rotation in a single operation. Depending upon the geometry selected, a joint type is automatically selected, but can be overridden. Joints use endpoints, midpoints and center points to locate the components. The Joint command can be used in conjunction with assembly constraints or alone.

Following are explanations of the types of joints that you can apply.

JOINT	DEGREES of FREEDOM	EXAMPLE
Rigid	Removes all degrees of freedom (DOF).	 Two plates that are welded together is an example of a rigid joint.
Rotational	Leaves one rotational degree of freedom.	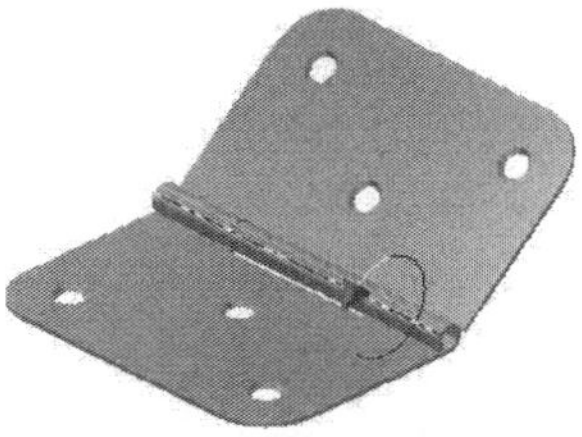 A door hinge is an example of a rotational joint.
Slider	Leaves one translational degree of freedom.	 A block sliding along a track is an example of a slider joint.
Cylindrical	Leaves one translational and one rotational degree of freedom.	 The shaft in a cylinder that can rotate and move in and out is an example of a cylindrical joint.
Planar	Leaves two translational and one rotational degree of freedom.	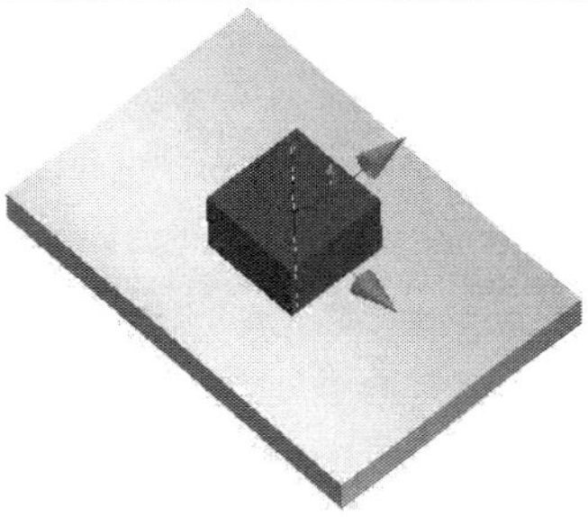 An example would be placing two plates on top of each other. One of the plates can move and rotate on top of the other plate.
Ball	Leaves three rotational degrees of freedom.	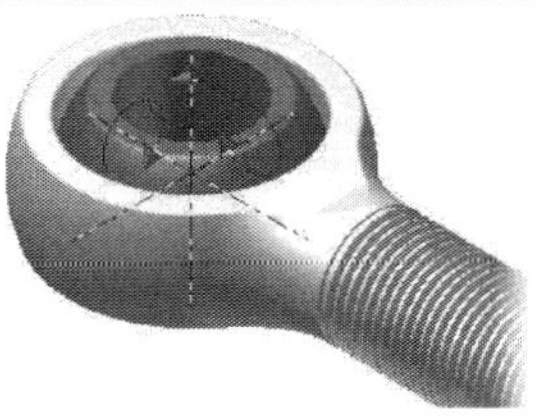 A ball joint is an example where the ball can rotate freely.

CREATING AN ASSEMBLY JOINT

Follow these steps to create a Joint.

1. Start the Joint command from the Assemble tab > Relationships panel as shown in Figure 6.62.

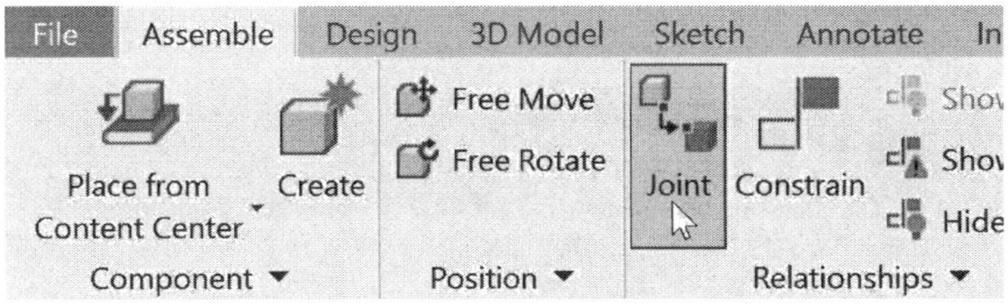

Figure 6.62

2. The Place Joint dialog box appears as shown in Figure 6.63.

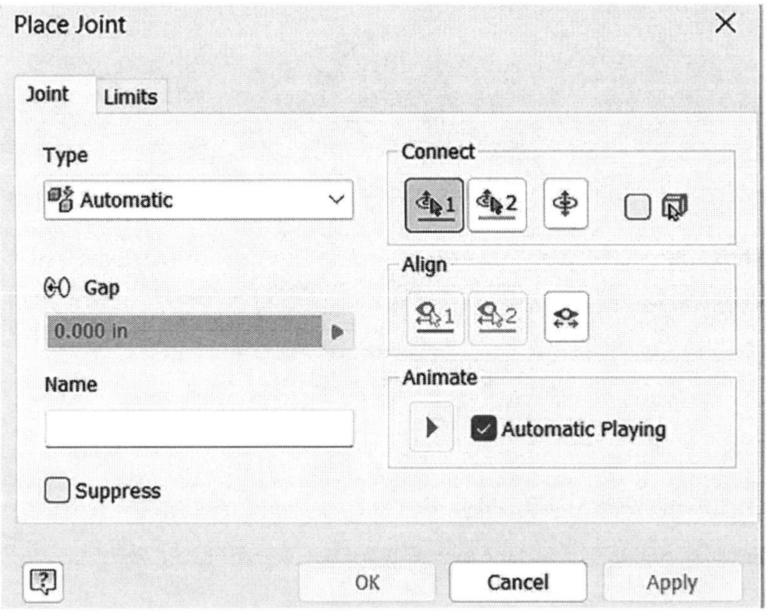

Figure 6.63

3. Select an endpoint, centerpoint, or midpoint on a face of the component that will move.
4. If you select a component that is grounded or has no degrees of freedom, a dialog box will appear alerting you that the component is grounded, and if you continue the component will be ungrounded.

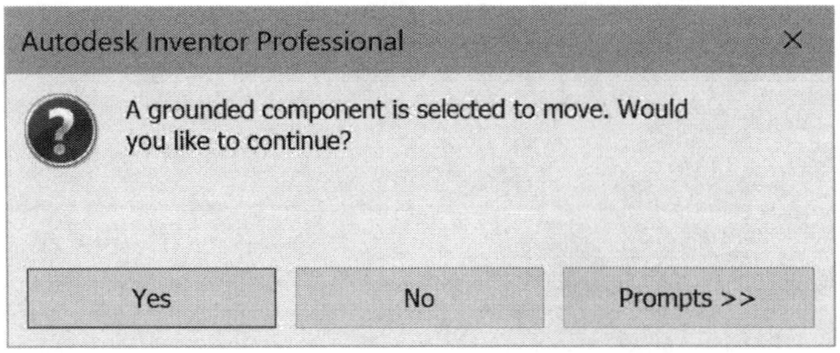

Figure 6.64

5. From the Place Joint dialog box select the desired Joint type as shown in Figure 6.65.
   Note that the Automatic option will change depending upon the geometry that is selected.

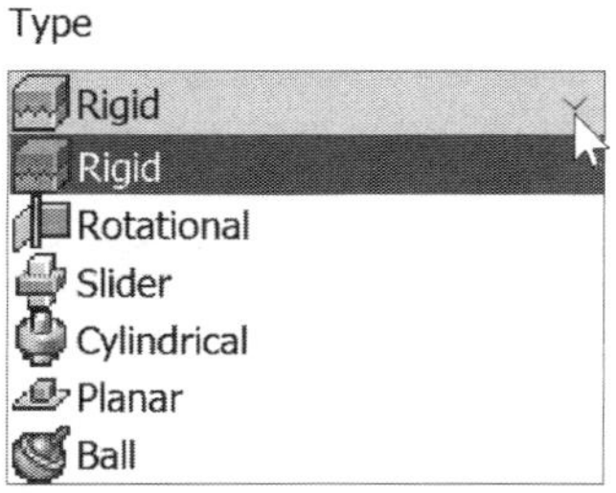

Figure 6.65

6. If a Rotational, Slider, or Cylindrical joint type is selected, green arrows will appear showing the direction of movement. Figure 6.66 left shows a cylindrical joint with two degrees of movement: one translational and one rotational.

7. To change the direction of the movement or to rotate the component, click the First alignment button in the dialog box, as shown in Figure 6.66, and select a face or an edge that you want aligned.

8. The Second Alignment option will be active; select an edge or face that the first selection will be aligned to.

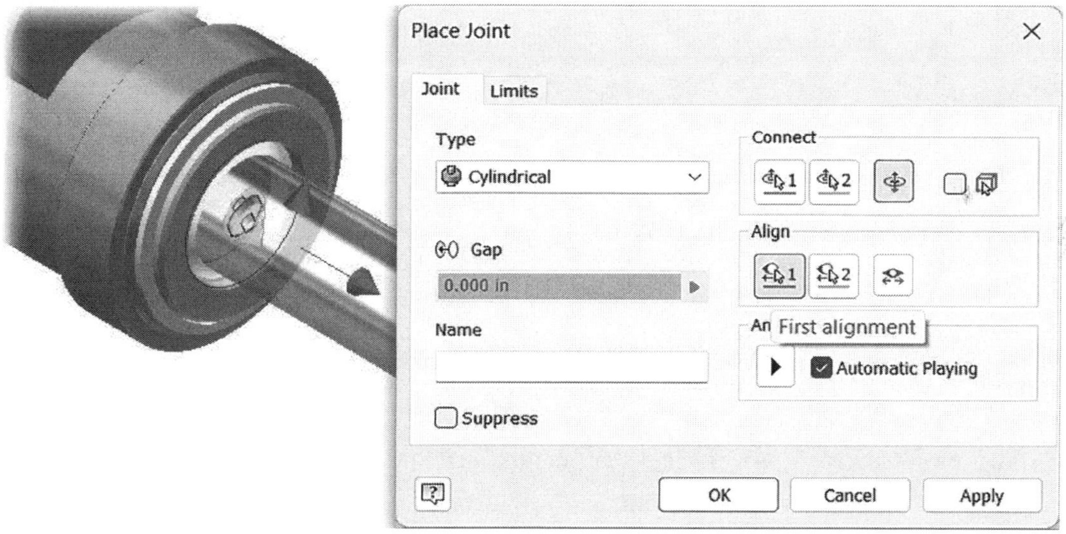

Figure 6.66

9. If needed, click the Invert alignment button to flip the direction.

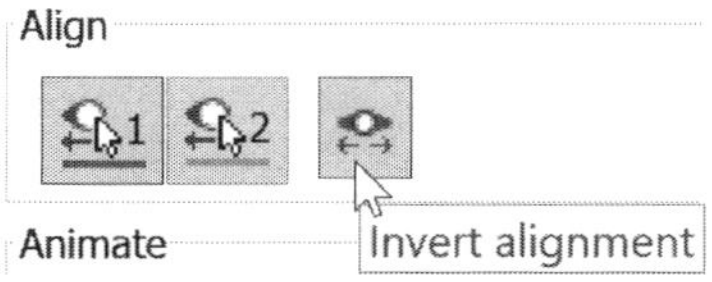

Figure 6.67

10. If desired, you can restrict the movement of the joint by clicking on the Limits tab and define a value in the start and end fields. This functionality is like limits for an assembly constraint.

11. Click the OK button to create the joint.

12. Edit the joint like you would an assembly constraint.

Tips for using the joint command

- The first component you select is the component that will move.
- When creating an assembly joint you can also define the offset or select a point that is between two faces by right-clicking and using the corresponding option as shown in the following image on the left.
- You can view or edit an existing joint or assembly constraints by expanding the Relationships folder or expanding the component in the browser as shown in the following image on the right (shows the same Joints in three locations), then use the Edit or Modify commands.

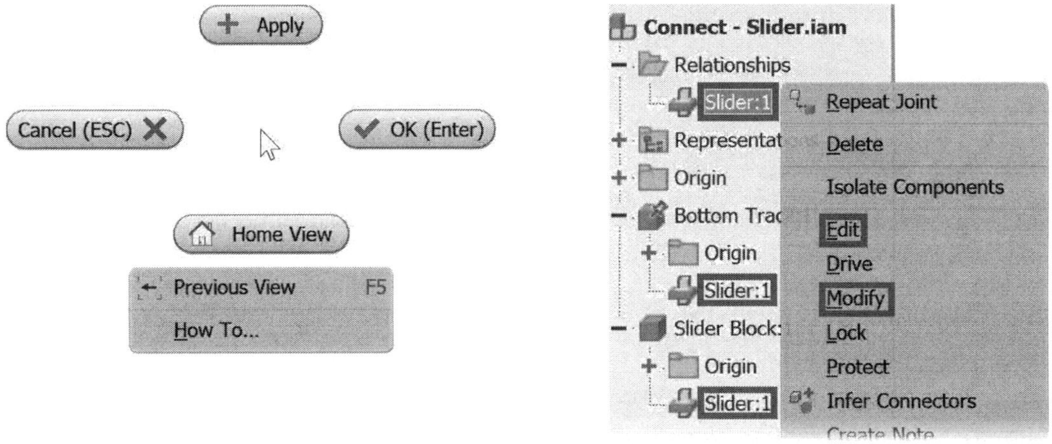

Figure 6.68

EXERCISE 6-2: ASSEMBLY JOINTS

In this exercise, you apply and edit an assembly joint between a block and a beam.

1 Open *ESS_E06_02.iam* in the Chapter 06 folder.

2 Start the Joint command from the Assemble tab > Relationships panel.

a. First select the midpoint on the inside face of the slider block as shown in Figure 6.69 left. This is the part that will move.

b. Select the midpoint on the top face of the track as shown in Figure 6.69 right. The second selection is on the component that will remain stationary.

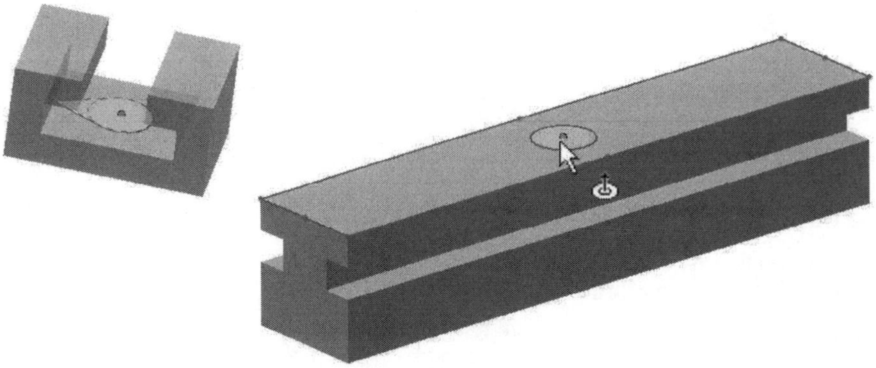

Figure 6.69

c. From the Place Joint dialog box click through the six different joint types, as shown in Figure 6.70 left, and watch the animated movement of each joint.

d. Select Rigid from the list and click OK to create the joint.

3 Try to click and drag on the slider block and notice that it cannot move because of the rigid joint.

4 Next you edit the Joint.

a. Start the Show Relationships command on the Assemble tab > Relationships panel.

b. In the graphics window, select the slider block (yellow part).

c. Right-click on the glyph of the joint and click Edit from the marking menu, as shown in Figure 6.70 right.

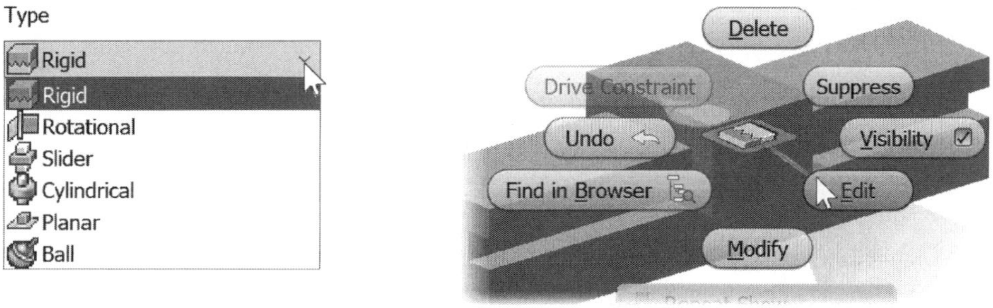

Figure 6.70

d. From the Edit Joint dialog box change the joint type to Slider.

e. Notice that the green arrow and animation shows the slider moving perpendicular to the length of the track.

f. To change the alignment direction, in the Align section of the dialog box click the First alignment button  and then select the bottom horizontal edge of the slider block, as shown in Figure 6.71 left.

g. The Second alignment button is automatically active. Select the bottom horizontal edge of the track as shown in the middle of Figure 6.71 (the first selected edge will slide in the direction of the second selected edge).

h. Notice the arrow is now parallel with the length of the track as shown in Figure 6.71 right.

i. Click OK to complete the edit.

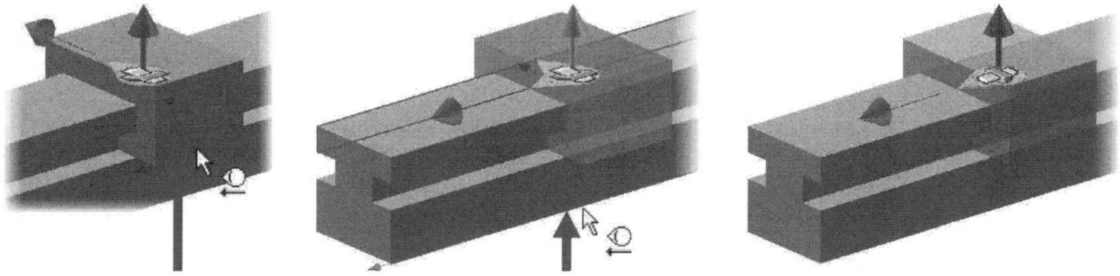

Figure 6.71

5 Click and drag on the slider block and notice that it can slide off the track.

6 Next you add limits to the joint.

a. Right-click on the glyph of the joint and click Edit from the marking menu.

b. In the Edit Joint dialog box click the Limits tab.

c. In the Linear area, check the Start option and change its value to **-2 inches.** Check the End option and change its value to **2 inches** as shown in Figure 6.72.

d. Click OK to complete the edit.

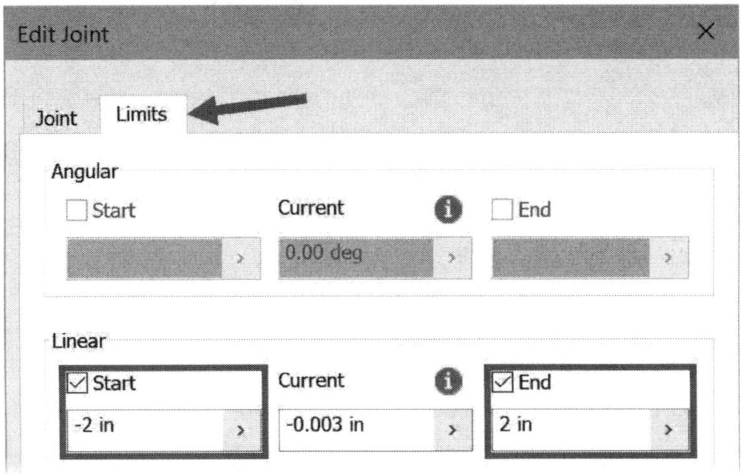

Figure 6.72

7 Click and drag on the slider block and notice that it is limited to 4 inches of travel along the length of the track.

8 Turn off the visibility of the glyph joint by clicking the Hide All Relationships command on the Assemble tab > Relationships panel.

9 Close all open files. Do not save changes.

End of exercise.

ADDITIONAL ASSEMBLY COMMANDS

Additional commands are available to help you control and manage data in an assembly file–for example, controlling how constraints are viewed in the browser, finding the other half of an assembly constraint, using tooltips to learn about an assembly constraint, isolating assembly components, controlling visibility of components and editing a component in the context of the assembly. The following sections describe these operations.

Browser Views

Two modes for viewing assembly information are in the top of the browser toolbar: Assembly View and Modeling View. Assembly View, the default mode, displays assembly constraint symbols nested below constrained components as well as under the Relationships folder, as shown in Figure 6.73 left. In this mode, the features used to create the part are not displayed.

When Modeling View is active, all assembly constraints in the assembly are in the Relationships folder at the top of the assembly tree, as shown in Figure 6.73 right. In this mode, the features used to create the part are displayed just as they are in the part file and can be edited by right-clicking on the feature in the browser and selecting Edit Feature from the menu.

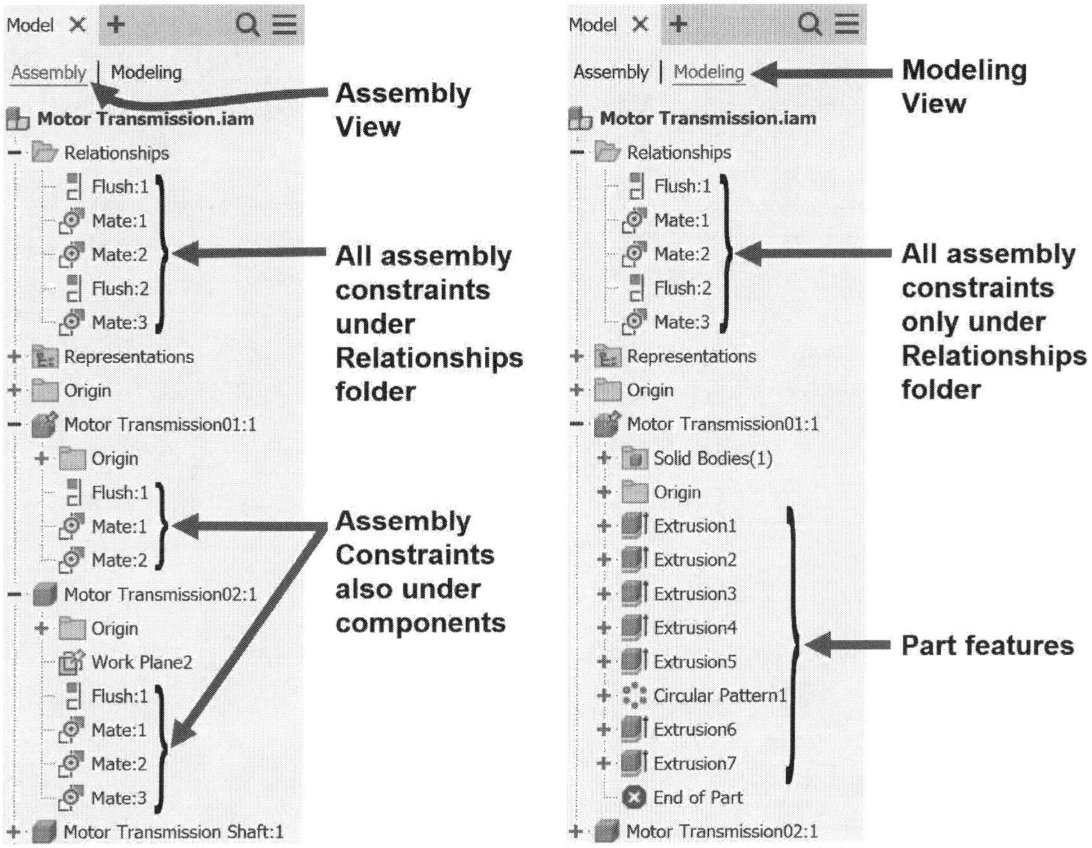

Figure 6.73

Other Half

You can use the Other Half command to find the matching part that participates in a constraint placed in an assembly. As you add parts, you may wish to highlight an assembly constraint and find the part to which it is constrained. To find the second part that the constraint is applied to, right-click on the constraint and click Other Half as shown in Figure 6.74 left. The constraint will highlight in the graphics window, as shown in the middle of the following image, and the browser will expand the second part and highlight the second half of the constraint, as shown in the image on the right.

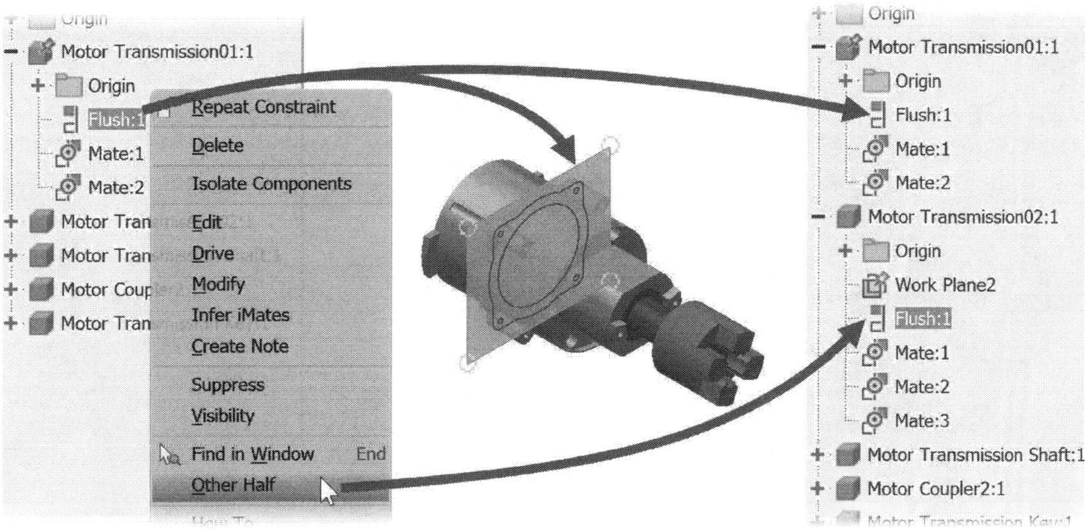

Figure 6.74

Constraint Tooltip

To display all property information for a specific constraint, move your cursor over the constraint icon, and a tooltip will appear, as shown in the following image. Although the constraint name is highlighted, you must hover your cursor over the constraint icon to view the tooltip.

The following information is displayed in the tooltip:

- Constraint and parameter names, applicable to offset and angle parameters
- Constrained components, that is, the two part names from the Assembly browser
- Constraint solution and type
- Constraint offset or angle value

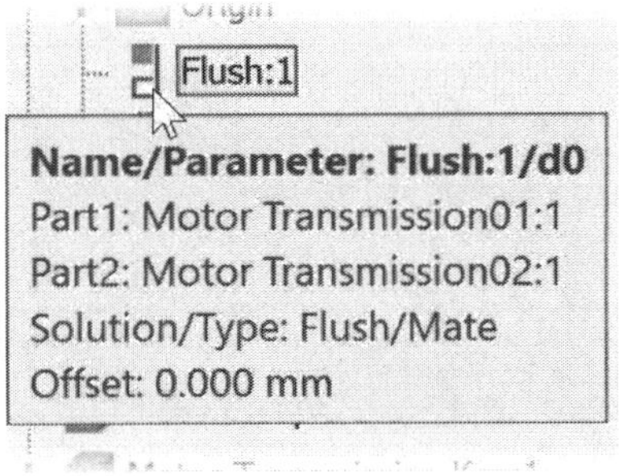

Figure 6.75

User-Defined Assembly Folders

User-defined folders in an assembly allow you to organize your work by grouping assembly components under a single folder name. This method allows you to simplify the appearance of an assembly in the browser. To create a user-defined assembly folder, select those components that you want to group, right-click and pick Add to new folder, as shown in Figure 6.76 left. You will be prompted in the browser to rename the default folder to something more meaningful, such as Fasteners, as shown in Figure 6.76 right. Notice also a unique icon that identifies the user-defined assembly folder.

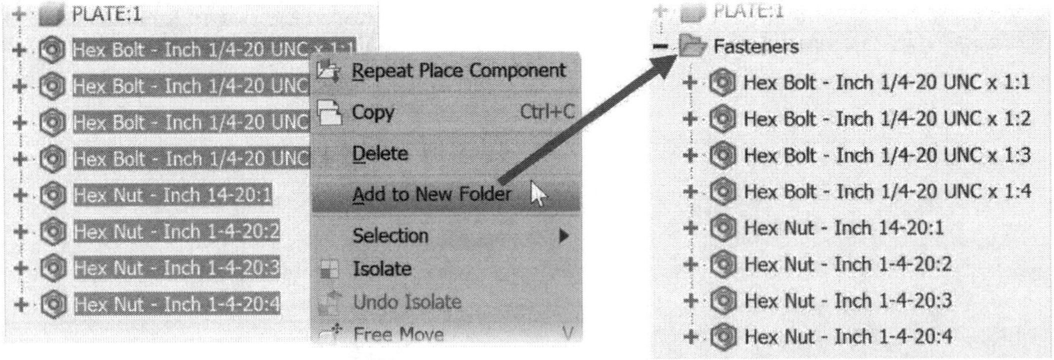

Figure 6.76

Isolating Assembly Components

Components can be isolated as a means of viewing smaller sets of components, especially in a large assembly. In the browser or in the graphics window select the component(s) to isolate, right-click and click Isolate from the menu as shown in the following image on the left. The visibility of all unselected components will be turned off.

To return the assembly to its previous assembled state, right-click inside the browser, or the graphics window, and click Undo Isolate from the menu, as shown in Figure 6.77 right.

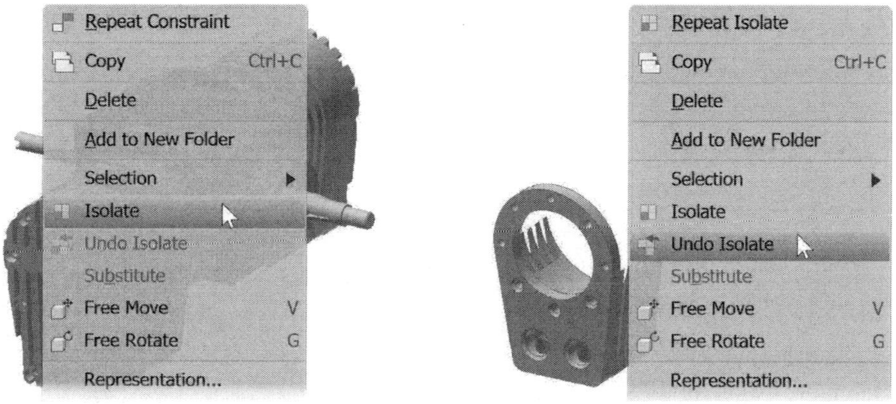

Figure 6.77

Visibility Control

Controlling the visibility of components is critical to managing assemblies. While working, there may be parts that are obscured by other components. You can turn off the visibility of components that are obstructing these components. You can control the visibility of any component in the active assembly even if the component is nested deep in the assembly hierarchy. To change the visibility of a component, expand the browser until the component occurrence is visible, right-click the occurrence and click Visibility from the menu. You can also right-click on a component in the graphics window and select Visibility from the marking menu, as shown in Figure 6.78 left.

After turning the visibility of a component off it will also be grayed out in the browser. To return the component to visible, right-click on it in the browser and choose Visibility from the menu, as shown in Figure 6.78 right.

Figure 6.78

ADAPTIVITY

Adaptivity is the function that allows the size of a part to be determined by setting up a relationship to another part in an assembly. Adaptivity allows under-constrained sketches—features that have undefined dimensions, angles or extents, and hole features—to adapt to changes. An example of adaptivity is determining the diameter of a pin from the size of a hole. Following are some key points to know about adaptivity:

- The adaptive relationship is defined by applying assembly constraints between an adaptive sketch or feature and another part.
- If a sketch is fully constrained, it cannot be made adaptive; however, the extruded length or revolved angle of the part can be adaptive.
- A part can only be adaptive in one assembly at a time.
- In an assembly that has multiple placements of the same part, only one occurrence can be adaptive.
- The other occurrences will reflect the size of the adaptive part.

Project Geometry in the Context of an Assembly

While working in the context of an assembly you can project geometry from one part onto another part creating an adaptive relationship. The projected geometry created is adaptive to the geometry it was projected from. To project geometry while in the context of an assembly follow these steps:

1. Open or create an assembly file.
2. Create or edit a part where you will project geometry.
3. Make a sketch active or create a sketch where the geometry will be projected onto.
4. Use the Project Geometry command on the Sketch tab > Create panel and project edges from another part.

To learn more about the Project Geometry command, refer to the Projecting Geometry section in Chapter 3.

Creating an Adaptive Sketch/Part Manually

Follow these steps to make a sketch and a part adaptive:

1. In a part file create a sketch and draw a sketch, but under constrain it in the direction(s) it will adapt. The following image on the left shows a rectangle that is under constrained in the horizontal and vertical direction (it does not contain dimensions).
2. Make the sketch adaptive by right-clicking on the sketch in the browser and click Adaptive in the menu, as shown in the middle of the following image. The Adaptive icon will appear to the left of the sketch, as shown in the following image on the right.

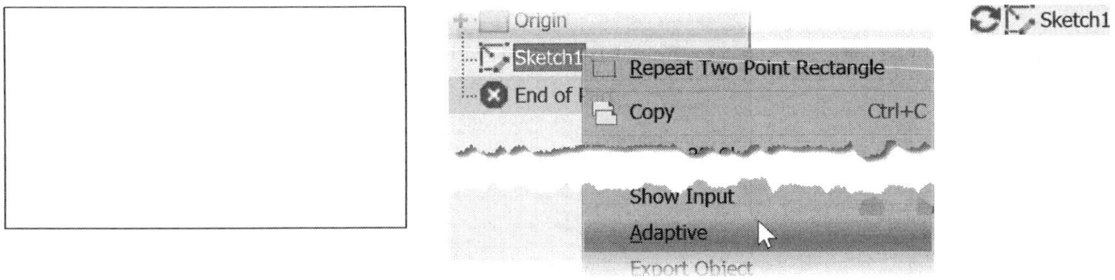

Figure 6.79

3. Extrude or revolve the sketch. The Adaptive icon is displayed to the left of the feature. 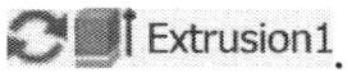
.
4. Save the file.
5. Open an assembly file that contains the part to which the adaptive part will be sized to.
6. Place the adaptive part into an assembly.
7. Make the part adaptive in the assembly by right-clicking on the part's name in the browser and click Adaptive in the menu, as shown in the following image on the left.
8. Define the size of the adaptive part by applying assembly constraints. The following image on the right shows the Adaptive Plate constrained to the top and outside faces of the Base component by applying a mate and four flush constraints. Notice how the plate matches the top size of the base.

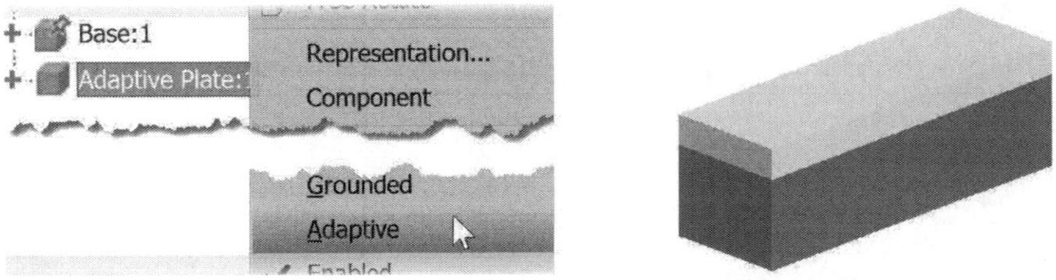

Figure 6.80

EXERCISE 6-3: DESIGNING A PART IN CONTEXT OF AN ASSEMBLY

In this exercise, you create a lid for a container, based on the geometry of the container, by projecting geometry in the context of the assembly. The projected geometry is adaptive and automatically updates to reflect design changes in the container.

1 Open *ESS_E06_03.iam* in the Chapter 06 folder and notice that the container component has already been placed into this assembly.

2 Begin the process of creating a new component in the context of an assembly.

a. Click the Create Component command on the Assemble tab > Component panel.

b. Enter ***ESS_E06_03-Lid*** for the New Component Name.

c. Click the Browse Templates button, and in the English tab, double-click Standard (in).ipt.

d. Change the location the file will be saved to by selecting *C:\Inv 2024 Ess Plus\Chapter 06*.

e. Ensure there is a checkmark beside the ☑ Constrain sketch plane to selected face or plane option.

f. Click the OK button to exit the dialog box.

g. Select the top planar face of the container, as shown in the following image on the left.

3 Create a new sketch by clicking the Start 2D Sketch command on the 3D Model tab > Sketch panel. The origin planes for the new part will be displayed in the graphics window. Select the XY origin plane, as shown in Figure 6.81. The selected plane becomes the active sketch plane for the new part.

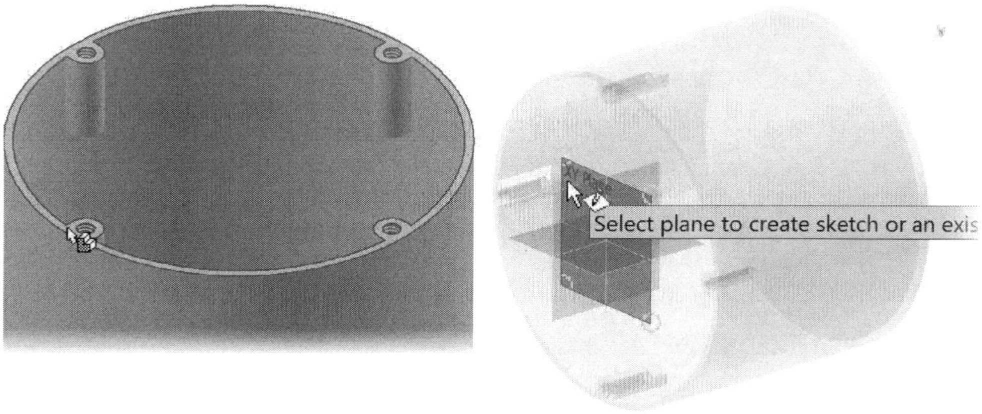

Figure 6.81

4 Next you project all geometry contained on this top face of the container.

a. Click the Project Geometry command from the Sketch tab > Create panel.

b. Move the cursor over the left face of the base part until the profile of the entire planar face is highlighted, as shown in the following image on the left.

c. Click to select all the edges on the selected plane. Your display should appear like Figure 6.82 right. Project the selected edges by right-clicking and click OK from the menu.

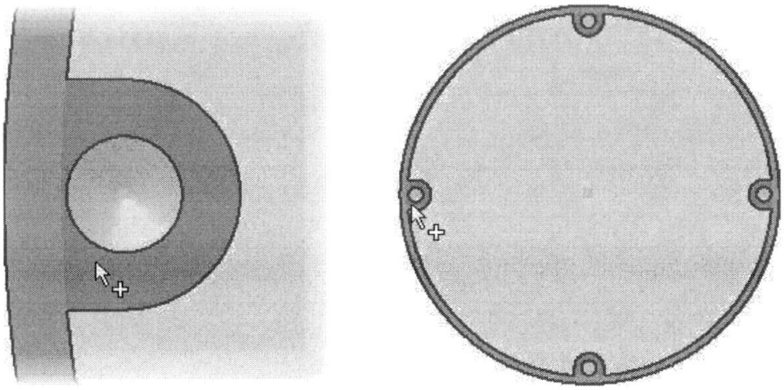

Figure 6.82

5 Next, you create a series of circles that will act as clearance holes. Zoom in to view a tapped hole.

6 Add four circles that are concentric to and larger than the four projected circles, as shown in Figure 6.83.

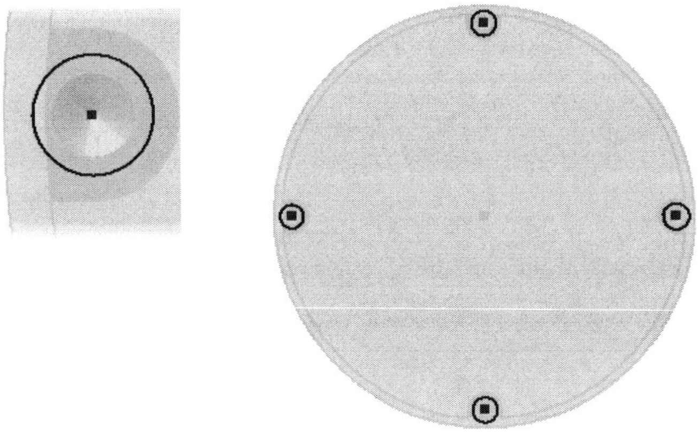

Figure 6.83

7 All four circles need to have the same diameter. You will add an equal constraint between the circles in the following steps.

a. Click the Equal constraint command on the Sketch tab > Constrain panel.

b. Click the left circle, and then click the top circle.

c. Click the left circle, and then click the circle on the right.

d. Click the left circle, and then click the bottom circle.

e. Next, fully constrain the circles by adding a **.15 inch** diameter dimension to the left most circle. When done, your sketch should resemble the following image.

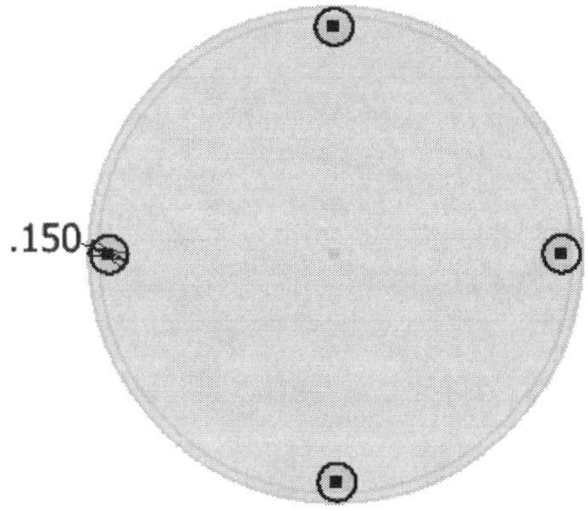

Figure 6.84

8 Exit the Sketch environment by right-clicking and click Finish 2D Sketch on the marking menu.

9 Next you extrude two profiles.

a. Select one of the edges on the sketch and click Extrude from the mini-toolbar.

b. To create the entire lid, select the two profiles as shown in the following image on the left. If only one profile is selected, the lid will be missing a section.

c. In the Extrude Properties panel, enter an extrude value of **.25 inches**.

d. Click OK to create the extrusion.

10 To make the top-level assembly active, click the Return command on the 3D Model tab > Return panel. When done, your display should appear like the following image on the right.

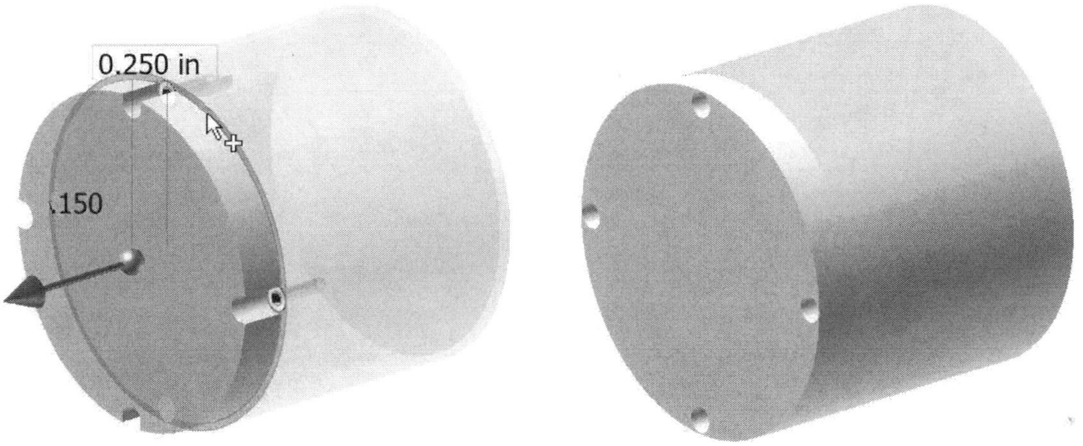

Figure 6.85

11 You will now modify the container and observe how this affects the lid. In the graphics window double-click on *ESS_E06_03-Container:1* (yellow component).

12 Next you change the diameter dimension of the sketch of the base extrusion.

a. Select the cylindrical face on the container and click Edit Sketch on the mini-toolbar, as shown in the following image on the left.

b. Double-click the 1.5000 diameter dimension, and change the value to **1.0 inches**.

c. Finish the sketch. Notice that the lid is now larger than the container as shown in Figure 6.86 right.

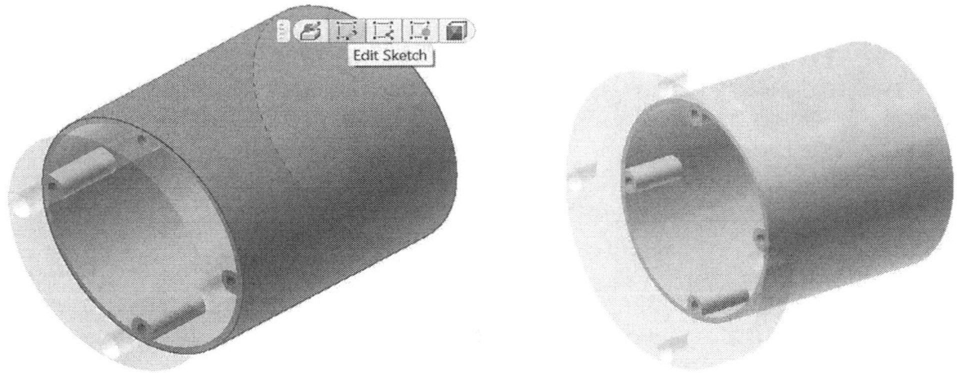

Figure 6.86

13 Click the Return command to return to the top-level assembly. Notice that the lid adapts to the modified dimensions of the container, as shown in the following image.

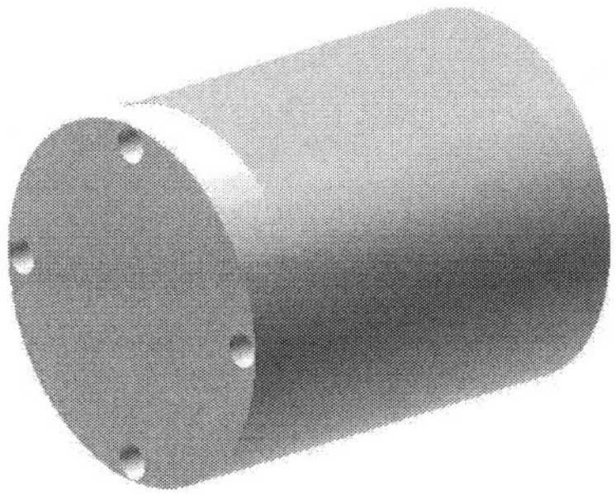

Figure 6.87

14 If desired, change the viewpoint to the Home View, as this will display the container from its original viewpoint.

15 Close all open files. Do not save changes.

End of exercise.

PATTERNING COMPONENTS

To speed the process of placing and constraining the same component or components that follow a set direction and have a fixed distance between them, you can use the Pattern Component command. This command has three options: pattern a component(s) about an existing feature pattern, or the components can be patterned in a rectangular or a circular pattern. Each of these options are described in the next sections. Start the Pattern Component command from the Assemble tab > Pattern panel, as shown in Figure 6.88 left. The Pattern Component dialog box will appear that has three tabs: Associative, Rectangular, and Circular, as shown in Figure 6.88 right, and are used to create the pattern in three different ways.

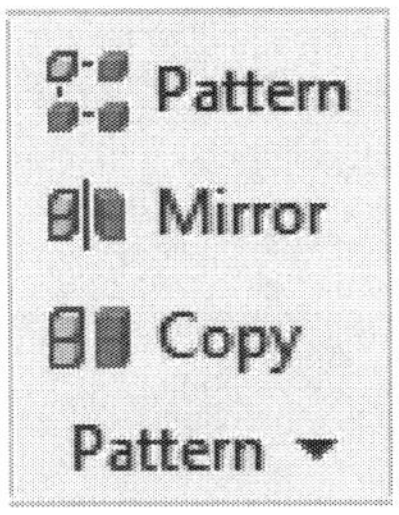

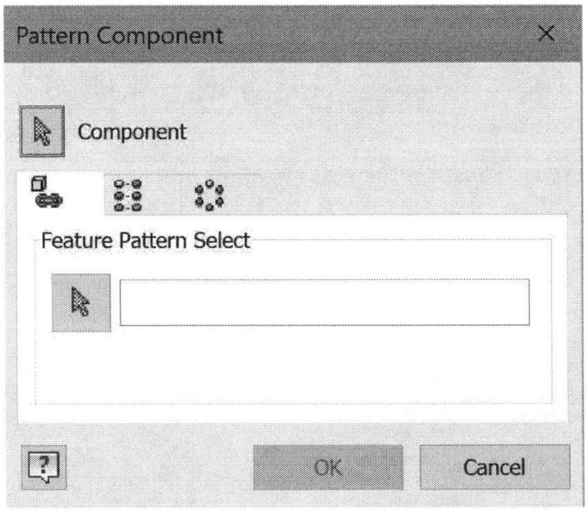

Figure 6.88

Associative Pattern

An associated component pattern will maintain a relationship to the feature pattern that you select. For example, a bolt is component-patterned to a part's bolt-hole circular pattern that consists of four holes. If the feature pattern, the bolt-hole, changes to six holes, the bolts will move to the new locations, and two new bolts will be added for the two new holes. To create an associative component pattern, there must be a feature-based rectangular or circular pattern, and the part that will be patterned should be constrained to the parent feature, that is, the original feature that was patterned in the component.

By default, the Component selection option is active. Select the component, such as the cap screw, or components to pattern. Next, click the Feature Pattern Select button in the dialog box and select a feature, such as a hole, that is part of the feature pattern. Do not select the parent feature. After selecting the pattern, it will highlight on the part and the pattern name will appear in the dialog box. When done, click OK to create the component pattern, as shown in the following image on the right.

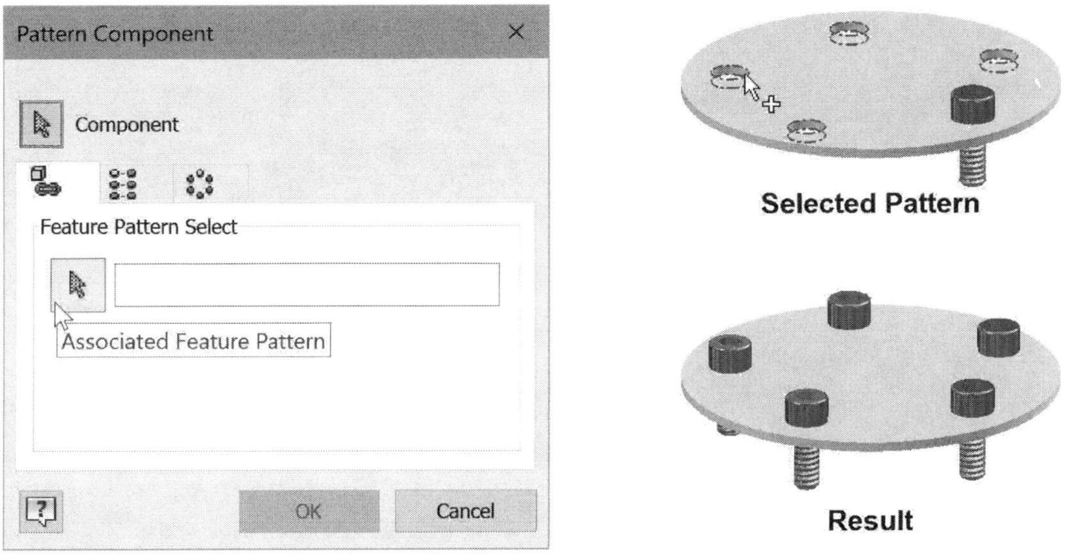

Figure 6.89

Rectangular Pattern

For a rectangular pattern operation, you select two edges to define the direction of the rows or columns. Note that a rectangular pattern can consist of a single row or column. When defining a pattern with a single row/column, you define a single direction. To define a direction, select an edge, work axis, origin axis or a face. After selecting the direction(s) enter the number of columns and rows and then the spacing or distance between the rows and columns. The resulting pattern acts like a feature pattern. After creation, you can edit the pattern to change its numbers, spacing, and so on.

Figure 6.90 illustrates the rectangular pattern option used to pattern a bolt by using the bottom horizontal edge and the right-angled edge of the plate to define the directions.

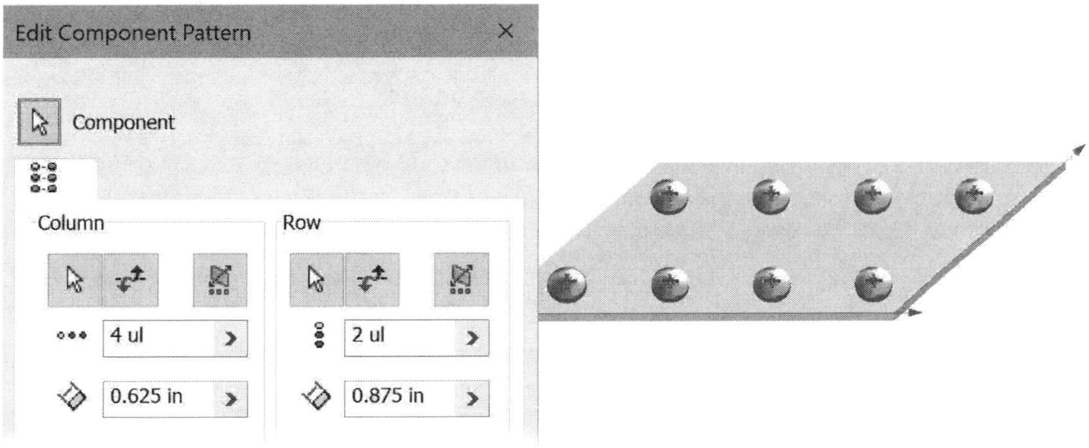

Figure 6.90

Circular Pattern

Circular patterns copy the selected components in a circular direction. After selecting the component or components to pattern, select an axis direction. In the following example, this element takes the form of the centerline of the circular plate and acts as a pivot point for the pattern. You then enter the number of occurrences or items that will make up the pattern and the circular angle. Figure 6.91 shows an example of a bolt patterned with a count of 12 and 30 degrees between them. If needed use the Orientation options to control if the patterned components should be rotated or keep their fixed position.

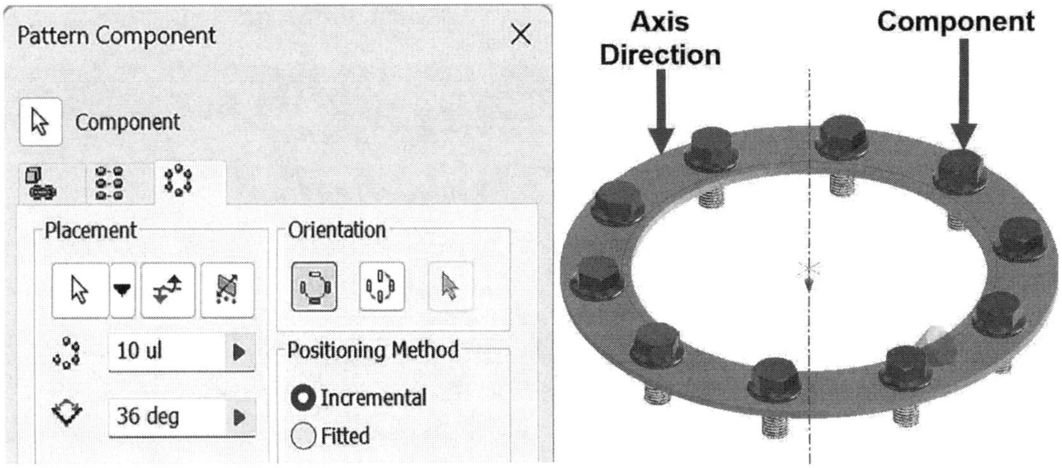

Figure 6.91

Editing a Pattern

When done patterning a component with any of the three methods, the patterned component is consumed into a Pattern entry in the browser. Expand the Pattern entry to see the elements, which represent the count. Expand the browser entry to view the component occurrences as shown in Figure 6.92.

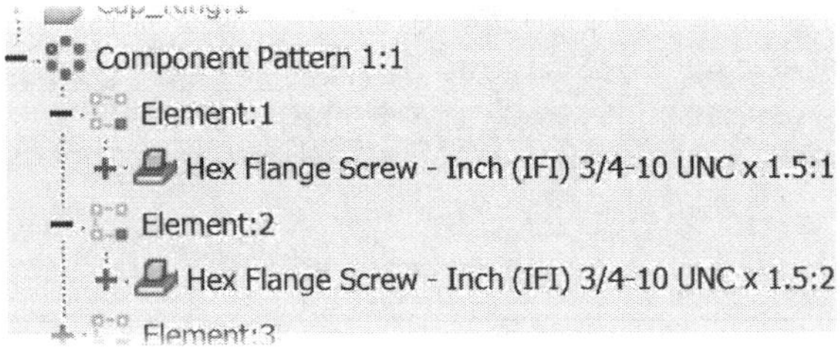

Figure 6.92

You can edit the original component pattern by right-clicking on the pattern in the graphics window and clicking Edit from the menu, or right-click on the pattern's name in the browser and click Edit from the menu, as shown in Figure 6.93 left.

You can also suppress an element (part occurrence) by right-clicking on the Element's name in the browser and clicking Suppress from the menu, as shown in the middle image. The suppressed component will be grayed out in the browser and a line will be struck through it, as shown in Figure 6.93 right.

You can remove an individual occurrence from the pattern by right-clicking on it and clicking Independent from the menu, as shown in the middle of Figure 6.93. Once a part is independent, an X will appear on the icon of the element in the browser as shown in Figure 6.93 right. An independent component has no relationship with the pattern. However, you can rejoin an element to the pattern by right-clicking on the element's name in the browser and clicking Independent from the menu to remove the checkmark next to the item in the menu.

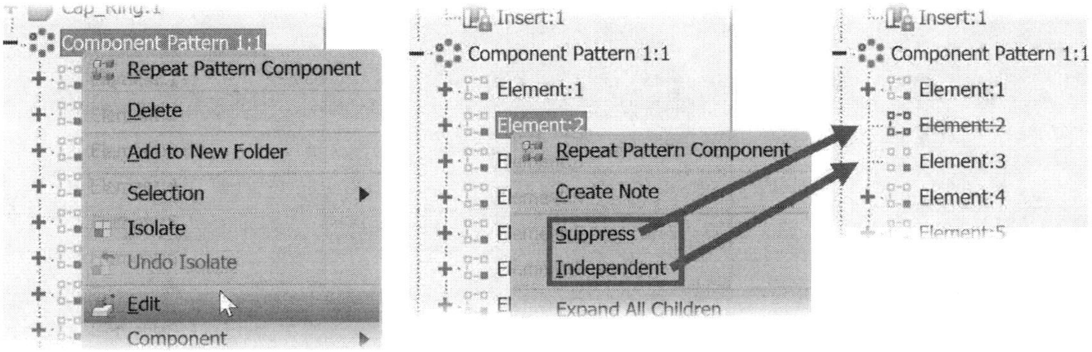

Figure 6.93

EXERCISE 6-4: PATTERNING COMPONENTS

In this exercise, you pattern a fastener to match an existing hole pattern and then change the count of the original hole pattern in a T-pipe assembly.

1 Open *ESS_E06_04.iam in* the Chapter 06 folder.

2 In the browser, expand the parts *ESS_E06_04-Cap_Bolt.ipt:1* and *ESS_E06_04-Cap_Nut.ipt:1*. Notice that the components are already constrained with an insert constraint, as shown in Figure 6.94.

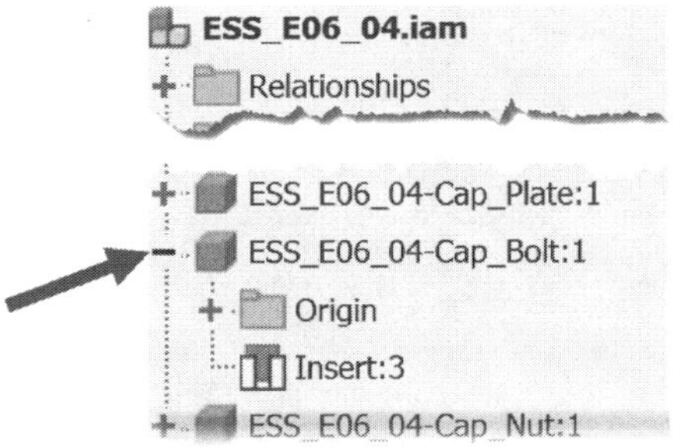

Figure 6.94

3 Next, you create a circular pattern consisting of a collection of nuts and bolts. Click the Pattern Component command from the Assemble tab > Pattern panel.

a. Select the components to pattern, press and hold the CTRL key and in the graphics window or in the browser select *ESS_E06_04-Cap_Bolt.ipt:1* and *ESS_E06_04-Cap_Nut.ipt:1* parts.

b. Click the Associated Feature Pattern button found under the Feature Pattern Select area of the Pattern Component dialog box. In the graphics window, move the cursor over a patterned hole in *ESS_E06_04-Cap_Plate:1*. When the circular pattern of holes in the cap plate is highlighted, select the edge of one of the holes, as shown in Figure 6.95 left.

c. You should see a preview of the component pattern, as shown in Figure 6.95 right.

d. Click the OK button to create the component pattern.

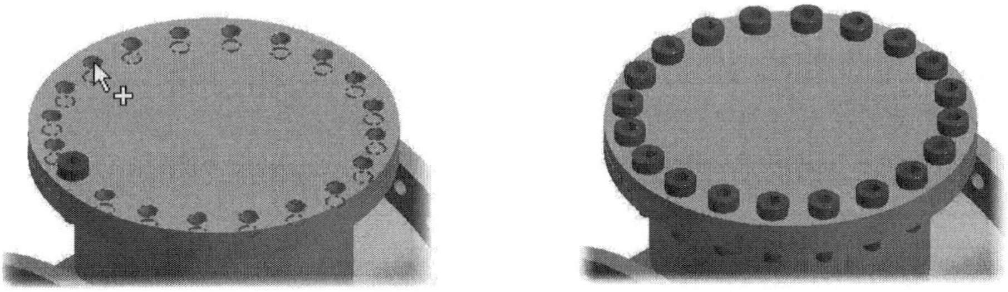

Figure 6.95

4 Next, you edit the original hole pattern on the Cap Plate.

a. Make the top plate current by double clicking on *ESS_E06_04-Cap_Plate:1* (the top circular plate) as shown in Figure 6.96 left.

b. Select a circular face of one of the patterned holes and select Edit Circular Pattern from the marking menu, as shown in Figure 6.96 right.

c. Change the Placement count to **10**.

d. Click OK to complete the edit.

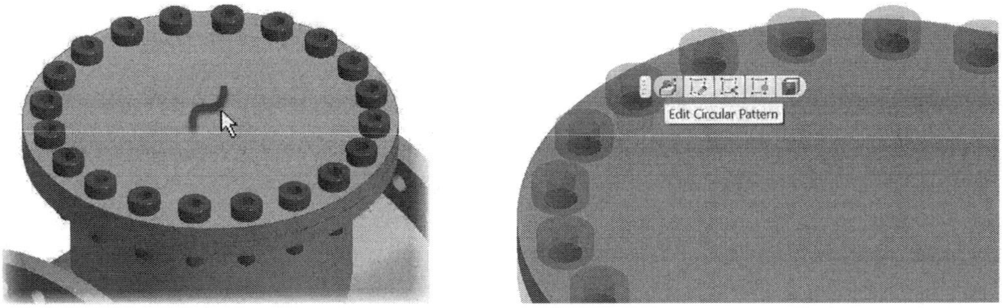

Figure 6.96

5 Click the Return command on the 3D Model tab > Return panel. When done, your screen should resemble the following image.

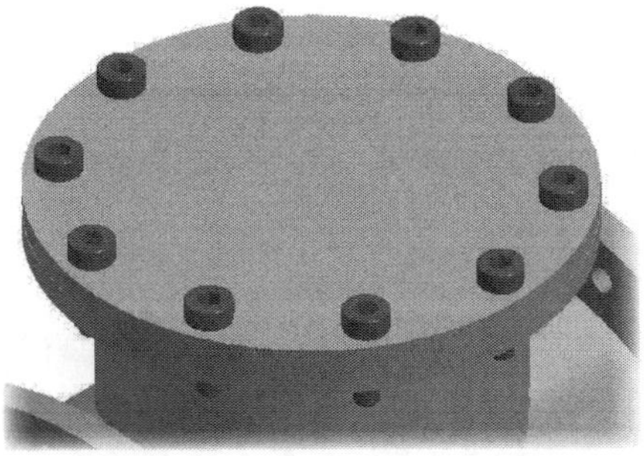

Figure 6.97

6 Close all open files. Do not save changes.

End of exercise.

ANALYSIS COMMANDS

Various commands are available to assist you in analyzing sketch, part, and assembly models. You can calculate the minimum distance between components, the center of gravity of parts and assemblies, and perform interference detection.

Measure Distance

In chapter 2 you learned about the Measure command; while in an assembly there is additional functionality to measure the distance, angle between components, faces and edges, determine the area of faces, and volume of solids. While in the assembly, select the Measure Distance command from the Inspect tab > Measure panel, as shown in Figure 6.98 left. Then in the Measure property panel change the selection priority, depending on what you want to measure. The three available selection modes are shown on the right.

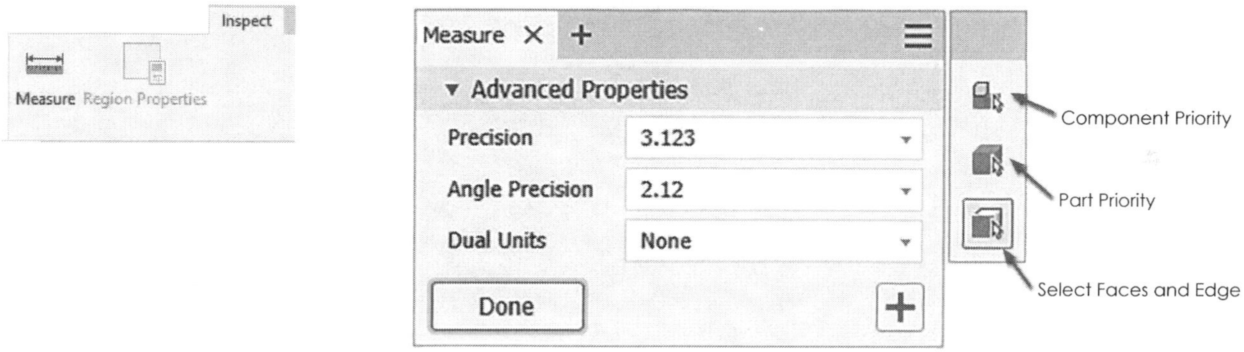

Figure 6.98

Identifying the Center of Gravity of an Assembly

Knowledge of where the center of gravity is in an assembly could be critical to the overall design of that assembly. In chapter 3 you learned how to change the material of a part and see the physical properties of the part through its iProperties. You can also use the iProperties command to see the center of gravity of an assembly. Start the iProperties command with one of the following techniques.

- Right-click the assembly's name in the browser and choose iProperties from the menu and click the Physical tab.
- Click the File tab > iProperties.

The image on the left shows the center of gravity as displayed in the iProperties for an assembly. You can also see where the center of gravity is in the graphics window by clicking on the View tab > Visibility panel > Center of Gravity as shown in the middle of the following image. The image on the right shows the center of gravity icon applied to an assembly model. This icon consists of a triad displaying the X, Y, and Z directions. Three selectable work planes and a selectable work point area are also available to measure distances and angles that reference the center of gravity.

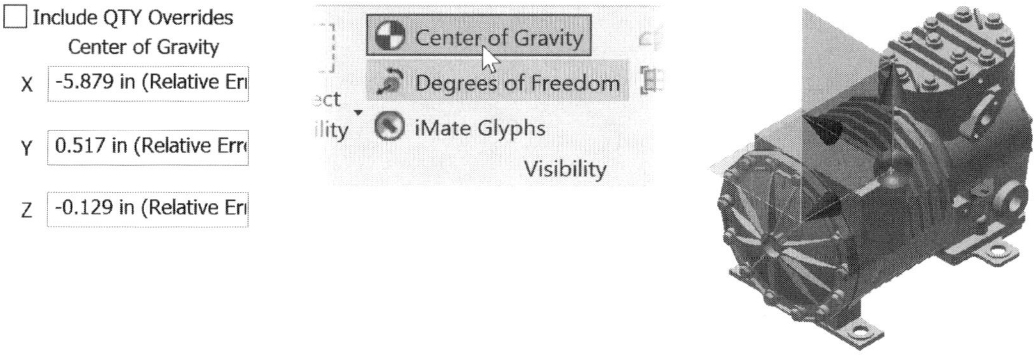

Figure 6.99

TIP: Before finding the center of gravity of an assembly, ensure that each part has the correct material applied to it.

Interference Checking

You can check for interference in an assembly using one or two sets of objects. To check the interference among sets of stationary components, make the assembly or subassembly in question active. Then click the Analyze Interference command on the Inspect tab > Interference panel, as shown in the following image on the left. The Interference Analysis dialog box will appear, as shown in Figure 6.100 right.

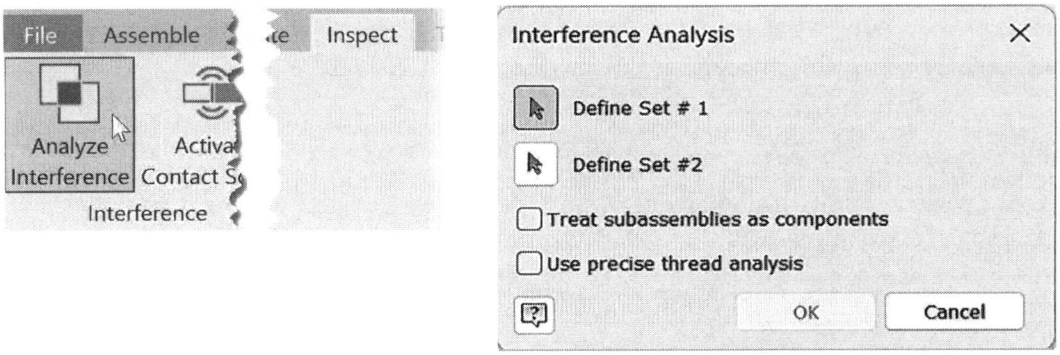

Figure 6.100

To run interference check, follow these steps.

1. Click on Define Set #1, and select the components that will define the first set.
2. Click on Define Set #2, and select the components that will define the second set. A component can exist in only one set.

TIP: Only use Define Set #1 if you want to check for interference against all selected components. Otherwise interference is only checked for set #1 vs. set #2.

3. To add or delete components from either set, select the button that defines the set that you want to edit. Click components to add to the set, or press the CTRL key while selecting components to remove from the set.
4. Once you have defined the sets, click the OK button. The order in which you selected the components has no significance.
5. If interference is found, the Interference Detected dialog box will appear, as shown in Figure 6.101.

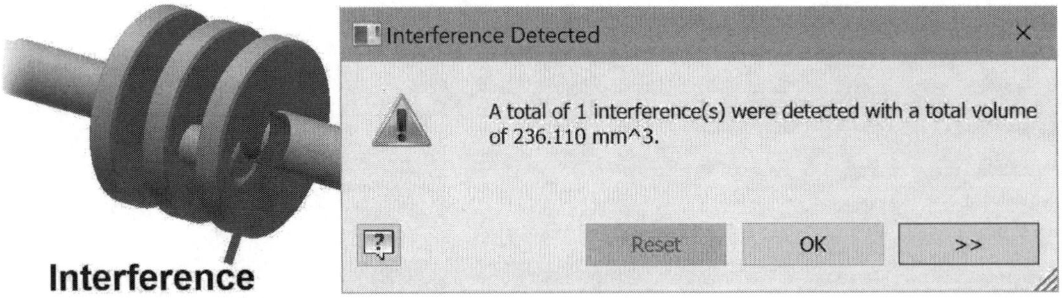

Figure 6.101

The information in the dialog box defines the X, Y, and Z coordinates of the centroid of the interfering volume. It also lists the volume of the interference and the components that interfere with one another. A temporary solid will also be created in the graphics window that represents the interference. Expand the Interference Detected dialog box by clicking the >> button and from the expanded dialog box you can see more information about the interference, copy the interference report to the clipboard or print it as shown in Figure 6.102. When the operation is complete, click the OK button, and the interfering solid will be removed from the screen.

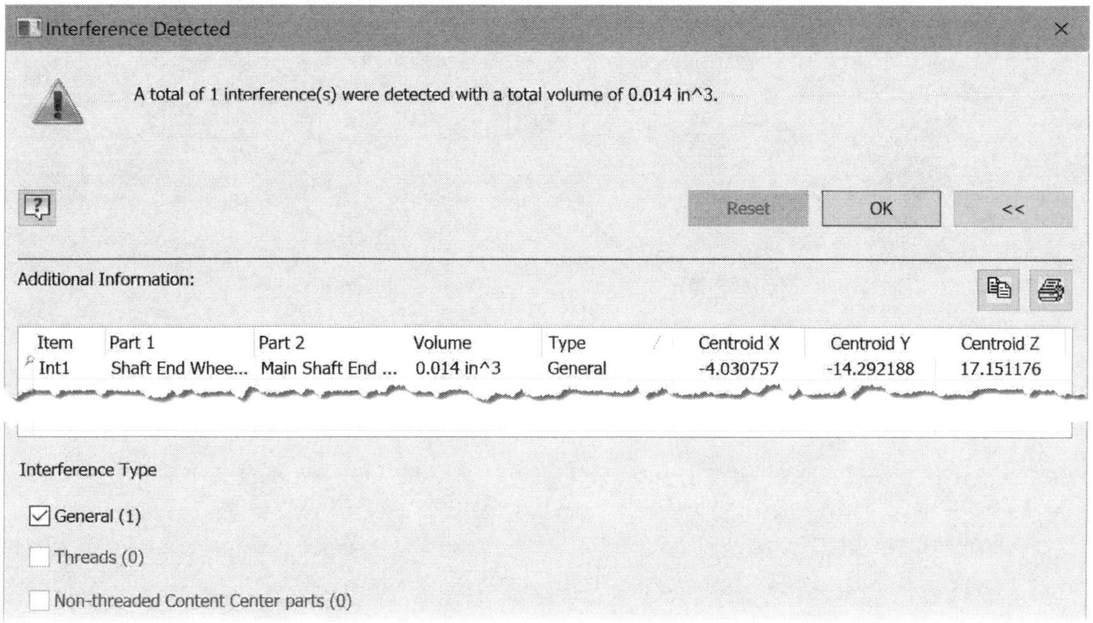

Figure 6.102

TIP: Performing an interference check does not fix the interfering problem; it only presents a graphical representation of the problem. After analyzing and finding interference, edit the assembly or components to remove the interference. You can also detect interference when driving constraints.

EXERCISE 6-5: ANALYZING AN ASSEMBLY

In this exercise, you analyze a partially completed assembly for interference between parts in a linkage, as shown in the following image on the left, and check the physical properties to verify design intent.

1. Open *ESS_E06_05.iam* in the Chapter 06 folder.
2. Begin the process of checking for interferences in the assembly by zooming into the linkages, as shown in the following image on the left.
3. Click the Analyze Interference command on the Inspect tab > Interference panel.
4. In the Analyze Interference dialog box, define Set #1 by selecting two components: *ESS_E06_01-Crank:1* and *ESS_E06_05-Claw1:1,* labeled (1) in the following image on the left.
5. Click the Define Set #2 button, then select the *ESS_E06_05-Spyder1:1*, labeled (2) in the following image on the left.
6. Click the OK button. Notice that the Interference Detected dialog box is displayed, as shown in the following image on the right, and the amount of interference displays on the parts of the assembly.

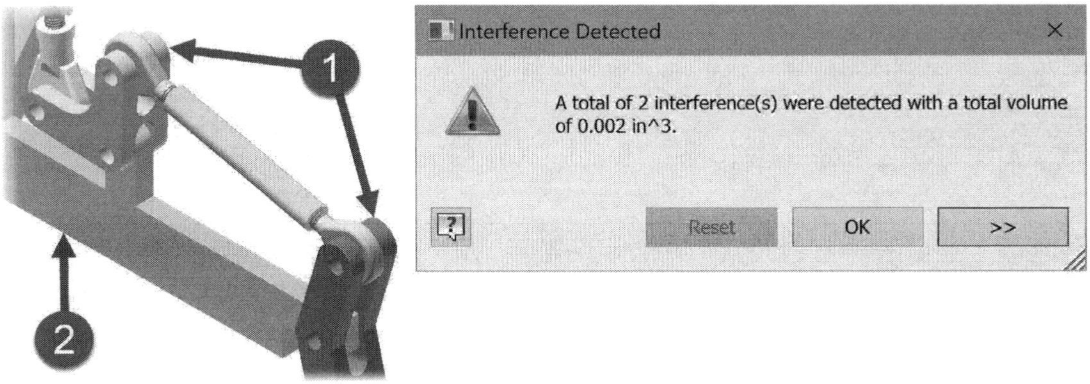

Figure 6.103

7 Click the More >> button to expand the dialog box and display additional information.

8 Click the OK button to close the Interference Detected dialog box.

9 The design intent of the lifting mechanism relies on knowing the center of gravity. You will now display the physical properties of the entire assembly. Change the viewpoint to the Home View.

10 In the browser, right-click on the top-level assembly *ESS_E06_05.iam*, and then click iProperties from the menu, as shown in the following image on the left.

11 When the iProperties dialog box displays, click the Physical tab and notice that the General Properties are displayed as N/A, noting that the properties are out of date.

12 Click the Update button, as shown in the following image on the right. Notice that the physical properties of the assembly are updated in the dialog box.

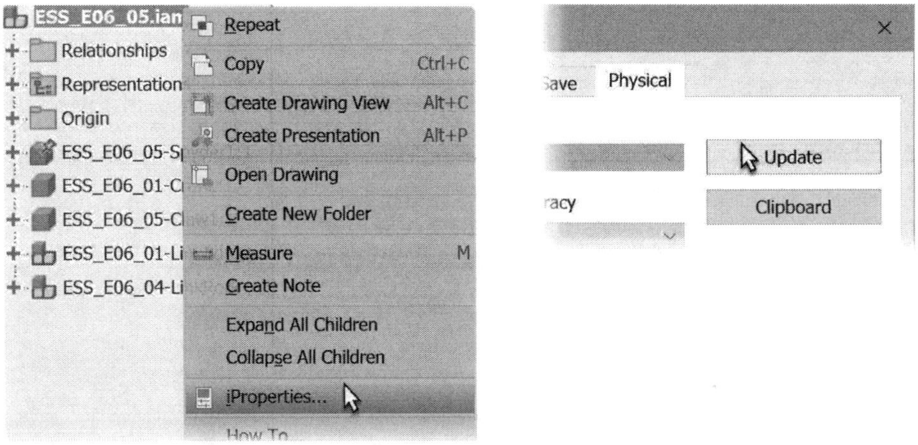

Figure 6.104

13 Click the Close button to dismiss this dialog box.

14 To display the center of gravity symbol on the parts click the Center of Gravity command on the View tab > Visibility panel.

15 Close all open files. Do not save changes.

End of exercise.

DRIVING A CONSTRAINT

You can animate mechanical motion by driving an angle, mate, tangent, or insert assembly constraint a set distance or angle. You can only drive one assembly constraint at a time, but you can use equations to create relationships to drive multiple assembly constraints simultaneously. To drive a constraint, right-click on the desired constraint in the browser and click Drive from the menu, as shown in Figure 6.105 left.

The Drive dialog box will appear, as shown in the following image on the right. Enter a Start value; the default value is the angle or offset for the constraint. Enter a value for End and a value for Pause Delay if you want a dwell time between the steps. In the dialog box, click the Record button  to save a video file (AVI or WMV) of the assembly motion. You can replay the recorded file without having Autodesk Inventor installed.

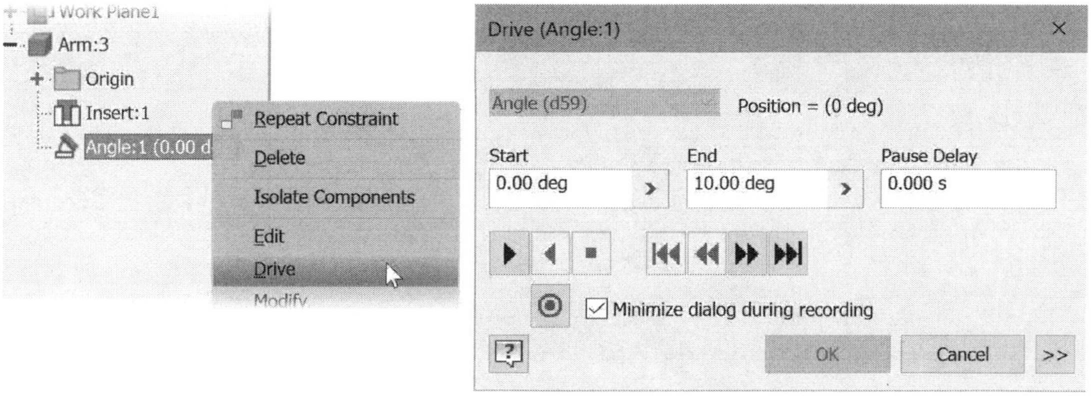

Figure 6.105

To set more conditions on how the motion will behave, click the More >> button. This action will display the expanded Drive Constraint dialog box, as shown in Figure 6.106. Following are descriptions of the options in this portion of the dialog box.

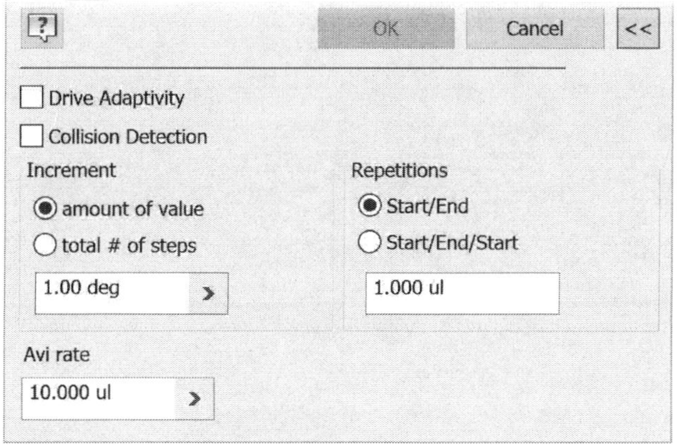

Figure 6.106

Drive Adaptivity. Click this option to adapt the component while the constraint is driven. It only applies to assembly components for which adaptivity has been defined and enabled.

Collision Detection. Click this option to drive the constraint until a collision is detected. When interference is detected, the drive constraint will stop and the components where the collision occurs will be highlighted. It also shows the constraint value at which the collision occurred.

Increment

The options and value in this area determine the method and value that the constraint will be incremented during the animation.

Amount of Value. Increments the offset or angle by this value for each step.

Total # of Steps. The constraint offset or angle is incremented by the same value per step based on the number of steps entered and the difference between the start and stop values.

Repetitions

Sets how the driven constraint will act when it completes a cycle and how many cycles will occur.

Start/End. Drives the constraint from the start value to the end value and resets at the start value.

Start/End/Start. Drives the constraint from the start value to the end value and then drives it in reverse from the end value to the start value.

AVI Rate. Specifies how many frames are skipped before a screen capture is taken of the motion that will become a frame in the completed AVI file.

TIP: If you drive a constraint and it fails, you may need to suppress or delete another assembly constraint to allow the required component to have degrees of freedom.
To reduce the size of a WMV or AVI movie file, reduce the screen size before creating the video and use a solid background color in the graphics window.

EXERCISE 6-6: DRIVING A CONSTRAINT

In this exercise, you drive an angle constraint to simulate motion in an assembly and then drive the constraint to determine if components interfere.

1 Open *ESS_E06_06.iam* in the Chapter 06 folder.

2 First you create an angle constraint between the pivot base and arm. Click the Constrain command on the Assemble tab > Relationships panel.

   a. In the Place Constraint dialog box, click the Angle button, labeled (1) in the following image.

   b. Click the Directed Angle solution, labeled (2) in the following image.

   c. Enter an angle of **45 deg**, labeled (3) in the following image.

   d. In order, select the front face on the Base labeled (4) and then front face on the Arm labeled (5) as shown in the following image on the right.

   e. Click the OK button.

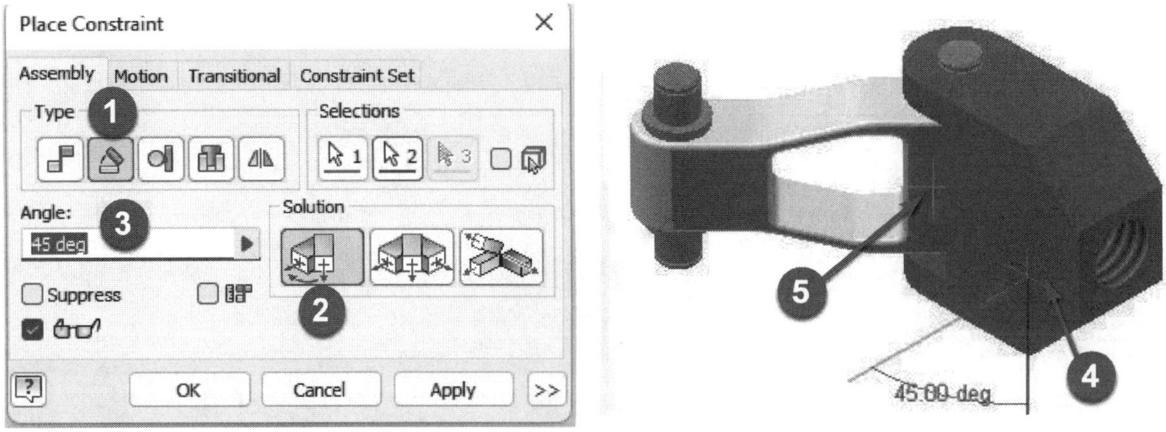

Figure 6.107

3 Now drive the angle constraint you just placed.

a. In the browser, expand *ESS_E06_06-Pivot_Base:1*.

b. Right-click on the Angle:1 constraint, and click Drive from the menu as shown in Figure 6.108 left.

c. In the Drive Constraint dialog box click the More >> button.

d. Verify that **45.00 deg** is the Start value, labeled (1) in the image on the right.

e. Change the End value to **120.00 deg**, labeled (2).

f. In the Increment section, enter **2 deg**, labeled (3).

g. Click the Forward button labeled (4) in the image. Notice that the arm subassembly interferes with the pivot base between 45.00° and 120.00° but Inventor continues to drive the constraint.

h. Click the Reverse button, labeled (5) in the image.

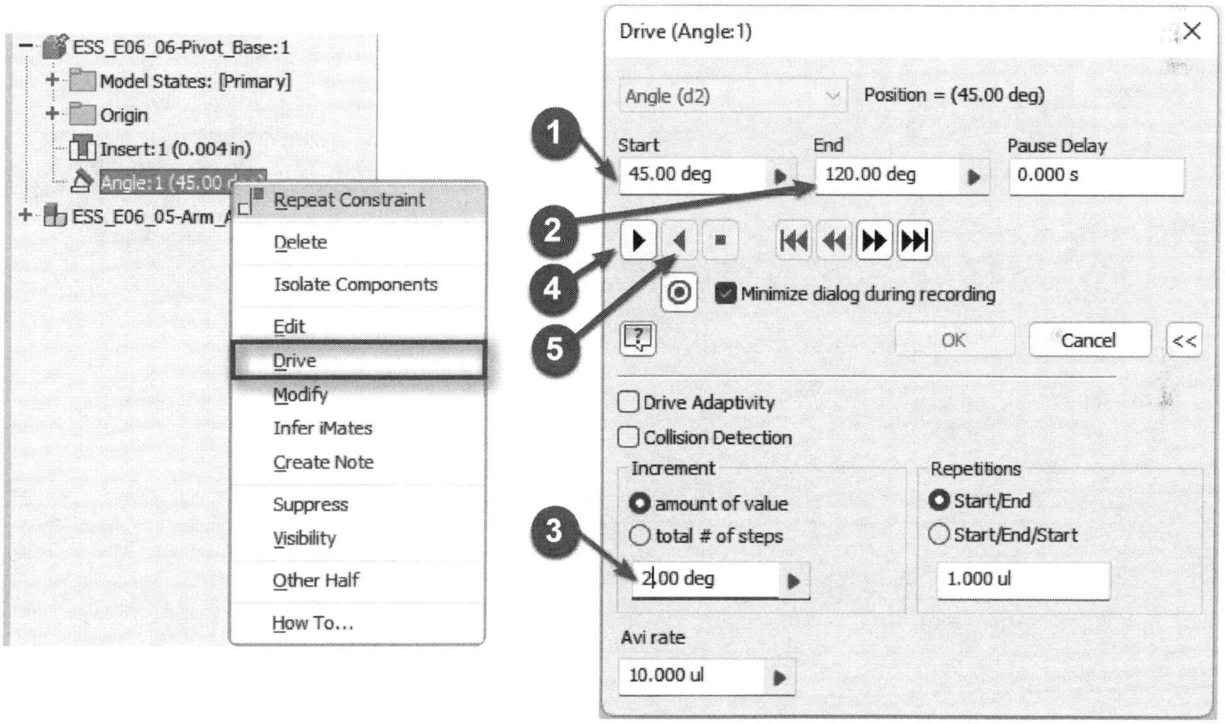

Figure 6.108

4 Next you run collision detection while driving the constraint.

a. In the Drive Constraint dialog box place a check in the Collision Detection section, labeled (1) in Figure 6.109.

b. Change the Increment value to **.7 deg**, labeled (2) in the following image.

c. Click the Forward button, labeled (3) in the following image.

d. Notice that a collision is detected at 81.7 degrees as displayed in the dialog box, labeled (4) in the following image.

e. In the graphics window, the parts that interfere are also highlighted, as shown in Figure 6.109.

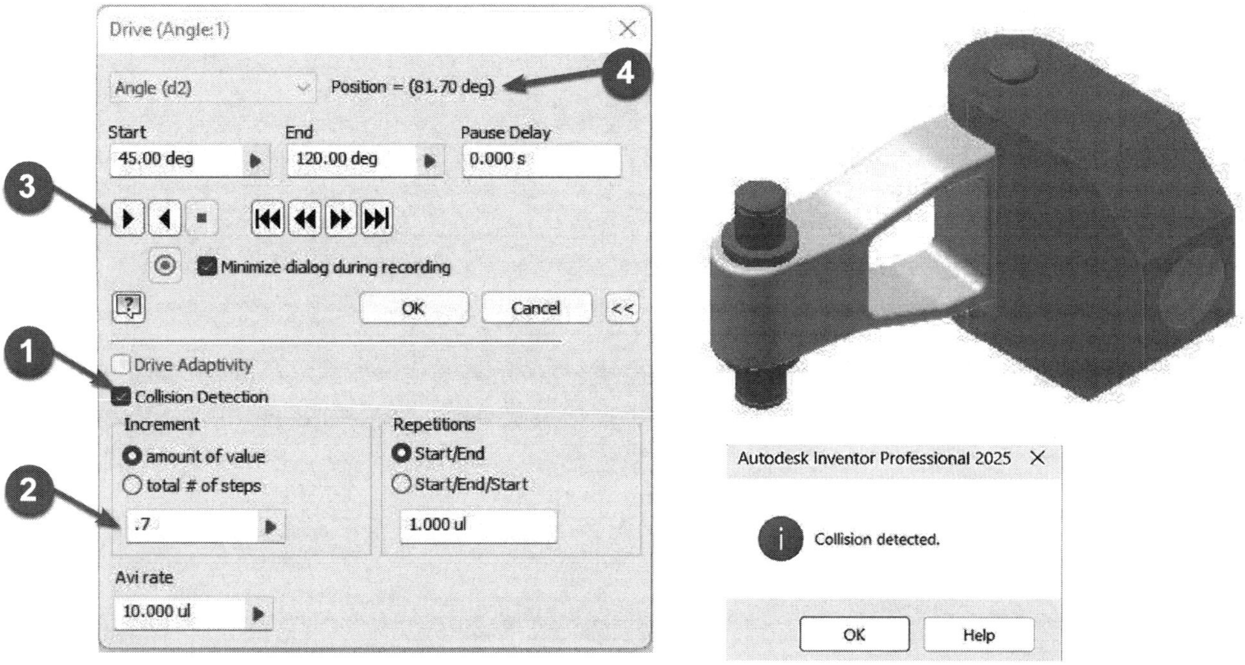

Figure 6.109

5 Click OK in the Collision detected dialog box to return to the Drive Constraint dialog box.

6 Now perform a check on the full range of motion for the arm.

a. In the Drive Constraint dialog box, enter **-80 deg** for the Start value, labeled (1) in the following image.

b. Change the End value to **80 deg**, labeled (2) in the following image.

c. Change the Increment value to **2 deg**, labeled (3) in the following image.

d. In the Repetitions area of the dialog box, select Start/End/Start, labeled (4) in the following image.

e. Change the number of repetitions to **4**, labeled (5) in the following image.

f. To test the full range of motion for interference, click the Reverse button labeled (6) in the following image.

g. The arm rotates four times and no collision is detected in this range of motion.

7 In the Drive Constraint dialog box, click the OK button. The 80 degrees represents the maximum angle before collision, and is applied to the Angle constraint.

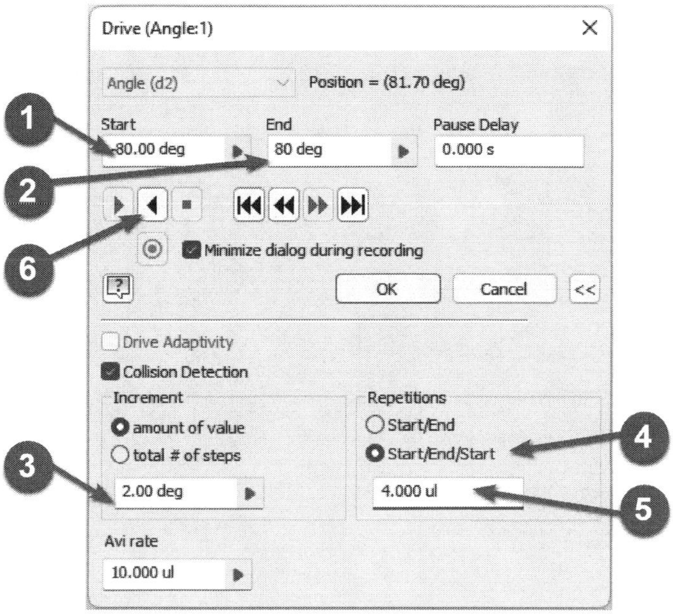

Figure 6.110

8 Close all open files. Do not save changes.

End of exercise.

CREATING A PRESENTATION FILE

After creating an assembly, you can create a presentation file to show how components are assembled or disassembled, or show their different positions like an exploded view. A presentation file can include a set of snapshot views and animation storyboards. In a presentation file, you can move and rotate (tweak) components to different positions, hide components, and create an animation that shows how to assemble and disassemble the components. A presentation file has a file extension of .ipn, and is associated with an assembly file. Changes made to the assembly file are reflected in the presentation file but note you cannot create components in a presentation file.

To create a new presentation file, follow one of these methods:

- In the Quick Access toolbar click the down arrow next to the New icon, and click Presentation, as shown in Figure 6.111 left.
- Expand the options under New and click Presentation on the Home page (middle image).
- From the File tab click New > Presentation, as shown in Figure 6.111 right.
- If desired, you can create a presentation file based on another template by using the New File command. This technique is described in the Creating Assemblies section at the beginning of this chapter.

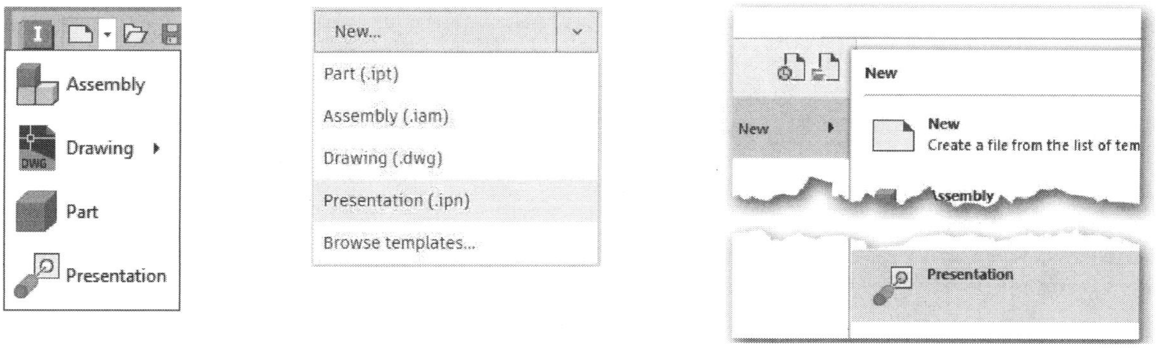

Figure 6.111

Creating Presentation Views and Snapshots

After starting a Presentation file, the Insert dialog box will appear; navigate to and select an assembly file as shown in the following image.

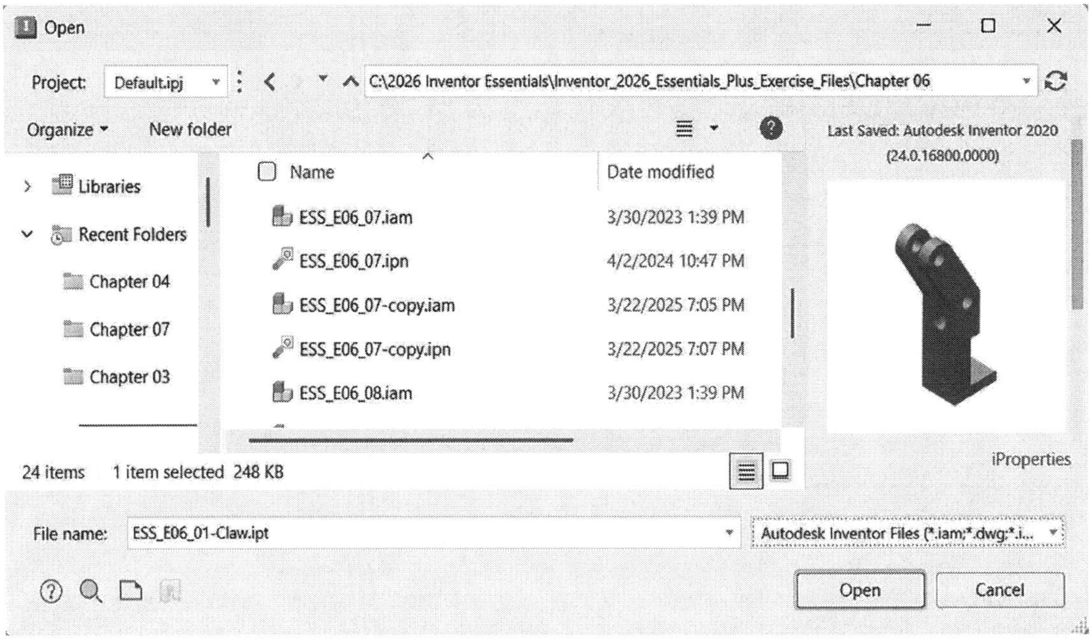

Figure 6.112

The Presentation environment will appear as shown in the following image. Following is a description of the main areas.

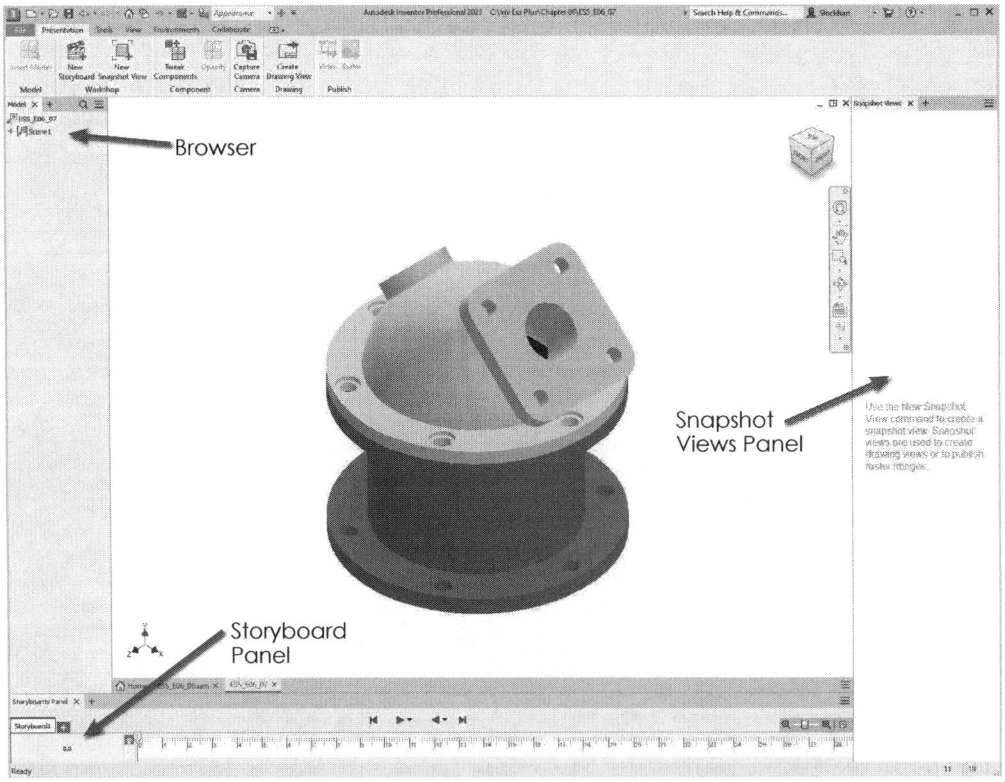

Figure 6.113

Snapshot Views Panel

Displays and manages snapshot views. To save a specific state and camera view create an independent snapshot view.

To associate a snapshot view to a specific time in the animation, create a snapshot view associated with the timeline. Independent snapshot views can be edited as needed. For associated views you can only change the camera.

Browser

Displays information about the presentation scene. The Model folder shows components and their visibility. The Tweaks folder lists component tweaks saved in the presentation file.

Storyboard Panel

Displays storyboard(s) that exist in the current document. A storyboard includes sequences of actions placed on the animation timeline. Storyboards are used to create animations and associative snapshot views. After creating an animation, the action is added to the storyboard as shown in Figure 6.114. Following is a description of the main sections in the storyboard.

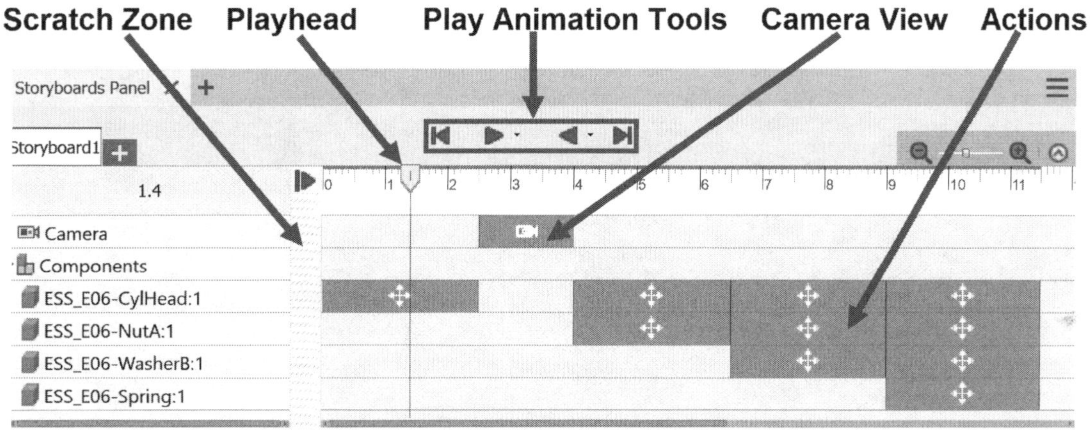

Figure 6.114

Scratch Zone

When the playhead is in this Scratch Zone, changes to the components and viewpoint are not saved as actions in the storyboard. Move the playhead to the Scratch Zone and position the components for an independent snapshot view, or to prepare the model before you start recording an animation.

Playhead

Marks the current instant on the timeline. The appearance of the components in the graphic window corresponds to the position of the playhead.

Play Animation Tools

After creating an animation, use these tools to play the animation.

Camera View

Represents a change in the camera's viewpoint.

Actions

Shows changes in the graphic window which can be components moving or rotating, change component visibility and opacity, and change of the camera position.

Inserting Models into a Presentation File

When creating a Presentation file and if you cancel the Insert command you can insert an assembly into the current scene by clicking the Insert Model command on the Presentation tab > Model tab as shown in the following image. Then navigate to and select an assembly file.

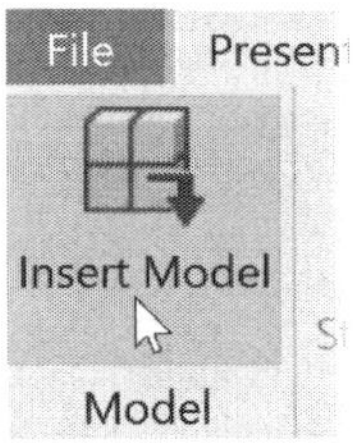

Figure 6.115

Tweak Components

After inserting an assembly into a presentation file, you move or rotate the components with the Tweak Components command on the Presentation tab > Component panel as shown in Figure 6.116 left. You can also right-click and choose Tweak Components from the marking menu. A triad and mini-toolbar appear, as shown in Figure 6.116 right. Following are explanations of the functionality.

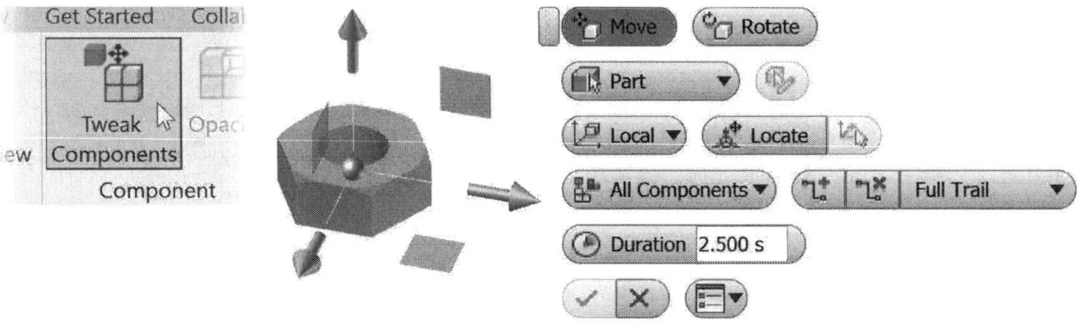

Figure 6.116

Tweak Types

Select how the components will be tweaked.

Move. When the Move option is active, the selection set is cleared when you click Apply. With the Move option you can create a series of tweaks with different selection sets.

Rotate. Click the Rotate option to rotate components.

Part or Component

Select the Part option to tweak parts, even if they are in a subassembly. Select the Component option to select subassemblies or individual parts that are not in a subassembly.

Locate

Position the triad by moving the cursor over a part or component.

Local or World

Select the Local option to align the triad to the selected geometry. Select the World option to snap the triad to a point or center point on the selected geometry when the triad is aligned to the 3D Origin indicator and not the selected component.

Default Trails

Set how the trails will be displayed.

No trail. Cancels the creation of trails.

All Components. Creates trails for all selected components.

All Parts. Creates trails for all selected parts; trails will not be created for subassemblies.

Single. Creates a single trail for the entire tweak regardless of the number of selected components. The trail starts from the triad location.

Add New Trails

Select the Add option to add a trail.

Delete Existing Trail

Select the Delete option to delete a trail.

Full Trail or Trail Segment

Select the Full Trail option to add or delete all trails on the selected component. Use the Trail Segment option to add or delete a trail for a single segment.

Steps to Tweak

To tweak a component, follow these steps:

1. Select the tweak type: Move or Rotate.
2. Select a part(s) or component(s) to tweak.
3. Select a Default Trails option.
4. Locate and align the triad.
5. Move or rotate the selected part(s) or component(s).
6. To continue to tweak the selected component and add another component to tweak, press the Ctrl key and select the component. Then move or rotate the selected components as needed.

Edit Tweaks in the Storyboard

After creating a tweak, you can move it to happen sooner or later in the storyboard by clicking and dragging on it in the storyboard as shown in Figure 6.117.

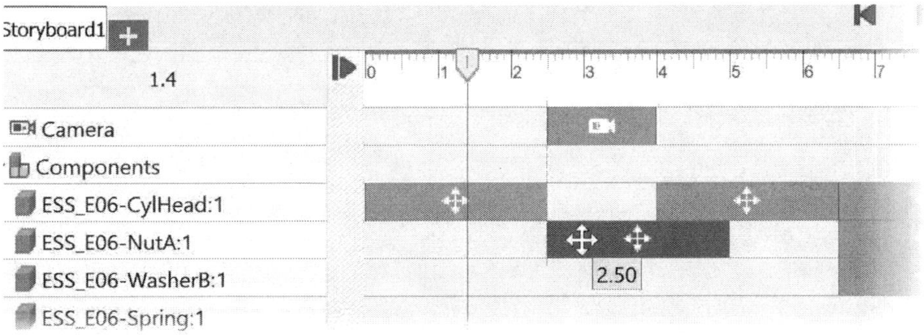

Figure 6.117

You can adjust the duration that the tweak happens by clicking and dragging on one of its vertical lines as shown in this image . Right-click on the tweak in the storyboard and choose Edit Time as shown in Figure 6.118 left, then in the mini-toolbar to adjust its duration as shown in the middle image.

To edit the tweak's distance, angle, or to add or delete a trail, right-click on the tweak in the storyboard and choose Edit Tweak from the menu. Adjust the tweak as needed via the mini-toolbars as shown in Figure 6.118 right.

To delete a tweak, click Delete from the menu.

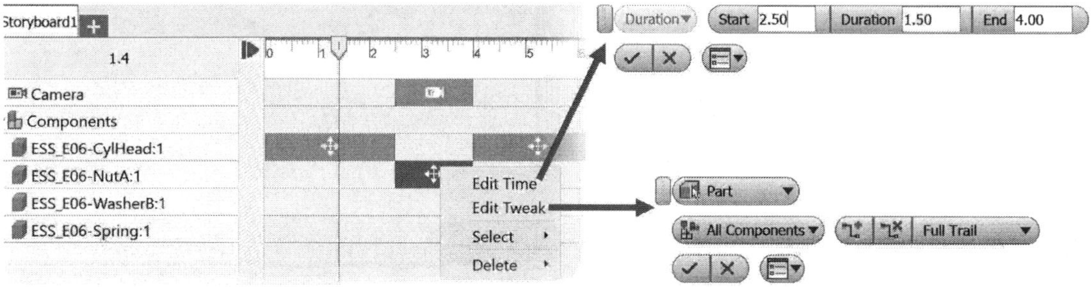

Figure 6.118

Edit Tweaks in Browser

You can also edit a tweak by expanding the Tweaks folder in the browser and right-click on the tweak to edit, then click the desired option from the menu as shown in Figure 6.119.

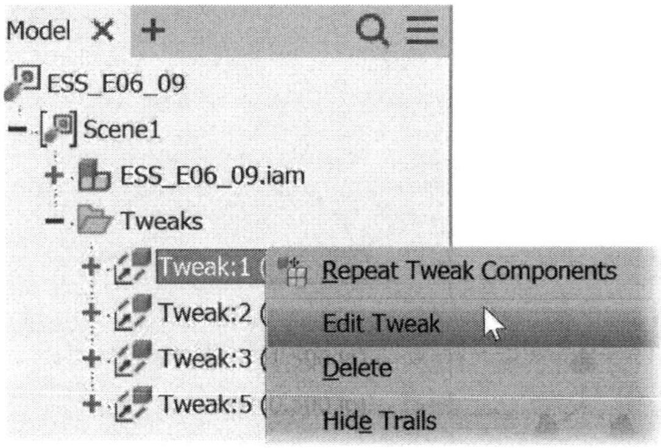

Figure 6.119

Capture Camera Viewpoint

After tweaking the components, you can add a camera viewpoint to the storyboard by positioning the Playhead in the storyboard's timeline at the time the viewpoint change is complete like what is shown in Figure 6.120 left. Click the Capture Camera command from the Presentation tab > Camera panel to complete the capture as shown in Figure 6.120 on the right.

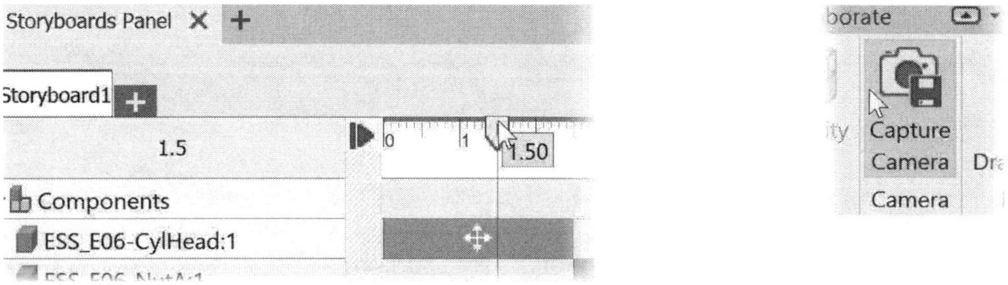

Figure 6.120

Position the cursor over the camera in the storyboard's timeline and click and drag to move the camera as shown in Figure 6.121.

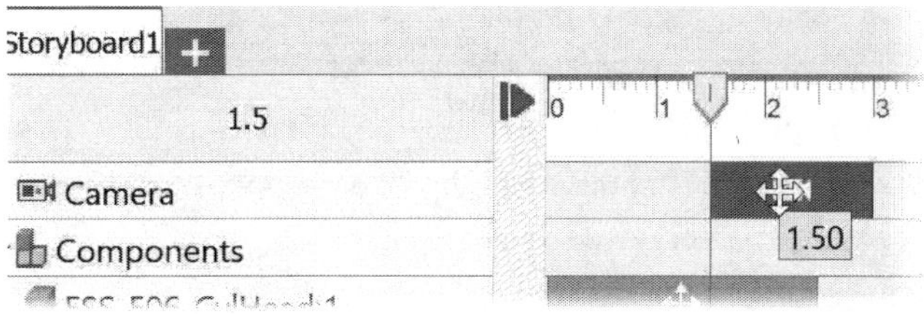

Figure 6.121

You can change the duration for the transition to the new viewpoint by moving the cursor over a vertical line of the camera in the storyboard and drag it. Edit the start and end time of the duration by right-clicking on the camera in the timeline and choose Edit Time from the menu as shown in Figure 6.122 left. Afterward you can edit the start and end time or you can change the Duration option to Instant.

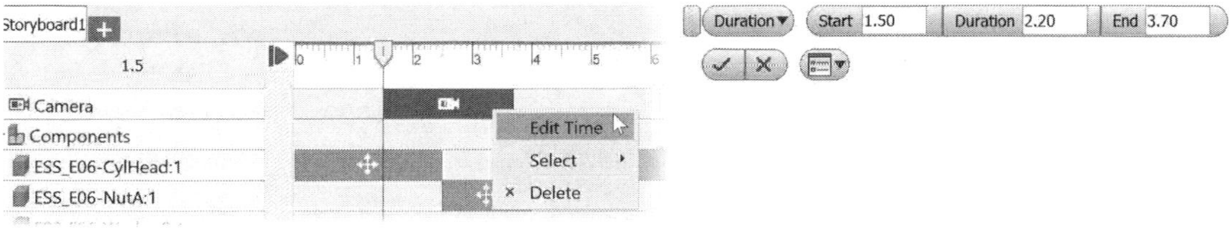

Figure 6.122

Play the Animation

While editing a storyboard it is helpful to play the animation. To play the animation use the Play Animation tools at the top of the Storyboard panel as shown in the following image.

Figure 6.123

Create a New Storyboard

While creating animations you may want to break the animations into separate animations/storyboards. For example, you may create a storyboard for instructions on how to assemble components and another storyboard to show how to disassemble the components. To create a new storyboard, follow these steps.

1. Click on the + icon from the left side of the Storyboard panel as shown in Figure 6.124 left or choose New Storyboard from the Presentation tab > Workshop panel as shown in the middle image.
2. In the New Storyboard dialog box, select the Clean option to create a storyboard that returns all components to their home position.
3. Click the Start from end of previous option to create a storyboard that has no actions but the components are in their last position.

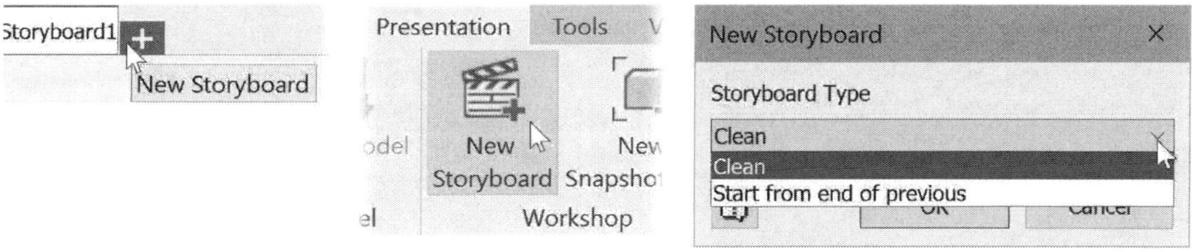

Figure 6.124

Publish a Video from a Storyboard

After creating and editing a storyboard you can publish the animation to a wmv or an avi video file by following these steps.

1. In the Storyboard panel, right-click on the storyboard from which to publish a video. Choose Publish to Video from the menu as shown in Figure 6.125 left. You can also choose Publish to Video from the Presentation tab > Publish panel as shown in the middle image. If you select Publish to Video from the Publish panel, select the desired storyboard before starting the command or if no storyboard is selected you can publish all storyboards.
2. The Publish to Video dialog box appears; select the desired options and click OK to publish the file.

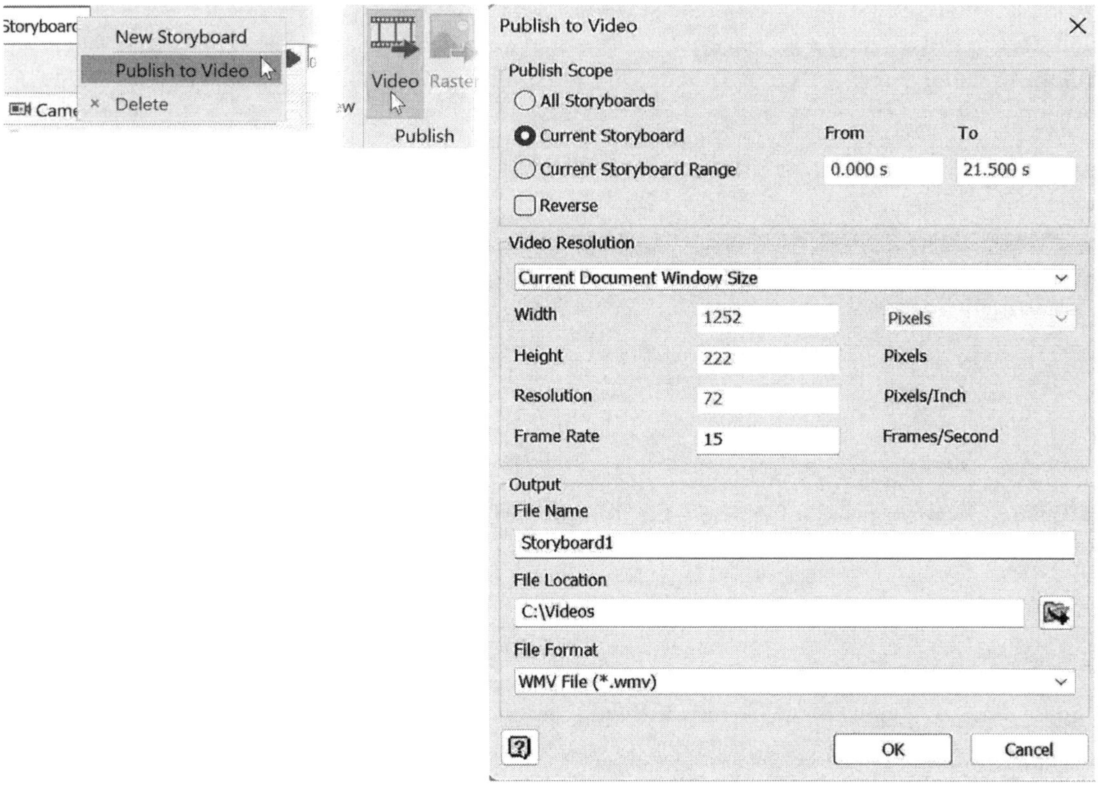

Figure 6.125

Snapshot View

A snapshot view stores the current viewpoint and the position of the components. Snapshot views can be independent or associated with the timeline. Snapshot views are used to create drawing views or raster images. You can create as many views as needed.

Create an Independent Snapshot View

An independent snapshot view has no association to animations or other snapshot views. You can edit an independent snapshot view and it will have no effect on the animation. To create an independent snapshot view, follow these steps.

1. In the Storyboard, move the Playhead to the Scratch Zone as shown in Figure 6.126.

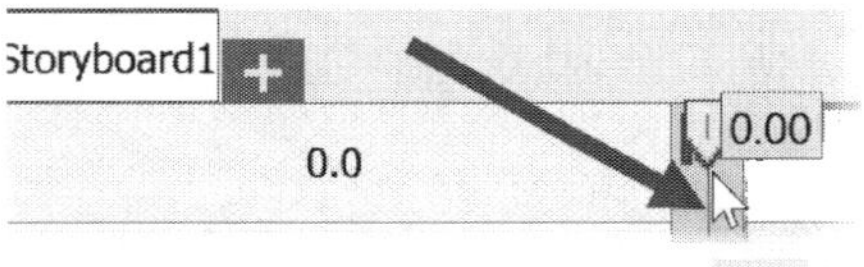

Figure 6.126

2. Control the visibility for components by right-clicking on the component in the graphic window or browser and click Visibility from the menu.
3. Tweak the components as desired.
4. Change the viewpoint as needed.
5. To create the independent snapshot view, click the New Snapshot View command on the Presentation tab > Workshop panel as shown in Figure 6.127 left.
6. The new view appears in the Snapshot Views panel as shown in Figure 6.127 right.

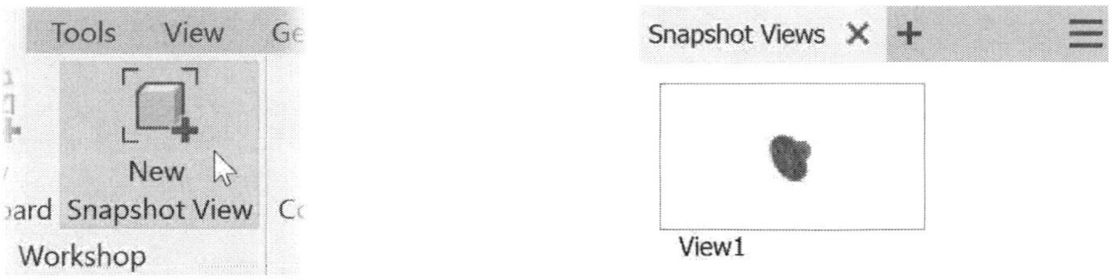

Figure 6.127

Create an Associated Snapshot View

An associated snapshot view is associated to the selected time in the storyboard. Associated snapshots are marked with the view mark on the timeline and in the Snapshot Views panel as shown in Figure 6.128.

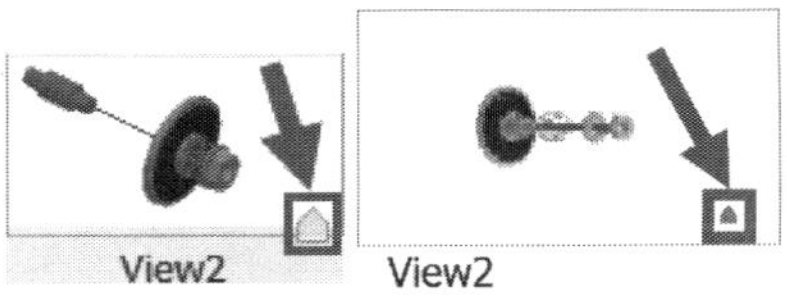

Figure 6.128

To create an associated snapshot view, follow these steps.

1. In the Storyboard move the playhead to the desired timeline position as shown in Figure 6.129 left. The graphic window displays the current state of the components appearance and camera position.
2. To create the associative snapshot view, click the New Snapshot View command on the Presentation tab > Workshop panel as shown in the middle image.
3. The new view will appear in the Snapshot Views panel as shown in Figure 6.129 right.

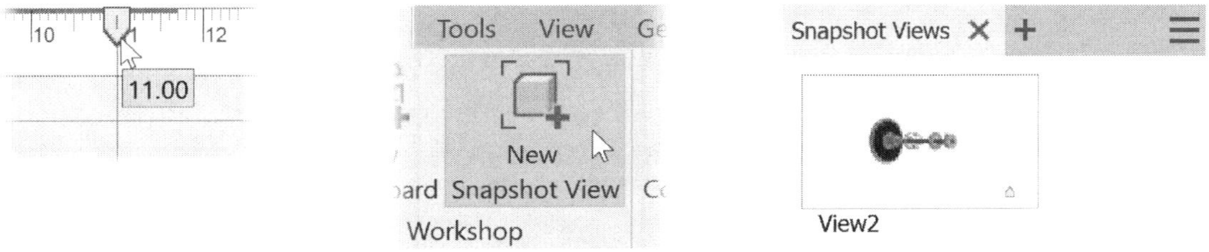

Figure 6.129

Create a Drawing from a Snapshot View

To create a drawing from a snapshot view, follow these steps.

1. In the Snapshot Views panel, right-click on the snapshot from which to create the drawing view. Choose Create Drawing View from the menu as shown in Figure 6.130 left or choose Create Drawing View from the Presentation tab > Drawing panel. If you select Create Drawing View from the Drawing panel, select the desired snapshot view before starting the command.
2. You are prompted to select a template for the drawing.
3. After selecting a drawing template, the Drawing View dialog box appears with the name of the Snapshot view as shown in Figure 6.130 right.
4. Create the drawing view as you normally would.

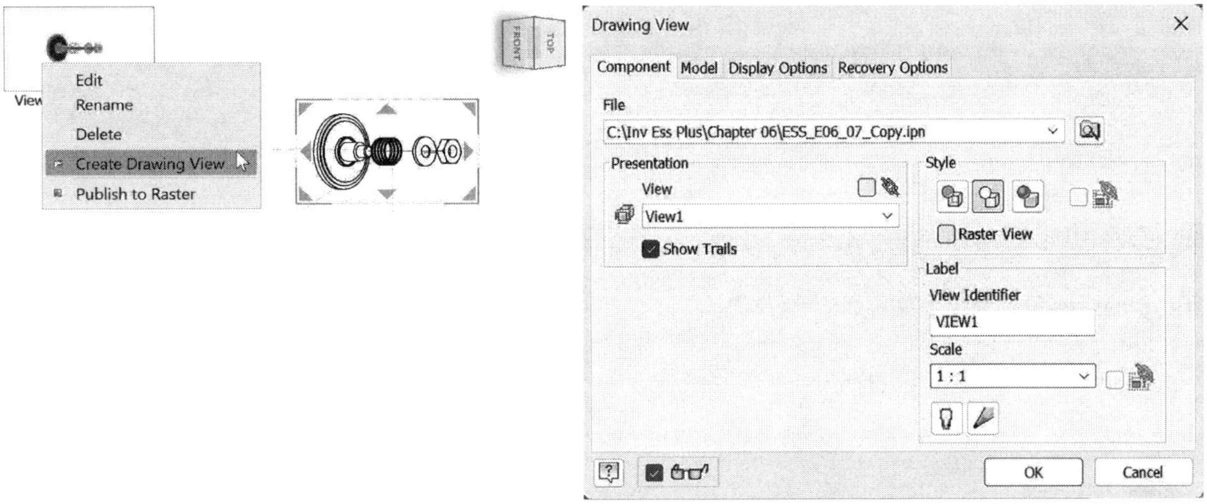

Figure 6.130

Publish a Raster Image from a Snapshot View

To create a raster image from a snapshot view, follow these steps.

1. From the Snapshot Views panel, right-click on the snapshot to use for the raster image. Choose Publish to Raster from the menu as shown in Figure 6.131 left or choose Publish to Raster from the Presentation tab > Publish panel. If you select the Publish to Raster command from the Publish panel, select the desired view before starting the command or if no snapshot view is selected you can publish all views.
2. From the Publish to Raster Images dialog box, select the desired option and click OK to publish the file.

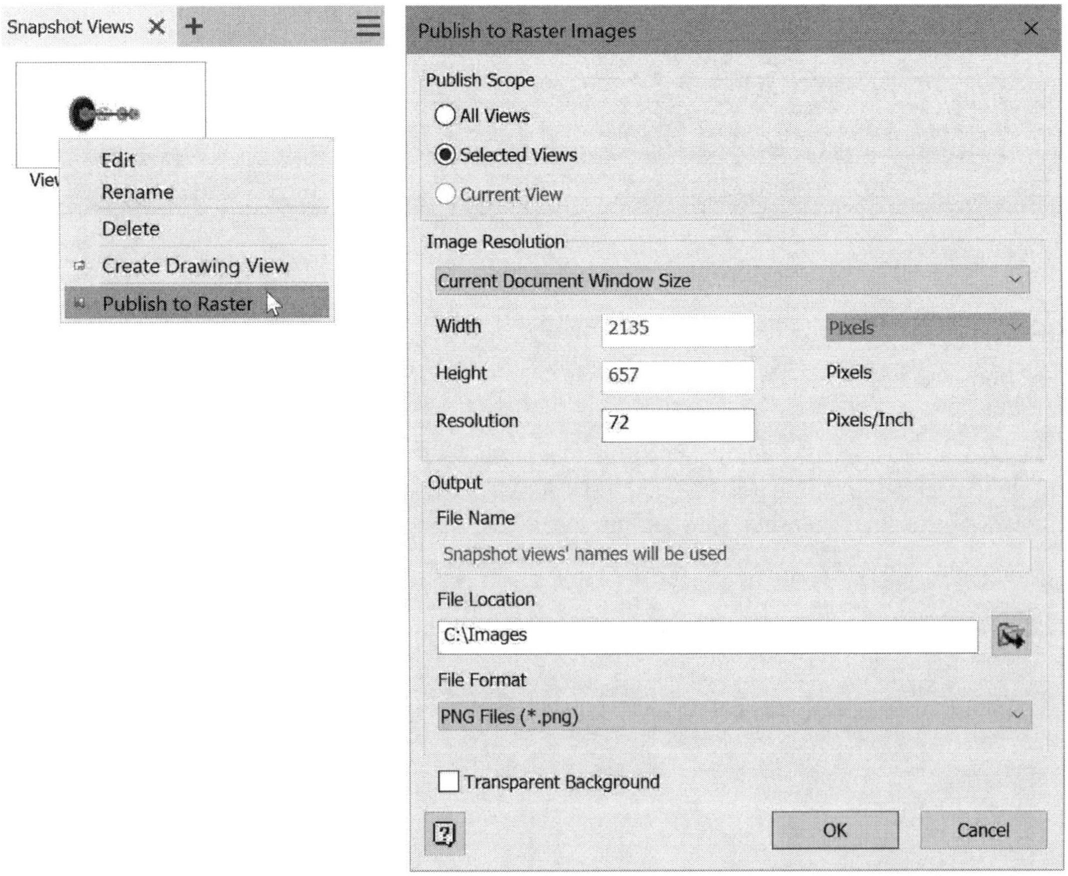

Figure 6.131

EXERCISE 6-7: CREATING A PRESENTATION STORYBOARD

In this exercise, you create a storyboard of an assembly and add tweaks to the components to create an exploded view.

1 Create a new presentation file.

a. Click the New command on the Quick Access toolbar.

b. Select the English template folder, and in the Presentation area double-click Standard (in).ipn.

c. In the Insert dialog box, navigate to the Chapter 06 folder, and select *ESS_E06_07.iam* and then click the Open button.

2 You will animate the internal components, so turn off the visibility of the two outside components.

a. Expand the browser so the components are displayed.

b. Press the Ctrl key and select the top two components: ESS_E06-Cylinder.ipt:1 and ESS_E06-CylHead.ipt:1

c. Right-click and click Visibility from the menu as shown in Figure 6.132.

d. Click Break from the dialog box that states the view is associative to a design view.

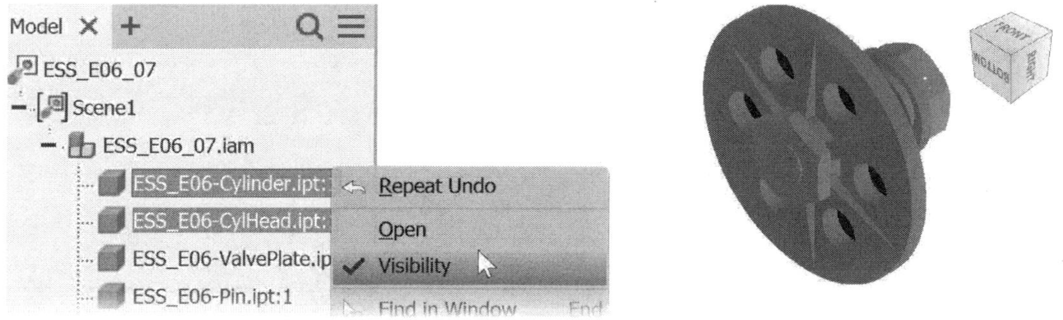

Figure 6.132

3 Zoom out so the tweaks have room to go toward you. Capture the camera using the Capture Camera command from the Presentation tab > Camera panel.

4 Next, tweak the location of the jamb nut ESS_E06-NutB.ipt.

a. Click the Tweak Components command on the Presentation tab > Component panel.

b. In the graphics window or the browser, select component ESS_E06-NutB.ipt (closest component).

c. In the Tweak Components mini-toolbar, ensure that Move and Part are both selected, and change the trail option to All Parts as shown in Figure 6.133 left.

d. Drag the nut toward you (Z arrow) and enter a value of **1.25 inches** as shown in Figure 6.133 right. Do not press Enter as this will exit the command.

Figure 6.133

5 Next tweak the Valve Plate. With the Move option active, press the Ctrl key and in the graphics window select the component *ESS_E06-ValvePlate.ipt.*

> **TIP:** With the Move option still active, press the Ctrl key and select another component, and it will be added to the selection set.

6 Again, select the Z arrow and either drag or enter a value of **1.75 inches** as shown in the following image. Do not press the Enter key as this will exit the command.

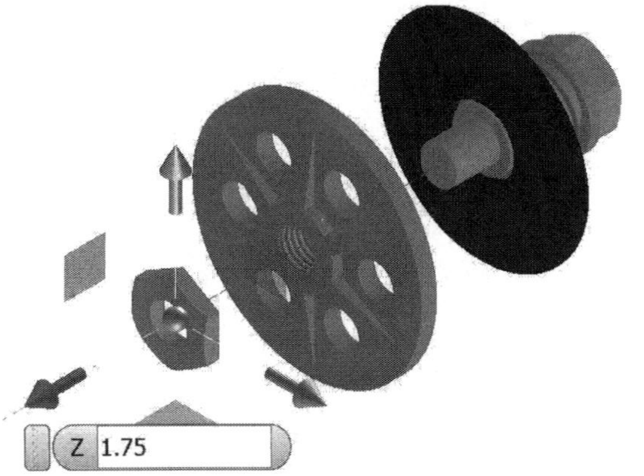

Figure 6.134

7 Complete this phase of the exercise by tweaking the remaining parts using the same steps described previously and entering input values as desired. Hold down the Ctrl key when adding components to tweak. When done, right-click and click OK from the menu. Your view should resemble the following image.

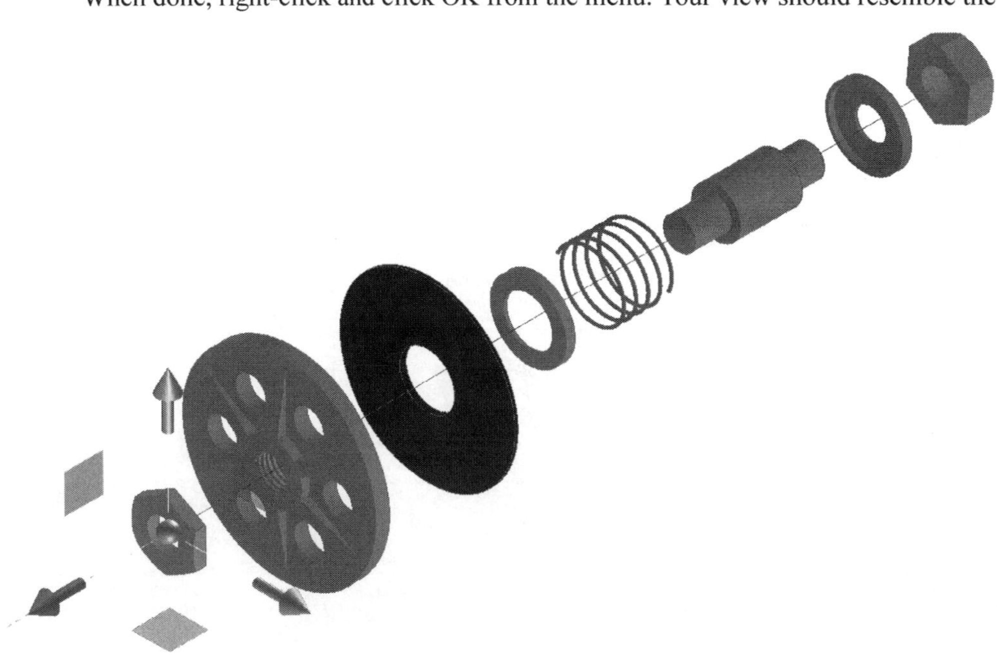

Figure 6.135

8 Expand the Storyboard panel which should be similar to Figure 6.136.

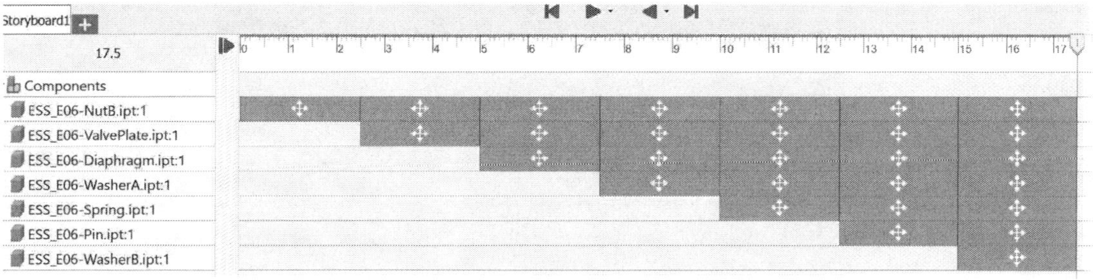

Figure 6.136

9 Next play the animation.

a. Move the Playhead to 0.00 by clicking the Back to Storyboard Beginning button on the top of the Storyboard panel.

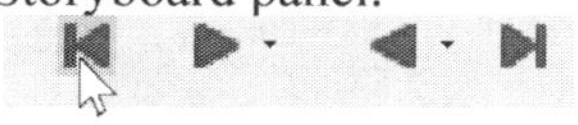

b. Play the animation by clicking the Play Current Storyboard on the top of the Storyboard panel.

10 Next you insert a new camera viewpoint.

a. Move the Playhead to 10.00 seconds.

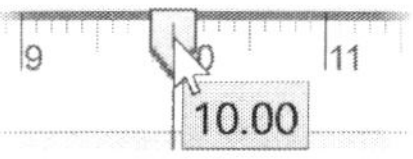

b. Change the viewpoint so it resembles the following image on the left.

c. Click the Capture Camera command on the Presentation tab > Camera panel.

d. A camera is added to the Storyboard panel as shown in the following image on the right.

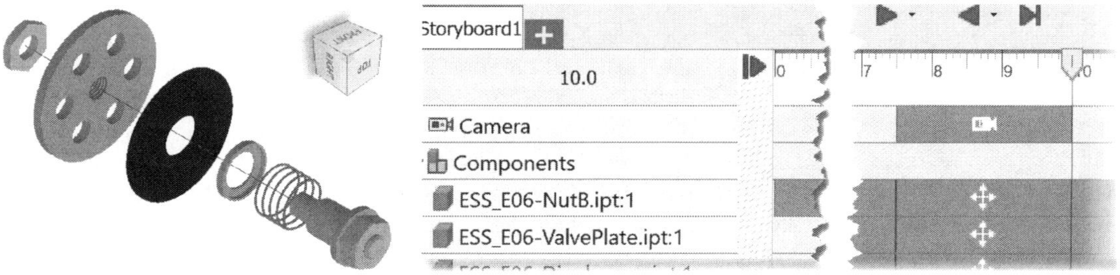

Figure 6.137

11 Move the playhead so it is before the new camera and play the animation; notice how the viewpoint changes.

12 Next edit the camera.

a. In the Storyboard click and drag the camera so it starts at 15.00 seconds as shown in Figure 6.138 left.

b. Reduce the duration for the camera to get to its viewpoint to 1.50 seconds. Right-click on the camera entry in the Storyboard and choose Edit Time from the menu. From the mini-toolbar change the end time to **16.50** as shown in Figure 6.138 right. Note that you could have also changed the start time in this mini-toolbar. Complete the change by clicking the green check mark in the mini-toolbar.

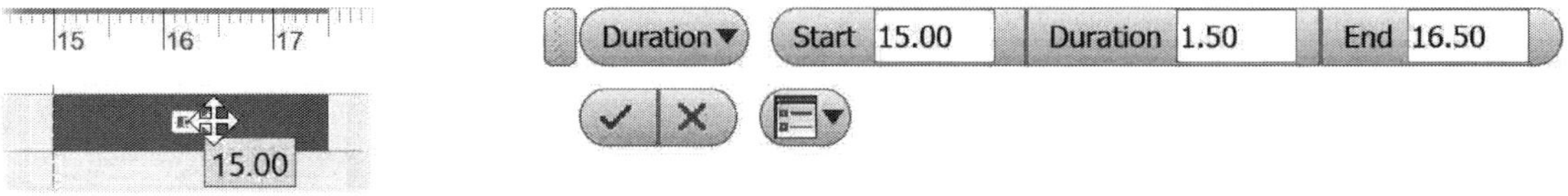

Figure 6.138

13 Notice that the camera changes its viewpoint at the same time as the last tweaks. In this step, you move the last tweaks so they happen after the camera.

a. Hold down the Ctrl key and select the seven tweaks on the far-right side.

b. Drag them so they start at 16.50 as shown in Figure 6.140.

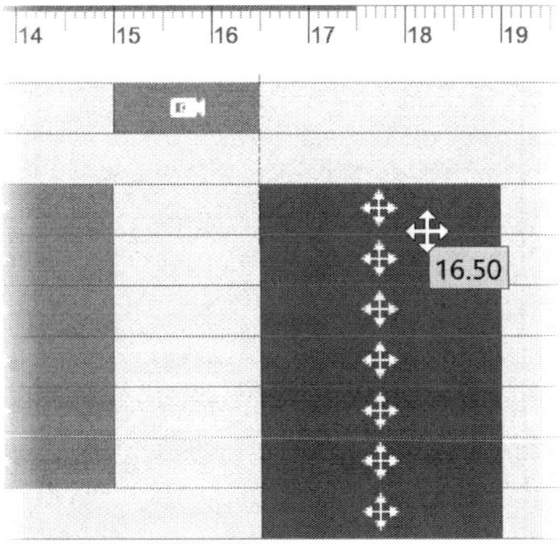

Figure 6.139

14 Next rotate the jamb nut so it rotates while it is moving out in the first animation.

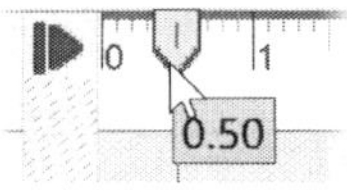

a. In the Storyboard panel move the Playhead to 0.50 .

b. Click the Tweak Components command on the Presentation tab > Component panel.

c. In the graphics window select on the outside circular edge on the component ESS_E06-NutB.ipt (the circular edge will align the coordinates to the center of the nut).

d. In the Tweak Components mini-toolbar, change the tweak type to Rotate.

e. Set the Duration to **2.000 seconds** as shown in Figure 6.140 left.

f. You want the nut to rotate counterclockwise for two full rotations. Click the sphere on the plane that is parallel to the outside of the nut and enter **-720** for the angle as shown in Figure 6.140 right.

g. Finish the command by clicking the green checkmark in the mini-toolbar.
Note: In the Storyboard panel, the actions for the nut are grouped together.

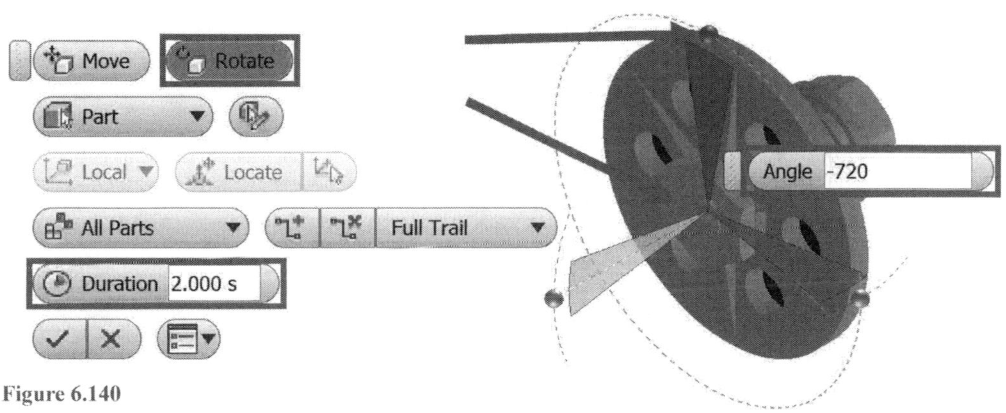

Figure 6.140

15 Play the animation from the beginning. In the first action, the nut should rotate and move at the same time.

16 Next publish a Video.

a. Click the Publish to Video command on the Presentation tab > Publish panel.

b. Change the options as desired and click OK to publish the video.

c. Navigate to the location where the video was saved and play the video.

17 Next create an associated Snapshot view and publish a raster image.

a. Move the Playhead to the end of the storyboard by 0.00 by clicking the Forward to Storyboard End on the storyboard.

b. Click the New Snapshot View command in the Presentation tab > Workshop panel.

c. To publish a raster image, right-click on View1 in the Snapshots View panel and click Publish to Raster from the menu as shown in Figure 6.141.

d. Change the options as desired and click OK to publish the raster image.

e. Navigate to the location where the raster file was saved and view the file.

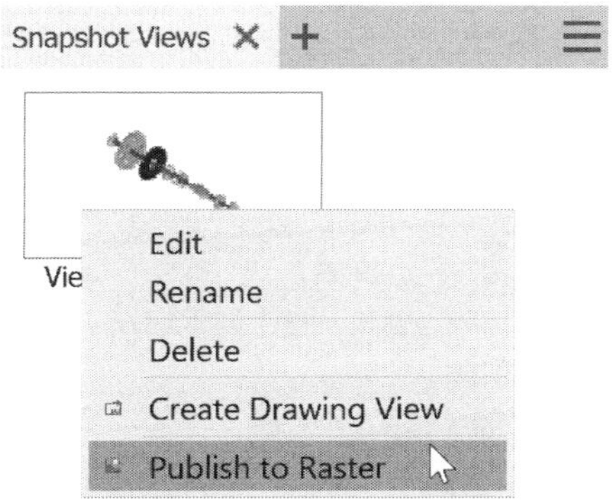

Figure 6.141

18 If desired, create a drawing view of the snapshot by right-clicking on View1 in the Snapshots View panel and choose Create Drawing View from the menu.

19 Practice tweaking components, editing tweaks, creating cameras and new videos, and raster images.

20 Close all open files. Do not save changes.

End of exercise.

CREATING DRAWING VIEWS FROM ASSEMBLIES & PRESENTATIONS

After creating an assembly or presentation file, you may want to create drawing views to document them. You can create drawing views from part files as described in Chapter 5. You can also create drawing views from assemblies or from any presentation view using the same method as you did for a part file. Start a new drawing file or open an existing drawing file. Use the Base View command and select the .iam or .ipn file from which to create the drawing. If needed, specify the design view or presentation view in the dialog box. Figure 6.142 shows a presentation view being selected after a presentation file was chosen.

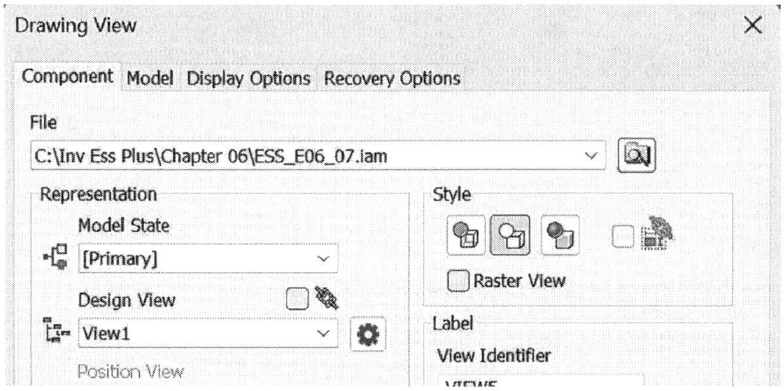

Figure 6.142

BILL OF MATERIAL (BOM)

A Bill of Materials, referred to as a BOM, is a table containing information about the components used in an assembly. A BOM can include item number, quantity, part number, description, vendor, and other information needed to describe the assembly. The BOM is associative to the assembly; when a component is added or removed from the assembly, the quantity field in the BOM updates to reflect this change in the assembly.

To facilitate managing item numbers, every item is automatically assigned a number, and the item number can be changed or reordered if necessary. Whenever these types of changes are made in the BOM, the same changes are updated automatically in the Parts List and balloons.

Values in the BOM can be changed in several ways. The first is to add information while inside the individual part file using the iProperties dialog box. Another way is to add the information directly in the BOM Editor; then, when the BOM is saved, the information supplied inside of the BOM is saved back to each part whose information changed. This is an efficient way to edit the iProperties of multiple parts in one operation.

To activate the BOM Editor, you must be inside an assembly file or editing a Parts List in a drawing. In an assembly click the Bill of Materials command on the Assemble tab > Manage panel, as shown in the following image on the left, and the Bill of Materials dialog box appears as shown in Figure 6.143 right.

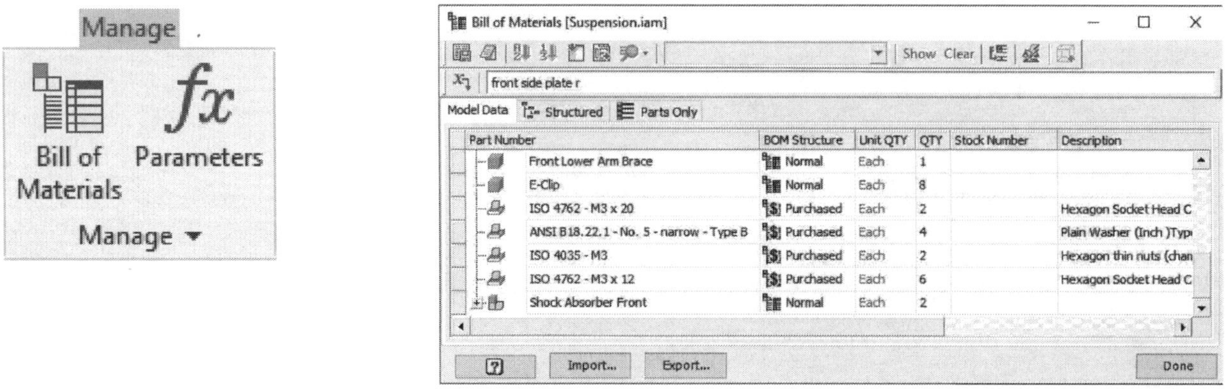

Figure 6.143

There are three tabs in the BOM Editor: Model Data, Structured, and Parts Only.

The Model Data Tab

On this tab, the components that make up the assembly are arranged in a format that resembles the assembly browser, as shown in Figure 6.144 left.

The Structured Tab

The data on this tab is the actual BOM data that is used to create a Parts List in a drawing. When arranging an assembly in Structured mode, all components are considered individual items, as shown in the middle of Figure 6.144.

Parts Only Tab

When this tab is current, all the parts that make up the assembly are listed, and no subassemblies are listed, as shown in Figure 6.144 right. This arrangement of parts is referred to as a flat list.

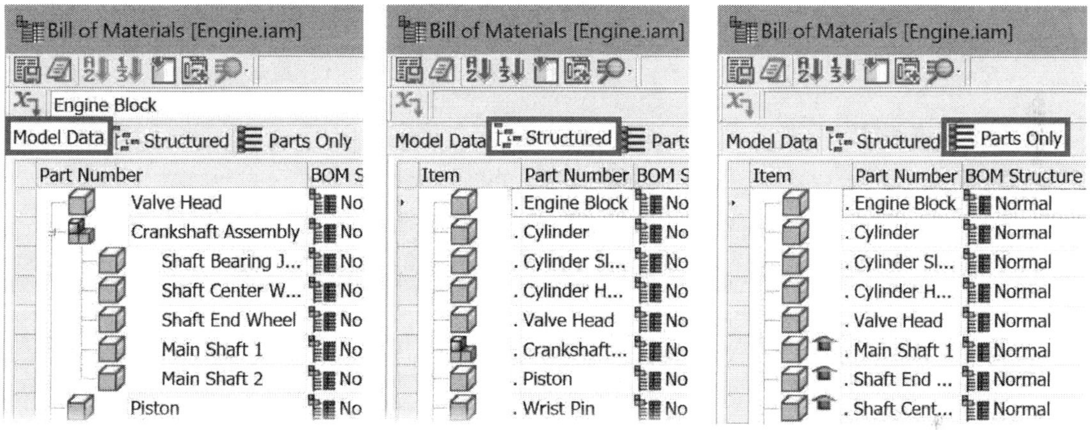

Figure 6.144

Following are common operations performed in the BOM:

Sorting Items in the BOM

Click the Sort button from the BOM dialog box to open the Sort dialog box, as shown in Figure 6.145. Sorting can be performed only when in the Structured and Parts Only tabs. In the Sort by area of the dialog box, select the first column to sort by, followed by selecting ascending or descending order. Columns can also be arranged by a secondary sort and, if desired, a tertiary sort.

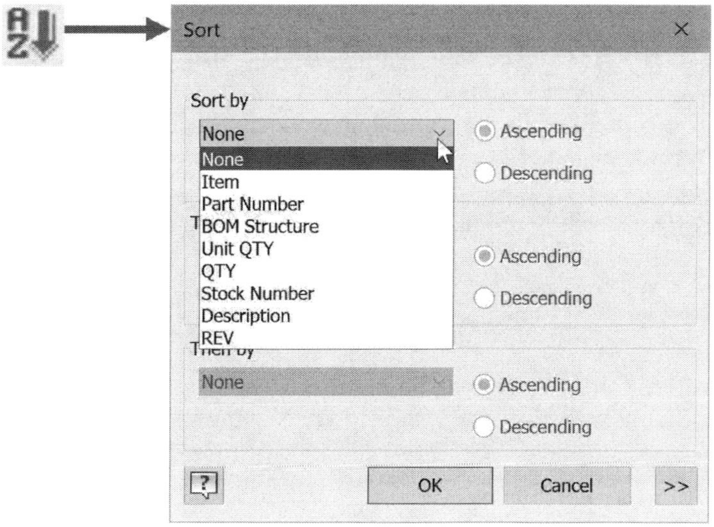

Figure 6.145

Renumbering Items in the BOM

Clicking on the Renumber Items button will display the Item Renumber dialog box, as shown in Figure 6.146. This operation is usually performed after sorting information in the BOM dialog box.

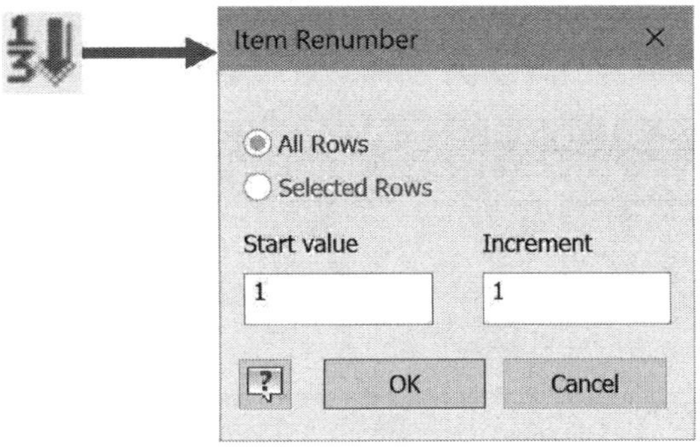

Figure 6.146

Adding Columns to the BOM

You can add columns to the BOM by clicking on the Choose Columns button and the Customization dialog box appears, as shown in Figure 6.147. Choose the desired property from the list, and then either double-click or drag and drop on the property so it is next to an existing column heading in the BOM dialog box. You can reorder the columns by dragging on its header and moving it to a new location. You can remove a column by dragging the column header and dropping it in the Customization dialog box.

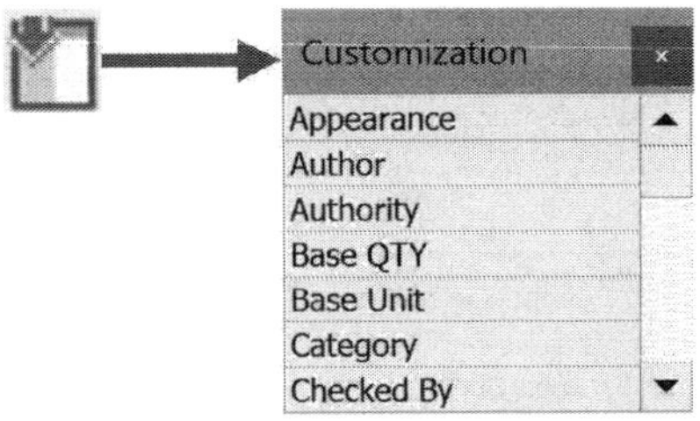

Figure 6.147

Changing the View Options of the BOM

When editing a BOM and the Structured tab or Parts only tab are disabled, you can enable them by clicking on the tab and selecting Enable BOM View, as shown in the following image on the left.

For a structured BOM, you can control how many levels are displayed by right-clicking on Structured tab and from the menu click View Properties, or click the View Options button as shown in the middle of the following image and from the menu click View Properties. The Structured Properties dialog box appears, as shown in Figure 6.148 right, where you can adjust the settings as desired.

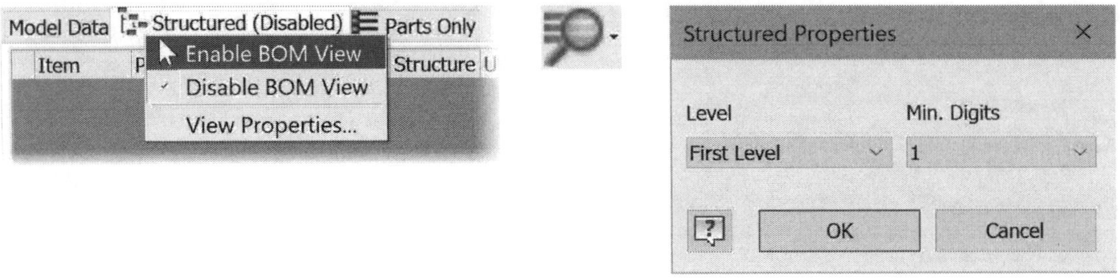

Figure 6.148

Opening a Part from the BOM Editor

The BOM editor can be used to open a part or assembly file. To perform this task, open the BOM editor, select the part from the list, and right-click and select Open from the menu, as shown in Figure 6.149.

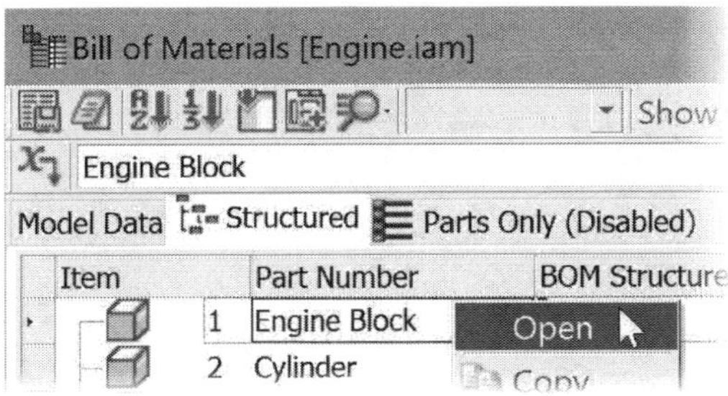

Figure 6.149

While editing a BOM, you can use the following tools, which are similar in functionality to Microsoft Excel:

- Use the SHIFT or CTRL keys to make selection sets of information while in the BOM editor.
- Use the grip point located in the lower-right corner of a cell to expand a cell or blocks of cells.
- BOM cells can be copied from the BOM Editor to another BOM view or to an external editor.
- Copy and paste information from the BOM editor into Microsoft Excel.
- Additional command options are available through a menu. These options include Open, Copy, Paste, Capitalize, Find, and Replace.

EXERCISE 6-8: EDITING A BILL OF MATERIAL (BOM)

In this exercise, you add descriptions, change the material and manipulate the number of items in the BOM.

1 Open *ESS_E06_08*.iam in the Chapter 06 folder.

2 Start the Bill of Materials command from the Assemble tab > Manage panel.

3 Examine the BOM and in the Model Data tab resize the columns as shown in the following image. Notice that the Description cells are blank.

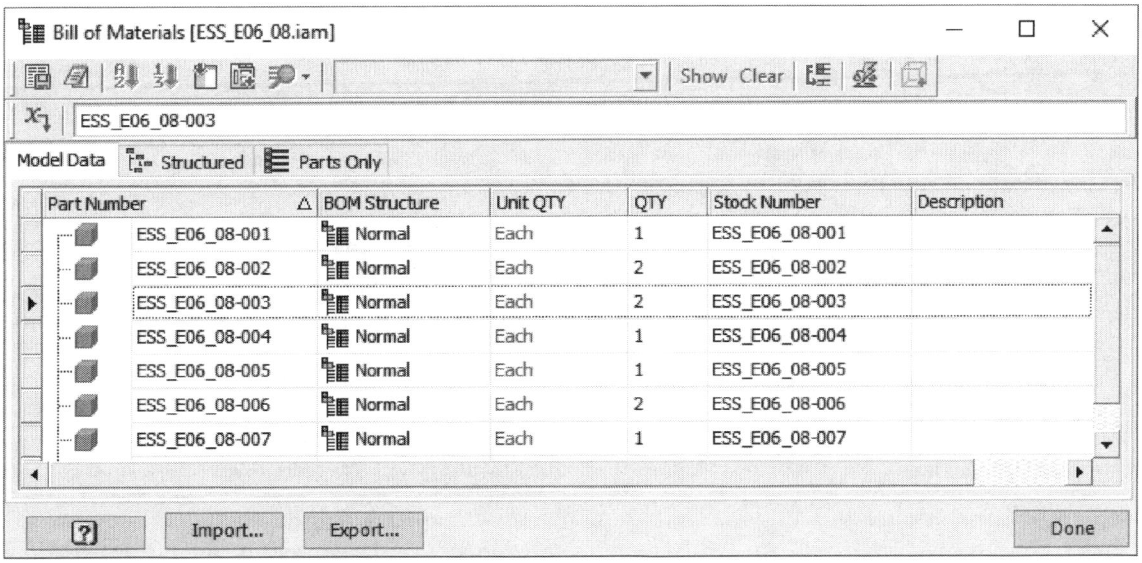

Figure 6.150

TIP: The individual descriptions of each part can easily be added through the BOM Editor. Using this method, you will NOT have to open each part file and add a description via the iProperties.

4 In the Bill of Materials dialog box, under the Model Data tab > Description column add a description to each part. Click in each Description cell and fill in the information, as shown in Figure 6.151. Only capitalize the first letter in each word.

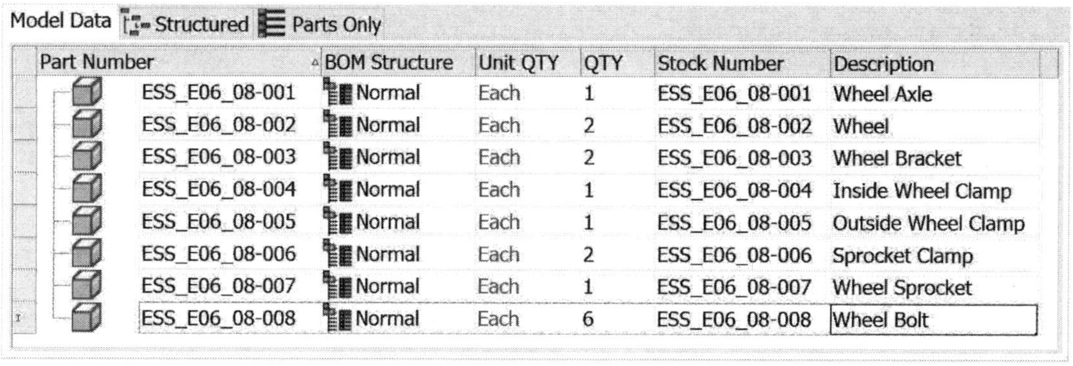

Model Data | Structured | Parts Only

Part Number	BOM Structure	Unit QTY	QTY	Stock Number	Description
ESS_E06_08-001	Normal	Each	1	ESS_E06_08-001	Wheel Axle
ESS_E06_08-002	Normal	Each	2	ESS_E06_08-002	Wheel
ESS_E06_08-003	Normal	Each	2	ESS_E06_08-003	Wheel Bracket
ESS_E06_08-004	Normal	Each	1	ESS_E06_08-004	Inside Wheel Clamp
ESS_E06_08-005	Normal	Each	1	ESS_E06_08-005	Outside Wheel Clamp
ESS_E06_08-006	Normal	Each	2	ESS_E06_08-006	Sprocket Clamp
ESS_E06_08-007	Normal	Each	1	ESS_E06_08-007	Wheel Sprocket
ESS_E06_08-008	Normal	Each	6	ESS_E06_08-008	Wheel Bolt

Figure 6.151

5 Most mechanical engineering / design drawings capitalize all letters. You can easily capitalize all the descriptions by doing the following.

a. Click and drag to select all the description cells.

b. Right-click and choose Capitalize from the menu as shown in Figure 6.152.

c. If needed, resize the Description column to fit the text and when done, all the descriptions should be capitalized, as shown in the following image on the right.

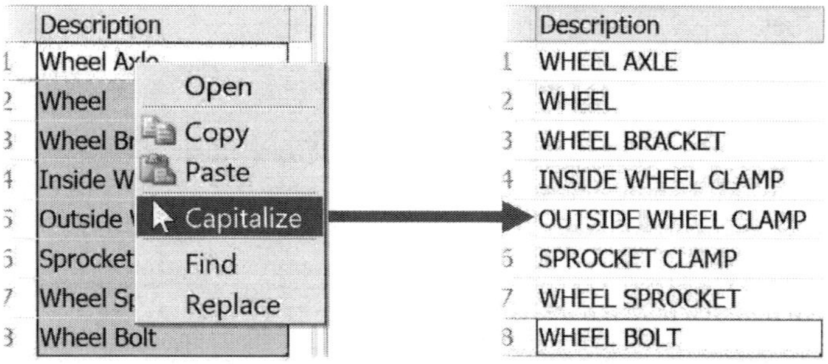

Figure 6.152

6 Next, add a Material column to the BOM. This allows you to edit or add the material type for each part.

a. To add a column called Material, click the Choose Columns button in the dialog box as shown in the following image.

b. The Customization box appears.

c. Scroll down in the Customization box and locate Material.

d. Double click on the Material entry or click and drag the Material property from the Customization box, and drop it to the right of the Description Column heading, as shown in the following image.

e. When finished, dismiss the Customization dialog box by clicking on the Close button (X) located in the upper right corner of the Customization box, not in the Bill of Materials dialog box.

f. If needed, resize the Material column to fit the text.

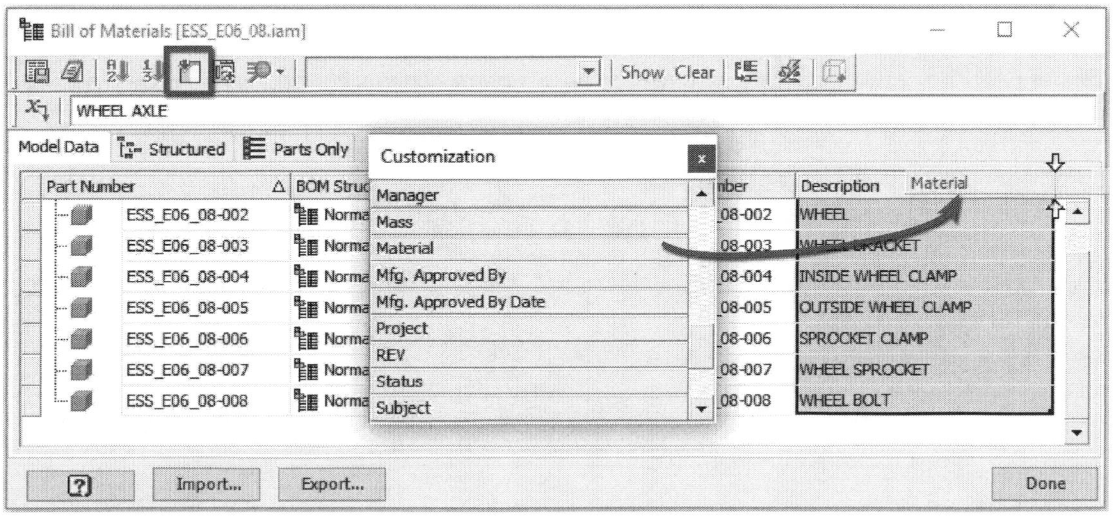

Figure 6.153

7 With the addition of the Material column added, notice that all Materials are listed as Generic. You will now change the material for each part.

a. Begin changing the material type by double-clicking in the material cell to activate the material drop-down list, as shown in the following image on the left.

b. Select the material from the list for each component, as shown in the image on the right. When finished, click the Done button in the Bill of Materials dialog box.

TIP: You can copy and paste data from one cell to another or click and drag on the square in the bottom-right corner of the cell and drag it up or down to copy the data to the highlighted cells.

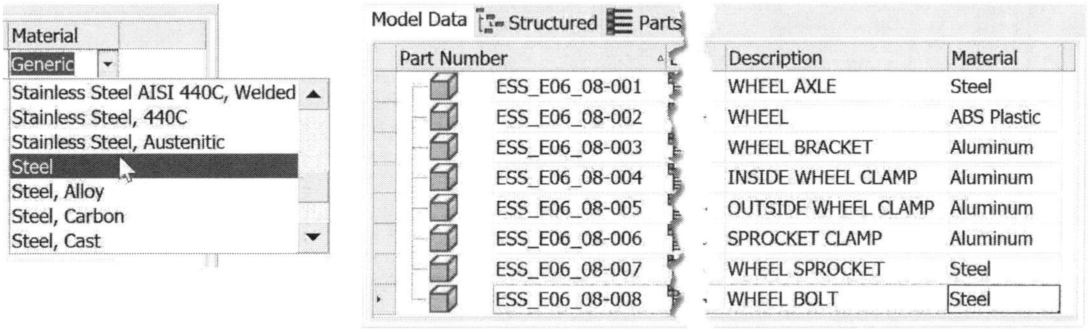

Figure 6.154

8 Usually, you would save the assembly. You would be prompted to save the changes back to the part files whose properties have changed, as shown in Figure 6.155. If you save the files for this exercise and you want to run through the exercise again, you will need to copy the original exercise files over the modified files.

Close the file. Do not save changes.

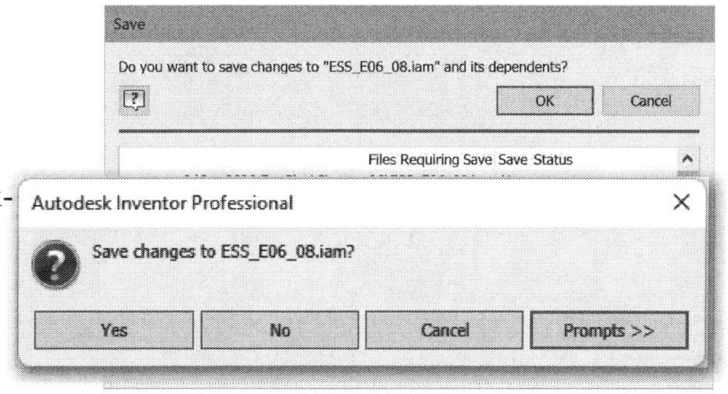

Figure 6.155

End of exercise.

CREATING BALLOONS

After you have created a drawing view of an assembly or a presentation file, you can add balloons to parts and/or subassemblies to document the component item numbers. The item numbers are received from the assembly's bill of material. Balloons and parts lists both reference the item numbers from the bill of material, so it does not matter if balloons or a parts list is placed first.

You can add balloons individually to components in a drawing, or you can use the Auto Balloon command to add balloons to selected components in one operation.

Placing Individual Balloons

To add individual balloons to a drawing, follow these steps:

1. In a drawing, on the Annotate tab > Table panel, click the Balloon command, as shown in the following image on the left, or use the shortcut key B.
2. Select a component to balloon.
3. If a balloon or a parts list has not been placed in the drawing, the BOM Properties dialog box will appear, as shown in the following image on the right. The options in this dialog box are explained in the next section. After selecting the BOM View setting click OK.

TIP: The BOM Properties dialog box does not appear if a parts list or balloon exists in the drawing.

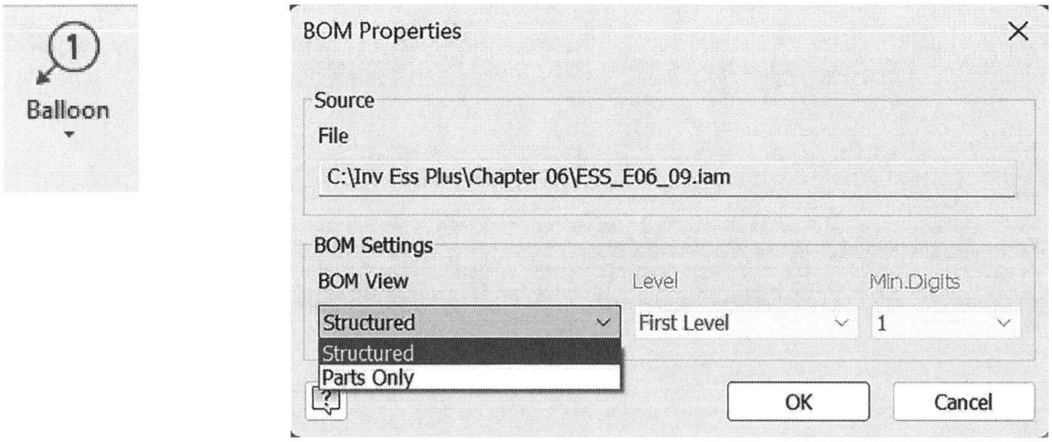

Figure 6.156

4. Position the cursor to place the second point for the leader, and press the left mouse button.
5. If desired, add segments to the balloon's leader by clicking additional points.
6. If you want to place balloons with single segments, right-click before clicking a point and click Single segment leader from the menu as shown in the following image. This setting will affect how future balloons are placed.
7. Or to place a balloon with a single segment click a point, right-click and choose Continue.

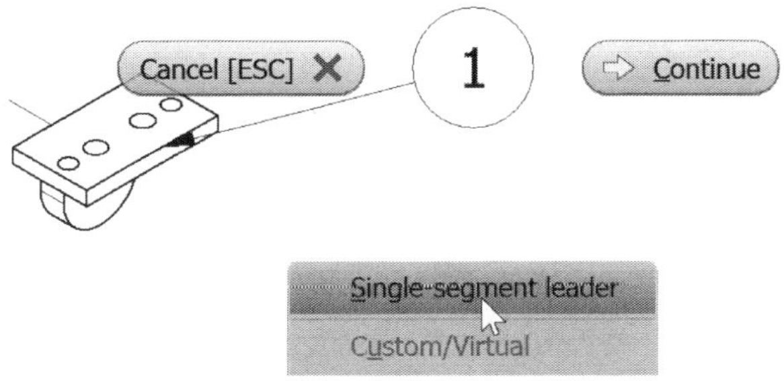

Figure 6.157

8. When finished adding segments to the leader, right-click, and select Continue from the marking menu to create the balloon. If the Single segment leader option is set, after clicking a second point, you will automatically continue to the next step.
9. Continue adding balloons by selecting components and placing balloons.
10. When done, right-click and click Cancel from the marking menu.

TIP: When placing balloons, the direction of the balloon leader line snaps to 15° increments; to freely move a balloon press the CTRL key while dragging.

A balloon can be aligned horizontally or vertically with another balloon by inferring a point.

BOM Properties Dialog Box Options

Source

Specify the source file on which the BOM will be based.

BOM Settings and BOM View

Structured. A Structured list refers to the top-level components of an assembly in the selected view. Subassemblies and parts that are NOT in a subassembly can be ballooned. Parts that are in a subassembly cannot be ballooned.

Parts Only View. This option allows you to select all the parts of the assembly regardless if they are in a subassembly. Subassemblies cannot be ballooned, but parts in the subassembly can be.

TIP: In a Parts Only View, components that are in a subassembly are not presented in the list unless they are considered inseparable or purchased.

Level

First Level. Select this option to determine how many levels of subassemblies deep can be selected when placing balloons or displayed in a parts list. In the Min. Digit list, you can select how many levels deep to display in a parts list, or how many levels can be selected.

All Levels. Select this option to be able to select all levels of subassemblies and display all subassembly levels in a parts list.

Min. Digits

Use this command to set the minimum number of digits displayed for item numbering. The range is fixed from 1 to 6 digits.

Auto Ballooning

In complex assembly drawings, it will become necessary to balloon many components. Rather than manually add a balloon to each component, you can use the Auto Balloon command to perform this operation on several components in a single operation automatically. Clicking on the Auto Balloon button on the Annotation tab > Table pane > under the Balloon command will display the Auto Balloon dialog box, as shown in the following image. The areas in this dialog box allow you to control how you place balloons.

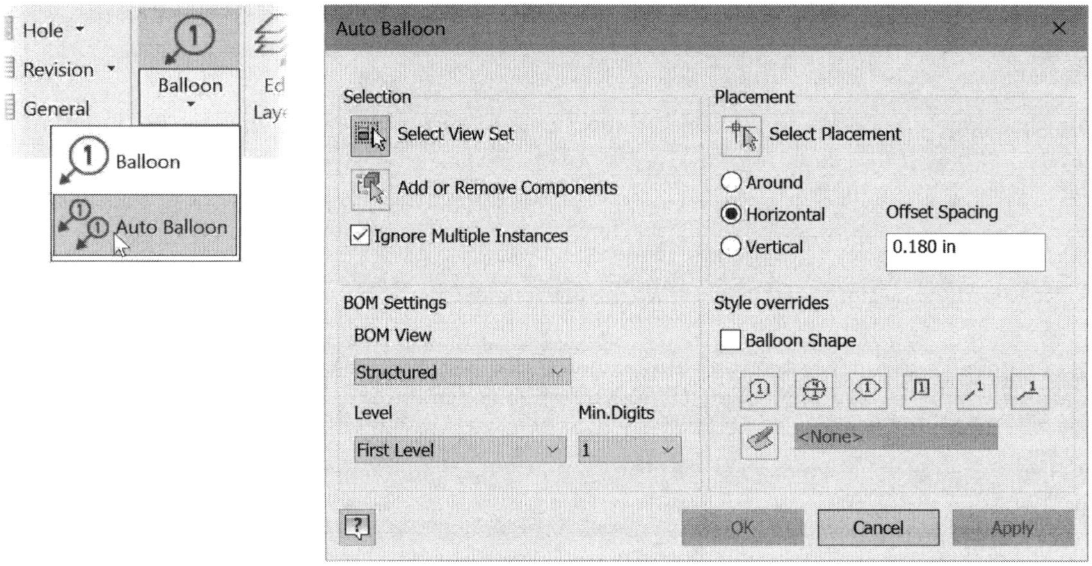

Figure 6.158

Selection

This area requires you to select where to apply the balloons. With the view selected, you then add or remove components to the balloon. The Ignore Multiple Instances option appears in this area, and when the box is checked, multiple instances of the same component will not be ballooned. This action greatly reduces the number of balloon callouts in the drawing. If your application requires multiple instances to each have a balloon, remove the check from this box.

Placement

Select the method to align the balloons: Around, Horizontal or Vertical.

Around

Places balloons for the selected components around the view's boundary. Move the cursor from the center of the view to readjust the balloons to different locations. The following image on the left shows an example where the balloons are located around the view boundary.

Horizontal

Align all balloons horizontally as shown in the middle of the following image; the Offset Spacing option determines the distance between the balloons.

Vertical

Aligns all balloons vertically, as shown in the following image on the right; the distance between the balloons is set with the Offset Spacing option.

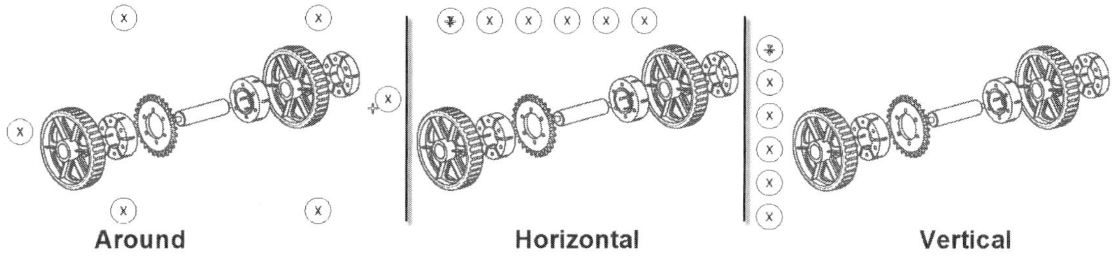

Figure 6.159

BOM Settings

This area allows you to control if balloons are applied to structured or parts only and determine how many levels deep can be ballooned.

Style Overrides

The Style overrides area allows you to change the shape of the balloon and assign a user-defined balloon shape. Click Balloon Shape to enable balloon shape style overrides.

To place balloons with the auto balloon command, follow these steps:

1. Select the view to which to add the balloons.
2. Select the components to which to add balloons. You will notice the color of all selected objects changes. Balloons will be applied to these selected objects. To remove components, hold down the SHIFT or CTRL key, and click a highlighted component. This action will deselect the component. You can also select components using a window or crossing window selection technique.
3. Select one of the three placement modes for the balloons: Around, Horizontal or Vertical.

Moving and Aligning Balloons

After you have placed balloons, you can drag a balloon to new locations, as well as reposition the arrows to a different location or point it to a different component by moving the arrow to an edge on a part. When a different component is selected the item number will change to reflect the item number of the new component. After moving the balloon, if a ● appears at the end of the leader line, the balloon is not attached to a component and can be repositioned so it is attached to a component.

To manually align two balloons vertically or horizontally, click and drag on the center of the balloon to move and drag it over the center of the balloon you want to align to; this infers the center point. Then drag the balloon vertically or horizontally; if dotted lines appear, the balloon is aligned. The following image on the left shows the balloon on the right being aligned horizontally to the balloon on the left.

To automate the alignment of multiple balloons, select the balloons, right-click and select an option from the menu as shown in the image on the right. Then click to locate the balloons (the offset option separates the balloons). With the To Edge option, select an edge to align to and then locate the balloons.

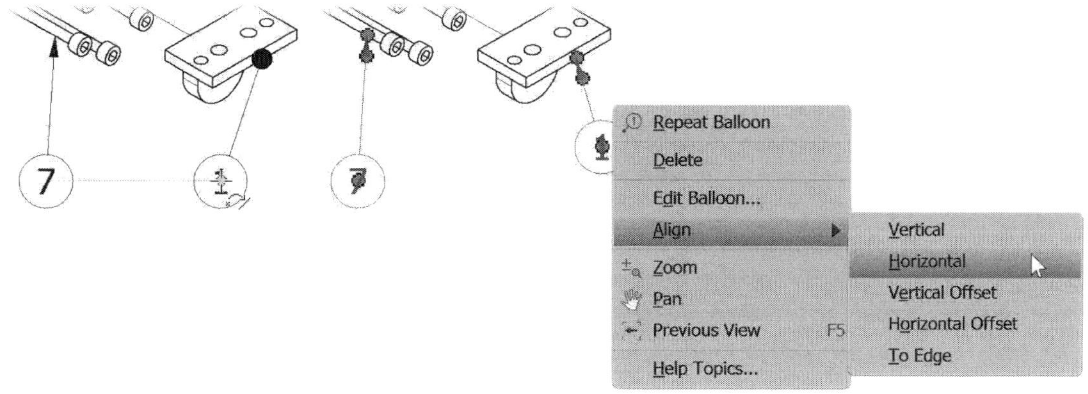

Figure 6.160

Editing a Balloon's Value

You can change the balloon's value/item number by either double-clicking on the balloon or right-clicking on the balloon and clicking Edit Balloon from the menu, as shown in the following image on the left. The Edit Balloon dialog box will appear, as shown in the following image on the right.

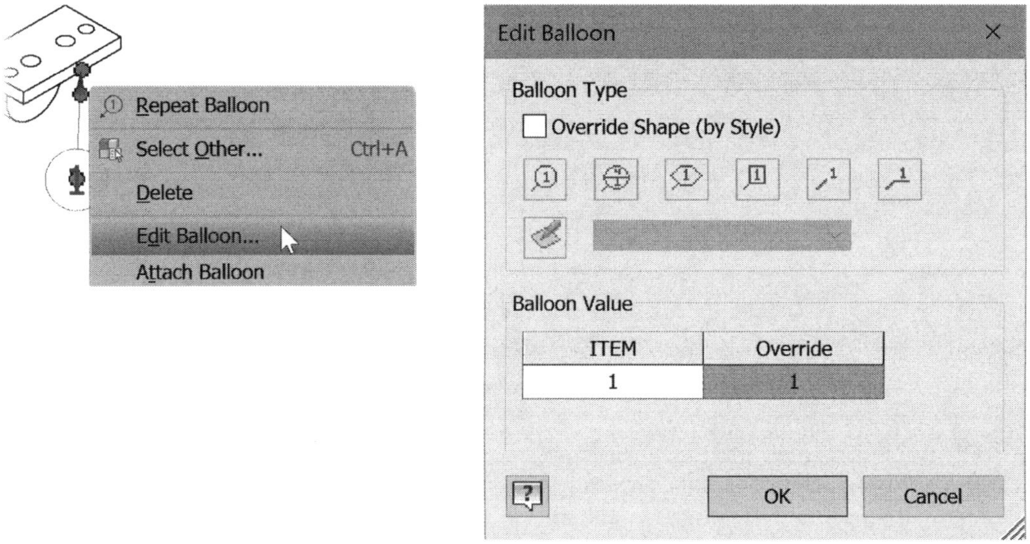

Figure 6.161

In the Balloon Value area, you can change the balloon's value by changing the Item or Override. When you edit the Item value in the Edit Balloon dialog box, the change will be made to the balloon and will be updated in the Parts List. In the following image on the left, the Item was changed to 2A. Notice that both the balloon and the Parts List are updated to the new value.

When you change the value in the Override column in the Edit Balloon dialog box, the balloon will update to reflect this change, but the Parts List remains unchanged. The following image on the right illustrates this; the balloon is overridden to a value of 2A, but the Parts List has the item Rod Cap still listed as 2.

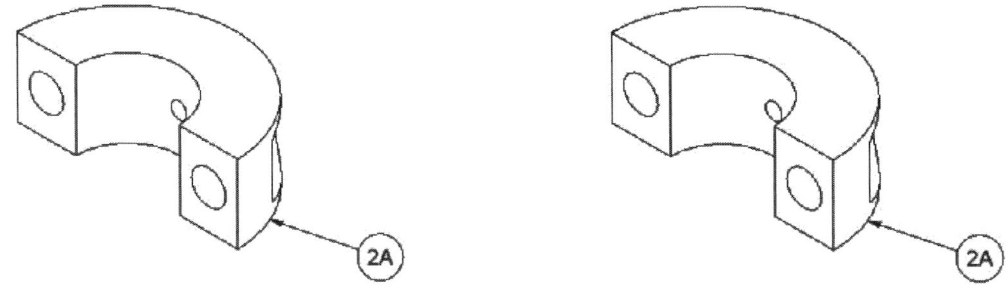

Parts			
ITEM	QTY	PART NUMBER	
1	1	Connecting Rod	
2A	1	Rod Cap	
3	2	Rod Cap Screw	
4	1	Piston	
5	1	Wrist Pin	

Parts			
ITEM	QTY	PART NUMBER	
1	1	Connecting Rod	
2	1	Rod Cap	
3	2	Rod Cap Screw	
4	1	Piston	
5	1	Wrist Pin	

Figure 6.162

PARTS LIST

After you have created the drawing views and placed balloons, you can also create a Parts List, as shown in the following image. As noted in the balloon section, components do not need to be ballooned before creating a Parts List.

Parts List

ITEM	QTY	PART NUMBER	DESCRIPTION
8	2	HANDLE CAP	HANDLE CAP
9	1	COLLAR	COLLAR
10	1	GIB PLATE	GIB PLATE
11	1	GROOVE PIN	GROOVE PIN
14	4	ANSI B18.3 - 1/4 - 20 UNC - 7/8	Hex Cap Screw
15	4	BS 4168 - M5 x 16	Hex Cap Screws
16	1	ISO 4766 - M5 x 5	Hex Cap Screws

Parts List

ITEM	QTY	PART NUMBER	DESCRIPTION
1	1	Arbor Press	ARBOR PRESS
2	1	FACE PLATE	FACE PLATE
3	1	PINION SHAFT	PINION SHAFT
4	1	LEVER ARM	LEVER ARM
5	1	THUMB SCREW	THUMB SCREW
6	1	TABLE PLATE	TABLE PLATE
7	1	RAM	RAM

Figure 6.163

To create a Parts List, select the Annotate tab >Table panel and click the Parts List command, as shown in the following image on the left. The Parts List dialog box will appear, as shown in the following image on the right. Select a view on which to base the Parts List. Specify the BOM Settings and Properties; see the Creating Balloons section for information about these settings as they are the same and are set upon placing a parts list or a balloon. After specifying your options, click the OK button, and a preview of the Parts List will appear attached to your cursor. Click a point in the graphics window to place the Parts List. The information in the list is extracted from the iProperties of each component.

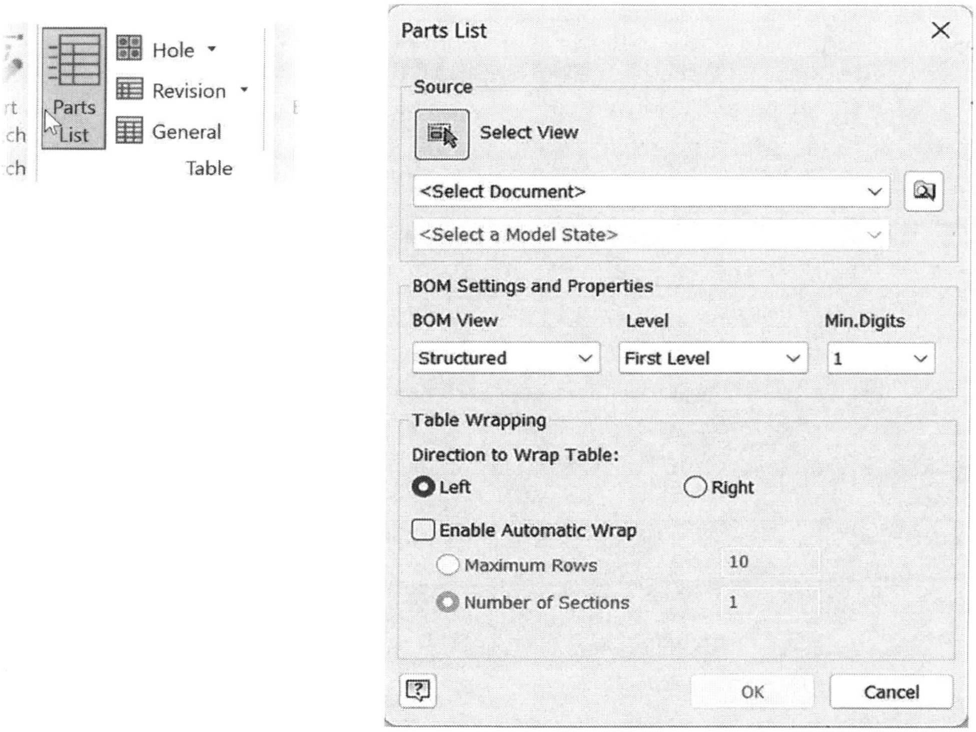

Figure 6.164

Editing a Parts List

After placing a parts list, you can edit the appearance and override values in the parts list. To edit a parts list, right-click on the Parts List in a drawing and select Edit Parts List from the menu, or double-click on the Parts List in the drawing.

This action displays the Edit Parts List dialog box, as shown in the following image. The row of icons along the top of the dialog box consists of operations that allow you to change the Parts List.

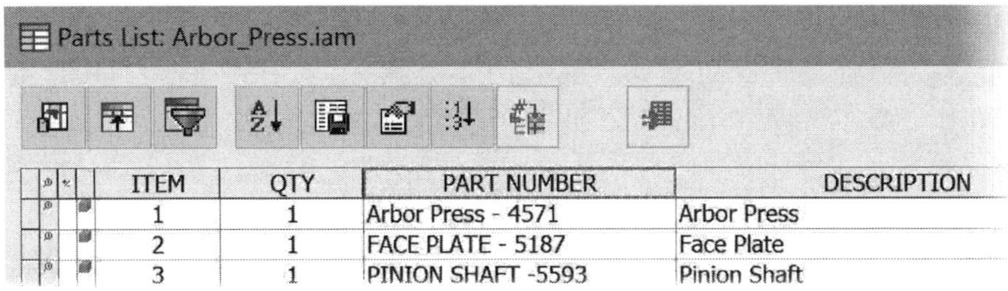

Figure 6.165

Following are descriptions of the icons in the Parts List dialog box:

Button	Function	Explanation
	Column Chooser	Opens the Parts List Column Chooser dialog box, where you can add, remove, or change the order of the columns for the selected Parts List. Data for these columns is populated from the properties of the component files.
	Group Settings	Opens the Group Settings dialog box, where you select Parts List columns to be used as a grouping key and group different components into one Parts List row. This button is active only when generating a Parts List using the Structured mode.
	Filter Settings	Opens the Filter Settings dialog box, where you can define such filter settings as Assembly View Representations, Ballooned Items Only, Item Number Range, Purchased Items, and Standard Content from which to filter the information. Once a filter is selected, you then filter out rows without changing the data in the Parts List. You can also add Parts List filters to a Parts List style.
	Sort	Opens the Sort Parts List dialog box, where you can change the sort order for items in the selected Parts List.
	Export	Opens the Export Parts List dialog box so that you can save the selected Parts List to an external file of the file type you choose.
	Table Layout	Opens the Parts List Table Layout dialog box where you can change the title text, spacing, or heading location for the selected Parts List.
	Renumber Items	Renumbers item numbers of parts in the Parts List consecutively.
	Save Item Overrides to BOM	Saves item overrides back to the assembly Bill of Materials.
	Member Selection	Used with iAssemblies. Clicking this button opens a dialog box where you can select which members of the iAssembly to include in the Parts List.

Editing Parts List BOM Data

Once a Parts List is created, and if there is missing or inaccurate data, you can open the BOM Editor directly from the drawing environment and edit the assembly BOM. You can save these changes back to the affected components by saving the drawing and accepting the prompt to also save the assembly and corresponding component files. To edit the BOM from a parts list, right-click on the Parts List in the drawing and then click Bill of Materials from the menu, as shown in the following image on the left. The Bill of Materials dialog box appears for the specific assembly file. The dialog box provides a convenient location to edit the iProperties and Bill of Material properties for all components in the assembly. Then edit the BOM as you learned in the previous Bill of Material (BOM) section. Figure 6.166 right illustrates the Part Number and Description for the Arbor Press entry capitalized. Complete the edits by clicking the Done button and then save the drawing, assembly and the affected assembly and component files.

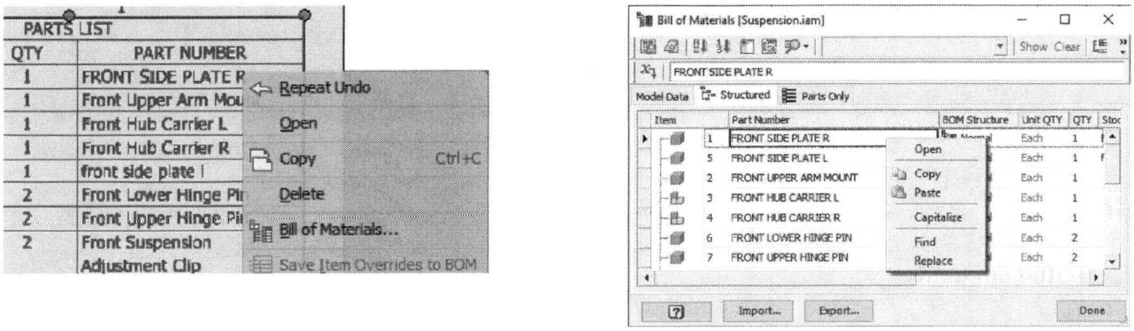

Figure 6.166

Creating Custom Parts

Custom parts are useful for parts in the Parts List that are not components or graphical data, such as paint or a finishing process. To add a custom part, right-click on an existing part in the Edit Parts List dialog box, as shown in Figure 6.167, and click Insert Custom Part. When a custom part no longer needs to be documented in the Parts List, you can right-click in the row of the custom part, and click Remove Custom Part from the menu.

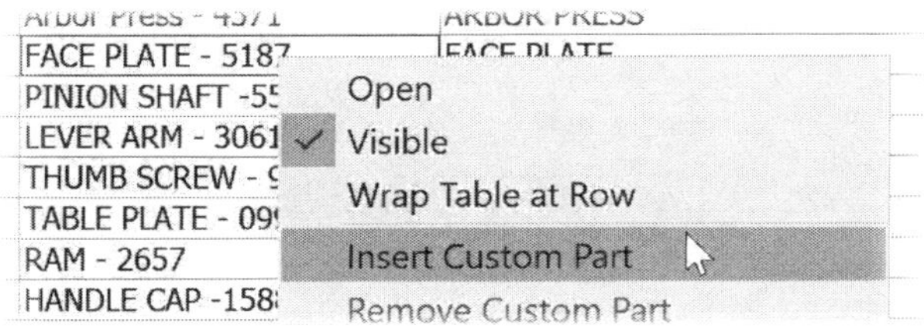

Figure 6.167

Wrapping a Parts List

When creating a parts list that is long, you can wrap it into multiple sections. In the Parts List dialog box place a check in the Enable Automatic Wrap box and then enter the number of sections to break the Parts List into or select the Maximum Rows option to specify the maximum number of rows for each section. You can also choose to have the sections added to the right or the left. Figure 6.168 shows the results of wrapping the Parts List into two sections and having it added to the left.

Parts List				Parts List			
ITEM	QTY	PART NUMBER	DESCRIPTION	ITEM	QTY	PART NUMBER	DESCRIPTION
8	2	HANDLE CAP - 2028	HANDLE CAP	1	1	Arbor Press - 4571	Arbor Press
9	1	COLLAR - 5339	COLLAR	2	1	FACE PLATE - 5187	FACE PLATE
10	1	GIB PLATE - 8783	GIB PLATE	3	1	SHAFT - 5593	PINION SHAFT
11	1	GROOVE PIN - 3597	GROOVE PIN	4	1	LEVER ARM - 3016	LEVER ARM
14	4	HHCS - 8840	1/4 - 20 UNC - 7/8 HS HCS	5	1	THUMB SCREW - 1166	THUMB SCREW
16	1	ISO 4766 - M5 x 5 - 1411	Slotted Headless Set Screw	6	1	TABLE PLATE - 9591	TABLE PLATE
15	4	ISO 4026 - M5 x 16 - 1422	Hexagon socket set screws	7	1	RAM - 4017	RAM

Figure 6.168

While editing a Parts List you can also manually split the table. To do so, right-click on the row after the one that you want to split and click Wrap Table at Row, as shown in the following image on the left. You can remove the wrap by right-clicking on the row and selecting the same option, to remove the checkmark next to the Wrap Table at Row option.

After wrapping a parts list, a thicker black line will appear in the Parts List dialog box representing where the wrap begins. You can remove the wrap by right-clicking on the row above where the wrap begins and clicking Wrap Text at Table, as shown in Figure 6.169 right.

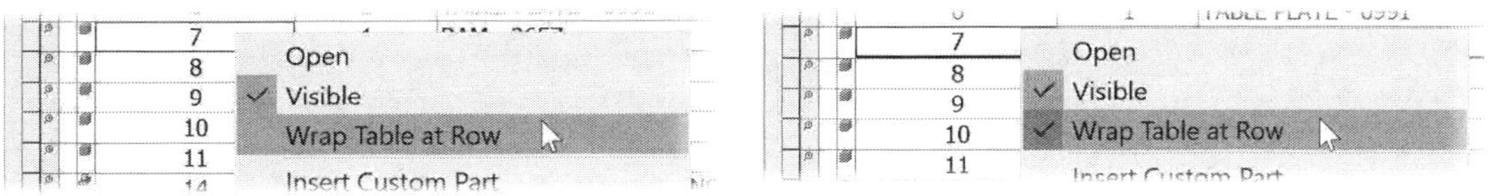

Figure 6.169

EXERCISE 6-9: CREATING A DRAWING FROM AN ASSEMBLY

In this exercise, you create drawing views of an assembly and then add a parts list and balloons.

1 Open the drawing *ESS_E06_09.idw* in the Chapter 06 folder. This drawing consists of one blank D size sheet with a border and title block.

2 First, you format the drawing sheet.

a. In the browser, right-click on Sheet:1 and click Edit Sheet from the menu.

b. In the Edit Sheet dialog box, rename the sheet by typing **Assembly** in the Name field.

c. In the Size list, select C as shown in the following image.

d. When finished, click the OK button.

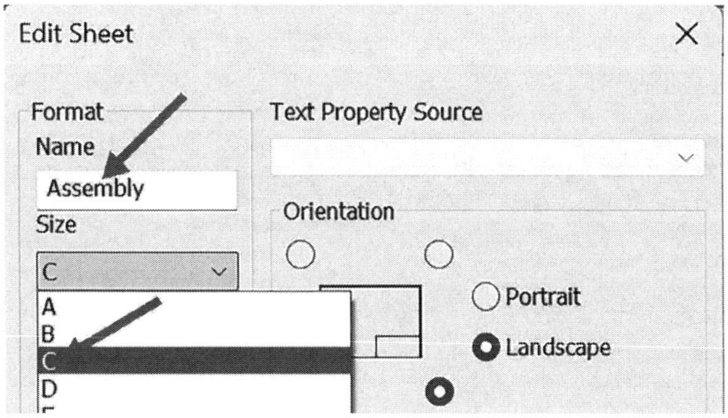

Figure 6.170

3 Next, you change the file properties to update the title block fields.

a. Click the File tab > iProperties.

b. Click on the Summary tab and make the following changes:

i. In the Title area, enter **PUMP ASSEMBLY**.

ii. In the Author area, enter your name as shown in Figure 6.171 left.

c. Click the Project tab and make the following changes:

i. In Part number, enter **123-456-789** as shown in Figure 6.171 right.

ii. In Creation Date, click the down arrow, and then select Today's date.

iii. Fill in the other fields as desired.

iv. When finished, click the OK button.

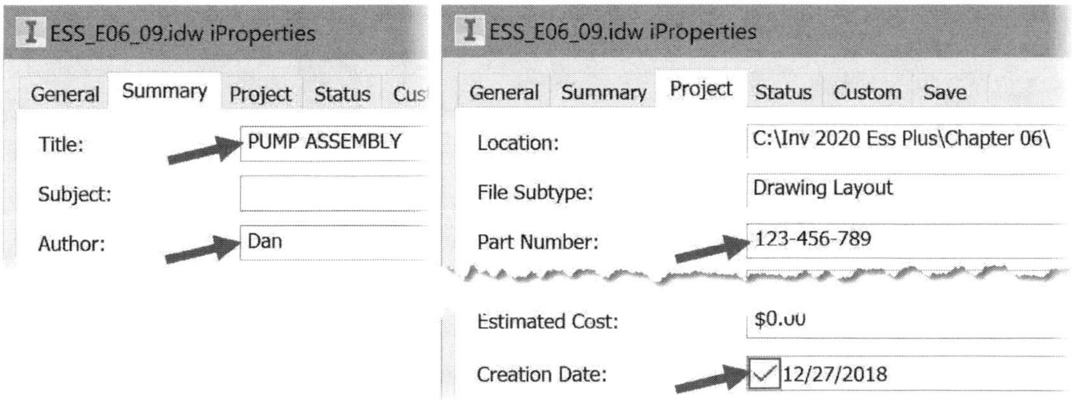

Figure 6.171

4 Zoom in to the title block and notice that the entries have updated, as shown in the following image.

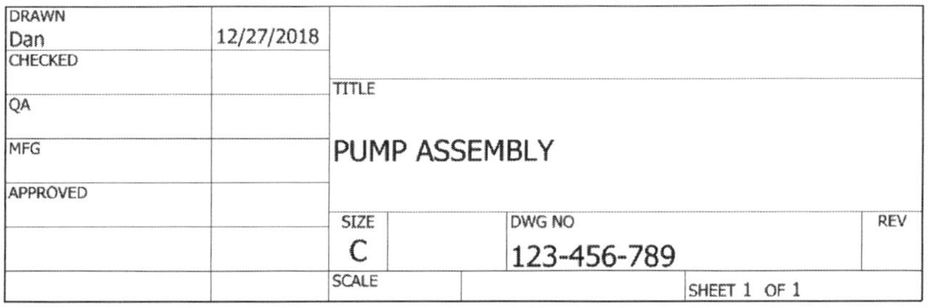

Figure 6.172

5 Next you create drawing views.

a. Zoom out to view the entire sheet.

b. Click the Base View command on the Place Views tab > Create panel. The Drawing View dialog box appears.

c. Click the Open an existing file button next to the File list and double-click on *ESS_E06_09.iam* in the Chapter 06 folder.

d. Change the view orientation from the ViewCube above and to the right of the view preview; change the view to the Top view .

e. Set the Style to Hidden Line, as shown in the following image.

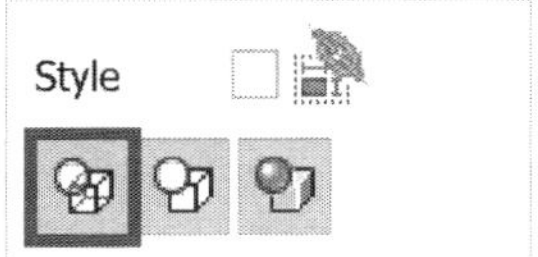

Figure 6.173

f. Ensure the view's Scale is 1:1.

g. Locate the view by clicking in the upper-left corner of the drawing sheet, similar to Figure 6.173. Note that the preview of the view is shaded until the view is created.

h. Create the view by right-clicking and click OK from the marking menu.

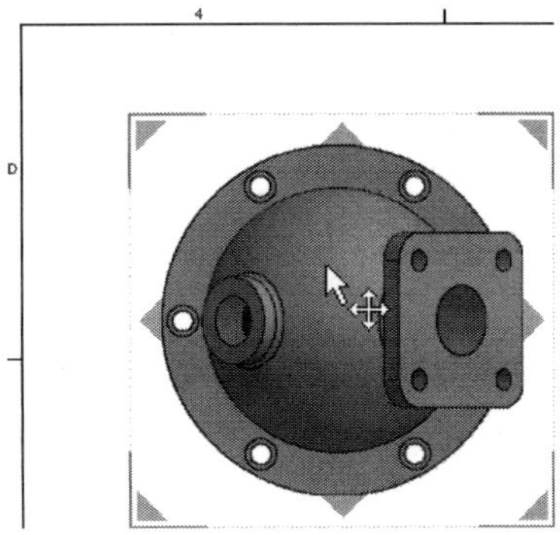

Figure 6.174

6 Before you create a section view, you turn off the Section Participation property for some components so that they are not sectioned. Follow these steps:

a. In the browser, expand Assembly:1 > *View1: ESS_E06_09.iam* > *ESS_E06_09.iam*.

b. Hold down the CTRL key and select:

   i. *ESS_E06-Pin.ipt:1*

   ii. *ESS_E06-Spring.ipt:1*

   iii. *ESS_E06-NutA.ipt:1*

   iv. *ESS_E06-NutB.ipt:1*

c. Right-click on one of the highlighted parts in the browser, and from the menu click Section Participation > None, as shown in Figure 6.175.

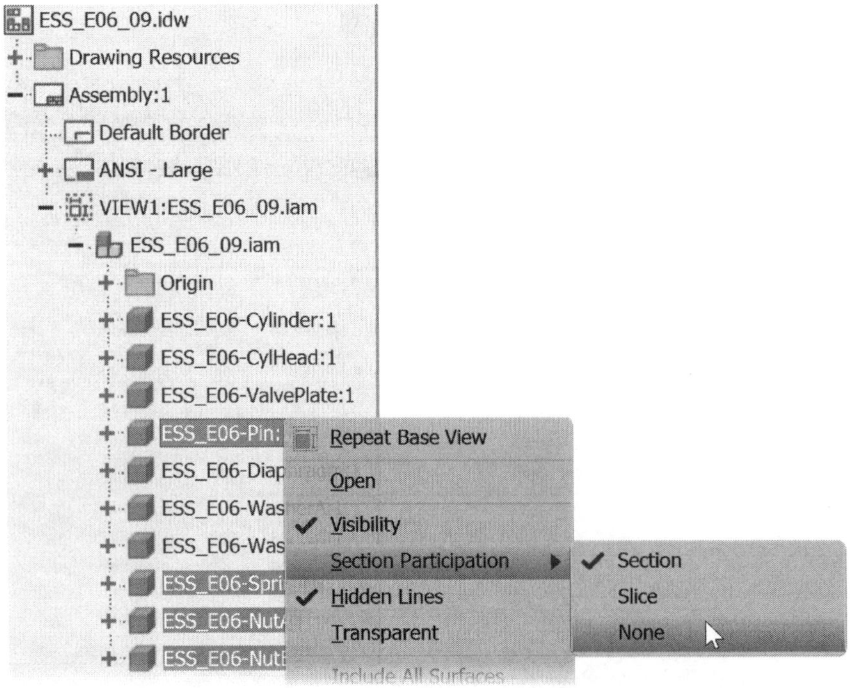

Figure 6.175

7 Next, you create a section view from the top view.

a. Right-click in the top view and click Section View from the marking menu.

b. Draw a horizontal section line through the center of the top view, as shown in Figure 6.176 left.

c. Right-click and click Continue from the marking menu.

d. Move the cursor down below the top view as shown in the middle of the following image.

e. Click to create the view.

f. Notice that the nuts, spring, and pin are visible in the sectioned view but are not sectioned, as shown in Figure 6.176 right.
Note: If the label for the section view does not appear, edit the section view and click the Toggle Label Visibility button so it's on .

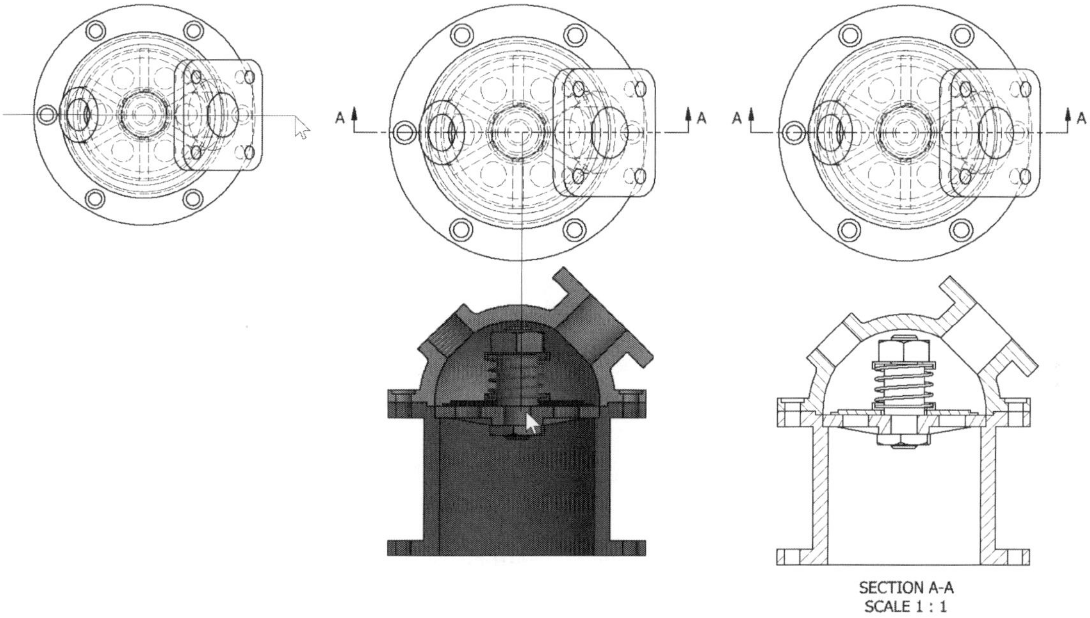

Figure 6.176

8 Now hide two components from displaying in the top view.

a. In the browser, under VIEW1:ESS_E06_09.iam > *ESS_E06_09.iam*, hold down the CTRL key and select:

i. *ESS_E06-CylHead:1*

ii. *ESS_E06-Spring:1*

b. Right-click, and clear the checkmark from Visibility as shown in the following image on the left.

c. Notice that the cylinder head and the spring are no longer displayed in the top view, as shown in Figure 6.177 right.

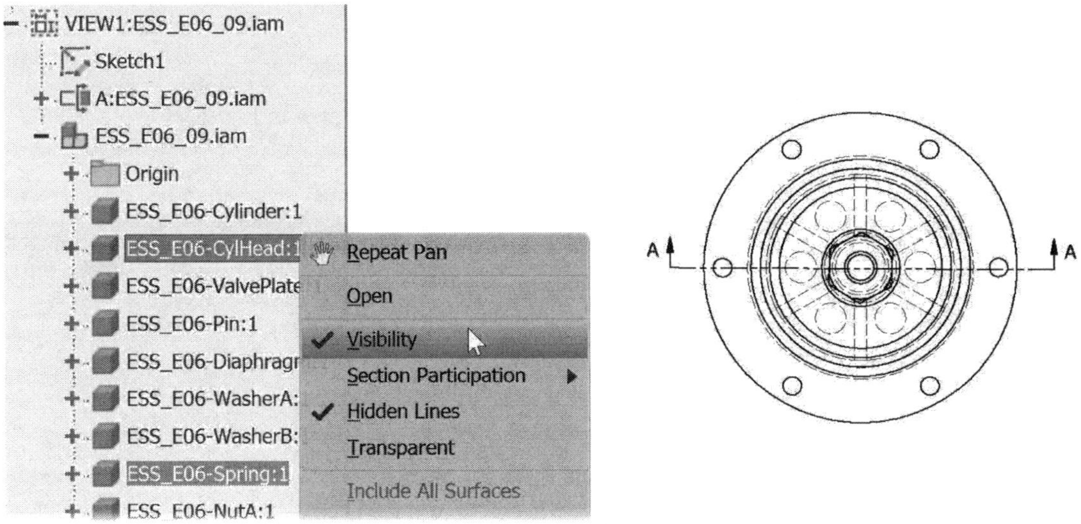

Figure 6.177

9 Next, add a text note stating that the cylinder head and spring are not visible.

a. Click the Text command in the Annotate tab > Text panel.

b. Select a point below and to the left of the top view.

c. Change the text justification to Center Justification.

d. Change the text size to **.25 inches**.

e. In the text entry area, type **FOR CLARITY CYLINDER HEAD AND** press ENTER to go to the next line, and type **SPRING REMOVED FROM VIEW**.
If needed, change the text to uppercase by selecting the text and use the Text Case option UPPERCASE.

f. In the Format Text dialog box, click OK and the text will be added to the drawing.

g. Right-click and click OK from the marking menu to exit the command.

h. Reposition and stretch the text box so the text is below the top view, like Figure 6.178.

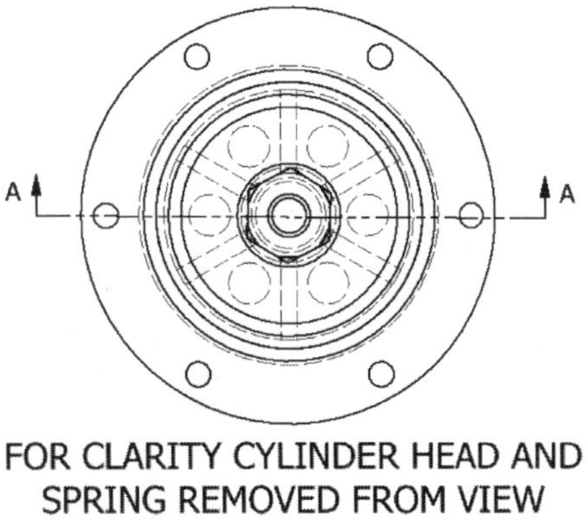

Figure 6.178

10 Next, create a Parts List that will list item numbers, quantity, part number, and descriptions.

a. Begin by clicking the Parts List command on the Annotate tab > Table panel.

b. The Parts List dialog box will display; select the sectioned view.

c. Verify that the BOM View is set to Structured, and click the OK button.

d. Move the Parts List until it connects to the upper-right corner of the title block and the border, and then click to place it. Zoom in to display the Parts List, as shown in Figure 6.179.

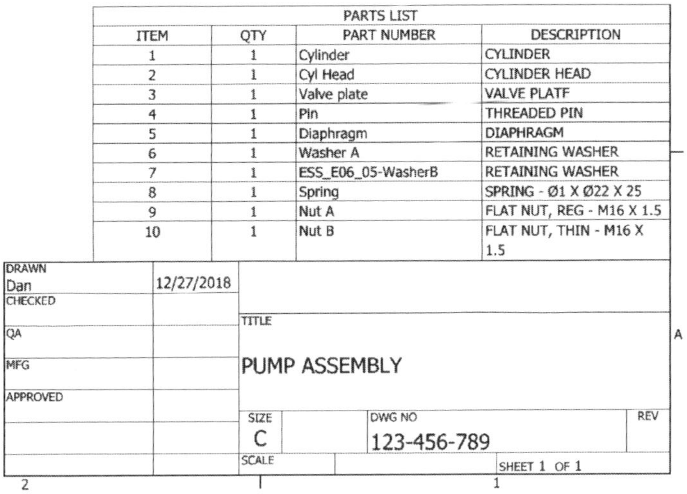

PARTS LIST			
ITEM	QTY	PART NUMBER	DESCRIPTION
1	1	Cylinder	CYLINDER
2	1	Cyl Head	CYLINDER HEAD
3	1	Valve plate	VALVE PLATF
4	1	Pin	THREADED PIN
5	1	Diaphragm	DIAPHRAGM
6	1	Washer A	RETAINING WASHER
7	1	ESS_E06_05-WasherB	RETAINING WASHER
8	1	Spring	SPRING - Ø1 X Ø22 X 25
9	1	Nut A	FLAT NUT, REG - M16 X 1.5
10	1	Nut B	FLAT NUT, THIN - M16 X 1.5

Figure 6.179

11 Next, change the layout of the parts list so its heading is at the bottom.

a. Double-click on the Parts List in the graphics window.

b. In the top of the Parts List dialog box, click the Table Layout button .

c. Change the header so it is at the bottom of the parts list table by clicking the Bottom option (middle option in the list) in the Heading area, as shown in Figure 6.180.

d. Click OK to close the Parts List Table Layout dialog box, but leave the Parts List dialog box open.

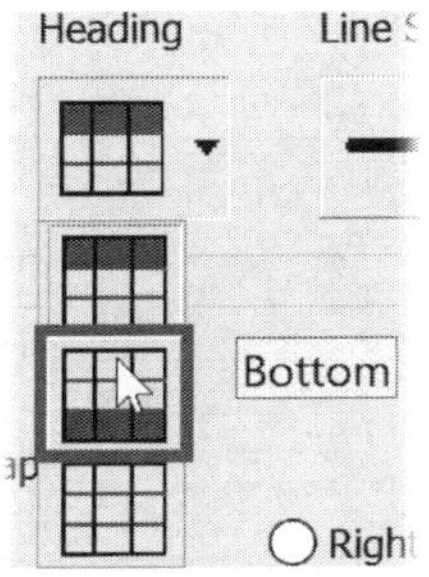

Figure 6.180

12 Next, add a column to the parts list and reorder a column.

a. While still in the Parts List dialog box, click the Column Chooser button  to display the Parts List Column Chooser dialog box.

b. Scroll down in the Available Properties list and select MATERIAL.

c. Click the Add button in the middle of the dialog box, as shown in the following image on the left. To add a property to the list you can also double-click on its name.

d. Move the Description Column up by selecting DESCRIPTION in the Selected Properties list and click the Move Up button twice, as shown in the image on the right.

e. Click OK to close the Parts List Column Chooser dialog box, but leave the Parts List dialog box open.

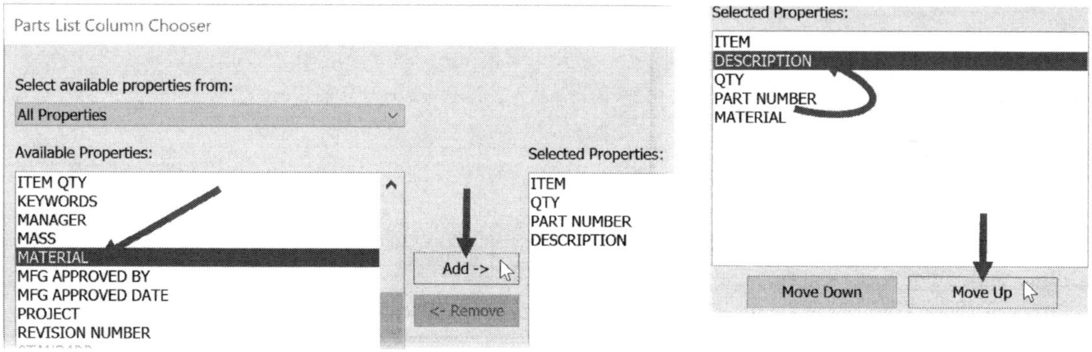

Figure 6.181

13 Next, you edit the column widths.

a. In the Parts List dialog box, right-click on the DESCRIPTION header and click Column Width from the menu, as shown in the following image on the left.

b. Enter **2.25** in the Column Width dialog box, as shown in the following image on the right.

c. Click OK to complete this edit.

Figure 6.182

14 Change the width of the following columns.

a. ITEM = **1**

b. PART NUMBER = **2**

c. MATERIAL = **1.5**

15 Next, you change the order in which the items are sorted.

a. Click the Sort button  to display the Sort Parts List dialog box.

b. Change the Sort by option to ITEM.

c. Change the sort order to Descending, as shown in the following image.

e. Click the OK button to close the Sort Parts List dialog box.

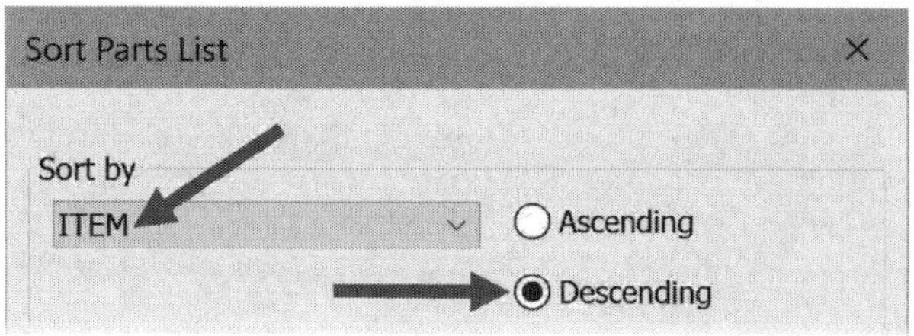

Figure 6.183

16 Click OK to make the changes and close the Parts List dialog box. Your display should look like the following image.

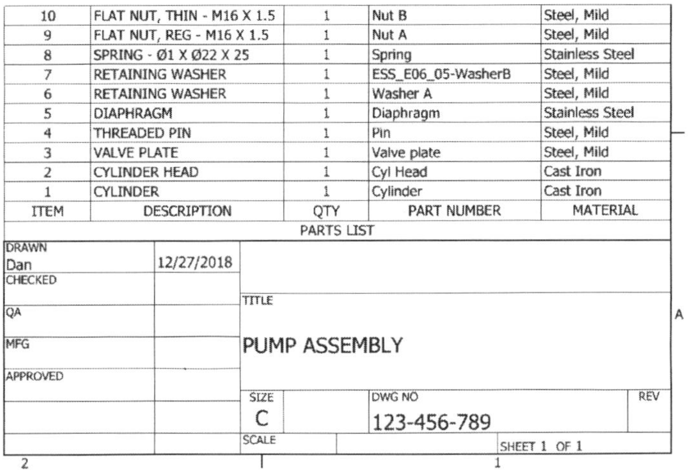

ITEM	DESCRIPTION	QTY	PART NUMBER	MATERIAL
10	FLAT NUT, THIN - M16 X 1.5	1	Nut B	Steel, Mild
9	FLAT NUT, REG - M16 X 1.5	1	Nut A	Steel, Mild
8	SPRING - Ø1 X Ø22 X 25	1	Spring	Stainless Steel
7	RETAINING WASHER	1	ESS_E06_05-WasherB	Steel, Mild
6	RETAINING WASHER	1	Washer A	Steel, Mild
5	DIAPHRAGM	1	Diaphragm	Stainless Steel
4	THREADED PIN	1	Pin	Steel, Mild
3	VALVE PLATE	1	Valve plate	Steel, Mild
2	CYLINDER HEAD	1	Cyl Head	Cast Iron
1	CYLINDER	1	Cylinder	Cast Iron

Figure 6.184

TIP: Within the graphics window, you can also resize the width of a column or height of a row by clicking and dragging on a border, like you would in Microsoft Excel.

17 Next, you add a balloon.

a. Zoom into the section view.

b. Click the Balloon command on the Annotate tab > Table panel.

c. At the start of the leader, select the middle of the right-outside edge of the cylinder.

d. Move the cursor up and to the right and click a point on the sheet to define the end of the first leader segment.

e. Right-click and select Continue from the marking menu to place the balloon, as shown in Figure 6.185 left.

f. Note that the item number in the balloon corresponds to the item number in the Parts List.

18 Next, you add balloons with the Auto Balloon command.

a. Start the Auto Balloon command from the Annotate tab > Table panel.

b. Select the sectioned assembly view as the view to place the balloons.

c. Select all the components in the view by dragging a window around all the components in the section view. The cylinder will not be selected because it has already been ballooned.

d. In the Auto Balloon dialog box click the Select Placement button.

e. Ensure Horizontal is the active placement option.

f. Change the Offset Spacing to **0.25 inches** as shown in the following image.

g. Locate the balloons by clicking a point above the cylinder as shown in Figure 6.185 right.

h. Click OK in the Auto Balloon dialog box to create the balloons.

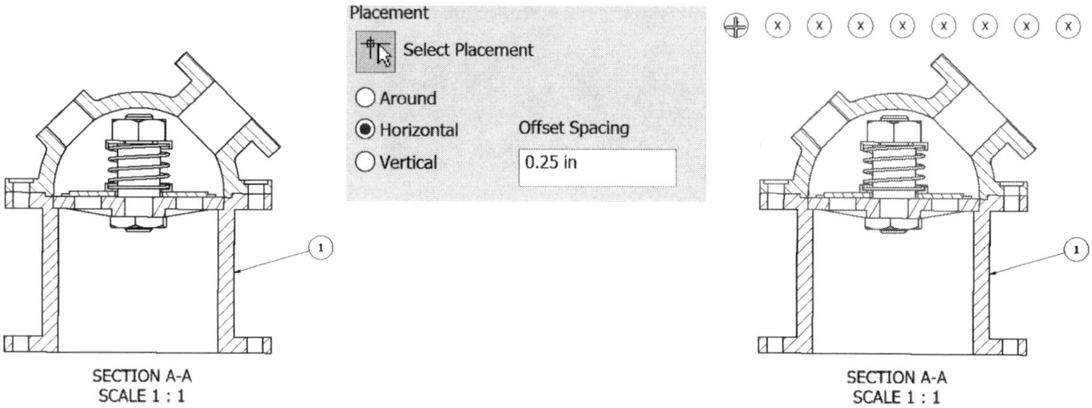

Figure 6.185

19 If needed you can reposition the balloons individually or after selecting the balloons you can right-click and use the Align options as shown in the following image.

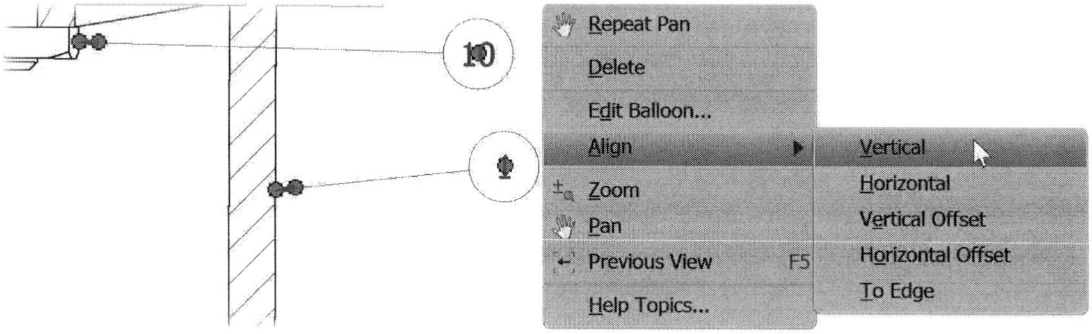

Figure 6.186

20 When done, your view should resemble the following image (view label not shown).

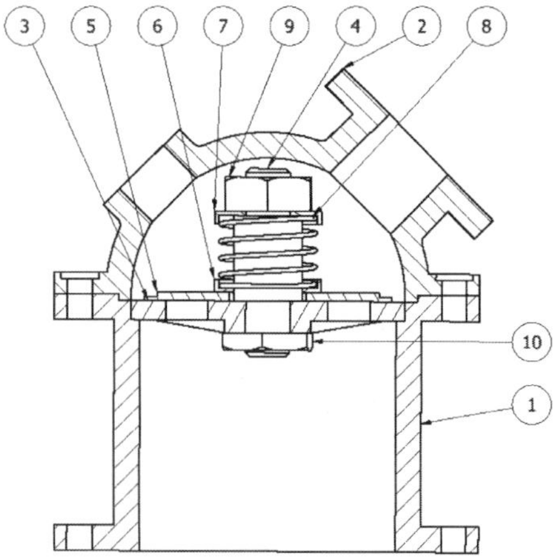

Figure 6.187

21 Close all open files. Do not save changes.

End of exercise.

APPLYING YOUR SKILLS

Skills Exercise 6-1

In this exercise, you create a new component for a charge pump and then assemble the pump.

1. Open ESS_E06_10.iam in the Chapter 06 folder.
2. From the Chapter 06 folder place the following components into the right side of the assembly as shown in the following image:
    - 2 occurrences of ESS_E06_10-M8x30.ipt
    - 1 occurrence of ESS_E06_10-Seal.ipt
    - 1 occurrence of ESS_E06_10-Union.ipt

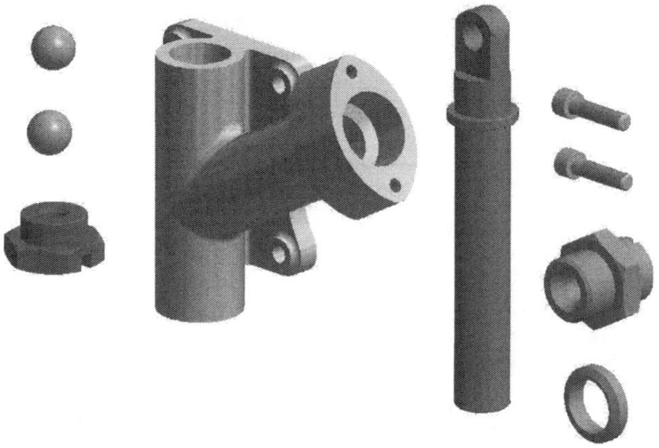

Figure 6.188

3. Next, create the Connector in the context of the assembly. Project edges from the *ESS_E06_10-Body.ipt:1* (pump body) to define the connector. Use the dimensions in the following image to model the connector. The missing dimensions and constraints are acquired from the projected edges.

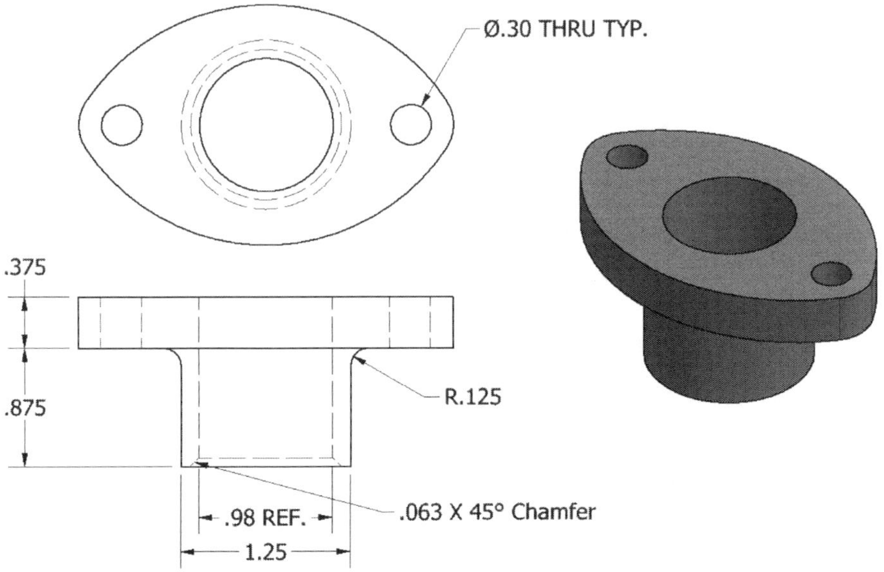

Figure 6.189

4. Use assembly constraints to build the assembly as shown in the following two images.

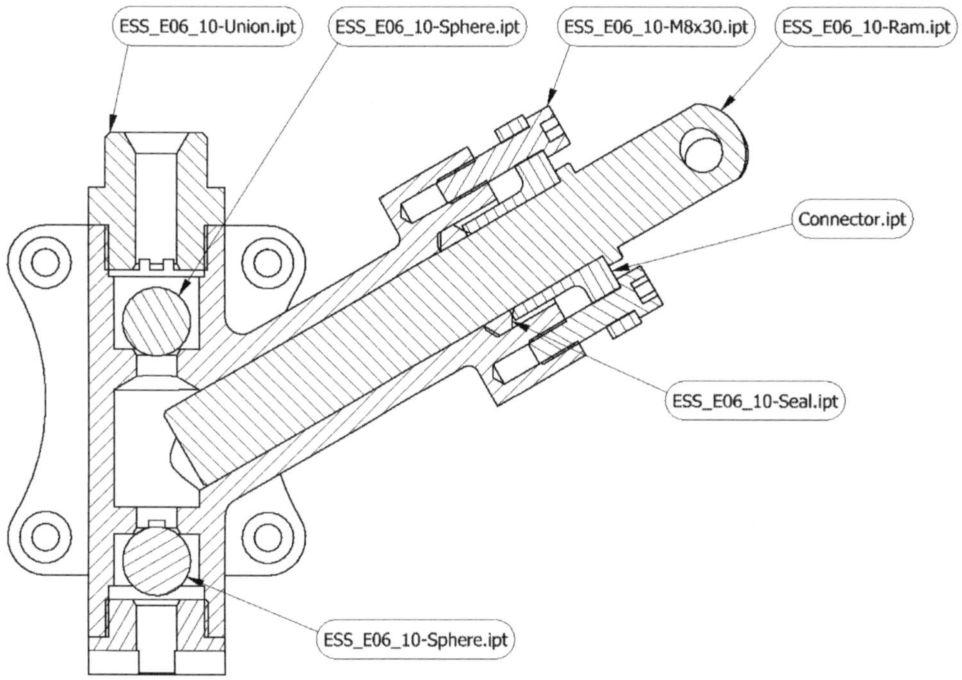

Figure 6.190

TIP: Use the origin axis in the Valve Sphere part file to align it to the center of the body and use a tangent constraint to assemble the Valve Sphere to the inside chamfered face.

Figure 6.191

5. Close all open files. Do not save changes.

End of exercise.

Skills Exercise 6-2

In this exercise, you create drawing views of an assembly, and add a parts list and balloons for a charge pump. The charge pump assembly is the completed version of the assembly that you created in skills exercise 6-1.

Create a new drawing using the ANSI (in) DWG or IDW template with a C size sheet named Assembly.

1. Use the Base View command to create a Top view of ESS_E06_11.iam with a scale of 1:1.
2. Use the top view to create a sectioned front view and exclude the following components from being sectioned:
    - ESS_E06_10-Sphere:1
    - ESS_E06_10-Sphere:2
    - ESS_E06_10-Ram:1
    - ESS_E06_10-M8x30:1
    - ESS_E06_10-M8x30:2
3. Use the Base View command to create an ISO Top Right view of ESS_E06_11.iam with a scale of 1:1. When done creating the views, your screen should resemble the following image.

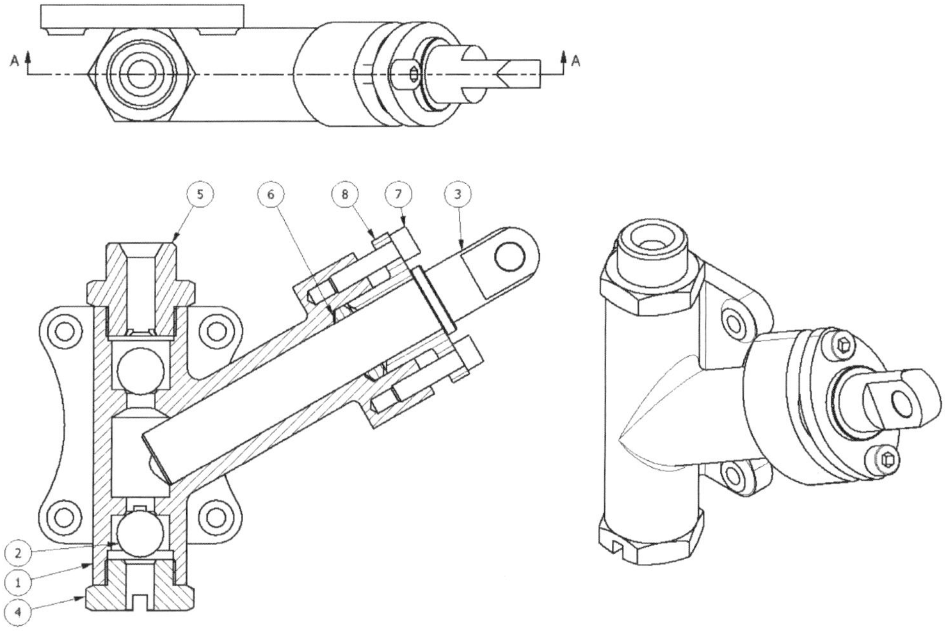

Figure 6.192

4. Insert a Parts List on top of the title block.
5. Add a Material column to the parts list and change the material of all components to steel.
6. Modify the parts list and title block to match what is shown in the following image. Note how the item numbers are sorted in descending order and the header row is at the bottom of the table.

ITEM	DESCRIPTION	QTY	PART NUMBER	MATERIAL
8	CONNECTOR	1	ESS_E06_11-CONNECTOR	STEEL
7	HHCS - PURCHASED	2	ESS_E06_10-M8x30	STEEL
6	SEAL	1	ESS_E06_10-Seal	STEEL
5	UNION	1	ESS_E06_10-Union	STEEL
4	SEAT	1	ESS_E06_10-Seat	STEEL
3	RAM	1	ESS_E06_10-Ram	STEEL
2	VALVE SPHERE	2	ESS_E06_10-Sphere	STEEL
1	BODY	1	ESS_E06_10-Body	STEEL

PARTS LIST

DRAWN DTB	12/27/2018			
CHECKED		TITLE		
QA		CHARGE PUMP ASSEMBLY		
MFG				
APPROVED				
		SIZE C	DWG NO ESS_E06_Skills 6-2	REV
		SCALE	SHEET 1 OF 1	

Figure 6.193

7. Add balloons in the section view to identify the components, as shown in the following image.

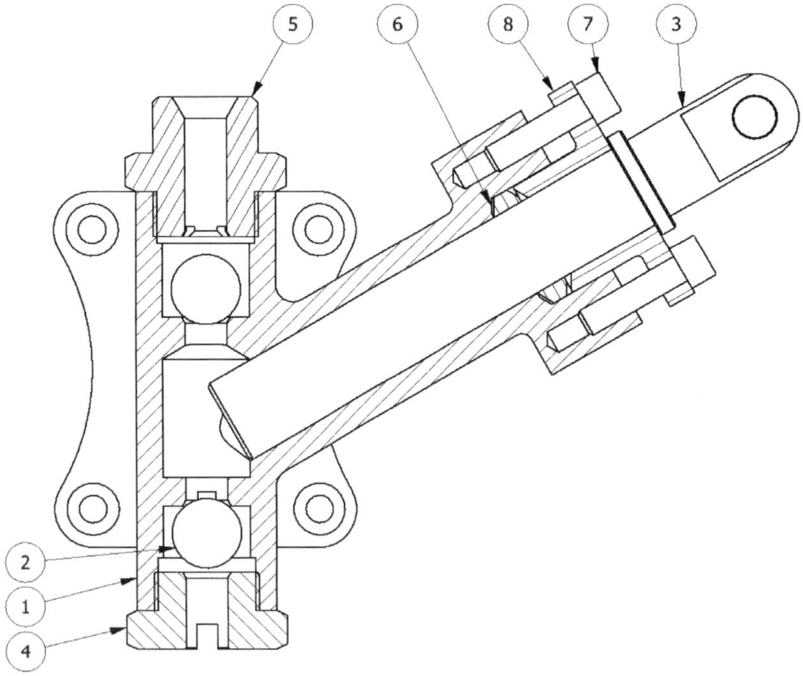

Figure 6.194

8. Close all open files. Do not save changes. End of exercise.

CHECKING YOUR SKILLS

Use these questions to test your knowledge of the material covered in this chapter.

1. True__ False__ The only way to create an assembly is by placing existing components into it.
2. Explain how to create a component in the context of an assembly.
3. True__ False__ An occurrence is a different version of an existing component.
4. True__ False__ Only one component can be grounded in an assembly.
5. True__ False__ Autodesk Inventor does not require components in an assembly to be fully constrained.
6. True__ False__ A sketch must be fully constrained before it can become adaptive.
7. What is the purpose of a presentation file?
8. True__ False__ Balloons can only be placed in a drawing after placing a Parts List.
9. True__ False__ When creating drawing views from a presentation file, you can create views from multiple presentation views.
10. True__ False__ A Bill of Materials only retrieves its data from a Parts List.
11. True__ False__ An animation of a presentation file can be exported to a video file.
12. True__ False__ A BOM Structured view shows all subassemblies and individual parts at the same assembly level.
13. True__ False__ An Associated Component Pattern will maintain a relationship to a feature pattern.
14. True__ False__ User-defined folders allow you to organize an assembly browser by grouping assembly constraints in a single folder.
15. True__ False__ In a Parts List, you can display custom parts that are not components or graphical data, e.g. paint.

7 Advanced Modeling Techniques

INTRODUCTION

In this chapter, you learn to use modeling techniques to be more productive and efficient with Inventor. You learn to control dimension display, set up relationships between dimensions, parameters, section a component, emboss text, sweep a profile along a path, create features that transition between profiles, create organic shapes, suppress features, reorder features, and validate your design by simulating stress in a part or assembly.

OBJECTIVES

After completing this chapter, you will be able to:

- ☐ Change the display of dimensions
- ☐ Create relationships between dimensions
- ☐ Create an equation to define a dimension's value
- ☐ Create parameters
- ☐ Create and place iParts
- ☐ Create and place iAssemblies
- ☐ Section a part or components in an assembly
- ☐ Create a design view representation
- ☐ Emboss text and close profiles
- ☐ Create sweep features
- ☐ Create 3D sketches
- ☐ Create coil features
- ☐ Create loft features
- ☐ Create and Edit Freeform shapes
- ☐ Split a part
- ☐ Mirror model features
- ☐ Suppress features of a part
- ☐ Reorder part features
- ☐ Place components from the Content Center
- ☐ Simulate stress on a part and an assembly
- ☐ Use the Frame Generator

DIMENSION DISPLAY, RELATIONSHIPS, AND EQUATIONS

When creating part features, you may set up relationships between features and/or sketch dimensions. For example, the length of a part may need to be twice that of its width, or a hole may always need to be in the middle of the part. In Inventor, you can use several different methods to create relationships between dimensions. The following sections cover these methods.

Dimension Display

When you create a dimension, it is automatically tagged with a label, or parameter name, that starts with the letter "d" and a number: for example, "d0" or "d27." The first dimension created for each part is given the label "d0." Each dimension that you place for subsequent part sketches and features are sequenced incrementally, one number at a time. If you erase a dimension, the next dimension does not go back and reuse the erased value. Instead, it continues the sequencing from the last value on the last dimension created. When creating dimensional relationships, you may want to view a dimension's display style to see the underlying parameter label of the dimension. Five options for displaying a dimension's display style are available.

Value. Use to display the actual value of the dimensions on the screen.

Name. Use to display the dimensions on the screen as the parameter name: for example, d12 or Length.

Expression. Use to display the dimensions on the screen in the format of parameter# value, showing each actual value: for example, d7 = 20in or Length = 5in.

Tolerance. Use to display the dimensions on the screen that have a tolerance style: for example, 40±.3.

Precise Value. Use to display the dimensions on the screen and ignore any precision settings that are specified: for example, 40.3563123344.

To change the dimension display style, while in a sketch, click the desired dimension display option from the status bar, as shown in Figure 7.1. After you select a dimension display style, all visible dimensions change to that style. As you create dimensions, they reflect the current dimension display style.

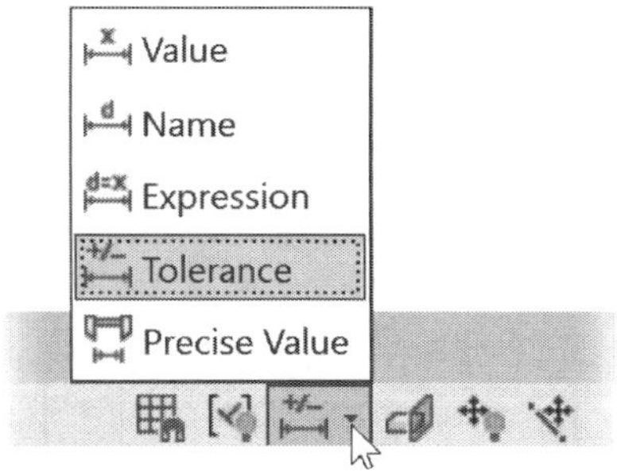

Figure 7.1

A Dimension's Relationship

To define a dimension to be equal to another dimension, instead of applying an equal constraint, you can set the value of one dimension to equal the value of another. Before creating a relationship between two dimensions of the same length, a dimension must exist to have the value to which the other dimension will be equal. Place or edit the dimension to be equal to the first dimension. Use the Edit Dimension dialog box to enter the parameter name (d#) of the other dimension. You can also choose the dimension with which to set the relationship from the graphics window. Figure 7.2 left shows the Edit Dimension dialog box for the vertical dimension after selecting the 1.25 inch horizontal dimension to which it will be equal. Note that "d0" in the Edit Dimension dialog box replaced the original value. After establishing a relationship to another dimension, the dimension has a prefix of fx: to indicate a function relationship as shown in Figure 7.2 right.

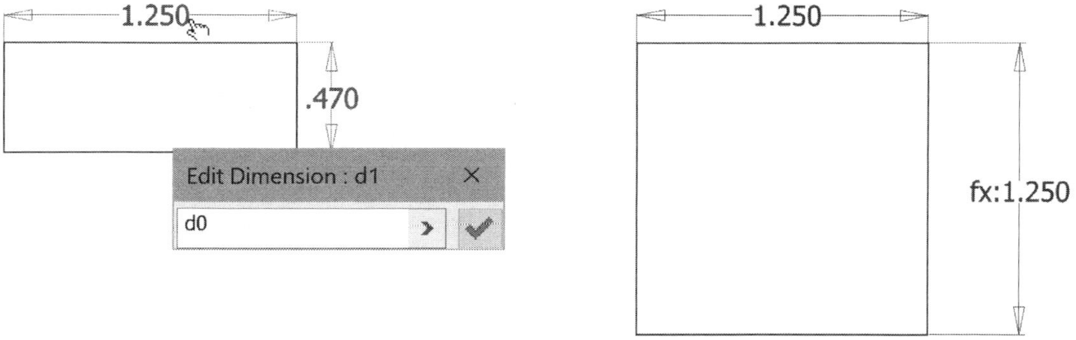

Figure 7.2

Equations

You can use an equation to set a value. Two examples of equations are (d9/4)*2 or 50 – 11.625. When creating equations, Inventor allows prefixes, precedence, operators, functions, syntax, and units. To see a complete listing of valid options, use the Help system and search for "Functions, Prefixes, and Algebraic Operators."

You can enter numbers with or without units; when no unit is entered, the default unit is assumed. As you enter an equation, Inventor evaluates it. An invalid expression will appear in red, and a valid expression appears in black. For best results include units for every term and factor in the equation.

To create an equation when a value is required, follow these steps:

1. Click in the cell where a value is required, for example, dimension, extrude length, number of feature to be patterned, etc.
2. Enter any valid combination of numbers, parameters, operators, or built-in functions. The following image shows an example of an equation that divides a number and then adds a number.
3. Press ENTER or click the green checkmark to accept the expression.

TIP: Use ul (unitless) where a number does not have a unit. For example, use a unitless number when dividing, multiplying, or specifying values for a pattern count.

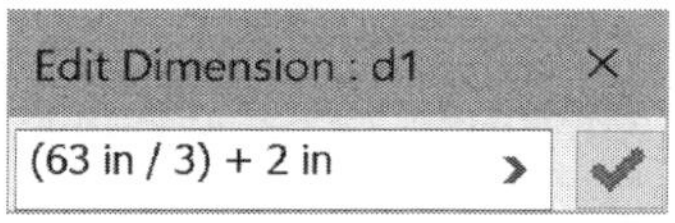

Figure 7.3

PARAMETERS

Another way to create relationships between dimensions is using parameters. A parameter is a user-defined name that is assigned a value, either explicitly or through equations. You can use multiple parameters in an equation, and you can use parameters to define another parameter such as *depth = length - width*. You can use a parameter anytime a numeric value is required. There are four types of parameters: model, user, reference, and linked.

TIP: Functions in Inventor such as sheet metal, stress analysis and design accelerator can automatically add parameter groups.

Creating Parameters

To create parameters via the dialog box, start the Parameters command from the Manage tab > Parameters panel, as shown in Figure 7.4 left, or by clicking the Parameters icon from the Quick Access toolbar. The Parameters dialog box is displayed, as shown in the following image on the right. Figure 7.4 shows an example of a part file with four types of parameters: model, user, reference and linked parameters.

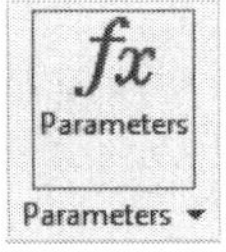

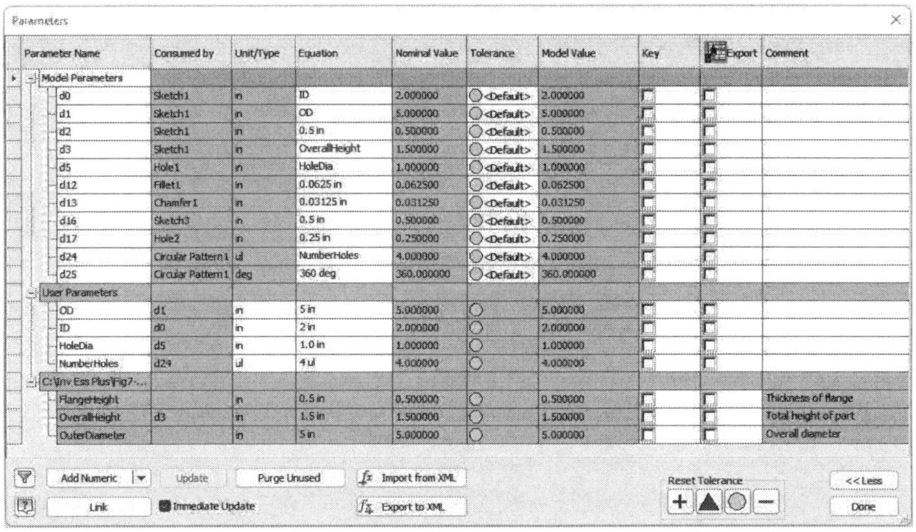

Figure 7.4

Parameter Dialog Box Sections

By default, the Parameters dialog box has two sections: Model Parameters and User Parameters. A section called Reference Parameters is displayed if driven dimensions exist and a section will be created for every spreadsheet you link to. Following are descriptions of the parameter types.

Model Parameter. This parameter type is created automatically and assigned a name when you create a sketch dimension, feature parameter such as an extrusion distance, draft angle, or coil pitch, and the offset, depth, or angle value of an assembly constraint. Inventor assigns a default name to each Model Parameter as you create it.

The default name format is a "d" followed by an integer incremented for each new parameter. You can rename Model Parameters via the Parameters dialog box or specify a name upon initial creation in the Edit Dimension dialog box. This method is called "creating a parameter on the fly" and can be done in any field that you can specify a parametric dimensional value. The parameter being defined by the field is renamed to the name specified at the beginning of the equation. For example, if you were defining a rectangular sketch and wanted to define a parameter named Length while dimensioning the sketch, enter "Length=10" (without the quotation marks) in the Edit Dimension dialog box. A parameter with the name of Length is automatically created.

User Parameter. This type of parameter is created manually by clicking the Add Numeric button in the Parameters dialog box. You can change the names and equations of both types of parameters, and you can add comments by double-clicking in the cell and entering the new information. To learn more about user parameters, see the Creating User Parameter section later in this chapter.

Reference Parameter. This parameter type is created automatically when you create a driven dimension. Inventor assigns a default name to each reference parameter as you create it. The default name format is a "d" followed by an integer incremented for each new parameter. You can rename reference parameters via the Parameters dialog box.

Linked Parameter. To use the same parameter in multiple part and/or assembly files, you can add the parameter data to a Microsoft Excel spreadsheet. You then embed or link the spreadsheet into a part or assembly file through the Parameters dialog box. When you embed a Microsoft Excel spreadsheet, there is no link between the spreadsheet and the parameters in the Inventor file, and any changes to the original spreadsheet are not reflected in the Autodesk Inventor file. When you link a Microsoft Excel spreadsheet to an Inventor part or assembly file, any changes in the original spreadsheet will update the parameters in the Inventor file. You can link more than one spreadsheet to an Inventor file. You can also link a spreadsheet to multiple part and assembly files. For more information, see the Creating Linked Parameters section later in this chapter.

Parameter Dialog Box Columns

The columns available for all four types of parameters are the same. The following sections define them.

Parameter Name. The name of the parameter appears in this cell. To change the name of an existing parameter, click in the box and enter a new name. When you update the model, all dependent parameters update to use the new name.

Consumed. Displays the name of the sketch or feature that uses this parameter.

Unit/Type. Enter a new unit of measurement for the parameter in this cell. With Inventor, you can build equations that include parameters of any unit type. All length parameters are stored internally in centimeters; angular parameters are stored internally in radians.

Equation. The equation that appears in this cell determines the value of the parameter. If the parameter is a discrete value, the value is rounded to match the precision setting for the active document. To change the equation, click on it and enter the new equation.

Nominal Value. The nominal tolerance result of the equation appears in this cell. It can only be modified by editing the equation.

Tol. (Tolerance). Click the pencil icon as shown in Figure 7.5 to open the Tolerance dialog box, and use it to set the tolerance condition to upper, median, nominal, or lower.

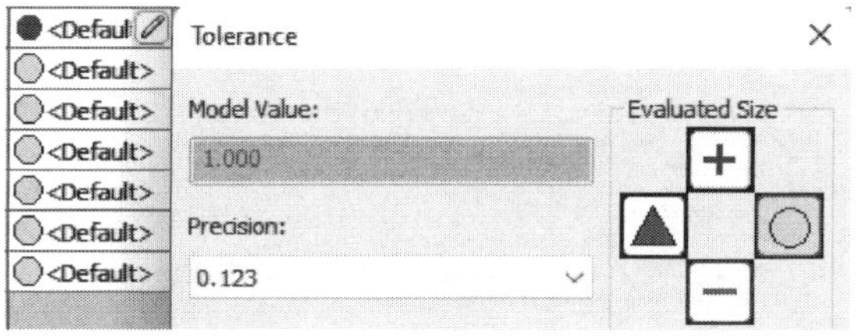

Figure 7.5

Model Value. The calculated model value of the equation, in full precision, appears in this cell. This value reflects the current tolerance condition of the parameter.

Key. Click to specify which defined parameters are identified as a key parameter. The display of parameters in the dialog box can be filtered based on this setting.

Export Parameters. Click to export the parameter to the Custom tab of the iProperties dialog box. The parameter will also be available when using the Derive command and in the Bill of Materials and Parts List Column Chooser dialog boxes.

Comment. You enter a comment for the parameter in this cell. Click in the cell, and type the comment.

Creating User Parameters

You define User Parameters in a part or an assembly file. Parameters defined in one file type are not directly accessible in the other file type. If parameters are to be used in both file types use linked parameters via a Microsoft Excel spreadsheet. When creating a User Parameter, follow these guidelines:

- Assign meaningful names to parameters, as other designers may edit the file and will need to understand your thought process. You may want to use the comments field for further clarification.
- The parameter name cannot include spaces, mathematical symbols, or special characters.
- Inventor detects capital letters and uses them as unique characters. Length, length, and LENGTH are each a different parameter name.
- When defining a parametric equation, you cannot use the parameter name to define itself; for example, Length = Length/2 is invalid.
- Duplicate parameter names are not allowed. Model, User, and Spreadsheet-driven parameters must have unique names.

To create and use a User Parameter, follow these steps:

1. Click the Parameters command from the Manage tab > Parameters panel.
2. Click the Add Numeric button located at the bottom of the Parameters dialog box.
3. A new row in the User Parameters section appears; enter data in the cells as required.
4. After creating the parameter(s), you can enter the parameter name(s) anywhere a value is required. When editing a dimension, click the arrow on the right, and click List Parameters from the menu, as shown in Figure 7.6 left. All the available User Parameters and any renamed Model Parameters will appear in a list like that in the following image on the right. Click the desired parameter from the list.

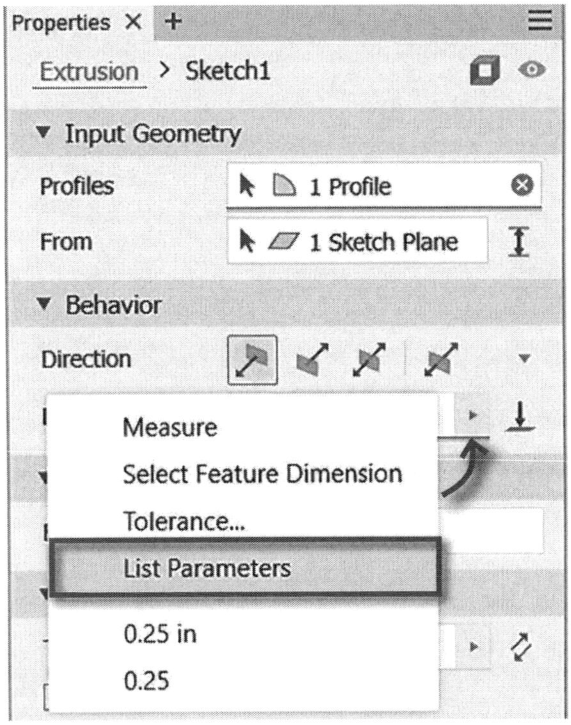

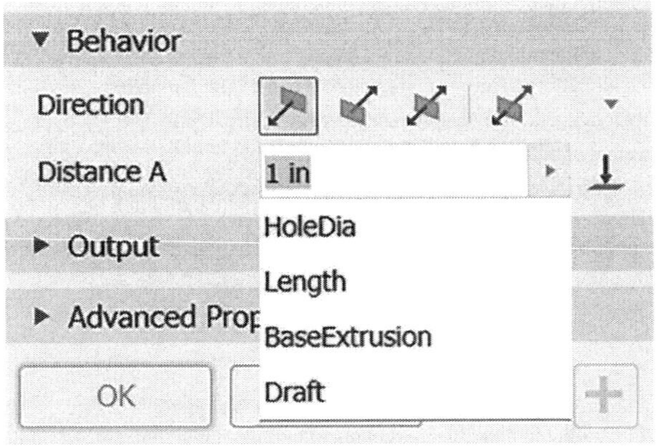

Figure 7.6

Creating Linked Parameters

When creating a Microsoft Excel spreadsheet with parameters, follow these guidelines:

- The data in the spreadsheet can start in any cell, but the cells must be specified when you link or embed the spreadsheet.
- The data can be in rows or columns, but they must be in this order: parameter name, value or equation, unit of measurement, and (if desired) a comment.
- The parameter name and value (equation) are required, but the other items are optional.
- The parameter name cannot include spaces, mathematical symbols, or special characters. You can use these to define the equation.
- Parameters in the spreadsheet must be in a continuous list. A blank row or column between parameter names eliminates all parameters after the blank row or column.
- If you do not specify a unit of measurement for a parameter, the default units for the document are used for the parameter. To create a parameter without units, enter "ul" (unitless) in the Units cell.
- Only parameters defined on the first worksheet of the spreadsheet are linked to the Inventor file.
- You can include column or row headings or other information in the spreadsheet, but they must be

outside the block of cells that contains the parameter definitions.

The following image on the left shows three parameters that were created in rows with a name, equation, unit, and comment. The image on the right shows the same parameters created in columns.

	A	B	C	D
1	Length	5	in	Length of plate
2	Width	Length/2	in	Width of plate
3	Depth	1	in	Depth of plate

	A	B	C
1	Length	Width	Depth
2	5	Length/2	1
3	in	in	in
4	Length of plate	Width of plate	Depth of plate

Figure 7.7

After you have created and saved the spreadsheet, create parameters from it by following these steps:

1. While in a part file click the Parameters command on the Manage tab > Parameters panel. From an assembly file click the Parameters command on the Manage tab > Manage panel.
2. Click the Link button Link on the bottom of the Parameters dialog box.
3. The Open dialog box appears as shown in the following image.
4. Navigate to and select the Microsoft Excel file to use.
5. In the lower-left corner of the Open dialog box, enter the start cell for the parameter data.
6. Select whether the spreadsheet will be linked or embedded.

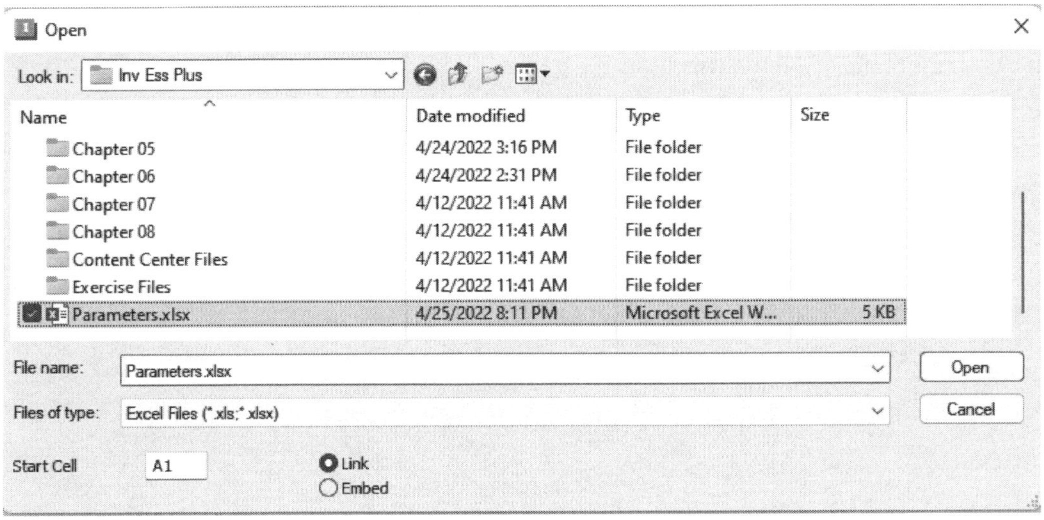

Figure 7.8

7. Click the Open button.
8. A new section showing the linked parameters is added to the Parameters dialog box, as shown in Figure 7.9. If you embedded the spreadsheet, the new section is titled Embedding #.
9. To complete the operation, click Done.

Length	Width	in	5 in	5.000000	○	5.000000	☐	☐	Length of plate
Width		in	Length / 2 ul	2.500000	○	2.500000	☐	☐	Width of place
Depth		in	1 in	1.000000	○	1.000000	☐	☐	Depth of place

Figure 7.9

To edit linked parameters, follow these steps:

1. Open the Microsoft Excel file.
2. Make the required changes.
3. Save the Microsoft Excel file.
4. Open the Inventor part or assembly file that uses the spreadsheet and if needed click the Update command on the Quick Access toolbar.

You can also use these steps when a spreadsheet has been linked. If you chose to embed the spreadsheet, the following steps must be used to edit the parameters:

1. Open the part or assembly file that uses the spreadsheet.
2. Expand the 3rd Party folder in the browser.
3. Double-click on the name of the spreadsheet or right-click on the name of the spreadsheet and click Edit from the menu, as shown in Figure 7.10.
4. The Microsoft Excel spreadsheet will open in a new window for editing.
5. Make the required changes.
6. Save the Microsoft Excel file.
7. Activate the Autodesk Inventor part or assembly file that uses the spreadsheet.
8. Click the Update command on the Quick Access toolbar.

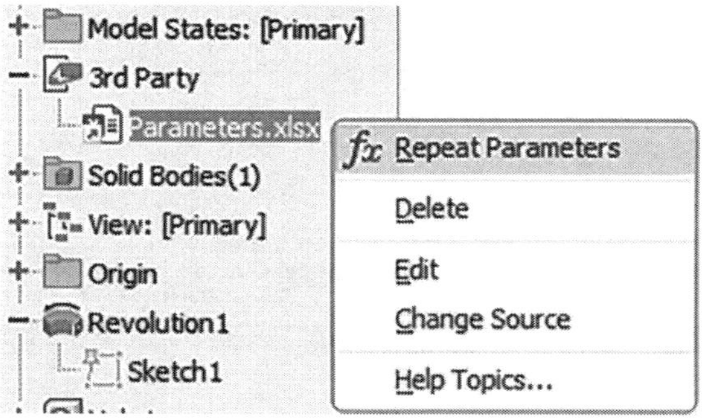

Figure 7.10

EXERCISE 7-1: RELATIONSHIPS AND PARAMETERS

In this exercise, you create a sketch and dimension it. You then set up relationships between dimensions, and define User Parameters in both the model and an external spreadsheet.

1 Create a new file based on the default Standard (in).ipt file.

2 Create a 2D Sketch on the XY origin plane and then draw and dimension the profile as shown in the following image on the left. The lower-left corner should be located at the origin point, and the arc is tangent to both lines.

3 Next, create a relation and an equation.

a. Double-click the **1.500 inch** radius dimension on the arc.

b. With the value in the Edit dimension dialog box highlighted, change it to relate to the horizontal dimension by selecting the **5.000 inch** dimension, as shown in Figure 7.11 right.

c. Next divide the horizontal dimension by four. In the Edit Dimension dialog box type **/4** after the d#. Figure 7.11 shows the value for the radius dimension equal to "**d0/4**."

d. Click the green checkmark to accept the dimension value in the Edit Dimension dialog box.

Your dimension my have a different name than the d0 in Figure 7.11 as the name depends upon the order the dimensions were placed.

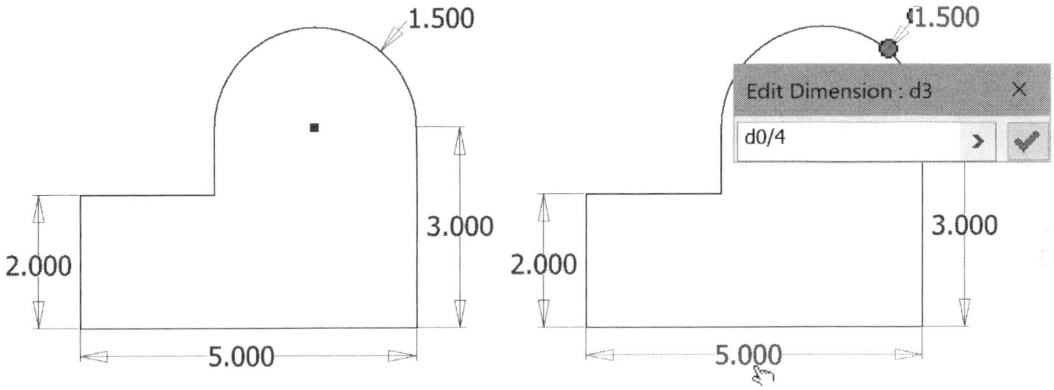

Figure 7.11

4 Add another relationship and equation.

a. Double-click the **2.000 inch** left vertical dimension.

b. Change it by selecting the **3.000 inch** right vertical dimension and type **/2** after the d#, i.e. **d#/2**.

c. Click the green checkmark to accept the dimension value in the Edit Dimension dialog box. The fx: text denotes that the dimensions are driven by an equation and reference another dimension.

5 Make the following change to these dimensions:

a. Change the **5 inch** horizontal dimension to **4 inches**.

b. Change the **3 inch** vertical dimension to **2.5 inches**.

c. When done, your sketch should resemble the following image on the left.

6 To see the underlying equations for the dimensions, click the Expression icon from the status bar on the bottom of the screen, and then click the Expression option on the menu. When done, the equations are displayed as shown in Figure 7.12.

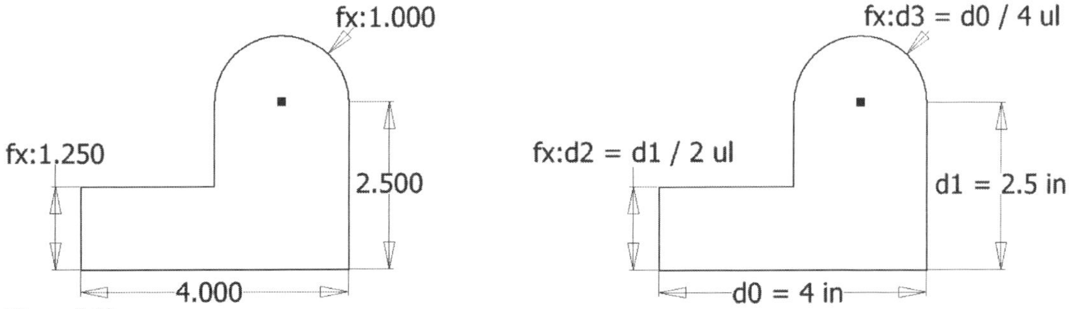

Figure 7.12

7 Next you create two user parameters to control the sketch.

a. Click Parameters from the Quick Access toolbar or from the Manage tab > Parameters panel.

b. In the Parameter dialog box, create a User Parameter by clicking Add Numeric.

c. Type the following information for the first parameter:

   i. Parameter Name = **Length**

   ii. Units = **in**

   iii. Equation = **6**

   iv. Comment = **Bottom length**

d. Create a second User Parameter by clicking the Add Numeric button. Use the following information for this parameter:

   i. Parameter Name = **Height**

   ii. Units = **in**

   iii. Equation = **Length/3**

   iv. Comment = **Height is 1/3 the length**

   v. When done, the User Parameter area should resemble Figure 7.13.

   vi. Close the Parameter dialog box by clicking Done.

User Parameters									
Length	Height	in	6 in	6.000000	○	6.000000	☐	☐	Bottom length
Height		in	Length / 3 ul	2.000000	○	2.000000	☐	☐	Height is 1/3 the length

Figure 7.13

8 Next, change two dimensions to use the parameters you just created.

a. Double-click the bottom-horizontal dimension. Click the arrow from the right side of the dialog box > List Parameters and choose Length from the list. When done click the checkmark from the Edit Dimension dialog box.

b. Double-click the right-vertical dimension. Click the arrow from the right side of the dialog box > List Parameters and choose Height from the list. Click the green checkmark to make the change to the dimension. When done, your screen should resemble Figure 7.14.

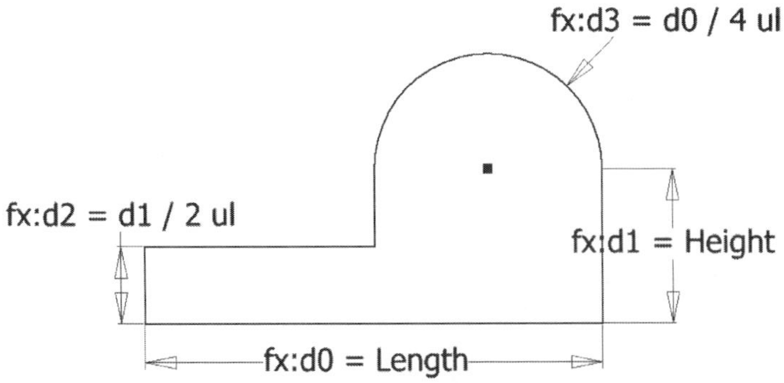

Figure 7.14

9 Next you edit the User Parameters. Start the Parameters command and from the User Parameters area, make the following edits.

a. Change the Equation value for the parameter "Length" to **4.75**.

b. Change the Equation value for the parameter "Height" to **2.75**.

c. Change the comment field of the Height parameter to "Height of the right side." When done making the changes, the User Parameter area should resemble Figure 7.15.

d. Notice that the sketch updates automatically when the values are changed in the Parameters dialog box (if the default option is selected).

e. Close the Parameter dialog box by clicking Done.

User Parameters									
Length	d0	in	4.75 in	4.750000		4.750000			Bottom length
Height	d1	in	2.75 in	2.750000		2.750000			Height of the right side

Figure 7.15

10 To see the values for the dimensions, click the Expression icon on the status bar on the bottom of the screen, and then click the Value option on the menu. The values of the dimensions will be displayed as shown in Figure 7.16.

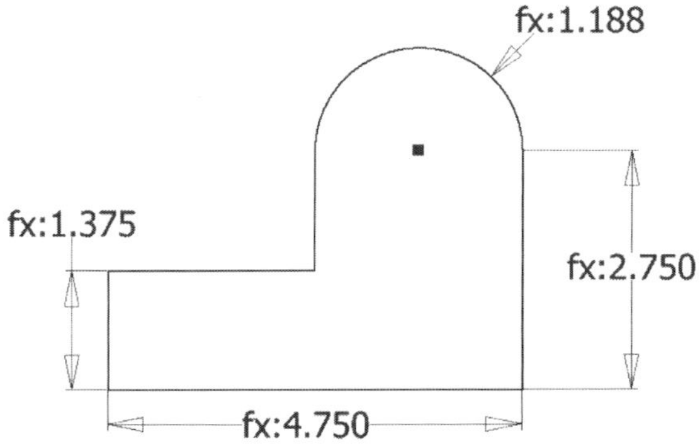

Figure 7.16

11 Finish the sketch.

12 Save the file as *ESS_E07_01.ipt* in the Chapter 07 folder.

13 Now create a spreadsheet that has two parameters.

a. Create a new Microsoft Excel spreadsheet.

b. In Sheet 1 create columns with the names Parameter Name, Equation, Unit, and Comment. Fill in the spreadsheet with the information as shown in Figure 7.17.

A	B	C	D
Parameter Name	**Equation**	**Unit**	**Comment**
BaseExtrusion	1.25	in	Base extrusion distance
Draft	5	deg	Draft angle

Figure 7.17

c. Save the spreadsheet as *ESS_07_Parameters.xlsx* in the *C:\Inv 2024 Ess Plus\Chapter 07* folder.

d. Close Excel.

14 Make Autodesk Inventor and the file *ESS_E07_01.ipt* current.

15 Next, link the parameters to the spreadsheet.

a. Start the Parameters command.

b. In the Parameter dialog box, click the Link button and select the *ESS_07_Parameters.xlsx* file in the Chapter 07 sub folder, but do not click Open yet.

c. For the Start Cell, change this to **A2** (lower-left corner of the Open dialog box). If you fail to do this, no parameters will be found. The first row contains the names of the columns and will not be imported.

d. Click the Open button, and a new spreadsheet area will appear in the Parameters dialog box, as shown in Figure 7.18.

e. Click the Done button to complete the operation.

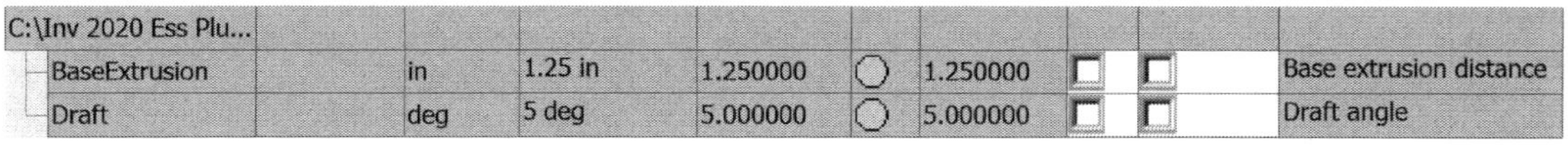

Figure 7.18

16 If needed, change to the Home View.

17 Next, extrude the profile using the linked parameters for the values.

a. Select an edge of the profile and click Create Extrude from the mini-toolbar.

b. In the Extrude Properties panel, for the value of the extrusion, enter the parameter name **Base-Extrusion** or click the arrow from the menu, click List Parameters, and then click the parameter **BaseExtrusion** from the list.

c. In the Advanced Properties area in the Extrude Properties panel, in the Taper A area, enter **Draft** or click the arrow from the menu, click List Parameters, and then click the parameter **Draft**. Figure 7.19 shows the values entered in the Extrude Properties panel and the preview of the extrusion.

d. To complete the operation, click the OK button.

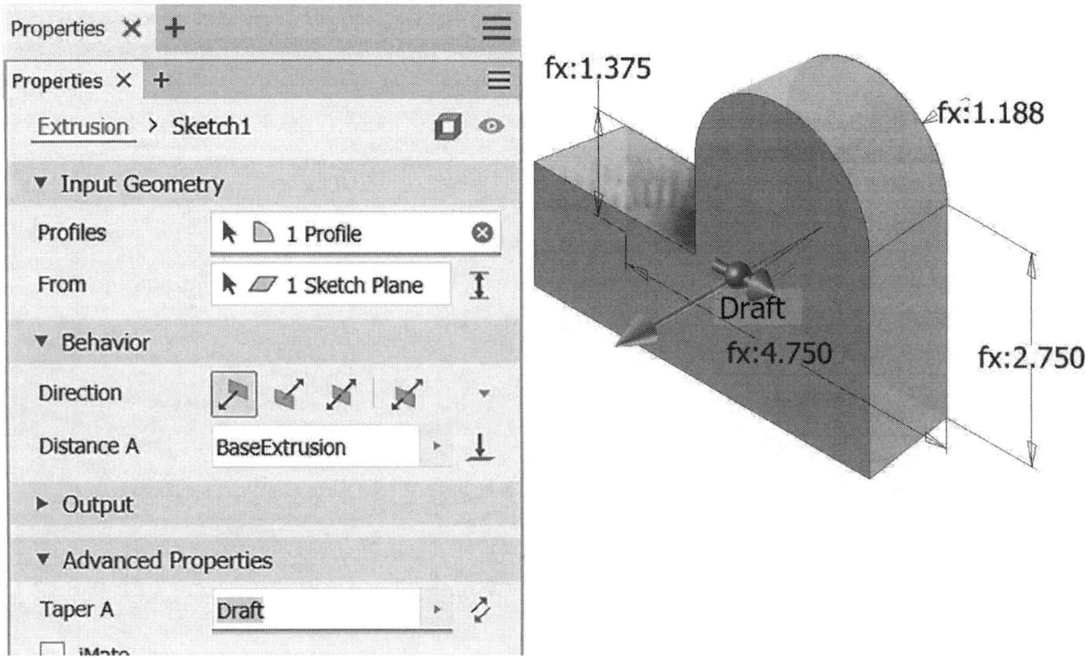

18 Next, edit the spreadsheet.

a. In the browser, expand the 3rd Party icon.

b. Double-click on the *ESS_07_Parameters.xlsx* or right-click on *ESS_07_Parameters.xlsx*, and click Edit from the menu.

c. In the Excel spreadsheet, make the following changes:

i. Change the Equation value for the parameter "BaseExtrusion" to **2**.

ii. Change the Equation value for the parameter "Draft" to **-10**.

d. Save the spreadsheet.

e. Close Excel.

19 The last step is to update the linked parameters.

a. Make Autodesk Inventor with the file *ESS_E07_01.ipt* current.

b. Click the Update command  from the Quick Access toolbar.

20 Change the viewpoint so it resembles Figure 7.20.

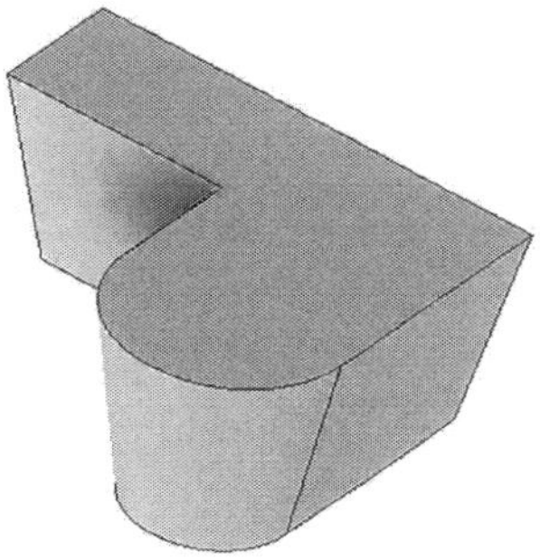

Figure 7.20

21 Close the file. Do not save changes.

End of exercise.

IPARTS

In the last section, you learned how to create parameters to drive the size and shape of a part. In this section you learn how to use parameters to create multiple parts based on configurations called iParts. In industry, configured parts are referred to as tabulated parts, charted parts, configurations, or a family of parts.

An iPart is generated from a standard Inventor part file (.ipt). When you activate the Create iPart command, the standard part file is converted into an iPart. You can add individual members or configurations to the iPart–it is then referred to as an iPart factory. Figure 7.21 shows an example of the dialog box that is used to author an iPart factory.

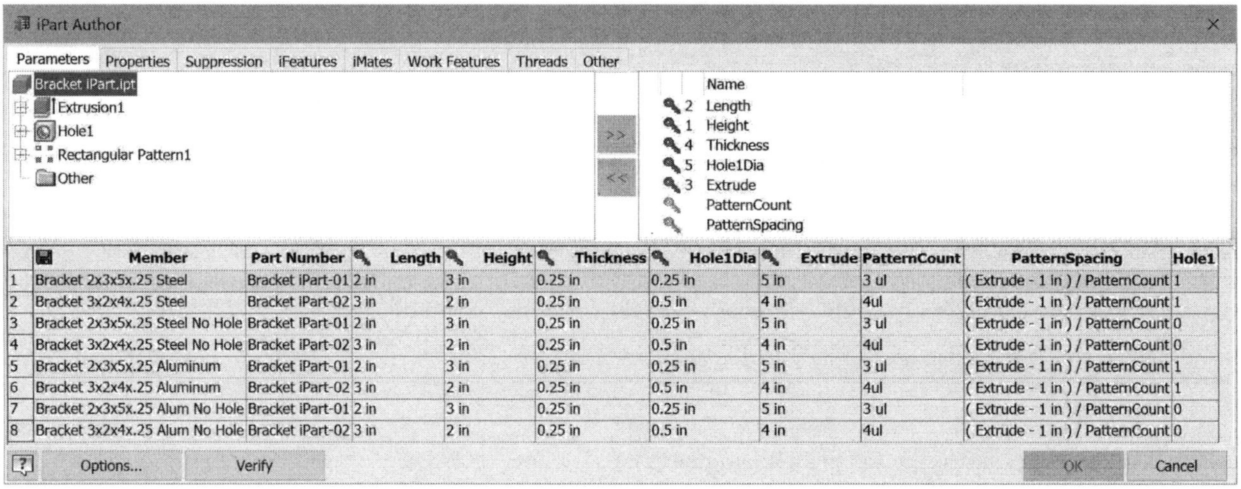

Figure 7.21

There are two steps to create an iPart: authoring the iPart, and placing an iPart into an assembly. In the authoring stage, you design the part and establish possible versions of the design by adding rows in a table with specific values, or ranges for the values. The rows in the table describe the members of the iPart factory.

In the placement stage, you insert a member from the iPart factory into an assembly.

Creating an iPart

You can create two types of iParts: standard and custom.

Standard iPart

A standard iPart is one that has all its sizes and values predefined, and you can only select a predefined member for placement into an assembly. After a standard iPart is placed into an assembly, you cannot add features to the placed iPart.

Custom iPart

A custom iPart has at least one value that the user defines when the iPart is placed into an assembly. After a custom iPart is placed in an assembly, you can add features to it.

To create an iPart, follow these steps:

1. Create or open a part, or a sheet metal part file.
2. Create or ensure that parameters used to define the iPart have descriptive names. User Parameter names are automatically added to the table when creating an iPart factory. If you do not use parameter names, you will need to determine what each parameter (d#) represents.

   TIP: Parameters with names other than d# are automatically added to the parameter table during the creation of the iPart factory.

3. Click the Create iPart command on the Manage tab > Author panel, as shown in Figure 7.22.

Figure 7.22

4. The iPart Author dialog box appears. Figure 7.23 shows a part with named User Parameters. Note: User or model parameters that are named something other than d# are automatically added to the Name section on the right side of the iPart Author dialog box.

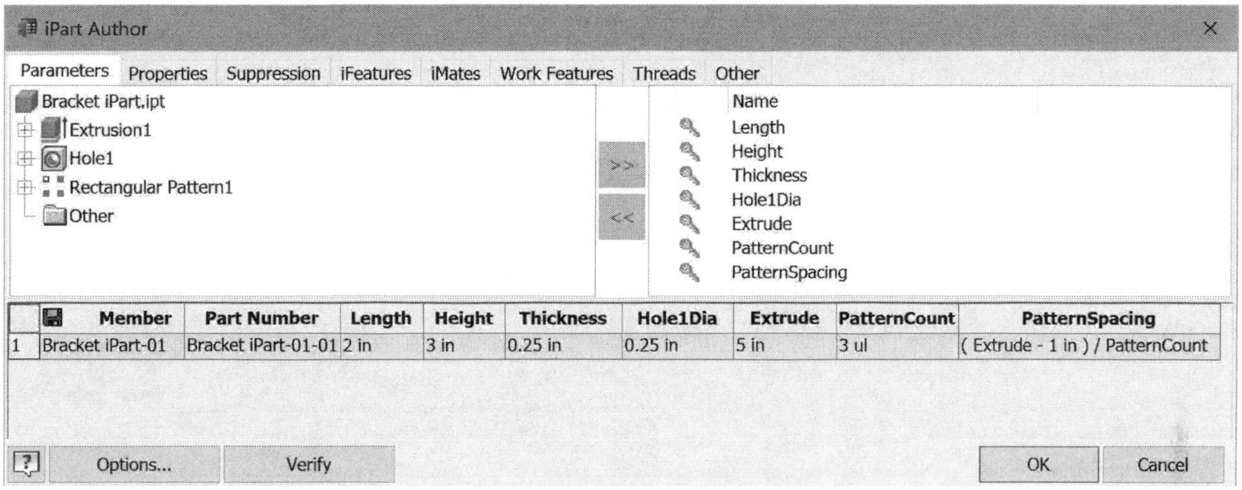

Figure 7.23

5. Add additional parameters to the Name section on the right side of the dialog box.

   a. To add a d# dimension, double-click on it in the left column, or click on the parameter then click the Add parameters (>>) button. If a feature is selected, all its parameters will be added.

   b. To remove a parameter from the Named section in the dialog box, select on it in the Name section, and click the Remove parameters (<<) button.

6. After you add parameters to the Named list, you define the Keys. Keys identify the column values used to define the iPart member when it is placed in an assembly. For example, setting a parameter as a primary key allows the user placing the iPart to choose from the allowed values for that parameter.

7. To specify the key order, from the Name section, click the key symbol of the item in the sequence that the parameter will be specified. You can also right-click on the item and select the key sequence number from the menu as shown in Figure 7.24 left. Selected keys are blue; items that are not defined as keys have dimmed key symbol. You don't have to define keys, but it is recommended, as they help the user place the correct iPart. Keys cannot be specified for custom table columns. Figure 7.24 right shows an example of the keys that were defined for all parameters.

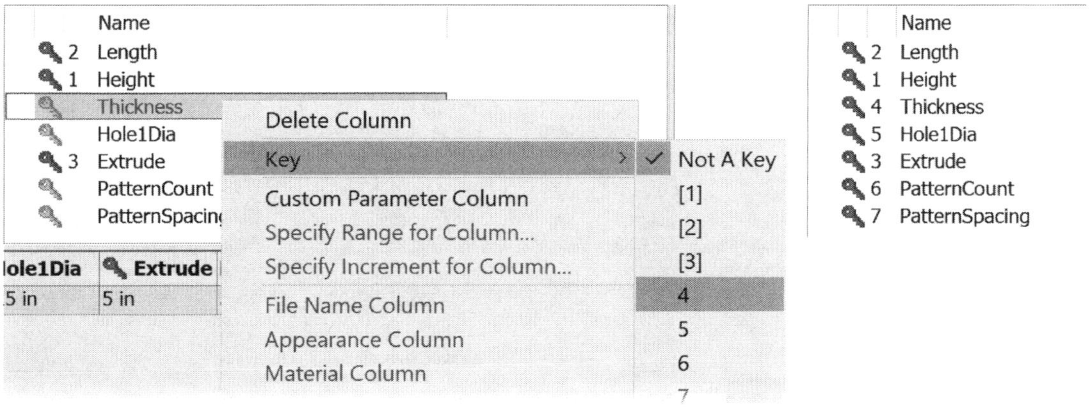

Figure 7.24

8. Add a row to the table by right-clicking on an existing row and click Insert Row as shown in the following image.

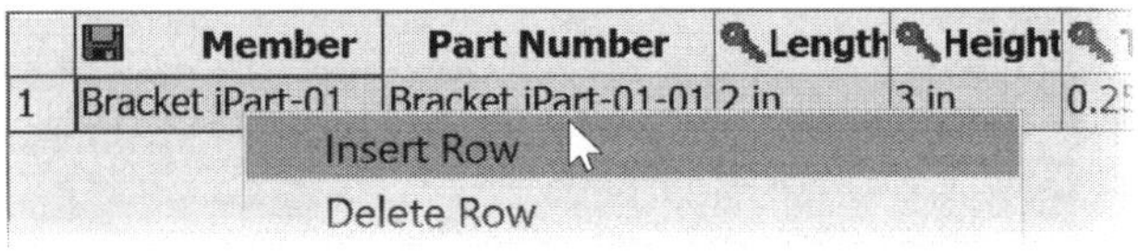

Figure 7.25

9. To allow the user placing the iPart to specify a value for a given column, first delete the Key that may be associated to the parameter by clicking on the blue key symbol in the Name section in the dialog box, then right-click on the column's name, and click Custom Parameter Column from the menu as shown in the following image on the left.

10. To make a specific cell custom, right-click in the cell, and click Custom Parameter Cell from the menu, Specify Range for Column, or Specify Increment for Column as shown in Figure 7.26 right.

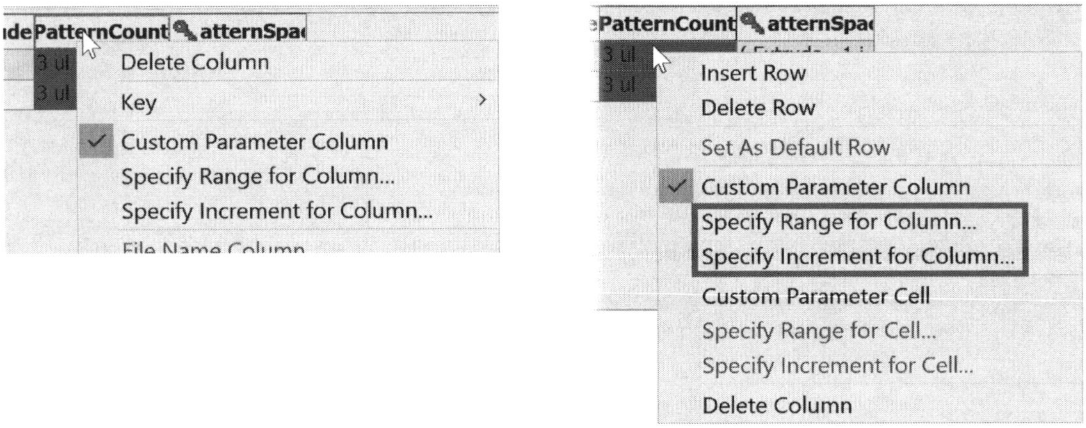

Figure 7.26

11. Edit a cell's content by clicking in the cell and typing new values needed. To delete a row, right-click the row number, and select Delete Row from the menu.

12. Click on the other tabs in the dialog box to add specific operations to the iPart. These items are not required to define the iPart factory.

13. To ensure all the settings can be generated, click the Verify button in the dialog box.

14. When finished defining the contents of the table Click OK in the iPart Author dialog box. The iPart Author dialog box closes, and the part is converted to an iPart factory. The table is saved, and a table icon appears in the browser, as shown in Figure 7.27. The member that is current has the check mark in front of it.

15. Confirm that the members work as desired by double-clicking on each member's name under the Table entry in the browser.

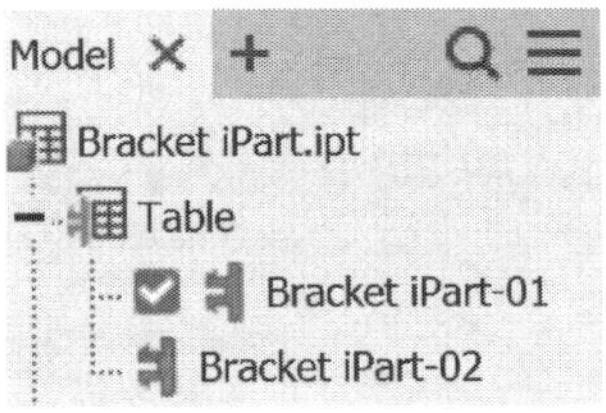

Figure 7.27

iPart Author Dialog Box – Tabs

Following are descriptions for the tabs in the iPart Author dialog box.

Properties Tab

The Properties tab lists all the summary, project, physical, and custom iProperties in the file. Use the material property on this tab to control the material of the iParts. For part properties to be used in drawings, Bill of Materials (BOMs), and other downstream purposes, they must include the properties in the table, even if their values do not vary between members.

Suppression Tab

The Suppression tab allows you to suppress features for specific members of the iPart factory. If you enter Suppress in a cell for a feature, the feature will be suppressed when the version is calculated. When the cell is entered as Compute, the feature is calculated. You can also use the value of 0 to Suppress, and 1 to Compute.

iFeatures Tab

On the iFeatures tab, select one or more table driven iFeatures included in the iPart. Control the iFeature member for each member of the iPart. They can also be suppressed as described above.

iMates Tab

On the iMates tab, select one or more iMates to include in the iPart. iMates are included in the iPart if their status is set to Compute. If their status is suppressed, they are not included. Control the iMate properties for each member using the iMates, Parameters, and Suppression tabs. On the iMates tab, you can perform the following actions:

- Define different offset values for different iPart members
- Suppress iMates or Composite iMates for different iPart members
- Change the matching name for different assembly configurations
- Change the sequence of the iMate

Work Features Tab

On the Work Features tab, select user-defined work features to include in the iPart. Origin work features are included automatically in an iPart factory, and iPart members, but they are not listed. The work features are managed in the iPart if its table status is Include. If the status is Exclude, the work features are not managed in the iPart.

Threads Tab

On the Threads tab, you can control thread features in the iPart factory to create table driven items for regular or tapered thread features. Use Type to identify the thread standard, Designation to control the size and pitch, and Class to define fit based upon the standard. You can also assign the thread Direction as right or left, and modify the pipe diameter in the table.

Other Tab

Use the Other tab to create custom table items. For example, you can add a column that represents the name of the iPart member. You can create a column named Version to identify the appropriate member when it's placed in an assembly. You can also create custom values such as Color if you want to control the display style of the iPart.

Editing iParts

After creating an iPart factory, you can edit the table by either double-clicking on the Table icon in the browser, or right-clicking on the Table icon in the browser and choosing Edit Table from the menu. The iPart Author dialog box appears; edit the iPart as needed.

You can also edit the table with Microsoft Excel by right-clicking on the Table icon in the browser and choosing Edit via Spreadsheet from the menu, as shown in Figure 7.28. Make changes as needed; you can incorporate spreadsheet formulas, and conditional statements. When done, save, and then close the file.

To delete the table and convert the iPart factory back to a normal part, right-click on the Table icon in the browser, and click Delete from the menu.

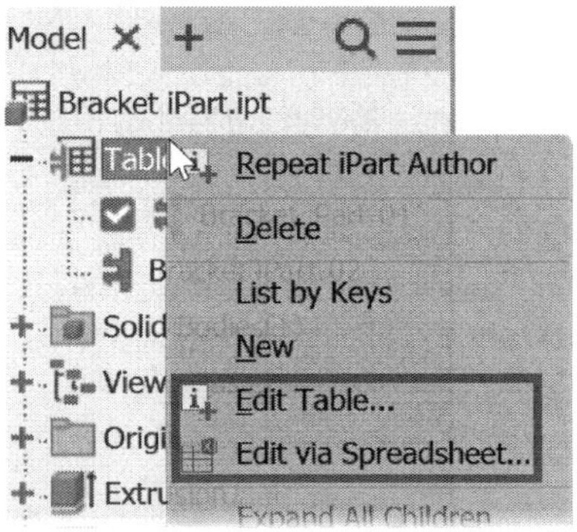

Figure 7.28

TIP: Changes made to iPart factories do not update automatically in members that have been placed in assemblies. To update the iPart members in an assembly, open the assembly, and choose the Update tool from the Quick Access toolbar.

Placing an iPart Member into an Assembly

To place an iPart into an assembly, follow these steps:

1. Start a new assembly or open an existing assembly where the iPart will be placed.
2. Click Place Component from the Assemble tab > Component panel. Use the Place Component dialog box to navigate to the iPart's name, double-click to select the iPart to place in the assembly (or select it and click the Open button.)
3. The Place Standard iPart dialog box appears as shown in Figure 7.29 left. If a custom cell(s) or column(s) exists, the Place Custom iPart dialog box appears as shown in Figure 7.29 right.

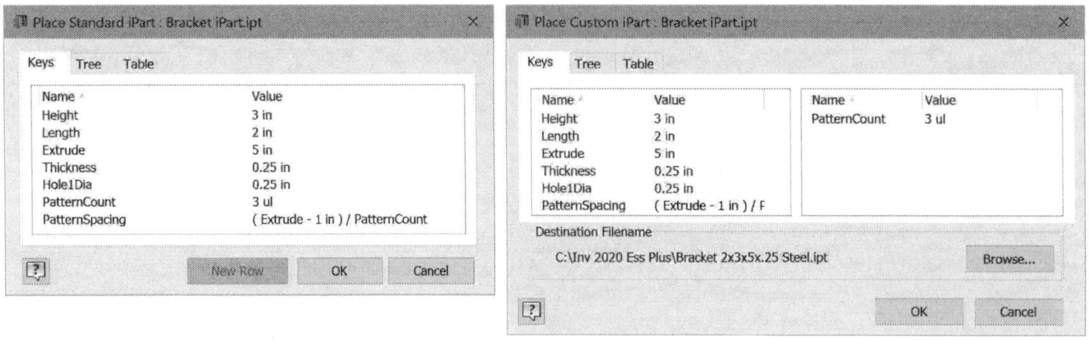

Figure 7.29

a. If you are placing a standard iPart, select from the member list, select the Keys, Tree, or Table tab, and then select the member that defines the part to place.

b. If you are placing a custom part, select the Keys, Tree, or Table tab, and enter a value in the right side of the dialog box, as shown in Figure 7.29 right. The value must fall within the limits set in the iPart factory; otherwise, an alert appears.

4. Place the iPart by clicking in the graphics window.
5. Continue placing instances of the iPart as needed, or select a different member from the Place iPart dialog box, and place it into the assembly.
6. When finished, right-click and click OK from the menu, or click Dismiss in the dialog box.
7. To change an iPart that has been placed in an assembly to a different configuration, expand the iPart in the browser, right-click on the table's name, and click Change Component. Then select a new member from the Keys, Tree, or Table tab.

When you place the first version of a standard iPart into an assembly, a folder is created with the same name as the iPart factory. By default, this folder is created in the same folder as the iPart factory file. As you place additional standard iParts into your assemblies, this folder is checked for existing iPart files prior to creating new iPart files.

TIP: Since each custom iPart is unique, you must provide different file names as you place different custom members.

Creating a Drawing for an iPart Factory

When creating a base drawing view of an iPart factory, you can select the member to document by clicking on the Model State tab of the Drawing View dialog, and select the standard member from the list, as shown in the following image. If a column in the iPart factory is set to Custom Parameter Column, the list will be blank.

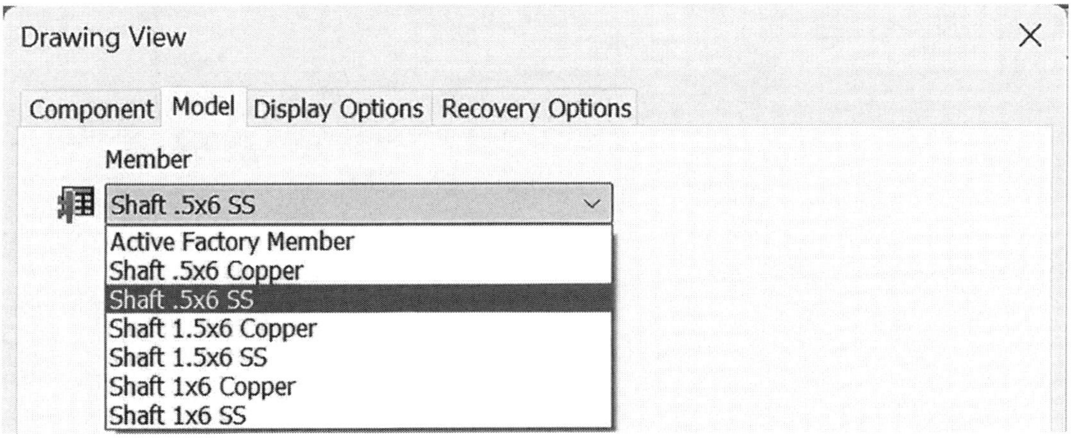

Figure 7.30

When documenting an iPart in a drawing, you can create a table that shows the available configurations. To create a table in a drawing for an iPart factory, follow these steps:

1. Ensure at least a base view for an iPart factory has been created.
2. Click the General Table command on the Annotate tab > Table panel as shown in the following image on the left.
3. The Table dialog box will appear; select which columns will appear in the table by clicking on the Column Chooser button as shown in Figure 7.31 right.

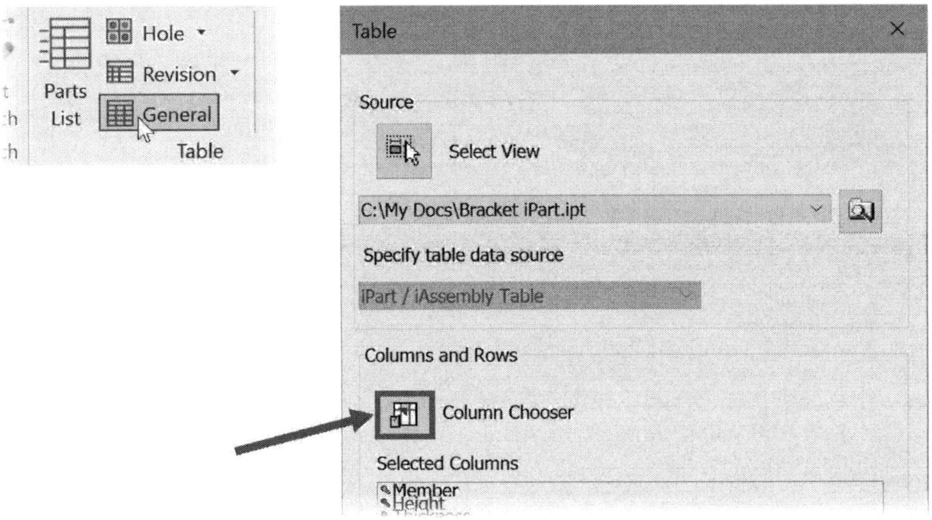

Figure 7.31

4. The Table Column Chooser dialog box appears as shown in Figure 7.32. Add a column from the Available Columns section by either double-clicking on the desired column, or by clicking on the desired column and click the Add -> button. Remove a column from the Selected Columns section by either double-clicking on the desired column, or by clicking on the desired column in the area and click the <- Remove button.

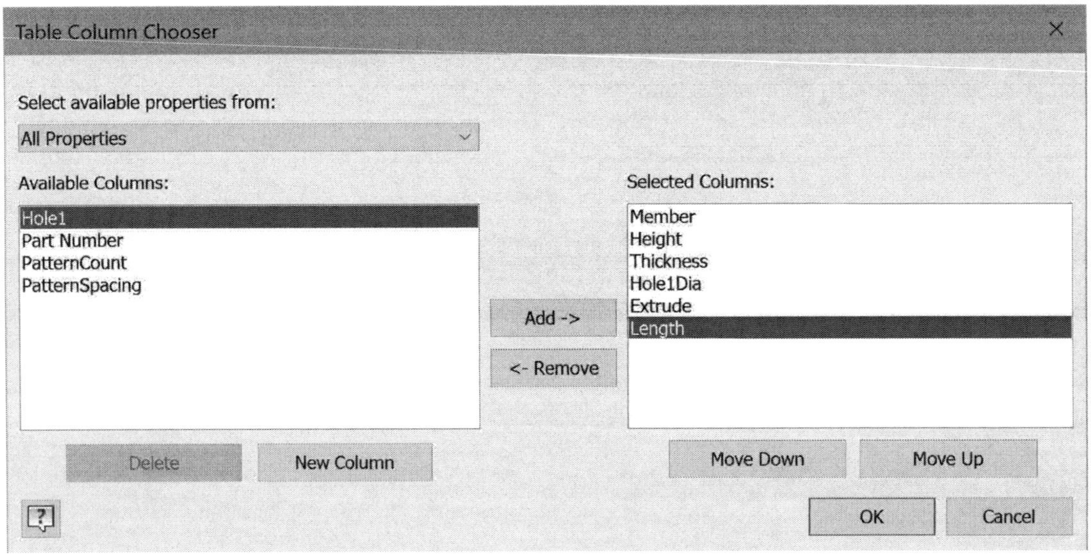

Figure 7.32

5. Click OK to create the table. Figure 7.33 shows an example of a table created from an iPart factory.

TABLE					
Member	Height	Thickness	Hole1Dia	Extrude	Length
Bracket 2x3x5x.25 Steel	3 in	0.25 in	0.25 in	5 in	2 in
Bracket 3x2x4x.25 Steel	2 in	0.25 in	0.5 in	4 in	3 in
Bracket 2x3x5x.25 Steel No Hole	3 in	0.25 in	0.25 in	5 in	2 in
Bracket 3x2x4x.25 Steel No Hole	2 in	0.25 in	0.5 in	4 in	3 in
Bracket 2x3x5x.25 Aluminum	3 in	0.25 in	0.25 in	5 in	2 in
Bracket 3x2x4x.25 Aluminum	2 in	0.25 in	0.5 in	4 in	3 in
Bracket 2x3x5x.25 Alum No Hole	3 in	0.25 in	0.25 in	5 in	2 in
Bracket 3x2x4x.25 Alum No Hole	2 in	0.25 in	0.5 in	4 in	3 in

Figure 7.33

EXERCISE 7-2: CREATING AND PLACING AN IPART

In this exercise, you convert a regular Inventor part to a Standard iPart factory, and then place a standard iPart member into an assembly.

1. Open *ESS_E07_02.ipt* from the Chapter 07 folder.
2. To prevent the original file being changed, create a copy of it by using the Save As command from the File tab, and type the name **Coupling iPart**.
3. Review the names of the five parameters using the Parameters command from the Quick Access toolbar. When finished reviewing the parameters, click Done at the bottom of the Parameters dialog box.
4. Begin an iPart factory by clicking on the Create iPart command from the Manage tab > Author panel.
5. The iPart Author dialog box appears. Note that five named parameters are automatically added to the Name section on the right side of the iPart Author dialog box.
6. Next add the Material property to the table. Click on the Properties tab > Physical folder > Material [Generic] and click the >> button as shown in Figure 7.34, or expand the Physical folder and double-click on Material [Generic].

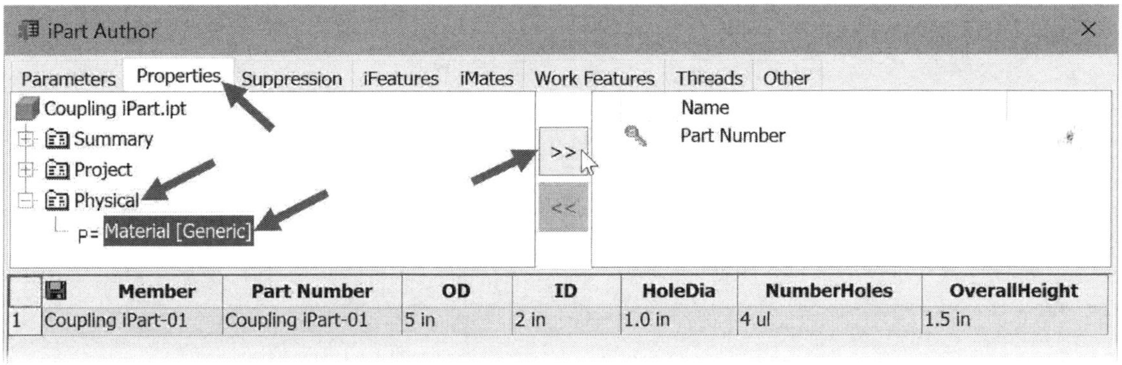

Figure 7.34

7. The Material column is added to the Name and the table area on the bottom of the dialog box. If you are not able to see the Material column in the table, expand the right side of the dialog box by clicking and dragging on the right-vertical edge of the dialog box.
8. Change the Material in the first row, click in the Material cell in the first row and enter Stainless Steel as shown in Figure 7.35.

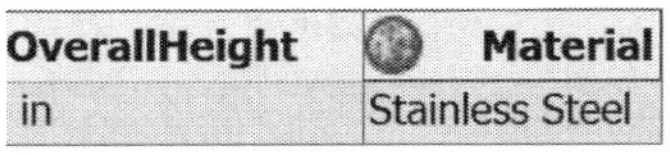

Figure 7.35

9 Next add an option to suppress the second hole. Click on the Suppression tab > Hole2 [Compute] and then click the >> button to add it to the list as shown in Figure 7.36, or double-click on Hole2 [Compute].

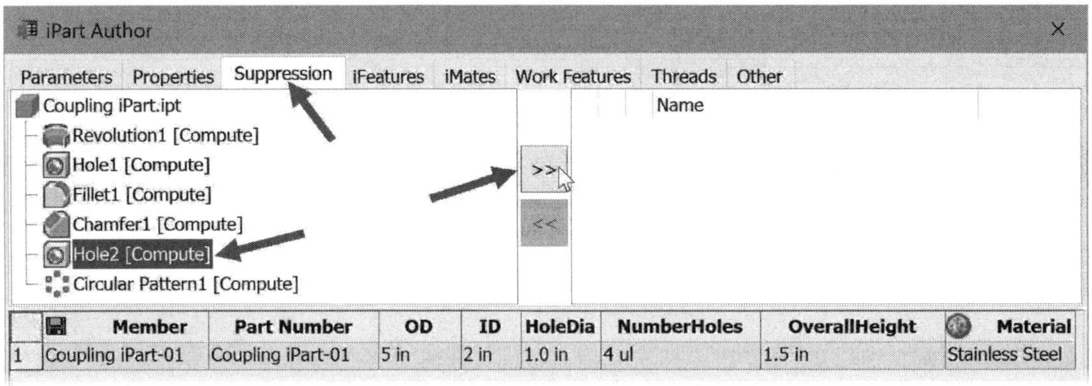

Figure 7.36

10 The Hole2 column is added to the Name and the table area. If you are not able to see the Hole2 column, Click and drag on the right-vertical edge to expand the dialog box.

11 Next add Keys to define the order in which the user will insert the members. Click the Parameters tab, then on the right side of the dialog box, click the Keys in order as shown in Figure 7.37.
Note the OverallHeight parameter is Key number 4, and the NumberHoles parameter is Key number 5.

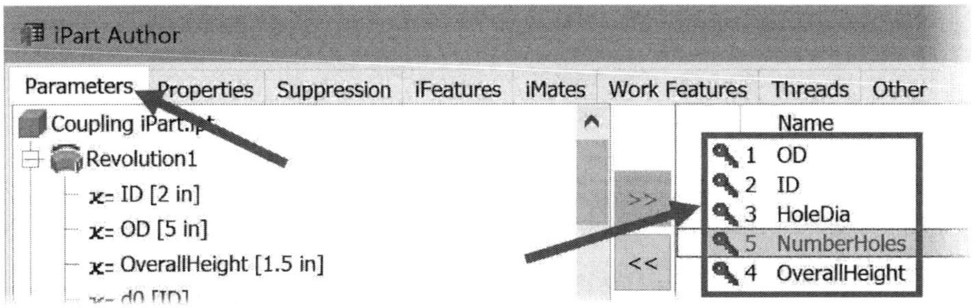

Figure 7.37

12 Next add a member to the table. Right-click on the only member row in the table, and choose Insert Row as shown in Figure 7.38.

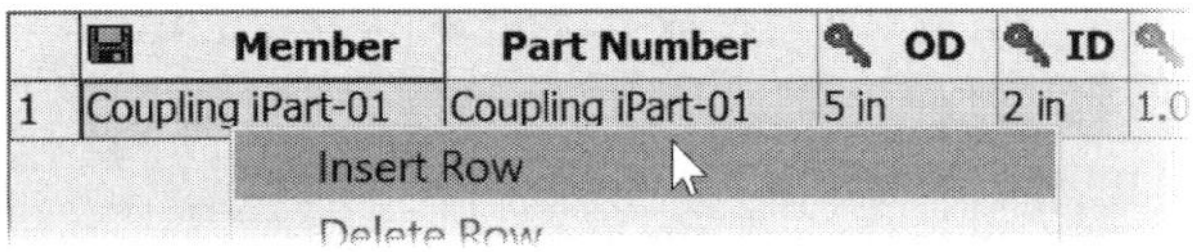

Figure 7.38

13 Rename the Member and Part Number cells in both rows, and change the parameter values for the second row in the table by clicking in each cell, and type the data as shown in Figure 7.39.

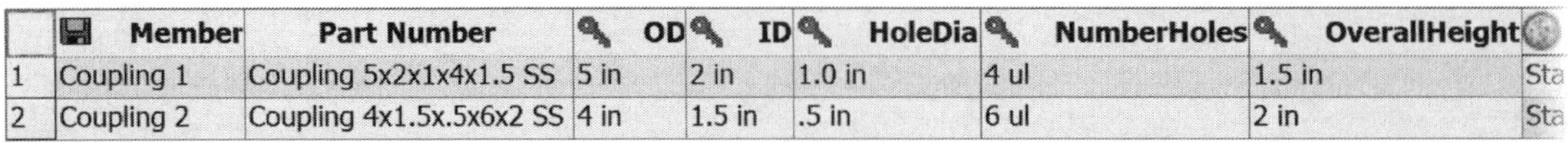

	Member	Part Number	OD	ID	HoleDia	NumberHoles	OverallHeight
1	Coupling 1	Coupling 5x2x1x4x1.5 SS	5 in	2 in	1.0 in	4 ul	1.5 in
2	Coupling 2	Coupling 4x1.5x.5x6x2 SS	4 in	1.5 in	.5 in	6 ul	2 in

Figure 7.39

14 Click OK to create the iPart factory.

15 A Table entry is created in the browser. Right-click on the Table entry in the browser and choose List by Member Name to expand the table as shown in Figure 7.40 left.

16 Review the iPart factory members and double-click on Coupling 2. Your screen should resemble Figure 7.40.

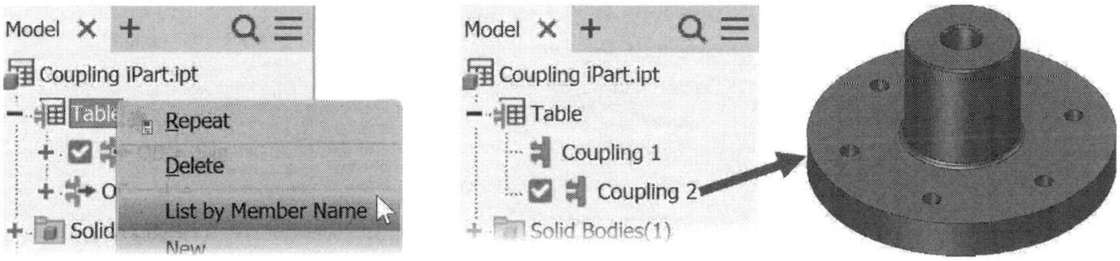

Figure 7.40

17 Next edit the iPart factory in Excel. In the browser, right-click on the Table entry and click Edit via Spreadsheet as shown in Figure 7.41. Click OK in the dialog box that states, The changes via Spreadsheet will take affect after the Excel process is closed.

Figure 7.41

18 Create additional members / rows in Excel; use the data that is shown in the following image. Where possible, utilize Excel's copy and paste functionality to create the new rows.

TIP: Instead of typing in Compute and Suppress, you can use the value 1 to Compute, and 0 to suppress.

Member<d	Part Number [Project]	OD<	ID<ke	HoleDia	NumberHoles	OverallHeight	Material	Hole2<ke
Coupling 1	Coupling 5x2x1x4x1.5 SS	5 in	2 in	1.0 in	4 ul	1.5 in	Stainless Steel	Compute
Coupling 2	Coupling 4x1.5x.5x6x2 SS	4 in	1.5 in	.5 in	6 ul	2 in	Stainless Steel	Compute
Coupling 3	Coupling 5x2x1x4x1.5 Copper	5 in	2 in	1.0 in	4 ul	1.5 in	Copper	Compute
Coupling 4	Coupling 4x1.5x.5x6x2 Copper	4 in	1.5 in	.5 in	6 ul	2 in	Copper	Compute
Coupling 5	Coupling 5x2x1x4x1.5 SS - No Hole	5 in	2 in	1.0 in	4 ul	1.5 in	Stainless Steel	Suppress
Coupling 6	Coupling 4x1.5x.5x6x2 SS - No Hole	4 in	1.5 in	.5 in	6 ul	2 in	Stainless Steel	Suppress
Coupling 7	Coupling 5x2x1x4x1.5 Copper - No Hole	5 in	2 in	1.0 in	4 ul	1.5 in	Copper	Suppress
Coupling 8	Coupling 4x1.5x.5x6x2 Copper - No Hole	4 in	1.5 in	.5 in	6 ul	2 in	Copper	Suppress

Figure 7.42

19 Save the Excel file, and then close Excel.

20 In *Coupling iPart.ipt* file, the Table entry will be updated in the browser, and the new members displayed below it. If the Table is listed by its Keys as shown in the following image on the left, you can display the members by their names by right-clicking on the Table entry in the browser, and click List by Member Name as shown in the following image on the right.

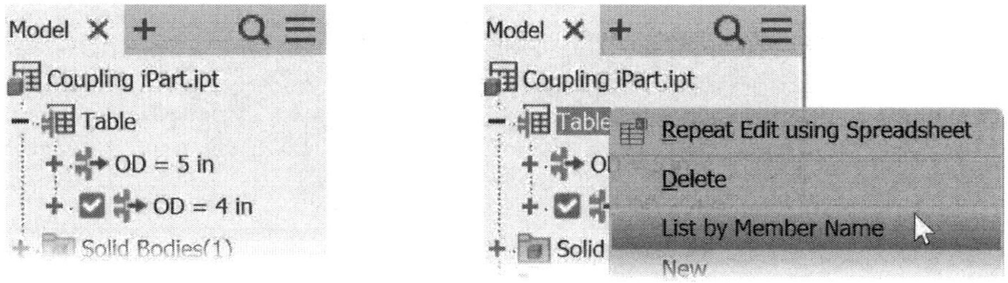

Figure 7.43

21 The following image shows Coupling 6 as the active member.

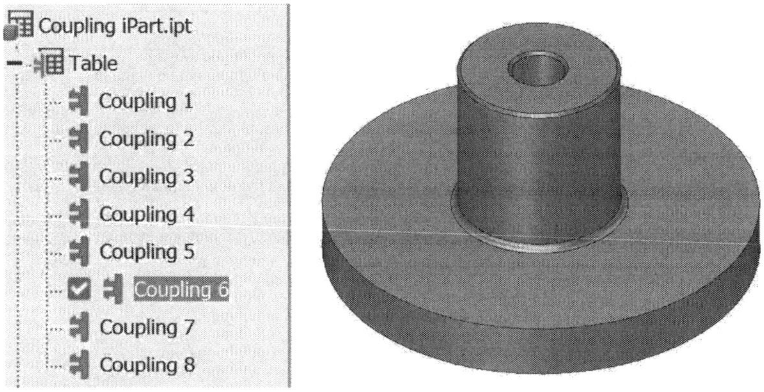

Figure 7.44

22 Before placing the iPart into an assembly, you define a Key for the Material and Hole2 columns. Edit the iPart by right-clicking on the Table entry in the browser and click Edit Table from the menu as shown in the following image.

Figure 7.45

a. Right-click on the Material column, and in the menu, click Key > 6 as shown in Figure 7.46.

b. Right-click on the Hole2 column, and in the menu, click Key > 7.

c. Then click OK to close the iPart Author dialog box.

23 Save the iPart factory file.

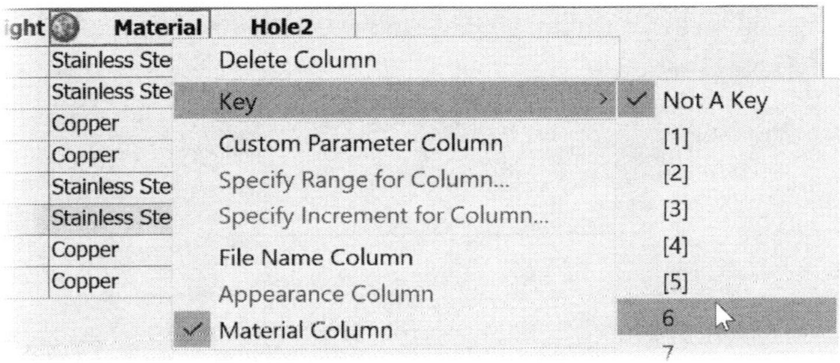

Figure 7.46

24 Create a new assembly file.

25 Click the Place Component command on the Assemble tab > Component panel, then double-click on the *Coupling iPart.ipt* file you just saved in the Chapter 07 folder.

26 The Place Standard iPart dialog box appears with the Key tab active. Practice placing an iPart member into the assembly by doing the following:

a. In the Key tab, click on the value for each parameter and select a value from the list. The following image shows the list for the parameter OD.

TIP: If all the values do not appear in the list, check the All Values option in the list as shown in the following image.

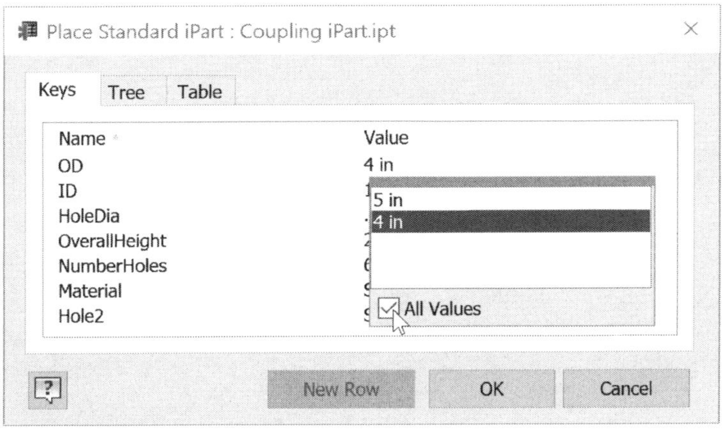

Figure 7.47

b. Make the Tree tab active, expand the parameters and then select the desired member. Figure 7.48 shows the expanded list for the parameter OD with a value of 5 inches.

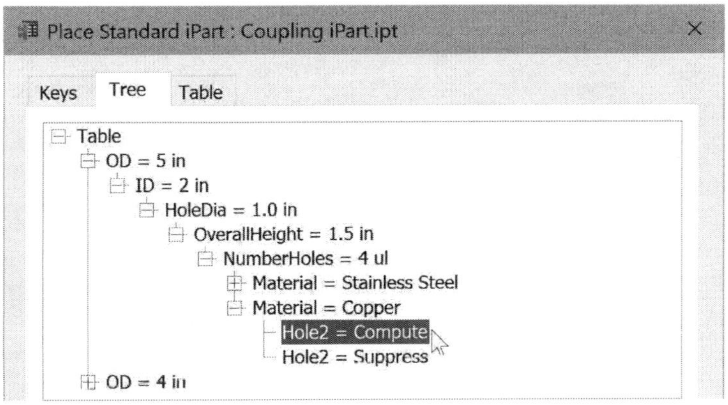

Figure 7.48

27 Make the Table tab active, and select the desired member from the list as shown in the following image. This is the same table that you created in the iPart factory.

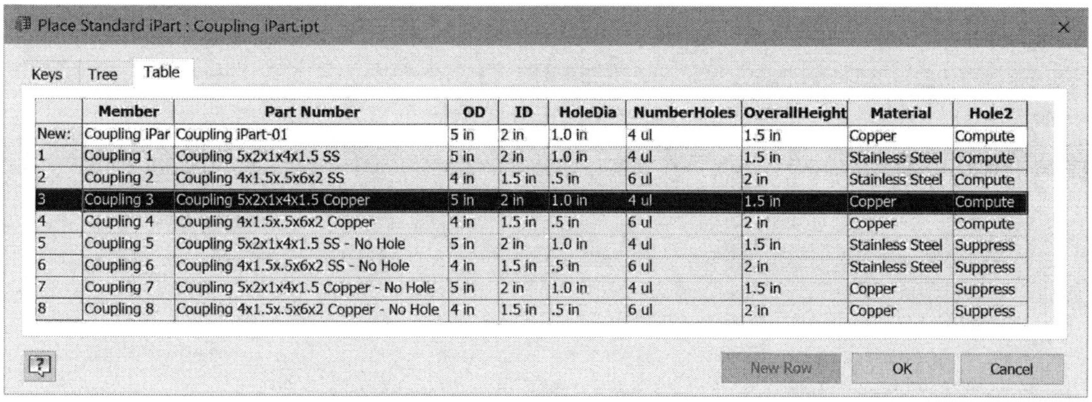

	Member	Part Number	OD	ID	HoleDia	NumberHoles	OverallHeight	Material	Hole2
New:	Coupling iPar	Coupling iPart-01	5 in	2 in	1.0 in	4 ul	1.5 in	Copper	Compute
1	Coupling 1	Coupling 5x2x1x4x1.5 SS	5 in	2 in	1.0 in	4 ul	1.5 in	Stainless Steel	Compute
2	Coupling 2	Coupling 4x1.5x.5x6x2 SS	4 in	1.5 in	.5 in	6 ul	2 in	Stainless Steel	Compute
3	Coupling 3	Coupling 5x2x1x4x1.5 Copper	5 in	2 in	1.0 in	4 ul	1.5 in	Copper	Compute
4	Coupling 4	Coupling 4x1.5x.5x6x2 Copper	4 in	1.5 in	.5 in	6 ul	2 in	Copper	Compute
5	Coupling 5	Coupling 5x2x1x4x1.5 SS - No Hole	5 in	2 in	1.0 in	4 ul	1.5 in	Stainless Steel	Suppress
6	Coupling 6	Coupling 4x1.5x.5x6x2 SS - No Hole	4 in	1.5 in	.5 in	6 ul	2 in	Stainless Steel	Suppress
7	Coupling 7	Coupling 5x2x1x4x1.5 Copper - No Hole	5 in	2 in	1.0 in	4 ul	1.5 in	Copper	Suppress
8	Coupling 8	Coupling 4x1.5x.5x6x2 Copper - No Hole	4 in	1.5 in	.5 in	6 ul	2 in	Copper	Suppress

Figure 7.49

28 After selecting the desired member from the dialog box, place the member by clicking in the graphics window.

29 Continue placing occurrences of the iPart as needed, or select a different member from the Place iPart dialog box, and place it into the assembly.

30 After placing an iPart member/component, you can change it to a different component by expanding the component in the browser and right-click on its table entry and click Change Component from the menu as shown in the following image. Practice changing the iPart member.

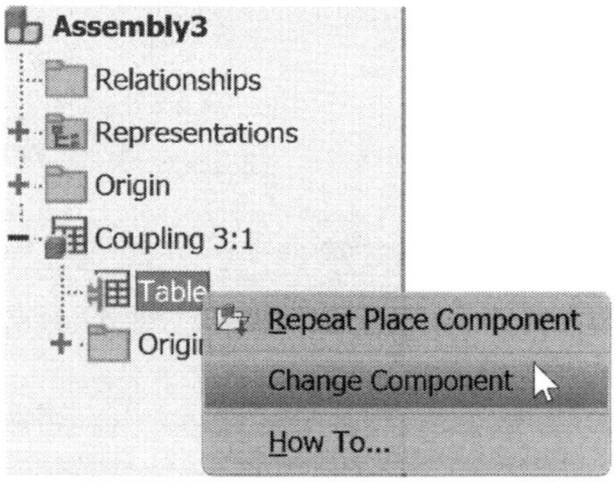

Figure 7.50

31 Experiment creating and editing the iPart factory - *Coupling iPart* and inserting the members into an assembly.

32 Save the file. Close the file.

End of exercise.

IASSEMBLIES

iAssemblies are used to group a set of similar parts/designs in a table format. These are also commonly referred to as configurations, and they function in a similar method to iParts. An iAssembly factory defines the family along with a table and a set of configurations of the same design. Each row of the table defines a unique product definition, which is called a configuration or member. When you use an iAssembly factory in another assembly, you can easily change from one configuration to another.

To Create an iAssembly Factory follow these steps:

1. Create iParts and normal parts that will be inserted into the iAssembly.
2. Create or open an assembly file.
3. Populate the assembly by inserting and constraining iParts and other components.
4. Click the Create iAssembly command on the Manage tab > Author panel, as shown in Figure 7.51.

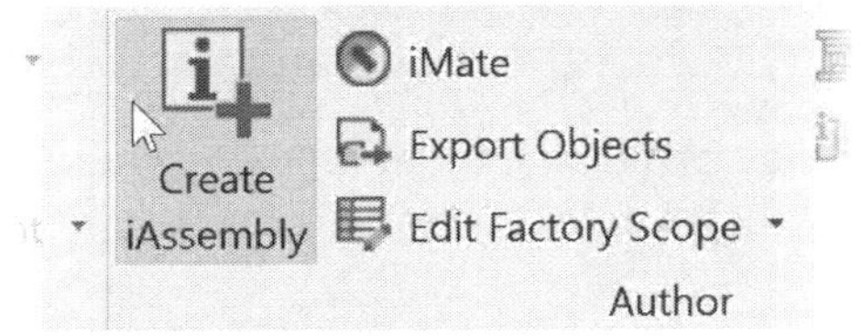

Figure 7.51

5. The iAssembly Author dialog box appears. Figure 7.52 shows an assembly that contains two iParts.

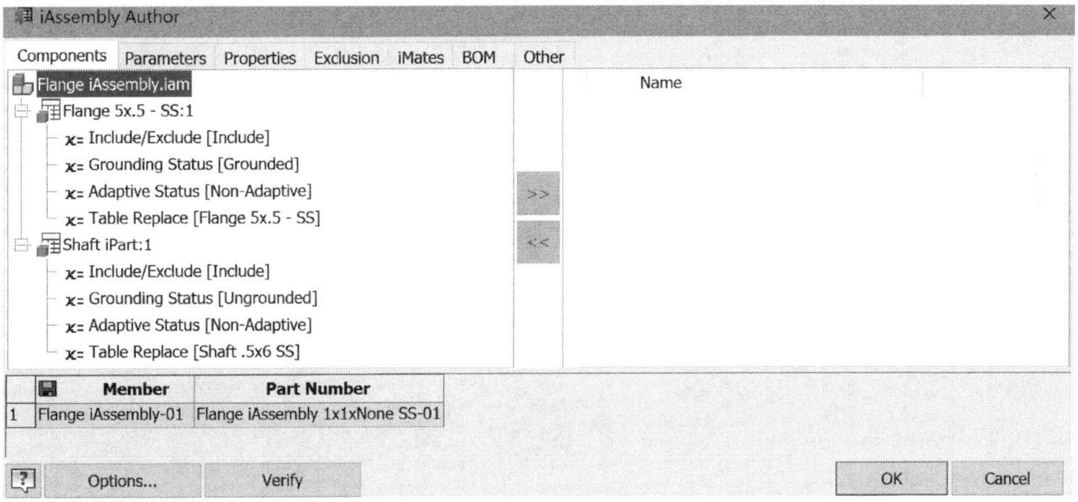

Figure 7.52

6. Determine which properties of the parts will be used to define the iAssembly members by double-clicking on the desired option on the left pane, or click on the desired option and then click the >> button in the dialog box. Descriptions for the options in the iAssembly Author dialog box are covered in the next section.
7. Add a row to the table by right-clicking on an existing row and click Insert row as shown in Figure 7.53.

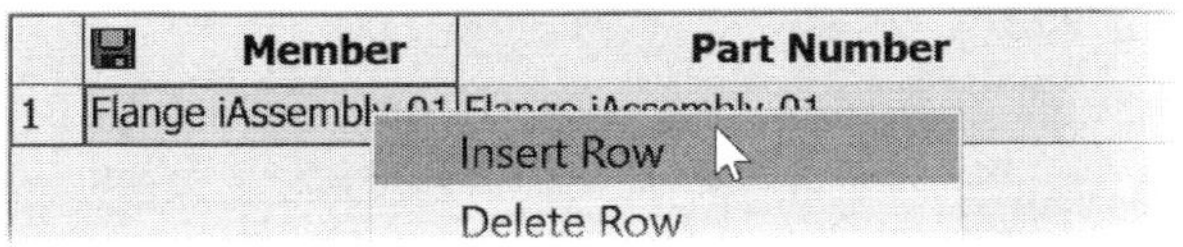

Figure 7.53

8. Change the configuration of each member by clicking in a cell and select an option from the menu. The following image shows an iPart member with the Table Replace option being changed.

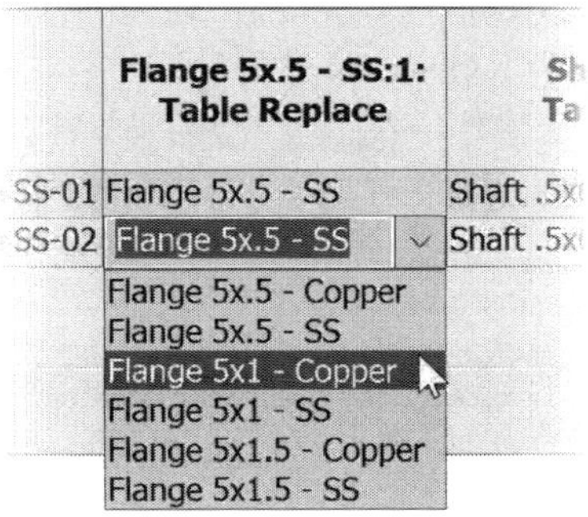

Figure 7.54

9. To ensure all the settings can be generated, click the Verify button in the dialog box.

10. When you have finished defining the contents of the table, click OK in the iAssembly Author dialog box. The iAssembly Author dialog box closes, and the assembly is converted to an iAssembly factory. The table is saved, and a table icon appears in the browser, as shown in the following image. The member that is current has the check mark in front of it.

11. Confirm that the members work as desired by double-clicking on each member's name under the Table entry in the browser.

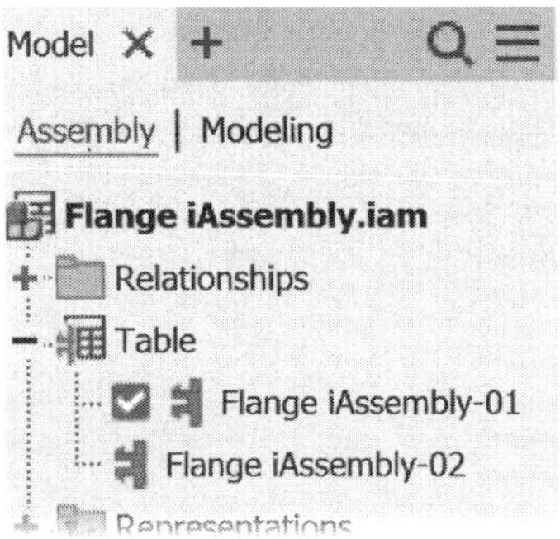

Figure 7.55

iAssembly Author Dialog Box Options

Following are descriptions for the tabs in the iAssembly Author dialog box.

Component Tab

The left pane lists all the components in the assembly. Click the + to the left of each component to see the following options.

- Included/Excluded: Select this option to include and exclude components from the iAssembly. The components can be iParts, iAssemblies, and normal parts and subassemblies.
- Grounded/Ungrounded: Select this option to be able to ground and unground the component.
- Non-adaptive/Adaptive/Flexible: Only one member can maintain adaptivity for a component. All other members containing the component show it as non-adaptive. However, each member can be flexible, if appropriate.
- Table Replace: Available only if the component is a member of an iPart or iAssembly factory. When editing the table, you can replace the member in the row by selecting a new member from the factory via a drop-down.

Parameters

Lists all parameters in the factory. Each parameter group lists parent objects, with their parameters nested below them. Parameter groups include:

- Relationships
- Assembly features
- Work features
- iMates
- Component patterns
- Other

Properties

Lists Summary, project, and physical properties. Properties can be used in drawings and bill of materials.

Exclusion

Specifies all configurable items in the factory that can be excluded, including components. Select Components, Relationships, Assembly features, Work Features, iMates, and Representations to exclude. A column is added in the table for each selected item.

iMates

Specifies individual iMates to include in a member. You can set four parameters: Offset Value, Include/Exclude, Matching Name, and Sequence Number.

BOM

Shows BOM-specific properties for child components of each instance of a component factory. BOM Structure is unique for each instance of the component. Each instance of a component shares the BOM Quantity.

Other

Creates custom column headings in the table, such as Description. May include text or numeric data; does not control size or placement of a member. For example, you can add a prompt to include instructions on how to place the component.

Editing iAssemblies

After creating an iAssembly factory, you can edit the table by either double-clicking on the Table icon in the browser, or right-clicking on the Table icon in the browser and clicking Edit Table from the menu. The iAssembly Author dialog box will appear; edit the iAssembly as needed.

You can also edit the table with Microsoft Excel by right-clicking on the Table icon in the browser and click Edit via Spreadsheet from the menu, as shown in the following image. Make changes as needed, save and then close the file. When editing the spreadsheet using Microsoft Excel, you can incorporate formulas and conditional statements.

To delete the table and convert the iAssembly factory back to a regular assembly, right-click on the Table icon in the browser, and click Delete from the menu.

Figure 7.56

Placing an iAssembly member into an Assembly

To place an iAssembly member into an Assembly, follow these steps:

1. Start a new assembly or open an existing assembly in which to place the iAssembly.
2. Click the Place Component command from the Assemble tab > Component panel, and in the Place Component dialog box navigate to and either double-click on the iAssembly's name, or select the iAssembly that will be placed in the assembly and click on the Open button.
3. Then the Place iAssembly dialog box appears. The following image shows an example with an iAssembly factory that has configurations defined.

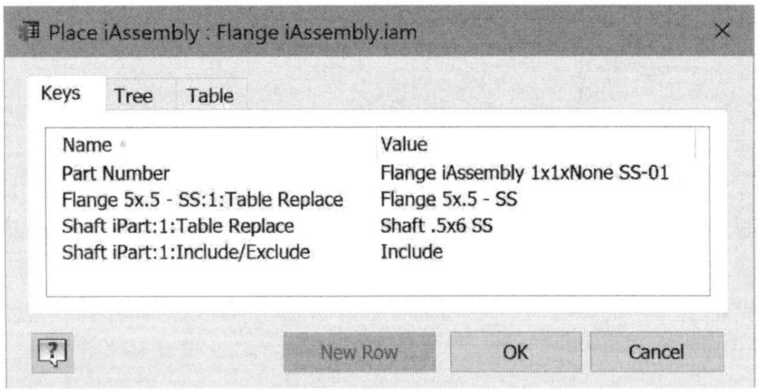

Figure 7.57

4. Select the desired configuration by selecting it from the Keys, Tree, or Table tab.
5. Place the iAssembly by clicking in the graphics window.
6. Continue to place multiple occurrences of the member, or select a different member from the Place iAssembly dialog box, and place it into the assembly.
7. When finished, right-click and click OK from the menu, or click Dismiss in the dialog box.

Creating a drawing for an iAssembly factory

When creating a drawing view of an iAssembly factory, you can select the member to document by clicking on the Model State tab of the Drawing View dialog, and select the member from the list, as shown in Figure 7.58.

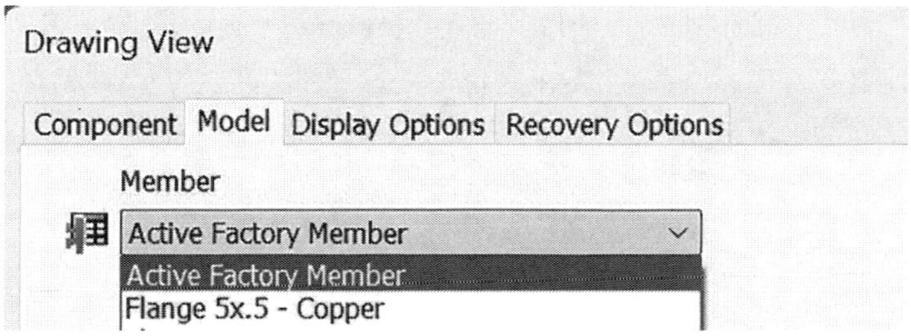

Figure 7.58

When documenting an iAssembly in a drawing, you can also create a table that shows the available configurations. To create a table in a drawing for an iAssembly factory, follow these steps:

1. Ensure that at least a base view for an iAssembly factory has been created.
2. Click the General Table command on the Annotate tab > Table panel as shown in Figure 7.59 left.
3. The Table dialog box will appear. To select which columns will appear in the table, click the Column Chooser button as shown in the following image on the right.

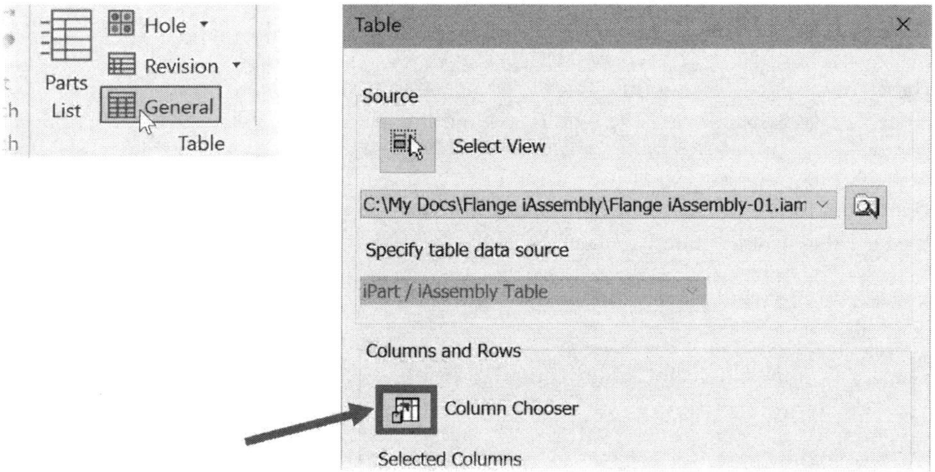

Figure 7.59

4. The Table Column Chooser dialog box appears as shown in Figure 7.60. Add a column from the Available Columns section by either double-clicking on it, or by clicking the desired column and choosing the Add -> button. Remove a column from the Selected Columns section by either double-clicking on it, or by clicking on the desired column in the area and choosing <- Remove.

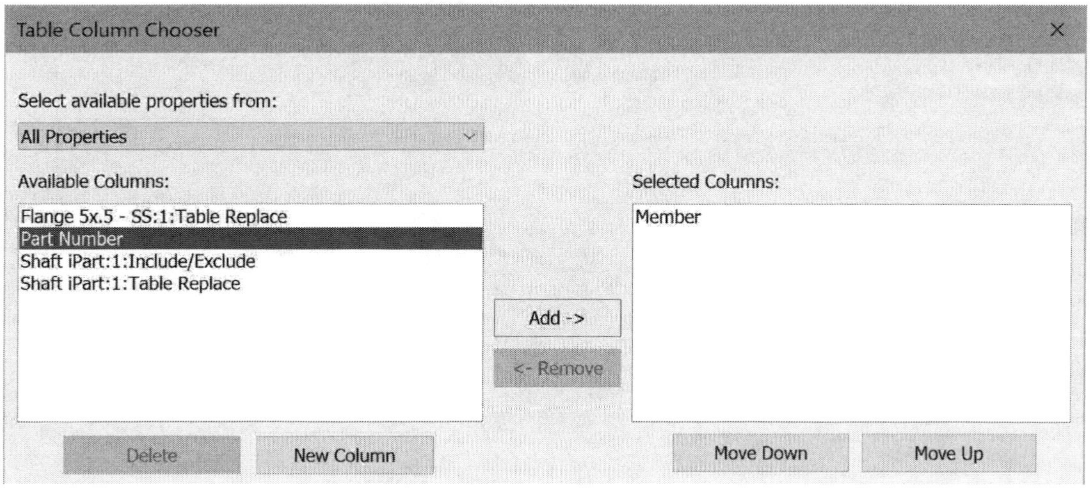

Figure 7.60

5. Click OK to create the table. Figure 7.61 shows an example of a table created from an iAssembly factory.

TABLE				
Member	Part Number	Flange 5x.5 - SS:1:Table Replace	Shaft iPart:1:Include/Exclude	Shaft iPart:1:Table Replace
Flange iAssembly-0 1	Flange iAssembly 1x1xNone SS-01	Flange 5x.5 - SS	Include	Shaft .5x6 SS
Flange iAssembly-0 2	Flange iAssembly 1x1xNone SS-02	Flange 5x.5 - SS	Include	Shaft .5x6 SS

Figure 7.61

EXERCISE 7-3: CREATING AND PLACING AN IASSEMBLY

In this exercise, you create an iAssembly factory, and then place an iAssembly member into an assembly.

1 Open *ESS_E07_03.iam* from the Chapter 07 folder.

2 So the original file does not get changed, create a copy of the file by clicking the Save As command from the File tab, and type the name **Flange iAssembly**.

3 Review the iParts by right-clicking on the two iParts in the browser and click Open from the menu. When done, close the iPart files without saving the changes. Note that *Flange iPart* has two occurrences in the assembly.

4 With the *Flange iAssembly.iam* file current, create an iAssembly by clicking the Create iAssembly command from the Manage tab > Author panel.

5 In the iAssembly Author dialog box, add the Table Replace option for the three iParts by doing the following:

a. On the Components tab, double-click the Table Replace option under *Flange 5x.5 - SS:1*. This adds the property to the right pane.

b. Repeat the process by expanding *Flange 5x.5 - SS:2,* and double-click its Table Replace option.

c. Expand *Shaft .5x6 - SS:1,* and double-click its Table Replace option.
The following image shows the Table Replace option added to the table for the three iParts.

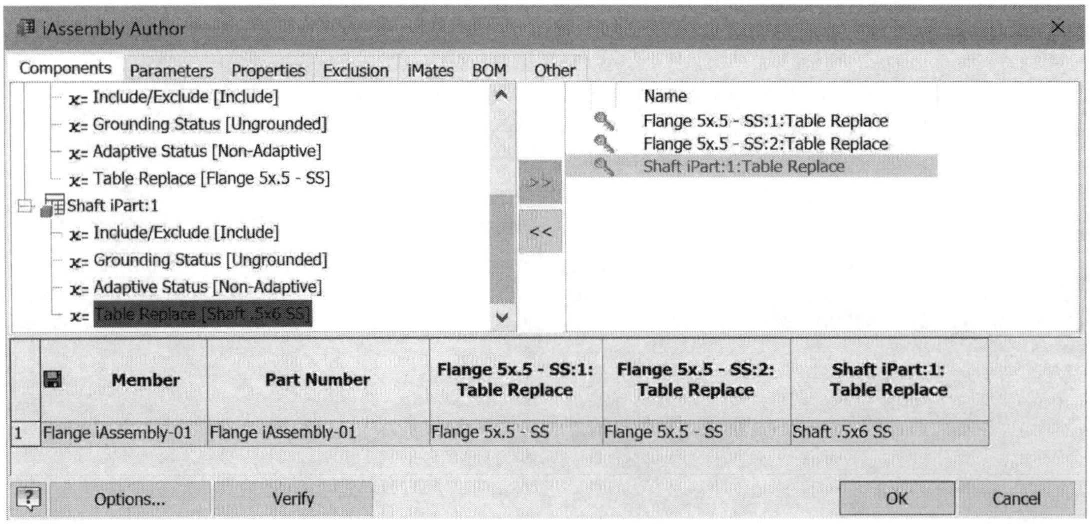

Figure 7.62

6 Next you add the option to include or exclude the second flange from the configuration. On the Components tab, under the options for *Flange 5x.5 - SS:2*, double-click on its Include/Exclude [Include] option as shown in the following image.

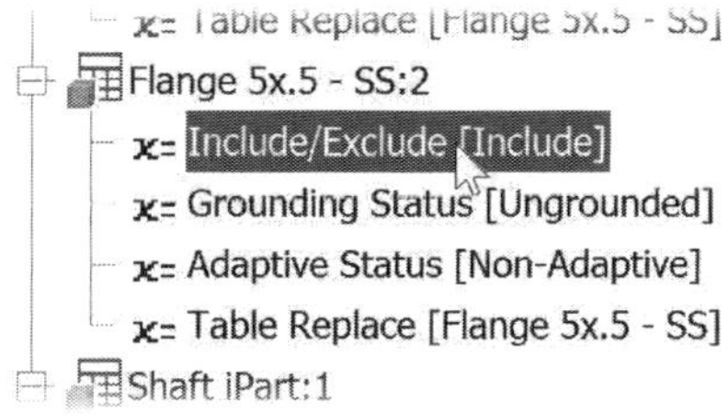

Figure 7.63

7 Next you add a member to the table. Right-click on the only member row in the table, and click Insert Row as shown in the following image.

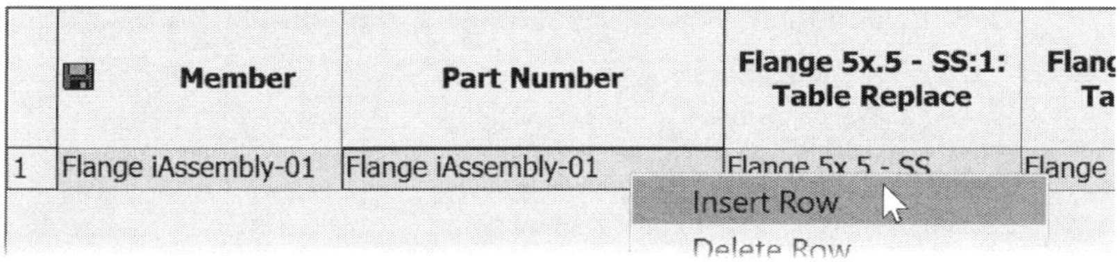

Figure 7.64

8 Rename the Part Number cells, and select the iPart member from the drop-down menu for the three iParts in both rows/members as shown in the following image.

If you are unable to see the full names of the iParts in the drop-down menu, expand the column by clicking and dragging on the right vertical line in the top row of that cell.

	Member	Part Number	Flange 5x.5 - SS:1: Table Replace	Flange 5x.5 - SS:2: Table Replace	Shaft iPart:1: Table Replace	Flange 5x.5 - SS:2: Include/Exclude
1	Flange iAssembly-01	Flange iAssembly .5x.5x.5 SS	Flange 5x.5 - SS	Flange 5x.5 - SS	Shaft .5x6 SS	Include
2	Flange iAssembly-02	Flange iAssembly 1x1x1 SS	Flange 5x1 - SS	Flange 5x1 - SS	Shaft 1x6 SS	Include

Figure 7.65

9 Insert two more rows, and change their options to match the following image.

	Member	Part Number	Flange 5x.5 - SS:1: Table Replace	Flange 5x.5 - SS:2: Table Replace	Shaft iPart:1: Table Replace	Flange 5x.5 - SS:2: Include/Exclude
1	Flange iAssembly-01	Flange iAssembly .5x.5x.5 SS	Flange 5x.5 - SS	Flange 5x.5 - SS	Shaft .5x6 SS	Include
2	Flange iAssembly-02	Flange iAssembly 1x1x1 SS	Flange 5x1 - SS	Flange 5x1 - SS	Shaft 1x6 SS	Include
3	Flange iAssembly-03	Flange iAssembly 1x1xNone SS	Flange 5x1 - SS	Flange 5x1 - SS	Shaft 1x6 SS	Exclude
4	Flange iAssembly-04	Flange iAssembly .5x.5x.5 Copper	Flange 5x.5 - Copper	Flange 5x.5 - Copper	Shaft 1.5x6 Copper	Include

Figure 7.66

10 Click OK to create the iAssembly factory.

A Table entry will be created in the browser, and the members are below it. Review the four iAssembly factory members by expanding the Table entry in the browser, and double-click on each member's name. Figure 7.68 shows Flange Assembly 3 active; notice that the second flange is excluded.

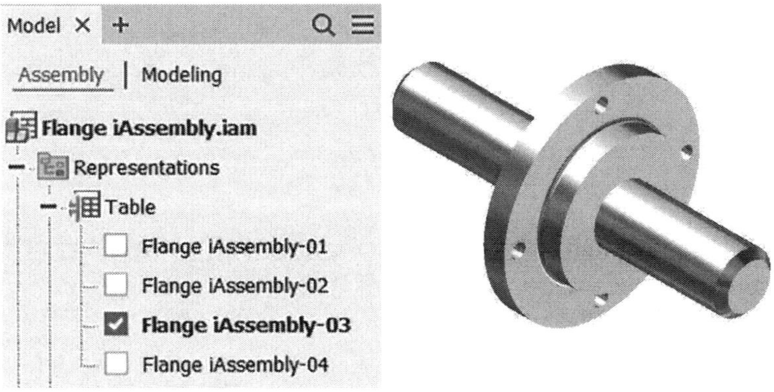

Figure 7.67

11 If desired add more rows to the table via Excel. In the browser, right-click on the Table entry and click Edit via Spreadsheet. When done, save the file, and then close Excel, and the table will update in the iAssembly factory.

12 Save the *Flange iAssembly* file, and click OK to save the member files.

13 Create a new assembly file.

14 Click Place Component command on the Assemble tab > Component panel, then double-click on the *Flange iAssembly.iam* file you just saved in the Chapter 07 folder.

15 The Place Standard iPart dialog box will appear, and the Key tab will be active. Practice placing an iAssembly members of your choice into the assembly by doing the following:

a. In the Key tab, click on the value for each option, and select a member from the list. Figure 7.68 shows the list for the Flange iPart.

TIP: If all the values are not displayed in the list, check the All Values option at the bottom of the menu.

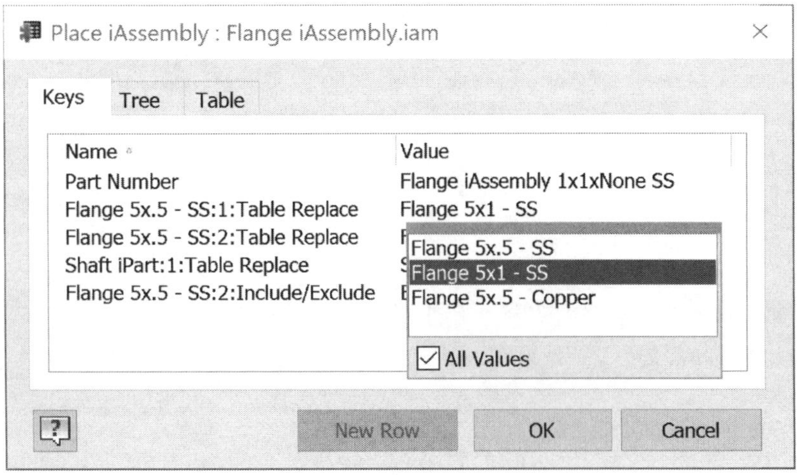

Figure 7.68

b. Make the Tree tab active, expand the options, and then select the desired member. The following image shows the expanded options for the first iAssembly member.

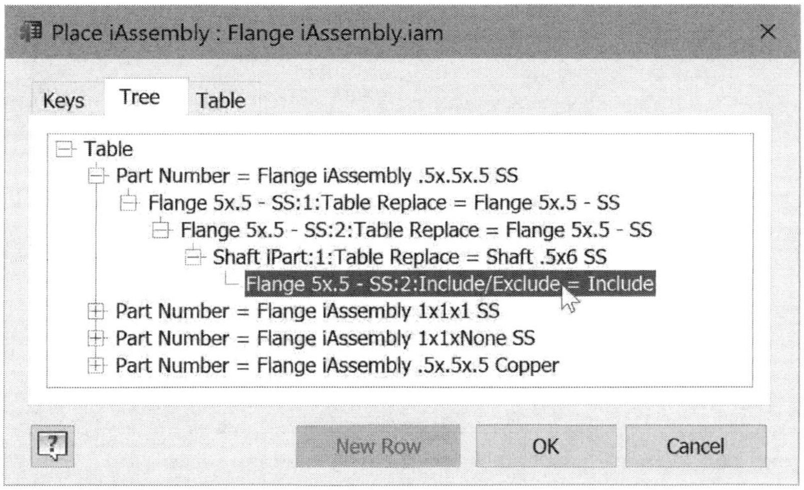

Figure 7.69

c. Make the Table tab active, and select the desired member from the list as shown in the following image. This is the same table that you created in the iAssembly factory.

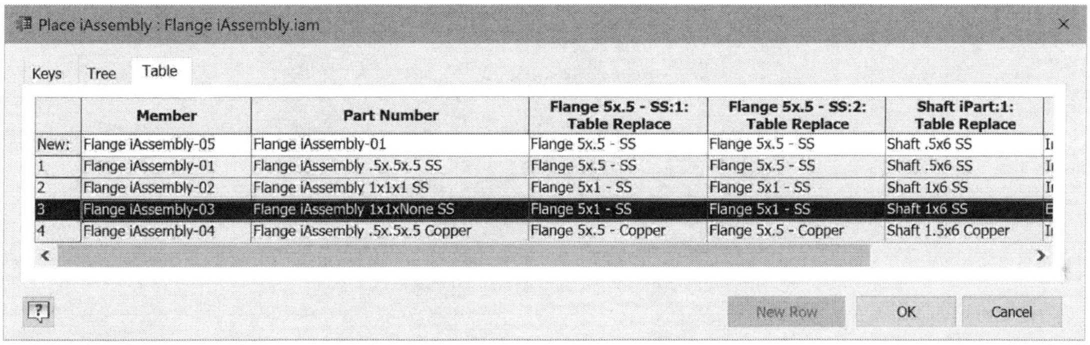

Figure 7.70

16 Place the member by clicking in the graphics window.

17 Continue placing occurrences of the iAssembly as needed, or select a different member from the Place iAssembly dialog box, and place it into the assembly.

18 After placing an iAssembly member/component, you can change it to a different component by expanding the component in the browser and right-click on its table entry and click Change Component from the menu as shown in Figure 7.71. Practice changing the iAssembly member.

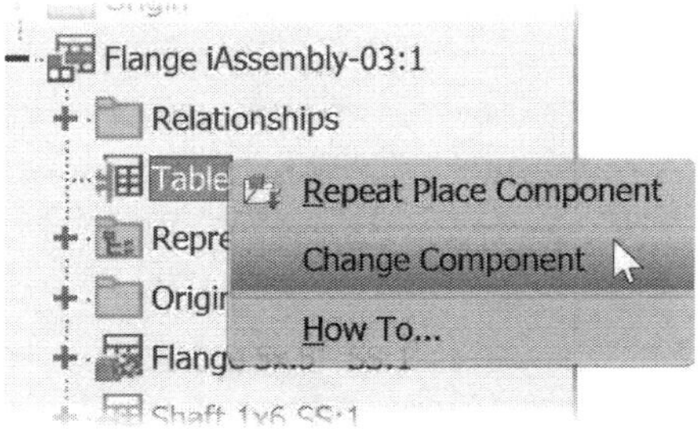

Figure 7.71

19 Continue to experiment creating and editing the iAssembly factory - *Flange iAssembly.iam* and inserting them into an assembly.

20 Save the file. Close the file.

End of exercise.

SECTIONING A PART OR COMPONENTS IN AN ASSEMBLY

While designing in a part or an assembly file you can visually remove portions of a part or components in an assembly. This is useful especially where features or components are hidden or obscured within chambers. While the part or assembly is sectioned, part and assembly tools can be used to create or modify parts within the assembly. You can use the exposed cut edges, along with the Project Cut Edges command, to assist in sketch construction for additional features that need to be created in a part.

TIP: A section in a part or assembly file is only visual and has no effect on the mass properties of the component and cannot be used to create drawing views from.

There are three types of sections you can create in a part or assembly file: quarter section, half section, or a three quarter section. To access the section commands, click the View tab > Visibility panel, as shown in Figure 7.72 left. The middle image shows a part before being sectioned and the right shows the part sectioned with the Three Quarter Section View command. Note that the sections are temporary and sketches cannot be actively placed on a sectioned plane. As always, you can create a sketch on an existing planar face or work plane.

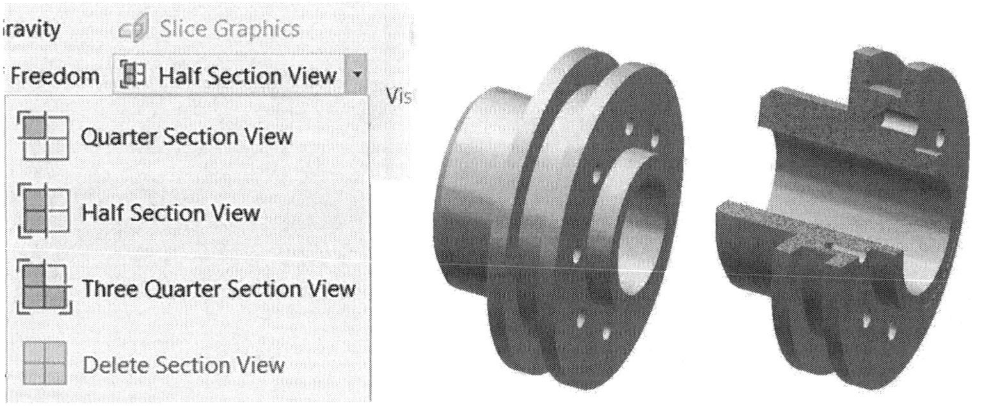

Figure 7.72

To section a part or components in an assembly, follow these steps:

1. Select the desired section command on the View tab > Visibility panel.
2. Select any existing planar face, work plane, or origin plane.
3. Adjust the section depth by clicking and dragging the visible arrow in the graphics window, as shown in the following image on the left, or enter a value in the Offset cell in the mini-toolbar. Another option to control the offset is to move the cursor over the Offset cell and scroll the wheel on the mouse. You can control the distance that each step takes when you scroll the wheel by right-clicking and click Scroll Step Size.
4. If you are creating a quarter or three quarter section, accept the current location by clicking the right-faced arrow in the mini-toolbar, as shown in the middle image, or right-click and click Continue in the marking menu and follow the same process to define a second plane.
5. While still in the command, you can specify which side of the section remains visible by right-clicking and click the Flip Section or change the section type if Quarter or Three Quarter section was selected, as shown in Figure 7.73 right.
6. To finish the command, click the green check mark in the mini-toolbar or right-click and choose OK.
7. To clear the section, click the Delete Section View command on the View tab > Appearance panel.

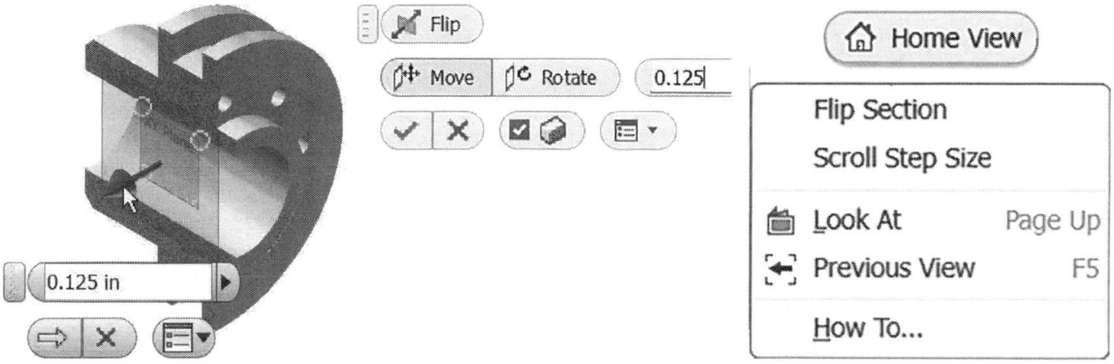

Figure 7.73

DESIGN VIEW REPRESENTATIONS IN A PART OR AN ASSEMBLY FILE

While working on a part or an assembly file, you can save a design view representation that stores information related to the current viewpoint, part color, sections, and work feature visibility. You can save as many design view representations as needed, but only one can be current. By default, there is a design view representation named Master; this master design view cannot be changed, deleted, or locked. To create a design view representation in a part file, follow these steps:

1. To create a Design View in a part file, expand the browser and right-click on View: Primary and click New from the menu, as shown in Figure 7.74 left.
2. To create a Design View in an assembly file, expand the Representation folder in the browser and right-click on View: Default, as shown in Figure 7.74 middle, and select New from the menu.
3. A design view will be created. You can rename the design view by slowly clicking twice on its name and type in a new name.
4. Once a design view is created, change the view point, section a part, part color, and work feature visibility. These changes will be captured in the current design view.
5. Since these changes are actively captured to the current design view, you can lock a design view so that any change to the current viewpoint, sectioning of a part, part color, or work feature visibility is not saved to the current design view. To lock a design view, right-click on its name in the browser and click Lock from the menu, as shown in Figure 7.74 right.
6. Create as many design views as needed.
7. Lock and unlock the current design view to capture the settings as needed.
8. To make a design view current, in the browser double-click on its name or right-click on its name and click Activate from the menu. The last settings in the design view or the locked settings in the design view will be displayed in the graphics window.

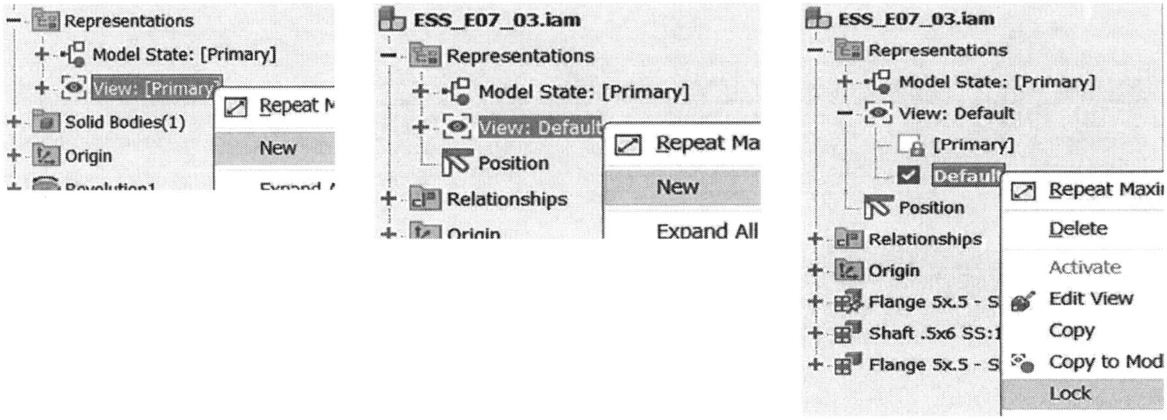

Figure 7.74

EMBOSS TEXT AND CLOSED PROFILES

You may need to have text or a profile embossed (raised) or engraved (cut) onto a part as shown in Figure 7.75. Next, you will learn how to emboss and engrave text or a closed profile onto a planar or curved face of a part. You can define a profile using the sketch commands on the Sketch tab but note that the profile must be closed. There are two steps to emboss or engrave: first create the text or profile and then you emboss the text or profile onto the part. The text can be oriented in a rectangle or about an arc, circle, or line.

Figure 7.75

Step 1: Creating Rectangular Text

To place rectangular text onto a sketch, follow these steps:

1. In a part file, create or activate a sketch.
2. Click the Text command on the Sketch tab > Create panel, as shown in Figure 7.76 left.
3. Pick a point or drag a rectangle where you will place the text. If a single point is selected, the text is fit on a single line. Grips can be used to resize the text bounding box. If a rectangle is used, it defines the width for text wrapping and can be modified.
4. The Format Text dialog box appears. Use the Format Text dialog box to specify the text font and format style, and enter the text to be placed in the sketch. Figure 7.76 right shows text entered in the bottom pane.

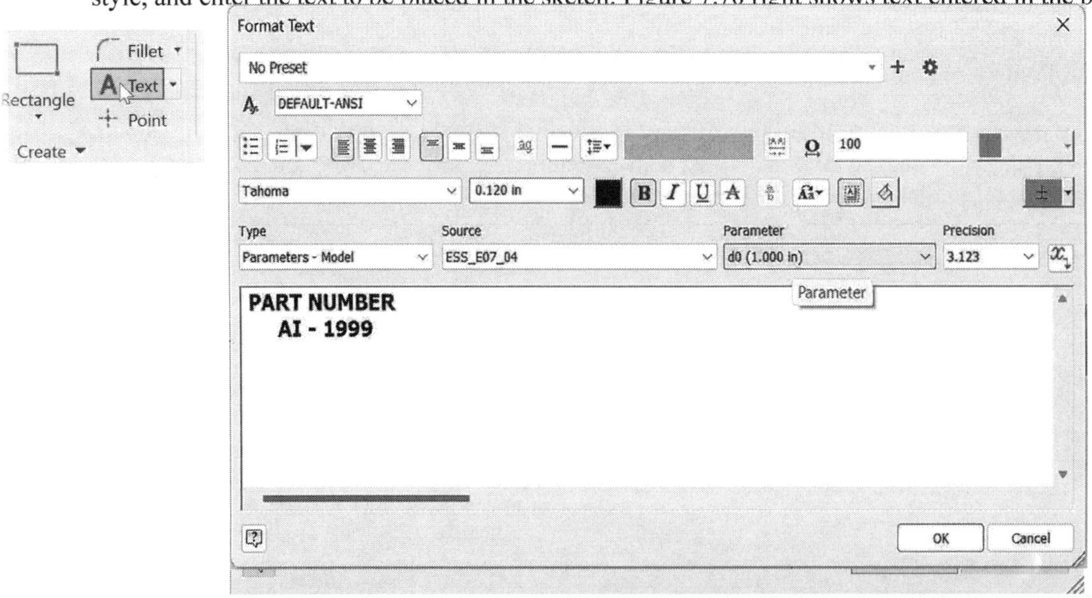

Figure 7.76

5. When you have finished typing the text, click OK to complete the text entry, then right-click and click OK in the marking menu to complete the command.
6. When the text is placed, a rectangular set of construction lines defines the perimeter of the text. You can click and drag a point on the rectangle and change the size of the rectangle. Dimensions or constraints can be added to these construction lines to refine the text's location and orientation, as shown in the following image.
7. To edit the text, double-click on the text or right-click on the text and click Edit text from the marking menu. The same Format Text dialog box that was used to create the text will appear.

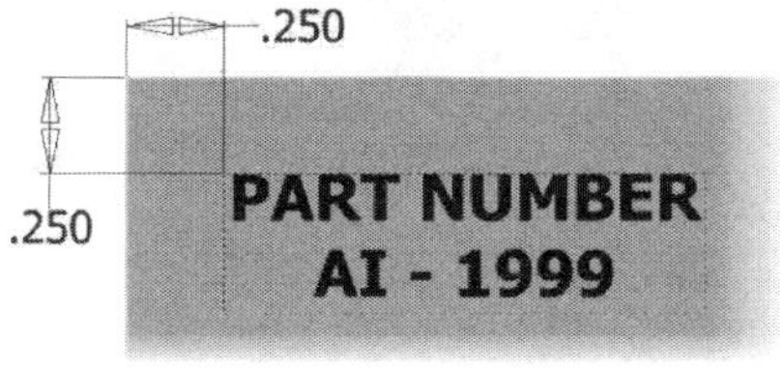

Figure 7.77

Step 2: Aligning Text to Geometry

To align text to an arc, circle, or line in a sketch, follow these steps:

1. In a part file, create a new sketch or make an existing sketch active.
2. Sketch and constrain an arc, circle, or line that the text will follow; it is recommended to use the construction geometry style for this so that it will not be used as the geometry for the profile.
3. Click the Geometry Text command on the Sketch tab > Create panel > Geometry Text command, as shown in Figure 7.78 left. The command may be under the Text command.
4. In the sketch, select an arc, circle, or line and the Geometry-Text dialog box will appear.
5. In the Geometry-Text dialog box, specify the text font and format style, direction, position, and start angle and enter the text to place in the sketch. The start angle is relative to the left quadrant point of a circle or the start point of the arc. Figure 7.78 right shows the dialog box with text entered in the bottom pane.

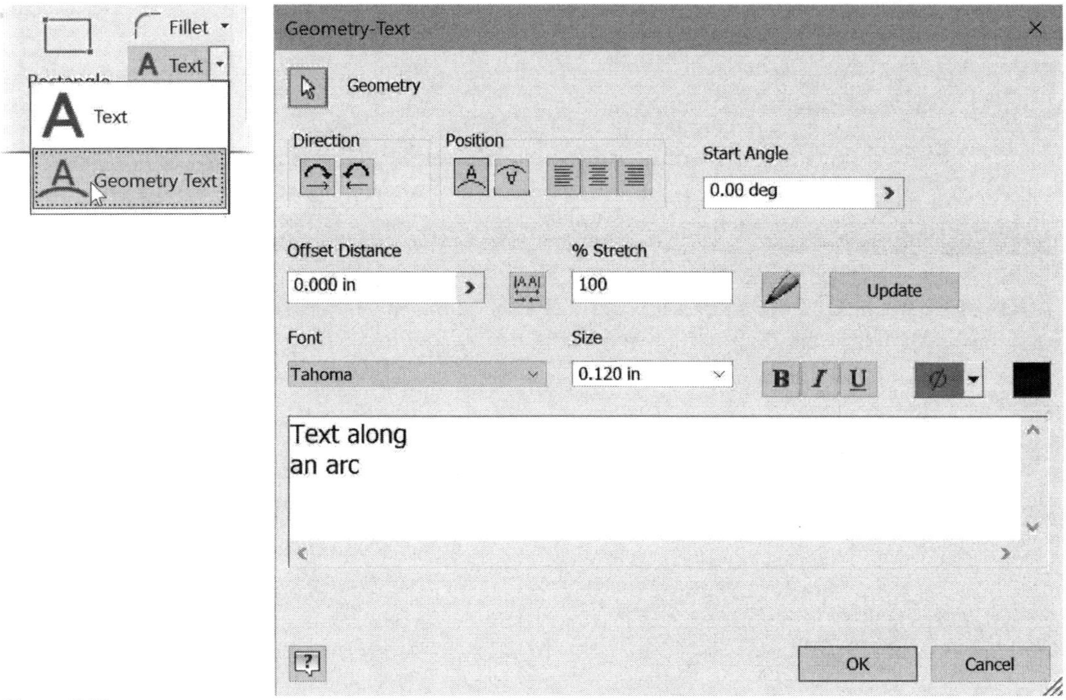

Figure 7.78

6. When you have finished entering text, click the Update button in the dialog box to see the preview in the graphics window.
7. Click OK to complete the operation. The following image shows text placed along an arc.
8. To edit the text, double-click on the text or right-click on the text and click Edit Geometry Text from the marking menu. The same Format Text dialog box that was used to create the text will appear.

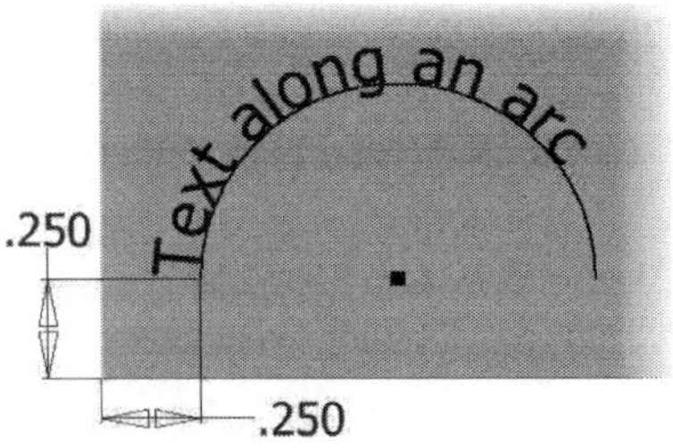

Figure 7.79

Step 3: Embossing Text or Closed Profile

To emboss text or a closed profile, click the Emboss command on the 3D Model tab > Create panel, as shown in the following image on the left. The Emboss dialog box will appear, as shown in Figure 7.80 right.

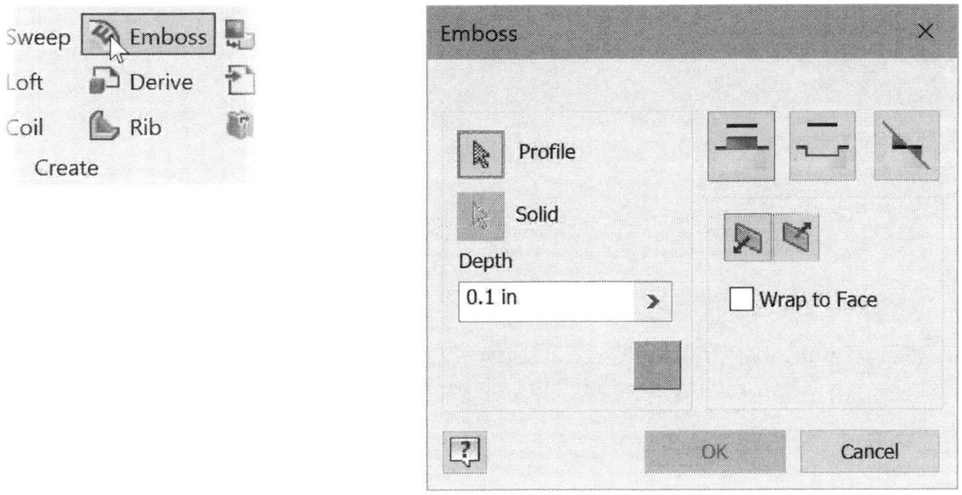

Figure 7.80

To emboss text or a closed shape, follow these steps:

1. Click the Emboss command on the 3D Model tab > Create panel (it may be under the sweep or another command in this group).
2. Define the profile by selecting a closed shape or text. If needed, use the Select Other—Face Cycling command to select geometry that is not selected immediately when the cursor is moved over it.
3. Select the type of emboss: Emboss from Face, Engrave from Face, or Emboss/ Engrave from Plane.
4. Specify the depth, color, direction, and face as needed. If you selected the Emboss/ Engrave from plane option, you can also add a taper angle to the created emboss/engrave feature.
5. You edit the emboss feature like any other feature.

TIP: You can use the Sheet Metal tab > Modify panel > Mark command to add text or other markings to sheet metal parts.

EXERCISE 7-4: CREATING TEXT AND EMBOSS FEATURES

In this exercise, you emboss and engrave closed profiles onto a razor and engrave text on the handle.

1 Open *ESS_E07_04.ipt* in the Chapter 07 folder.

2 First, you emboss a closed profile. From the browser turn on visibility of Sketch11 by right-clicking on Sketch11 and click Visibility from the menu.

3 Change the viewpoint so it resembles the following image.

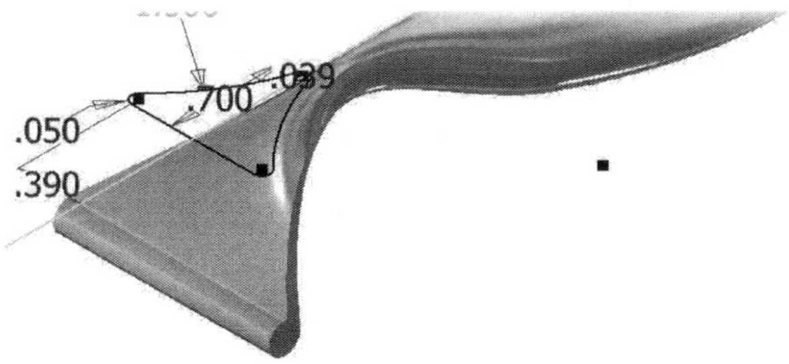

Figure 7.81

4 Click the Emboss command on the 3D Model tab > Create panel to engrave a closed sketch profile. Notice that the only visible closed profile is automatically selected.

a. In the Emboss dialog box, click the Engrave from Face option (middle button).

b. Change the Depth to **0.03125 in**.

c. Click the Top Face Appearance button, and choose Aluminum (Flat) from the Appearance dialog box drop-down list, as shown in the following image on the left.

d. Click OK twice to close both dialog boxes and create the engraved feature, as shown in Figure 7.82 right.

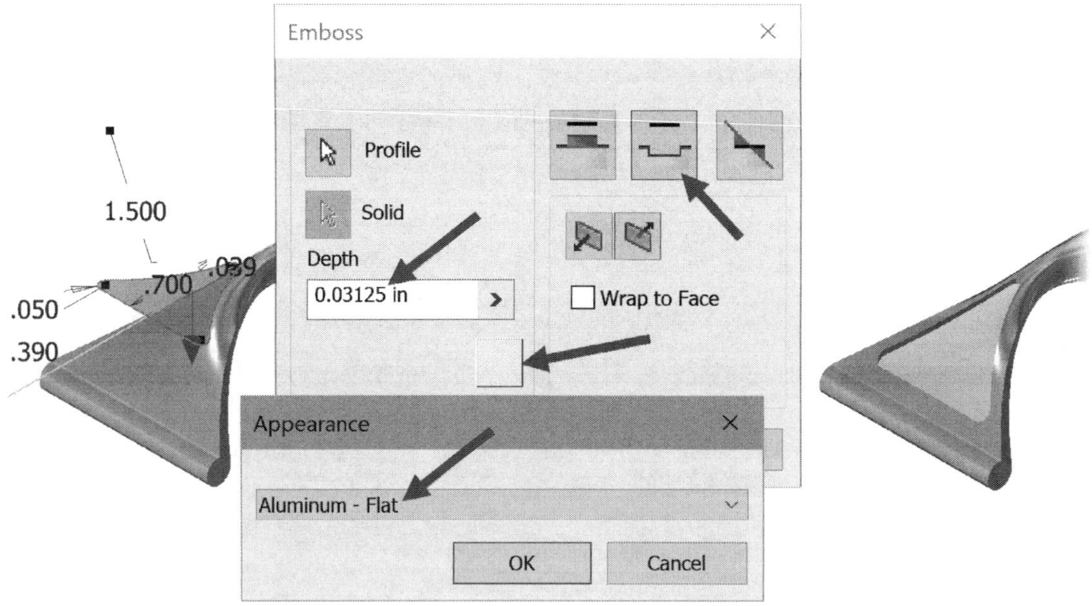

Figure 7.82

5 Change to the home view by clicking the Home View icon above the ViewCube.

6 From the browser turn on the visibility of Sketch10.

7 Click the Emboss command from the 3D Model tab > Create panel.

a. For the profile, click the oblong (near the front) and the five closed profiles. Be sure to select the left and right halves of the herringbone profiles, as shown in the following image on the left.

b. In the Emboss dialog box, click the Emboss from Face option (left button).

c. Change the Depth to **0.015625 in**.

d. Click the Top Face Appearance button, and select Black from the Appearance dialog box drop-down list, as shown in the following image on the right.

e. Click OK twice to close both dialog boxes and create the emboss feature.

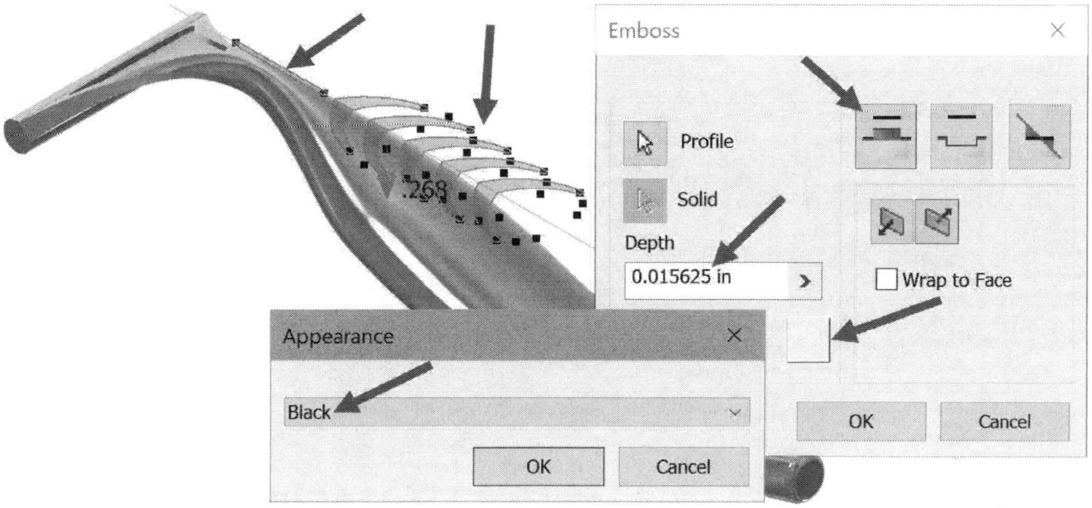

Figure 7.83

8 Use the Free Orbit command to examine the engraved and embossed features.

9 Next, you create text on a sketch. From the browser turn on the visibility of Sketch9.

10 In the browser, double-click Sketch9 to make it the active sketch.

11 If needed, use the rotate tools on the ViewCube as shown in the following image on the left, so your viewpoint resembles the image on the right.

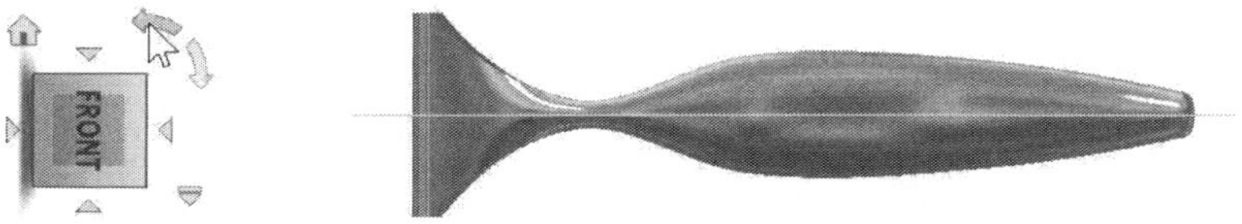

Figure 7.84

12 Click the Text command in the Sketch tab > Create panel.

a. Place the insertion point of the text in an open area below the part, and the Format Text dialog box appears.

b. In the dialog box, click the Center and Middle Justification buttons labeled (1) and (2) in the following image.

c. Click the Italic option labeled (3).

d. Change the % Stretch value to **120** labeled (4).

e. Click in the text field, and type **The SHARP EDGE** labeled (5).

f. In the text field, double-click on the word "The" and change the text size from .120 in to **0.09 in** labeled (6).

g. Click OK to create the text.

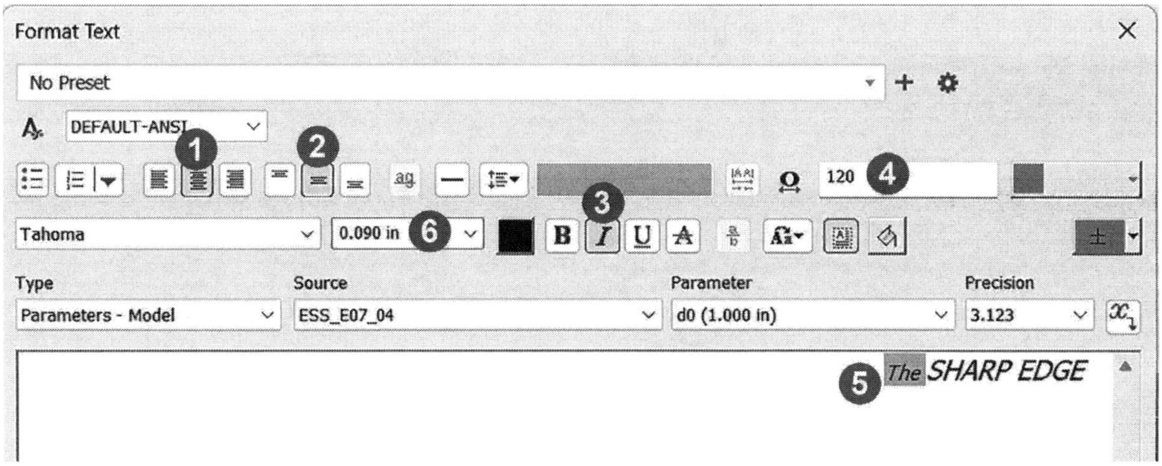

Figure 7.85

13 Draw a diagonal construction line coincident in the text bounding box from the top-left to bottom-right corners, as shown in the following image.

Figure 7.86

14 Place a coincident constraint between the midpoint of the diagonal construction line on the text box and Sketch9, as shown in the following image on the left. The completed operation is shown in the following image on the right.

15 Move the text so it is positioned near the right side of the part. If desired a dimension could be added.

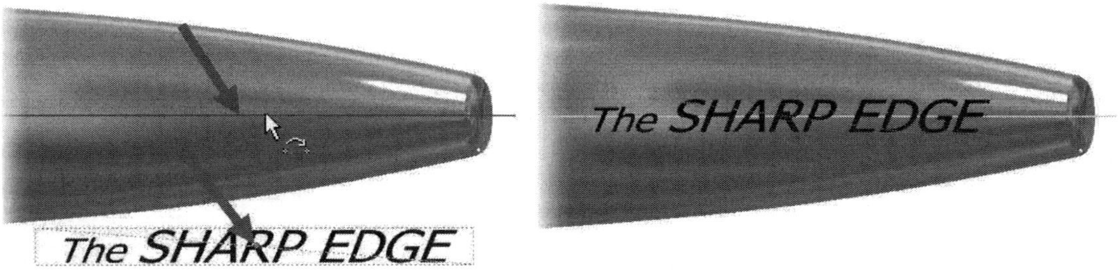

Figure 7.87

16 Finish the sketch by right-clicking and click Finish 2D Sketch from the marking menu.

17 Start the Emboss command from the 3D Model tab > Create panel.

a. For the profile, select the text.

b. In the Emboss dialog box, click the Engrave from Face option.

c. Change the Depth to **0.015625 in**.

d. For the direction, click the right button to change the direction down.

e. Click the Top Face Color button, and click Black from the Appearance drop-down list, as shown in Figure 7.88 left.

f. Click OK twice to close both dialog boxes and create the emboss feature as shown in the following image on the right (for clarity the sketch's visibility is off).

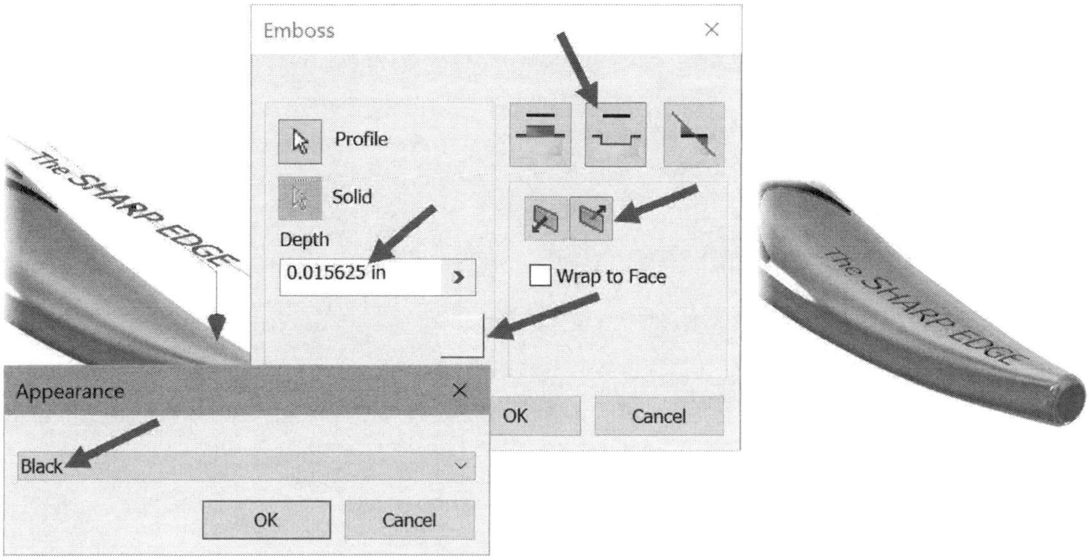

Figure 7.88

18 Use the Free Orbit command to examine the engraved text.

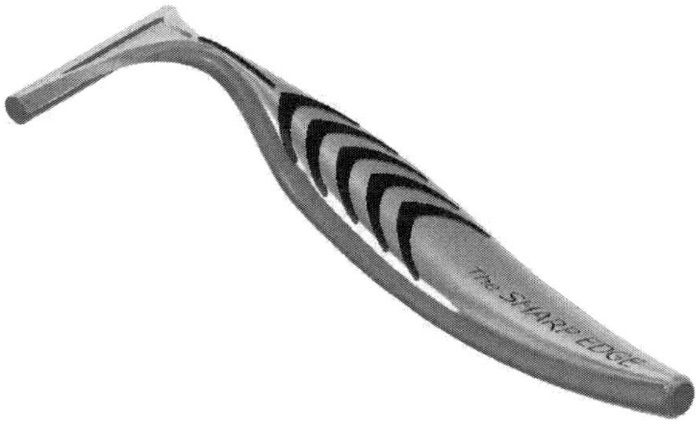

Figure 7.89

19 If desired, edit the sketches and features and try different dimensions and options, such as placing text along an arc.

20 Close the file. Do not save changes.

End of exercise.

SWEEP FEATURES

Sweep features are used to create more complex geometry beyond extruded or revolved features. In this section, you learn how to create sweep features such as a hook, cabling, or piping. A sweep feature can be a base or a secondary feature.

Unlike extrude and revolve features, a sweep feature usually uses two unconsumed sketches: a profile to be swept and a path the profile follows; or edges of a part the path follows. Additional profiles can be used as guide rails or a surface can also be used to help shape the feature.

The profile sketch and the path sketch cannot lie on the same or parallel planes. The path can be an irregular shape or use part edges by projecting and including the edges onto the active sketch. The path can be either an open or closed profile and can lie in a plane or lie in multiple planes (3D Sketch).

To create a sweep feature, use the Sweep command on the 3D Model tab > Create panel, as shown in Figure 7.87 left, and the Sweep Property panel will appear, as shown in Figure 7.87 right.

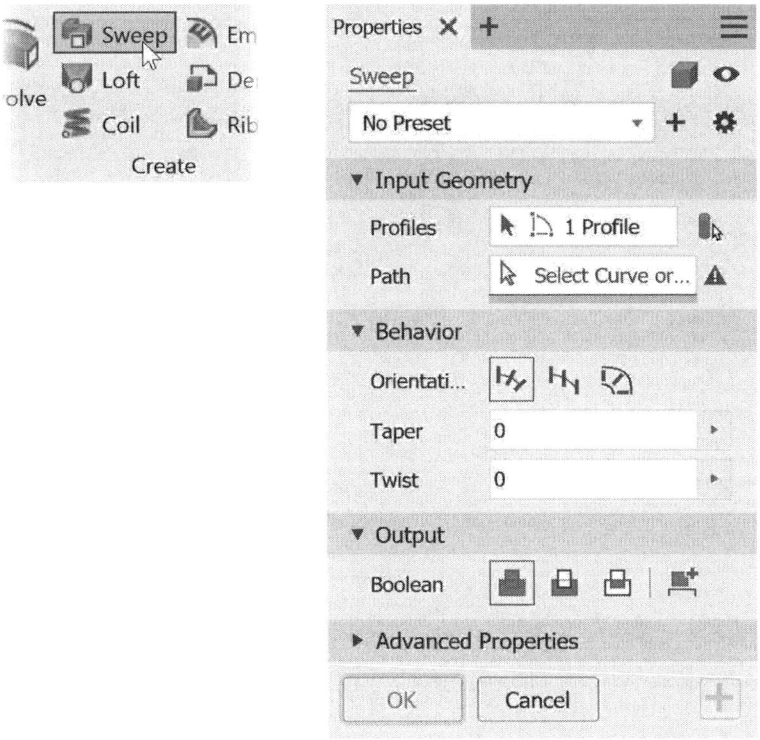

Figure 7.90

The Sweep Dialog Box

The following sections describe the options in the Sweep dialog box.

Profile. Click this button to select the sketch profile to sweep. If the Profile button is depressed and red, it means you need to select a sketch or sketch area. If there are multiple closed profiles, you need to select the profile to sweep. If there is only one possible profile, Inventor selects it for you. If you selected the wrong profile or sketch area, choose the Profile button, and deselect the incorrect sketch by clicking it while holding down the CTRL key. Release the CTRL key, and select the desired sketch profile.

Path. Click this button to select the path along which to sweep the profile. The path can be an open or a closed profile on its own sketch, but must pierce the sketch the profile is drawn on. You can also use edge(s) of a part as a path. The profile is typically perpendicular to and intersects with the start point of the path. The start point of the path is often projected into the profile sketch to provide a reference point.

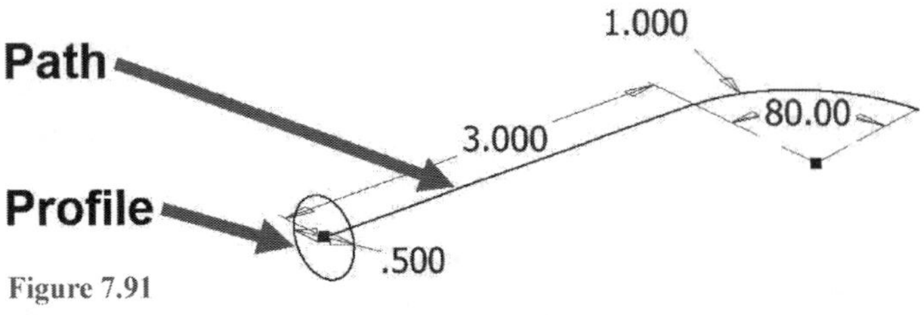

Figure 7.91

Solids. If there are multiple solid bodies, use this button to choose the solid body(ies) for the operation.

Output Buttons. From the Output section, choose between Solid and Surface for the type of feature to create.

Join. Adds material to the part.

Cut. Removes material from the part.

Intersect. Creates a new feature from the shared volume of the sweep feature and existing part volume. Material not included in the shared volume is deleted.

Operation

Creates a sweep feature by sweeping a profile along a path.

Follow Path. Holds the swept profile constant to the sweep path. All sweep sections maintain the original profile relationship to the path, as shown in Figure 7.92 left.

Fixed. Holds the swept profile parallel to the original profile, as shown in the following image on the right.

Guide. Select a rail or surface to guide and scale the sweep profile.

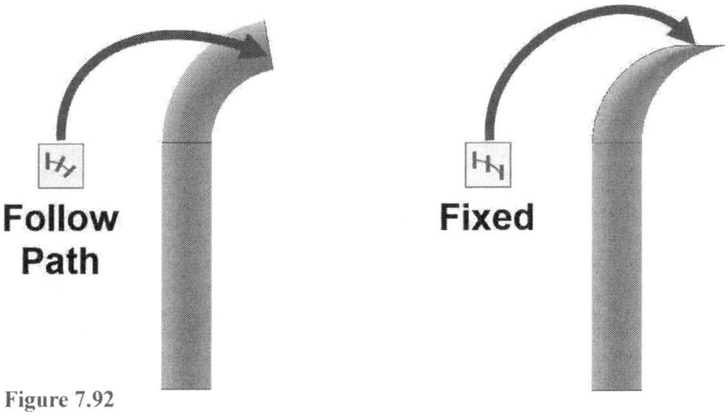

Figure 7.92

Options for Follow Path

When using the Follow Path option you can control the taper and twist of the sweep feature.

Taper. Enter a value for the angle at which you want the profile to be drafted. By default, the taper angle is zero.

Twist. Use the twist option to spiral the profile along the path the number of degrees or radians you specify. For example, the following image on the left shows a triangle profile and a line as the path, the middle image shows the triangle swept along the line with no twist, and the image on the right shows the same sweep but 360 degrees of twist was added.

Figure 7.93

Guide Type

Creates a sweep feature by sweeping a profile along a path and a guide rail to control scale and twist of the swept profile. Figure 7.94 shows the options for the path and guide rail type.

Guide. Select a guide curve or rail that controls the scaling and twist of the swept profile. The guide rail must touch the profile plane. If you project an edge to position the guide rail and the projected edge is not the path, change the projected edge to a construction line.

Profile Scaling. Specify how the swept section scales to meet the guide rail. Figure 7.94 shows the path, the profile, and guide rail along with the three different profile-scaling options.

- X & Y. Scales the profile in both the X and Y directions as the sweep progresses.
- X. Scales the profile in the X direction as the sweep progresses. The profile is not scaled in the Y direction.
- None. Keeps the profile at a constant shape and size as the sweep progresses. Using this option, the rail controls only the profile twist and is not scaled in the X or Y direction.

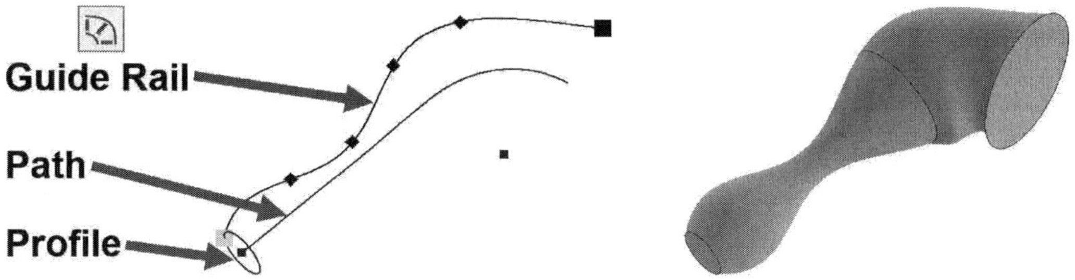

Figure 7.94

Solid Sweep

If needed you can also sweep a solid instead of a profile. For example, Figure 7.95 left shows two solids: a cylinder and a sphere (the sphere will be swept around the cylinder with a twist). The middle image shows the Sweep Properties panel with the Solid Sweep option on and the cells filled in. The image on the right shows the sphere swept around the cylinder with the twist option.

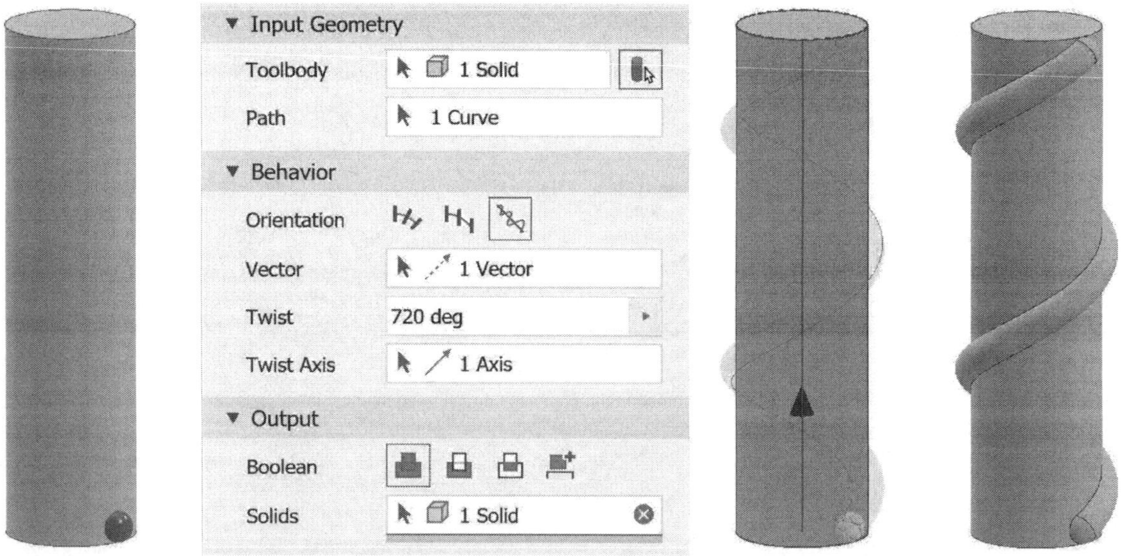

Figure 7.95

Creating a Sweep Feature Using a Profile

To create a sweep feature, follow these steps:

1. Create an unconsumed sketch for the profile to sweep. Create another sketch for the path. If you are going to use existing part edges for the path, you don't need to create a second sketch for the path.

The profile and path sketches must lie on separate nonparallel planes. It is recommended that the profile intersect the path. Use work planes to place the location of the sketches, if required. The path sketch can be open or closed. Add dimensions and constraints to both sketches as needed. If needed, create a third sketch that will be used as a rail, or create a surface to be used as a guide surface.

2. Click the Sweep command on the 3D Model tab > Create panel.
3. The Sweep Properties panel appears. If two unconsumed sketches do not exist, Inventor notifies you that two unconsumed sketches are required.
4. Click in the Profile cell, and select the sketch to sweep from the graphics window. If only one closed profile exists, this step is automated for you.
5. If it is not already selected (appears depressed), click in the Path cell, and then select the sketch to be used as the path from the graphics window.
6. Select the type of sweep to create: Follow Path, Fixed, or Guide.
7. Define if the resulting sweep will create a solid or a surface by clicking the appropriate button in the Output area. Select the required options, as outlined in the previous descriptions.
8. If this is a secondary feature, click the operation to specify whether material will be added or removed or retain what is common between the existing part and the new sweep feature.
9. Define the taper and twist as needed.
10. Click the OK button to complete the operation.

EXERCISE 7-5: CREATING SWEEP FEATURES

In the first part of this exercise, you create a swept component. You will create a profile, a path, and a guide rail sketch. In the second part of the exercise you sweep a profile using edges on a part as the path.

1 Start a new part file based on the Standard (in).ipt template file.

2 Create a 2D Sketch on the XY origin plane and draw, constrain, and dimension the geometry as shown in the following image. Place the lower left endpoint of the sketch on the origin point. The arcs are tangent to the adjacent lines and the three arcs have the same radius.

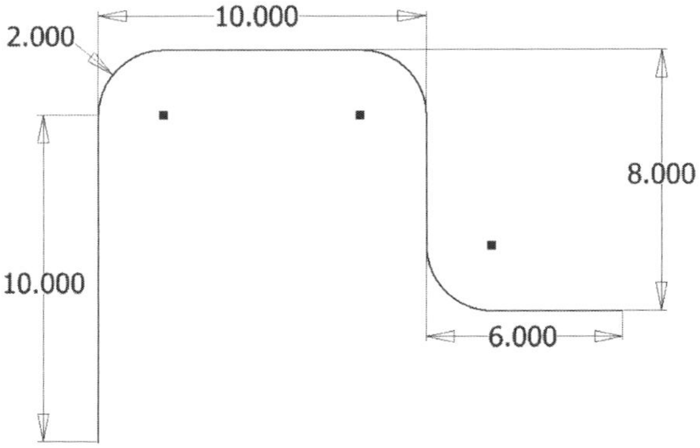

Figure 7.96

3 Finish the sketch.

4 Create another 2D Sketch on the XZ origin plane.

5 Change the viewpoint by selecting the Home icon.

6 Zoom into the lower-left corner of the first section. Draw two circles centered on the origin point (lower-left endpoint of the first sketch). Add a **1.5 inch diameter** dimension for the outside circle and a **.125 inch** dimension between the two circles to define the thickness, as shown in Figure 7.97 left.

7 Finish the sketch.

8 Click the Sweep command on the 3D Model tab > Create panel.

a. For the profile click in-between the two circles, as shown in Figure 7.97 right.

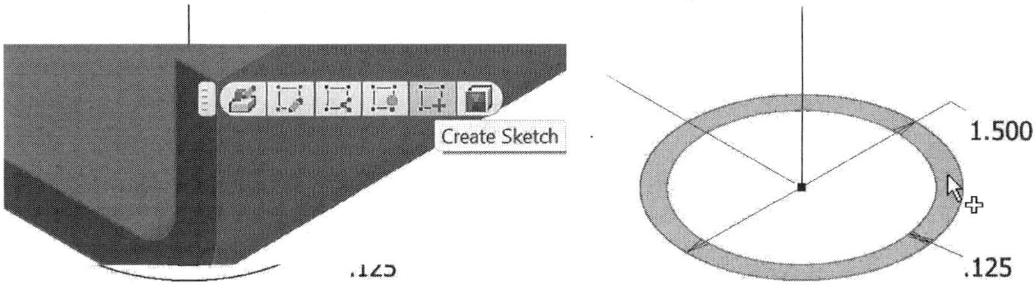

Figure 7.97

b. Next, define the path; select in the Path cell and then select one line or arc of the open profile.

c. Click OK to create the sweep feature.

9 Change your viewpoint so you can see the front of the sweep as shown in Figure 7.98 right.

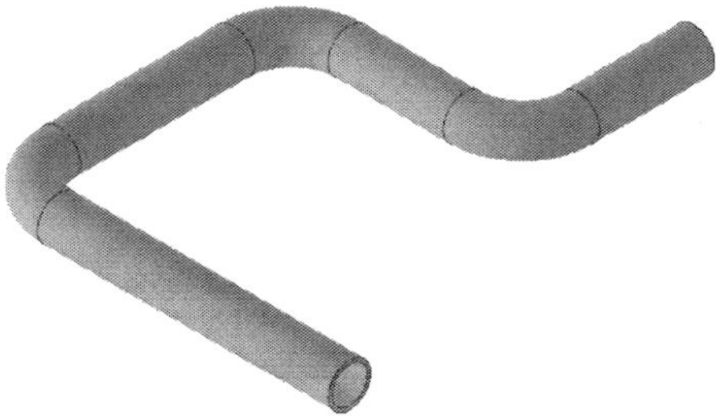

Figure 7.98

10 Close the file. Do not save changes.

In the next portion of the exercise you create a sketch and sweep the profile along the edges of a part.

11 Open *ESS_E07_ 05-2.ipt* in the Chapter 07 folder.

12 Zoom into the front-right side of the part and create a sketch on the front face of the part as shown in Figure 7.99 left.

13 The viewpoint changes so you are looking directly at the sketch. However, it is difficult to see where you are sketching. To return to your previous view, press the F5 key.

14 Sketch a **.0625 in** circle on the projected top-right end point as shown in Figure 7.99 right. By default, the dimension rounds to three decimal places in the sketch.

Figure 7.99

15 Click the Finish Sketch command on the Sketch tab > Exit panel.

16 Next, sweep the circle along edges of the part.

a. Click Sweep from the 3D Model tab > Create panel.

b. For the profile select the circle.

c. Click in the Path cell and then select the top-right outside edge of the part for the path, as shown in Figure 7.100 left. Inventor automatically selects all the edges that are tangent to the selected edge.

d. To remove material, in the Sweep Properties panel, Operation section, click the Cut button.

e. Click the OK button in the Sweep Properties panel. Figure 7.100 right shows the completed part.

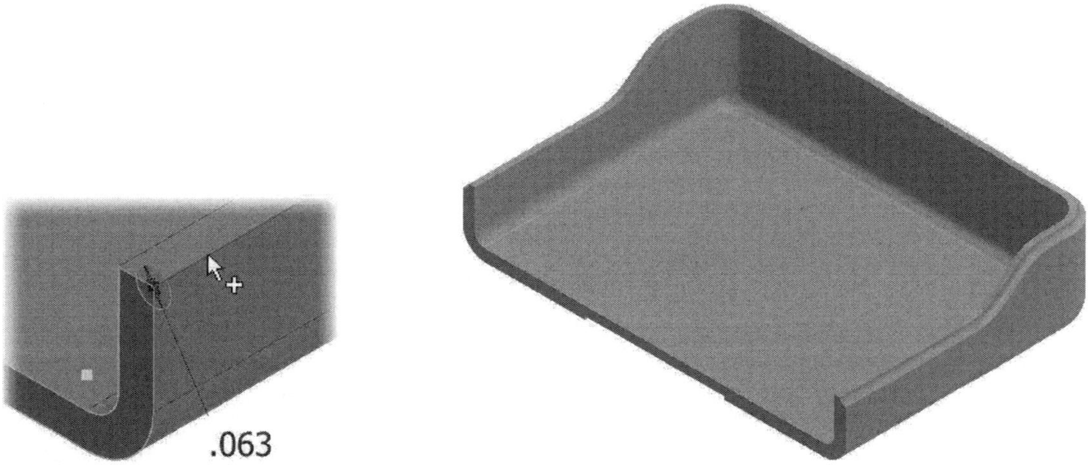

Figure 7.100

17 Use the Zoom and Free Orbit commands to examine the Sweep feature.

18 Close the file. Do not save changes.

End of exercise.

3D SKETCHING

To create a sweep feature whose path does not lie on a single plane, you can create a 3D sketch that will be used for the path. You can use a 3D sketch to define the routing path for an assembly component, such as a pipe or duct work that crosses multiple faces on different planes. You need to define a 3D sketch in the part environment, and you can do this within an assembly or in its own part file. You can use the Autodesk Inventor adaptive technology during 3D sketch creation to create a path that updates automatically to reflect changes to referenced assembly components. In this section, you will learn strategies for creating 3D sketches.

3D Sketch Overview

When creating a 3D sketch, you use many of the sketching techniques that you have already learned with the addition of a few commands. 3D sketches use work points and model edges or vertices to define the shape of the 3D sketch by creating line or spline segments between them. You can also create bends between line segments. When creating a 3D sketch, you use a combination of lines, splines, fillet features, work features, constraints, and existing edges and vertices.

3D Sketch Environment

The 3D sketch environment is used to create 3D or a combination of 2D and 3D curves. Before creating a 3D sketch, change the environment to the 3D sketch environment by clicking the Start 3D Sketch command on the 3D Model tab > Sketch panel which may be beneath the Start 2D Sketch command, as shown in Figure 7.101.

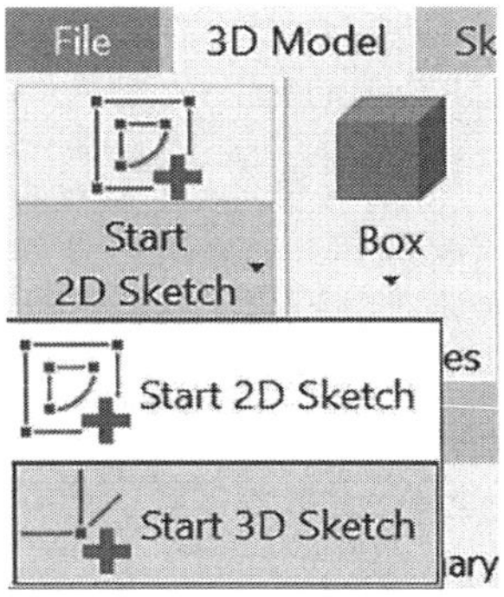

Figure 7.101

The ribbon changes to show the 3D Sketch tab and commands, as shown in Figure 7.102. The commonly used 3D sketch commands are explained in the next sections.

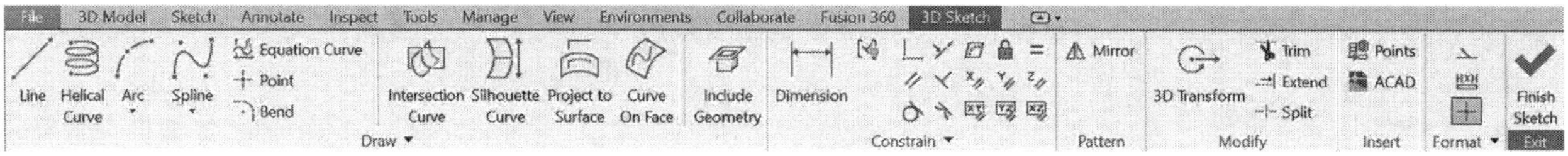

Figure 7.102

3D Sketch from Intersecting Geometry

One method to create a 3D path is to use geometry or features that intersect with faces on the part. If the intersecting geometry defines the 3D path, you can use it. The intersection can be defined by a combination of any of the following: planar or nonplanar part faces, surface faces, a quilt, or work planes.

To create a 3D path from an intersection, follow these steps:

1. Create the intersecting features that describe the desired path.
2. Create a 3D Sketch by clicking the Start 3D Sketch command on the 3D Model tab > Sketch panel. It may be under the Start 2D Sketch command.
3. Click Intersection Curve from the 3D Sketch tab > Draw panel, as shown in Figure 7.103 left.
4. The 3D Intersection Curve dialog box appears, as shown in the middle of the following image.

5. Select the two intersecting geometries or features.
6. To complete the operation, click OK.
7. The following image on the right shows geometry created from where a work plane intersects the cylindrical face.

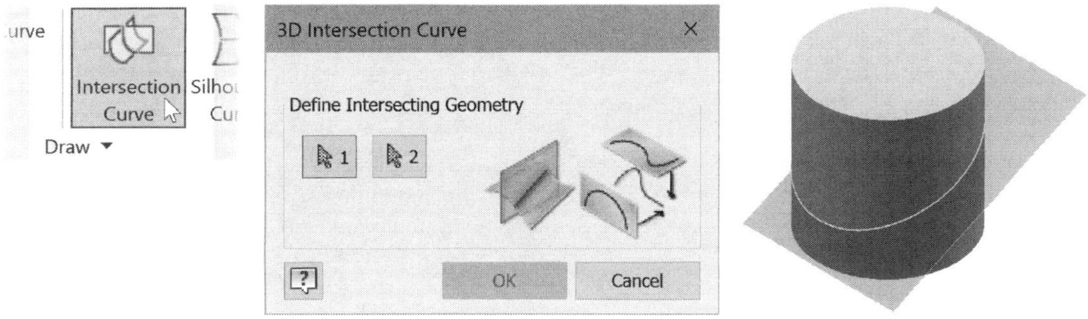

Figure 7.103

Project to Surface

While in a 3D sketch, you can project curves, 2D or 3D geometry, part edges, and points onto a face or selected faces of a part. To project curves onto a face, follow these steps:

1. Create or open a part onto which to project the curves.
2. Create the curves to project onto the part's face(s).
3. Click the Start 3D Sketch command on the 3D Model tab > Sketch panel.
4. Click Project to Surface from the 3D Sketch tab > Draw panel, as shown in Figure 7.104 left.
5. The Project Curve to Surface dialog box appears, as shown in the middle of Figure 7.104. The Faces button is active. In the graphics window, select the face(s) onto which to project the curves.
6. Click the Curves Button, and then from the graphics window, select the individual objects to project. Figure 7.104 right shows an ob-round projected onto a cylindrical face.

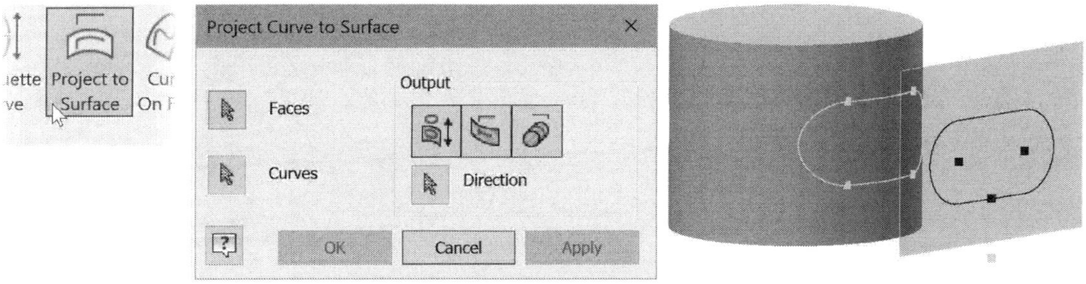

Figure 7.104

7. In the Output area, select one of the following options. Click OK to complete the action.

	Project along vector	Specify the vector by clicking the Direction button and selecting a plane, edge, or axis. If a plane is selected, the vector will be normal (90°) from the plane. The curves will be projected in the direction of the vector.

	Project to closest point	Projects the curves onto the surface normal to the closest point.
	Wrap to Surface	The curves are wrapped around the curvature of the selected face or faces.

By default, the projected curves are linked to the original curve. If the original curves change size, they are updated. To break the link, move the cursor into the browser over the name of the Projected to Surface entry, right-click, and choose Break Link from the menu, as shown in Figure 7.105.

You can display the sketch constraints and delete the reference constraints. You can also change how curves are projected by right-clicking on the Project to Surface entry in the browser and choosing Edit Projection Curve from the menu to show the Project Curve to Surface dialog box. Use it to make changes as required.

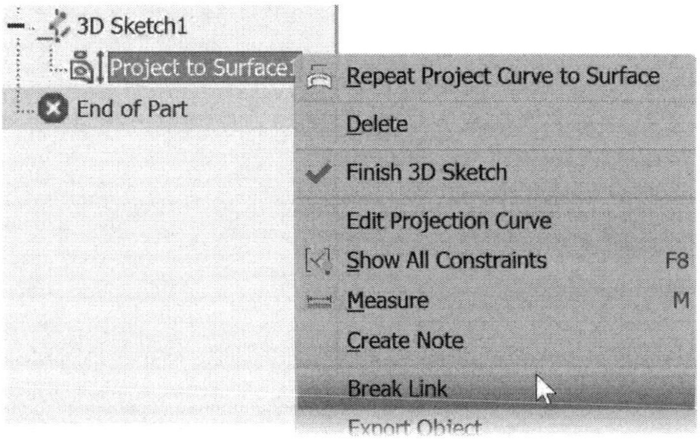

Figure 7.105

Project to a 3D Sketch

Another method to project 2D geometry onto a non-planar face is the Project to 3D Sketch command. While in a 2D Sketch, using Project to 3D Sketch projects all the geometry of the active 2D sketch geometry onto a nonplanar face and automatically creates a 3D sketch.

To use the Project to 3D Sketch command, follow these steps.

1. First, create a 2D sketch and sketch and constrain geometry as needed or make an existing sketch active that contains the geometry that you want to project.
2. Click Project to 3D Sketch from the Sketch tab > Create panel, as shown in Figure 7.106 left.
3. If needed, in the Project to 3D Sketch dialog box check the Project option.
4. Select a face or faces that the geometry will be projected onto. All the geometry on the active sketch shows in the preview how it will be projected on the selected face(s), as shown in Figure 7.106 right.
5. Click OK to create a 3D sketch and project all the geometry of the active sketch to the selected face(s).

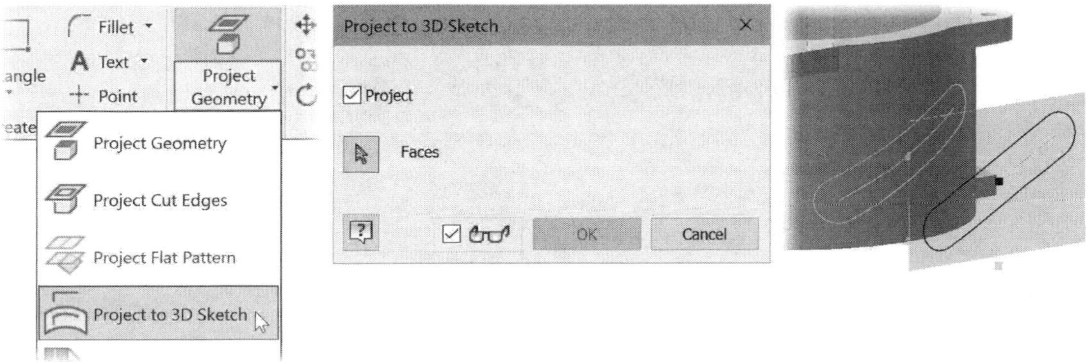

Figure 7.106

3D LINES

Another option to create a 3D path is to draw 3D lines that are nonplanar. To draw 3D lines, follow these steps:

1. While in a 3D Sketch, use Line from the 3D Sketch tab > Draw panel, as shown in Figure 7.107 left.
2. Once in this 3D line command, you can create lines by selecting a plane on the triad, as shown in Figure 7.107 right, and drawing line segments. After creating the lines, you can add dimensions.

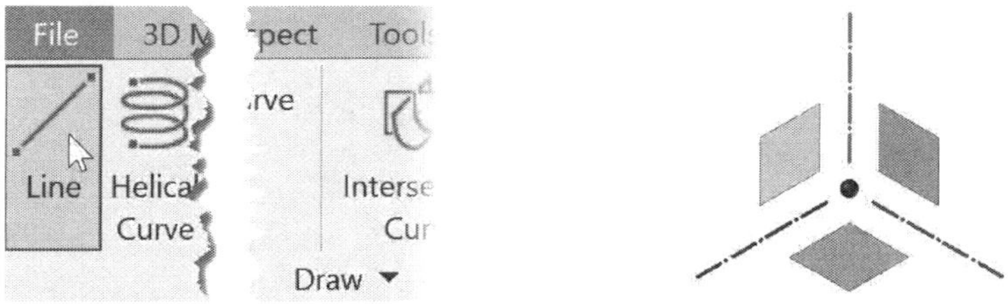

Figure 7.107

3. Or while in the line command, you can turn on the Precise Input mini-toolbar. While in the line command click the down arrow on the 3D Sketch tab > Draw panel, as shown in Figure 7.108 left, and from the expanded panel click Precise Input as shown in the middle image. Use the Precise Input mini-toolbar to enter X, Y, and Z point data, as shown in the image on the right. Values can be relative to the last point or absolute to 0, 0, 0. Consult the help system "Create a 3D Line" for more information about using Precise Input.

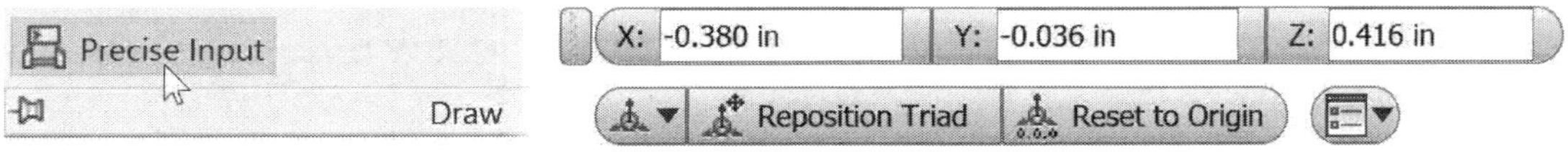

Figure 7.108

4. Dimensions and constraints can be applied to the lines.
5. By default, a bend is not applied between 3D line segments, but this option can be toggled on and off by right-clicking while in the 3D line command and selecting or deselecting Auto-Bend on the menu, as shown in Figure 7.109 left.
6. To set the default radius of the bend, Click Document Settings from the Tools tab > Options panel, and change the 3D Sketch Auto-Bend Radius setting from the Sketch tab.
7. To manually add a bend between two 3D lines, use the Bend command on the 3D Sketch tab > Draw panel, as shown in Figure 7.109 middle, or right-click in a blank area of the graphics window and choose Bend from the marking menu, as shown in the image on the right.

8. The Bend dialog box appears; use it to enter a value for the bend, and then select two 3D lines or the endpoint where they meet. To complete the command, right-click and click OK from the marking menu. You can edit the bend by double-clicking on the dimension and entering a new value.

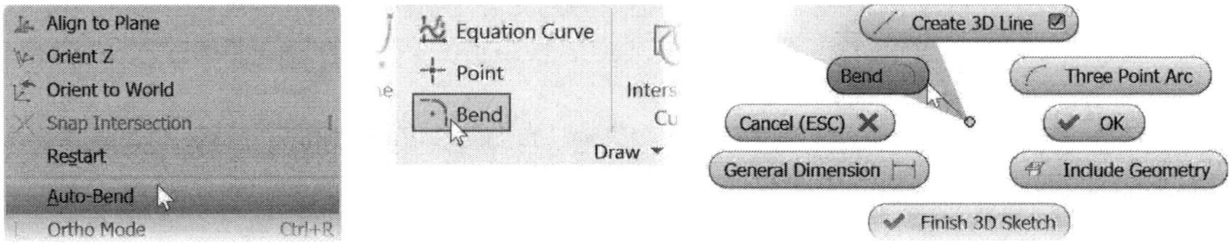

Figure 7.109

CREATE A 3D SWEEP

To create a 3D sweep, follow these steps:

1. Click the Start 3D sketch command from the 3D Model tab > Sketch panel.
2. Create a 3D path using any of the previously discussed techniques.
3. After creating a 3D path, next you create a 2D sketch, draw, and constrain a closed profile.
4. Start the Sweep command on the 3D Model tab > Create panel and use the closed sketch as the profile and 3D lines as the path.

Figure 7.110 left shows a part with 3D lines, two bends, and dimensions. The image on the right shows the completed sweep.

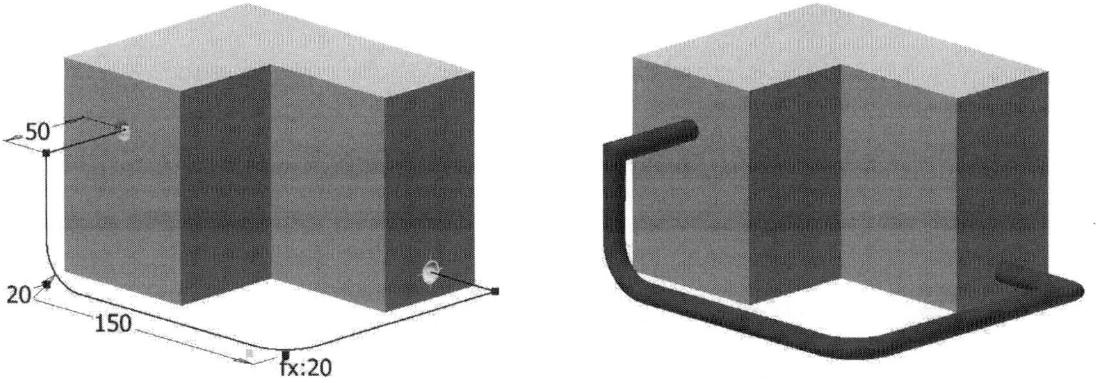

Figure 7.110

IMPORT POINTS

Another option to create geometry is to import X, Y, Z point data from a Microsoft Excel spreadsheet. The imported points can be connected by lines or splines or left as points. The imported points are not associated back to the spreadsheet: if the data in the spreadsheet changes, the imported point data do not update.

While in a 2D sketch, only X and Y values are imported. While in a 3D sketch, X, Y and Z values are imported. The data in the spreadsheet must start in cell A1 and be in the first worksheet. Cell A1 can define a unit; if no unit is defined in cell A1 the document unit will be used. Rows can include an X, Y and Z heading if desired, but it is not required. The columns must be in the following order and the C column is not required.

Column A = X value   Column B = Y value   Column C = Z value

Figure 7.111 left shows an example of a spreadsheet with the unit and column heading data. The image on the right shows an example of only the required fields.

	A	B	C
1	**in**		
2	**X**	**Y**	**Z**
3	0	0	0
4	5	0	0
5	5	5	0
6	5	5	5
7	10	5	5

	A	B	C
1	0	0	0
2	5	0	0
3	5	5	0
4	5	5	5
5	10	5	5

Figure 7.111

To import points from a spreadsheet, follow these steps:

1. Create a Microsoft Excel spreadsheet with X, Y and Z point data. The Z data is optional.
2. Create or make a 2D or 3D sketch active.
3. Click the Import Point data command using either of the following methods:
   a. In a 2D sketch, click the 2D Sketch tab > Insert Panel > Import Points, as shown in Figure 7.112 left.
   b. *Or,* in a 3D sketch, click the 3D Sketch tab > Insert Panel > Import Points, as shown at right.

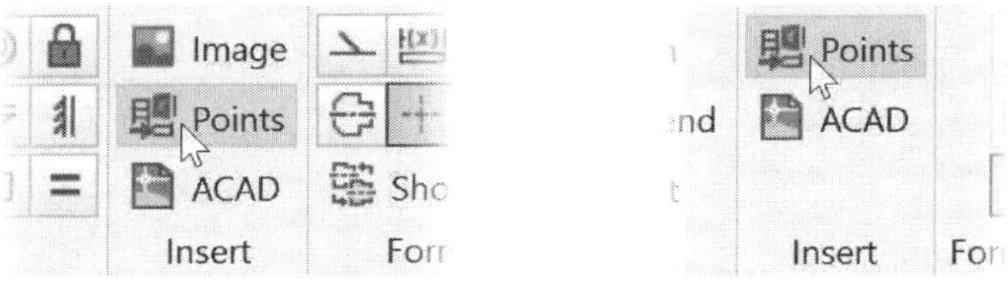

Figure 7.112

4. The Open dialog box appears; navigate to and select the Excel spreadsheet.
5. On the bottom-right corner of the Open dialog box click the Options button and the File Open Options dialog box appears, as shown in the following image. Select the option to Create points (the default), connect the points by clicking Create lines, or connect the points with a spline with the Create spline option.

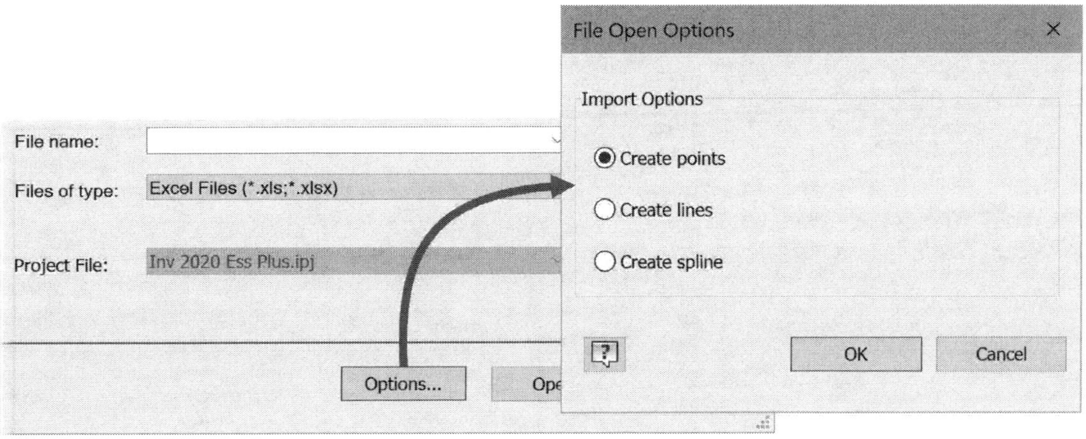

Figure 7.113

6. Click OK to accept the options and then click Open to create the data.
7. Add dimensions and constraints as needed.

EXERCISE 7-6: 3D SKETCH & SWEEP FEATURES

In this exercise, you create geometry in a 3D sketch using different methods.

1 Start a new part file based on the *Standard (in).ipt* template file.

2 Create a 3D Sketch by clicking Start 3D Sketch from the 3D Model tab > Sketch panel.

3 Press the F6 key to change to the Home View.

4 Next create four lines.

a. Click the Line command on the 3D Sketch tab > Draw panel.

b. From the graphics window, select the XY plane (the red and green axis) of the 3D coordinator.

c. For the first point click on the origin. Place the second point in the X direction so the X parallel symbol X// appears as shown in Figure 7.114 labeled (1).

d. Draw another line parallel in the Y direction Y// (green axis) as shown in Figure 7.114 labeled (2).

e. From the graphics window select the YZ plane (the blue and green axis) on the 3D coordinator.

f. Draw a line parallel to the Z axis Z// (blue axis) as labeled (3) in Figure 7.114.

g. From the graphics window select the XZ plane (the red and blue axis) on the 3D coordinator.

h. From the graphics window draw a line in the X direction with the X parallel X// constraint as shown in Figure 7.114 labeled (4).

i. Finish the command by right-clicking and choose OK from the marking menu.

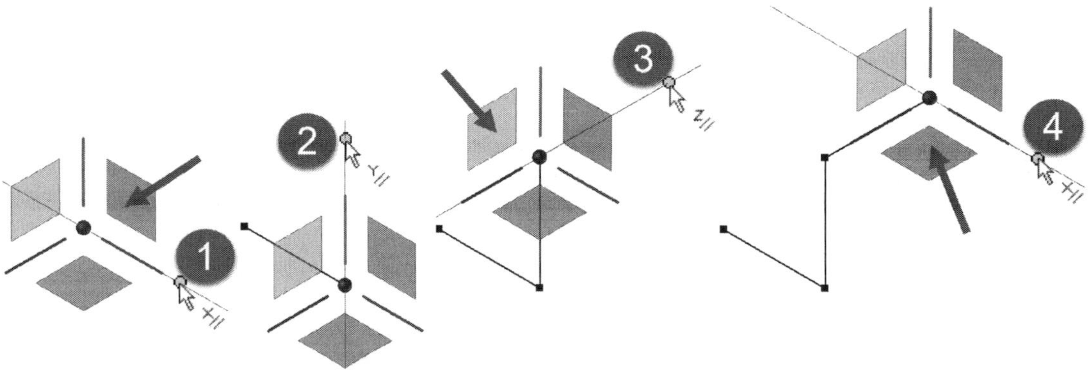

Figure 7.114

5 Click Dimension from the 3D Sketch tab > Constrain panel and add four dimensions to the lines, as shown in Figure 7.115. If needed, you can add perpendicular and parallel constraints between the lines.

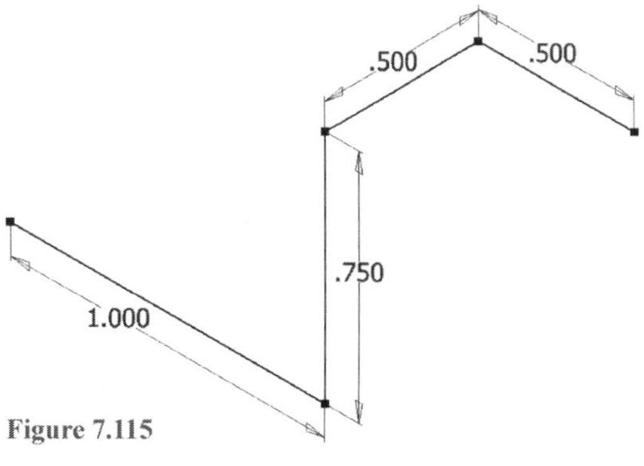

Figure 7.115

6 Delete the lines you just created.

7 Next, you place lines via the Precise Input mini-toolbar.

a. Click the Line command on the 3D Sketch tab > Draw panel.

b. If the Precise Input mini-toolbar does not appear, click the down arrow on the 3D Sketch tab > Draw panel and click the Precise Input command. Practice creating lines by entering point data in the Precise Input mini-toolbar.

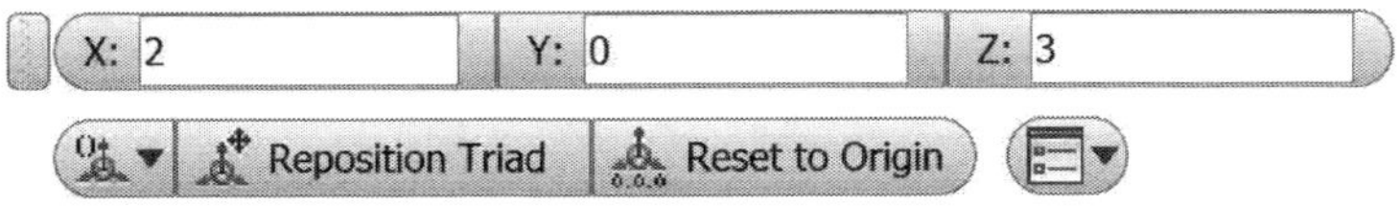

Figure 7.116

c. Add dimensions and constraints as desired.

d. Delete the lines you just created.

8 Next you import point data from an Excel spreadsheet.

a. In Excel, navigate to and open the spreadsheet *XYZ - Point Data.xlsx* in the Chapter 07 folder and review the point data. Then close the file.

b. While still in a 3D sketch, click Import Points from the 3D Sketch tab > Insert Panel.

c. The Open dialog box appears; navigate to and single click on the Excel spreadsheet *XYZ - Point Data.xlsx* in the Chapter 07 folder.

d. From the bottom-right of the Open dialog box, choose the Options button. In the File Open Option dialog box click the Create lines option, as shown in Figure 7.117.

e. Click OK to accept the options. Click the Open button in the Open dialog box to create the data.

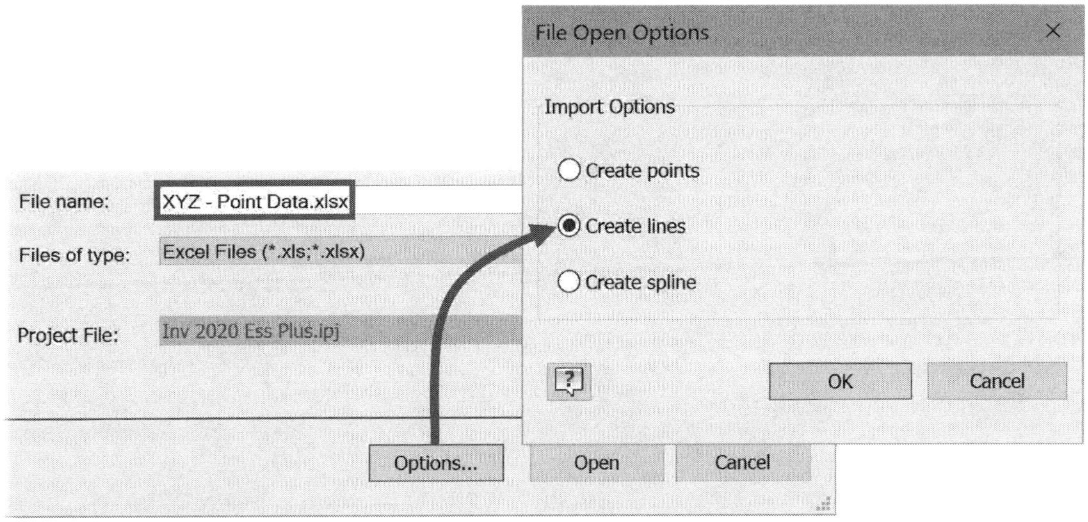

Figure 7.117

9 The points with the connecting lines will appear as shown in Figure 7.118. Examine the geometry by rotating the viewpoint.

10 If desired add dimensions and a perpendicular and parallel constraint between the lines.

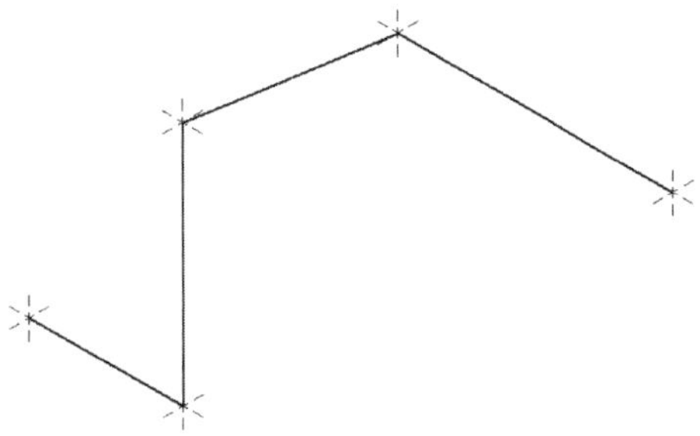

Figure 7.118

11 Close the file. Do not save changes.

In the next portion of this exercise, you create geometry on a 3D sketch from intersecting geometry.

12 Open *ESS_E07_06-2.ipt* from the Chapter 07 folder. The part contains a cylinder and an angled work plane.

13 Create a 3D Sketch by clicking the Start 3D Sketch command on the 3D Model tab > Sketch panel.

14 Click the Intersection Curve command from the 3D Sketch tab > Draw panel.

a. Select the visible work plane and the circular face of the part.

b. Click OK to create the curve.

15 Turn off the visibility of the work plane by right-clicking on an edge of the work plane and click Visibility from the menu. When done, your screen should resemble the following image.

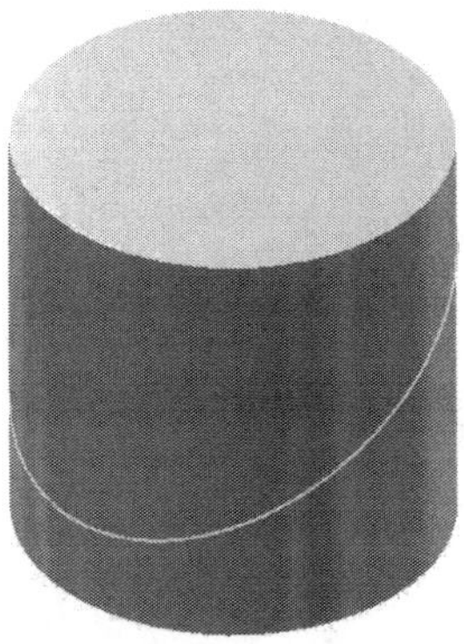

Figure 7.119

16 Finish the 3D sketch by clicking on Finish Sketch on the 3D Sketch tab > Exit panel.

17 Next you create a 2D sketch that the sweep command will use as the profile.

a. Create a 2D Sketch by right-clicking and click New Sketch from the marking menu.

b. In the browser, expand the Origin folder and click the YZ Plane, as shown in Figure 7.120 left.

18 Change to the Home View by right-clicking and click Home View from the menu.

19 Slice the graphics by pressing the F7 key.

20 Use the Project Geometry command on the Sketch tab > Create panel and project the intersecting geometry that you created.

21 Change the projected line to a construction line. This will prevent the projected line from being used when you create the sweep feature.

22 Create a **.5 inch** diameter circle on the bottom left endpoint of the projected geometry as shown in the middle of the following image.

23 Finish the Sketch.

24 Start the Sweep command on the 3D Model tab > Create panel and use the following options:

a. The circle should be automatically selected as the profile; if not, select the circle as the profile.

b. In the Sweep Properties panel click in the Path cell and then select the geometry that was created from the intersecting plane and face for the sweep path.

c. Change the Output to cut.

d. Click OK to create the sweep. When complete, your screen should resemble the following image on the right.

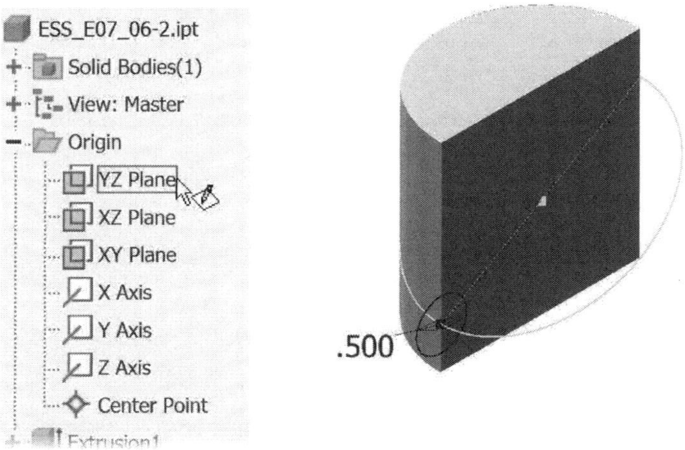

Figure 7.120

25 Close the file. Do not save changes.

In the next section of this exercise, you use the Project to 3D Sketch command to project geometry onto a cylindrical face.

26 Open *ESS_E07_06-3.ipt* from the Chapter 07 folder. The part contains an extrusion and a constrained sketch.

27 Make Sketch2 active by clicking on one of the arcs or lines on the sketch, and click Edit Sketch from the mini-toolbar or by double-clicking on Sketch2 in the browser.

28 To better see the part, change to the Home View.

29 Click the Project to 3D Sketch command on the Sketch tab > Create panel. The command may be under the Project Geometry command.

a. In the Project to 3D Sketch dialog box, ensure the Project option is checked. This option automatically projects geometry on the 2D sketch to the selected face.

b. In the graphics window select the inside circular face on the part to project the sketch onto, as shown in the following image on the left.

A

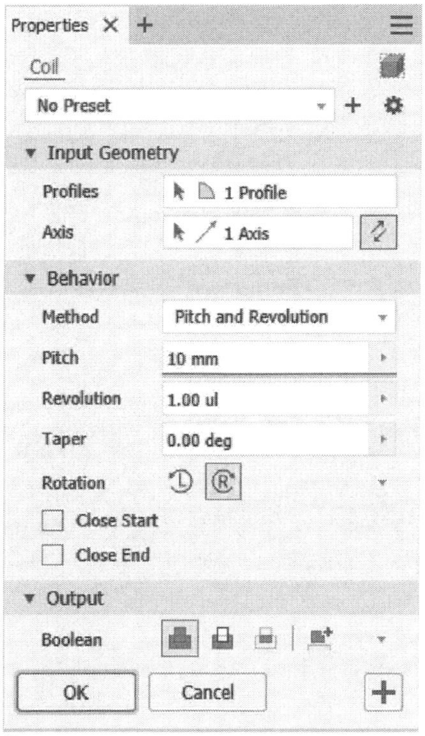

B

c. Click OK to complete the operation.

30 Finish the sketch.

31 In the browser, turn off the visibility of Sketch2. When complete, your screen should resemble the following image on the right.

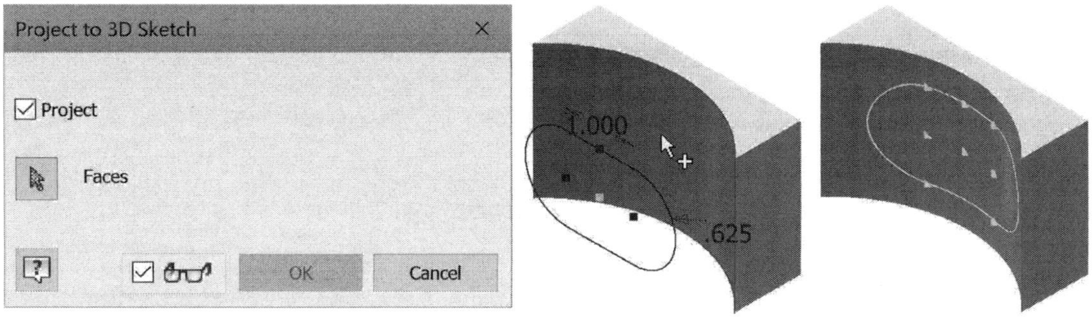

Figure 7.121

32 If desired, you can create a profile and sweep it along the projected geometry as you did earlier in this exercise. Note that you could have also used the Project to Surface command in the 3D sketch tab to project the geometry onto the face.

33 Close the file. Do not save changes.

End of exercise.

COIL FEATURES

The Coil feature makes it easy to create many types of helical, coil, spring, or spiral/spring geometry. Various springs can be created using different settings in the Coil dialog box. You can also use the Coil feature to remove or add a helical shape around the outside of a cylindrical part to represent a thread profile.

To create a coil, you need to have at least one unconsumed or shared sketch available in the part. This sketch describes the profile or shape of the coil feature. The sketch can also contain a line used as the coil's axis of revolution. For the axis of revolution, you can also use an edge on the part or a work axis. If no unconsumed sketch is available, Inventor prompts you with a message stating, "No unconsumed visible sketches on the part."

After creating a sketch, choose the Coil command from the 3D Model tab > Create panel, as shown in Figure 7.122A. The Coil dialog box appears, as shown in Figure 7.122B, with options as described next.

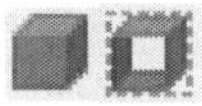

Solid/Surface: Click this button at the upper right of the dialog box to toggle between creating a solid or surface for the result of the coil feature.

Input Geometry

Profiles. Click to select the sketch that you will use as the profile shape of the coil feature. By default, the Profile button is shown selected; this tells you that you need to select a sketch or sketch area. If there are multiple closed profiles, select the profile to revolve. If there is only one possible profile, Inventor selects it for you, and you can skip this step. If you select the wrong profile or sketch area, click the Profile button and select a new profile or sketch area.

Figure 7.122

Axis. Click to select a sketched line or centerline, a projected straight edge, or an axis about which to revolve the profile sketch. If selecting an edge or sketched centerline, it must be part of the sketch. If selecting a work axis, it cannot intersect the profile.

Flip. Click to change the direction in which the coil will be created along the axis. The direction changes on either the positive or negative X or Y axis, depending upon the edge or axis you selected. A preview shows the direction in which the coil will be created.

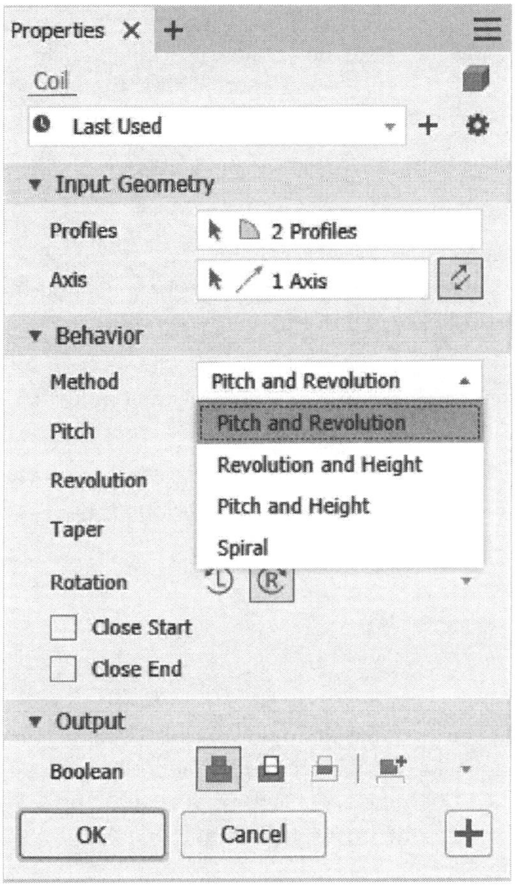

Figure 7.123

Behavior

The options for the coil listed in the Behavior area change to be appropriate for the method you select.

Method. Expand the options by clicking the down arrow and select from:

- Pitch and Revolution
- Revolution and Height
- Pitch and Height
- Spiral.

Pitch. Type in the value for the height to which you want the helix to elevate with each revolution.

Revolution. Specify the number of revolutions for the coil. A coil cannot have zero revolutions. Fractions can be used in this field, for example, you can create a coil that contains 2.5 turns. If end conditions are specified, as using the Coil Ends tab section, these end conditions are included in the number of revolutions.

Rotation. Click to specify the direction in which the coil will rotate, either clockwise or counterclockwise.

Height. Specify the height of the coil. This is the total coil height as measured from the center of the profile at the start to the center of the profile at the end.

Taper. Specify an angle at which you want the coil to be tapered along its axis.

TIP: A spiral coil type cannot be tapered so this option will not appear for that method.

Use the coil ends options, as shown in Figure 7.125 right, to specify the start and end conditions of the coil. When selecting the Flat option, the helix, not the profile you selected for the coil, is flattened. Coil features can have unique conditions that are not consistent between the start and the end of the coil.

Close Start. Check box to show additional options for either Natural or Flat for the start of the helix.

Close End. Check box to show additional options for either Natural or Flat for the end of the helix.

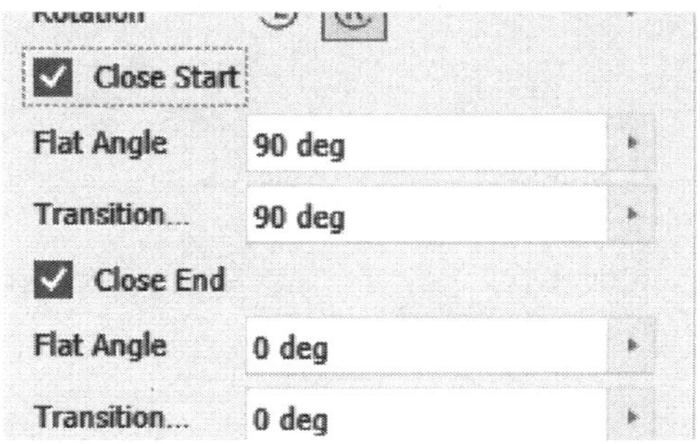

Figure 7.124

Flat Angle. This is the rotational angle, specified in degrees, that describes the amount of flat coil that extends after the transition. It specifies the transition from the end of the revolved profile into a flattened end.

Transition Angle. This is the rotational angle, specified in degrees, in which the coil achieves the coil start or end transition. It normally occurs in less than one revolution.

Click the right arrow to show the drop down menu for Measure or Select Feature Dimension.

Output

This area provides the options for how the coil is added to the part.

Boolean. The operation buttons are the column of buttons in the center of the dialog box. These buttons are only available if a feature exists before creating the Coil feature. By default, the Join operation is selected. Select to add or remove material from the part using the Join or Cut options or to keep what is common between the existing part volume and the completed coil feature using the Intersect option.

- Join: Adds material to the part.
- Cut: Removes material from the part.
- Intersect: Keeps what is common to the part and the coil feature.
- New solid: Creates a new solid body. The first solid feature that is created uses this option by default. Select it to create a new body in a part file with an existing solid body.

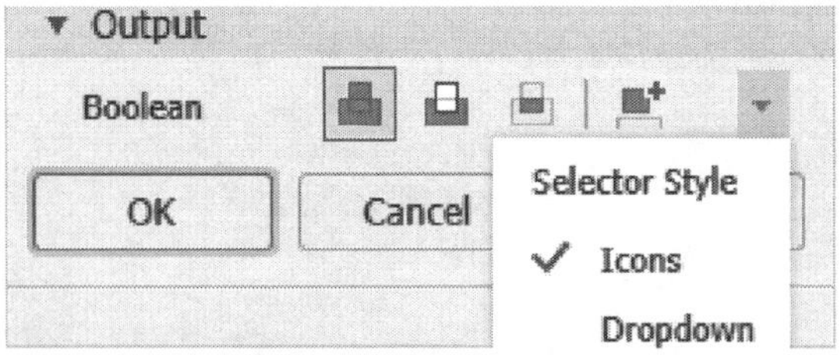

Figure 7.125

Figure 7.126 shows a coil created as the base feature. The image on the left shows the coil in its sketch stage; it contains a rectangle that will be used as the profile, and a centerline that will be used as the axis of rotation. The finished part, on the right of Figure 7.126, shows the coil with flat ends.

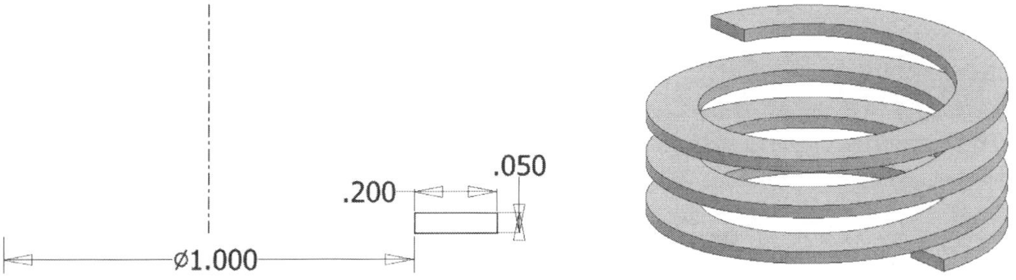

Figure 7.126

Figure 7.127 shows a coil created as a secondary feature. The image on the left shows the coil in its sketch stage with slice graphics on. The sketch, a rectangle, is drawn on the end of the cylinder with a centerline. The rectangle will be used as the profile, and the centerline will be used as the axis of rotation (a work axis could also be used for the axis of rotation). The finished part on the right shows the coil with flat ends on the bottom and top.

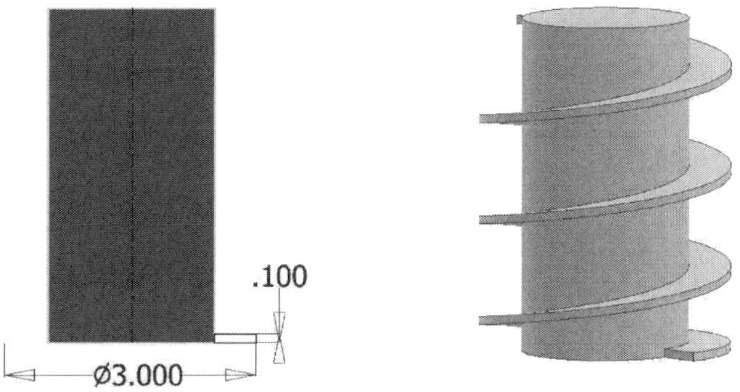

Figure 7.127

LOFT FEATURES

The Loft command creates a feature that blends a shape between two or more different sections or profiles. A point can be used to define the beginning and ending section of the loft. Lofts are used frequently when creating plastic or molded parts which have complex shapes that are difficult to create using other modeling techniques. A loft is like a sweep, but can have multiple sections and rails. You can create loft features that blend between two or more cross-section profiles that reside on different planes. You can also control the area of a specific section in the loft. You can use a rail, multiple rails, or a centerline to define a path(s) that the loft will follow. Four types of geometry are used to create a loft: sections, rails, centerlines, and points. The following sections describe these types of geometry.

Create a Loft

To create a loft feature, follow these steps:

1. Create the profiles or points that will be used as the sections to define the loft. If required, use work features, sketches, or projected geometry to position the profiles.
2. Create rails or a centerline that will be used to define the direction or control of the shape between sections.
3. Click the Loft command on the 3D Model tab > Create panel, as shown in Figure 7.128 left.
4. On the Curves tab of the Loft dialog box, the Sections option will be active. In the graphics window, select the sketches, face loops, or points in the order in which the loft sections will blend.
5. If rails are to be used to define the loft, click the "Click to add" text in the Rails area on the Curves

tab, and then in the graphics window select the rail or rails.

6. If needed, change the options for the loft on the Conditions and Transition tabs.

TIP: The loft options are also available by right-clicking in a blank area in the graphics window and clicking an option on the menu.

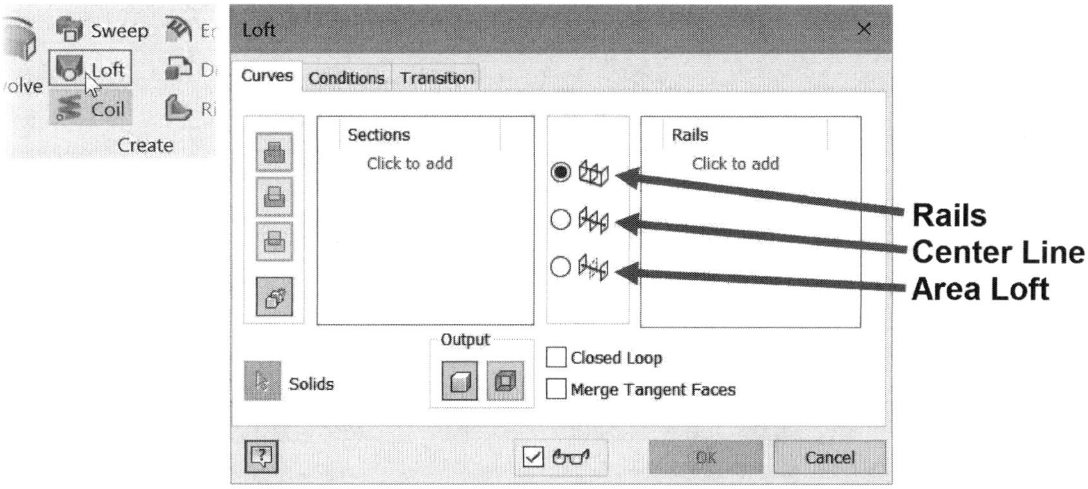

Figure 7.128

Figure 7.129 shows a loft created from two sections and a centerline rail. Figure 7.129 left shows two sections and a centerline in their sketch stages, shown in the top view. Figure 7.129 right shows the completed loft.

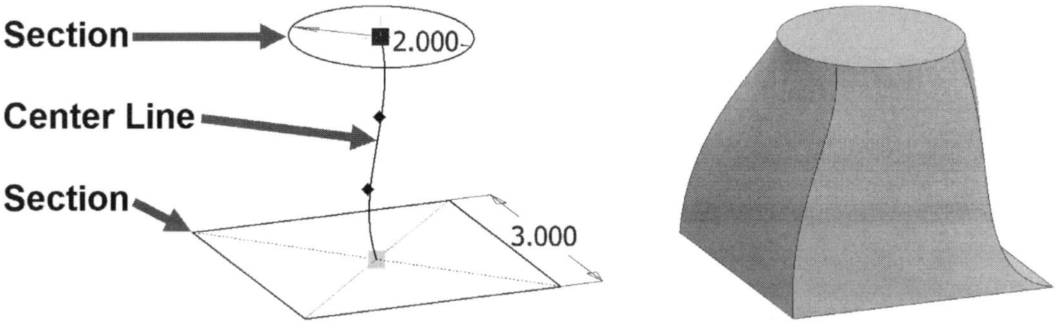

Figure 7.129

The following sections explain the options in the Loft dialog box and on its tabs.

Curves Tab

The Curves tab, as shown in the previous image, allows you to select which sketches, part edges, part faces, or points will be used as sections to select if a rail or centerline will be used and to determine the output condition.

Sections

You can define the shape(s) between which the loft will blend. The following rules apply to sections:

1. There is no limit to the number of sections that you can include in the loft feature.
2. Sections do not have to be sketched on parallel planes.
3. You can define sections with 2D sketches (planar), 3D sketches (nonplanar), and planar or nonplanar faces, edges on a part, or points.
4. All sections must be either open or closed. You cannot mix open and closed profile types within the

same loft operation. Open profiles result in a lofted surface.

Points

A point can be used as a section to help define the loft. The following rules apply to points used for a loft profile:

1. A point can be used to define the start or end of the loft.
2. An origin point, sketch point, center point, edge point, or work point can be used.

Rails

You can define rails using the following elements: 2D sketches (planar), 3D sketches (nonplanar), or part faces and edges. The following rules apply to rails:

1. There is no limit to the number of rails that you can create or include in the loft feature.
2. Rails must not cross each other and must not cross mapping curves.
3. Rails affect all the sections—not just faces or sections that they intersect. Section vertices without defined rails are influenced by neighboring rails.
4. All rail curves must be open or closed.
5. Closed rail curves define a closed loft, meaning that the first section is also the last section.
6. No two rails can have identical guide points, even though the curves themselves may be different.
7. Rails can extend beyond the first and last sections. Any part of a rail that comes before the first section or after the last is ignored.
8. You can apply a 2D or 3D sketch tangency or smooth constraint between the rail and the existing geometry on the model.

Center Line

A center line is treated like a rail, and the loft sections are held normal to the center line. When a centerline is used, it acts like a path used in the sweep feature, and it maintains a consistent transition between sections. The same rules apply to center lines as to rails except that the centerline does not need to intersect sections and only one centerline can be used.

Area Loft

Select a sketch to be used as the centerline, and then click on the centerline to define the area of the profile at the selected point. After picking a point, the Section Dimensions dialog box appears, as shown in the following image. You define its position either as a proportional or absolute distance, and you can define the section's size by area or scale factor related to the area of the original profile.

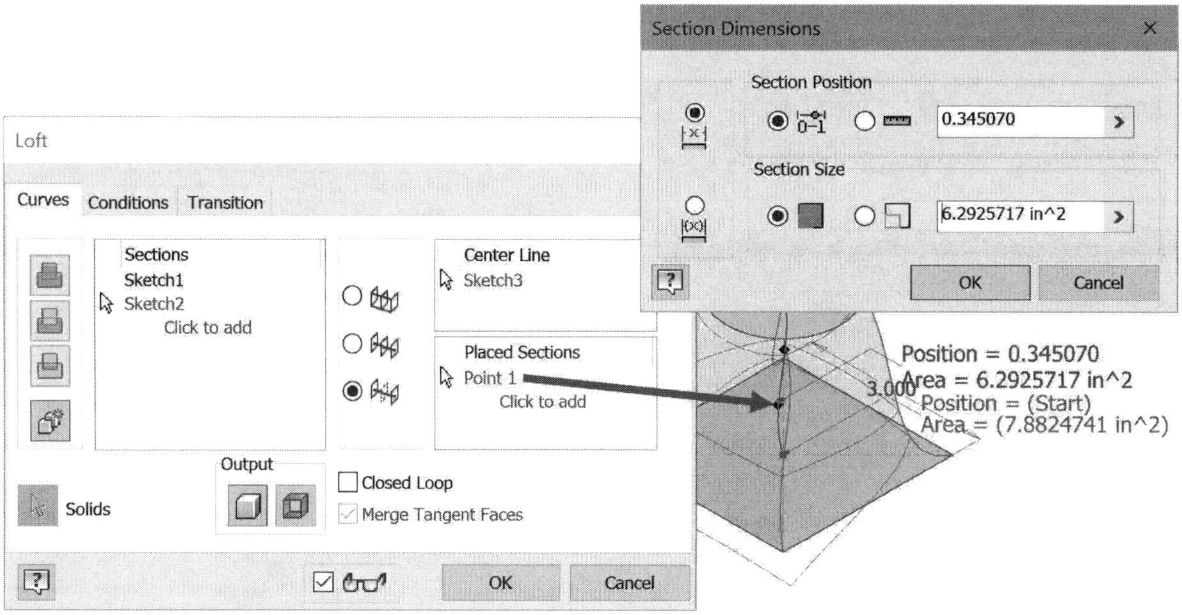

Figure 7.130

Output

Select whether the loft feature will be a solid or a surface. Open sketch profiles selected as loft sections will define a lofted surface.

Operation

Select an operation button to add or remove material from the part, using the Join or Cut options, or to keep what is common between the existing part and the completed loft using the Intersect option. By default, the Join operation is selected.

Closed Loop

Click to join the first and last sections of the loft feature to create a closed loop.

Merge Tangent Faces

This option is available when one of the sketches of the loft is on an existing face of a feature. This option joins the loft feature so it is tangent to the existing feature.

Conditions Tab

The Conditions tab, as shown in the following image, allows you to control the tangency condition, boundary angle, and weight condition of the loft feature. These settings affect how the faces on the loft feature relate to geometry at the start and end profiles of the loft. This may be existing part geometry or the plane or work plane containing the loft section sketch.

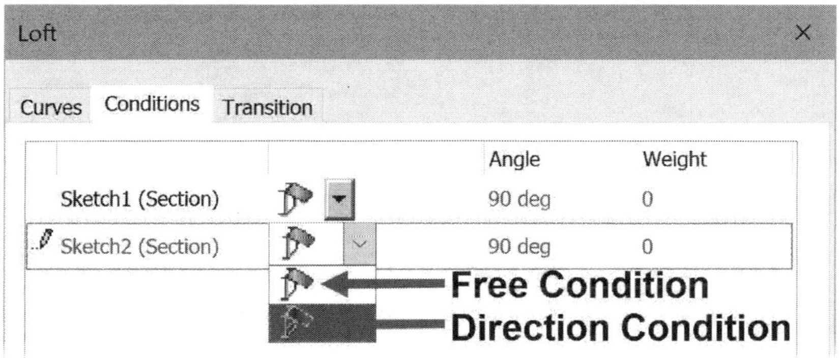

Figure 7.131

The column on the left lists the sketches and points specified for the sections. To change a sketch's condition, click on its name, and then select a condition option.

Conditions - Sketch

Two boundary conditions are available when the first or last section is a sketch.

- **Free Condition** (top button): With this option, there is no boundary condition, and the loft will blend between the sections in the most direct fashion.
- **Direction Condition** (bottom button): This option is only available when the curve is a 2D sketch. When selected, you specify the angle at which the loft will intersect or transition from the section.

Conditions - Edge

When a face loop or edges from a part are used to form a section, as shown in Figure 7.132 left, three conditions are available, as shown in Figure 7.132 right.

- Free Condition (top button): With this option, there is no boundary condition, and the loft will blend between the sections in the most direct fashion.
- Tangent Condition (middle button): With this option, the loft will be tangent to the adjacent section of the face.
- Smooth (G2) Condition (bottom button): With this option, the loft will have curvature continuity to the adjacent section of the face.

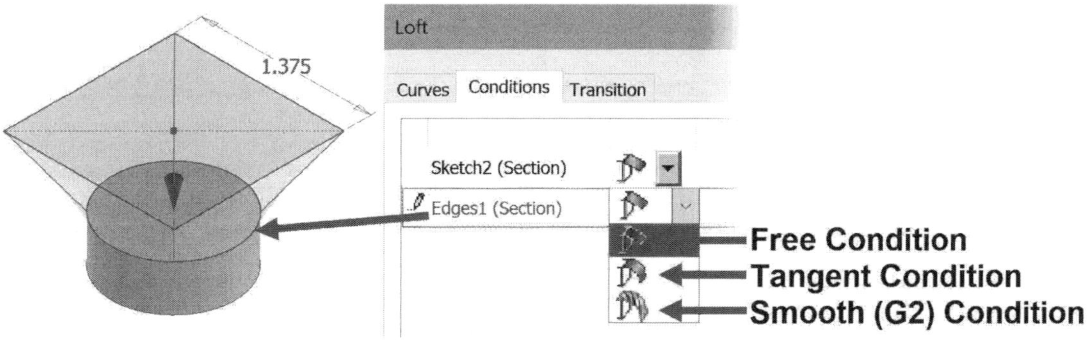

Figure 7.132

Conditions – Point

Three conditions are available when a point is selected for a loft profile, as shown in Figure 7.133.

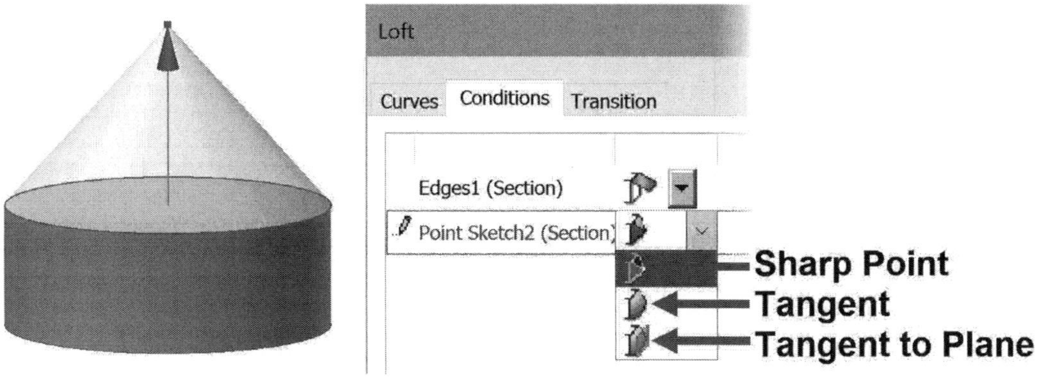

Figure 7.133

- Sharp Point (top button): With this option, there is no boundary condition, and the loft will blend from the previous section to the point in the most direct fashion.
- Tangent (middle button): When selected, the loft transitions to a rounded or domed shape at the point.
- Tangent to Plane (bottom button): When selected, the loft transitions to a rounded dome shape. You select a planar face or work plane to be tangent to. This option is not available when a centerline loft is used.

Angle

This option is enabled for a section only when the boundary condition is Tangent or Direction. The default is set to 90° and is measured relative to the profile plane. The option sets the value for an angle formed between the plane that the profile is on and the direction to the next cross-section of the loft feature. Valid entries range from 0.0000001° to 179.99999°.

Weight

The default is set to 0. The weight value controls the angle's influence on the tangency of the loft shape to the normal of the starting and ending profile. A small value creates an abrupt transition, and a large value creates a more gradual transition. High weight values may result in twisting the loft and cause a self-intersecting shape.

Transition Tab

The Transition tab, as shown in the following image with the Automatic mapping box unchecked, allows you to specify point sets. A point set is used to define section point relationships and control how segments blend from one section to the segments of the adjacent sections. Points are reoriented or added on to adjacent sections.

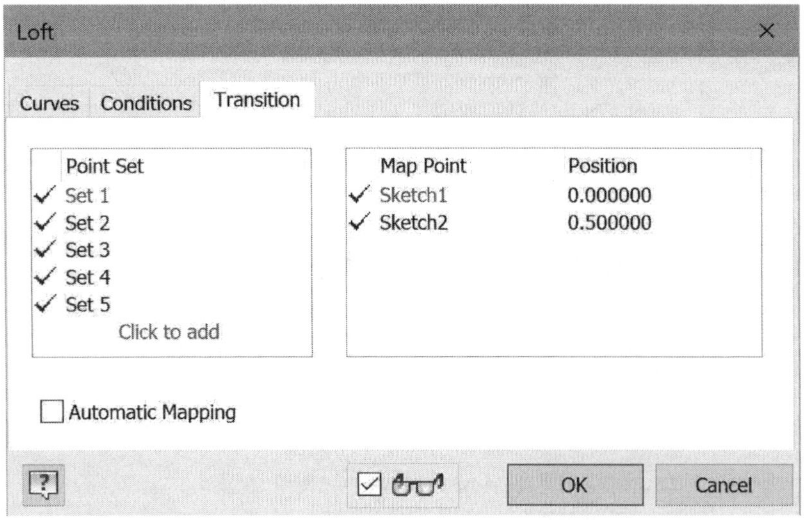

Figure 7.134

EXERCISE 7-7: CREATING A LOFT FEATURE

In this exercise, you use the loft command to define the shape of a razor handle.

1 Open *ESS_E07_07.ipt* in the Chapter 07 folder.

2 Edit Sketch1 by double-clicking on Sketch1 in the browser.

3 Click the Point, Center Point command on the Sketch tab > Create panel.

4 Place a point so it is coincident on the spline near the bottom-right of the curve, making sure that the coincident glyph is displayed as you place the point, as shown in Figure 7.135.

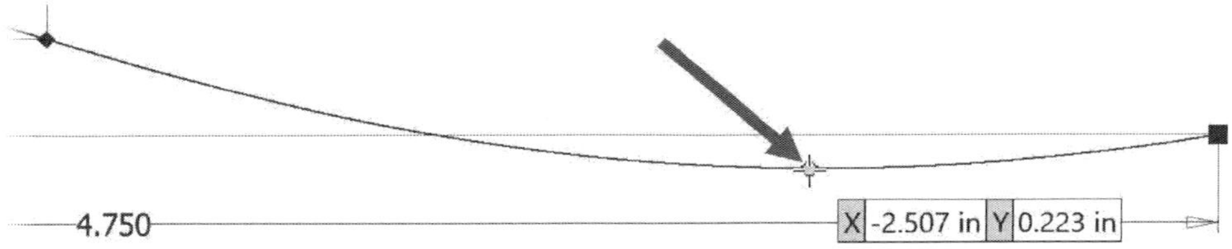

Figure 7.135

5 Draw two construction lines that are coincident with the sketched point you just placed and the nearest spline points on both sides of the point, as shown in Figure 7.136.

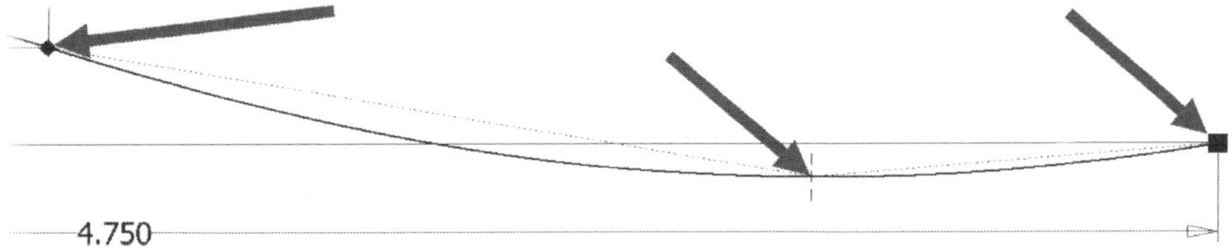

Figure 7.136

6 To parametrically position the sketched point midway between the spline points, place an equal constraint between the two construction lines you just created.

7 Change the 4.750 horizontal dimension to **5 inches**, and verify that the sketched point moves along the spline to maintain its position on the spline.

8 Finish the sketch.

9 Next, you create a work plane that a new profile will be located on. The work plane will be perpendicular to the spline and located at the point you just created.

a. Click the Work Plane Normal to Curve at Point command on the 3D Model tab > Work Features panel.

b. Locate the work plane by selecting the point you just created.

c. Position the work plane so it is perpendicular to the spline by selecting the spline, as shown in Figure 7.137 left (do not select the construction line).

10 Next, create a profile for a loft section on the new work plane.

a. Create a new sketch on the work plane you just created.

b. To see the sketch better, change to the Home View.

c. Use the Project Geometry command on the Sketch tab > Create panel and select the Point, Center Point you created in step 4.

d. Draw an ellipse with its center coincident with the projected point and the second point so the ellipse's axis is horizontal to the point. Click a third point, so the ellipse's axis is vertical to the point. The Ellipse command is under the Circle command on the Create panel.

e. Place **.1875 inch** and **.375 inch** dimensions to control the size of the ellipse, as shown in the image on the right.

f. Finish the sketch.

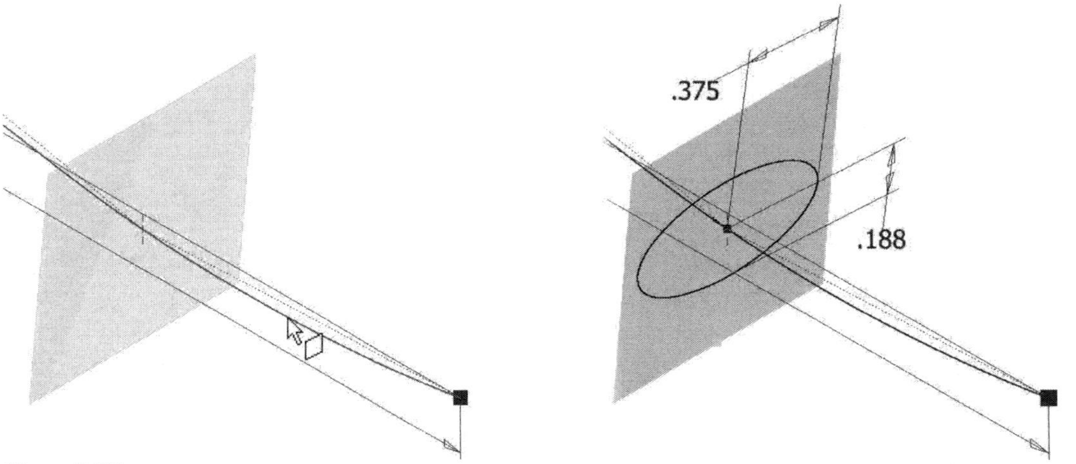

Figure 7.137

11 Click the Loft command on the 3D Model tab > Create panel.

a. For the first section, click the edge of the concave 3D face, as shown in Figure 7.138 left.

b. To define the remaining sections, select the other three profile sections from left to right, and then click the point on the right end of the spline as shown in Figure 7.138 right.

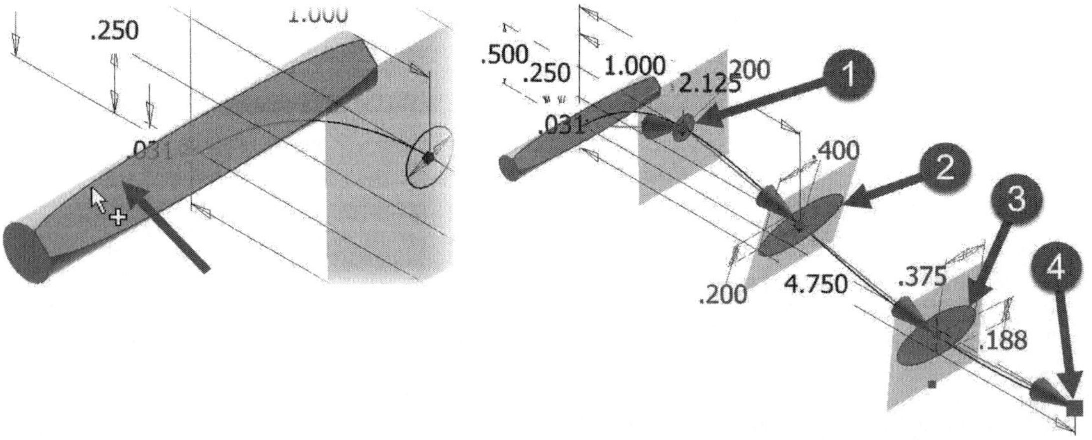

Figure 7.138

c. Click the Center Line option in the Loft dialog box, as shown in the following image on the left, and select the spline. When finished, the preview should resemble Figure 7.139 right.

d. Click OK to create the loft.

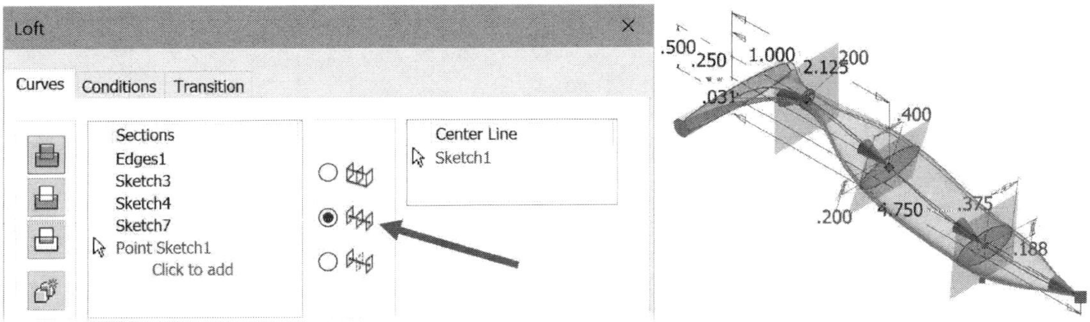

Figure 7.139

12 Turn off the visibility of all the work planes by pressing the ALT and the] keys and then use the Free Orbit command to examine the loft. Notice that the end of the loft on the right side is sharp.

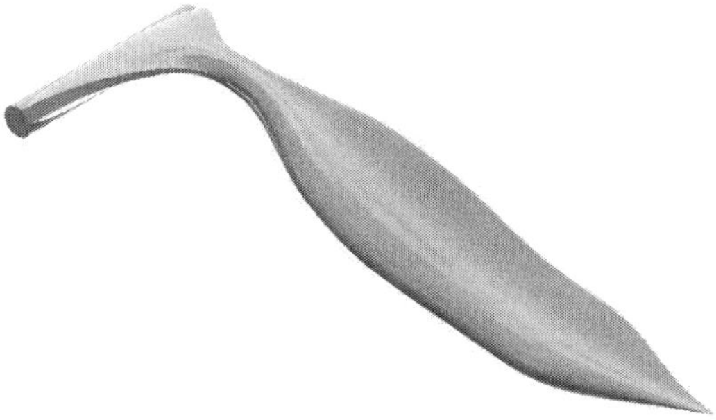

Figure 7.140

13 Next you edit the loft.

a. Edit the loft, select a face on the Loft in the graphics window, and click Edit Loft from the mini-toolbar that is displayed.

b. Click the Conditions tab, and change the Point Sketch1 entry to a Tangent condition, as shown in the following image on the left.

c. Click OK to update the loft. The bottom of the handle should resemble Figure 7.141 right.

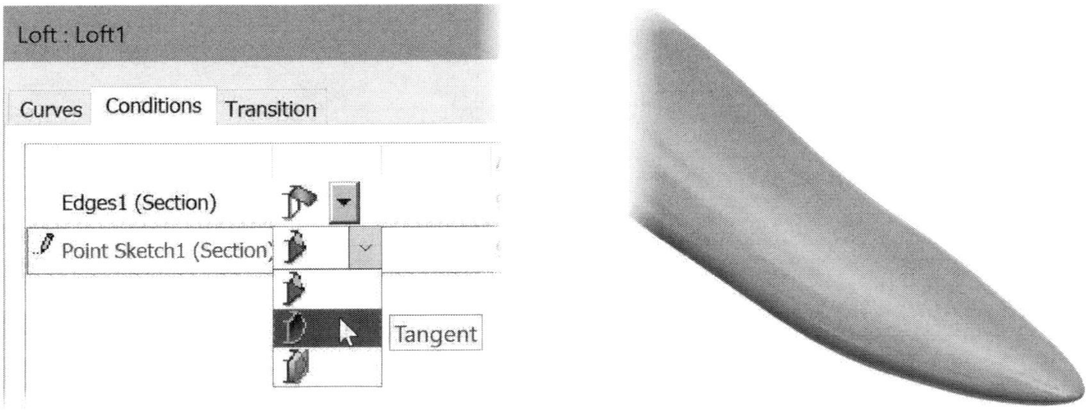

Figure 7.141

14 Next, edit the loft again and define a new section by area.

a. In the browser, double-click on the entry Loft1.

b. From the Curves tab, click the Area Loft option labeled (1) in Figure 7.141.

c. In the Placed Sections area, select Click to add, labeled (2).

d. In the graphics window, click a point near the middle of the spline labeled (3).

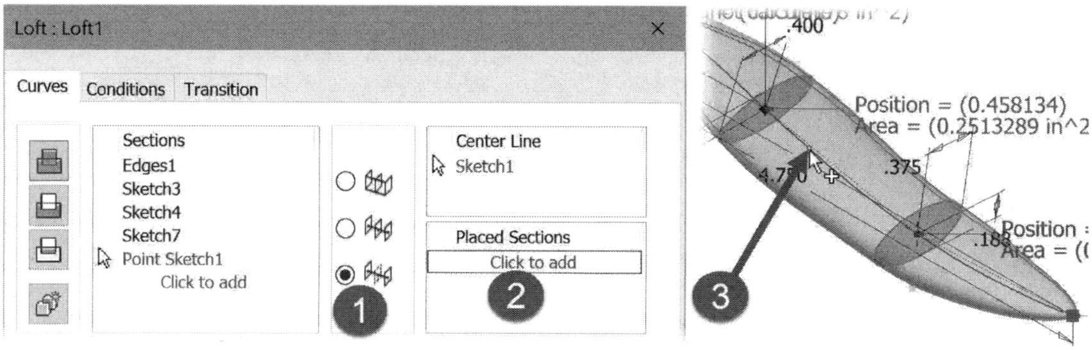

Figure 7.142

e. The Section Dimensions dialog box appears; in the Section Position area change the Proportional Distance option to **0.5** labeled (1) in the following image.

f. In the Section Size area change the Area option to **0.375 in^2** labeled (2).

g. Click OK in the Section Dimensions dialog box.

h. In the graphics window, notice the values of each section and new section half way through the loft.

i. Complete the edit by clicking OK. The image on the right shows the updated loft.

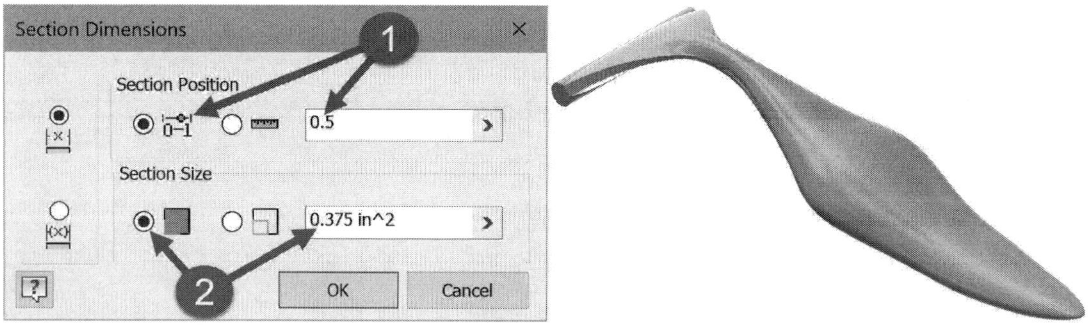

Figure 7.143

15 Close the file. Do not save changes.

End of exercise.

FREEFORM MODELING

If you need to model freeform or organic shapes that are difficult to create with traditional modeling commands, you can use the freeform tools that allow you to edit a model by simply clicking and dragging on an edge(s), face(s) or point(s). To create a freeform shape, follow these basic steps.

1. Start a command on the 3D Model tab > Create Freeform panel as shown in the following image. This will enable the Freeform environment.

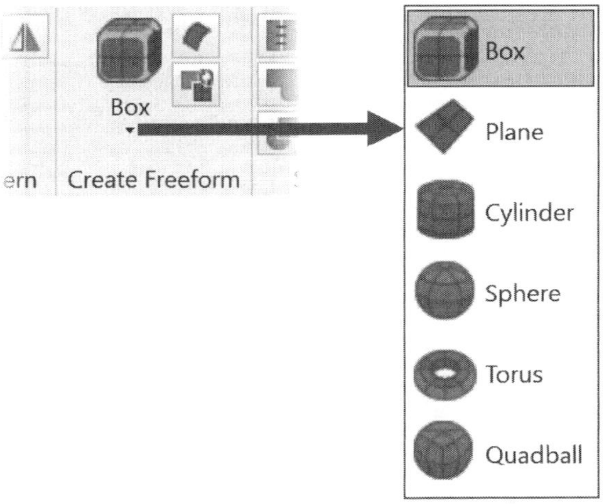

Figure 7.144

2. Select a planar face and start point.
3. Define the size of the shape by entering values in the dialog box or by dragging the arrows on the model. If needed you can also turn on symmetry. The following image shows a freeform box being created with faces added.

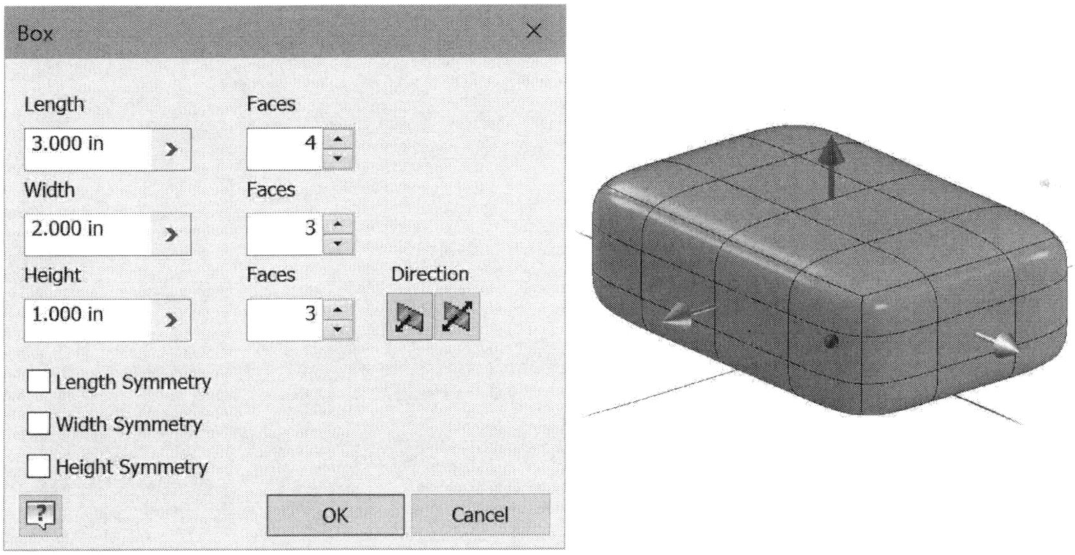

Figure 7.145

4. Select the geometry to edit. Hold down the CTRL or Shift key to add or remove objects, and double-click on an edge to select all the edges that go around the freeform shape.
5. To edit the selected geometry, right-click and click Edit Form from the menu as shown in the following image on the left or start the Edit Form command from the Freeform tab > Edit panel. The Edit panel appears when in the Freeform environment.
6. An Edit Form manipulator will appear. The image on the right shows the back-right face moved up. Use the following options on the manipulator to change the shape.
   a. **Move in a Direction**. Click and drag on an arrow 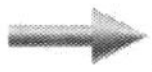.
   b. **Move in a Plane**. Click and drag on a square face 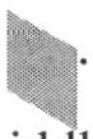.
   c. **Rotate**. Click and drag on a sphere that is in the middle of an arc with dashed lines .

d. **Scale Globally**. Click and drag on the sphere that is in the middle of the manipulator.

e. **Scale Along an Axis**. Click and drag in the cylinder that is in the middle of the center sphere and arrow.

f. **Scale Along a Plane**. Click and drag on the quarter-circle.

g. **Add or Remove Material**. Hold down the ALT key when moving and scaling faces.

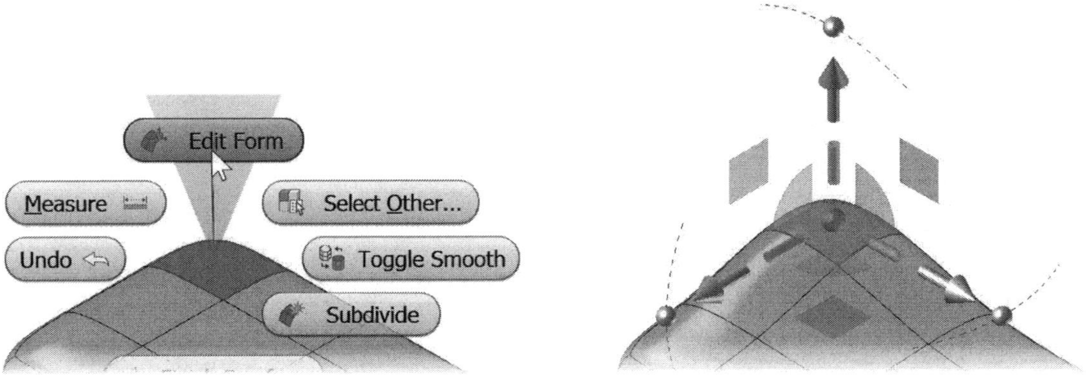

Figure 7.146

7. Use the Edit Form dialog box controls to edit the form as shown in Figure 7.147.

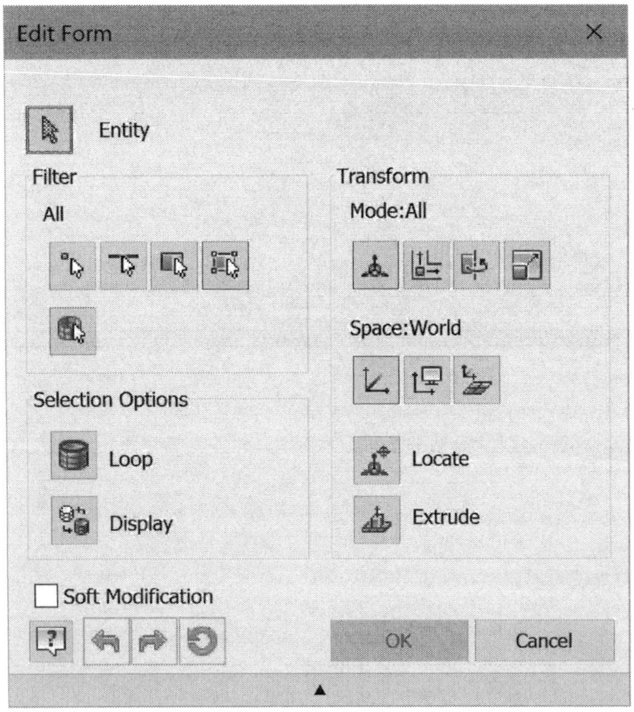

Figure 7.147

8. Continue to edit the model using the tools in the Freeform tab as shown in the following image.

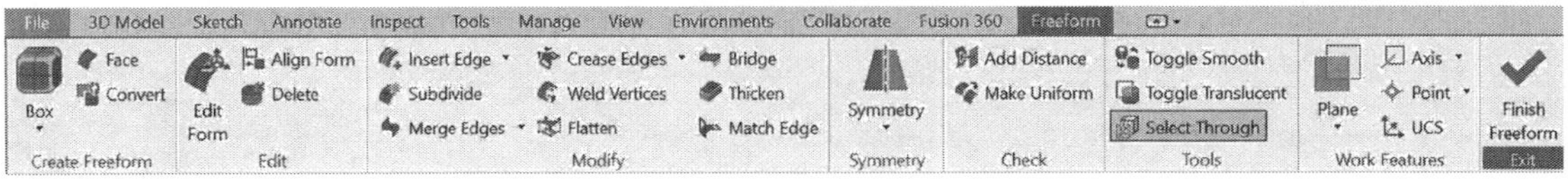

Figure 7.148

9. Click the Finish Freeform command on the Freeform tab > Exit panel to finish.

10. Add regular features to the model as needed.
11. To edit a Freeform body, double-click on the Form entry in the browser or right-click the feature in the browser and choose Edit Freeform as shown in Figure 7.149.

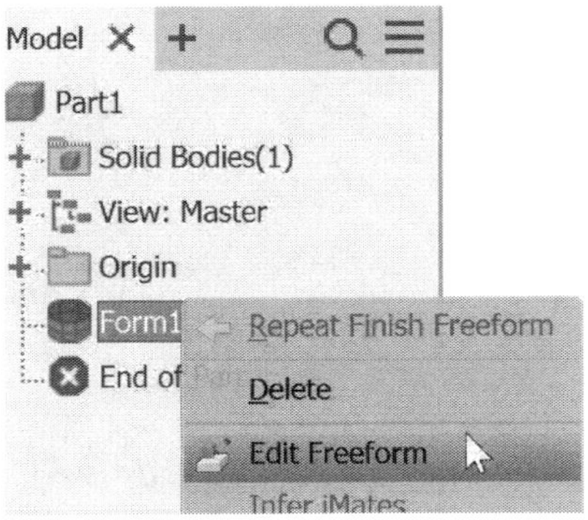

Figure 7.149

SPLIT A PART OR FACE

The Split command allows you to split a part into two solids, split the solid by removing one portion of the part, to split individual faces, or to split all faces. A typical application to split a face is to allow the creation of face drafts to the split faces of a part. You can use the Split command to perform the following:

- Split a solid into multiple solid bodies.
- Remove a section of the part by using a surface, planar face, or a work plane and to cut material from the part in the direction you specify. The side that is removed is suppressed rather than deleted. To create a part with the other side removed, edit the split feature, redefine it to keep the other side, save the other half of the part to its own file using the Save Copy As option, or create a derived part.
- Split individual faces using a surface, sketching a parting line, or placing a work plane, and then selecting the faces to split. You can edit the split feature to add or remove the desired part faces to be split.

The Split command is located on the 3D Model tab > Modify panel, as shown in Figure 7.150 left. Once selected, the Split dialog box appears as shown in Figure 7.150 right.

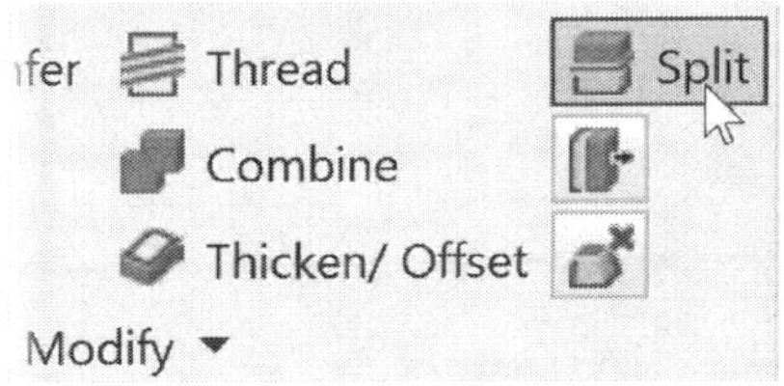

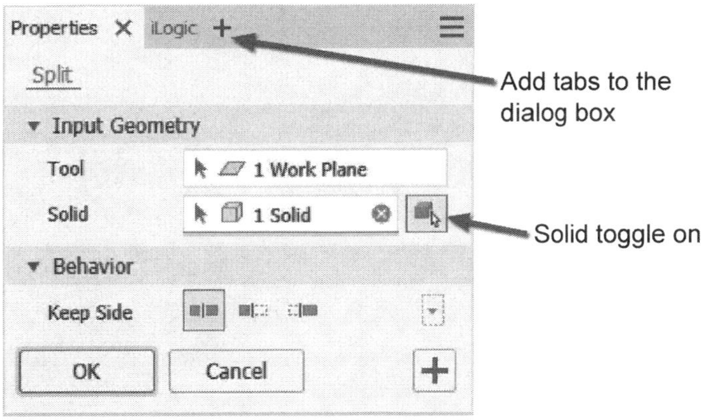

Figure 7.150

The Split dialog box contains the following sections:

Input Geometry

Tool. The button is preselected. Select a surface, work plane, or sketch to use to split the part.

Faces. The Faces selection is available only when you have the Solid toggle off. See Figure 7.150 left. Select the specific faces that you want to split as shown in Figure 7.151. Use the check box to select all faces to save time when appropriate.

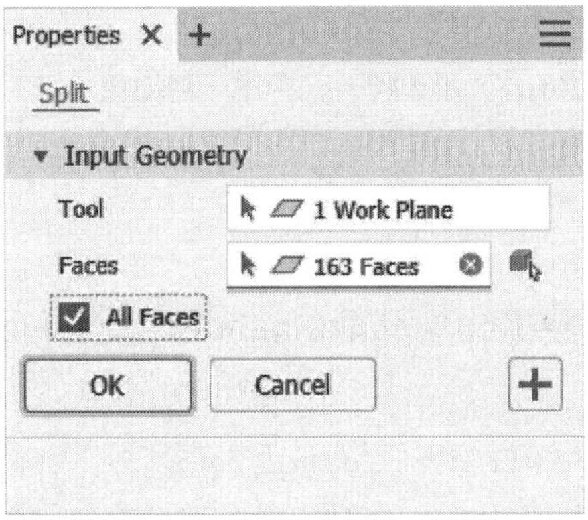

Figure 7.151

Solid. Click the Solid toggle icon as shown in Figure 7.150 left. When selected all faces of the solid body that intersect the tool selected are sliced.

Behavior

Keep Side. Click to keep both sides, keep the default side, or the opposite side. These icons or drop down menu only shows when the Solid toggle is on.

EXERCISE 7-8: SPLITTING A PART INTO MULTIPLE SOLID BODIES

In this exercise, you split a part into multiple solid bodies and export them to an assembly. For more information about working with multi-body parts consult the help system.

1 Open *ESS_E07_08.ipt* in the Chapter 07 folder.

2 First, you use an origin plane to split the part into two solids.

a. Start the Split command from the 3D Model tab > Modify panel.

b. Click to toggle the Solid method button on as shown in the following image.

c. For the Tool, click XY Plane from the Origin folder in the browser.

d. Click OK to split the part into two solids.

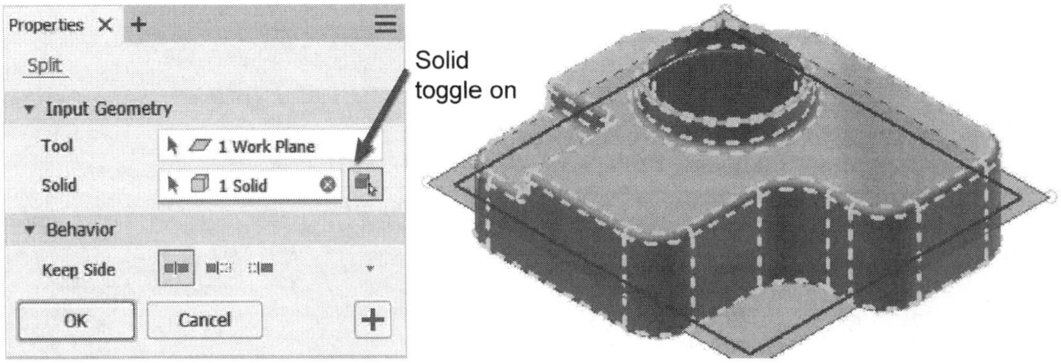

Figure 7.152

3 In the browser, expand the Solid Bodies folder and click on Solid2 and then Solid3 to verify that another solid has been created.

4 Next, you turn off the visibility of a solid. In the browser, in the Solid Bodies folder, right-click Solid2 and click Visibility on the menu.

5 Use the Zoom and Free Orbit commands to examine the solid.

6 Next, you switch the visibility of the solids. In the browser, in the Solid Bodies folder, right-click Solid2 and on the menu, click Hide Others.

7 Use the Zoom and Free Orbit commands to examine the solid.

8 Turn the visibility of Solid3 back on, under the Solid Bodies folder, right-click Solid3 and click Visibility or Show All on the menu.

9 Next, you create a hole feature.

   a. From the 3D Model tab > Modify panel click the Hole command.

   b. Add a **.25 inch** diameter simple hole that is concentric to the top right circular edge of the top solid, and the termination set to Through All, as shown in the following image.

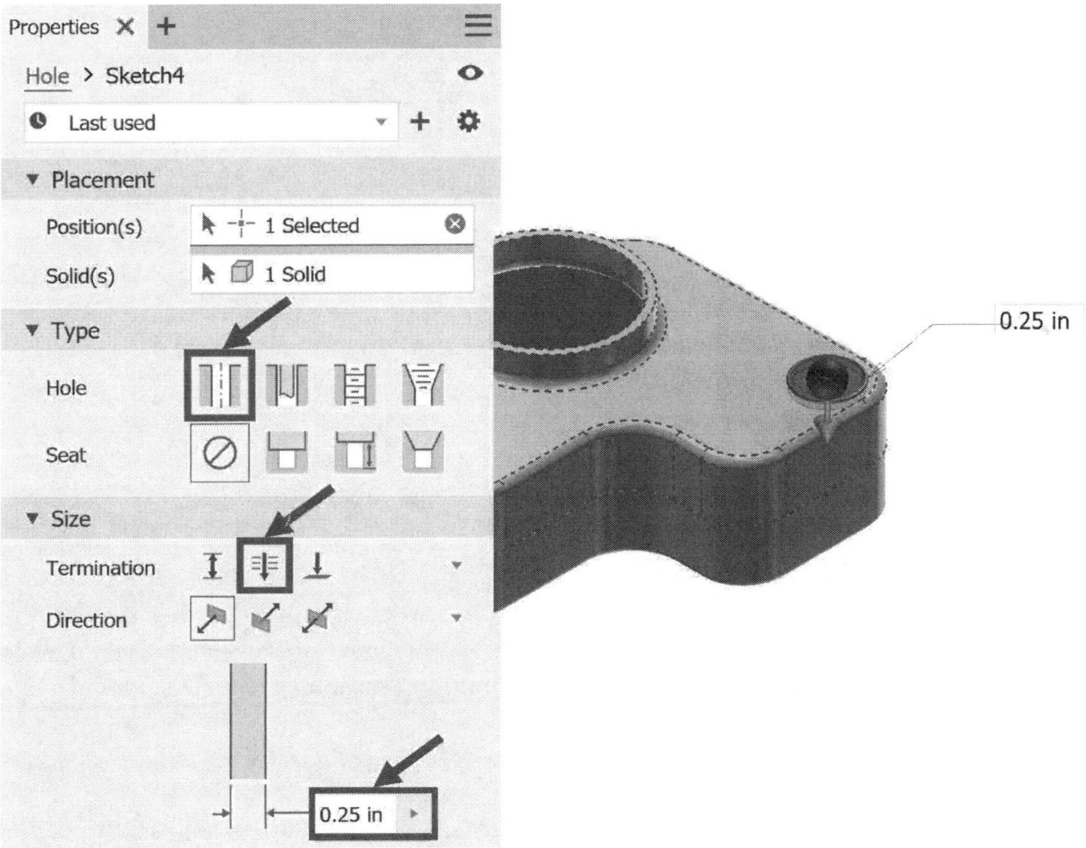

Figure 7.153

10 After creating the hole, use the Free Orbit command to examine the bottom solid and notice that the hole does not go through the bottom solid.

11 Change to the Home View.

12 In the browser, expand the Solid Bodies folder and expand the Solid2 and Solid3 entries. Notice that

Hole1 is only under Solid2, as shown in Figure 7.153 left.

13 Edit the Hole1 feature.

a. In the Hole Properties panel click in the Solids cell, as shown in the middle image.

b. Select the bottom solid (Solid3) in the graphics window, as shown in the image on the right.

c. Click OK to complete the edit.

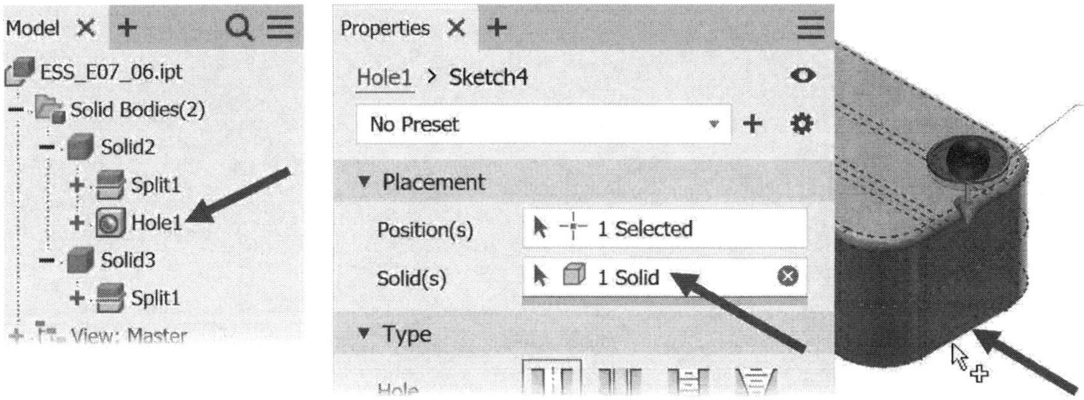

Figure 7.154

14 Use the Free Orbit command to ensure that the hole now goes through both solids.

15 In the browser, also notice that the Hole1 feature is now listed under both Solid2 and Solid3.

16 If desired use the Make Part command from the Manage tab > Layout panel to place a solid into a part file. Consult the Help system for more information about this procedure.

17 Close the files. Do not save changes.

End of exercise.

MIRROR FEATURES

When creating a part that has features that are mirror images of one another, you can use the Mirror Feature command to mirror a feature(s) about a planar face or work plane instead of recreating the features from scratch. Before mirroring a feature, a work plane or planar face to use as the mirror plane must exist. The mirror plane can be a planar face, a work plane on the part, or an origin plane.

The mirrored feature(s) are dependent on the parent feature—if the parent feature(s) change, the resulting mirror feature will also update to reflect the change. To mirror a feature or group of features, use the Mirror command from the 3D Model tab > Pattern panel, as shown in Figure 7.155 left. The Mirror dialog box will appear, as shown in the following image on the right. The following sections explain the options in the Mirror dialog box.

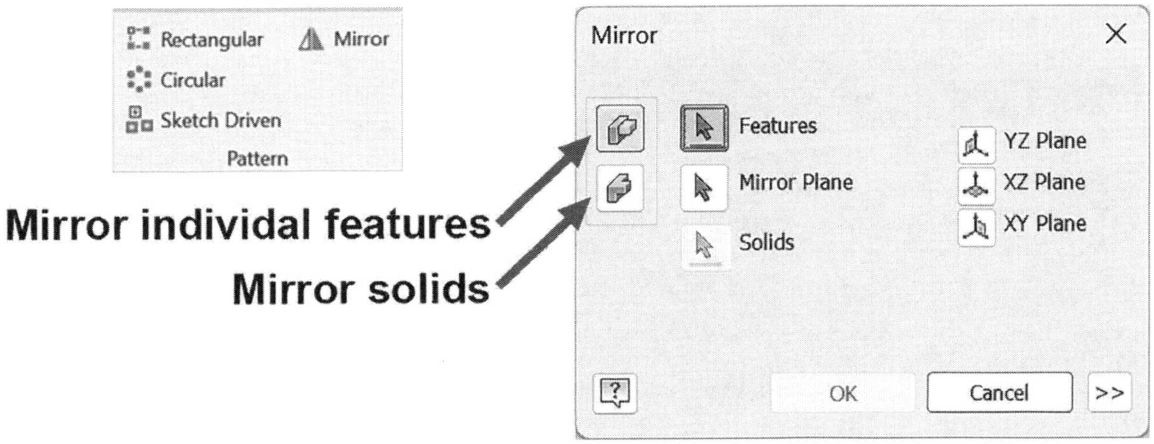

Figure 7.155

Mirror Individual Features. Click this button to mirror a feature or features.

Mirror a Solid. Click this button to mirror the solid body.

Features. Click the button to select the feature or features to mirror.

Mirror Plane. Click this button and then select a planar face or work plane to mirror the feature(s) about. You can also click one of the origin plane icons in the Mirror dialog box YZ , XZ , or XY .

Solid. If there are multiple solid bodies, click this button to choose the solid body(ies) to receive the mirrored feature.

To mirror a feature or features, follow these steps:

1. Click Mirror Feature from the 3D Model tab > Pattern panel to show the Mirror dialog box.
2. Click the Mirror individual features option or the Mirror solids option.
3. Select the feature(s) or solid body to mirror.
4. Click the Mirror Plane button, and select the plane on which the feature(s) will be mirrored about.
5. Click the OK to complete the operation. Figure 7.156 left shows a part with the features to be mirrored, the middle image shows the part with a work plane the features will be mirrored about. The image at right shows the left-vertical and the bottom slot features mirrored about the work plane.

Figure 7.156

SUPPRESSING FEATURES

You can suppress a model's feature or features to temporarily turn off their display and not calculate them. Feature suppression can be used to simplify parts, which may increase system performance. This capability can also be used to access faces and edges that you would not otherwise be able to access. If you need to dimension to a theoretical intersection of an edge that was filleted, for example, you could suppress the fillet and add the dimension and then unsuppress the fillet feature. If the feature you suppress is a parent feature for other dependent features, the child features will also be suppressed. Features that are suppressed appear gray in the browser and have a line drawn through them, as shown in the following image on the right. A suppressed feature will remain suppressed until it is unsuppressed, which will also return the features to their unsuppressed state. The following image on the left shows a part with no features suppressed and the following image on the right shows the suppressed extrusion and its dependent features suppressed.

To suppress and unsuppress a feature in a part, use one of these methods:

1. Right-click on the feature in the browser, and select Suppress Features from the menu as shown in Figure 7.157 left.
2. To unsuppress a feature, right-click on the suppressed feature's name in the browser and click Unsuppress Features from the menu.

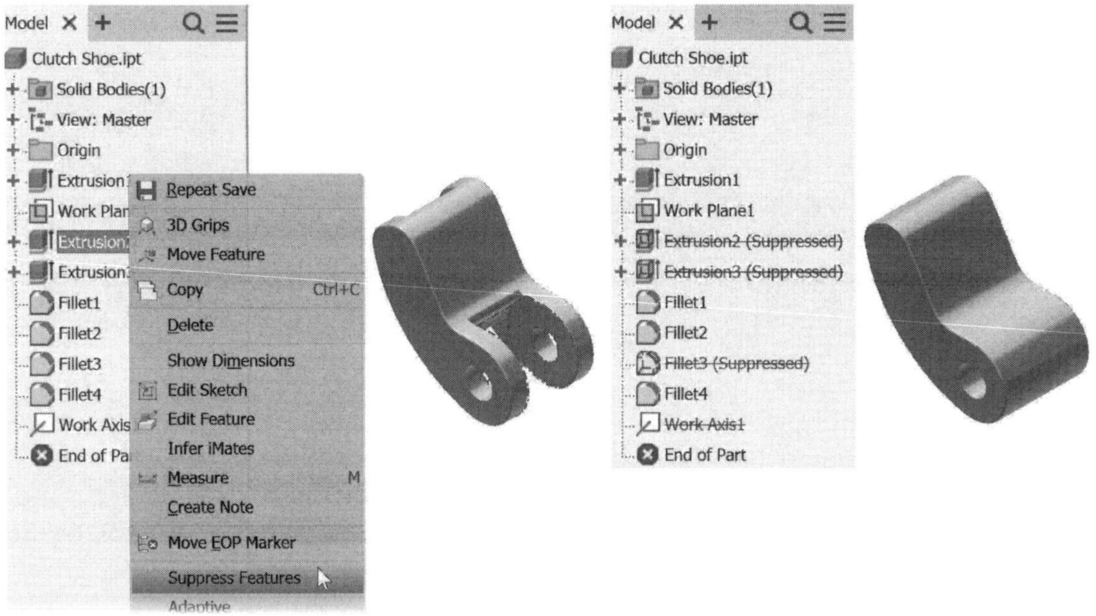

Figure 7.157

REORDERING A FEATURE

While designing, you may create features in an order that you would like to change. Inventor allows you to reorder a feature in the browser if there are no dependent features. For example, if you created a fillet feature using the All Fillets or All Rounds option and then created an additional extruded feature, such as a boss, you can reorder the fillet feature so it is below the boss in the browser. By reordering the fillet feature, so it is after the extrusion of the boss, the edges of the boss will be filleted. To reorder features, follow these steps:

1. Click the feature's name or icon in the browser. Hold down the left mouse button and click and drag the feature to the desired location in the browser. A horizontal line appears in the browser to show you the feature's relative location while it is being reordered. Figure 7.158 left shows a hole feature being reordered in the browser.
2. Release the mouse button and the model features will be recalculated in their new sequence. Figure 7.158 right shows the browser and the reordered hole feature.

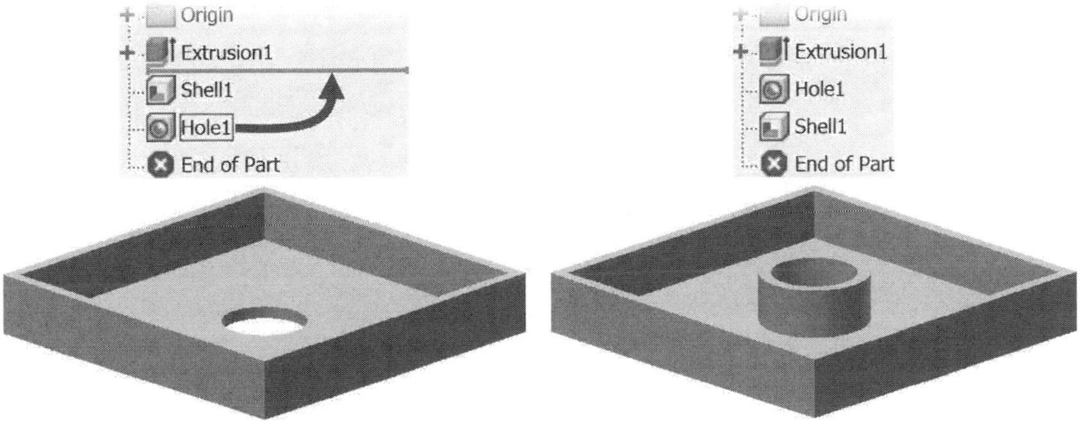

Figure 7.158

If you cannot move the feature due to parent-child relationships with other features, Autodesk Inventor will not allow you to drag the feature to the new position. In the browser, the cursor will change to a No symbol instead of a horizontal line, as shown in Figure 7.159.

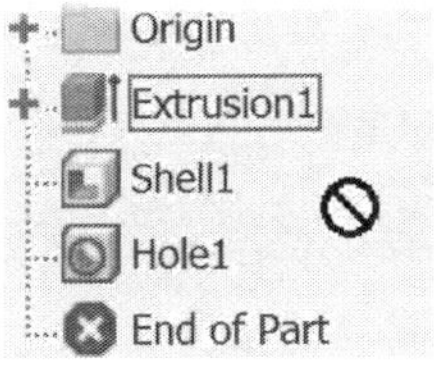

Figure 7.159

FEATURE ROLLBACK

While designing, you may not always place features in the order that your design later needs. In the last section, you learned how to reorder features, but reordering features will not always allow you to create the desired results. To help solve this issue, you can roll back the design to an earlier state and then place the additional new features. To roll back a design, follow these steps:

Note that Figure 7.159 left shows the original part.

1. Move the cursor over the End of Part marker in the browser.
2. With the left mouse button depressed, drag the End of Part Marker to the new location in the browser. While dragging the marker, a line will appear, as shown in the middle of Figure 7.160.
3. Release the mouse button, and the features below the End of Part marker are temporarily removed from calculation of the part. The End of Part marker is moved to its new location in the browser.
4. Another method is to right-click on a feature in the browser and click Move EOP Marker from the menu, as shown in Figure 7.159 right. The End of Part Marker moves below the selected feature.

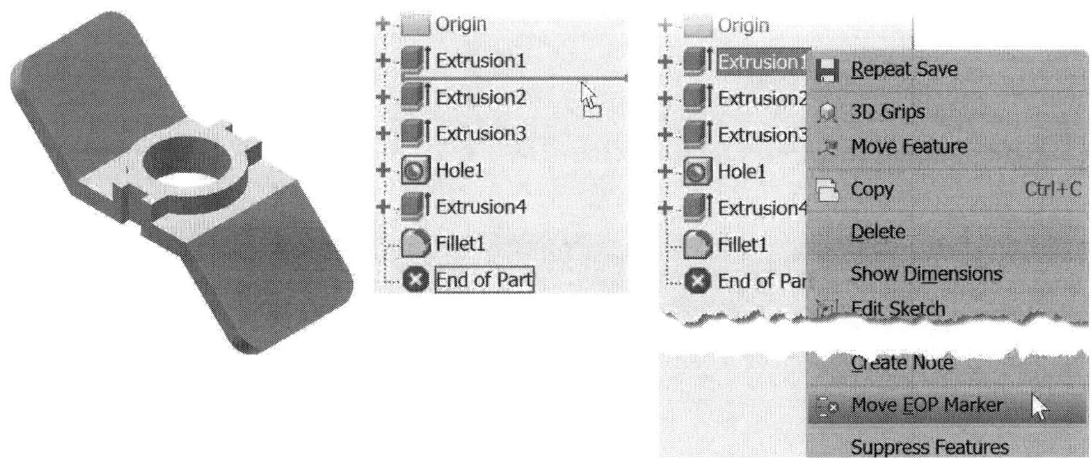

Figure 7.160

1. Figure 7.161 left shows the browser and part after moving the End of Part Marker below Extrusion 1. Then you can create new features as needed. The new features will appear in the browser above the End of Part Marker.
2. To return the part to its original state, including the new features, drag the End of Part marker below the last feature in the browser or right-click on the End of Part marker and click Move EOP to End.
3. If needed, you can delete all features below the End of Part marker by right-clicking on the End of Part marker and click Delete All Features Below EOP, as shown in Figure 7.161 right.

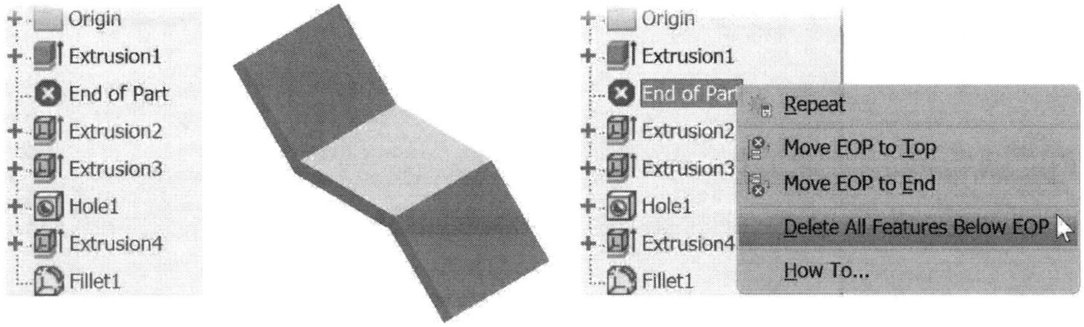

Figure 7.161

CONTENT CENTER

Autodesk Inventor's Content Center contains many standard components. It contains thousands of parts such as screws, nuts, bolts, washers, pins, and so on. You can place these standard components into an assembly file using the Place from Content Center command that is found on the Assemble tab > Component panel, as shown in the following image on the left.

After the command is selected, the Place from Content Center dialog box opens, as shown in the following image on the right, and you can navigate between items that are either included or published to the Content Center. Select the item you want to place, click OK, and place the part as you would any other assembly component.

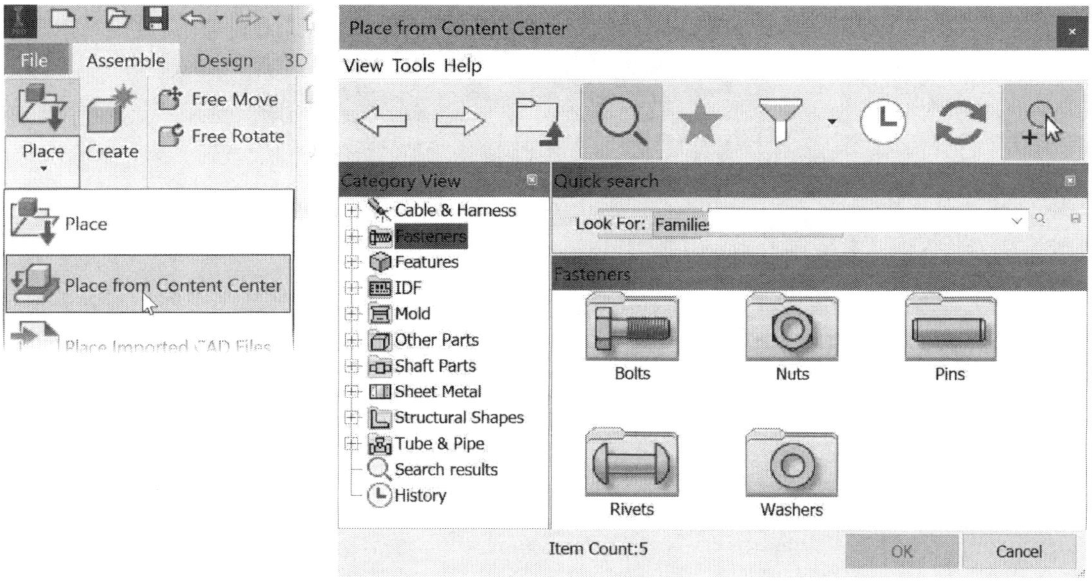

Figure 7.162

In addition to using the default items in the Content Center, you can publish your own features and parts to the Content Center. You publish parts and features using the Editor or Batch Publish commands, found on the Manage tab > Content Center panel as shown in the following image.

Figure 7.163

To publish content to the Content Center, a read/write library must exist and be added to the active project file.

TIP: For more information about the Content Center, publishing features, publishing parts, or configuring libraries, refer to Autodesk Inventor's Help system.

INTRODUCTION TO STRESS ANALYSIS

Autodesk Inventor Professional has the functionality to simulate the stress that will be applied to a part or an assembly which it will encounter when used in a real-world environment. Before a simulation can be performed, a material must be assigned, constraints established, and a load applied, as shown in the following image on the left. Autodesk Inventor uses Finite Element Analysis (FEA) to create a grid across the surface, called a mesh, as shown in the middle of the following image. The mesh divides the object into a continuous set of elements. Using properties of the assigned materials, Inventor calculates the stress in each element as shown in the following image on the right. Note that Inventor allows you to create multiple simulations on the same part or assembly, allowing you to test different designs and determine which design is best.

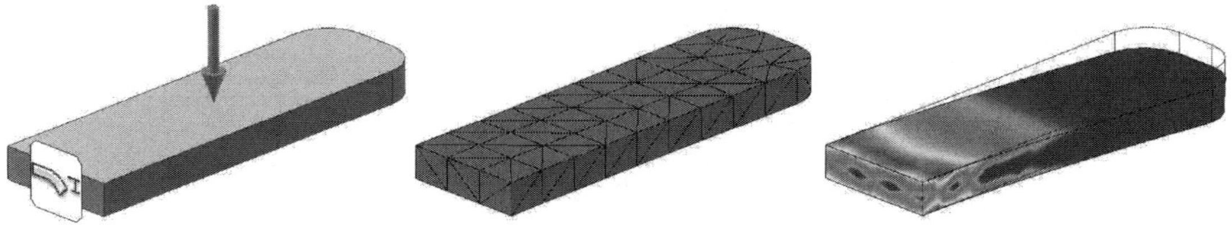

Figure 7.164

The following sections review the steps to run a stress analysis / simulation in Inventor Professional.

Create a Simulation Study – Step 1

The first step when performing a Stress Analysis is to create a simulation. Follow these steps to create a simulation in a part or an assembly file.

1. Start the simulation environment by clicking Stress Analysis from the 3D Model tab > Simulation panel, as shown in Figure 7.165 left. The Stress Analysis tab will become current.
2. Start a simulation study by clicking the Create Study command on the Analysis tab > Manage panel, as shown in the middle image.
3. The Create New Simulation dialog will appear, as shown in the image on the right.
4. Before a study can be created, you must select the type of simulation that will be created: Static or Modal Analysis. After selecting the type of simulation, click OK.

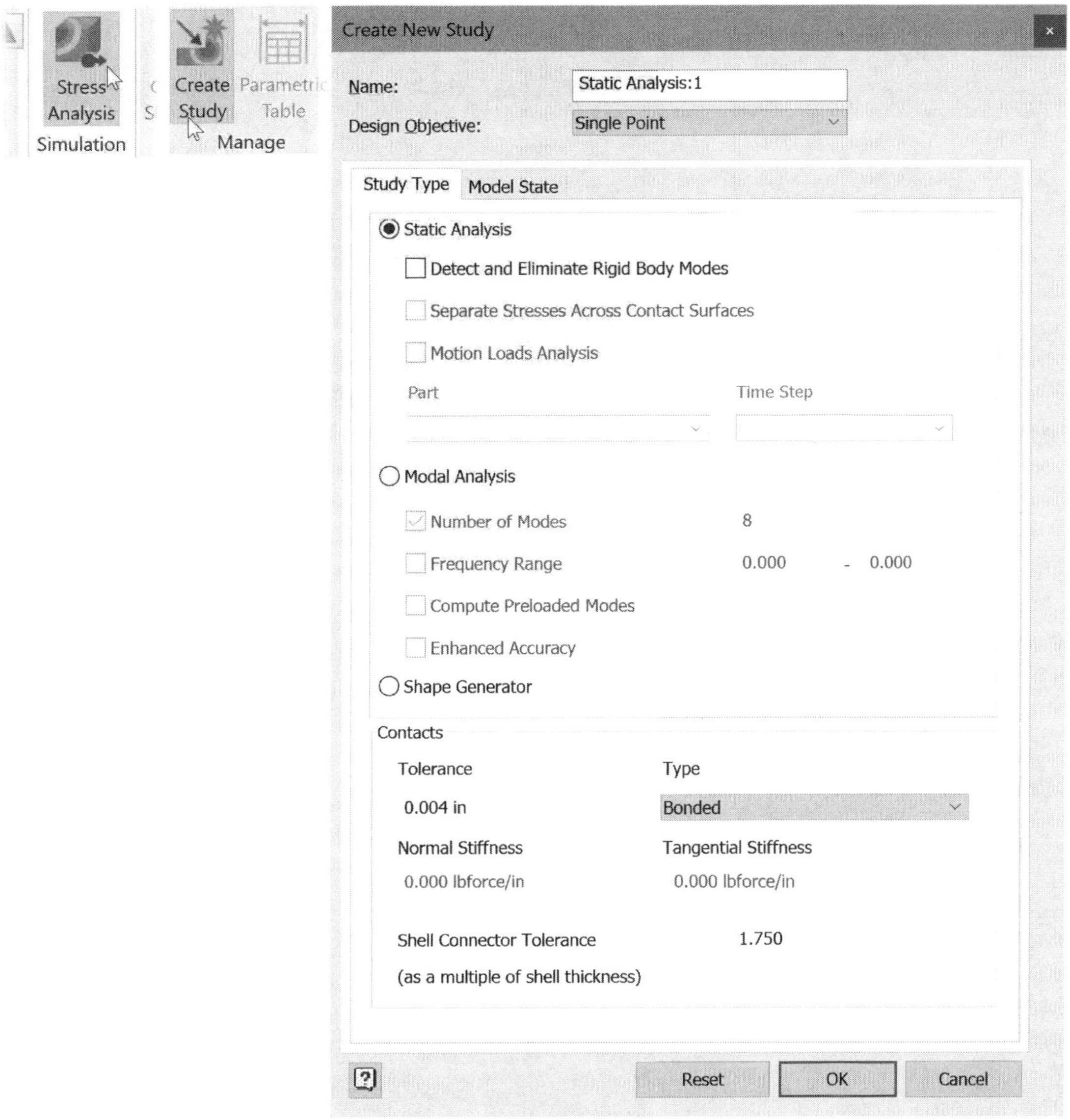

Figure 7.165

The following description explains the two types of simulations.

Static Analysis

Static analysis performs a simulation that calculates stresses and displacement. The simulation helps you determine the simple structural loading conditions of the part without motion and the load is constant (no impact or changing of the load values).

Modal Analysis

Modal Analysis performs a simulation that calculates the dynamic properties of a model with different frequencies of vibration. Rigid body movements are also a consideration within this type of analysis.

Assign Material – Step 2

The second step is to define the material. Before an analysis can be performed, a material must be assigned to each part. The material can be applied at the part level as was discussed in Chapter 3 Part Material, Properties and Appearance section, or you can override the material in the simulation. To assign a material while in a simulation, click the Assign command on the Analysis tab > Material panel, as shown in the following image on the left. The Assign Materials dialog box will appear, as shown in Figure 7.166 right. From the Override Material drop down menu, select the desired material. The material selected in the override will not change the material set in the part's iProperties.

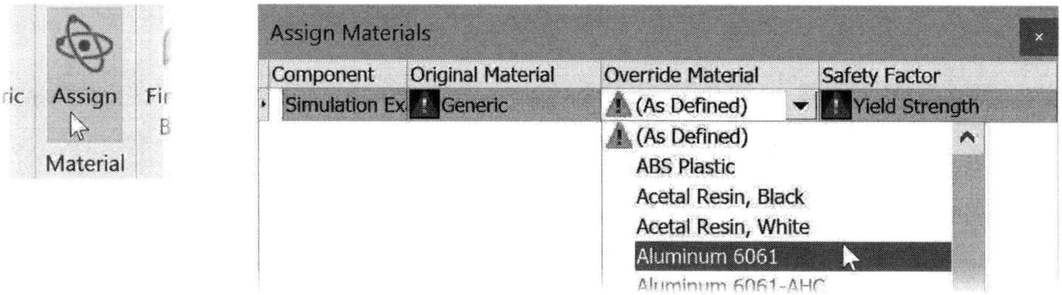

Figure 7.166

Applying Constraints – Step 3

The third step after the material(s) is assigned is to constrain the model. These constraints represent the conditions the part or assembly will experience in the real world. Inventor has three constraints within the Stress Analysis environment: fixed, pin and frictionless. These constraints are in the Analysis tab > Constraints panel, as shown in the following image.

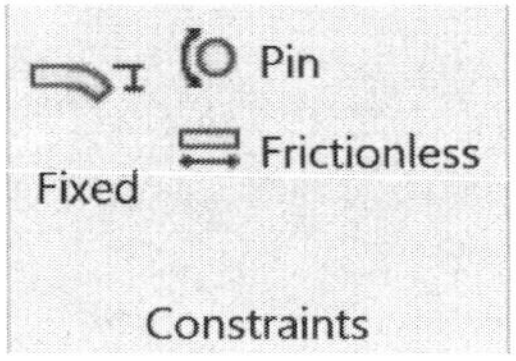

Figure 7.167

Fixed, Pin, and Frictionless constraints constrain the part or assembly to assure that the model reacts to an applied force in a realistic manner. Before the proper constraint can be applied, you must determine how the constraints will limit the movement of the model. You need to understand the three constraints that you can apply to the model that emulate the real-world scenario that the simulation represents. A brief description of each constraint follows.

Fixed

A fixed constraint removes all the degrees of freedom on the selected face(s) or edge(s). You apply a fix constraint when no translational or rotational movements are permitted. You may apply multiple fix constraints on different faces or edges on the same model.

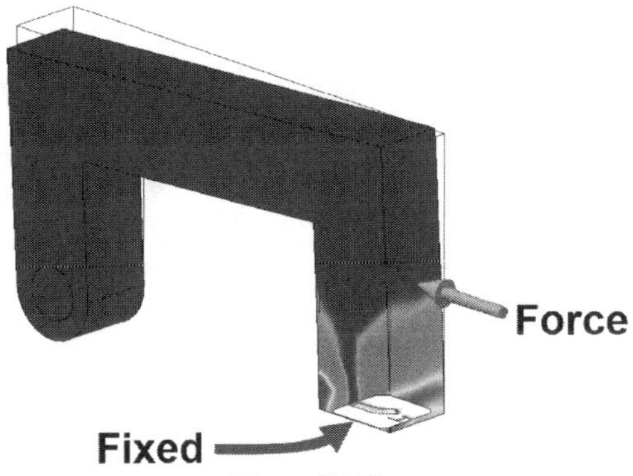

Figure 7.168

Pin

A Pin constraint removes translation degrees of freedom along the axes of a cylindrical face (no linear or sliding movement is allowed) but allows rotation. Apply this constraint if rotational movement is desired. A pin constraint can only be applied to a cylindrical face. A Pin constraint can work in conjunction with the other constraints.

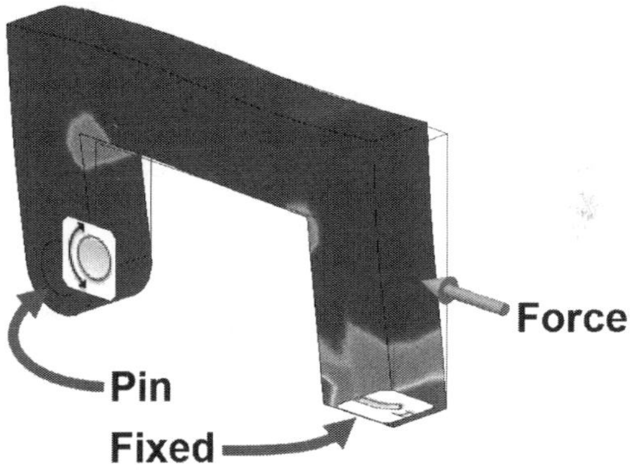

Figure 7.169

Frictionless

A Frictionless constraint limits a face to a rotation or translation movement (slide) along a plane. The face is prevented from moving or deforming perpendicular to the plane. A Frictionless constraint can be used in conjunction with the other constraints.

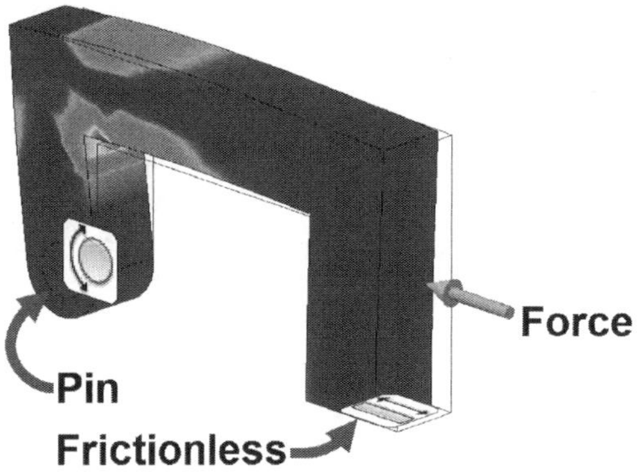

Figure 7.170

Applying a Load – Step 4

The next step to set up a stress analysis is to apply a load to the model. Inventor has different loads that can be applied to the model: force, pressure, bearing, moment and gravity. These constraints are located on the Analysis tab > Loads panel, as shown in the following image. Review the following descriptions to determine the type of load that closely resembles the desired load you need to apply.

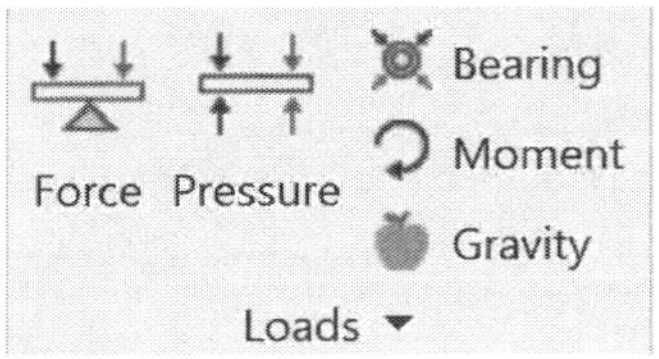

Figure 7.171

Force

Applies a user-defined load to a face or edge. A Force load is perpendicular to the surface, and parallel to an edge.

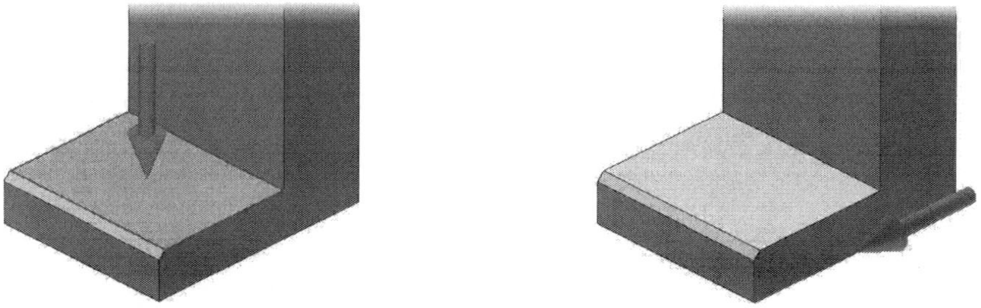

Figure 7.172

Pressure

Applies a user-defined pressure to a face. Pressure is a uniform load applied perpendicular to the whole surface.

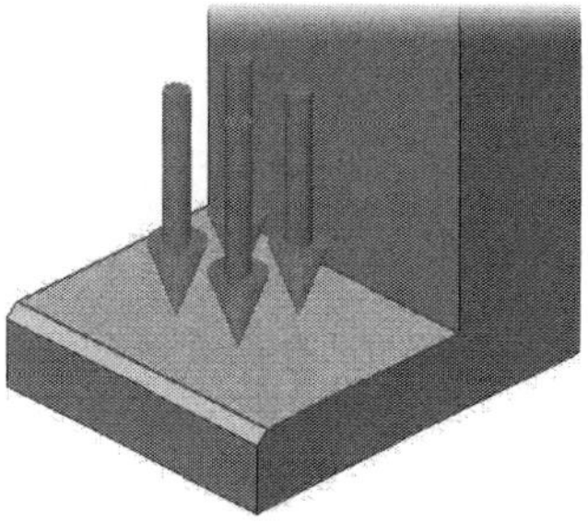

Figure 7.173

Bearing

Applies a user-defined axial or a radial load to a cylindrical face.

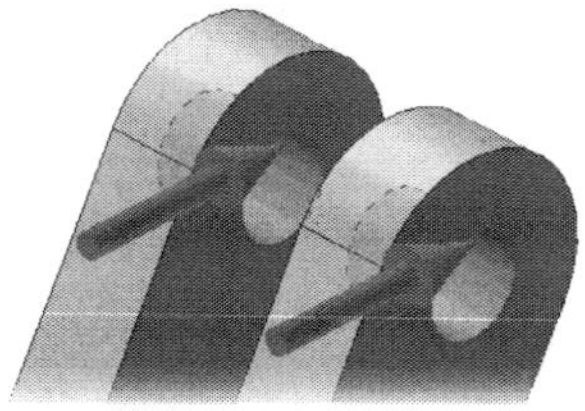

Figure 7.174

Moment

Applies a user-defined rotational load around an axis. If applied to a face, the momentary load is perpendicular to the surface.

Figure 7.175

Gravity

Applies a user-defined gravitational pull on a part. The gravitational pull is perpendicular to the surface, or parallel to an edge.

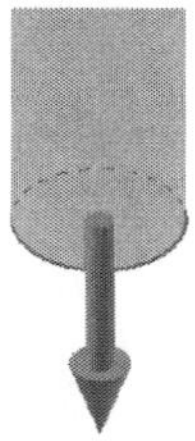

Figure 7.176

Mesh Options – Step 5

This step is optional. If you skip this step, Inventor will automatically mesh the part(s). As you gain experience performing simulations you may want to adjust the mesh to fine tune the results. There are four mesh commands that you can use from the Analysis tab > Mesh panel, as shown in Figure 7.177.

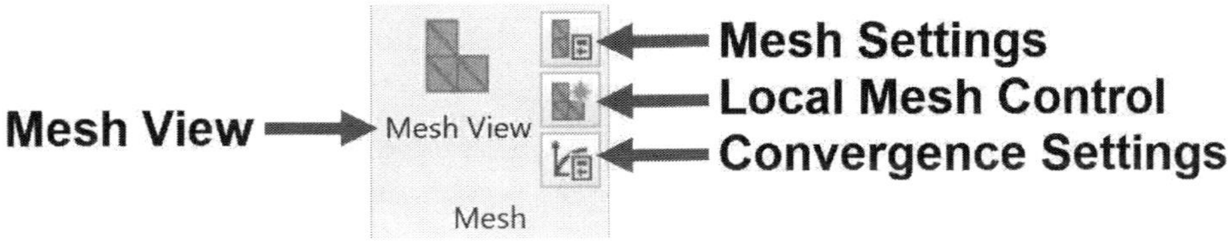

Figure 7.177

Mesh View

Use the Mesh View command to compute the mesh of the model. If this step is not done before a simulation is performed, Inventor will mesh the model automatically before doing the simulation. After the model has been meshed, click the Mesh View command to toggle the display of the mesh on and off.

Mesh Settings

Use the Mesh Settings command to adjust the average element size, minimum element size, grading factor, maximum turn angle, and if a curved mesh should be generated.

Local Mesh Control

Use the Local Mesh Control command to set the average element size for a face or an edge.

Convergence Settings

Use the Convergence Settings command to specify the maximum number of refinements, stop criteria, h refinement threshold, select the results to converge (Von Misses Stress, 1st Principal Stress, 3rd Principal Stress or Displacement) and select the geometry that will be utilized.

Run Simulation – Step 6

After materials are assigned, loads and constraints applied, and mesh adjusted, the last step is to run the simulation. To run a simulation, click the Simulate command from the Analysis tab > Solve panel, as shown in Figure 7.178 left. The Simulate dialog box will appear. Click the Run button as shown in Figure 7.178 right.

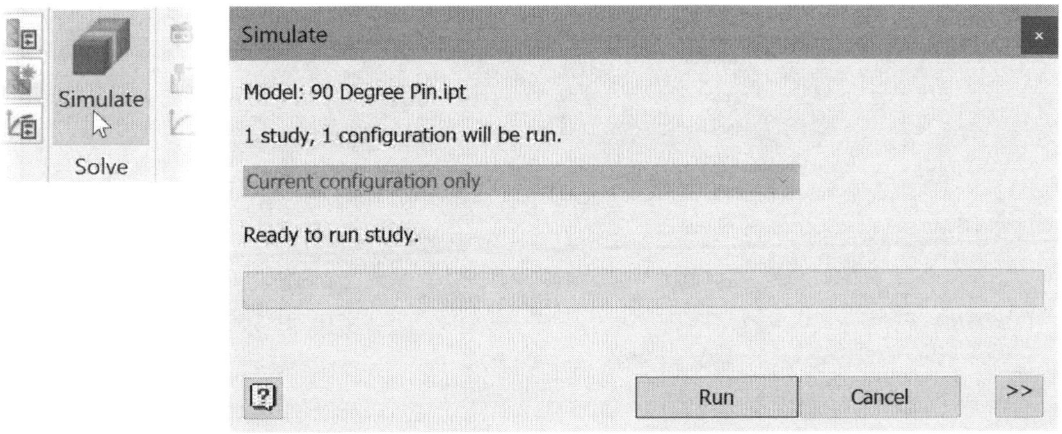

Figure 7.178

Review Simulation Results – Step 7

After a simulation is performed, several results can be displayed. To view the results from a different simulation, activate the desired test result in the browser, under the results section, as seen in Figure 7.179.

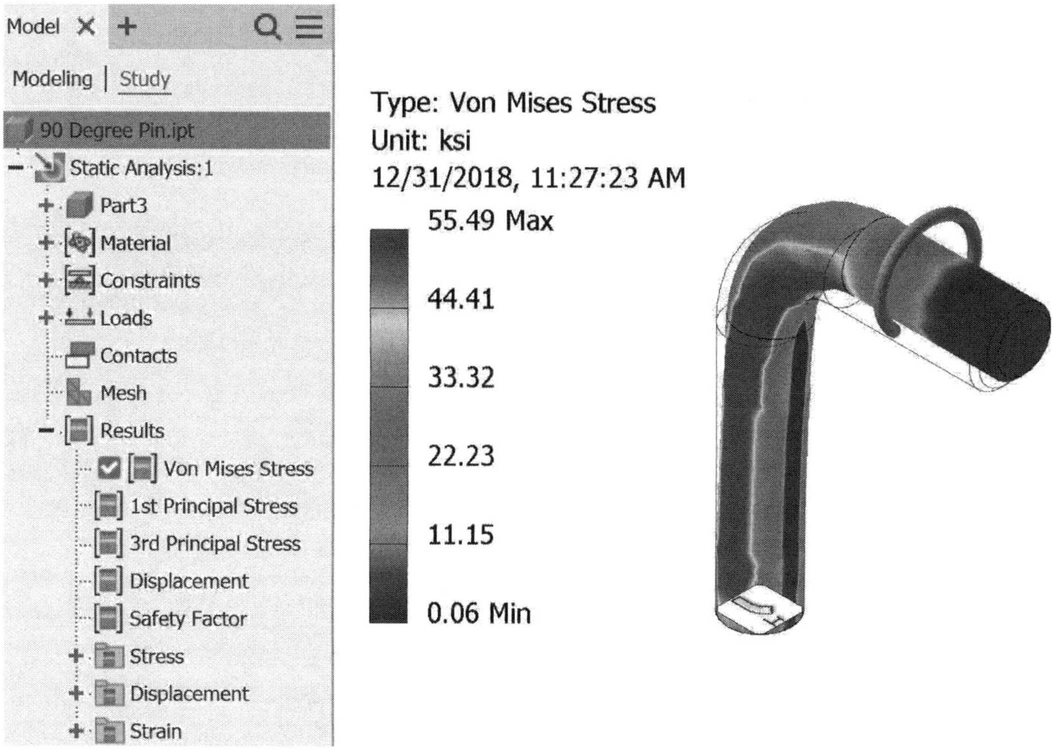

Figure 7.179

Display Results – Step 8

The results of the simulation can be analyzed by comparing the colors on the part to the color bar on the left side of the Graphics Window. How the simulation is displayed can be changed by selecting one of the options under the Analysis tab > Display panel, as shown in Figure 7.180.

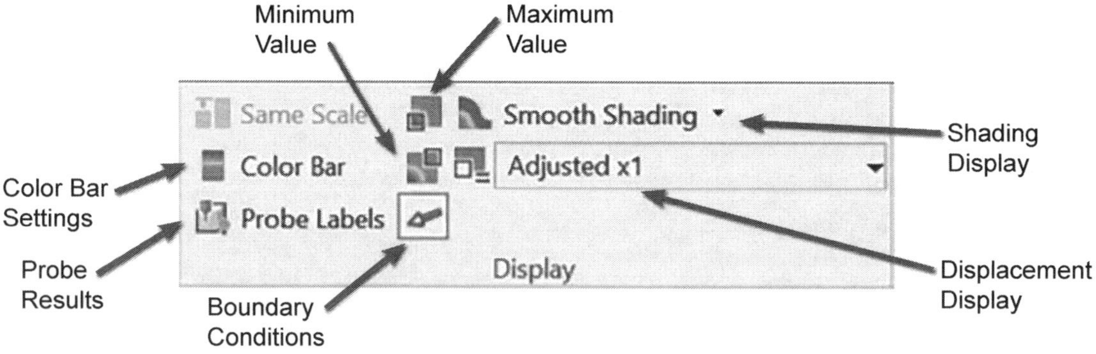

Figure 7.180

Color Bar Settings

Opens the Color Bar Settings Menu.

Minimum Value

Displays the location of the minimum value.

Maximum Value

Displays the location of the maximum value.

Shading Display

Allows the user to change the quality of the shading.

Displacement Display

Adjusts the level of model deformation.

Boundary Conditions

Toggles the display of glyphs for the loads applied.

Probe Results

Toggles the display of user probes.

Animate or Probe Results

To further analyze the results, you can run an animation of the simulation or probe a specific point(s) for information. These commands are in the Analysis tab > Result panel, as shown in the following image.

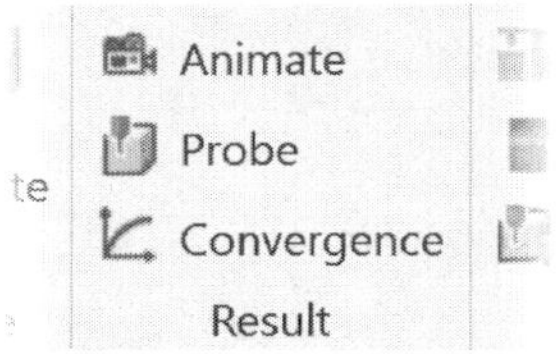

Figure 7.181

Animate

The Animate command plays an animation of the results. Once the Animate Results dialog opens, click the play button to view animation. If desired, an animation can be recorded.

Probe

The Probe command creates a probe to measure the simulation results. The probe will measure the exact value of the simulation at the selected point. To place a probe, select the area of the model where the exact measurement is to be displayed.

Figure 7.182

Report

Once the Simulation is complete, you can generate a report. The Report command is located under the Analysis tab > Report panel, as shown in Figure 7.183 left. To determine what simulation results to include in the report, click the options desired on the Studies tab in the report dialog, as shown in the image on the right. From the Format tab select the file type to export: html, mhtml or rtf. Click the OK button to create the report.

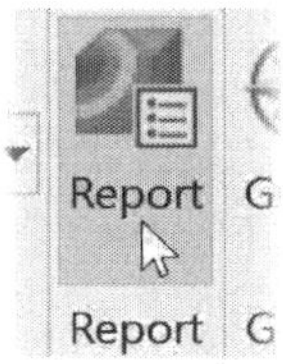

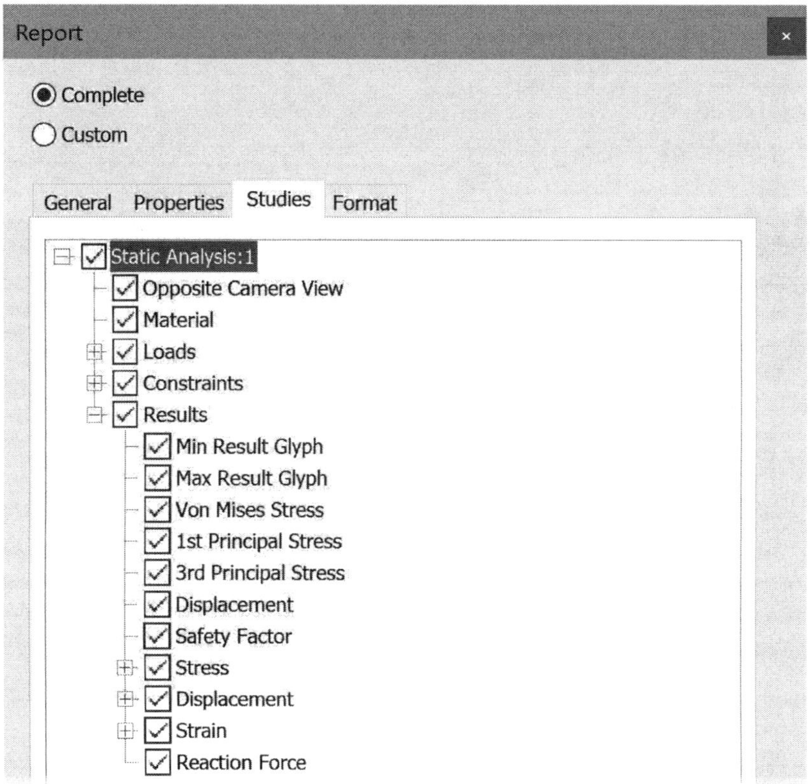

Figure 7.183

EXERCISE 7-9: RUN A STRESS ANALYSIS ON A PART

In this exercise, you will use the Stress Analysis commands to create a simulation, add constraints, loads, simulate, animate and create a report of a lifting claw that was used in the assembly constraint exercise in chapter 6.

1 Open *ESS_E07_09.ipt* in the Chapter 07 folder.

2 Create a new simulation study.

a. Begin a simulation by clicking the Stress Analysis command on the 3D Model tab > Simulation panel, as shown in the following image on the left.

b. Create a new simulation by clicking the Create Study command on the Analysis tab > Manage panel, as shown in the following image on the right.

c. Create a new static analysis by clicking OK in the Create New Simulation dialog box.

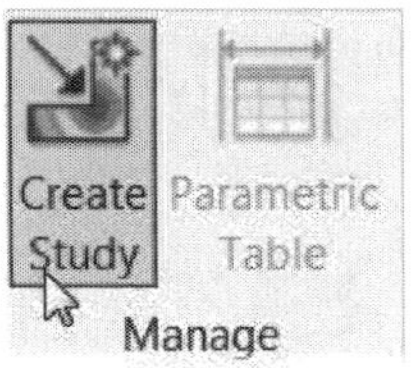

Figure 7.184

3 Verify that a material has been applied to the part by clicking the Assign Materials command on the Analysis tab > Material panel. Verify that the Original Material assigned to the part via the iProperties is Steel, Mild, as shown in the following image and then click OK.

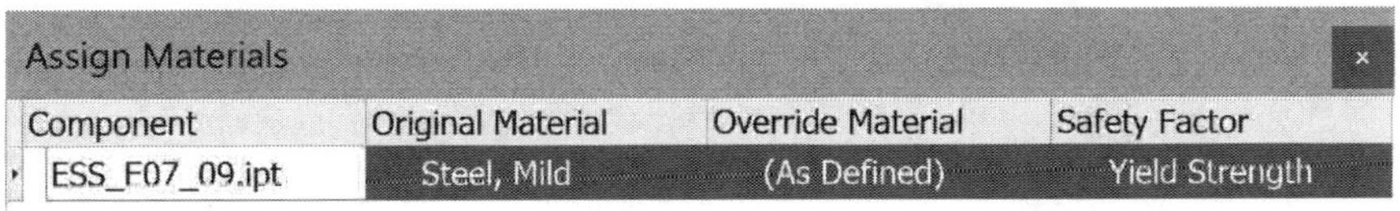

Figure 7.185

4 Next, you constrain the part by applying a pin constraint to the four holes.

a. Click the Pin Constraint command from the Analysis tab > Constraints panel.

b. Select the two circular faces of the bottom holes, labeled (1) in Figure 7.186, and click Apply.

c. While still in the Pin Constraint dialog box, select the two circular faces of the top holes labeled (2) in the following image.

d. Then click OK to create the pin constraint.

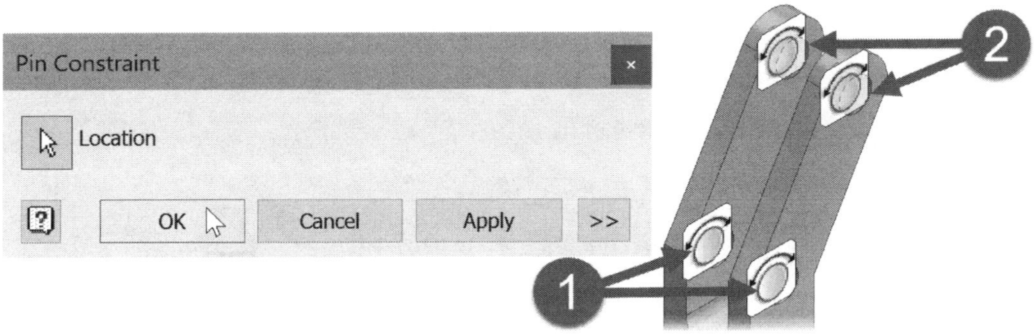

Figure 7.186

5 Next, you apply a load to the part.

a. Click the Force command on the Analysis tab > Loads panel.

b. Select the bottom horizontal face and change the Magnitude to **200 N,** as shown in the following image. The direction of the force should be directed down.

c. Click OK to create the force.

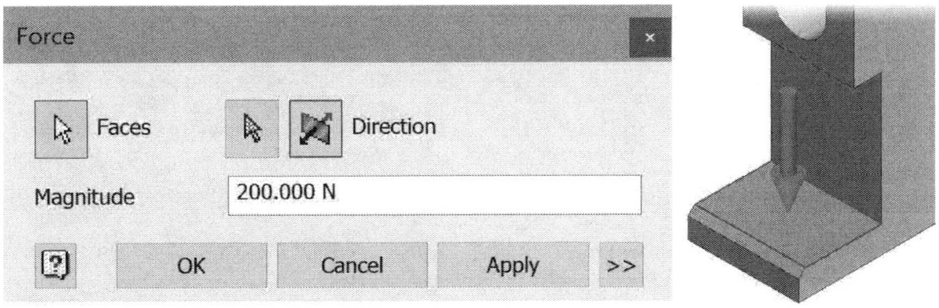

Figure 7.187

6 Next, run the simulation.

a. Click the Simulate command from the Analysis tab > Solve panel.

b. In the Simulate dialog box click Run.

c. The result of your simulation should resemble the following image.

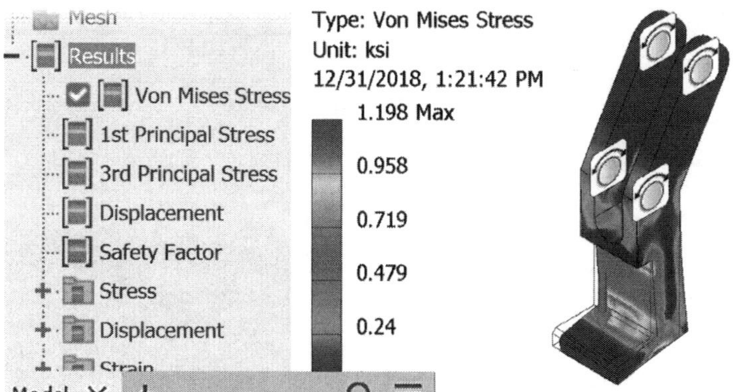

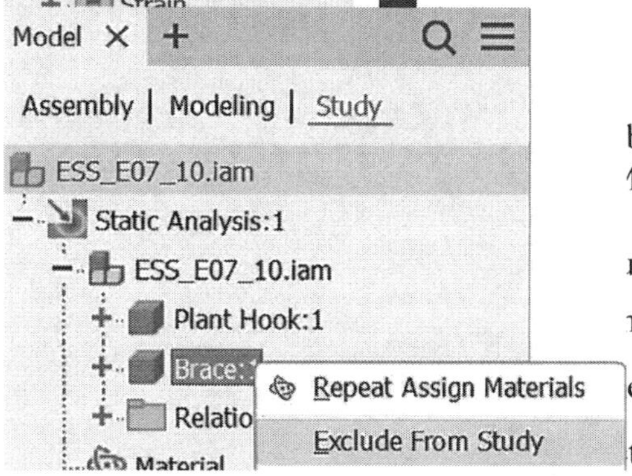

by double-clicking different options under the Results entry in the
lises Stress result current.

lts.

the Analysis tab > Result panel.

e Play button in the Animate Results dialog.

ting different options from the drop list.

d. Click OK to finish the animation.

9 Next, you copy the simulation study, override the material of the part, and run the simulation.

a. Copy the simulation study by right-clicking on Static Analysis:1 in the browser and click Copy Study, as shown in the following image.

b. Static Analysis:2 (the new simulation) will be current.

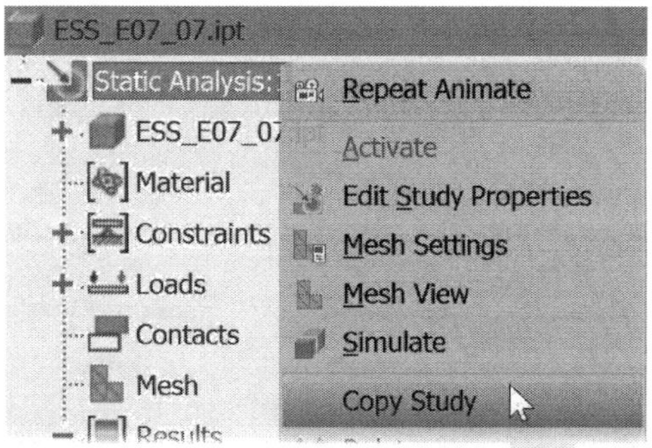

Figure 7.189

10 Next, you override the material.

a. Click the Assign Materials command on the Analysis tab > Material panel.

b. In the Override Material cell click in the cell and from the list of materials click Aluminum 6061 as shown in Figure 7.190.

c. To complete the operation, click OK.

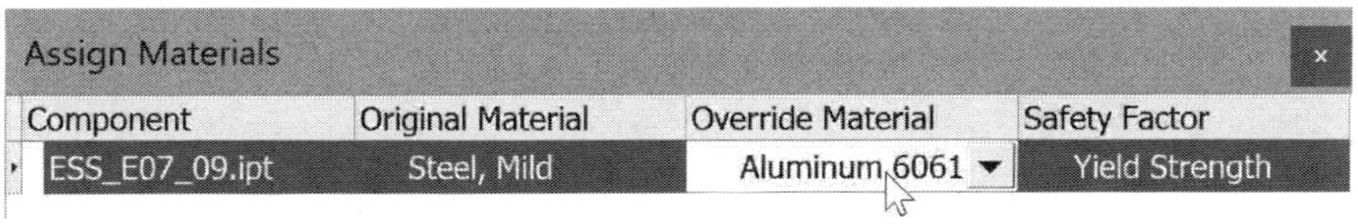

Figure 7.190

11 The constraints and the force were copied from the first study and do not need to be reapplied. Verify this by expanding the Constraints and Loads in the browser and double-click on the entries.

12 Next run the new simulation to reflect the change in material. Click the Simulate command from the Analysis tab > Solve panel and then click the Run button in the Simulate dialog box.

13 From the browser analyze the results of the simulation by selecting the different results method.

14 Lastly, generate a report. Click the Report command from the Stress Analysis tab> Report panel, from the Format tab change the report format to Rich Text Format (*.rtf) and then click OK.

15 To return to the part file environment, click the Finish Stress Analysis command on the Analysis tab > Exit panel.

16 Close the file. Do not save changes.

End of exercise.

EXERCISE 7-10: RUN A STRESS ANALYSIS ON AN ASSEMBLY

In this exercise, you will use the Stress Analysis commands to analyze a design of a plant hook to determine how much displacement there will be with and without a brace.

1 Open *ESS_E07_010.iam* in the Chapter 07 folder.

2 Create a new simulation.

a. Click the Environments tab.

b. Begin a simulation by clicking the Stress Analysis command on the Environments tab > Begin panel.

c. Create a new study by clicking the Create Study command on the Analysis tab > Manage panel.

d. The Create New Simulation dialog box appears; click OK to create a new static analysis.

3 Verify that a material has been applied to both parts.

a. Click the Assign Materials command on the Analysis tab > Material panel.

b. Verify that the Original Material assigned to both parts is Steel, Mild.

c. Dismiss the dialog box by clicking the Cancel button.

For the first simulation, you simulate how far the plant hook will displace without the brace.

4 In the browser, expand *ESS_E07_10.iam* and exclude the Brace part by right-clicking on Brace:1 in the browser and click Exclude From Study from the menu, as shown in Figure 7.191 left.

5 Next, you constrain the plant hook.

a. Click the Fixed Constraint command on the Analysis tab > Constraints panel.

b. Select the two circular faces of the holes in Figure 7.191 right.

c. Create the constraints by clicking OK.

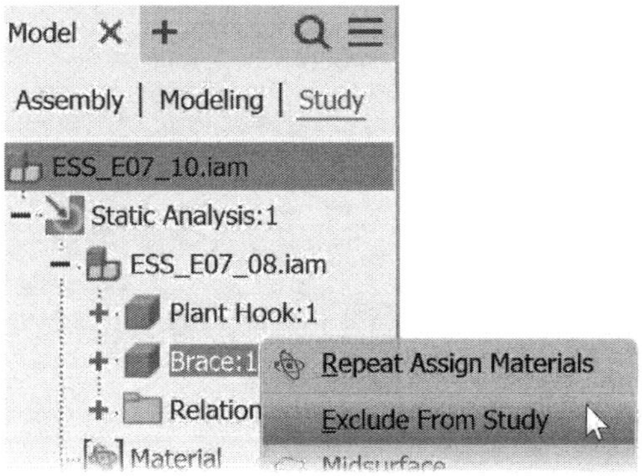

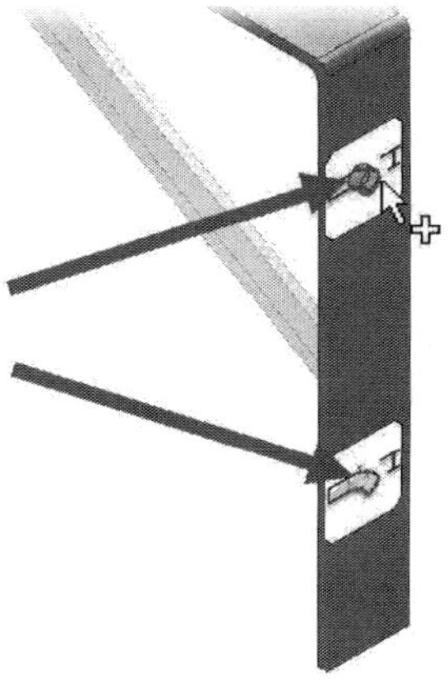

Figure 7.191

6 Next, you apply a load to the part. Click the Force command from the Analysis tab > Loads panel. To create the force, make the following changes:

a. For the location, select the inside circular face on the left end of the plant hook labeled (1) in the image on the right.

b. Notice that the force is at an angle. To adjust the direction of the force, click the direction selection button labeled (2) and in the graphics window, select the top horizontal face of the part labeled (3).

c. Enter **5.000 lbforce** labeled (4) and verify the direction of the force is pointing down.

d. Create the force by clicking OK.

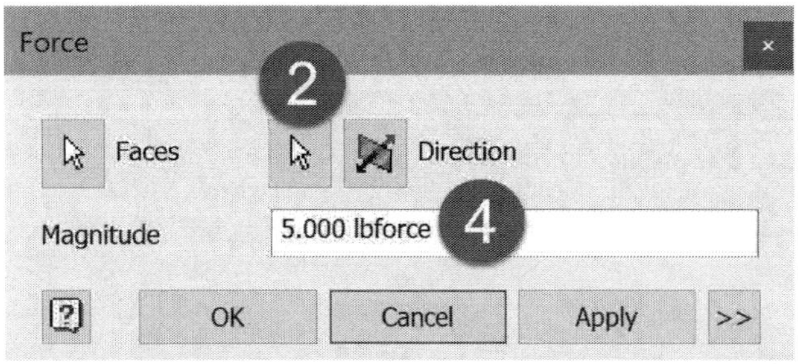

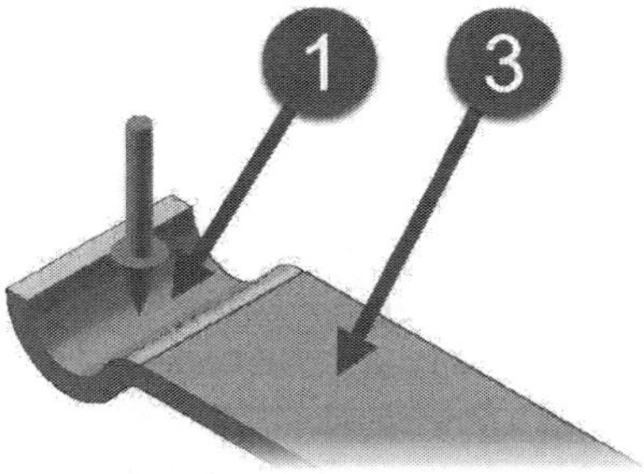

Figure 7.192

7 Next, you run the simulation.

a. Click the Simulate command from the Analysis tab > Solve panel.

b. Click the Run button in the Simulate dialog box.

8 From the Results entry in the browser, double-click Displacement and the results of your simulation should resemble Figure 7.193.

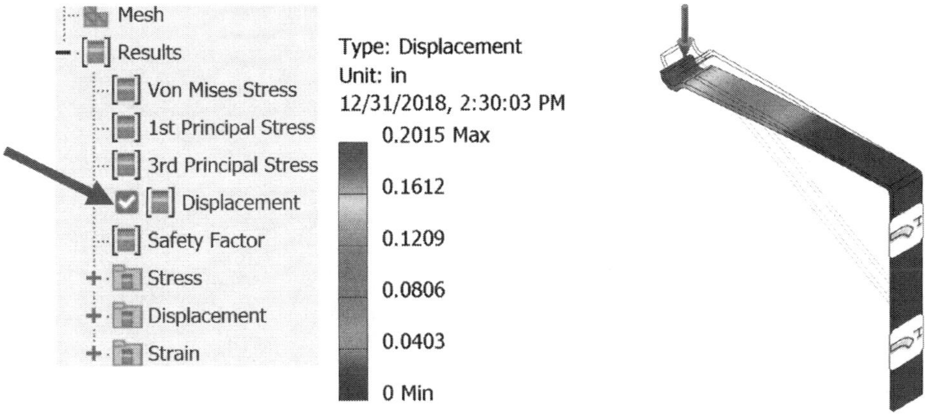

Figure 7.193

9 Next, you animate the simulation results.

a. Click the Animate command from the Analysis tab > Result panel.

b. Start the animation by clicking the Play button in the Animate Results dialog.

c. Adjust the speed by selecting different options from the drop list.

d. Click OK to finish the animation.

10 Next, you copy the study to include the brace and run the simulation.

a. Copy the study by right-clicking on Static Analysis:1 in the browser and click Copy Study, as shown in the following image on the left.

b. The new study, Static Analysis:2, will be active.

11 In the browser, expand *ESS_E07_10.iam* and include the Brace part by right-clicking on Brace:1 in the browser and click (uncheck) Exclude From Simulation from the menu, as shown in the following image on the right.

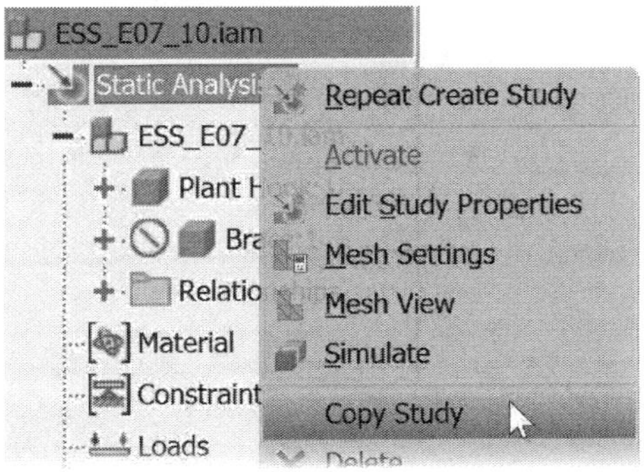

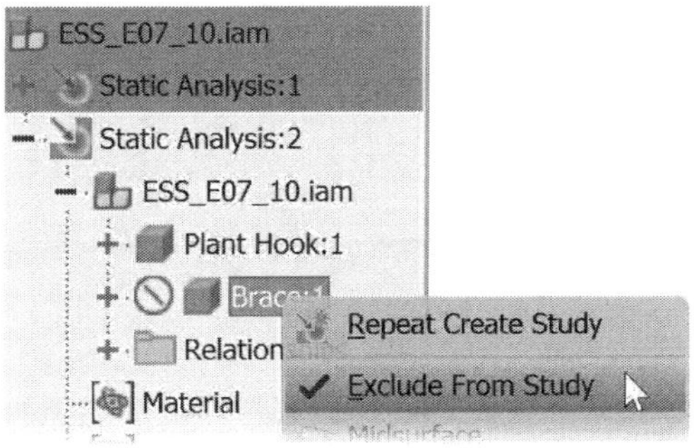

Figure 7.194

12 The constraints and the load were copied from the first study and do not need to be reapplied, but you need to run the simulation to reflect the inclusion of the brace. Run the simulation by clicking the Simulate command from the Analysis tab > Solve panel and then click the Run button in the Simulate dialog box.

13 From the Results entry in the browser, click Displacement and your results of your simulation should resemble the following image. As you can see, the displacement went from approximately 2 inches without the brace to approximately 0.003 inches with the brace.

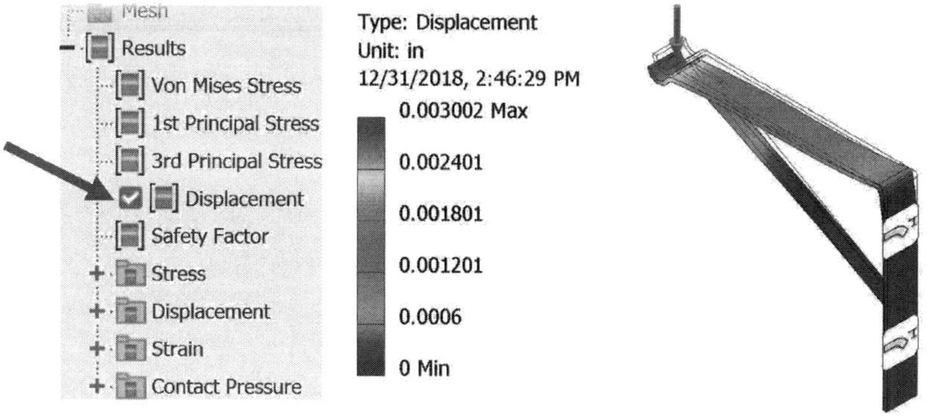

Figure 7.195

14 To see the maximum and minimum value of the active simulation click the Maximum Value and the Minimum Value command from the Analysis tab > Display panel, as shown in the following image on the left.

15 To see a leader, point to the location of the result and click and drag the result away from the part. The following image on the right shows the maximum and minimum results moved away from the part.

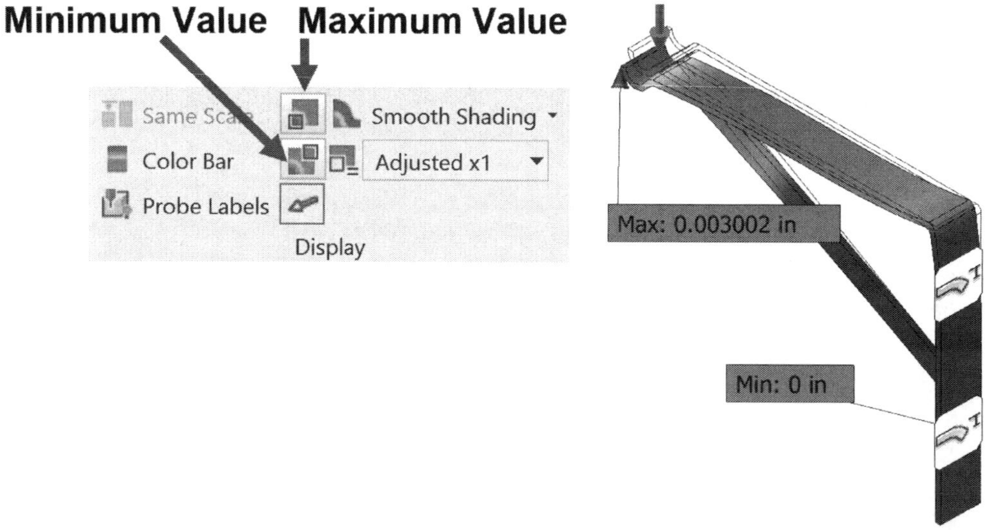

Figure 7.196

16 If desired, copy the simulation, assign different materials to the parts and edit the location of the holes to see if you can further minimize the displacement.

17 To return to the part file environment click the Finish Stress Analysis command on the Analysis tab > Exit panel.

18 Close the file. Do not save changes.

End of exercise.

THE FRAME GENERATOR

The Frame Generator tools allow you to design platforms, equipment racks, and other types of structural frames. The Frame Generator has its own set of tools which give you the ability to easily create and edit structural members. These tools can be accessed from the Design tab > Frame panel, as shown in Figure 7.197.

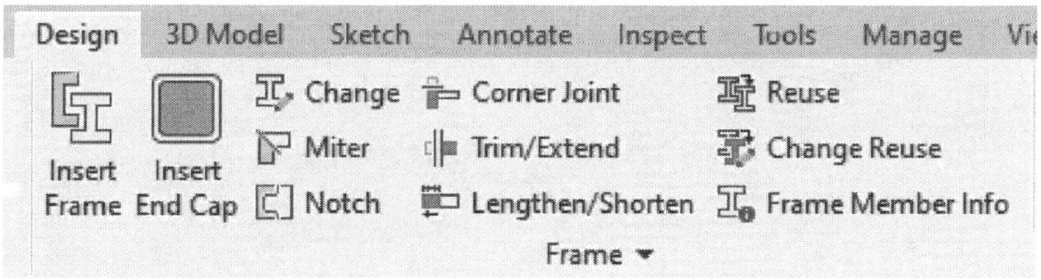

Figure 7.197

Before creating a frame, you first create a model that is referred to as a skeleton model because it often only contains lines, work points, and other "bones" of the structure. The geometry of the skeleton model is used to place the frame members.

Figure 7.198 left shows a skeleton model that consists of lines (wireframe). The image on the right shows the frame generated from the skeleton model (for clarity the skeleton model is turned off).

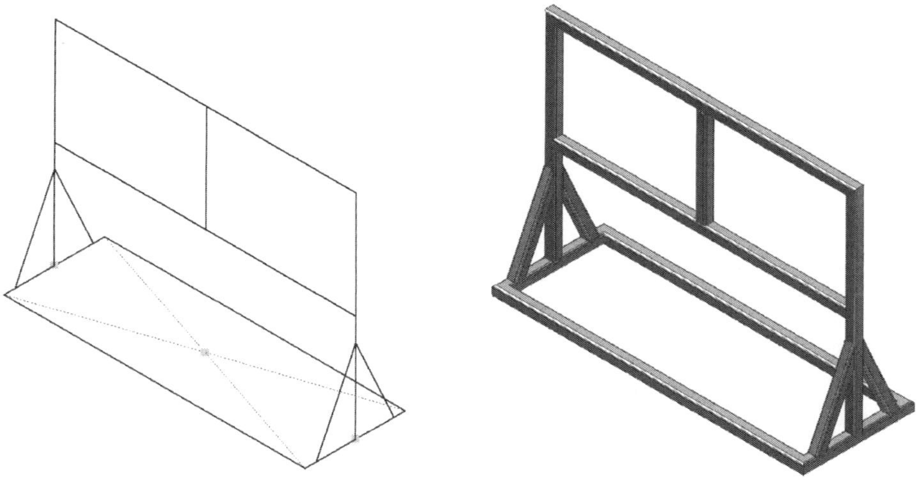

Figure 7.198

Figure 7.199 left shows a skeleton model that consists of a solid body with a sketch created on the front face, and a vertical line in the middle of the sketch. The image on the right shows the frame generated from the edges on the solid skeleton model and the vertical line (for clarity the skeleton model is turned off).

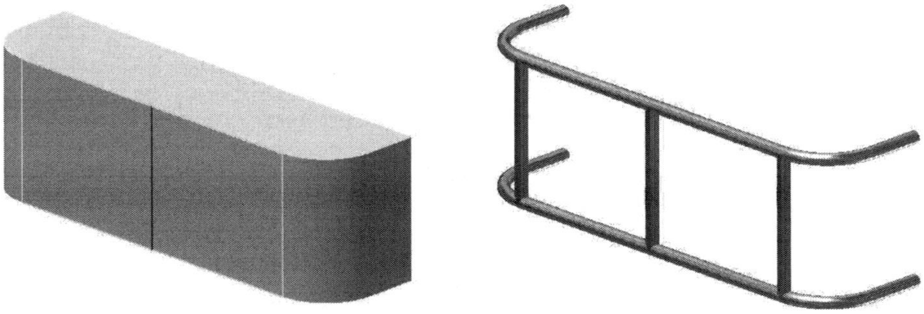

Figure 7.199

To create a frame, follow these steps:

1. Create or open a part that will be used as the skeleton model. The skeleton model can consist of 2D and 3D sketches (referred to as a wireframe), a solid and surface body(s), and work geometry.
2. Create or open an assembly file in which the frame will be created.
3. Use the Place Component command to place the skeleton model in the assembly.

4. Save the assembly file.
5. Click Insert Frame from the Design tab > Frame panel. The Insert Frame dialog box appears similar to Figure 7.200. The Standard set to ANSI; Family set to ANSI AISC (Rectangular), and the size at 2 x 1 x 1/8.

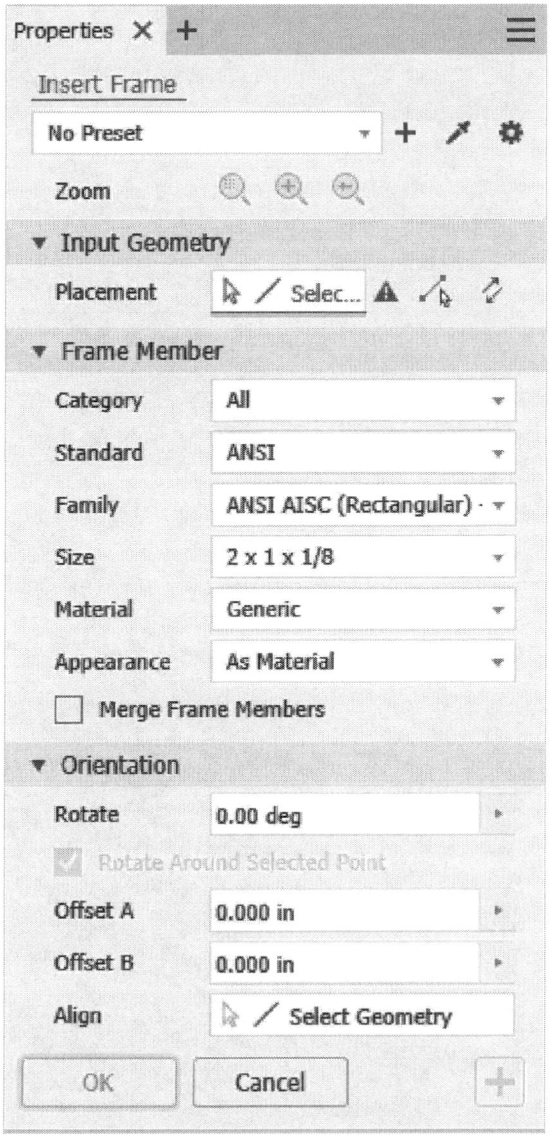

Figure 7.200

6. Use the Input Geometry area of the dialog box to select the desired option(s) as shown in Figure 7.201. Note the Specify frame location by two points is a toggle button. When it is active you can only draw points to locate the frame and you cannot select your existing geometry.
7. Locate where you want the frame members to be placed by selecting edges, sketch objects, or points on the skeleton model.

TIP When selecting multiple edges, you can use the window or crossing selection technique.

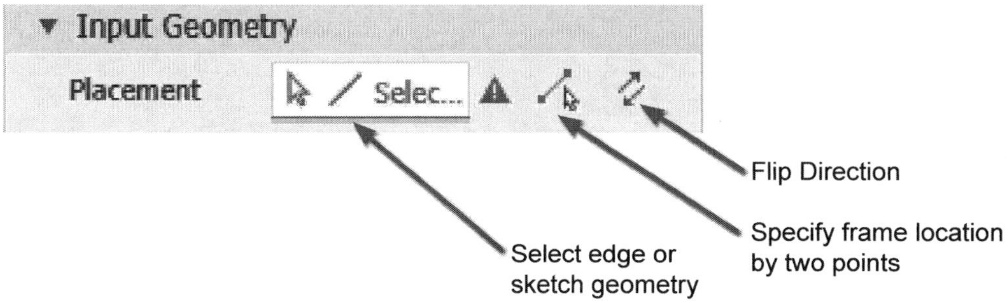

Figure 7.201

TIP Use the Merge Frame Members check box to create a single member from multiple selected edges.

1. The frame members preview in the graphics window. If needed, use the Flip Direction option. From the Orientation area you can set the offset distance, and rotation angle.
2. When the members are in the correct location, click the OK or Apply button, and the Create New Frame dialog box will appear as shown in Figure 7.202 left (your directory names will be different than what is shown). Here you define a new file location that will hold all the frame members.
3. Click OK in this dialog box to activate the Frame Member Naming dialog box, as shown in the following image on the right. You can keep the default Display Names and File Names, or edit them as needed; the names will be reflected in the browser.

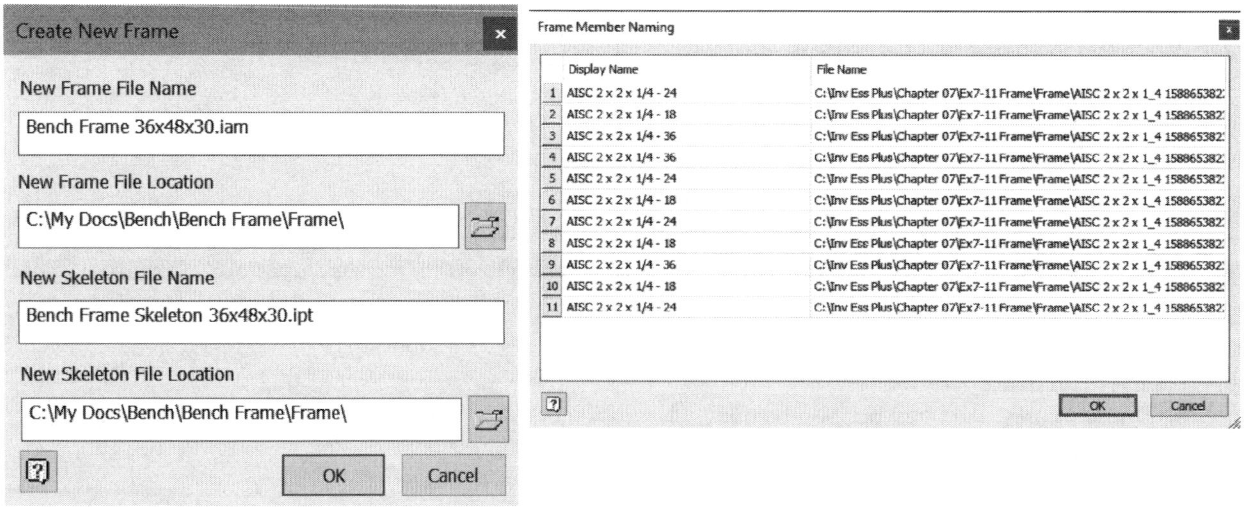

Figure 7.202

4. Edit the frame members by using the frame edit tools on the Design tab > Frame panel as shown in the following image.

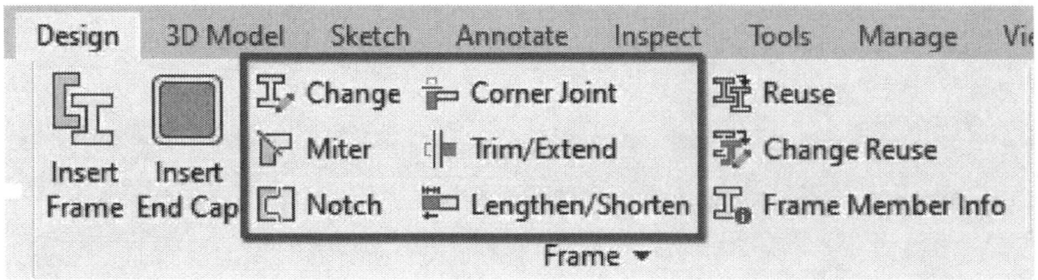

Figure 7.203

Following is a description of the frame editing tools.

- **Change:** Edits the properties of frame members, and controls the position of frame members in relation to the skeleton model.
- **Miter:** Applies miter cuts as end treatments between two frame members.
- **Notch:** Notches one frame member to fit another, using one profile to cut the other.
- **Corner Joint:** Trims and extends two frame members at their ends. Provides options to create your own presets to save time for frequently used settings. Select Corner Joint to show the Corner Joint dialog box as shown in Figure 7.204 left. Click to edit the dimension values for the corner joint as needed as shown in the middle image. Click the + near the upper right of the dialog box to create a new preset from the current settings. Many Inventor commands have time saving options to save a Preset that is easily reused. To select a Preset expand the options using the down arrow as shown in the image at right. Once the frame members are selected, click OK to apply trim/extend the frame members.
- **Trim/Extend:** Trims or extends multiple frame members to a model's planar face or work plane.
- **Lengthen/Shorten:** Extends or retracts a frame member.

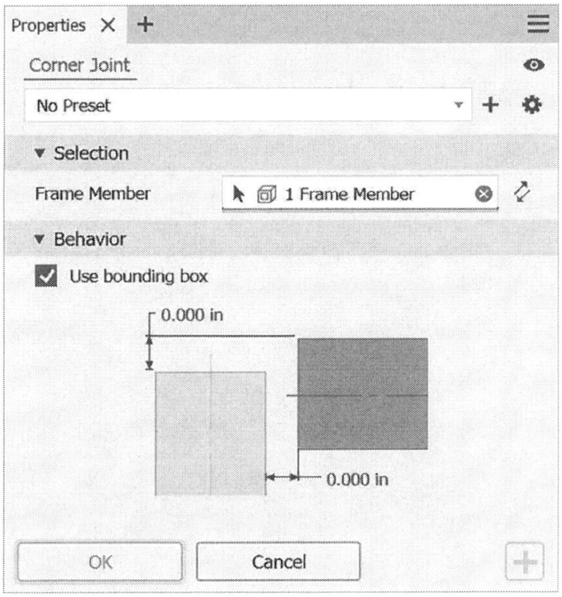

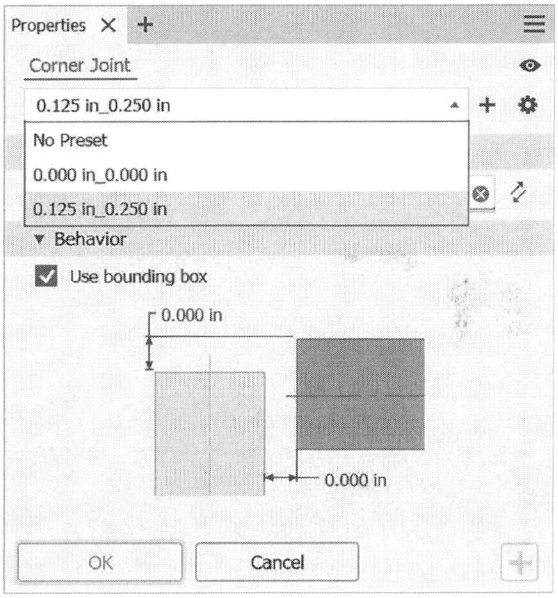

Figure 7.204

EXERCISE 7-11: CREATING A FRAME

In this exercise, you create a rectangular frame using the frame generator tools. The skeleton model will consist of a solid body and a sketch.

1. Create a new part file based on the Standard (in).ipt template file.
2. Draw a **36** x **18** inch rectangle on the XZ plane and place the lower-left corner on the origin point, and then extrude it up **24** inches. The **36 inch** dimension should run parallel with the X axis and the **18 inch** dimension should run parallel to the Z axis as shown in the following image.

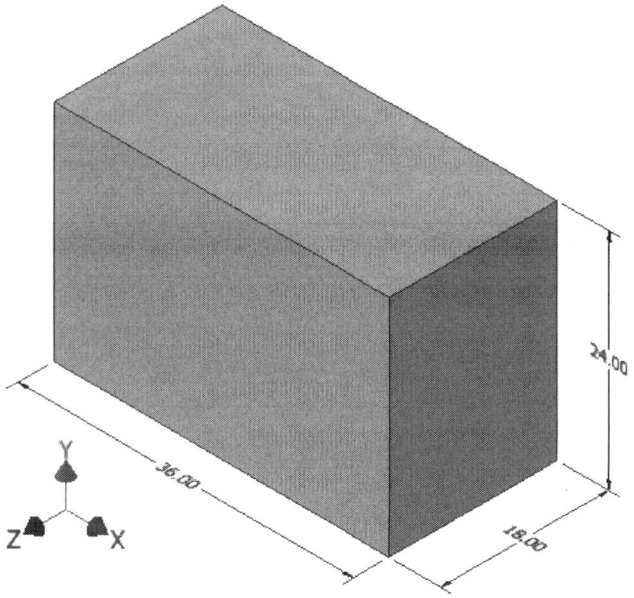

Figure 7.205

3 Save the file with the name *Ex7-11 Solid Skeleton.ipt* in the Chapter 07 folder.

4 Create a new assembly file based on Standard (in).iam template file.

5 Save the file with the name *Ex7-11 Frame.iam* in the Chapter 07 folder.

6 Place and ground one occurrence of the file *Ex7-11 Solid Skeleton.ipt*. You can easily do this by selecting the file, and before placing the part, right-click and click Place Grounded at Origin from the menu as shown in the following image.

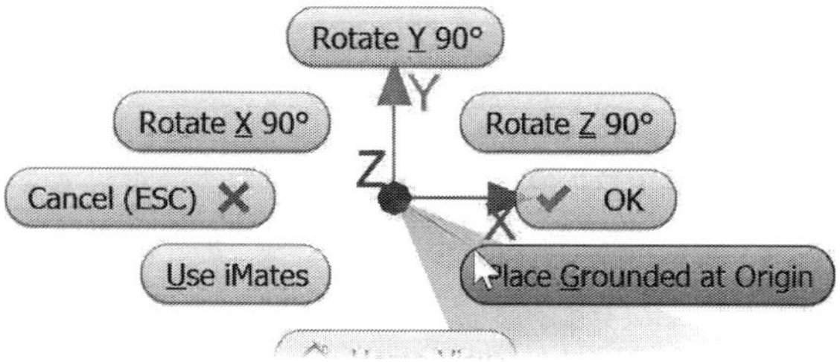

Figure 7.206

7 Next you place frame members on the edges on the solid body.

a. Click Insert Frame from the Design tab > Frame panel and in the Insert Frame dialog box change the Standard, Family, and Size as shown in the following image on the left.

b. With the Insert Frame dialog box still open, select all the edges except the front-bottom edge as shown in the following image on the right. You can drag a window around the part and then press the Ctrl key and deselect the front edge.

c. Click OK to create the frame members in the default location, and with the default names.

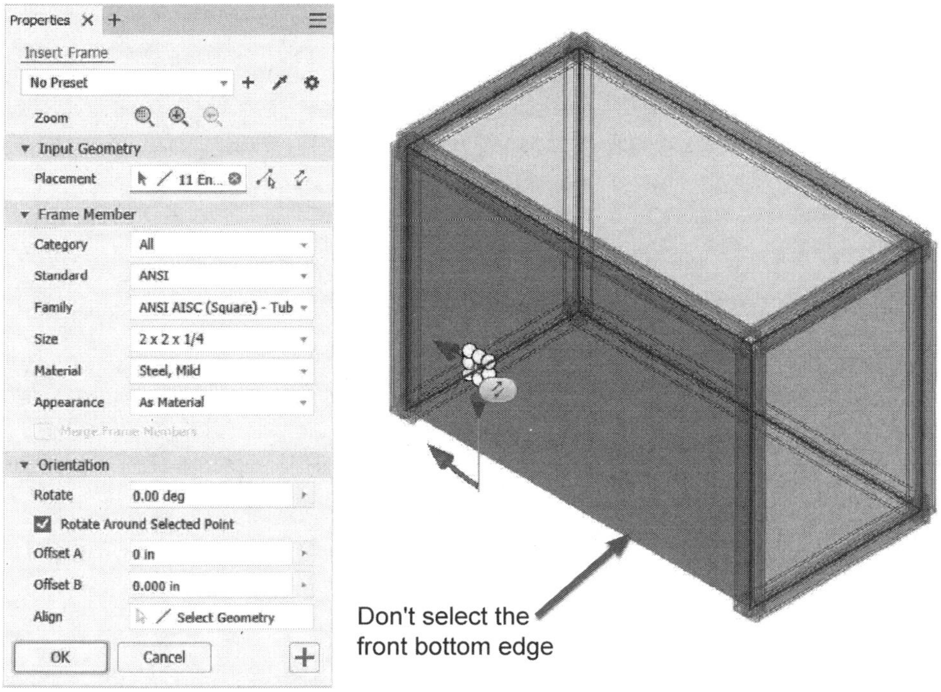

Figure 7.207

8 For clarity, turn off the visibility of the skeleton by right-clicking on *Ex7-11 Solid Skeleton:1* in the browser, and click Visibility from the menu. When done, your frame should resemble the following image on the left.

9 Zoom in and use the Free Orbit command and observe that the members interfere. Figure 7.208 right shows the interference on the top-right side of the frame.

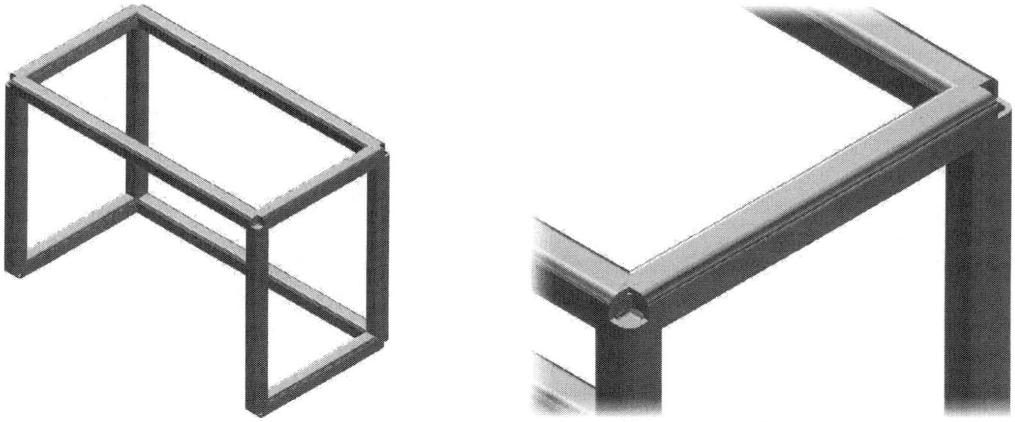

Figure 7.208

10 Next you miter the frame members to remove the interference.

a. Click the Miter command on the Design tab > Frame panel and the Miter Properties dialog box appears as shown in Figure 7.209.

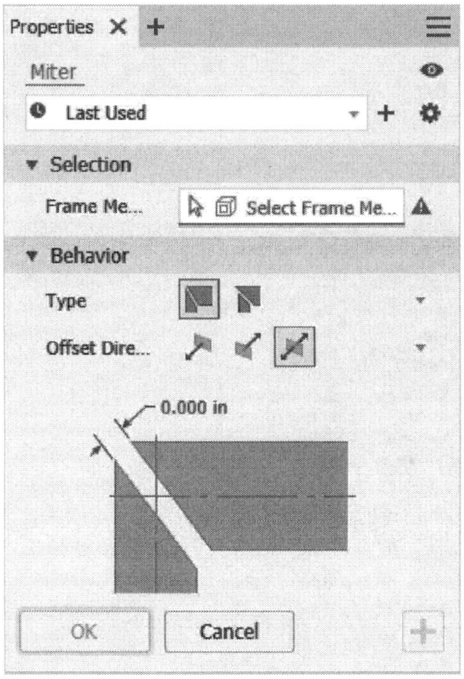

Figure 7.209

b. Click on the top-front and top-left members in the graphic window, labeled (1) in Figure 7.210 left. Use the + sign in the lower right of the dialog box to Apply the miter and keep the dialog box open.

c. Create seven more miters (2) through (8) by selecting two members at a time and clicking the + to Apply the miter after each pair. When done, click OK from the Properties panel. Your frame should look like Figure 7.210 right.

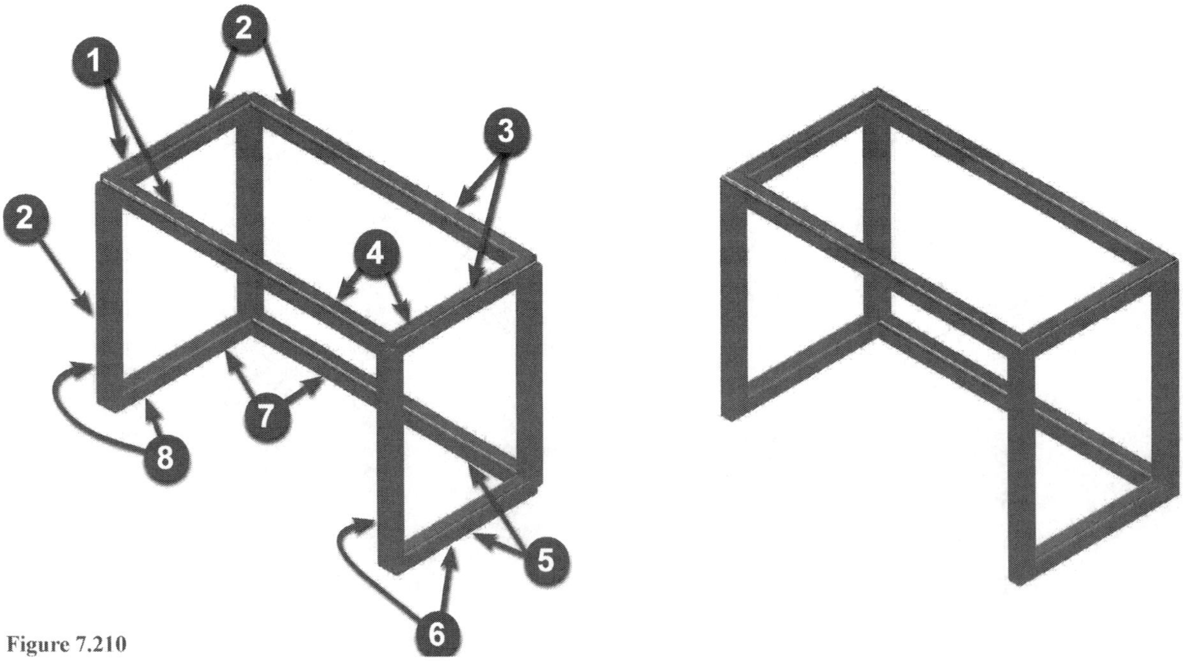

Figure 7.210

11 Next you remove the remaining interference by trimming the vertical frame members.

a. Click Trim/Extend from the Design tab > Frame panel to show the Trim – Extend To Face dialog box.

b. For the Tool select the top horizontal plane on the bottom member labeled (1). Note that you can select any planar face to define the plane to which the members will be trimmed.

c. Select the back two vertical members labeled (2) in the Figure 7.211.

d. Click the + to Apply the trim and keep the dialog box open.

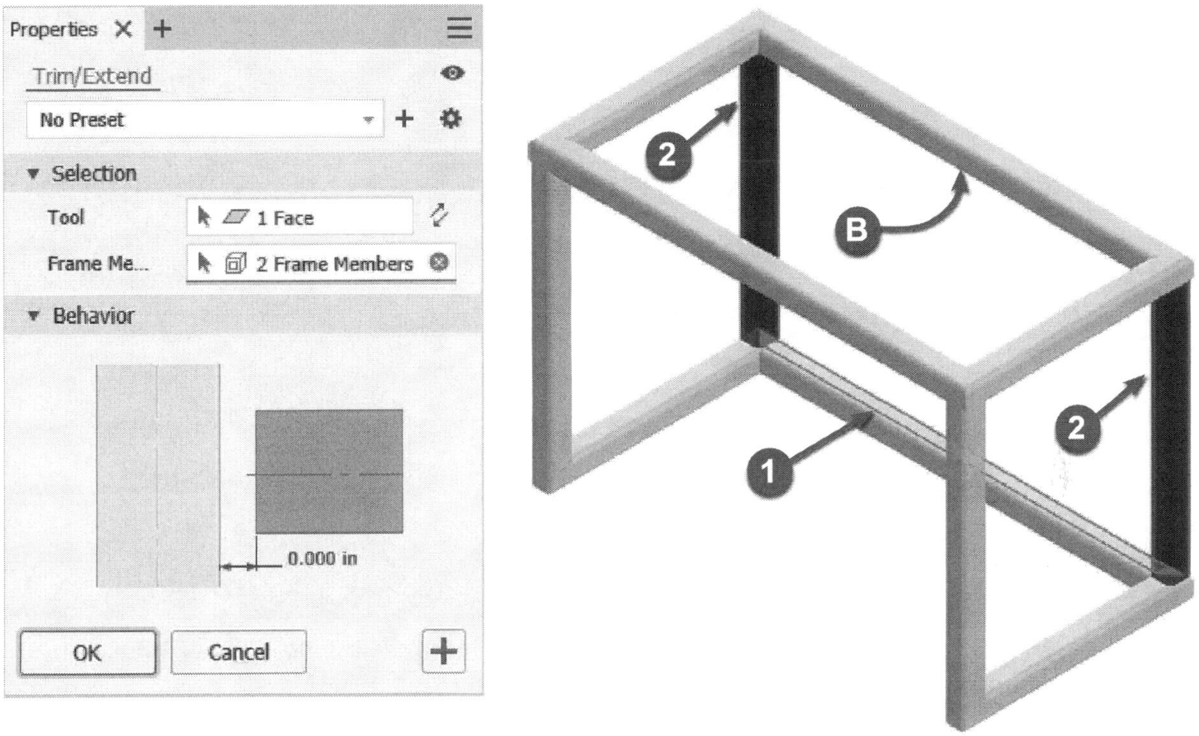

Figure 7.211

Figure 7.212

e. Click the vertex on the ViewCube as shown in Figure 7.212 to change the viewpoint so you can see the bottom faces.

f. In the Trim – Extend To Face dialog box, for the Tool select the underside of the top horizontal member labeled (B). With your viewpoint rotated this surface is easy to select.

g. Select the four vertical members – the ones labeled (2) and the other two parallel to them.

h. Trim the members and close the dialog box by clicking the OK button.

12 Zoom in and use the Free Orbit command to see that the members no longer interfere.

13 Change to the Home view. Your frame should resemble Figure 7.213 with the mitered and trimmed members.

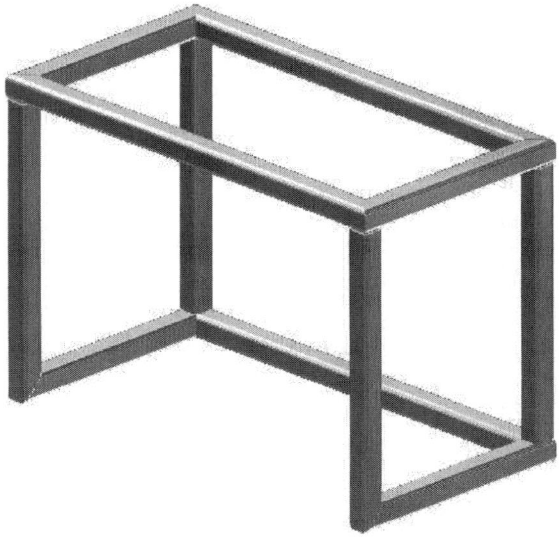

Figure 7.213

14 Save the assembly file.

Next you edit the skeleton part.

15 Make the part file *Ex7-11 Solid Skeleton.ipt* active by clicking on its tab near the bottom of the screen.

16 Edit the rectangle so its dimensions are **48** x **30 x 30** inches as shown in the following image.

17 Create a sketch on the front face.

18 Draw a horizontal line that is centered between the vertical lines as shown in the following image.

19 Save the file.

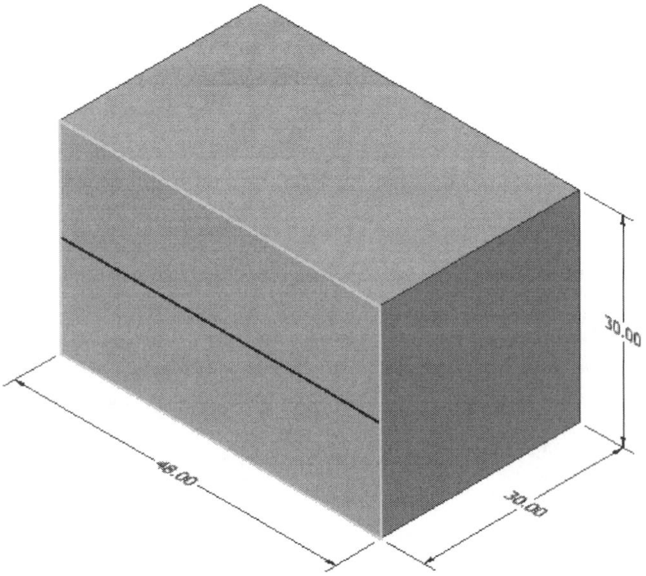

Figure 7.214

20 Make the assembly file *Ex7-11 Frame.iam* active by clicking on its tab near the bottom of the screen.

21 Update the frame by clicking the Local Update icon  from the Quick Access toolbar.

22 Next you insert a new member in the middle of the frame.

a. Turn **on** the visibility for Ex7-11 *Solid Skeleton:1* by right-clicking on it in the browser and clicking Visibility from the menu. (The image shows it off to make it easier to see the shapes. It is easier to make the selections with the solid skeleton visible.)

b. Click the Insert Frame command on the Design tab > Frame panel and change the Family and Size as shown in Figure 7.215.

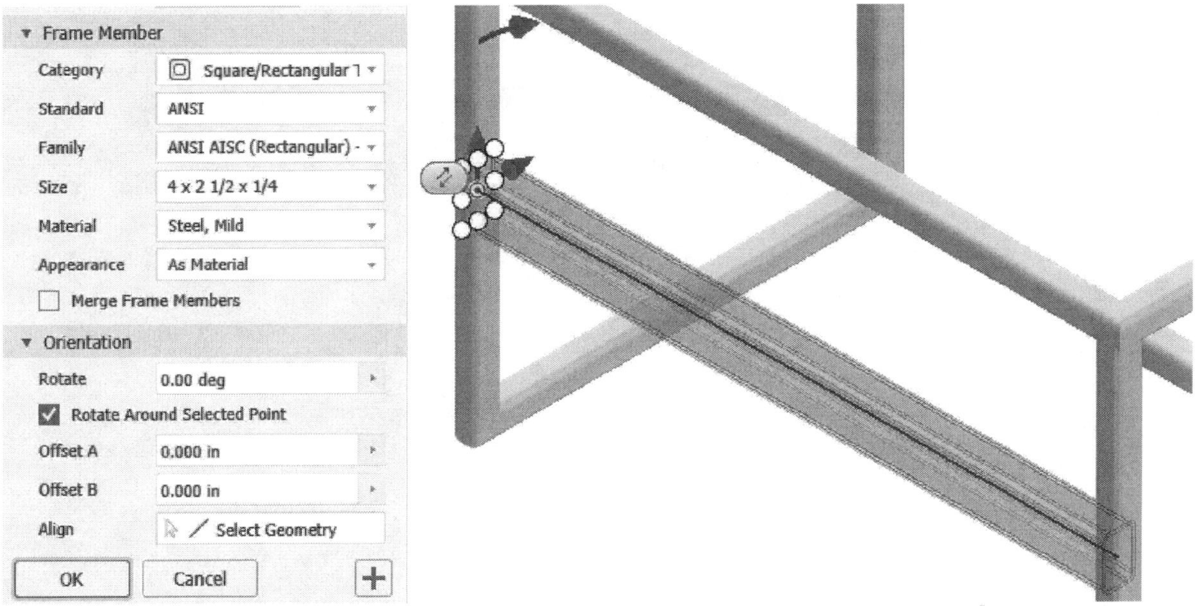

Figure 7.215

c. With the Insert dialog box still open, select the horizontal line you just added to the sketch in the skeleton part.

d. If the rectangle is outside the original frame, use the rotate option in the dialog box to correct it.

e. Click OK twice to create the member.

23 For clarity, turn off the visibility of *Ex7-11 Solid Skeleton:1* by right-clicking on it in the browser, and click Visibility from the menu.

24 Next remove the interference between the new 4 x 2 member and the two vertical members by notching the 4 x 2 member back on both sides.

a. Click Notch from the Design tab > Frame panel to show the Notch Properties panel.

b. In the Selection area, the Frame Member option is automatically active. From the graphics window select the horizontal member labeled (1) in Figure 7.216.

c. Click the Notch Tool selection to make it active, labeled (2). From the graphics window select the front-left and front-right vertical members labeled (3).

d. Notch the members by clicking the OK button.

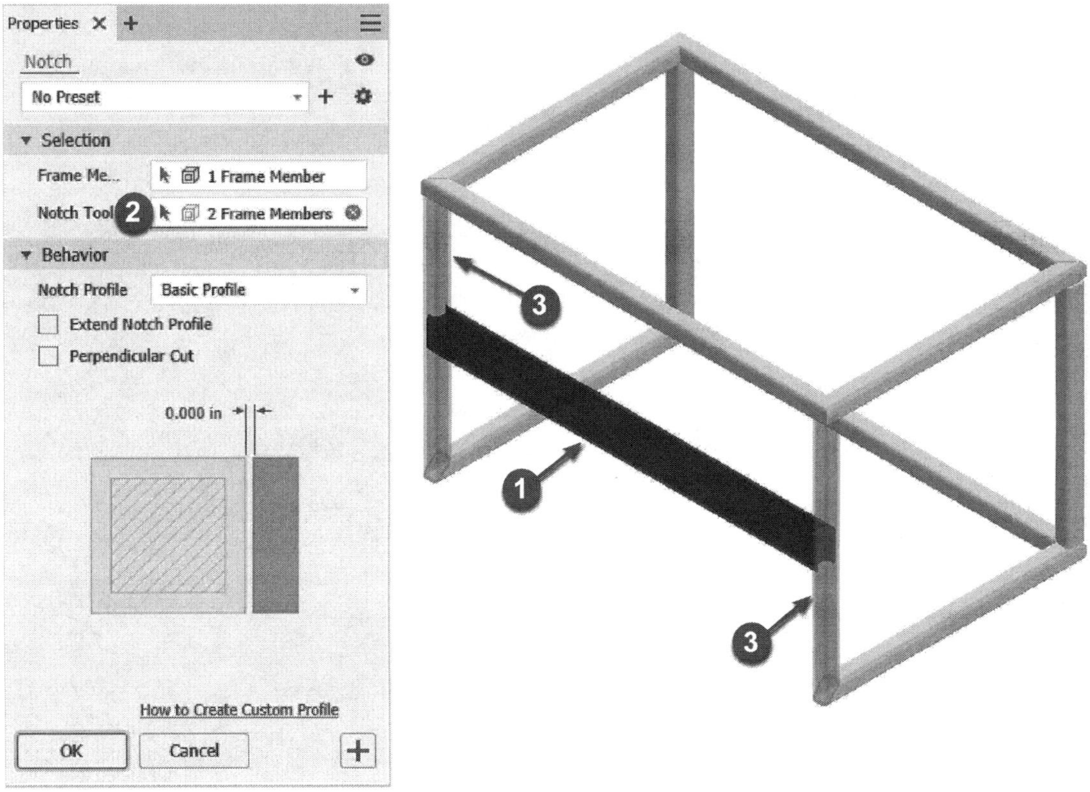

Figure 7.216

25 When done, your frame should resemble the following image.

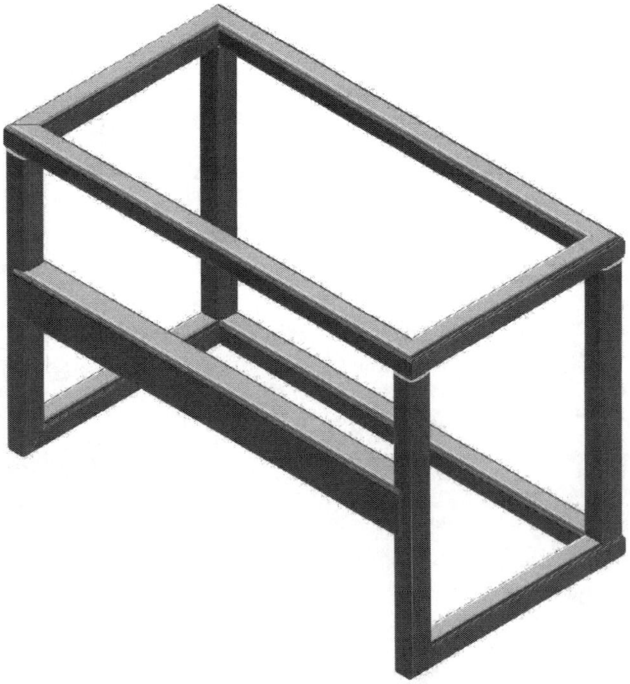

Figure 7.217

26 Save the file. Close the file.

End of exercise.

APPLYING YOUR SKILLS

Skill Exercise 7-1

In this exercise, you use the skills gained in this chapter to create a joystick handle using the loft, split, and a few plastic part commands. The file you will use has the profiles and rails created.

1. Open ESS_Skills_7-1.ipt from the Chapter 07 folder.
2. Create a loft, use the top and middle elliptical profiles, and the bottom circle profile for the sections, and use the two splines for the rails. Your preview should resemble the following image on the left.
3. Create a 0.1875 inch fillet around the top edge. The middle image shows the completed loft and fillet features.
4. Shell the part towards the inside with a thickness of 0.0625 inches. Do not remove any faces.
5. Next, create a 0.75 inch diameter hole that is concentric to the bottom circular face and has a depth of 0.0625 inches. The image on the right shows the completed part with the hole feature.

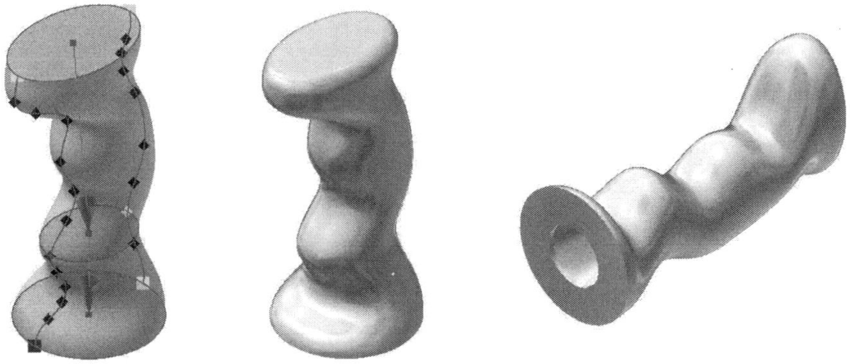

Figure 7.218

6. Use the Split command and the XY Origin plane to split the part into two solid bodies, as shown in Figure 7.219 left.
7. Use the Move Bodies command (found by clicking the down arrow on the 3D Model tab > Modify panel) to move the back solid body 1.5 inches in the Z direction. The middle image shows the moved body.
8. Turn off the visibility of the front solid and your screen should resemble the image on the right.
9. If desired, use the Make Components command to export the solids into an assembly.
10. Close the file. Do not save changes. End of exercise.

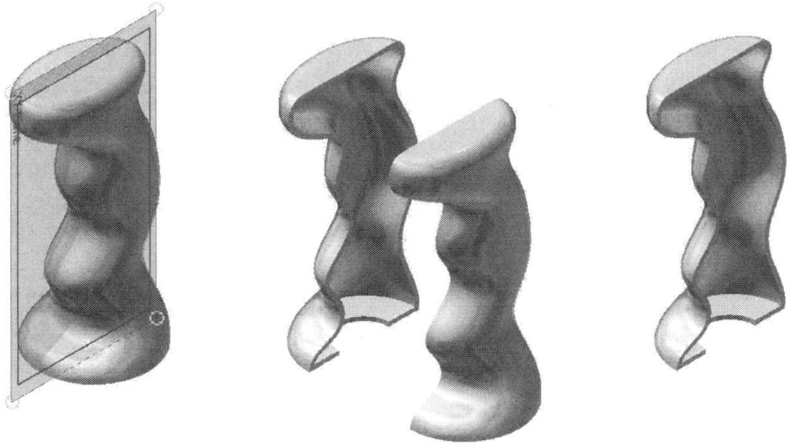

Figure 7.219

CHECKING YOUR SKILLS

Use these questions to test your knowledge of the material covered in this chapter.

1. When creating an equation, what does it signify when the equation is red in color?
2. What is the difference between a Model Parameter and a User Parameter?
3. True__ False__ User or parameters that have been renamed to something other than d# are automatically added to the Name section on the right side of the iPart Author dialog box.
4. True__ False__ You can only emboss text on a planar face.
5. True__ False__ A sweep feature requires three unconsumed sketches.
6. True__ False__ In a 3D Sketch, the Project Geometry command is used to project curves onto a circular face.
7. True__ False__ You can create a 3D curve with a combination of both 2D and 3D curves.
8. Explain how to create a 3D path using geometry that intersects with a part.
9. True__ False__ The easiest way to create a helical feature is to manually create a 3D path and then loft a profile along this path.
10. True__ False__ You can control the twisting of profiles in a loft by defining point sets.
11. Explain how to split a part into two solid bodies and then save them to their own part file.
12. Explain the difference between suppressing and deleting a feature.
13. True__ False__ After mirroring a feature, the mirrored feature is always independent from the parent feature. If the parent feature changes, the mirrored feature will NOT reflect this change.
14. Explain how to reorder features.
15. When creating a sweep with the Path & Guide Surface option, what does the surface control?
16. True__ False__ With Inventor Professional you can only run a stress analysis on a single part.
17. True__ False__ When using the frame generator tools to create a frame, you can locate frame members by selecting edges, sketch objects, or points on the skeleton model.

8 Introduction to Sheet Metal Design

INTRODUCTION

This chapter describes how to model basic sheet metal parts, create a flat pattern, and document sheet metal parts.

A sheet metal part is folded from a flat blank or sheet of material (usually metal) to form the finished shape. Sheet metal parts are relatively inexpensive to manufacture and are common components in mechanical assemblies. A sheet metal part has a uniform thickness. Examples of sheet metal parts include enclosures, guards, simple-to-complex brackets, and structural members. Although the term sheet metal is often associated with these components, heavy plates can also be formed using similar methods.

OBJECTIVES

After completing this chapter, you will be able to:

- ☐ Create a new sheet metal part
- ☐ Modify settings for sheet metal parts
- ☐ Create basic sheet metal parts
- ☐ Create a sheet metal flat pattern
- ☐ Create drawing views of a sheet metal part

SHEET METAL DESIGN

Sheet metal parts are often modeled in the folded state, as shown in Figure 8.1 left. A common practice to model a folded sheet metal part is to first extrude a profile of the thickness of the material and then add flanges, bends, hems and cutouts to the part as required. The model is then unfolded into a flat pattern, as shown on the right.

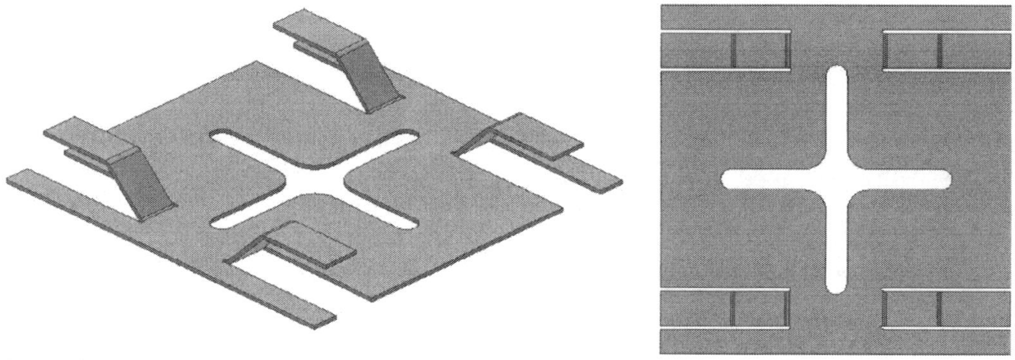

Figure 8.1

CREATING A SHEET METAL PART

The first step in creating a sheet metal part is to select a sheet metal template with which to work. Inventor comes with a sheet metal template named *Sheet Metal.ipt*. Start the New file command and use the Create New File dialog box to select the desired template folder and from the Part – Create 2D and 3D objects area, choose the Sheet Metal template file. Figure 8.2 left shows the English Sheet Metal template.

Convert a Regular Part File to a Sheet Metal Part

From a regular part file a part that has a uniform thickness can be converted to a sheet metal part. To do this click Convert to Sheet Metal from the 3D Model tab > Convert panel or Environments tab > Convert panel as shown in Figure 8.3 right.

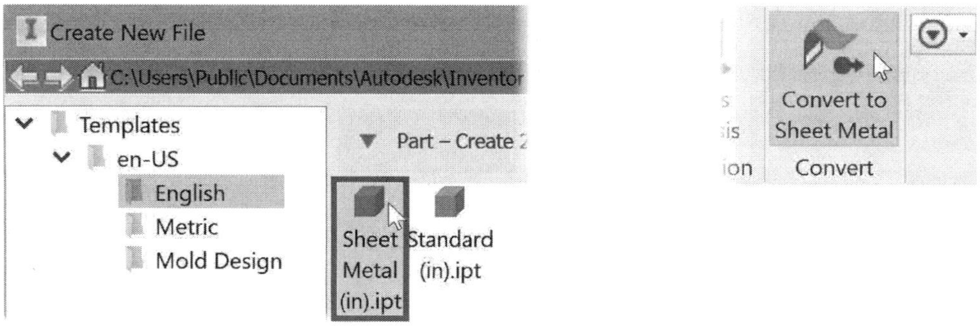

Figure 8.2

SHEET METAL ENVIRONMENT

When working on sheet metal parts, you can use Inventor's general modeling tools to add standard part features by switching to the standard modeling environment by clicking the 3D Model tab.

After creating the new sheet metal part from a sheet metal template file, the Sheet Metal tab will be active as shown in Figure 8.3. The sketch commands are common to both sheet metal and standard parts. The first sketch of a sheet metal part must be either a closed profile extruded to the sheet thickness to create a sheet metal face or an open profile that is thickened and extruded as a contour flange.

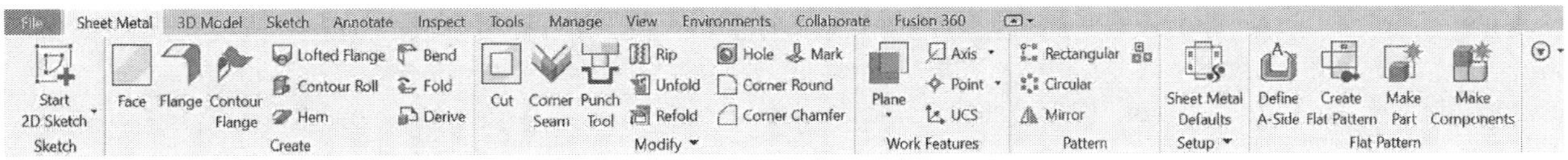

Figure 8.3

Adding Features from the 3D Model Tab

Use general modeling commands on the 3D Model tab to add features, such as holes. If you add a feature that adds thickness, the thickness must be equal to the value of the Thickness parameter. Otherwise the feature may not unfold when a flat pattern is generated from the folded model.

Sheet Metal Parameters

When you create a sheet metal part, parameters with numeric values, such as Thickness and Bend Radius, are automatically created, saved, and can be accessed in the Sheet Metal Parameters section in the Parameters dialog box, as shown in Figure 8.4. You can access the Parameter command on the Manage Tab > Parameters panel. The parameter values are updated when you modify the sheet metal defaults that will be discussed in the next section. Other model parameters and user-defined parameters can reference these parameters in equations.

TIP: Change the Sheet Metal parameter values via the Sheet Metal defaults and styles (covered in the following section). Do NOT manually change the values in the Parameters dialog box.

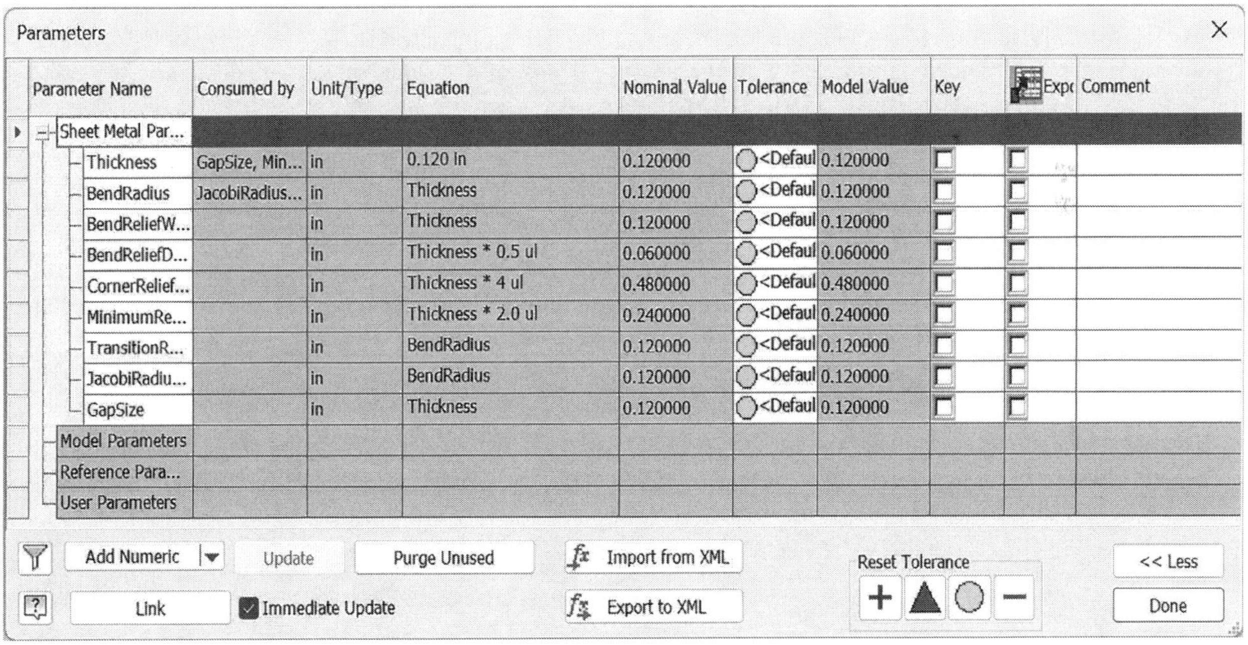

Parameter Name	Consumed by	Unit/Type	Equation	Nominal Value	Tolerance	Model Value	Key	Expo	Comment
Sheet Metal Par...									
Thickness	GapSize, Min...	in	0.120 in	0.120000	<Defaul	0.120000			
BendRadius	JacobiRadius...	in	Thickness	0.120000	<Defaul	0.120000			
BendReliefW...		in	Thickness	0.120000	<Defaul	0.120000			
BendReliefD...		in	Thickness * 0.5 ul	0.060000	<Defaul	0.060000			
CornerRelief...		in	Thickness * 4 ul	0.480000	<Defaul	0.480000			
MinimumRe...		in	Thickness * 2.0 ul	0.240000	<Defaul	0.240000			
TransitionR...		in	BendRadius	0.120000	<Defaul	0.120000			
JacobiRadiu...		in	BendRadius	0.120000	<Defaul	0.120000			
GapSize		in	Thickness	0.120000	<Defaul	0.120000			
Model Parameters									
Reference Para...									
User Parameters									

Figure 8.4

SHEET METAL DEFAULTS, RULES AND STYLES

The creation of a sheet metal part usually begins by specifying the sheet metal settings such as sheet thickness, material, bend radius and additional bend settings.

Sheet Metal Defaults

Through the Sheet Metal Defaults command, you can override the material thickness, the sheet thickness, and set rules to be current. Rules define how the sheet metal features will be created and are defined through Sheet Metal Styles that will be covered in the next section.

Start the Sheet Metal Defaults command from the Sheet Metal tab > Setup panel as shown in Figure 8.5 left. The Sheet Metal Defaults dialog box is shown in the middle image. Use it to select a material, uncheck the Use Thickness from Rule option to override the material thickness. From the drop list, you can change the current rule as shown in Figure 8.5 right.

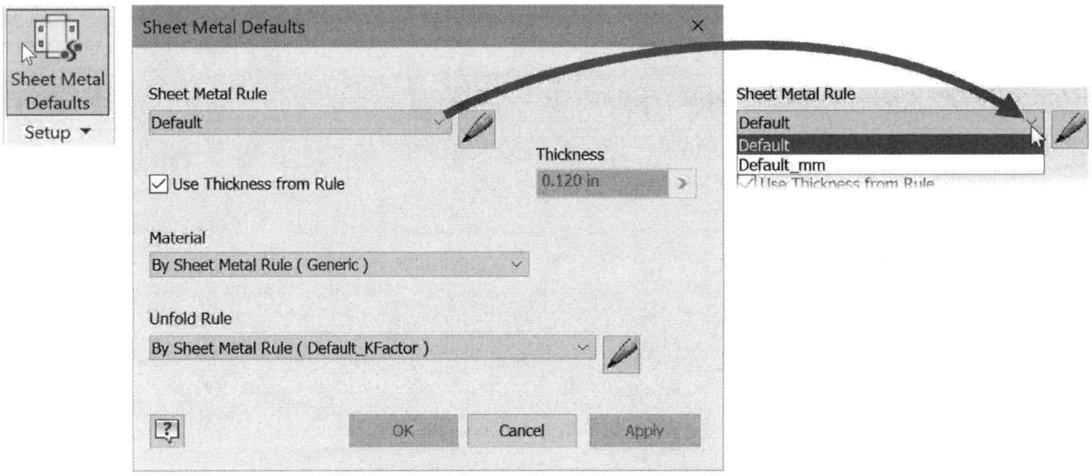

Figure 8.5

Sheet Metal Rules and Styles

Sheet metal specific parameters include the thickness and material of the sheet metal stock, a bend allowance factor to account for metal stretching during the creation of bends, and various parameters dealing with sheet metal bends and corners. The sheet metal specific parameters of a part are stored in a sheet metal rule. You can create additional sheet metal rules to account for various materials and manufacturing processes or material types. If you create sheet metal rules in a template file, they are available in all sheet metal parts based on that template. Sheet metal rules can also be created and managed using style libraries, which is the recommended workflow. The thickness of the sheet metal stock is the key parameter in a sheet metal part. Sheet metal commands such as Face, Flange, and Cut automatically use the Thickness parameter to ensure that all features are the same wall thickness, a requirement for unfolding a model. In the default sheet metal rule, all other parameters, such as Bend Radius, are initially based on the Thickness value.

To create and edit a sheet metal rule, follow these steps:

1. Click the Edit Rule icon  in the Sheet Metal Defaults dialog box or click the Styles Editor command on the Manage tab > Styles and Standards panel. The Style and Standard Editor dialog box will appear as shown in the following image.
2. In the left side of the dialog box, select an existing Sheet Metal Rule and click the New button, or right-click on the rule and click New Style from the menu. This creates a copy of the rule that is currently selected in the Sheet Metal Rule list.
3. In the New Local Style dialog box, enter a name for the rule and click OK.
4. Edit the values and settings on the Sheet, Bend, and Corner tabs to define the default feature properties of parts created with this sheet metal rule. These settings will be described in the next section.
5. Click the Save button.
6. Set the active sheet metal rule by selecting it in the Sheet Metal Defaults dialog box.
7. Changing to a different sheet metal rule or making changes to the active rule in the Style and Standard Editor updates the sheet metal part to match the new settings.

Following is a description of the settings for a sheet metal rule.

Sheet Tab

The Sheet tab in the Style and Standards Editor of the Sheet Metal Rule settings, as shown in Figure 8.6, contains settings for the sheet metal material, thickness, unfold rule, miter, rip, or seam gap, flat pattern bend angle, and the flat pattern punch representation.

Material. Specify a material from the list of defined materials in the active material library. The material appearance setting is applied to the part. You can define new materials using the Material Browser.

Thickness. Specify the thickness of the flat stock that will be used to create the sheet metal part.

Miter/Rip/Seam Gap. Specify a value to be used when creating features that require a miter, rip, or corner seam. The default gap is set to equal the Thickness parameter.

Flat Pattern Bend Angle. Specify the type of bending angle that you want to be reported. There are two options: report bending angle or report open angle. Based on the selection you choose, the appropriate angle will be used per the values designated in the Style and Standards Editor dialog box.

Flat Pattern Punch Representation. Choose from one of four options to specify how you want a sheet metal punch to appear when the model is displayed as a flat pattern. These options allow you to display the punch as a formed punch feature, a 2D sketch representation, a 2D sketch representation with a center mark, or as a center mark only.

Unfold Rule. Specify the rule or method used to calculate bend allowance, which accounts for material stretching during bending. The drop-down list provides access to the unfolding rules that are defined in the Sheet Metal Unfold node of the Style and Standard Editor.

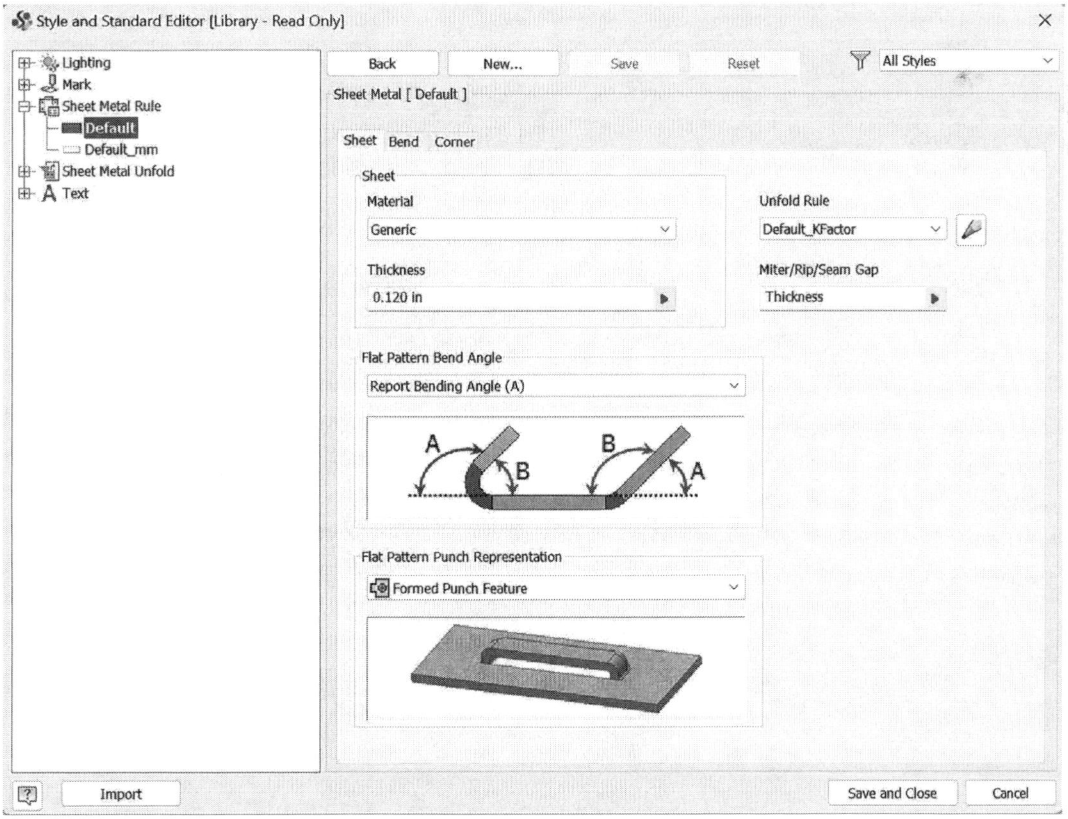

Figure 8.6

Bend Tab

The Bend tab contains settings for sheet metal bends, as shown in Figure 8.7. Bend settings are typically defined as a function of the sheet thickness. Bend Radius refers to the inside radius of the completed bend. This setting is the default for all bends, but you can override it while creating a bend. Specify bend reliefs to use when a bend zone (the area deformed during a bend) does not extend completely beyond a face. If bend reliefs are used, they are incorporated in the flat blank prior to folding. Bend reliefs are manufactured by punching or with laser, water-jet, or other cutting methods. Adding bend reliefs may increase production costs. It is common practice with thin deformable materials, such as mild steel, to add bends without bend reliefs. However, the material can tear or deform where the bend zone intersects the adjacent face.

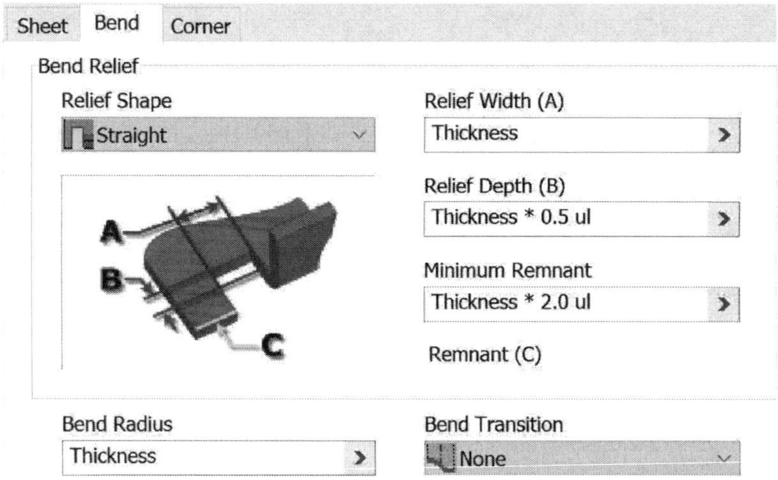

Figure 8.7

Relief Shape. If a bend does not extend the full width of an edge, a small notch is cut next to the end of the edge to keep the metal from tearing at the edge of the bend. Select from Straight, Round, or Tear for the shape of the relief. This setting displays as the default value in the sheet metal feature creation dialog boxes. As you select from one of the three relief shapes, a preview image appears, showing the shape of the relief and how the settings for Relief Width, Relief Depth, and Minimum Remnant apply to the selected shape. Figure 8.8 shows the preview of the Straight, Tear, and Round reliefs.

Figure 8.8

Relief Width. Specifies the width of the bend relief.

Relief Depth. Specifies the distance a relief is set back from an edge. Round relief shapes require this distance to be at least one-half of the Relief Width value.

Minimum Remnant. Specifies the distance from the edge at which, if a bend relief cut is made, the small tab of remaining material is also removed.

Bend Radius. Defines the inside bend radius value between adjacent, connected faces.

Bend Transition. Controls the intersection of edges across a bend in the flattened sheet. For bends without bend relief, the unfolded shape is a complex surface. Transition settings simplify the results, creating straight lines or arcs, which can be cut in the flat sheet before bending. Figure 8.9 shows the five transition types.

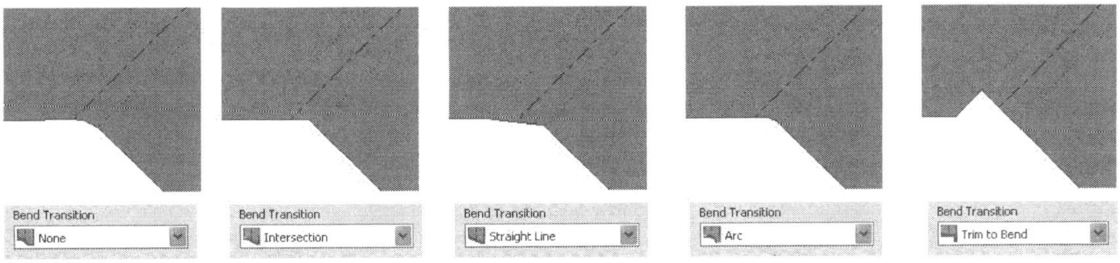

Figure 8.9

Corner Tab

A corner occurs where three faces meet. The corner seam feature controls the gap between the open faces and the relief shape at the intersection. As with bend reliefs, corner reliefs are added to the flat sheet prior to bending.

Use the options in the Corner tab, as shown in Figure 8.10, to set how corner reliefs will be applied to the model. You can designate the corner relief size, shape, and radius.

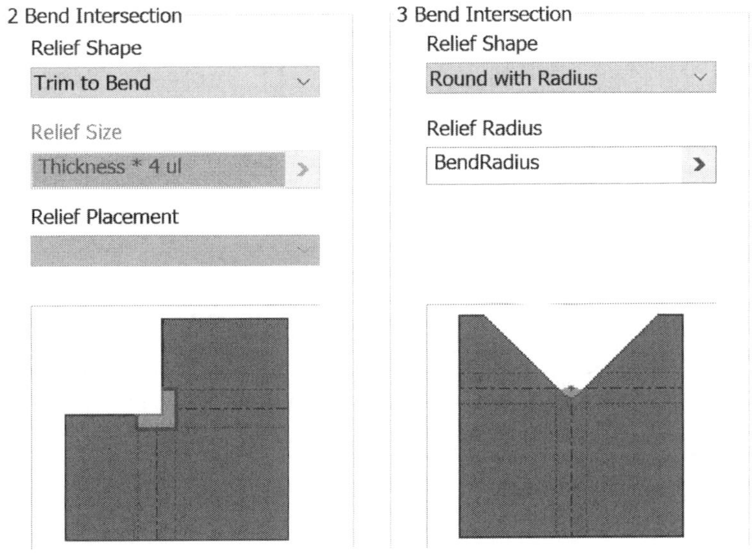

Figure 8.10

Relief Shape. Specify the shape of the corner relief for either 2 or 3 bend intersections. When a 2 bend intersection is formed in your model, you can select from one of seven corner relief options as shown in the following two figures.

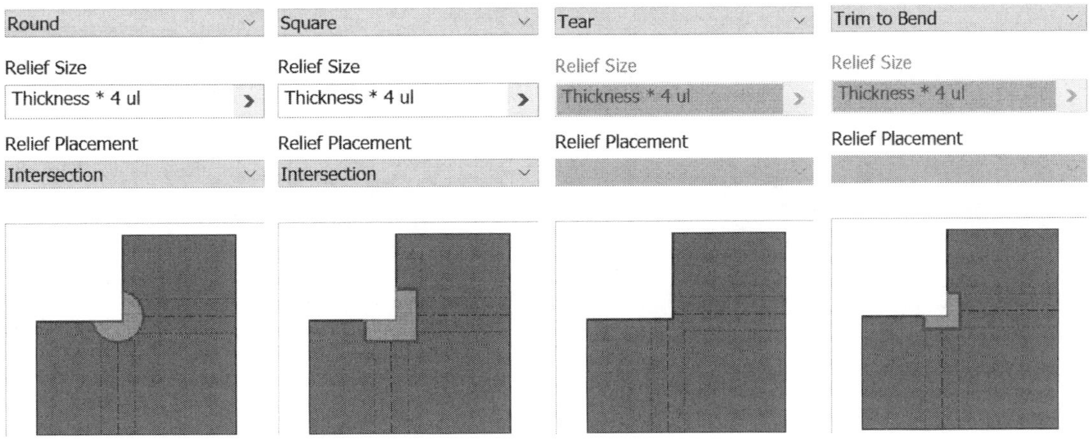

Figure 8.11

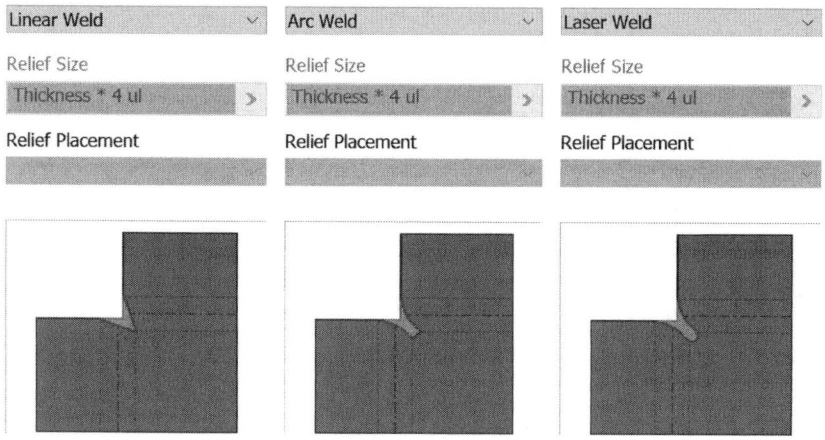

Figure 8.12

When a 3 bend intersection, also referred to as a Jacobi corner, is formed in your model, four relief shapes can be used, as shown in Figure 8.13.

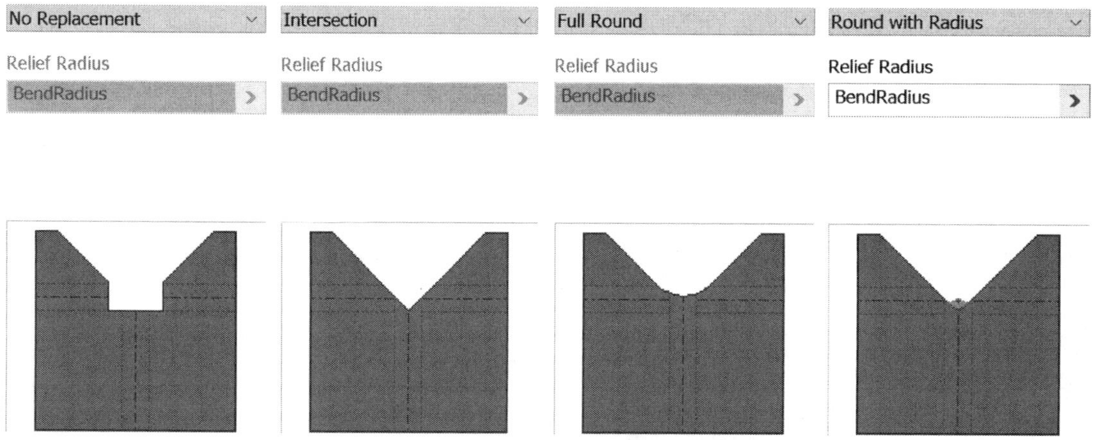

Figure 8.13

Relief Size. Sets the size of the corner relief for two-bend intersections when either round or square relief shapes are selected.

Relief Radius. Sets the radius of the corner relief for three-bend intersections when the round with radius relief shape is selected.

Sheet Metal Unfold Rule

In addition to defining sheet metal rules for creating sheet metal features you can define sheet metal unfold rules by clicking the Edit Unfold Rule in the Style and Standard Editor > Sheet tab, as shown in Figure 8.14. Additional unfolding rules are created the same way as sheet metal rules. There are three options available for the Unfold Method: Linear, Bend Table, or Custom Equation for more complex or precise requirements.

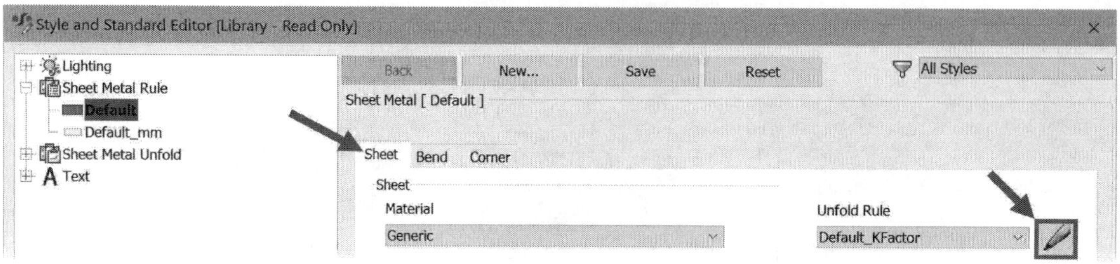

Figure 8.14

Linear Unfold Method. When Linear is selected, you specify the default KFactor value used for calculating bend allowances. A KFactor is a value between 0 and 1 that indicates the relative distance from the inside of the bend to the neutral axis of the bend. A KFactor of 0.5 specifies that the neutral axis lies at the center of the material thickness. The default value is 0.5, which can be adjusted up or down to represent your manufacturing requirements. The Unfold Method is combined with the Unfolding Rule specified on the Sheet tab of the sheet metal rule.

Custom Equation Method. The Custom Equation should be used when you need additional control over the unfolding of your model.

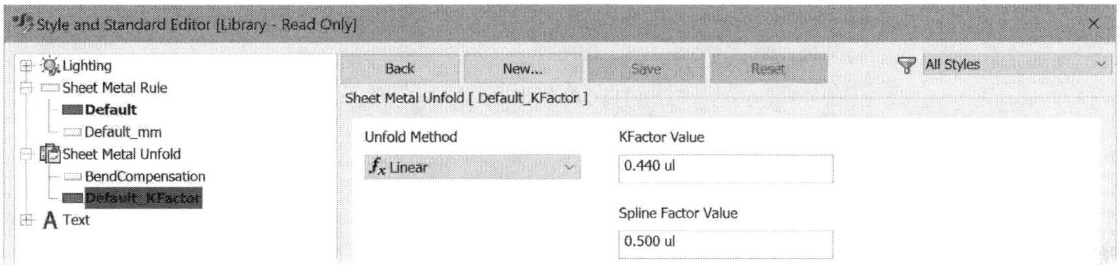

Figure 8.15

EXERCISE 8-1: EDITING A SHEET METAL STYLE AND RULE

In this exercise, you edit a sheet metal style and change sheet metal rules on a sheet metal part.

1 Open *ESS_E08_01.ipt* in the Chapter 08 folder. Zoom out and use the Free Orbit command to review the sheet metal features as shown in Figure 8.16. When done, click the Home View icon.

Figure 8.16

2 To edit the sheet metal styles, click Sheet Metal Defaults from the Sheet Metal tab > Setup panel.

3 The thickness can be changed by unchecking the Use Thickness from Rule option in the Sheet Metal Defaults dialog box, but for this exercise you will create a new rule. In the Sheet Metal Defaults dialog box, click Edit Sheet Metal Rule as shown in Figure 8.17.

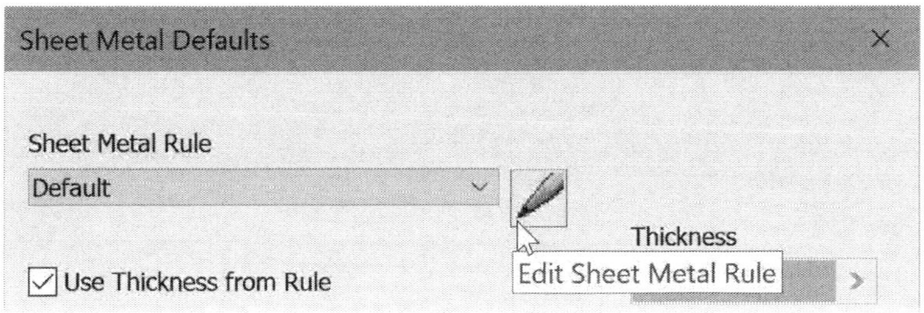

Figure 8.17

The new style will be based on the Default style. In the Style and Standard Editor dialog box expand Sheet Metal Rule, right-click on Default and click New Style from the menu as shown in Figure 8.18 left.

4 From the New Local Style dialog box, enter **16_Gauge_Aluminum** for the name of the new style. Click OK to create the style.

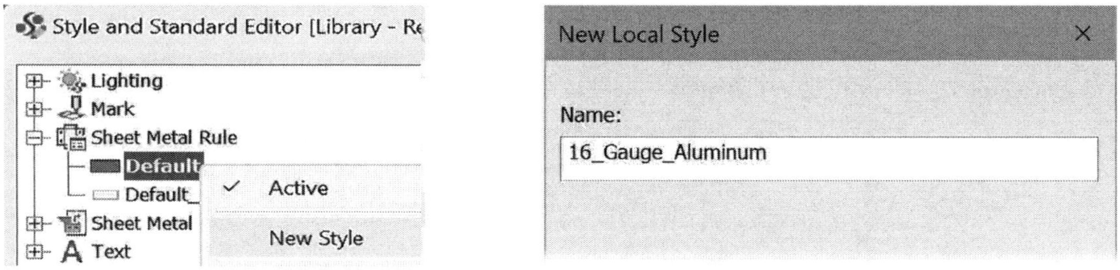

Figure 8.18

5 Next edit the new style.

a. Ensure the new style is current by double-clicking on **16_Gauge_Aluminum.** The active style will be bold and highlighted as shown in Figure 8.19.

b. From the Material list, change the material to Aluminum 6061.

c. Change the Thickness to **0.0508 inches** as shown in Figure 8.19.

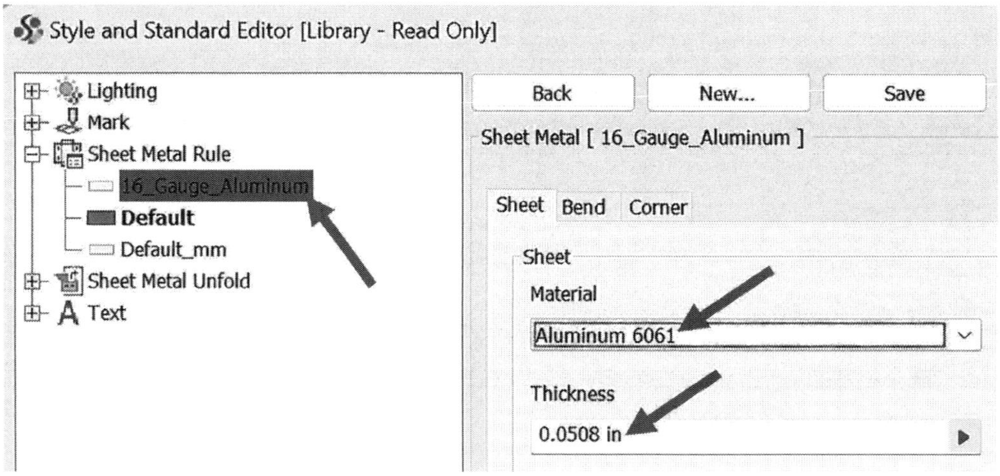

Figure 8.19

d. Click the Bend tab.

e. Change the 2 Bend Intersection - Relief Shape to Round as shown in the following image.

f. Change the Relief Width (A) to **Thickness/2** as shown in Figure 8.20.

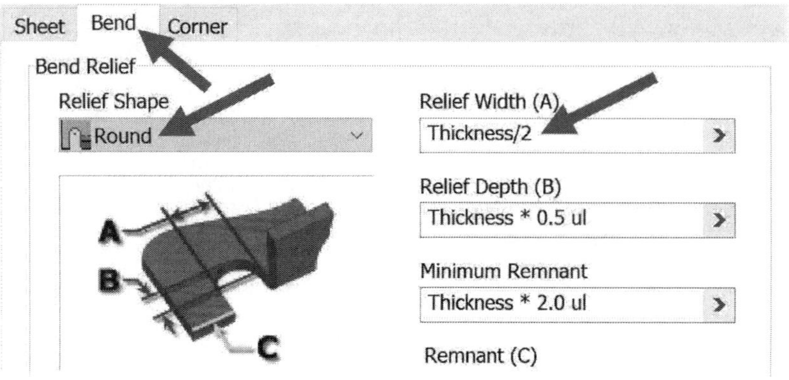

Figure 8.20

g. Click the Corner tab.

h. Change the 2 Bend Intersection – Relief Shape to Tear as shown in Figure 8.21.

i. Save the style by clicking Save and Close on the bottom-right of the Style and Standard Editor dialog box.

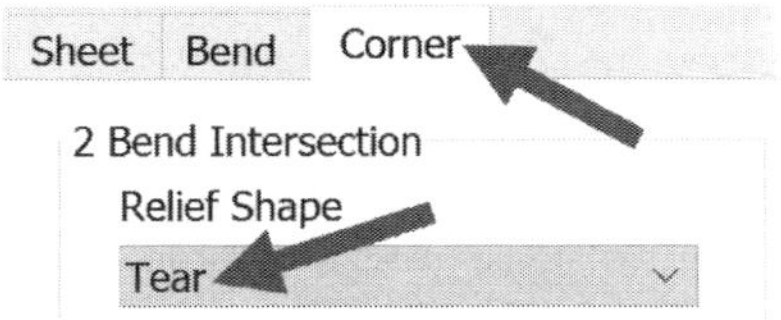

Figure 8.21

6 In the Sheet Metal Defaults dialog box, change the current Sheet Metal Rule to 16_Gauge_Aluminum as shown in Figure 8.22 left.

7 Close the Sheet Metal Defaults dialog box by clicking OK. The part will update to reflect the changes in the new rule. Your part should resemble Figure 8.22 right.

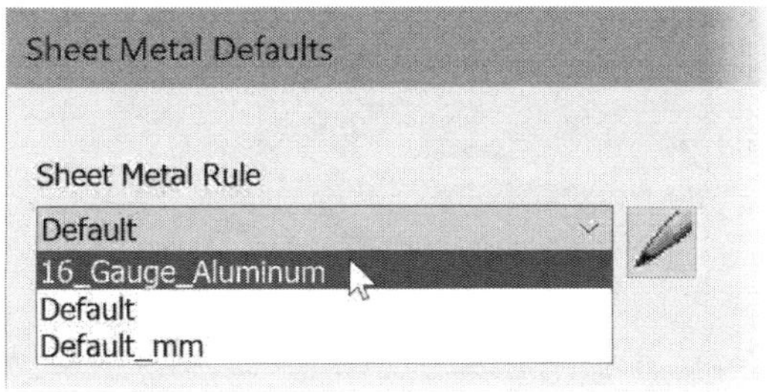

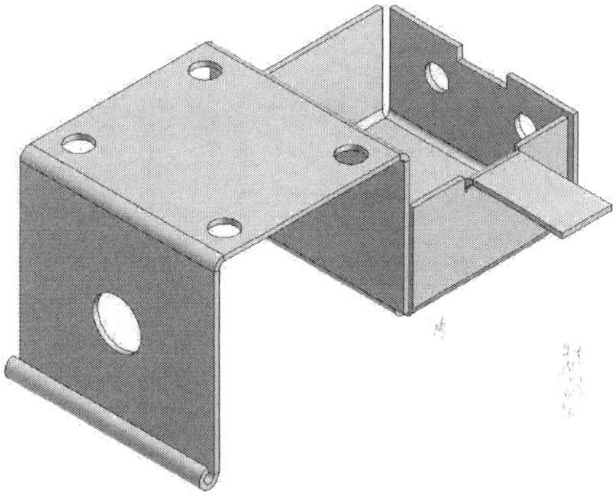

Figure 8.22

8 Zoom in and use the Free Orbit command to review the sheet metal features.

9 Edit the Sheet Metal Defaults and Change the Sheet Metal Rule back to the original Default rule.

10 Experiment with different style settings.

11 Close the file. Do not save changes.

End of exercise.

FACE

The Face command extrudes a closed profile for a distance equal to the sheet metal thickness parameter. If the face is the first feature in a sheet metal part, you can flip the direction of the extrusion. Figure 8.23 left shows a closed profile. The image on the right shows the feature after using the Face command. During the design process, you can connect an adjacent face to the previous face feature. If the sketch shares an edge with an existing feature, a bend is added automatically. As an option, you can select a parallel edge on a disjointed face. This action will extend or trim the attached face to meet the new face, with a bend created between the two faces.

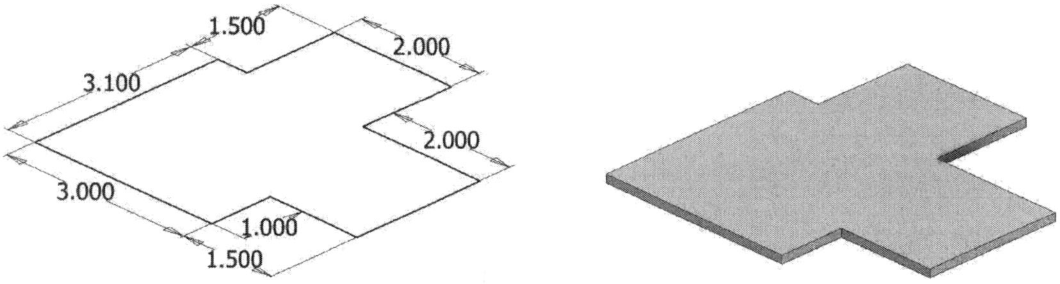

Figure 8.23

To create a sheet metal face, follow these steps:

1. Create a sketch with a single closed profile or a closed profile containing islands. The sketch is most often on a work plane created at either a specific orientation to other part features or by selecting a face on another part in an assembly.

TIP: You can create a face feature by selecting multiple closed profiles.

2. Click the Face command on the Sheet Metal tab > Create panel as shown in the following image on the left.
3. The Face dialog box appears, as shown in Figure 8.24 right. If a single closed profile is available, it is selected automatically. If multiple closed profiles are available, you must select in the area(s) to give a thickness to.

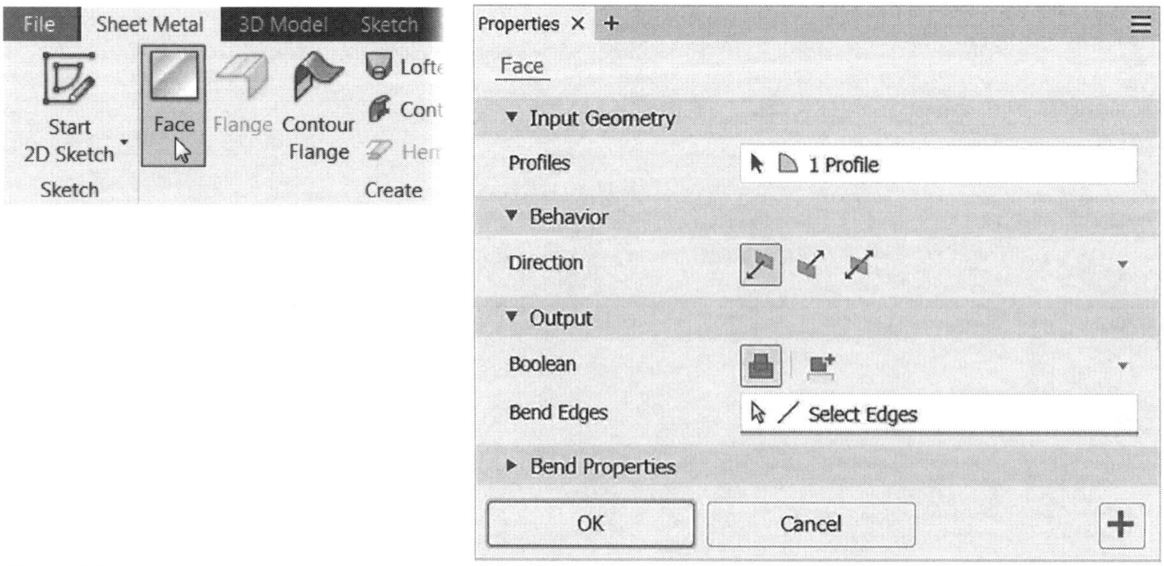

Figure 8.24

4. If required, in the Offset Direction area flip the direction of the thickness, or select Both Sides to use the selected profile as the mid-plane for the extrusion.
5. If the face is not the first feature and the sketch does not share an edge with an existing feature, click the Edges button to select an edge the face connects with. The two faces are extended or trimmed as required, meeting at a bend. The bend is listed as a child of the new face feature in the browser. If the face attached to the selected edge is parallel to but not coplanar with the new face, the dialog box expands to show the double-bend options. An additional face is added to connect the two parallel faces.
6. If needed, click the Unfold Options or Bend tab to override these options for this feature.
7. Click OK.

Unfold Options and Bend Tabs

The options available on the Unfold Options and Bend tabs are used to override the active rule.

CONTOUR FLANGE

Use the Contour Flange command to extrude an open sketch profile perpendicular to the sketch plane. The open profile is thickened to match the sheet metal thickness parameter. The profile does not require a sketched radius between line segments—bends are added at sharp intersections. Arc or spline segments are offset by the sheet metal thickness. A contour flange can be the first feature in a sheet metal part, or it can be added to existing features. Figure 8.25 left shows a profile consisting of lines; the image on the right shows the results after the Contour Flange command extrudes the profile, and adds bends and thickness.

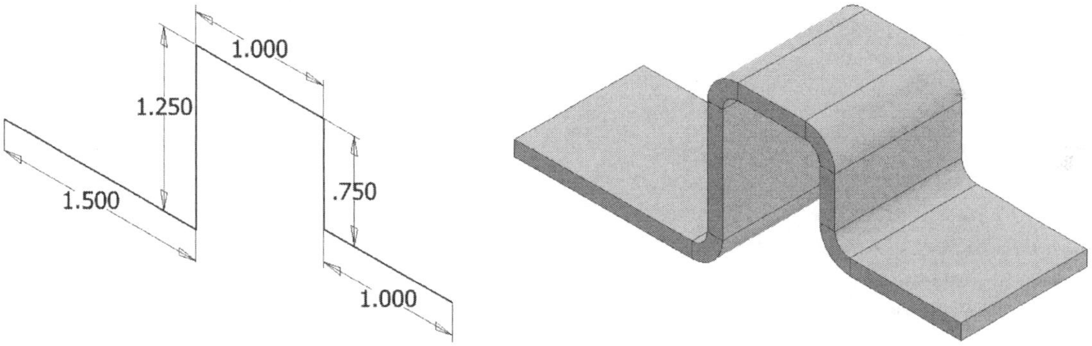

Figure 8.25

To create a contour flange, follow these steps:

1. Create a sketch with a single open profile. The sketch can contain line, arc, and spline segments, and it can be constrained to projected reference geometry to define a common edge between the contour flange and an existing face.
2. Click the Contour Flange command on the Sheet Metal tab > Create panel as shown in the following image on the left.
3. The Contour Flange dialog box appears, as shown in Figure 8.26 right.

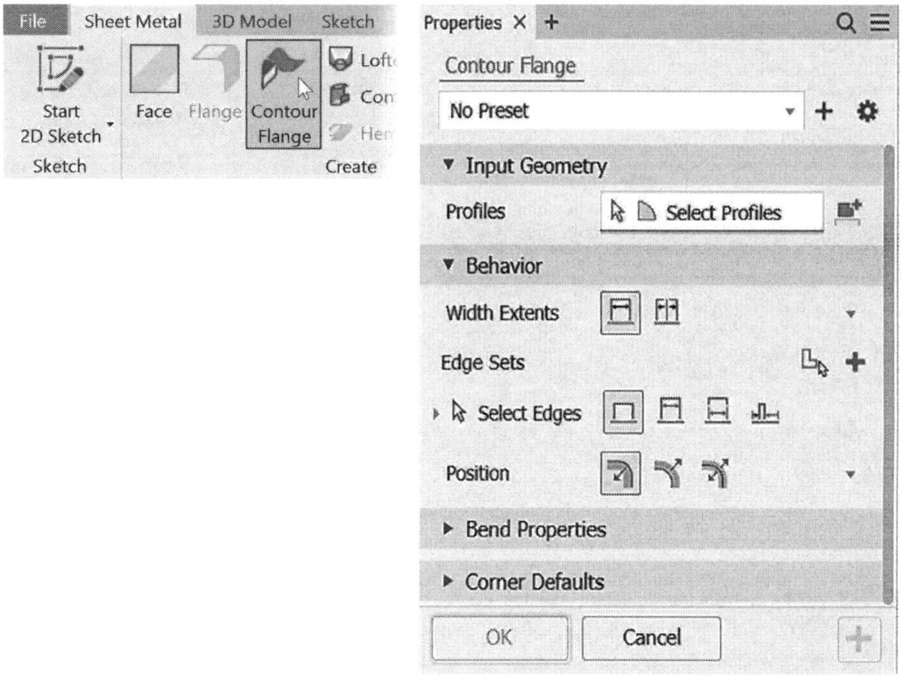

Figure 8.26

1. Select the open sketch profile to define the shape of the contour flange.
2. If required, flip the side to offset the profile or select the Symmetric icon to use the selected profile as the mid-plane for the contour flange.
3. If the contour flange is the first feature in the part, enter an extrusion distance and direction. If the contour flange is not the first feature, you can select an edge perpendicular to the sketch plane to define the extrusion extents. If the sketch is attached to an existing edge, select that edge. If the sketch is not attached to any projected geometry, the contour flange is extended or trimmed to meet the selected edge. The selected edge is typically the closest edge to the end of the sketch. You can also use Loop Select Mode to have the flange created along all edges of a selected loop.
4. Apply any unfold, bend, or corner overrides in the Bend Properties and Corner tabs.
5. If you select an edge to define the edge sets, you can select from four options to further refine the length of the contour flange. The options are described next.
6. Click OK.

Width Extents

Use the Width Extents area to select the extents type for the feature. If this is the first feature, From Sketch Plane is the only extent type that is available as shown in Figure 8.27.

Figure 8.27

From Sketch Plane

From. Defines the starting plane for the contour flange. By default, the profile sketch plane is used. This is inactive unless the Between option is selected.

Between. Use to specify both a start and end location for the flange.

From. Select where the flange starts.

To. Select the feature where the flange ends.

Direction. Choose the flange direction from four options: Default, Flipped, Symmetric, Asymmetric.

Distance A. Enter the length of the contour flange.

Bend Edges. Select additional edges where the flange will be applied.

Position. Defines the side of the profile where the flange is created.

Default. Places the flange on the interior side of the profile.

Flipped. Places the flange on the exterior side.

Symmetric. Centers the flange equally on both sides of the profile.

Edge Sets

After the first feature is created, the Edge Sets options are available as shown in Figures 8.28 and 8.29. These edge types are also available for flange and hem features. However, the hem feature does not include the Distance option because the distance is specified in the main body of the dialog box.

Edge. Select so that the contour flange extends the full length of the selected edge.

Width. Select so that a point defines one extent of the contour flange. The flange can be offset from this point and extended a fixed distance. You define the starting point by selecting an endpoint on the selected edge, a work point on a line defined by the edge, or a work plane perpendicular to the selected edge.

Between. Select so that the width of the feature is defined from two selected part features. You can select work points, work planes, vertices, or planar faces that intersect the selected edge.

Offset. Like the Width option, select so that two selected points define both extents of the flange. You can specify an offset distance from each point.

The following two figures show a sheet metal part with an open profile that was used with the contour flange command and the Edge Sets.

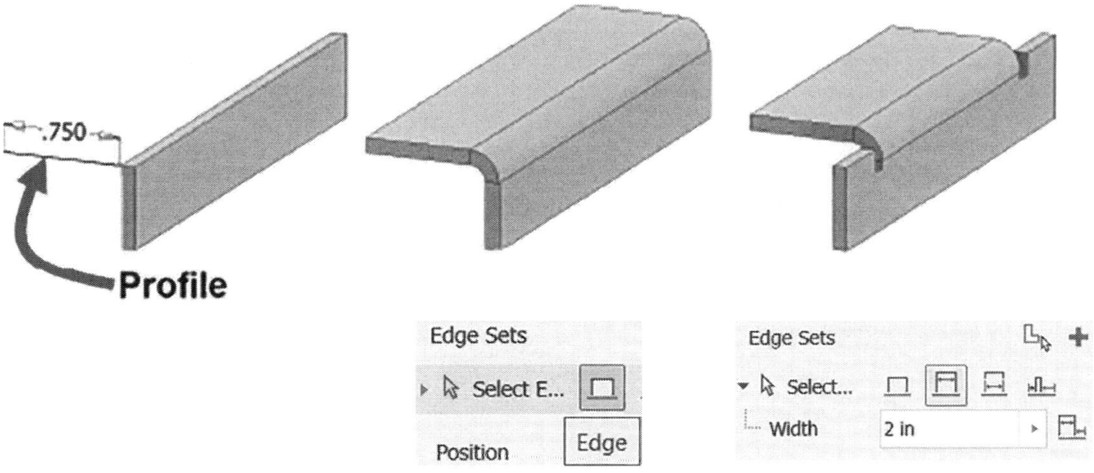

Figure 8.28

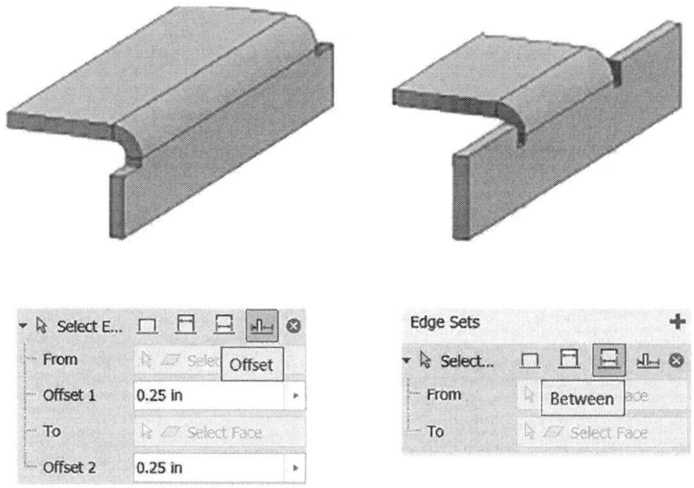

Figure 8.29

Bend Properties and Corner Defaults

The options available on the Bend Properties and Corner Defaults tabs are used to override the active rule.

FLANGE

A sheet metal flange is a simple rectangular face created from existing face edges. A sketch is not required when creating a flange. The flange can extend the full length of the selected edge and can create automatic corner and miter features. The Flange command adds a new sheet metal face and bend to an existing face. If multiple edges are selected for the flange feature, corner seams and miter features are also created. You set the flange length and the angle relative to the adjacent face in the Flange dialog box. The selected edge or edges define the bend location between the selected faces.

TIP: A minimum of one sheet metal face or contour flange must exist before creating a flange feature.

To create a flange, follow these steps:

1. Click the Flange command on the Sheet Metal tab > Create panel as shown in Figure 8.30 left.
2. The Flange dialog box appears, as shown in Figure 8.30 right.

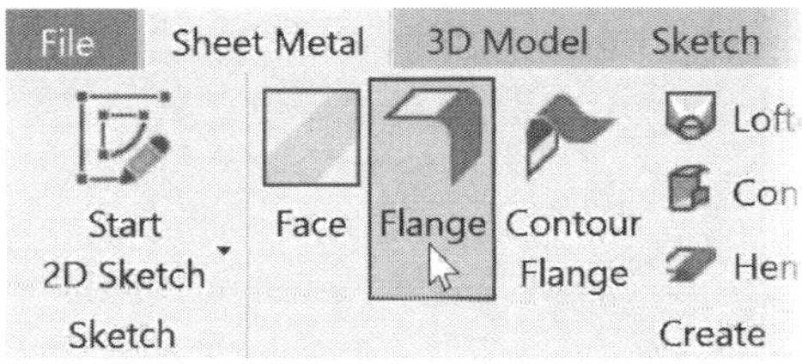

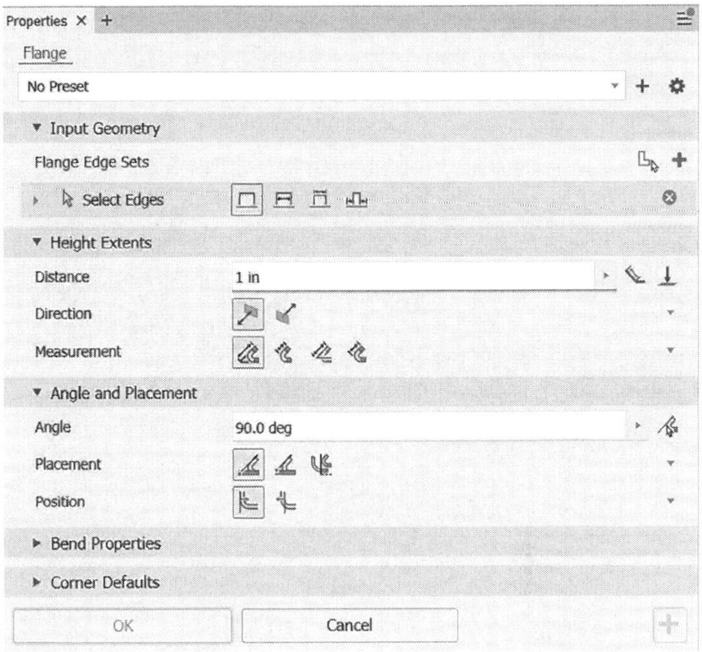

Figure 8.30

3. Select an edge or edges of an existing face.
4. Enter the distance and angle for the flange in the Flange dialog box. The flange preview updates to match the current values.
5. If required, flip the direction for the flange, offset the thickness, and modify the datum used for the height measurement or the position of the bend for the feature.
6. Expand the dialog box, and select the appropriate extent type. See the previous Width Extents section for additional information on extents.
7. Click Apply to continue creating flanges, or click OK to apply the flange and exit the command.

Flange Dialogue Box

The options in the Flange Dialogue Box allow you to select the edge, height extent, angle, and bend properties for the flange. Following are descriptions of the main options in the Flange Dialogue Box.

Input Geometry. Select the Flange Edge Sets on which to create flanges. Sets consist of one or more edges. To select an edge for a flange, click on an edge in the graphics window. For each flange edge set, determine the width of the flange edge by specifying an edge option to use, choosing from the Edge, Width, Between, or Offset options. Feature manipulators (arrows displayed at the top of a flange in the graphics window) may be used to adjust the flange angle and height. View a flange's manipulators by clicking to highlight its flange edge set in the Flange Dialogue Box. You may click on the manipulator arrow to open a field and enter a new value. You can also click and drag the arrow to manipulate the feature.

Height Extents. Select an option for specifying the height of the flange(s). Click the Orthogonal icon to enter a value specifying an orthogonal height, or select the To icon to specify a termination point.

Angle and Placement. Enter a value for the angle for the flange(s). The value must be less than 180°. Alternatively, by selecting the By reference plane icon located to the right of the angle input field, you may select a reference plane for the angle. Use the Placement icons to plane through a partnered edge, selected edge, or plane tangent to side faces. The Position icons allow you to position the angle plane on the outer face or inner face.

Bend Properties. This group includes controls for Bend Radius, Unfold Rule, Relief Shape, and Bend Transition. Relief Shape options include Default, Tear, Round, and Straight.

Corner Defaults. Set the Auto-Mitering option, Miter Gap size, Corner Shape, and Corner Relief.

If you create or edit a flange or contour flange feature with multiple edges converging in a corner, a glyph is displayed at that corner. Clicking the glyph opens the Corner Overrides tab, as shown in the following image. The Corner Overrides tab provides control over the geometry of the corner. These overrides can be set during feature creation or edit and control the type and size of relief, the overlap condition, and the miter gap.

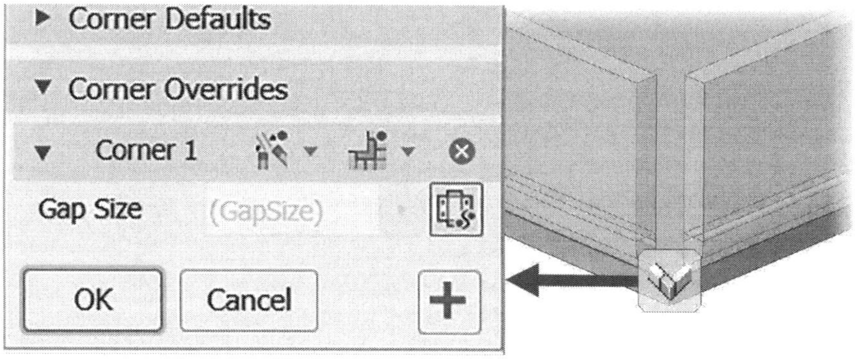

Figure 8.31

EXERCISE 8-2: CREATING SHEET METAL PARTS

In this exercise, you create a sheet metal face, contour flange, and flanges to build sheet metal parts.

1 Create a new sheet metal part by clicking the New file command in the Quick Access toolbar. In the Create New File dialog box click the English template folder and in the Part – Create 2D and 3D objects area click Sheet Metal (in).ipt.

2 Create a sketch on the XZ plane.

3 Draw and dimension a **6 inch x 4 inch rectangle**, with the lower-left corner on the origin point.

4 Exit the sketch and your screen should resemble the following image on the left.

5 Click the Face command on the Sheet Metal tab > Create panel. Since there is only one closed profile and the extrusion distance is set by the current sheet metal rule, all you need to do is click OK. When done, your screen should resemble the image on the right.

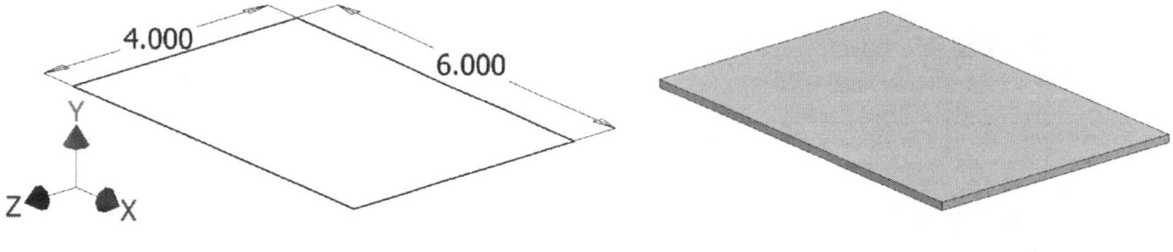

Figure 8.32

6 Next you create flanges. Start the Flange command on the Sheet Metal tab > Create panel.

a. By default, you select individual edges. Select the top four edges.

b. Try different values for the height and angle.

c. Enter a height of **1.5 inches** and an angle of **45 degrees** as shown in Figure 8.33. When an angle besides 90 degrees is used, the flanges are automatically mitered.

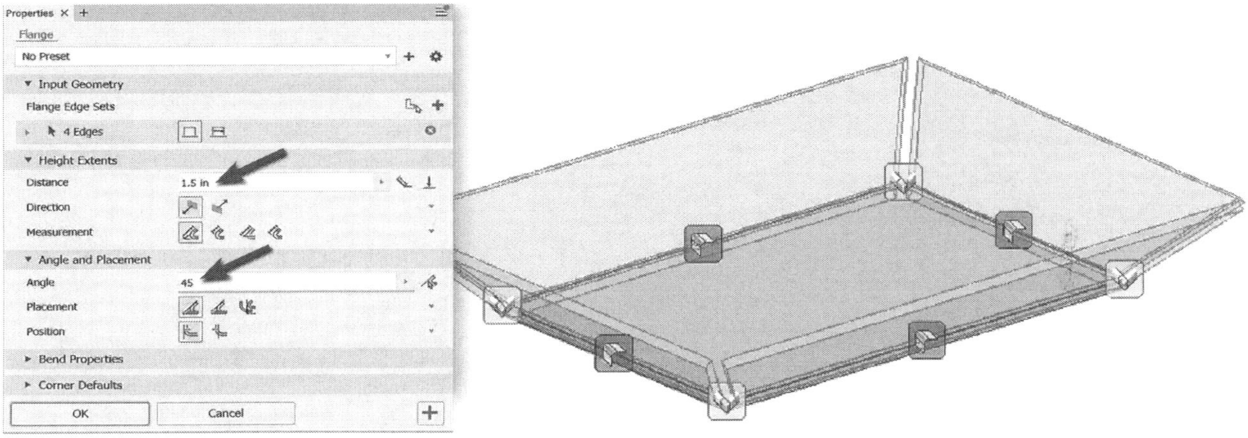

Figure 8.33

d. To turn off the automatic mitering, click Corner Defaults and uncheck Apply Auto-Mitering as shown in Figure 8.34. Notice how the flanges now have a gap.

e. Turn Apply Auto-Mitering back on.

f. From the Shape tab change the angle to **90 degrees** and click OK to create the flanges.

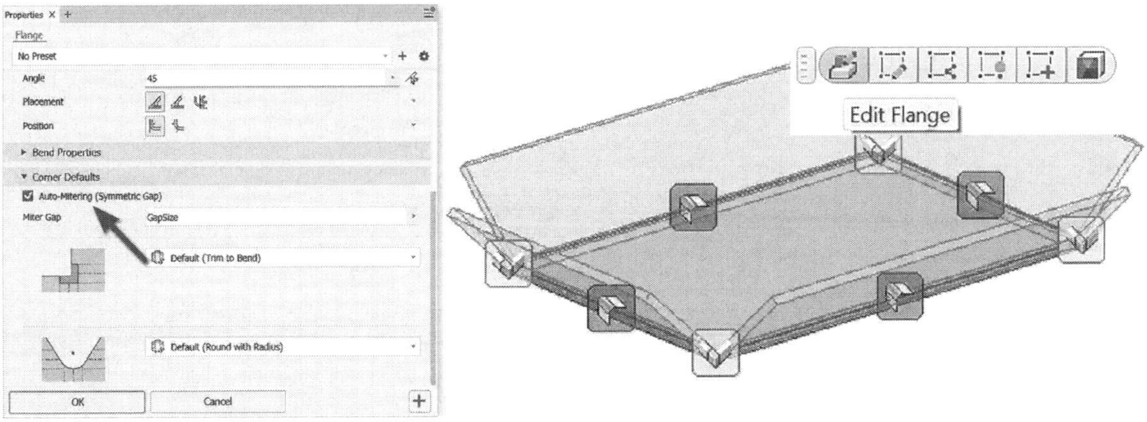

Figure 8.34

7 Next edit the corners.

a. Zoom into the front corner and notice the gap and the relief shape as shown in Figure 8.35 left.

b. Since the four flanges were created in one operation, you can edit all of them with one edit. Edit the flanges by selecting the front vertical face of the flange and click Edit Flange from the mini-toolbar as shown in Figure 8.35 right.

c. In the Flange dialog box click Bend Properties and change the Relief Shape to **Tear**. Click Corner Defaults and reduce the Miter Gap to **GapSize/2** as shown in Figure 8.36 left.

d. Click OK and the corners should resemble Figure 8.36 right.

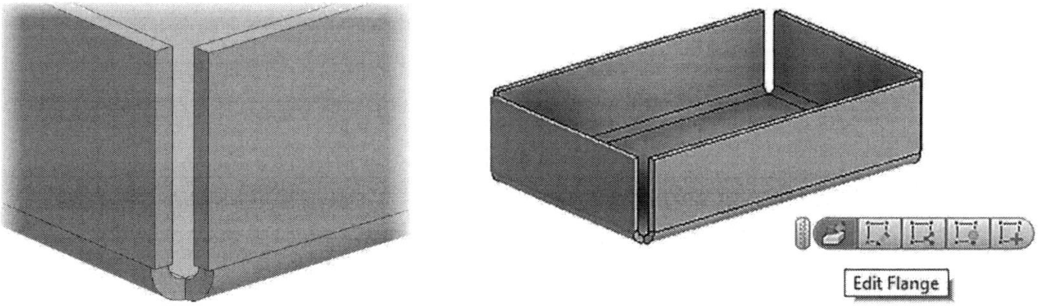

Figure 8.35

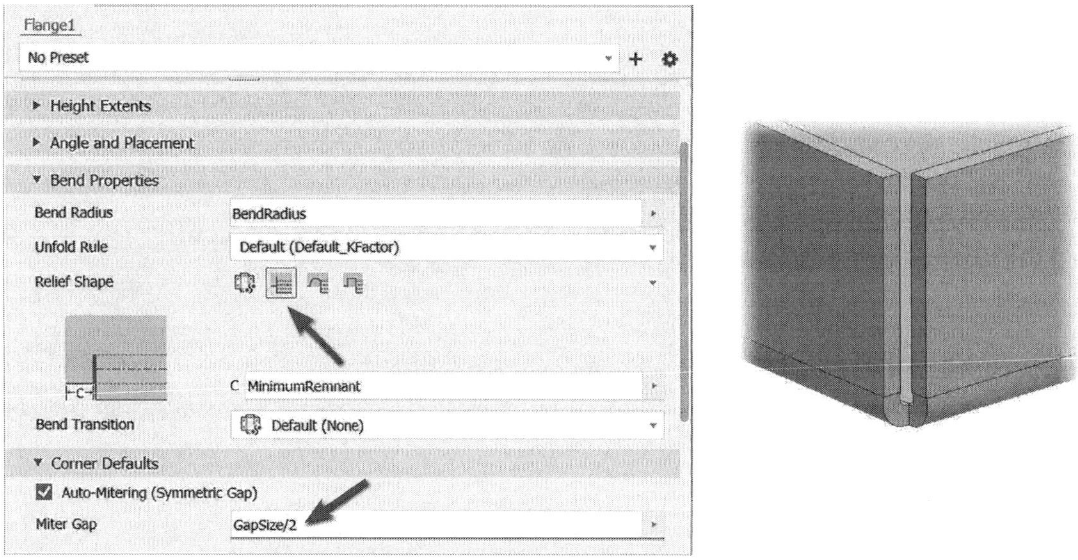

Figure 8.36

8 Next use the Loop option to create flanges.

a. In the browser, right-click on Flange1 and click Delete from the menu.

b. Start the Flange command on the Sheet Metal tab > Create panel.

c. To easily select all edges on a face you can use the Loop option. In the Flange dialog box click the Loop button as shown in Figure 8.37.

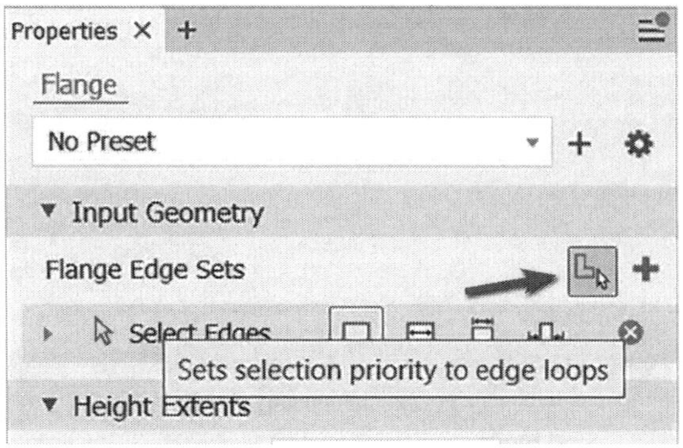

Figure 8.37

d. In the graphics window, select the top edge and the four flanges will be previewed as shown in Figure 8.38.

e. Change the height to **1 inch**. Your screen should resemble the image on the right.

f. When done, click OK to create the flange feature.

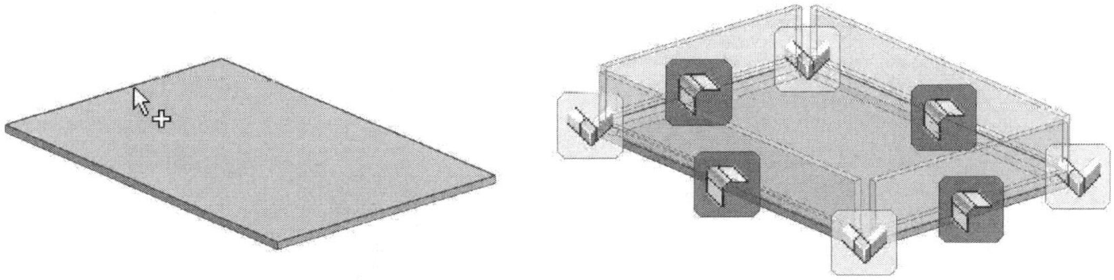

Figure 8.38

9 Next you use a 2D open profile to create a contour flange. You start by deleting the two features.

a. In the browser, right-click on Face1 and choose Delete from the menu. Click OK to delete the features and the sketch. The flange is deleted because it was dependent on the face feature.

b. Create a sketch on the YZ plane.

c. Sketch and dimension the four lines as shown in Figure 8.39. Place the lower-left corner of the profile on the origin point, and the left and right sides are equal in size.

d. Finish the sketch.

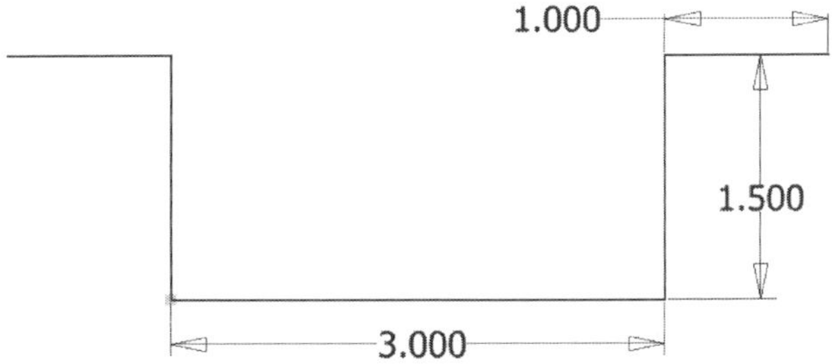

Figure 8.39

10 Click the Contour Flange command on the Sheet Metal tab > Create panel.

11 In the Contour Flange dialog box change the Distance to **3 inches** and click OK to create the feature. When done, your screen should resemble Figure 8.40.

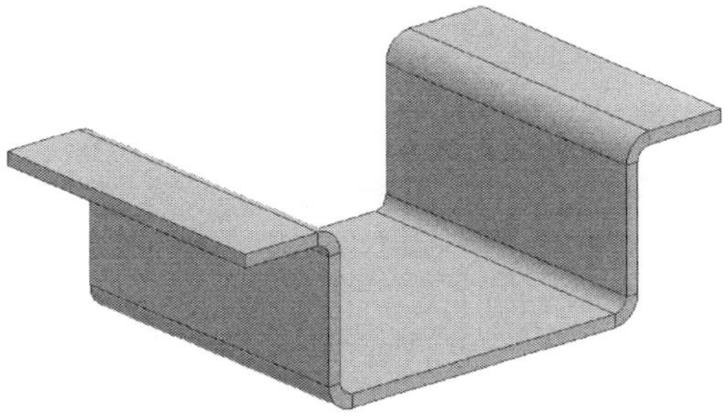

Figure 8.40

12 Next you practice placing a flange with different Width Extent options.

a. Start the Flange command on the Sheet Metal tab > Create panel.

b. Select the top back edge.

c. Change the Edge icon to Width and the Width distance to **1 inch** as shown in Figure 8.41.

d. Experiment with different Width options and applying additional flanges.

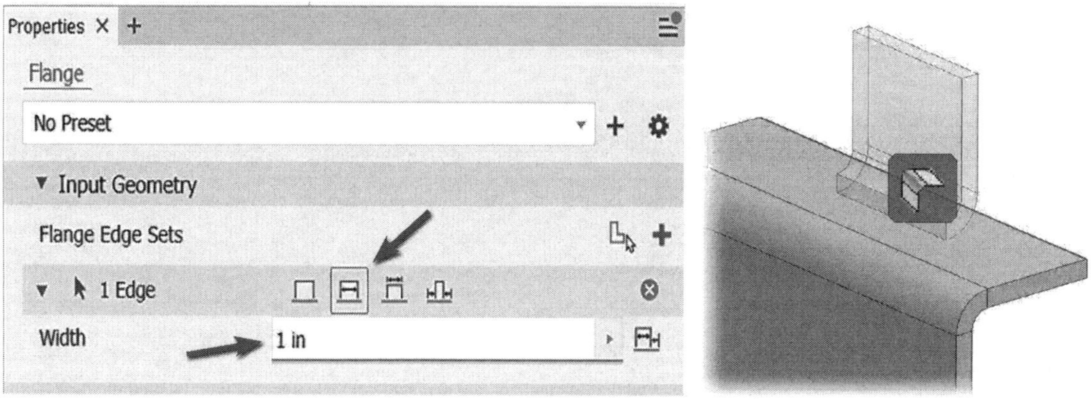

Figure 8.41

13 When done, close the file. Do not save changes.

End of exercise.

HEM

Hems eliminate sharp edges or strengthen an open edge of a face. Material is folded back over the face with a small gap between the face and the hem. A hem does not change the length of the sheet metal part; the face is trimmed so that the hem is tangent to the original length of the face.

To create a hem, follow these steps:

1. Open or create a sheet metal part that has at least one face or contour flange.
2. Click the Hem command on the Sheet Metal tab > Create panel as shown in Figure 8.42 left.
3. The Hem dialog box appears, as shown in Figure 8.42.

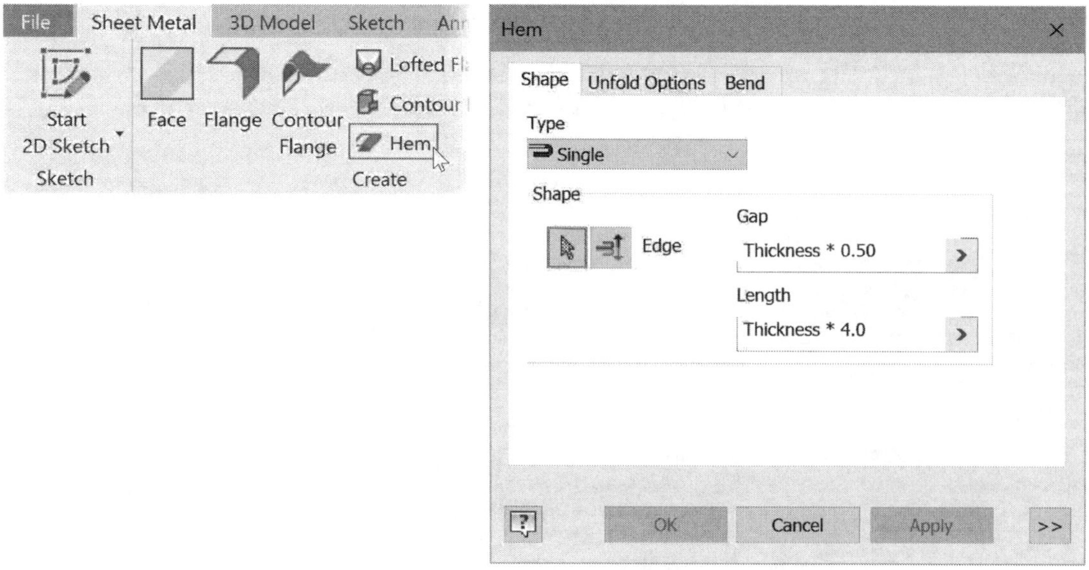

Figure 8.42

4. Select an open edge on a sheet metal face.

5. Select the hem type. Examples are shown in Figure 8.43:
   a. Single: A 180° flange
   b. Teardrop: A single hem in a teardrop shape
   c. Rolled: A cylindrical hem
   d. Double: Single hem folded 180° resulting in a double-thickness hem
6. Enter values for the hem. Teardrop and rolled hems require radius and angle values, while single and double hems require gap and length values. The hem preview changes to match the current values.
7. Expand the dialog box by clicking the More (>>) button and selecting Edge, Width, Offset, or From To for the hem extents.
8. Click Apply to continue creating hems, or click OK to complete the hem and exit the dialog box.

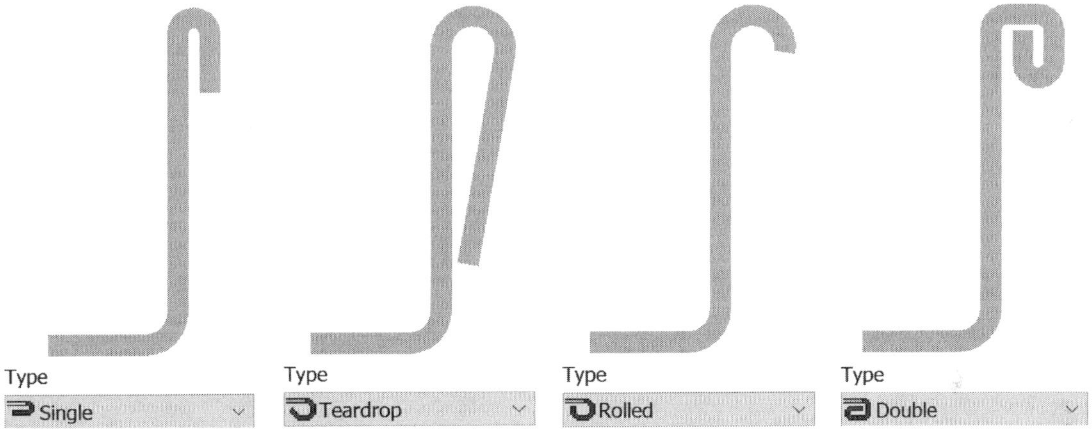

Figure 8.43

Shape Tab

The Shape tab allows you to select the edge and type of hem to create. Based on the type of hem that you want to create, different options will become active in the Hem dialog box. The following options are available on the Shape tab.

Select Edge. Select the edge along which the hem will be created.

Flip Direction. Toggle the direction in which the hem will be created.

Gap. Specify the distance between the inside faces of the hem. This feature is available when you select a single or double hem type.

Length. Specify the length of the hem. This feature is available when you select a single or double hem type.

Radius. Specify the bend radius to apply at the bend. This feature is available when you select a rolled or tear-drop hem type.

Angle. Specify the angle applied to the hem. This option is available when you select a rolled or teardrop hem type.

The options available on the Unfold Options and Bend tabs are used to override the active rule.

Unfold Options and Bend Tabs

The options available on the Unfold Options and Bend tabs are used to override the active rule.

BEND

You create bends to join faces. For example, Figure 8.44 left shows two disjointed faces. The image on the right shows the results after using the bend command to connect the faces.

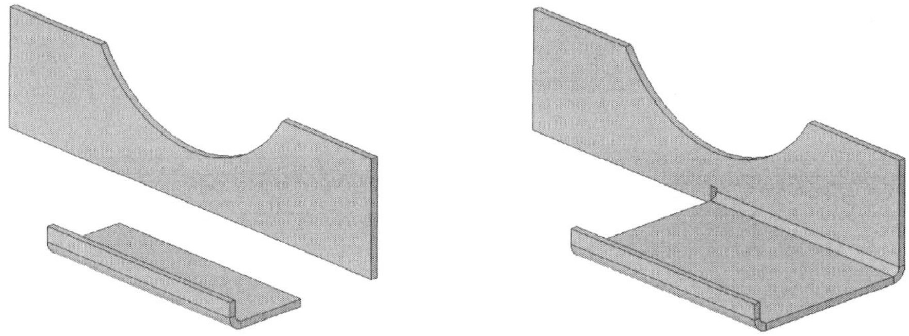

Figure 8.44

When you select parallel faces for the bend feature, a joggle or z-bend will be created, depending on the edge location. Joggles are often used to allow overlapping material. A sample of a joggle feature is shown in the following image. You can also create double bends when the two faces are parallel and the selected edges face the same direction.

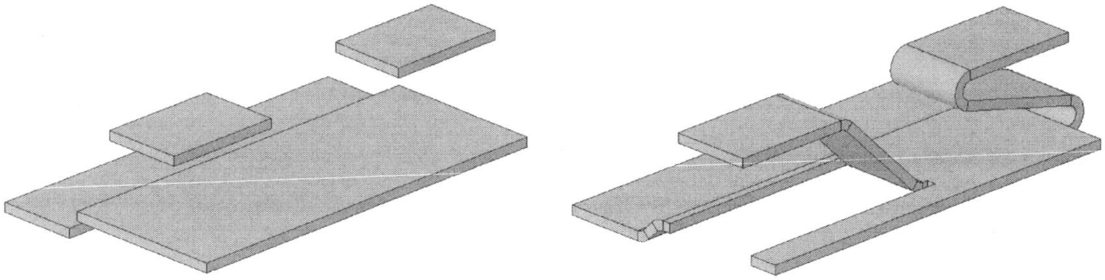

Figure 8.45

You can create the bend as two 90° bends with a tangent face between them, as shown in the following image on the left, or a single full-radius bend between the two faces, as shown in the image on the right.

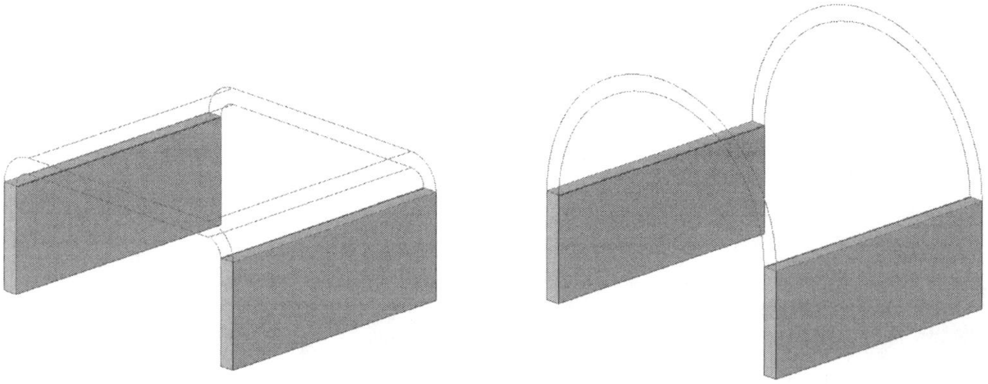

Figure 8.46

TIP: To apply a bend, the sheet metal part must have two disjointed or sharp-cornered intersecting faces.

To create a bend, follow these steps:

1. Click the Bend command on the Sheet Metal tab > Create panel as shown in Figure 8.47 left.
2. The Bend dialog box appears, as shown in the image on the right.

3. Select the common edge of two intersecting faces, or select two parallel edges on disjointed, non-co-planar faces. If the two faces are parallel, the Double Bend options are available for selection. Depending on the position of the two faces, the allowed Double Bend options will be Fix Edges and 45 Degree or Full Radius and 90 Degree.

4. Make any changes on the Unfold Options or Bend tabs.

5. Click Apply to continue creating bends, or click OK to complete the bend and exit the dialog box.

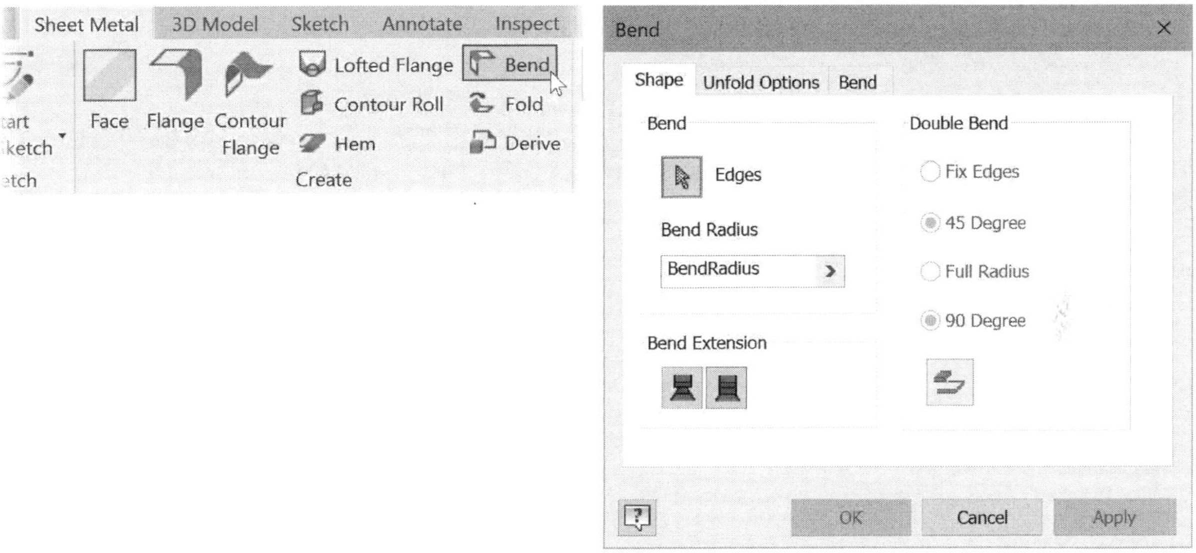

Figure 8.47

Shape Tab

The Shape tab allows you to select the edges between which to create the bend and specify the type of bend that you want to create. You can also override the bend radius specified in the sheet metal style. The following options are available on the Shape tab.

Edges. Select the edges where the bend will be applied. The selected edges will be trimmed or extended as needed to create the bend feature.

Bend Radius. Override the default bend radius set by the active sheet metal style.

Bend Extension. Controls how bend extensions are calculated when a profile has edges that are not coincident to the selected edge. Extend Bend Aligned to Side Faces: Aligns the bend of the shorter edge when the edges are not equal. Extend Bend Perpendicular to Side Faces: Extends the bend from the longer side when the edges are not equal.

Fix Edges. Click to create equal bends to the selected sheet metal edges.

45 Degree. Click to create 45° bends on the selected edges.

Full Radius. Click to create a single semi-circular bend between the selected edges.

90 Degree. Click to create 90° bends between the selected edges.

Flip Fixed Edge. Click to reverse the order of the edges being fixed. Normally, the first edge selected is fixed by default, and the second edge will be trimmed or extended. This button reverses the order.

Unfold Options and Bend Tabs

The options available on the Unfold Options and Bend tabs are used to override the active rule.

Edit a Bend

You can edit a bend that is listed under a feature in the browser at any time, allowing you to reselect the edges that define the bend. You can even edit the bend to connect two faces that were not initially joined when the bend feature was created.

CUT

The Cut command is a sheet metal specific implementation of the standard Extrude command. The Cut command always performs an extrude cut. The default extents are a blind cut at a distance equal to the sheet metal Thickness parameter. This action ensures that the cut extends only through the face containing the sketch and not through other faces that may be folded under the sketch face. You create a cut that does not go all the way through the material by entering a value that is less than the thickness of the part.

Most cuts are manufactured on the flat sheet stock before the sheet is bent to form the folded part, and cuts often cross bend lines. Because the part is modeled in the folded state, representing cuts that cross bend boundaries requires the bend be unfolded to represent the flat sheet. When the bend is refolded, the cut deforms around the bend to ensure that the extrusion remains perpendicular to the sheet metal faces on both sides of the bend and deforms throughout the bend, if required.

When sketching a profile that will cut across a bend that you will need to dimension to an edge that is on a face which is not in the same plane as the sketch, use the Project Flat Pattern command on the Sketch tab > Create panel as shown in Figure 8.48.

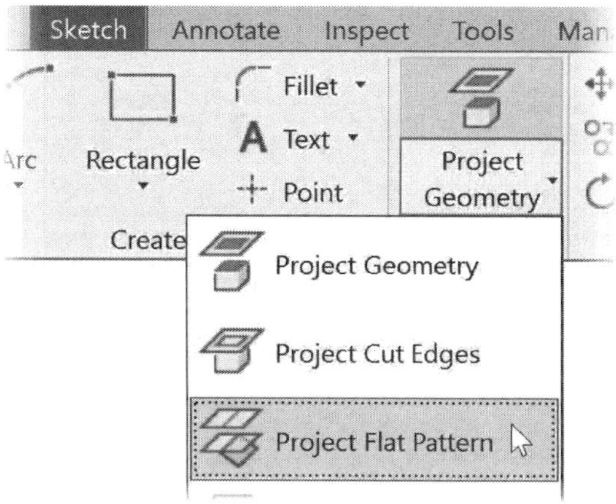

Figure 8.48

Then select faces on the folded sheet metal part to project the unfolded flat pattern geometry onto the sketch, as shown in Figure 8.49.

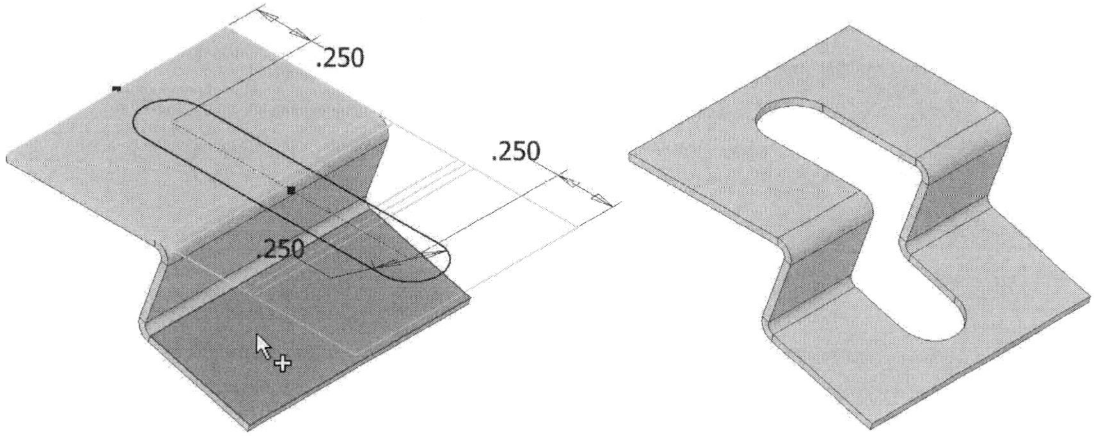

Figure 8.49

To create a cut feature, follow these steps:

1. Create a sketch on a sheet metal face that includes one or more closed profiles representing the area(s) to cut. If required, use the Project Flat Pattern command to project the unfolded geometry of connected faces onto the sketch.
2. Click the Cut command on the Sheet Metal tab > Modify panel as shown in Figure 8.50 left.
3. The Cut dialog box appears, as shown in the image on the right.

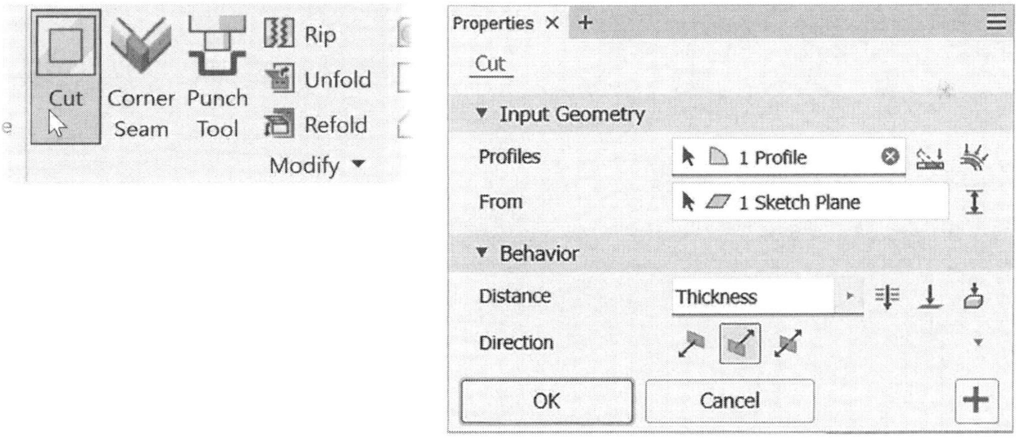

Figure 8.50

4. Select the profiles to cut.
5. If a profile crosses a bend, click Cut Across Bend.
6. Select the Cut Normal option to project the profile to the surface and cut perpendicular to the faces that the projection intersects.
7. Enter a distance for the extents of the cut.
8. Click OK to complete the cut and exit the dialog box.

The following commands are available in the Cut dialog box:

Profile. Select the profile(s) to be cut into the sheet metal part.

Cut Across Bend. Click to project the profile across faces that are bent in the sheet metal part.

Cut Normal. Projects the selected profiles (sketch, and so on) onto the surface and then cuts perpendicular to the faces that the projection intersects.

Extents. Choose one of the typical extrusion options:

Distance, To Next, To, From To, or All. If distance is selected, enter a distance for the extents of the cut or accept the default value of thickness to cut the feature using the thickness parameter defined in the sheet metal rule. The cut does not need to go through the part.

Direction. Specify the direction for the cut to be created in the part.

FOLD

An alternate method for creating sheet metal features is to start with a flat pattern shape and then fold the part to form a sheet metal part. The Fold command can also be used to create sheet metal shapes that are difficult to create using the Face or Flange commands. Figure 8.51 shows a corner being folded up.

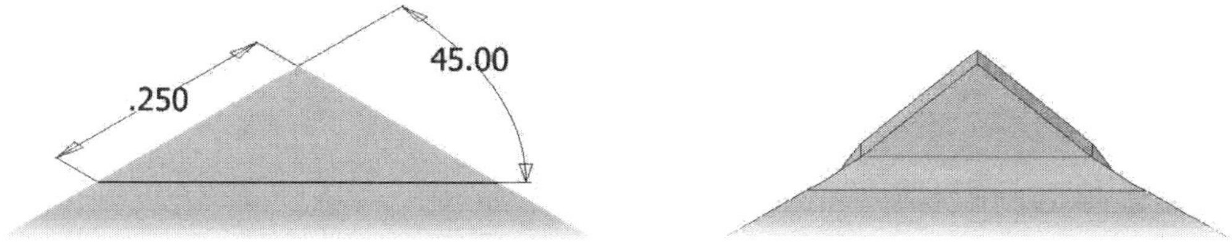

Figure 8.51

To create a fold, follow these steps:

1. Create a sketch on an existing face. Sketch a line between two open edges on the face. The sketched line endpoints must be coincident to the face edges.
2. Click the Fold command on the Sheet Metal tab > Create panel as shown in the following image on the left.
3. The Fold dialog box appears, as shown in Figure 8.52.

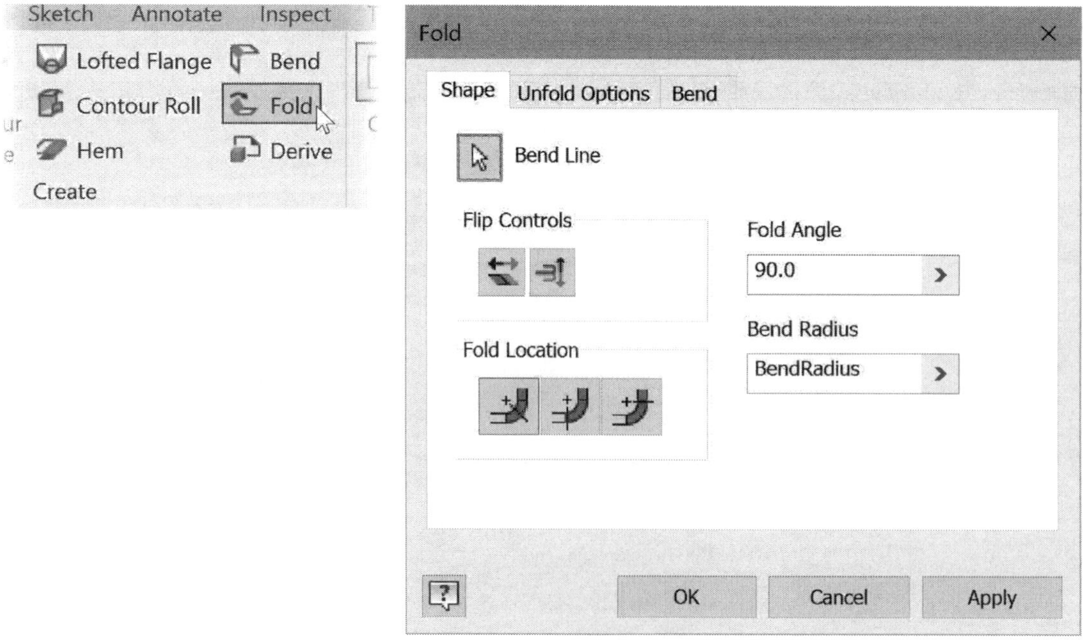

Figure 8.52

4. Select the sketch or bend line. The fold direction and angle are previewed in the graphics window. The fold arrows extend from the face that will remain fixed. The face on the other side of the bend

line will fold around the bend line.

5. If required, flip the fold direction and side.
6. Enter the angle of the fold.
7. Select the positioning of the fold with respect to the sketched line. The line can define the centerline, start, or end of the bend. The fold preview updates to match the current settings.
8. Make any needed changes on the Unfold Options or Bend tabs.

Shape Tab

The Shape tab allows you to select the bend line for the fold to be created. You can set the location of the selected bend line relative to the fold feature that determines the folded shape. The angle for the fold and direction are also specified on this tab, and you can also override the bend radius that is specified in the sheet metal rule. The following options are available on the Shape tab:

Bend Line. Select a sketch line to use as the fold line. The sketched line endpoints must be coincident to the face edges.

Flip Controls. The first button flips the side used for the angle of the fold. The second button toggles the direction the fold will be created.

Fold Location. Determine the centerline of the fold from the selected sketch line.

Fold Angle. Specify the angle to apply to the fold.

Bend Radius. Override the default bend radius set by the active sheet metal rule.

Unfold Options and Bend Tabs

The options available on the Unfold Options and Bend tabs are used to override the active rule.

CORNER ROUND

The Corner Round command is available when at least one face or contour flange feature exists and you are in the Sheet Metal environment. This command is only available when you are working on a sheet metal part and only edges that run perpendicular to the material thickness. It enables you to select these small edges easily without zooming in on the part.

To create a corner round, follow these steps:

1. Click the Corner Round command on the Sheet Metal tab > Modify panel as shown in Figure 8.53 left.
2. The Corner Round dialog box appears, as shown in Figure 8.53 right.

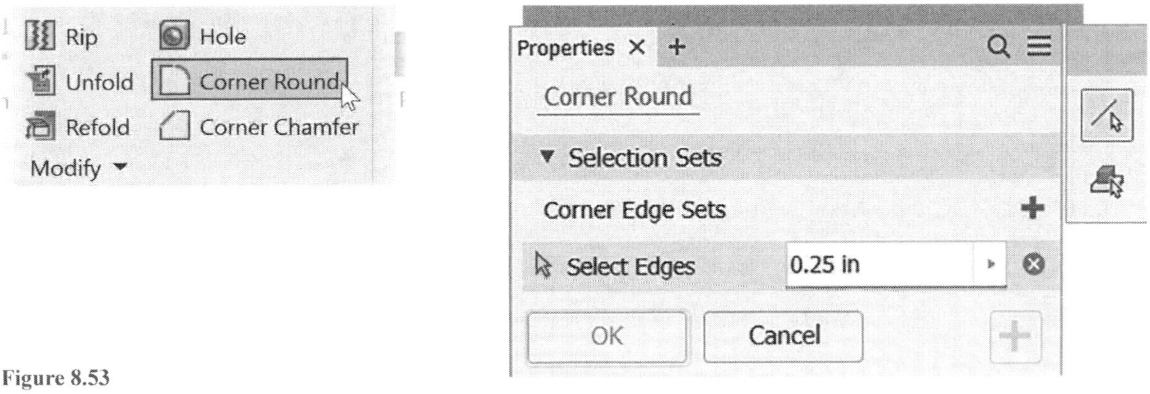

Figure 8.53

3. Select Edge Priority or Feature Priority from the upper right-hand tool palette.
4. Enter a radius for the corner round.
5. Select the corners or features to include.
6. Add additional corner rounds with different radii, and select corners or features for the additional corner rounds.
7. Click OK to complete the corner round and exit the dialog box.

Corner Chamfer

The Corner Chamfer command is a sheet metal specific chamfer tool. As with the Corner Round command, all edges other than those at open corners of faces are filtered out. Because the edges are always discontinuous, the Edge Chain and Setback settings available with the standard Chamfer command are not available.

To create a corner chamfer, follow these steps:

1. Click the Corner Chamfer command on the Sheet Metal tab > Modify panel as shown in Figure 8.54 left.
2. The Corner Chamfer dialog box appears, as shown in Figure 8.54 right.

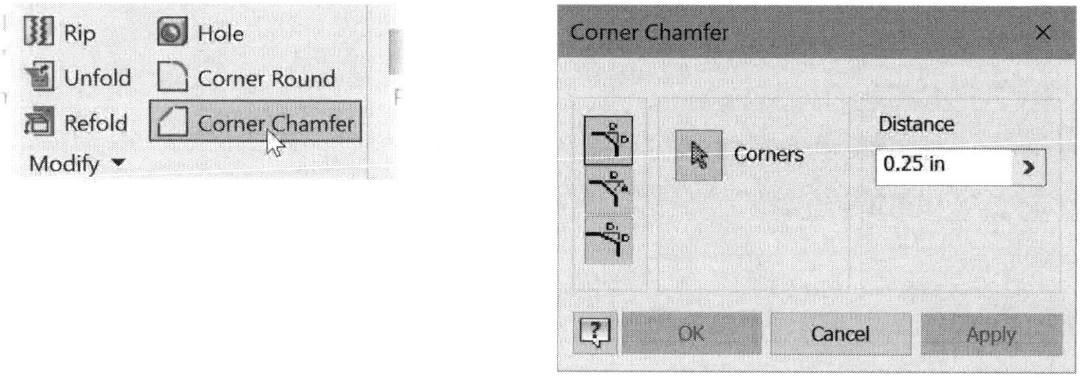

Figure 8.54

3. Select the chamfer style: One Distance, Distance and Angle, or Two Distances. See the following list for explanations of these styles.
4. Select the corners or edge and corner for Distance and Angle that you wish to include.
5. Enter the chamfer values.
6. Click OK to complete the corner chamfer and exit the dialog box.

EXERCISE 8-3: CREATING BEND, CUT, HEM AND FOLD FEATURES

In this exercise, you modify an existing sheet metal part by adding bend, cut, hem, and fold features.

1 Open *ESS_E08_03.ipt* in the Chapter 08 folder.

2 Zoom in and use the Free Orbit command to review the sheet metal features.

3 Change the viewpoint to the Home View.

4 First you create a bend between two disjointed faces (faces that do not touch).

a. Click the Bend command on the Sheet Metal tab > Create panel.

b. Select an edge on both disjointed faces (Face2 and Face 3), labeled (1) and (2) in Figure 8.55; it does not matter if you select the top or bottom edges as the selection defines the faces to join.

c. In the Bend dialog box, notice that Full Radius and 90 Degree options are grayed out; this is because the selected edges are not in the same plane.

d. Select the Fix Edges option labeled (3) in the following image. Notice how the faces would be joined together.

e. Then select the 45 Degree option labeled (4) as shown in Figure 8.55.

f. Click Apply to create the bend feature, and keep the dialog box open.

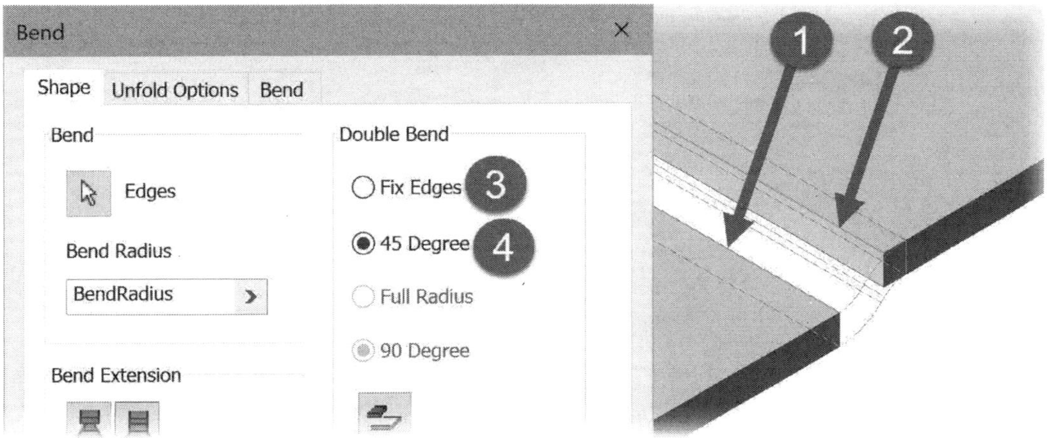

Figure 8.55

5 Next you create another bend feature.

a. Zoom into the top-right side of the part.

b. Select the Edges button in the Bend dialog box.

c. Select an edge on both disjointed faces that are facing outward labeled (1) and (2) in Figure 8.56; it does not matter if you select the top or bottom edges.

d. In the Bend dialog box, only the Full Radius and 90 Degree options are available because the selected edges are on the same plane.

e. Select the Full Radius option labeled (3) in the following image and notice how the faces would be joined together.

f. Then select the 90 Degree option labeled (4) in the following image.

g. Click Apply to create the bend feature, and keep the dialog box open.

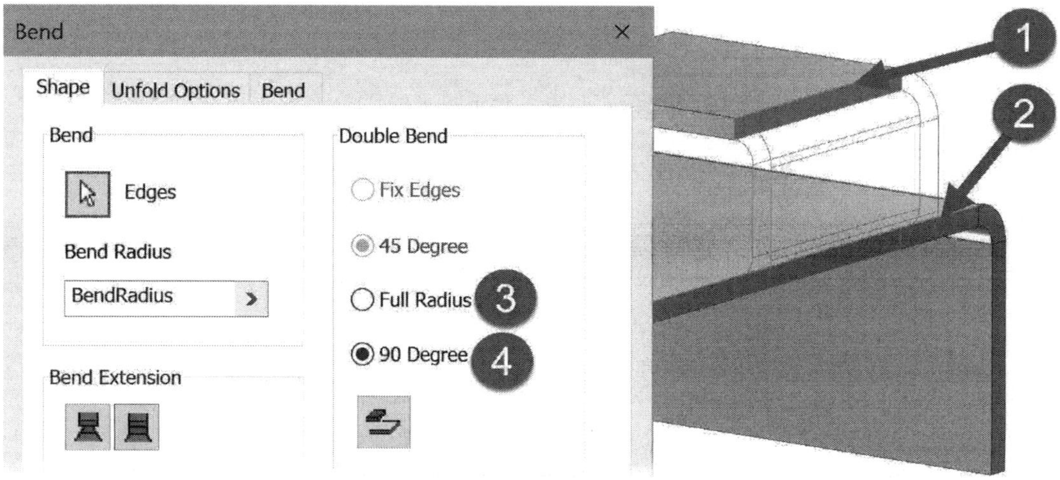

Figure 8.56

6 Now you create another bend on the last disjointed face.

a. Zoom into the top-left side of the part.

b. Select the Edges button in the Bend dialog box.

c. On the top disjointed face select the bottom-inside edge labeled (1) and the top edge of the bottom horizontal face labeled (2) in Figure 8.57.

d. Select the Fix Edges option labeled (3) in the following image and notice how the face would be joined together.

e. Then select the 45 Degree option labeled (4) in the image.

f. Click OK to create the bend feature and close the dialog box.

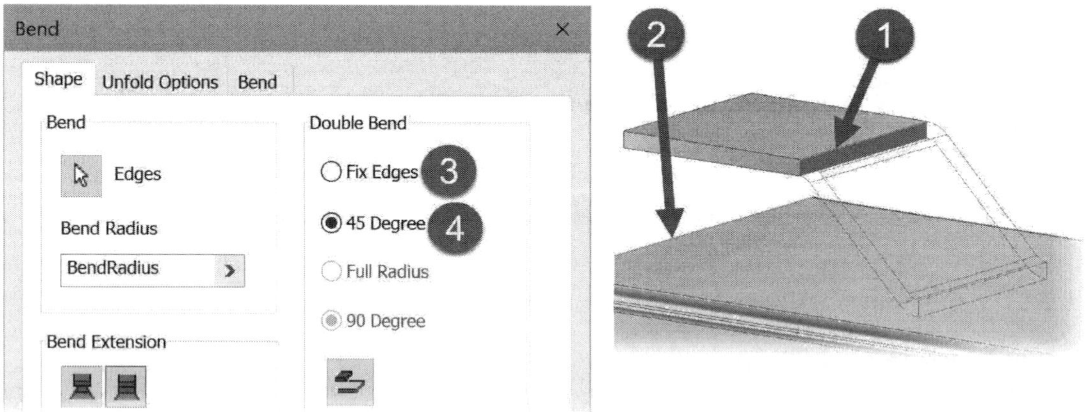

Figure 8.57

7 Next you create a cut feature that wraps around multiple faces.

a. Select the horizontal plane and click Create Sketch from the mini-toolbar as shown in the following image on the left.

b. Draw and dimension a rectangle as shown in the following image on the right. The last dimension needs to be added to an edge on the bottom flange.

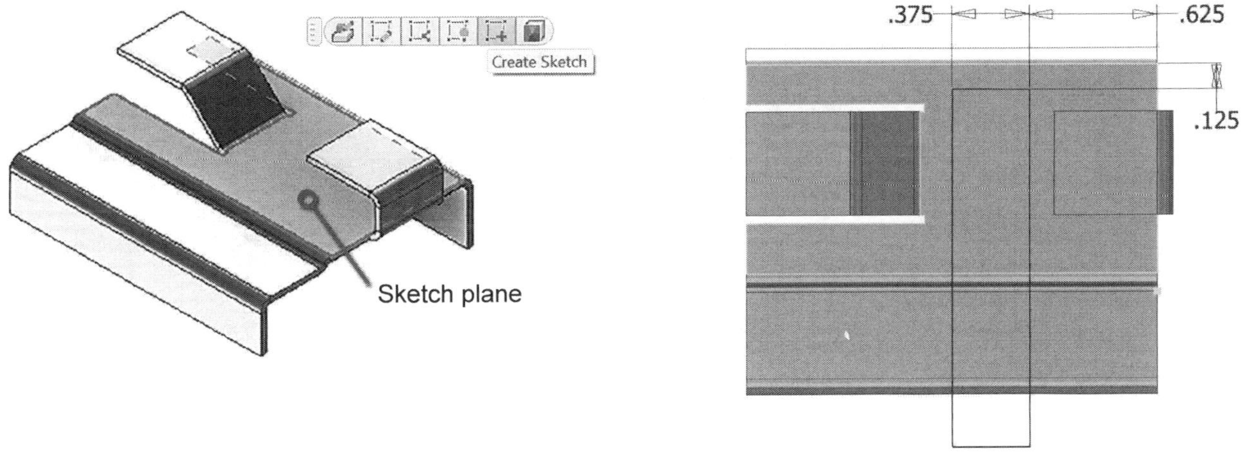

Figure 8.58

c. To make it easier to visualize the next steps, change to the Home View.

d. Next you project the face on the front flange. Click the Project Flat Pattern command on the Sketch tab > Create panel. The command may be under the Project Geometry command.

e. Select the front face of the flange as shown in the following image on the left.

f. Add a **.125 inch** dimension between the left projected edge and the edge of the rectangle as shown in Figure 8.59 right.

g. Finish the sketch.

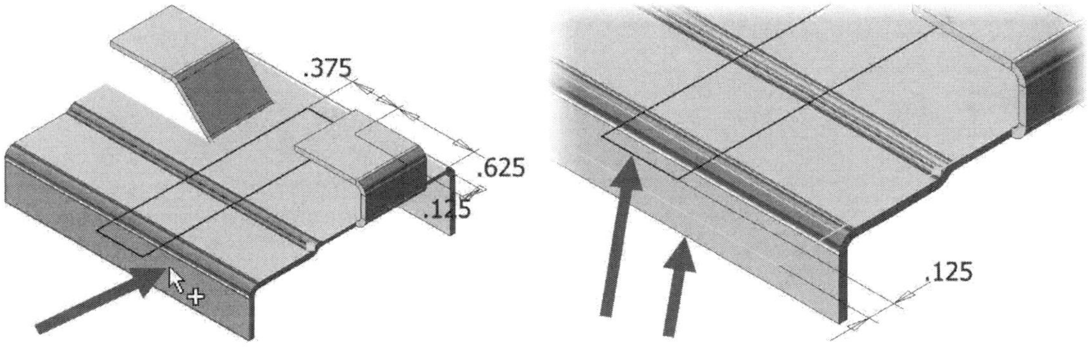

Figure 8.59

8 Start the Cut command on the Sheet Metal tab > Modify panel.

a. For the Profile select the rectangle you just sketched and in the Cut dialog box check the Cut Across Bend option as shown in Figure 8.60.

b. Click OK to create the cut feature.

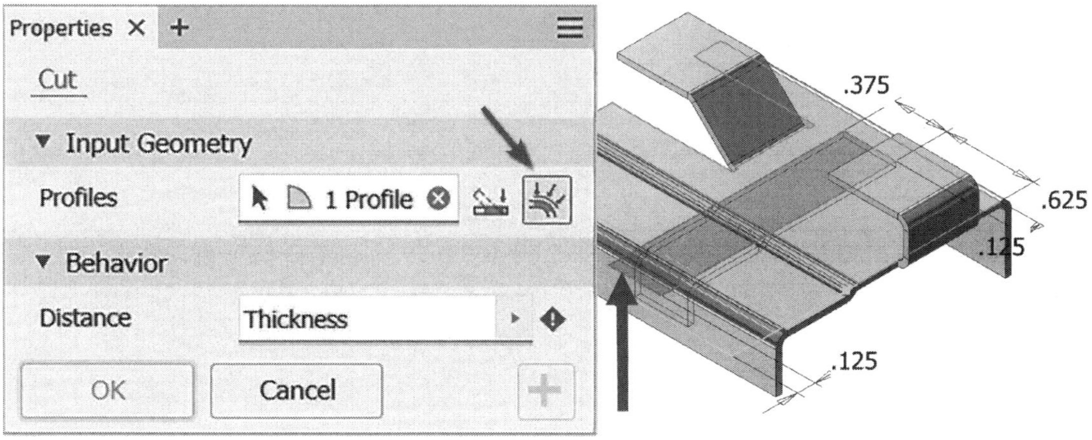

Figure 8.60

9 Next you create a fold feature.

a. Select the top-right horizontal face and click Create Sketch from the mini-toolbar as shown in the following image on the left.

b. Draw and dimension two lines as shown in the following image on the right.

c. Finish the sketch.

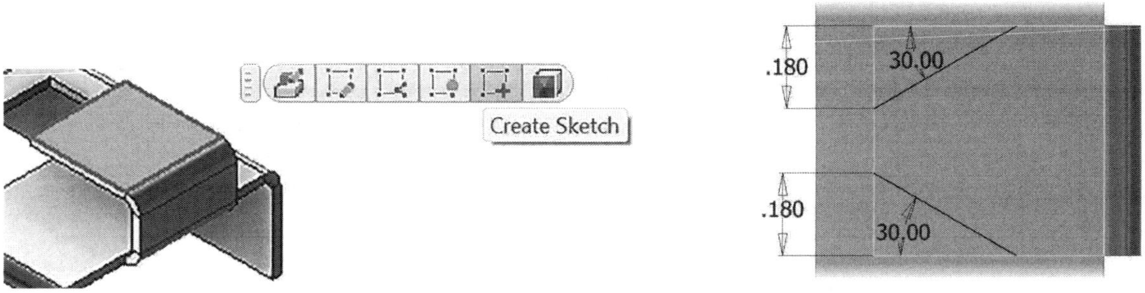

Figure 8.61

d. Click the Fold command on the Sheet Metal tab > Create panel.

e. Select the top angled line on the top-right face you just sketched. Note that you cannot select both lines in one operation. You'll use the other line to fold the bottom corner.

f. Notice that the preview shows the small angled area will fold up as shown in Figure 8.62 left. This is the correct direction.

g. Click OK to create the fold.

10 Notice that the second angled line you sketched is not visible; this is because the sketch was consumed by the fold feature. You can make this sketch visible by selecting the fold feature in the graphics window and click Make Sketch Visible from the mini-toolbar as shown in Figure 8.62 right.

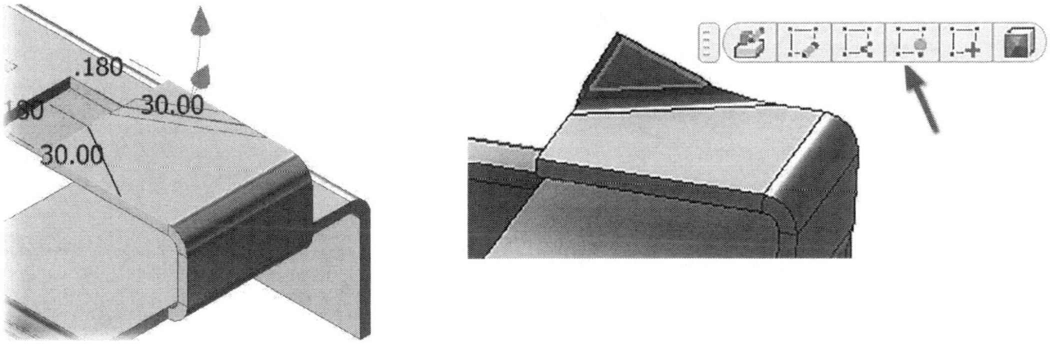

Figure 8.62

11 Next you fold the bottom corner. Click the Fold command on the Sheet Metal tab > Create panel.

a. Select the line on the bottom-side of the top face.

b. Ensure that the fold will fold up; if needed, reverse the direction.

c. Click OK to create the fold.

12 Turn off the visibility of the shared sketch by right-clicking on the sketch under one of the fold features in the graphics window and click Make Sketch Invisible from the mini-toolbar. When done, your screen should resemble Figure 8.63 right.

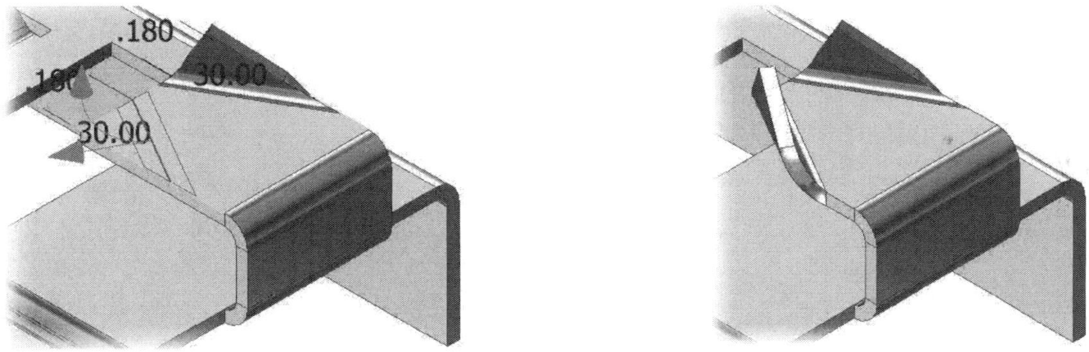

Figure 8.63

13 Next you round the corners of the folded features.

a. Click the Corner Round command on the Sheet Metal tab > Modify panel.

b. Select the two top edges on the fold features and change the radius to **0.125 inches** as shown in Figure 8.64.

c. Click OK to create the corner rounds.

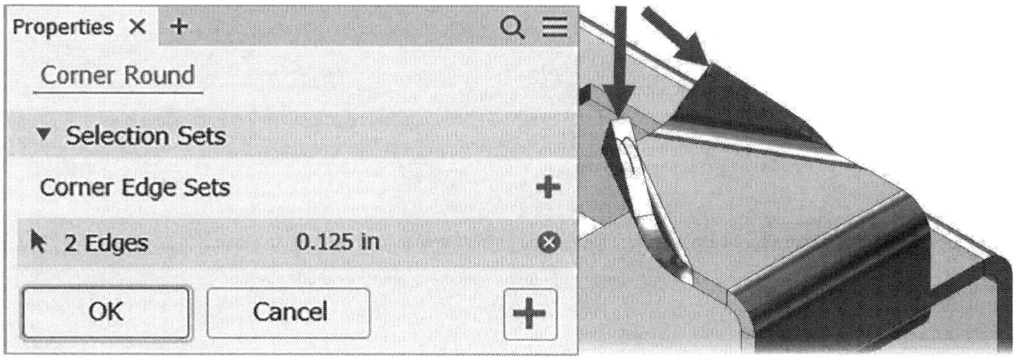

Figure 8.64

14 The last feature you create is a hem.

a. Change the viewpoint so it resembles Figure 8.65 and zoom into the back flange.

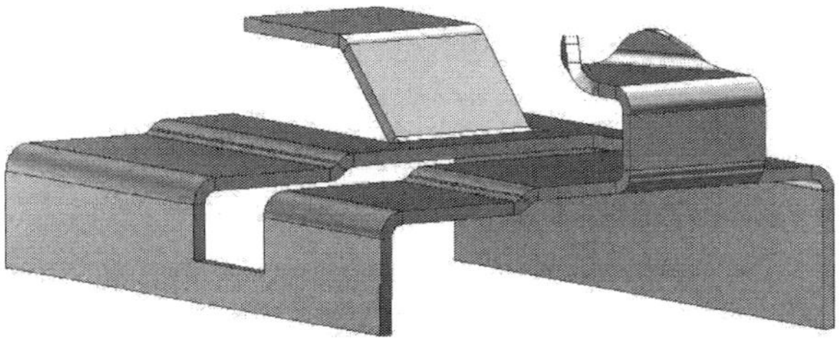

Figure 8.65

b. Click the Hem command on the Sheet Metal tab > Create panel.

c. Select the bottom-inside edge of the back flange labeled (1) in the following image on the right.

d. In the Hem dialog box change the Type to Teardrop labeled (2) and the Angle to **290 degrees** labeled (3) in the following image. If needed flip the direction of the hem labeled (4) so its preview is bending into the part as shown in Figure 8.66 right.

e. Click OK to create the hem. Your part should resemble Figure 8.67.

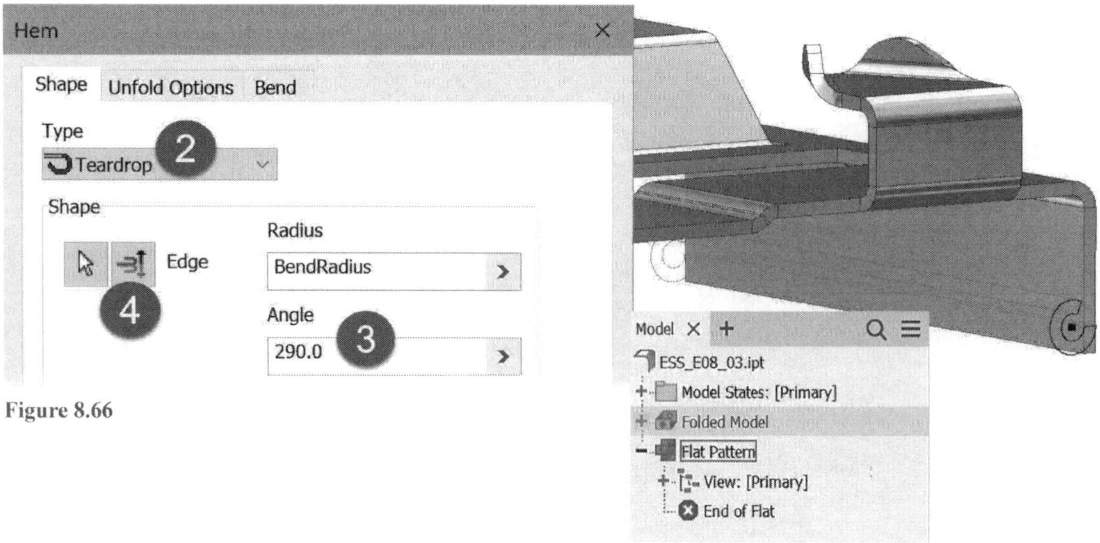

Figure 8.66

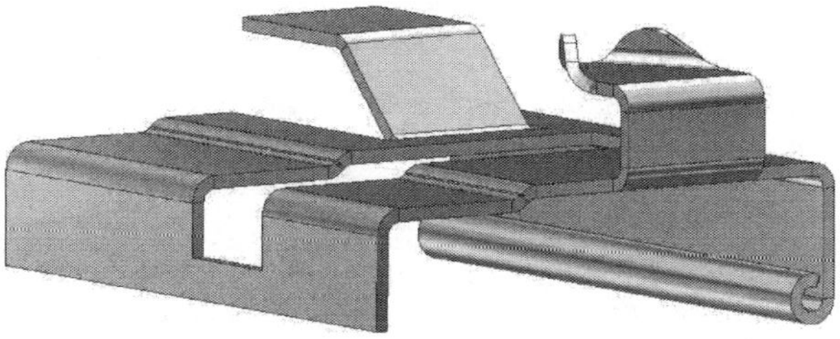

Figure 8.67

15 Practice editing the features and creating new features.

16 When done, close the file. Do not save changes.

End of exercise.

FLAT PATTERN

You can create a sheet metal flat pattern that unfolds all the sheet metal features. The flat pattern represents the starting point for the manufacturing of the sheet metal part. The flat pattern can also be used in a 2D drawing view, complete with lines indicating bend centerlines.

Define A-Side

Before creating a flat pattern, define the side of the part that will be up when manufactured. Use the Define A-Side command on the Sheet Metal tab > Flat Pattern panel as shown in the following image and then select a face on the part that is on the up side of the part.

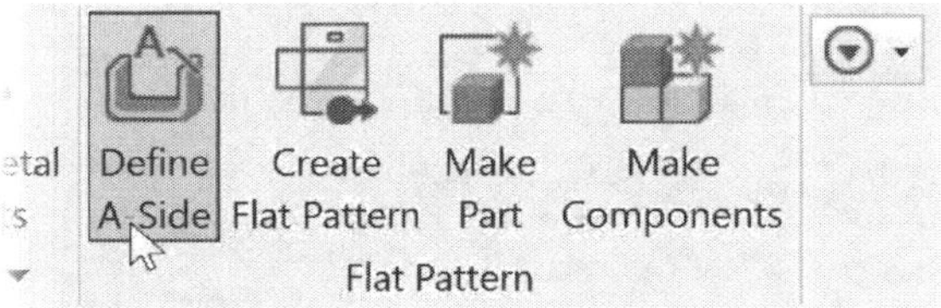

Figure 8.68

Create Flat Pattern

After the up side of the part has been defined with the Define A-Side command you can create a flat pattern by clicking the Create Flat Pattern command on the Sheet Metal tab > Flat Pattern panel, as shown in the following image on the left (for clarity the feature tree for the part has been collapsed). A 3D model of the unfolded part is generated in a Flat Pattern environment and a new Flat Pattern entry is added to the browser as shown in the image on the right.

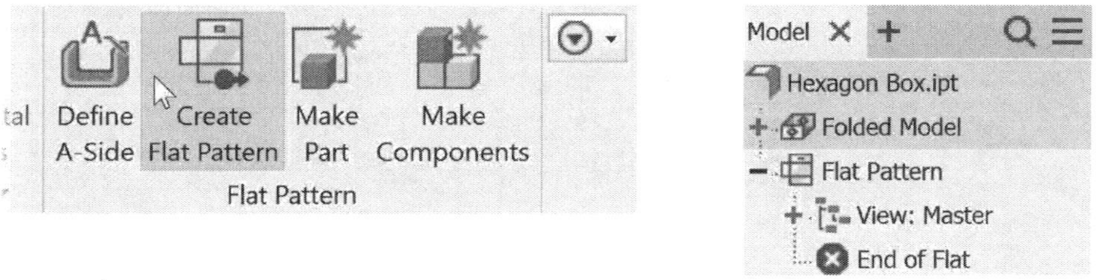

Figure 8.69

Flat Pattern Environment

To return to the Sheet Metal modeling environment click the Go to Folded Part command on the Flat Pattern tab > Folded Part panel as shown in the image on the left. When returning to the flat pattern, any changes done to the sheet metal part will be updated in the flat pattern.

While working on the folded part, you can return to the Flat Pattern environment by double-clicking on the Flat Pattern entry in the browser or click the Go to Flat Pattern command on the Sheet Metal tab > Flat Pattern panel as shown in the image on the right.

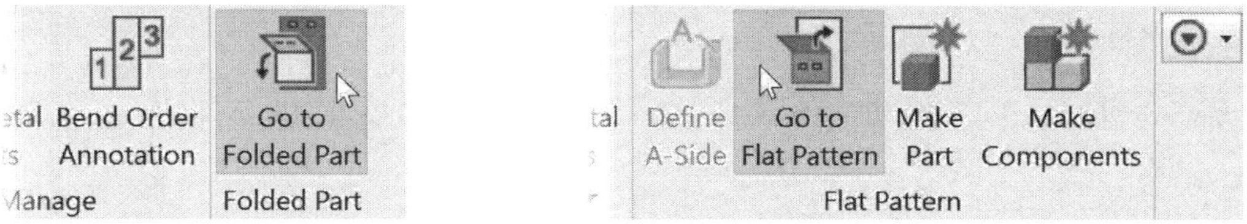

Figure 8.70

TIP: If you delete the flat pattern model from the browser, you will not be able to create a drawing view of the flat pattern. If you delete the flat pattern model after the creation of a flat pattern drawing view, the drawing view will be deleted.

Flat Pattern Definition and Extents

If you need to change the alignment of the flat pattern, while in the flat pattern environment right-click on Flat Pattern and click Edit Flat Pattern Definition from the menu as shown in the following image on the left. Use the options in the Flat Pattern dialog box as shown on the right of the following image to change the alignment.

If you need to determine the extents of the flat pattern, click Extents from the menu and in the Flat Pattern Extents dialog box the width, length and area of the flat pattern will display as shown in the image on the right.

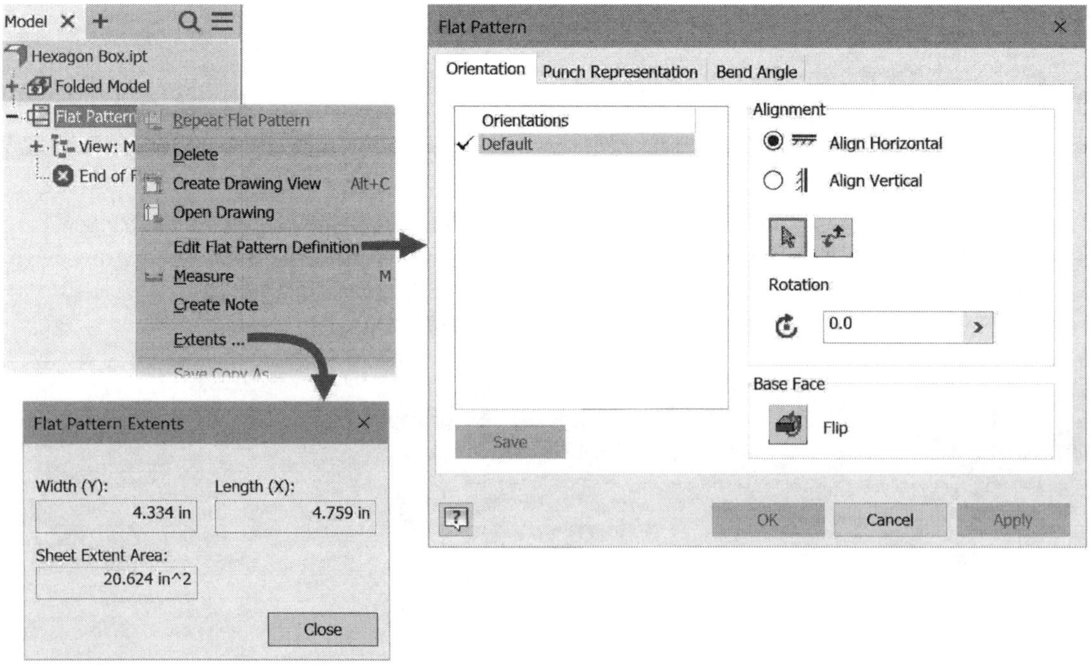

Figure 8.71

Export a Flat Pattern

If your manufacturing process requires a 3D (SAT file) or 2D (.dxf or .dwg) file format, you can export the flat pattern that has already been generated in a sheet metal part file. Right-click on Flat Pattern in the browser and click Save Copy As, as shown in the following image on the left, then in the Save Copy As dialog box, select the desired format as shown in the image on the right.

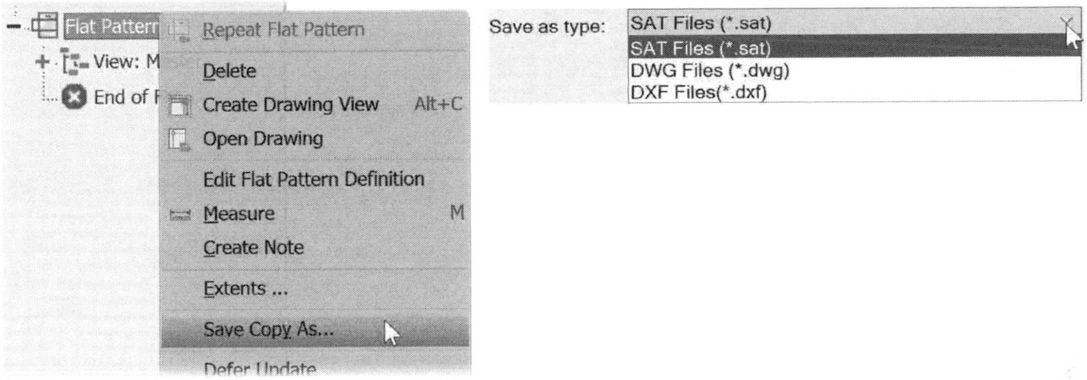

Figure 8.72

DETAILING SHEET METAL PARTS

You can create drawing views of the folded 3D sheet metal part and the flat pattern. The flat pattern drawing view enables all bend locations, bend extents, bend and corner reliefs, and cutouts to be located and sized with dimensions. The 3D model views describe the folded shape of the part and allow the forming tool operator to validate the folded part, as shown in the following image.

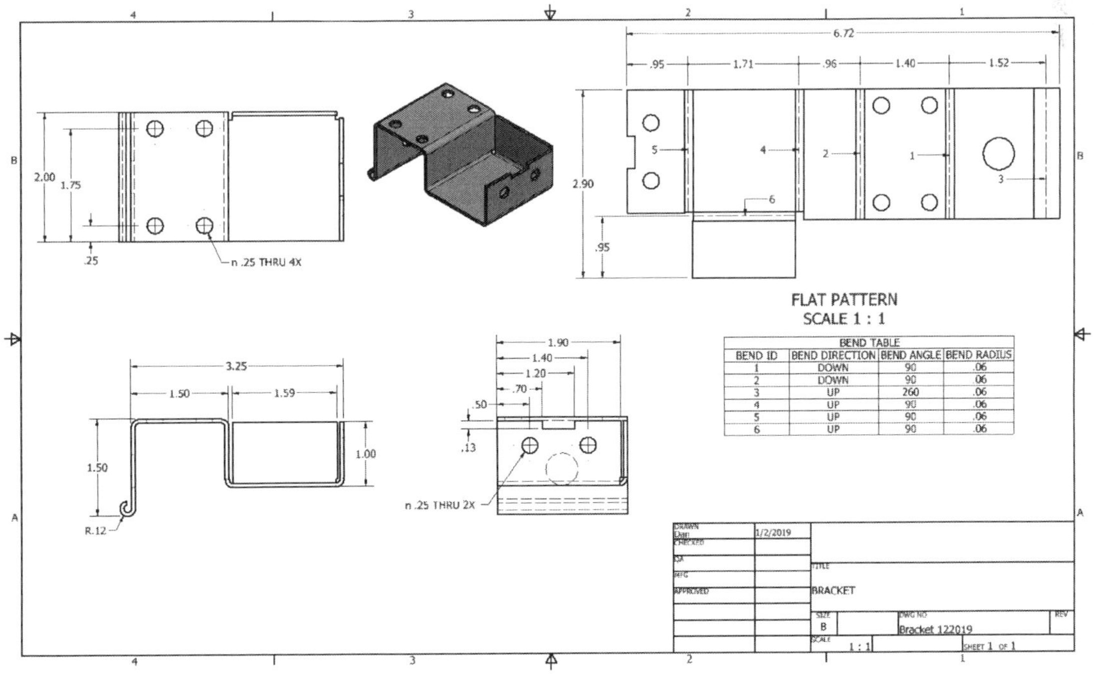

Figure 8.73

To document a sheet metal part, follow these steps.

1. Create a sheet metal part, and generate a flat pattern. If a flat pattern is not generated, you won't be able to document the flat pattern in a drawing.

2. Start a new drawing.
3. Create a base drawing view, and select Folded Model or Flat Pattern from the Sheet Metal View list in the Drawing View dialog box, as shown in the following image. When Flat Pattern is selected you have an option to also generate the Bend Extents and Punch Centers.
4. If desired use the Base View command to create the other sheet metal view type (folded or flat).

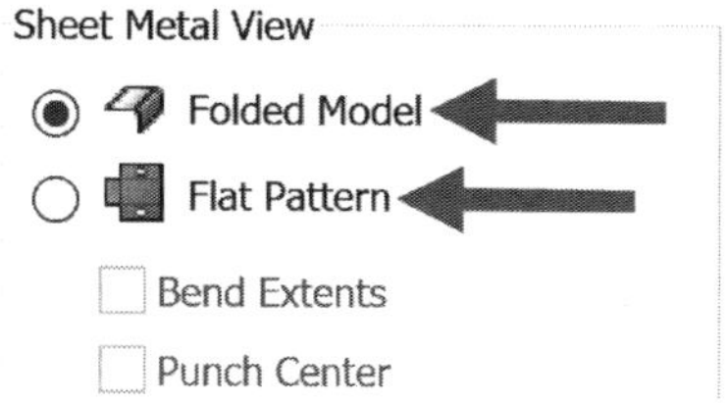

Figure 8.74

5. From the Annotate tab, add annotations as needed:
   a. Bend, Hole, and Punch Notes: Annotate tab > Feature Notes panel as shown in the following image on the left.
   b. Bend Table: Annotate tab > Table panel > General Table command as shown in Figure 8.75 right.

Figure 8.75

EXERCISE 8-4: FLATTENING & DOCUMENTING SHEET METAL PARTS

In this exercise, you create a flat pattern model of a sheet metal part, obtain the flat pattern extents and create drawing views of the folded sheet metal part and flat pattern.

1 Open *ESS_E08_04.ipt* in the Chapter 08 folder.

2 First you set which side of the part is up. Click the Define A-Side command on the Sheet Metal tab > Flat Pattern panel and click the top face as shown in the following image on the left.

3 Create a flat pattern by clicking the Create Flat Pattern command on the Sheet Metal tab > Flat Pattern panel. The flat pattern model of the sheet metal part is displayed in the graphics window, as shown in the image on the right.

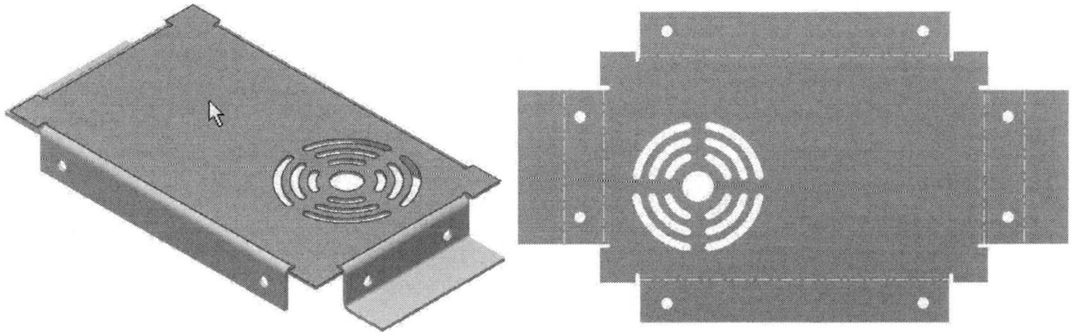

Figure 8.76

4 Determine the extents of the flat pattern by right-clicking on the Flat Pattern entry in the browser and click Extents from the menu as shown in the following image on the left. The extents are displayed in the Flat Pattern Extents dialog box as shown in Figure 8.77 right. Click Close to exit the dialog box.

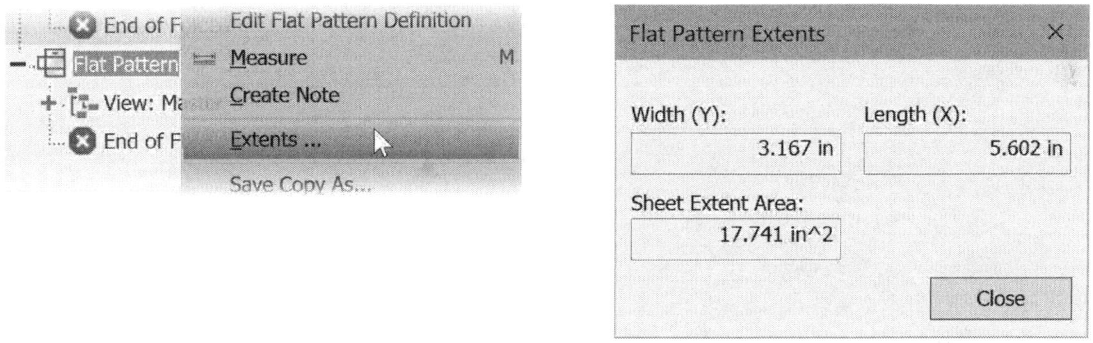

Figure 8.77

5 Exit the flat pattern environment by clicking the Go to Folded Part command on the Sheet Metal tab > Folded Part panel.

6 To keep the original file in its original state, click the Save As command from the File tab and enter the name *ESS_E08_04_Flat_Pattern.ipt*.

7 Create a new drawing based on the ANSI (in).dwg template.

8 Next you change the sheet size. In the browser right-click on Sheet:1 and click Edit Sheet from the menu. In the Edit Sheet dialog box change the Size to "C."

9 Use the Base View command to create the 1:1 drawing views of the folded model near the lower-left corner of the sheet, so they resemble Figure 8.78.

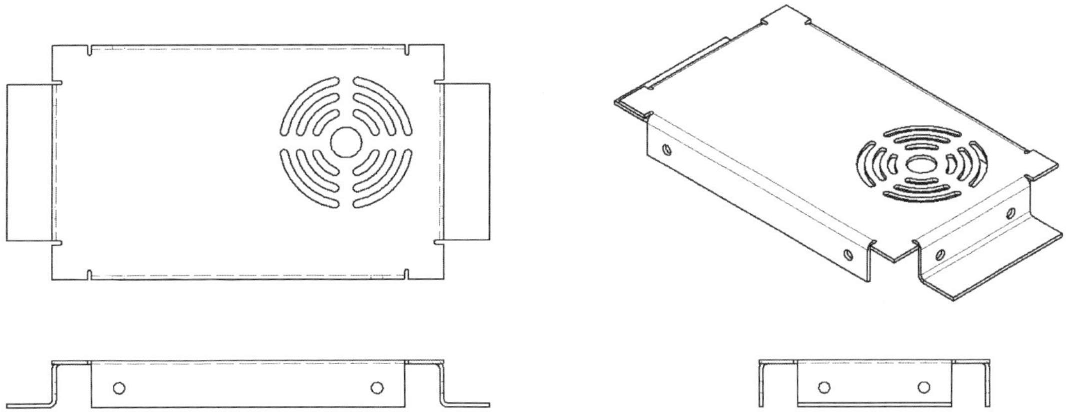

Figure 8.78

10 Next you add a drawing view of the flat pattern.

a. Start the Base View command.

b. In the Drawing View dialog box, in the Sheet Metal View section, select the Flat Pattern option, and the Bend Extents option as shown in Figure 8.79 left.

c. Place a 1:1 drawing view above the top view and click OK. The flat pattern should resemble the image on the right.

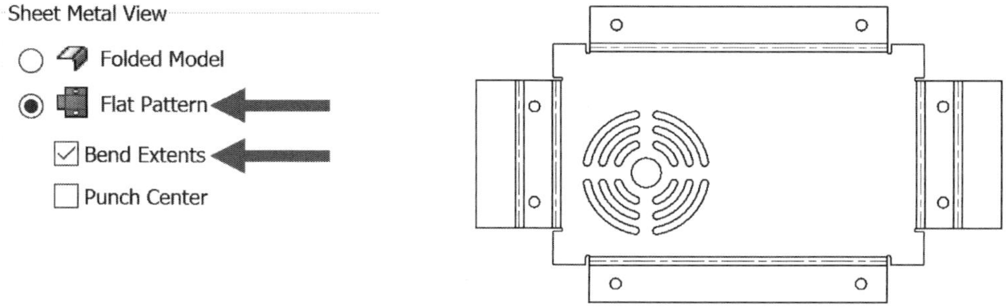

Figure 8.79

11 Next you add bend notes.

a. Zoom into the view of the flat pattern.

b. Click the Bend Notes command (labeled Bend) on the Annotate tab > Feature Notes panel.

c. Select the bend lines by selecting the six bend lines (represented with centerlines) individually or click and drag a window around the flat pattern as shown in Figure 8.80 left.

d. Add the bend lines by right-clicking and click OK from the marking menu. Figure 8.80 right shows the bend notes added to the drawing view.

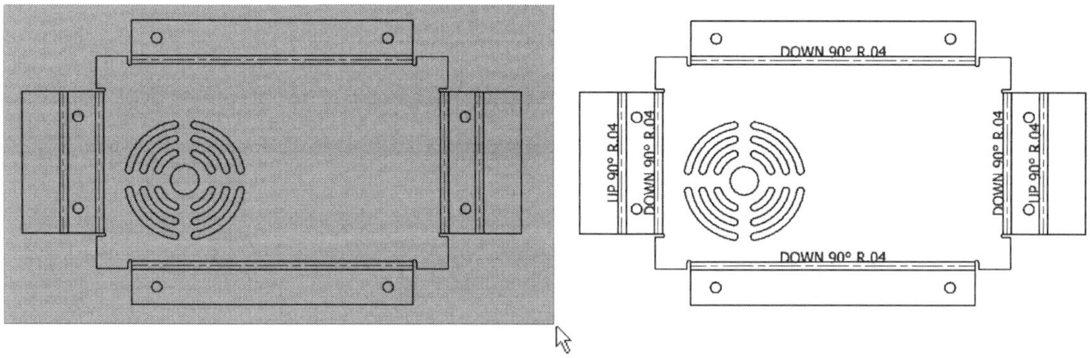

Figure 8.80

e. If desired, you can move the bend note away from the bend line by moving the cursor over the bend note and click and drag on the green circle as shown in Figure 8.81 left. As the bend note is repositioned a leader is added as shown on the right.

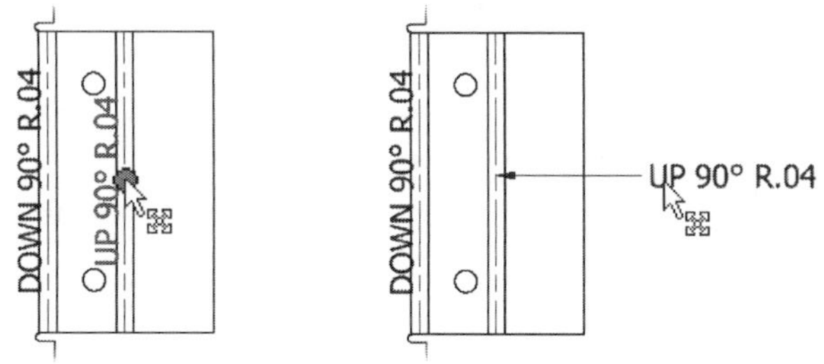

Figure 8.81

12 Next you add a bend table.

a. Use the Undo command to remove the bend notes you just placed.

b. Click the Table command which is labeled General on the Annotate tab > Table panel.

c. Select inside the flat pattern view as the view to generate the bend table from.

d. Click OK in the Table dialog box to accept the default settings.

e. Position the table to the right of the flat pattern view. When done, your screen should resemble Figure 8.82.

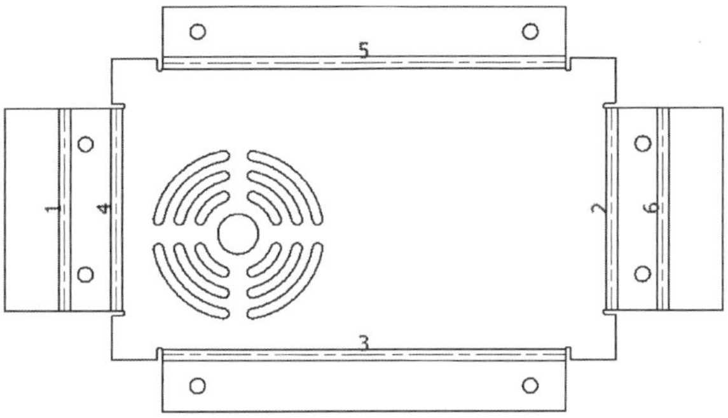

TABLE			
BEND ID	BEND DIRECTION	BEND ANGLE	BEND RADIUS
1	UP	90	.04
2	DOWN	90	.04
3	DOWN	90	.04
4	DOWN	90	.04
5	DOWN	90	.04
6	UP	90	.04

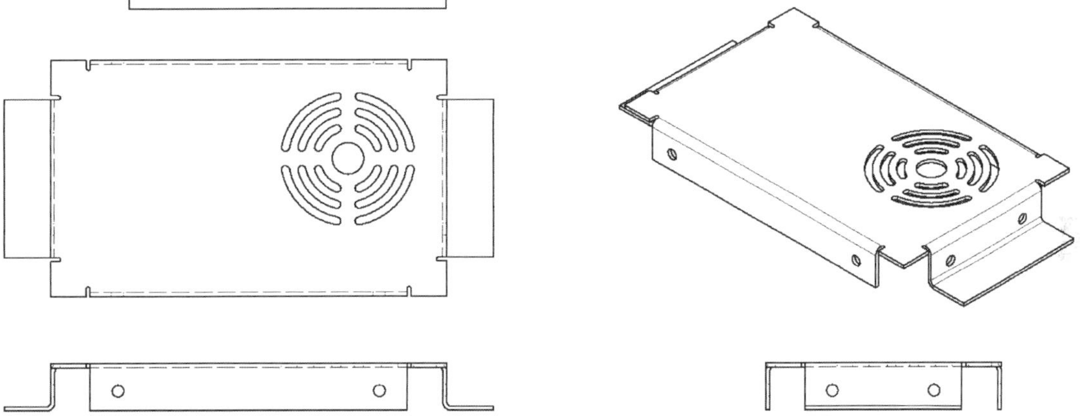

Figure 8.82

13 If desired add dimensions to the drawing, and make edits or add features to the sheet metal part. After making changes to the part, ensure the flat pattern is updated and then make the drawing current. The drawing views will automatically update.

14 When done, close the file. Do not save changes.

End of exercise.

APPLYING YOUR SKILLS

Skill Exercise 8-1

In this exercise, use the knowledge you gained in this chapter to create the sheet metal part shown in the following image. As part of the exercise, create a drawing that details both the flat and folded states of the part as shown.

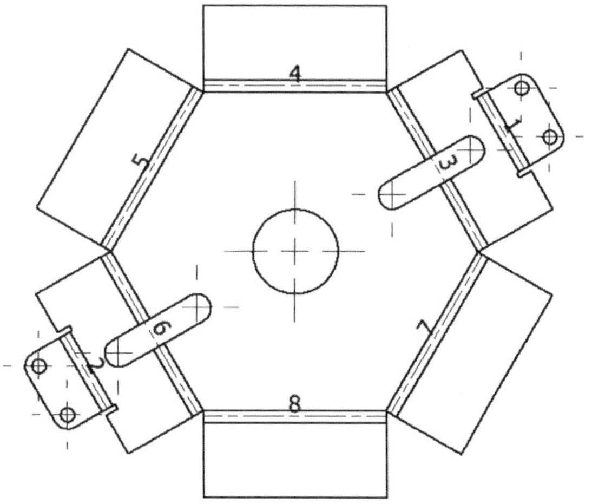

BEND TABLE			
BEND ID	BEND DIRECT	BEND AN	BEND RAD
1	DOWN	90	.048
2	DOWN	90	.048
3	UP	90	.048
4	UP	90	.048
5	UP	90	.048
6	UP	90	.048
7	UP	90	.048
8	UP	90	.048

NOTES:
Thickness = 0.0478 inches
Material = Steel, Mild
Corner 2 Bend Intersection = Tear

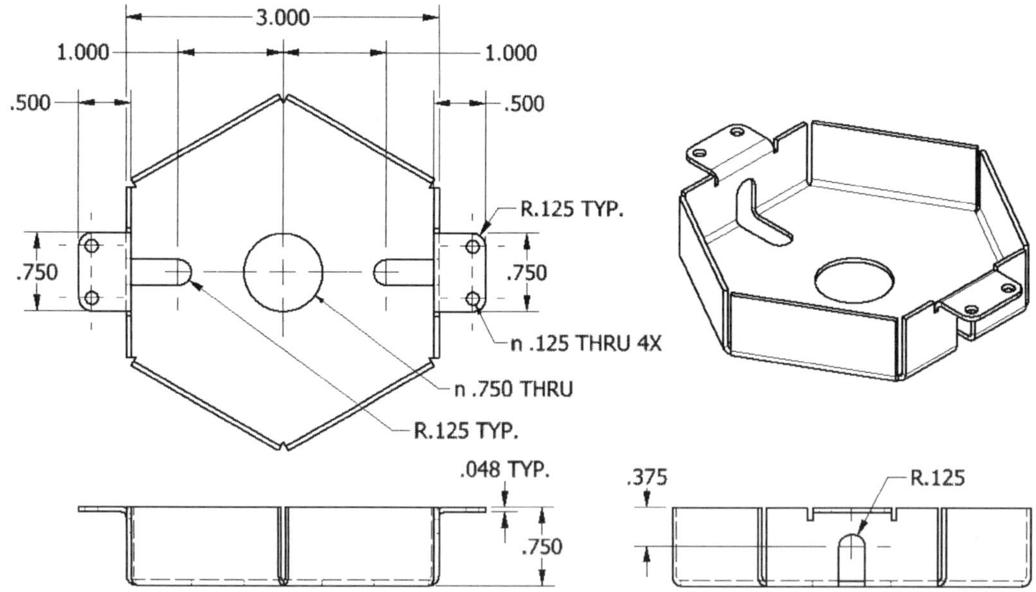

Figure 8.83

CHECKING YOUR SKILLS

Use these questions to test your knowledge of the material covered in this chapter.

1. True__ False__ While working on a sheet metal part, you should change the value of the Thickness parameter via the Parameters command.
2. Explain how to create a new sheet metal style.
3. True__ False__ In a sheet metal part, use the Base Feature command to extrude a closed profile the value of Thickness.
4. True__ False__ The Contour Flange command is used to extrude a closed profile the value of the thickness parameter.
5. Explain how to create a 1-inch-wide flange feature in the middle of an edge that is wider than 1 inch.
6. True__ False__ The Bend command is used to join two disjointed faces.
7. When dimensioning a profile that will be used to cut across a bend, which Project Geometry command is used to copy the required edge(s) to the current sketch?
   a. Project Geometry
   b. Project Cut Edges
   c. Project Flat Pattern
   d. Project to 3D Sketch
8. True__ False__ Use the Fold command to create a flat pattern.
9. Explain how to export a flat pattern from a sheet metal part file.
10. Explain how to create a drawing view of a flat pattern.

Index

Symbols

A

B

C

D

E

F

G

H

I

J

K

L

M

N

O

P

Q

R

S

T

U

V

W

X

Y

Z